Nonlinear Dynamics of Time Delay Systems

Jian Xu

Nonlinear Dynamics of Time Delay Systems

Methods and Applications

Springer

Jian Xu
School of Aerospace Engineering
and Applied Mechanics
Tongji University
Shanghai, China

ISBN 978-981-99-9906-4 ISBN 978-981-99-9907-1 (eBook)
https://doi.org/10.1007/978-981-99-9907-1

This Springer imprint is published by the registered company Springer Nature Singapore Pte Ltd.
The registered company address is: 152 Beach Road, #21-01/04 Gateway East, Singapore 189721, Singapore

Preface

Last century has witnessed the significant progress of control science. In order to achieve promising performance, the feedback control, which refers to the control mechanism that monitors the system output and adjusts the input or parameters based on the difference between the desired state and the actual state of the system, is usually employed especially in the applications which demand high control accuracy. Due to the finite speed of signal generation, transmission and processing, there exists time lag for a control system to react to changes in input or environment. Such time delay is a critical parameter that can significantly affect the stability and dynamic behaviour of control systems. For instance, traditional views hold that the time delay can destabilize the control system which is stable for some values of delay and unstable as time delay increases over some threshold for stability switch. Time delay can also degrade the performance of control systems in resulting in slow responses, low accuracy in tracking reference signals, and even oscillatory response with large amplitude. Consequently, control strategies must account for time delay to maintain the control performance, especially for the systems which are sensitive to variations of time delay.

However, researches in recent decades have shown that the influence of time delay is a twofold problem. In spite of the aforementioned negative effect on the system performance, time delay can stabilize the control system for some cases or serve as a trigger for the emergence of the desired dynamic response. This calls for the elaborate investigation on the system responses of time delay, particularly on the response prediction when the system loses the stability due to the existence of time delay. As a result, one may have to take into account the nonlinearity in addition to the time delay because the linearized system may not capture the features of system dynamics when the response moves far away from the unstable equilibrium. In other words, it is necessary to consider the joint effect of time delay and nonlinearity on the dynamics of the control system for a more promising design of control.

Nonlinear dynamics of systems with time delay play important role not only in artificial systems but also in natural sciences. For example, in the context of neuroscience, time delay refers to a certain level of time lag of response in neuronal activity and can manifest itself in various aspects of neurons, such as synaptic transmissions and generations of action potentials. Time delay within the nervous system can potentially impact information transmission and processing as it can alter the temporal relationships and coordination between neurons. In the field of neuroscience, the study of time delay in neuronal activity has become a significant topic, as it contributes to a full understanding of the complexity of the neuronal system during its information processing. The mathematical way to approach this target involves the analysis of dynamical model, expressed in terms of a group of time-delayed differential equations which have incorporated with nonlinearity since the 1950s. That is to say, the study of nonlinear dynamics of systems with time delay provides an opportunity of revealing the complete process of neuronal activity and the regulatory mechanism within neural networks, as well as an insight towards the excitatory and inhibitory behaviours of neurons and various patterns of synchronization among neurons.

The challenges of investigating nonlinear systems with time delay originate from the following facts: (i) the system of concern is nonlinear, implying the possibility of complicated dynamics when the equilibrium is unstable and consequently the dynamic behaviour of such system may be unpredictable due to lack of closed-form analytical solutions, and (ii) systems with time delay are of infinite dimension, which results in the difficulty in resorting to the well-developed methods for ordinary differential equations. Besides of difficulties in solving nonlinear systems with time delay, the inverse problem of systems with time delay, or more specifically, the parameter identification of time delay systems, requires special treatment due to the sensitivity of such system to the variation of time delay.

This text is aiming to provide a comprehensive introduction of the qualitative and quantitative studies on nonlinear dynamics of time delay systems from both natural sciences and engineering applications by the author and his former Ph.D. students or current colleagues in the past 20 years, containing the main background of systems with time delay and challenges of studying nonlinear dynamics of such systems, their nonlinear phenomena, novel methods of quantitative study of time delay systems, parameter identification in the presence of delay effect, vibration suppression utilizing time delay, and applications of nonlinear dynamics of time delay systems from various fields, such as manufacturing, communication networks, biology, public transportation. Following these contents, one can gradually enter this research area and can find the significant and powerful ability of the wavelet-based solution method with both high accuracy and low calculation.

My research on nonlinear dynamics of systems with time delay dated back to 1998 when I was a post-doctoral researcher at Huazhong University of Science and Technology. After I joined Tongji University in 1998, I started to systematically investigate the stability, bifurcation, chaos and other complex nonlinear dynamics of systems with feedback delay and coupling delay, under the support of general projects, key projects and national outstanding young fund of National Natural Science Foundation of China (NSFC). The author deeply appreciates his early Ph.D. student and current colleague, Prof. Xiuting Sun, now a professor of Tongji University, who was awarded the National Excellent Youth Science Fund of NSFC for her hard work in developing time-delay-based vibration suppression method together with the author. At the same time, I also thank my former Ph.D. students, Dr. Shu Zhang from Tongji

University, Dr. Zigen Song from Tongji University, Dr. Xiaoxu Zhang from Fudan University, and Dr. Yao Yan from University of Electronic Science and Technology of China, for their consistent work in the area of nonlinear dynamics of systems with time delay. Furthermore, my former Ph.D. students, Dr. Jvhong Ge, Dr. Ronggai Xu, Dr. Yueli Chen, Dr. Bin Zhen, Dr. Wanyong Wang, Dr. Lijun Pei, Dr. Huilin Shang, Dr. Caihong Wang, Dr. Dong Zhang, Dr. Gaoxiang Yang, Dr. Yanying Zhao, Dr. Yixia Sun, Dr. Shanying Jiang, Dr. Yanyan Zhang, contributed a lot to my previous work which shaped a solid foundation of the research content covered by this book. Without their efforts in the researches in this area, it is impossible for me to write this textbook with the present relatively mature version. In addition, I have to give my deep appreciation to my wife, Ms. Yu Huang, for her lasting encouragement and accompanying in my career development. Finally, I sincerely appreciate the Springer Nature to give me a chance for publishing this textbook and Ms. Ella Nan Zhang for selecting the topic of this textbook and her help during the publication program.

Shanghai, China
November 2023

Dr. Prof. Jian Xu
xujian@tongji.edu.cn

Acknowledgments

I am deeply indebted to the lasting supports of the Natural Science Foundation of China with the general programs, key programs, and a grant of the National Outstanding Young Fund (Nos. 10472083, 10532050, 10625211, 11032009, 11272236, 11572224, 11772229, 11932015, 12372022), the Shanghai Outstanding Academic Leaders Plan (08XD1404400), the Shanghai Special Fund for Leading Talents Team Construction, and the funds of Tongji University to the relevant researches of this textbook during past 20 years. I am most grateful to my former students, Associate Prof. Zigen Song (Tongji University), Prof. Shu Zhang (Tongji University), Associate Xiao Xu Zhang (Fudan University), Associate Prof. Yao Yan (University of Electronic Science and Technology of China), and Prof. Xiuting Sun (Tongji University), as well as to the graduate students in my research group for their help in revising this manuscript. Without their dedication, the organization of the content of this book and its production would have taken much longer.

Contents

About the Author

Dr. Prof. Jian Xu received the Ph.D. degree in dynamics and control from Tianjin University, Tianjin, China, in 1994. He is currently a distinguished Professor and Director of the Institute on Dynamics and Control, Tongji University, Shanghai, China. He was Postdoctoral Researcher with the Beihang University, Beijing, China, from 1994 to 1996, and with the Huazhong University of Science and Technology, Wuhan, China, from 1996 to 1998. He joined Tongji University, in 1998, as Associate Professor, and has been Professor since 2000. He now serves as the director of Institute on Dynamics and Control, Tongji University. He is Member of the Expert Advisory Group for the NSFC Major Research Program "Fundamental Theories and Key Technologies of Coexisting-Cooperative-Cognitive Robots", Executive Council Member of the Chinese Society of Theoretical and Applied Mechanics, chief supervisor of the Shanghai Society of Theoretical and Applied Mechanics, Council Member of the Chinese Society of Vibration Engineering, Associate Editor of the journal *Theoretical and Applied Mechanics Letters*, and Editorial Board Member of six domestic and international journals, including *Cognitive Neurodynamics.*

His research interests are mainly concentrated in the areas of time delay-induced dynamics, locomotion robotics, nonlinear dynamics and control, vibration control, bifurcation and chaos in nature and engineering systems, neural networks with delayed coupling, and biological systems. He has published over 200 peer-reviewed journal papers, two academic books, and one textbook. He received the National Science Fund for

Outstanding Young Scholars (2006) and was awarded the Leading Talent of Shanghai (2008), the Shanghai Outstanding Academic Leader (2007), and the Model Teacher of Shanghai (2009). His research contributions were awarded the Provincial Natural Science Prize (first grade) and the Ministerial Science and Technology Development Prize (second grade). He was invited multiple times to deliver keynote speeches and take on the roles of chair, vice-chair or session chair at domestic and international academic conferences. Three Ph.D. students supervised by him won the award of Outstanding Ph.D. Dissertations of Shanghai in 2009, 2013, and 2015, respectively.

Chapter 1
Introduction

Due to the ubiquity of time delay in real systems, it has been an important topic of concern to scientists in many fields such as mathematics, mechanics, high-precision mechanical manufacturing engineering, biology, etc. Investigations of systems with time delays are essential due to their significant impact on the stability, performance, and dynamical behaviour across various domains. Time delays are inherent in numerous real-world processes ranging from natural sciences to engineering applications, and understanding the effect induced by time delays is crucial for understating underlying mechanisms or designing effective control strategies to ensure the reliability of complicated artificial systems.

1.1 Brief Review of Nonlinear Ordinary Differential Equations

The real world is governed by physical laws which are typically formulated in terms of differential equations, such as Newtonian equations of motion, heat equation, Maxwell's equations, Schrödinger equation, etc. Among the models given in the form of differential equations, we are particularly interested in evolution equations where the unknown functions only depend on time, namely, ordinary differential equations (ODEs). Such equations can be used to model the motion of finitely many mass points or rigid bodies, the interaction of group of neurons, temporal evolution of control systems, dynamics of Internet congestion control algorithm, and dynamic behaviour of car-following. From the perspective of mechanics, ODEs describe the motion of a system with finite degrees of freedom (DOF). Note that partial differential equations are sometimes converted to ODEs for simplicity, thus the method developed for ODEs is not restricted to the study of ODEs.

J. Xu, *Nonlinear Dynamics of Time Delay Systems*,
https://doi.org/10.1007/978-981-99-9907-1_1

Generally, ODEs are classified into linear equations and nonlinear equations. Linear ODEs can be written as [1]

$$\frac{\mathrm{d}\mathbf{x}(t)}{\mathrm{d}t} = \mathbf{A}(t)\mathbf{x}(t), \tag{1.1}$$

where $t \in I$, $I = [a, b] \subset \mathbb{R}$, $\mathbf{x}(t) \in C^1(I, \mathbb{R}^n)$ and $C^1(I, \mathbb{R}^n)$ represent the set of all continuously differentiable functions from I to $\mathbb{R}^n$, $\mathbf{A}(t) \in C(I, \mathbb{R}^{n\times n})$. It is clear that once $\mathbf{x}_1(t)$ and $\mathbf{x}_2(t)$ solve Eq. (1.1), then $\alpha\mathbf{x}_1(t) + \beta\mathbf{x}_2(t)$ also solves Eq. (1.1) for α, $\beta \in \mathbb{R}$.

Nonlinear differential equations, however, do not satisfy the above principle of linear superposition, which induces difficulties in formulating general solutions of such equations. For example, based on Kepler's work, Isaac Newton formulated the following differential equations that govern the motion of a planet orbiting a star [2]:

$$\begin{cases} \dfrac{d^2r}{dt^2} - r\left(\dfrac{d\theta}{dt}\right)^2 = -\dfrac{GM}{r^2} \\ \dfrac{1}{r}\dfrac{d}{dt}\left(r^2\dfrac{d\theta}{dt}\right) = 0 \end{cases}, \tag{1.2}$$

where $r(t)$ represents the position of the planet in its orbital plane, measured from the star, $\theta(t)$ is the polar angle, M is the mass of the star, G is the gravitational constant. Clearly, the equations are nonlinear due to the existence of terms $r\dot{\theta}^2$, $r^2\dot{\theta}$, and $-GM/r^2$, and consequently difficult to solve in an analytical manner. Compared with linear systems, nonlinear systems exhibit qualitatively different phenomena such as multi-stability, bifurcations, chaos, and super/subharmonics, which impose challenges to analysis and response prediction of real systems [3–5].

1.2 Brief Review of Differential Equations with Time Delay

Due to the ubiquity of time delay in real world systems [6], mathematical models expressed in terms of differential equations with time delay have become an important topic of concern to scientists and engineers in many fields such as mathematics [7–10], control technology [11–15], structural/mechanical engineering [16–20], biological science [21–25], and telecommunications [26–30].

Time delay, according to its origination, can be categorized as (i) the feedback delay and (ii) the coupling delay, both of which are direct consequences of the finite speed of signal transmission and processing. Feedback delay in control refers to the time needed for the entire loop of feedback control to actuate. It includes the time taken for the control system to process the information, compute the necessary adjustments, and implement those adjustments in the controlled process. Delays can occur at various stages of the feedback loop, including sensor measurement,

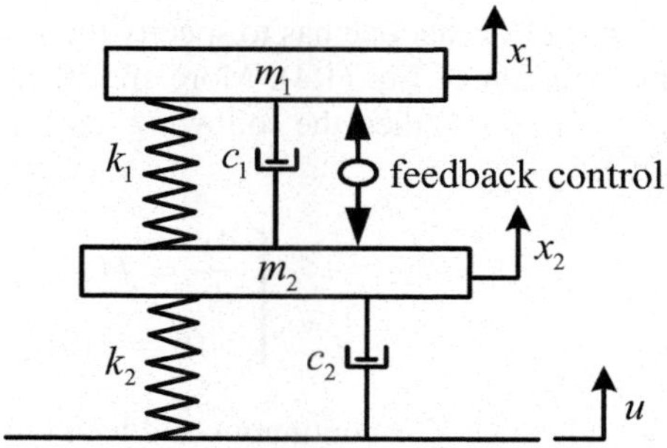

Fig. 1.1 Mechanical model of the 2-DOF vibration system where an actuator is placed between m_1 and m_2

signal processing, controller computation, and actuator response. Coupling delay typically refers to the time lag that arises when two or more interconnected or coupled systems communicate or interact with each other. This type of delay can occur in various fields and applications, including nervous systems, control systems, signal processing, information technology, and telecommunications. Figure 1.1 illustrates a vibration absorber with feedback control whose mathematical model is given by

$$\begin{cases} m_1\ddot{x}_1 + k_1(x_1 - x_2) + c_1(\dot{x}_1 - \dot{x}_2) - kx_1(t-\tau) = 0, \\ m_2\ddot{x}_2 + k_2x_2 + c_2\dot{x}_2 + k_1(x_2 - x_1) + c_1(\dot{x}_2 - \dot{x}_1) + kx_1(t-\tau) = f\sin(\omega t), \end{cases}$$

where x_1 and x_2 represent the displacements of the absorber and the primary mass, respectively, m_1 is the mass of the absorber, m_2 is the primary mass, c_1 and c_2 are the damping coefficients of m_1 and m_2, respectively, k_1 and k_2 are the stiffness's of the springs, k represents the control gain, $f\sin(\omega t)$ is the external excitation. Here τ represents the time delay, and is considered as the feedback delay from the perspective of m_1, and coupling delay from the view of m_2 [31].

For most real applications with discrete constant delays, the mathematical model appears in the form similar like a group of ordinary differential equations:

$$\frac{\mathrm{d}\mathbf{x}}{\mathrm{d}t} = \mathbf{F}(\mathbf{x}, \mathbf{x}_\tau, \boldsymbol{\mu}), \tag{1.3}$$

where $\mathbf{x}$ represents the vector of the quantity under consideration, $\mathbf{F}$ is the function (probably nonlinear) that models the law governing the evolution of $\mathbf{x}$, $\mathbf{x}_\tau = \mathbf{x}(t-\tau)$ is the delay term of $\mathbf{x}$ and τ the time delay, and $\boldsymbol{\mu}$ is the vector of parameters. Though Eq. (1.1) looks like ordinary differential equations, its characteristics are more like partial differential equations. More specifically, Eq. (1.3) is classified as functional differential equations [32] as will be explained in the following. Let us consider the scalar case for simplicity, then Eq. (1.3) can be written as

$$\frac{\mathrm{d}x}{\mathrm{d}t} = f(x, x_\tau, \boldsymbol{\mu}).$$

It is clear that one has to specify the values of x on the interval $[-\tau, 0]$ to obtain the solution of Eq. (1.4). More specifically, if we are given the Cauchy data φ : $[-\tau, 0] \to \mathbb{R}$, then the following equation can be solved in the same way as an ODE:

$$\begin{cases} \dfrac{\mathrm{d}x}{\mathrm{d}t} = f(x, x_\tau, \mu), \ t \in [0, \tau], \\ x(0) = \varphi(0), \end{cases} \tag{1.5}$$

provided that f is continuous and locally Lipschitz with respect to its first variable. Based on the solution on the interval $[-\tau, 0]$, the following equation can also be solved

$$\frac{\mathrm{d}x}{\mathrm{d}t} = f(x, x_\tau, \mu), \ t \in [\tau, 2\tau], \tag{1.6}$$

for which the initial condition $x(\tau)$ and right-hand term x_τ are obtained by evaluating the solution of Eq. (1.5) on $[-\tau, 0]$. Continuing such process will provide us with a solution of Eq. (1.4).

The above simple example clearly shows the difference between delayed differential equations and ordinary differential equations: the Cauchy data of an ODE is a vector in $\mathbb{R}^n$ for some n while that of a delayed differential equation is a function. Such difference can be mathematically clarified as follows. Let C_τ denote the Banach space of continuous functions which map from $[-\tau, 0]$ to $\mathbb{R}$, namely, $C_\tau = C([-\tau, 0], \mathbb{R})$ and is equipped with the supremum norm $\|\varphi\| = \sup\limits_{-\tau \le \theta \le 0} |\varphi(\theta)|$. Based on these notions, a functional differential equation or an infinite-dimensional dynamical system related with Eq. (1.4) can be defined by introducing $x_t(\theta) = x(t + \theta)$ for $\theta \in [-\tau, 0]$ [33].

We use a specific example to address the difference between differential equations with time delay and ODEs. Assume that in Eq. (1.4), $f(x, x_\tau, \mu) = kx_\tau = kx(t-\tau)$, where k is a positive gain parameter. Namely, the equation of concern is given by

$$\frac{\mathrm{d}x}{\mathrm{d}t} = kx(t - \tau). \tag{1.7}$$

In order to solve Eq. (1.7), we substitute $x(t) = Ce^{\lambda t}$ into Eq. (1.7) and obtain that

$$C\lambda e^{\lambda t} = Cke^{\lambda t - \lambda\tau} \Rightarrow \lambda = e^{-\lambda\tau}. \tag{1.8}$$

Let us consider the possibility of emergence of periodic solutions in Eq. (1.7). To this end, we substitute $\lambda = \omega \mathrm{i}$ into (1.8) and separate the real and imaginary parts to yield

$$\begin{cases} 0 = k\cos(\omega\tau), \\ \omega = -k\sin(\omega\tau), \end{cases}$$

and thus

$$\omega = k,\ \tau = \frac{1}{k}\left(\frac{\pi}{2} + (2m-1)\pi\right), \tag{1.9}$$

where $m \in \mathbb{N}$. It is then clear that as long as the time delay assumes the value in Eqs. (1.7), (1.9) admits periodic solutions. This reveals again that differential equations with time delay differs greatly from ODEs, which do not possess periodic solutions for the scalar case.

1.3 Main Challenge of Study on Nonlinear Dynamics of Systems with Time Delay

Traditional views hold that the time delay can destabilize the control system which is stable for small values of delay and unstable as time delay increases over some threshold for stability switch. It was also widely accepted that time delay can degrade the performance of control systems in resulting in slow responses, low accuracy in tracking reference signals, and even oscillatory response with large amplitude. Consequently, control strategies must account for time delay to maintain the control performance especially for the systems which are sensitive to variations of time delay.

However, researches in recent decades have shown that the influence of time delay is a two-fold issue. In spite of the aforementioned negative effect on the system performance, time delay can stabilize the control system for some cases or serve as a trigger for the emergence of the desired dynamic response. This calls for the elaborate investigation on the system responses in the presence of time delay, particularly on the response prediction when the system loses the stability due to the existence of time delay. As a result, one may have to take into account the nonlinearity in addition to the time delay because the linearized system may not capture the characteristics of system dynamics when the response moves far away from the unstable equilibrium. In other words, it is necessary to consider the joint effect of time delay and nonlinearity on the dynamics of the control system for a more promising design of control.

In this subsection, we will address the main challenges of investigation from the following aspects.

1.3.1 Nonlinear Dynamics of Time Delay Systems

Since the 1970s, with the development of computer technology and numerical methods, researchers have gradually recognized that parameter perturbations can lead to complex dynamic behaviours in nonlinear dynamic systems. Based on the typical nonlinear examples commonly found in mechanics and physics, numerous new nonlinear phenomena, including chaos and fractals, have been discovered. The corresponding fundamental theories and methods have been proposed and established.

The main questions relating to nonlinear dynamics of differential equations with time delay are listed as follows [34]. Firstly, it has been clear that the time delay can trigger the stability switch when focusing on the equilibrium. Thus it is natural to ask what effect time delay can have on oscillatory solutions, especially synchronously oscillatory solutions which characterize nonlinear systems with high dimension. Second, from the perspective of double Hopf bifurcation theory, the resonance with low order, namely the strongly resonant case, is of great interest since such case does not have the normal form as simple as the weakly resonant case or the non-resonant case. Considering the existence of time delay, the computation of the normal form of double Hopf bifurcation with strong resonance will be more difficult than the case without time delay and worth effort to study. Thirdly, as the variation of time delay can lead to switches between simple solutions, such as the equilibrium solution and periodic solutions of various types, one may ask if it is possible that change in time delay induces more complicated solutions of nonlinear systems such as quasi-periodic motion and chaotic motion.

1.3.2 Methods of Solving Nonlinear Systems with Time Delay

Asymptotic method is typically used to obtain the analytic form of the solution of a nonlinear dynamical system, which is of great importance in understanding the mechanism of certain nonlinear dynamics. For nonlinear dynamical systems with time delay, commonly used asymptotic methods include the center manifold reduction with normal form computation, the method of multiple scales, the Lindstedt–Poincaré method, the method of harmonic balance, etc., among which the center manifold reduction and the method of multiple scales are of major concern as they represent the prevailing methodologies when solving nonlinear systems with time delay. This subsection is devoted to providing examples to demonstrate the basic steps of applying such methods to the study of the Hopf bifurcation for nonlinear systems with time delay, showing the complexity of these methods.

1.3.2.1 Center Manifold Reduction

Let's consider the following delayed differential equations which describe a class of nervous system and whose nonlinear dynamics will be discussed in detail in Chap. 7:

$$\begin{cases} \dfrac{dx(t)}{dt} = -x(t) + a_1 S(x(t)) + a_2 S(y(t-\tau)) + P, \\ \dfrac{dy(t)}{dt} = -y(t) + a_3 S(x(t-\tau)) + a_4 S(y(t)) + Q, \end{cases} \tag{1.10}$$

where τ is the delay of synaptic junction, $x(t)$ and $y(t)$ represent the activity potential of the neuron at time t, and a_i, $(i = 1, \ldots, 4)$ represents the strength of the synaptic junction. When $a_i < 0$, we say that the connection is inhibitory, otherwise we call it excitatory. P and Q indicate the external stimulation received by the sensor organs from the environment. S(u) is the neuronal activation function. In this example, we assume $S(u) = 1/\left(1 + e^{-u}\right)$.

By the transformation $x \to x - x_0$ and $y \to y - y_0$, we can get the linearized system corresponding to Eq. (1.10)

$$\begin{cases} \dfrac{dx(t)}{dt} = \left(-1 + \dfrac{a_1 e^{-x_0}}{(1+e^{-x_0})^2}\right) x(t) + \dfrac{a_2 e^{-y_0}}{(1+e^{-y_0})^2} y(t-\tau), \\ \dfrac{dy(t)}{dt} = \dfrac{a_3 e^{-x_0}}{(1+e^{-x_0})^2} x(t-\tau) + \left(-1 + \dfrac{a_4 e^{-y_0}}{(1+e^{-y_0})^2}\right) y(t). \end{cases} \tag{1.11}$$

When the parameter P crosses the critical value, Eq. (1.10) will experience a Hopf bifurcation at the equilibrium point (x_0, y_0). In the following discussion, the center manifold reduction and normal form computation [35] are applied to study the direction of the Hopf bifurcation of Eq. (1.10) and the stability of the bifurcated periodic solution. To this end, we apply the following coordinate transformation and unfolding of parameters, that is

$$P = P_0 + \mu, \quad v_1(t) = x(t) - x_0, \quad v_2(t) = y(t) - y_0. \tag{1.12}$$

Then Eq. (1.10) can be rewritten as follows

$$\begin{cases} \dot{v}_1(t) = a_{10} v_1(t) + a_{01} v_2(t-\tau_0) + a_{20} v_1^2(t) \\ \quad + a_{02} v_2^2(t-\tau_0) + a_{30} v_1^3(t) + a_{03} v_2^3(t-\tau_0) + \cdots, \\ \dot{v}_2(t) = b_{10} v_1(t-\tau_0) + b_{01} v_2(t) + b_{20} v_1^2(t-\tau_0) \\ \quad + b_{02} v_2^2(t) + b_{30} v_1^3(t-\tau_0) + b_{03} v_2^3(t) + \cdots, \end{cases} \tag{1.13}$$

where

$$a_{10} = -1 + \frac{7e^{x_0}}{(1+e^{x_0})^2}, \quad a_{01} = \frac{-7e^{y_0}}{(1+e^{y_0})^2},$$

$$b_{10} = \frac{7e^{x_0}}{(1+e^{x_0})^2},\ b_{01} = -1 + \frac{2e^{y_0}}{(1+e^{y_0})^2},$$
$$a_{20} = \frac{7e^{x_0}(1-e^{x_0})}{2(1+e^{x_0})^3},\ a_{02} = \frac{-7e^{y_0}(1-e^{y_0})}{2(1+e^{y_0})^3},$$
$$b_{20} = \frac{7e^{x_0}(1-e^{x_0})}{2(1+e^{x_0})^3},\ b_{02} = \frac{e^{y_0}(1-e^{y_0})}{(1+e^{y_0})^3},$$
$$a_{30} = \frac{7e^{x_0}(1-4e^{x_0}+e^{2x_0})}{6(1+e^{x_0})^4},$$
$$a_{03} = \frac{-7e^{y_0}(1-4e^{y_0}+e^{2y_0})}{6(1+e^{y_0})^4},$$
$$b_{30} = \frac{7e^{x_0}(1-4e^{x_0}+e^{2x_0})}{6(1+e^{x_0})^4},$$
$$b_{03} = \frac{e^{y_0}(1-4e^{y_0}+e^{2y_0})}{3(1+e^{y_0})^4}.$$

To apply the center manifold reduction, we first need to rewrite Eq. (1.13) as a functional differential equation. For any $\varphi(\theta) = (\varphi_1(\theta), \varphi_2(\theta)) \in C$, we define an operator as follows:

$$L(\mu)\varphi = \begin{pmatrix} a_{10} & 0 \\ 0 & b_{01} \end{pmatrix} \begin{pmatrix} \varphi_1(0) \\ \varphi_2(0) \end{pmatrix} + \begin{pmatrix} 0 & a_{01} \\ b_{10} & 0 \end{pmatrix} \begin{pmatrix} \varphi_1(-\tau_0) \\ \varphi_2(-\tau_0) \end{pmatrix}, \tag{1.14}$$

where $C = C([-\tau_0, 0], \mathbb{R}^2)$. According to Riesz representation theorem, there must be a matrix function with bounded elements $\eta(\theta, \mu)$, $(\theta \in [-\tau_0, 0])$, satisfying

$$L(\mu)\varphi = \int_{-\tau_0}^{0} d\eta(\theta, \mu)\varphi(\theta), \varphi \in C[-\tau_0, 0], \tag{1.15}$$

where

$$\eta(\theta, \mu) = \begin{pmatrix} a_{10} & 0 \\ 0 & b_{01} \end{pmatrix} \delta(\theta) + \begin{pmatrix} 0 & a_{01} \\ b_{10} & 0 \end{pmatrix} \delta(\theta + \tau_0), \tag{1.16}$$

and $\delta(\theta)$ is a Dirac delta function. It is easy to see that Eq. (1.14) can be satisfied. Now we define

$$A(\mu)\varphi = \begin{cases} \dfrac{d\varphi(\theta)}{d\theta}, & \theta \in [-\tau_0, 0), \\ \displaystyle\int_{-\tau_0}^{0} d\eta(\xi, \mu)\varphi(\xi), & \theta = 0, \end{cases}$$

and

$$R(\mu)\varphi = \begin{cases} 0 & \theta \in [-\tau_0, 0), \\ F(\mu, \varphi) & \theta = 0, \end{cases}$$

where

$$F(\mu, \varphi) = \begin{pmatrix} a_{20}\varphi^2(0) + a_{02}\varphi^2(-\tau_0) \\ b_{02}\varphi^2(0) + b_{20}\varphi^2(-\tau_0) \end{pmatrix} + \begin{pmatrix} a_{30}\varphi^3(0) + a_{03}\varphi^3(-\tau_0) \\ b_{03}\varphi^3(0) + b_{30}\varphi^3(-\tau_0) \end{pmatrix} + \cdots.$$

Then Eq. (1.13) is equivalent to the following functional differential equation

$$\dot{u}_t = A(\mu)\, u_t + R(\mu)\, u_t, \tag{1.17}$$

where $u(t) = (v_1(t), v_2(t))^{\mathrm{T}}$, $u_t = u(t+\theta)$ for any $\theta \in [-\tau_0, 0]$.

We define A* , the conjugate operator of A, as

$$A^*(\mu)\psi = \begin{cases} -\dfrac{d\psi(s)}{ds}, & s \in [-\tau_0, 0), \\ \displaystyle\int_{-\tau_0}^{0} d\eta^{\mathrm{T}}(s, \mu)\psi(-s), & s = 0, \end{cases} \tag{1.18}$$

where η^T is the transpose matrix of η. For any $\varphi \in C\big([-\tau_0, 0], \mathbb{R}^2\big)$ and $\psi \in C\big([0, \tau_0], \mathbb{R}^2\big)$, we define the following bilinear function

$$\langle \psi(s), \varphi(\theta) \rangle = \overline{\psi}(0)\varphi(0) - \int_{-\tau_0}^{0}\int_{\xi=0}^{\theta} \overline{\psi}(\xi - \theta)d\eta(\theta)\varphi(\xi)d\xi, \tag{1.19}$$

where $\eta(\theta) = \eta(\theta, 0)$. Using Eq. (1.19), we can prove

$$\langle \psi,\ A\varphi \rangle = \langle A^*\psi, \varphi \rangle. \tag{1.20}$$

Next we calculate the eigenvector of A, i.e., q which corresponds to the eigenvalue $i\omega_0$, and the eigenvector of A^*, i.e., q^* which corresponds to the eigenvalue $-i\omega_0$. Through a direct calculation, we have

$$q(\theta) = \begin{pmatrix} 1 & \alpha \end{pmatrix}^{\mathrm{T}} e^{i\omega_0\theta},$$

and

$$q^*(s) = \overline{K}\begin{pmatrix} 1 & \beta \end{pmatrix} e^{i\omega_0 s},$$

where

$$\alpha = \frac{b_{10}e^{-i\omega_0\tau_0}}{-b_{01} + i\omega_0}, \quad \beta = \frac{a_{01}e^{i\omega_0\tau_0}}{-b_{01} - i\omega_0},$$

$$\frac{1}{K} = 1 - \frac{a_{01}b_{10}(2i\omega_0\tau_0 - 2\tau_0 b_{01} + 1)}{e^{2i\omega_0\tau_0}(ib_{01} + \omega_0)^2}.$$

Furthermore, we can verify that they satisfy $\langle q^*, q\rangle = 1$ as well $\langle q^*, \overline{q}\rangle = 0$.

The bifurcated periodic solution $z(t, \mu(\varepsilon))$ has non-zero Floquet multiplier $\beta(\varepsilon)$ which satisfies $\beta(0) = 0$, where $\mu(\varepsilon)$ and $\beta(\varepsilon)$ can be expanded into the following form.

$$\mu(\varepsilon) = \mu_2\varepsilon^2 + \mu_4\varepsilon^4 + \cdots,$$
$$\beta(\varepsilon) = \beta_2\varepsilon^2 + \beta_4\varepsilon^4 + \cdots.$$

Therefore the sign of μ_2 determines the direction of the bifurcation, and the sign of β_2 determines the stability of the periodic solution. Now we derive the expressions of μ_2 and β_2. To this end, we define

$$z(t) = \langle q^*, u_t\rangle, \tag{1.21}$$

and

$$W(t, \theta) = u_t(\theta) - 2\text{Re}[z(t)q(\theta)]. \tag{1.22}$$

On the center manifold C_0, we have

$$W(t, \theta) = W(z(t), \overline{z}(t), \theta), \tag{1.23}$$

where

$$W(z, \overline{z}, \theta) = W_{20}(\theta)\frac{z^2}{2} + W_{11}(\theta)z\overline{z} + W_{02}(\theta)\frac{\overline{z}^2}{2} + W_{30}(\theta)\frac{z^3}{6} + \cdots, \tag{1.24}$$

where z and $\overline{z}$ are local coordinates on the center manifold in terms of q^* and $\overline{q}^*$. Note that if u_t is real, so is W. In this example, only real solutions are considered.

From Eqs. (1.21) and (1.22), we get

$$\begin{aligned}\langle q^*, W\rangle &= \langle q^*, u_t(\theta) - 2\text{Re}[z(t)q(\theta)]\rangle \\ &= \langle q^*, u_t(\theta) - z(t)q(\theta) - \overline{z}(t)\overline{q}(\theta)\rangle \\ &= \langle q^*, u_t(\theta)\rangle - z(t)\langle q^*, q\rangle - \overline{z}(t)\langle q^*, \overline{q}\rangle \\ &= 0. \end{aligned} \tag{1.25}$$

For the solution of Eq. (1.25) $u_t \in C_0$, through Eqs. (1.19) and (1.20), we have

$$\begin{aligned} \dot{z}(t) &= \langle q^*, \dot{u}_t \rangle = \langle q^*, Au_t + Ru_t \rangle \\ &= \langle q^*, Au_t \rangle + \langle q^*, Ru_t \rangle = \langle A^* q^*, u_t \rangle + \langle q^*, Ru_t \rangle \\ &= i\omega_0 z + \overline{q}^*(0) F(0, W(t, 0) + 2\mathrm{Re}[z(t)q(0)]). \end{aligned} \tag{1.26}$$

That is,

$$\dot{z}(t) = i\omega_0 z(t) + g(z, \overline{z}), \tag{1.27}$$

where

$$g(z, \overline{z}) = g_{20}\frac{z^2}{2} + g_{11} z\overline{z} + g_{02}\frac{\overline{z}^2}{2} + g_{21}\frac{z^2\overline{z}}{6} + \cdots. \tag{1.28}$$

Furthermore, we have

$$\begin{aligned} g(z, \overline{z}) &= \overline{q}^*(0) \cdot F(z, \overline{z}) \\ &= K\begin{pmatrix} 1 & \overline{\beta} \end{pmatrix} \cdot \begin{pmatrix} a_{20}u_{1t}^2(0) + a_{02}u_{2t}^2(-\tau_0) + \\ a_{30}u_{1t}^3(0) + a_{03}u_{2t}^3(-\tau_0) + \cdots \\ b_{20}u_{1t}^2(-\tau_0) + b_{02}u_{2t}^2(0) + \\ b_{30}u_{1t}^3(-\tau_0) + b_{03}u_{2t}^3(0) + \cdots \end{pmatrix}, \end{aligned} \tag{1.29}$$

where $u_t(\theta) = (u_{1t}(\theta), u_{2t}(\theta))^T = W(t, \theta) + zq(\theta) + \overline{z}\overline{q}(\theta)$. Thus we have

$$\begin{aligned} u_{1t}(0) &= z + \overline{z} + W_{20}^{(1)}(0)\frac{z^2}{2} + W_{11}^{(1)}(0) z\overline{z} + W_{02}^{(1)}(0)\frac{\overline{z}^2}{2} + \cdots, \\ u_{2t}(0) &= \alpha z + \overline{\alpha}\overline{z} + W_{20}^{(2)}(0)\frac{z^2}{2} + W_{11}^{(2)}(0) z\overline{z} + W_{02}^{(2)}(0)\frac{\overline{z}^2}{2} + \cdots. \end{aligned} \tag{1.30}$$

Substituting Eq. (1.30) into Eq. (1.29) and comparing the coefficients with Eq. (1.28), we have:

$$\begin{aligned} g_{20} &= 2K\begin{bmatrix} (\alpha^2 a_{02} + \overline{\beta} b_{20}) e^{-2i\omega_0\tau_0} + \\ (\alpha a_{11} + \alpha\overline{\beta} b_{11}) e^{-i\omega_0\tau_0} + \alpha^2\overline{\beta} b_{02} + a_{20} \end{bmatrix}, \\ g_{11} &= K\begin{bmatrix} \alpha(a_{11} + \overline{\alpha}\overline{\beta} b_{11}) e^{-i\omega_0\tau_0} + (\overline{\alpha} a_{11} + \alpha\overline{\beta} b_{11}) e^{i\omega_0\tau_0} + \\ 2(a_{20} + \alpha\overline{\alpha}(a_{02} + \overline{\beta} b_{02}) + \overline{\beta} b_{20}) \end{bmatrix}, \\ g_{02} &= 2K\begin{bmatrix} (\overline{\alpha}^2 a_{02} + \overline{\beta} b_{20}) e^{2i\omega_0\tau_0} + \overline{\alpha}(a_{11} + \overline{\beta} b_{11}) e^{i\omega_0\tau_0} + \\ \overline{\alpha}^2\overline{\beta} b_{02} + a_{20} \end{bmatrix}, \\ g_{21} &= K\left[2\mathrm{M}_1 e^{-2i\omega_0\tau_0} + 2\mathrm{M}_2 e^{-i\omega_0\tau_0} + \mathrm{M}_3 + \mathrm{M}_4 e^{i\omega_0\tau_0}\right], \end{aligned} \tag{1.31}$$

where

$$
\begin{aligned}
\mathrm{M}_1 &= \alpha^2 a_{12} + \overline{\alpha}\overline{\beta} b_{21},\\
\mathrm{M}_2 &= \alpha\overline{\alpha}\left(3\alpha a_{03} + 2\overline{\beta} b_{12}\right) + \overline{\beta}\left(3b_{30} + 2b_{20} W_{11}^{(1)}(-\tau_0) + b_{11} W_{11}^{(2)}(0)\right)\\
&\quad + \alpha\left(2a_{21} + a_{11} W_{11}^{(1)}(0) + 2a_{02} W_{11}^{(2)}(-\tau_0)\right),\\
\mathrm{M}_3 &= 6a_{30} + 2a_{20}\left(2W_{11}^{(1)}(0) + W_{20}^{(1)}(0)\right)\\
&\quad + 2\alpha\overline{\beta}\left(2b_{21} + b_{11} W_{11}^{(1)}(-\tau_0) + 2b_{02} W_{11}^{(2)}(0)\right)\\
&\quad + \overline{\alpha}\left(4\alpha a_{12} + \overline{\beta}\left(6\alpha^2 b_{03} + b_{11} W_{20}^{(1)}(-\tau_0) + 2b_{02} W_{20}^{(2)}(0)\right)\right)\\
&\quad + a_{11}\left(2W_{11}^{(2)}(-\tau_0) + W_{20}^{(2)}(-\tau_0)\right),\\
\mathrm{M}_4 &= \overline{\beta}\left(2\alpha^2 b_{12} + 2b_{20} W_{20}^{(1)}(-\tau_0) + b_{11} W_{20}^{(2)}(0)\right)\\
&\quad + \overline{\alpha}\left(2a_{21} + a_{11} W_{20}^{(1)}(0) + 2a_{02} W_{20}^{(2)}(-\tau_0)\right).
\end{aligned}
$$

Then we can give the expression of μ_2 and β_2 as follows

$$
\begin{aligned}
&\dot{z}(t) = cz(t) + dz(t-\tau_c) + f(z(t), z(t-\tau_c-\varepsilon\tau_\varepsilon), z(t-\tau_c), \dot{z}(t), \varepsilon),\\
&\mu_2 = -\frac{\mathrm{Re}[C_1(0)]}{\mathrm{Re}[d\lambda/dP]},\ \beta_2 = 2\mathrm{Re}[C_1(0)],
\end{aligned}
\tag{1.32}
$$

where $C_1(0) = \frac{i}{2\omega}\left(g_{20}g_{11} - 2|g_{11}|^2 - \frac{1}{3}|g_{02}|^2\right) + \frac{g_{21}}{2}$.

According to the Hopf bifurcation theorem, we know that if $\mu_2 > 0(\mu_2 < 0)$, then the Hopf bifurcation is supercritical (subcritical), and if $\beta_2 < 0(\beta_2 > 0)$, then the bifurcated periodic solution on the center manifold is stable (unstable), correspondingly. When the parameters satisfy $P > P_0$ ($P < P_0$), the bifurcated periodic solution exists. It can be seen that μ_2 and β_2 are related to g_{20}, g_{11}, g_{02} and g_{21}. From (1.32), we know that g_{21} is determined by $W_{20}(\theta)$ and $W_{11}(\theta)$ at $\theta = 0$ and $\theta = -\tau_0$.

1.3.2.2 Method of Multiple Scales

Another method of theoretically solving delayed differential equations is the method of multiple scales. Assume that the equation under investigation is given by

$$
\dot{z}(t) = cz(t) + dz(t-\tau) + \tilde{f}(z(t), \dot{z}(t)), \tag{1.33}
$$

where $f(.)$ represents a nonlinear function containing quadratic terms. When we study the Hopf bifurcation, we need to perturb the bifurcation parameter in Eq. (1.33) near its critical value, that is, let $\tau = \tau_c + \varepsilon\,\tau_\varepsilon$, then we can get

$$\dot{z}(t) = cz(t) + dz(t - \tau_c) + f(z(t), \dot{z}(t), \varepsilon\tau_\epsilon),$$

which ε represents a small positive parameter. Then according to the method of multiple scales [36, 37], the solution of is expanded into a power series of ε, that is

$$\begin{aligned} z(t) &= Z(T_0,\ T_1,\ T_2,\ \cdots) \\ &= \varepsilon\, Z_1(T_0,\ T_1,\ T_2,\ \cdots) + \varepsilon^2 Z_2(T_0,\ T_1,\ T_2,\ \cdots) + \cdots, \end{aligned} \tag{1.34}$$

where $T_i = \varepsilon^i\,t, \quad i = 0,\ 1,\ 2,\ \cdots$. Substituting (1.34) into (1.33), expanding $Z_i(t - \tau_c - \varepsilon\,\tau_\varepsilon, \varepsilon\,(t - \tau_c - \varepsilon\,\tau_\varepsilon), \cdots)$ into a power series of ε at $(T_0 - \tau_c, T_1, \cdots)$, and considering the equation at the lowest order of ε, we have

$$\mathrm{D}_0\, Z_1(T_0,\ T_1,\ T_2,\ \cdots) + \omega\, Z_1(T_0 - \tau_c,\ T_1,\ T_2,\ \cdots) = 0, \tag{1.35}$$

where $\mathrm{D}_0 = \frac{\partial}{\partial T_0}$ and ω represent the frequency of the periodic solution of the system for $\tau = \tau_c$. Assume that the solution of Eq. (1.35) has the following form

$$Z_1(T_0,\ T_1,\ T_2,\ \cdots) = A_{11}\, sin(\omega\, T_0) + B_{11}\, cos(\omega\, T_0), \tag{1.36}$$

where $A_{11} = A_{11}(T_1,\ T_2,\ \cdots)$, $B_{11} = B_{11}(T_1,\ T_2,\ \cdots)$. For simplicity, we will refer $A_{11}(T_1,\ T_2,\ \cdots)$ to A_{11} and $B_{11}(T_1,\ T_2,\ \cdots)$ to B_{11}, respectively. Substitute Eq. (1.36) into Eq. (1.34), then at the second order of ε, we can get the following equation

$$\begin{aligned} &\mathrm{D}_0\, Z_2(T_0,\ T_1,\ T_2,\ \cdots) + \omega\, Z_2(T_0 - \tau_c,\ T_1,\ T_2,\ \cdots) + F_2 \\ &\quad + P_{21}\, sin(\omega\, T_0) + Q_{21}\, cos(\omega\, T_0) + P_{22}\, sin(2\omega\, T_0) + Q_{22}\, cos(2\omega\, T_0) = 0, \end{aligned} \tag{1.37}$$

where $P_{2i} = P_{2i}(D_1 A_{11}, D_1 B_{11}, A_{11}, B_{11})$, $Q_{2i} = Q_{2i}(D_1 A_{11}, D_1 B_{11}, A_{11}, B_{11})$, $\mathrm{D}_1 = \frac{\partial}{\partial T_1}$, $i = 1, 2$ and $F_2 = F_2(A_{11}, B_{11})$. Note that the appearance of F_2 is due to the presence of the squared nonlinear terms in $f(.)$. In order to avoid secular terms in the solution of Eq. (1.33), the first-order harmonic term should not exist. According to this, we can obtain a set of equations about $\mathrm{D}_1 A_{11}$ and $\mathrm{D}_1 B_{11}$. Solving these equations we have

$$\mathrm{D}_1 A_{11} = M_1(A_{11},\ B_{11}), \mathrm{D}_1 B_{11} = N_1(A_{11},\ B_{11}).$$

Therefore, the solution of Eq. (1.37) can be written as

$$Z_2(T_0,\ T_1,\ T_2,\ \ldots) = C_2 + A_{22}\, sin(2\omega\, T_0) + B_{22}\, cos(2\omega\, T_0). \tag{1.38}$$

Substituting (1.38) into (1.37), we get

$$A_{22} = A_{22}(A_{11}, B_{11}), B_{22} = B_{22}(A_{11}, B_{11}), C_2 = C_2(A_{11}, B_{11}). \quad (1.39)$$

Repeating the above procedure for equations of higher orders of ε, we can get $\mathrm{D}_i A_{11}$ and $\mathrm{D}_i B_{11}$ expressed as functions of A_{11} and B_{11}, where $\mathrm{D}_i = \frac{\partial}{\partial T_i}$, $i = 1, 2, \cdots$. Finally we have

$$\frac{\mathrm{d}A_{11}}{\mathrm{d}t} = \varepsilon \mathrm{D}_1 A_{11} + \varepsilon^2 \mathrm{D}_2 A_{11} + \cdots, \quad \frac{\mathrm{d}B_{11}}{\mathrm{d}t} = \varepsilon \mathrm{D}_1 B_{11} + \varepsilon^2 \mathrm{D}_2 B_{11} + \cdots. \quad (1.40)$$

Using the following transformation

$$A_{11} = R(t) cos(\phi(t)), B_{11} = R(t) sin(\phi(t)), \quad (1.41)$$

then Eq. (1.40) in polar coordinate can be expressed as

$$\begin{aligned} \dot{R}(t) &= r_1(\varepsilon, \tau_\varepsilon) R(t) + r_3(\varepsilon, \tau_\varepsilon) R(t)^3 + r_5(\varepsilon, \tau_\varepsilon) R(t)^5 + \cdots, \\ \dot{\phi}(t) &= f_0(\varepsilon, \tau_\varepsilon) + f_2(\varepsilon, \tau_\varepsilon) R(t)^2 + f_4(\varepsilon, \tau_\varepsilon) R(t)^4 + \cdots. \end{aligned} \quad (1.42)$$

The differential equation governing the evolution of the amplitude of the oscillation is independent of the phase variable, and consequently can be directly solved. Meantime, the stability of the solution can also be determined from Eq. (1.42).

From the above examples, it can be seen that the center manifold reduction and the method of multiple scales, as the most commonly used methods for approximately solving nonlinear systems with time delay, are complicated. Undoubtedly, it is significant, though challenging at the same time, to seek efficient methods for quantitative analysis of nonlinear systems with time delay.

1.3.3 Identification of Nonlinear Systems with Time Delay

In the presence of time delay, it has been accepted that the dynamic system's smartness and stability are sensitive to time delay. Therefore, identifying the time delay is necessary if one wants to match the theoretical prediction with real dynamic responses accurately. Generally, there are two factors challenging the identification of systems with time delay.

First, sensitivity analysis of time delay in the time domain is intricate because the system's response is variational to delay. Although auxiliary functions [38–42] can be constructed to calculate parameter sensitivity about time delay, which is required by the gradient-based identification procedure, the dimension of the state-space equation will be significantly expanded. Moreover, when the initial estimation of time delay is badly set, the expanded system will be unstable so that the identification cannot be continued. On the contrary, the representation of time delay in the frequency domain is a complex exponential function, which is algebraic and periodic. Therefore, there is no need to expand the differential state-space equation, and one can take advantage of the periodicity to qualitatively or quantitatively estimate the time delay.

Second, nonlinearity-noise coupling usually misleads parameter identification. Conventional identification approaches always consider irrelative noise, e.g., Gaussian white noise, in measurements. In accordance, Kalman filters [43, 44] or moving window filters [45, 46] are widely employed to enhance the accuracy of parameter identification. However, when the system has asymmetric nonlinearity, e.g., quadratic restoring force, the system's response center will deviate from its static equilibrium. Such deviation increases with the amplitude of oscillation and is called "streaming" or "drifting" [47]. Because the streaming is a low frequency component, it cannot be accurately measured by dynamic sensors. In other words, such distortion couples with the system's nonlinearity. Apparently, the existing filters cannot distinguish the streaming from low frequency noise.

1.3.4 Utilization of Time Delay in Vibration Suppression

In various fields such as aerospace, automotive engineering, ship engineering, and precision instrument manufacturing, engineers deploy elastic components or control devices to create effective systems for vibration isolation. These systems are placed between the source of vibration and the protected object to dissipate vibrational energy and ultimately suppress the vibration and noise. In engineering applications, there is an increasing demand for vibration isolation structures that exhibit enhanced load-bearing capacity, a broader range of frequency for effective vibration isolation, and smartness in deploying the controller. The contemporary trend toward larger and more complex structures leads to designs with significant flexibility, resulting in a substantial decrease in the fundamental frequency of the structure. In comparison with high-frequency vibrations, low-frequency vibrations often come with large amplitudes. On one hand, factors such as clearances and frictions may induce low-frequency or even ultra-low-frequency vibrations across a wide band of low frequency. On the other hand, structures are frequently exposed to low-frequency excitations due to external time-varying disturbances over a broad bandwidth. As a result, the research that focuses on low-frequency vibration isolation has become a key area of interest in almost every field involving vibration.

When suppressing low-frequency or ultra-low-frequency vibrations, engineers commonly employ isolators to isolate the protected object from the vibrational environment. However, isolators designed based on classical theory of linear isolation are no longer adequate to meet the demands of modern vibration isolation. Firstly, unlike those structures with relatively high fundamental frequencies, structures subject to excitations with ultra-low-frequency or low-frequency are more prone to exhibit nonlinear vibrations with extremely large amplitudes. Secondly, according to the linear vibration theory, the frequency band for effective isolation of linear isolation structures starts at $\sqrt{2}$ times natural frequency of the structure. To facilitate the vibration isolation in low frequency band, lowering the stiffness of the isolation structure becomes inevitable, leading to a decrease in its load-bearing capacity, and thus contradicting the design requirements for engineering structures. Currently, there are two challenges in applying classical linear theory to isolator design: firstly, reducing the stiffness of the elastic component to broaden the effective band of frequency for vibration isolation will decrease the system's load-bearing capacity; secondly, increasing damping to reduce resonance peak values will sacrifice the isolation effectiveness in the high frequency band. Needless to say, tackling these issues will require innovative designs of vibration isolation schemes.

Suffered from limitations and shortcomings of linear isolation structure, researchers have begun to introduce nonlinearity in designing and constructing novel vibration isolators. They take into account the negative stiffness design to reduce the dynamic natural frequency and use nonlinearities to compensate for the loss of load-bearing capacity due to the negative stiffness. This leads to the development of isolator with the feature of so called "high static, low dynamic" [48]. However, compared with linear systems, nonlinear systems exhibit qualitatively different phenomena such as multiple equilibria, bifurcations, and chaos, presenting new challenges for vibration suppression design [49, 50]. The stationary response of nonlinear structures with large bandwidth and multi-stability characteristics is closely related to the initial conditions of vibration. When subjected to short-duration impact loads, the amplitude of structural response may experience a sudden change, potentially causing safety issue. Moreover, in high-tech fields such as aerospace, advanced CNC machine tools, lithography machines, etc., the sources of vibration that induce structural responses are usually complicated, which poses new requirements for effective vibration suppression in a wide band. Consequently, to further enhance the performance of vibration isolators and achieve wide band and effective vibration suppression, active control is introduced to supplement the traditional design of vibration design.

Indeed, in applications such as aerospace and automobile engineering, the demand for intelligent isolators often leads to the use of active control to suppress vibration responses or excitations, and generally the time delay in the control loop cannot be neglected. Time delay has a profound impact on both the stability and vibration characteristics of the system. The dynamic properties of the system may change with variations in parameters such as time delay and control strength [51, 52]. Adjusting the value of time delay can improve the system performance without altering the energy consumption. Therefore, introducing the control with adjustable time-delay and establishing the relationship between control parameters and control effects becomes essential.

My group's previous work has validated the existence of time delay in control loops through parameter identification, confirming the feasibility of both theoretical modeling and practical application of time-delay feedback control devices [53]. Olgac and Holm-Hansen [54] introduced time-delay absorbers to the primary system and found that the time delay could regulate the anti-resonance frequency to achieve vibration suppression. Alhazza and Majeed [55] studied the suppression of free vibrations of a cantilever beam using time-delay acceleration feedback. Hamdi and Belhaq [56], with a simply supported beam subjected to axial excitation of high-frequency as the control object, investigated the effect of time-delay displacement feedback for suppressing the self-excited vibration. Professor Cai Guoping's group introduced time delay feedback control to continuous structures, and experimental results indicated that time delay in control could improve the performance of vibration suppression. However, they also identified numerous issues in time delay feedback control that required further investigation, including design of the control law, the robustness of time delay control, time delay control of nonlinear structures, and experimental methods for time delay control [57, 58]. Xu and Sun [31, 59] implemented a linear time delay absorber using motion controllers, servo motors, and elastic coupling elements to achieve vibration suppression on a linear spring oscillator based on the concept of time-delay absorbers. Subsequently, Sun XT et al. [60, 61] built a nonlinear isolator with time delay control based on the qualitative and quantitative analysis, and validated the impact of time delay on the stability, frequency, and other dynamic characteristics of nonlinear systems. They provided optimal values for the time delay in different frequency bands, demonstrating the applicability of time delay control in the field of nonlinear isolation.

It should be noted that nonlinear isolators with time delay control utilize multi-cell components and quasi-zero-stiffness structures as basic elements. Considering the engineering requirements for load-bearing capacity, the effective isolation band, and the intelligent vibration isolation demands at various frequency bands, adjustable time delay should be introduced to these isolators in a simple control manner to improve the system performance. The main difficulty is that the study on such systems will have to face the challenges of nonlinear structural design and time delay control of vibration systems, which requires a lot of fundamental work on the linear and nonlinear analysis of time delay systems, both autonomous and nonautonomous, optimization of time delay system, smart yet simple control and the relative experimental techniques.

1.3.5 Other Applications in the Presence of Time Delay

A great amount of references can be found in the field of nonlinear dynamics of systems with time delay. However, there still exists strong demand for studying the effect of time delay on the stability and dynamics of the model abstracted from certain applications on which my research group has been concentrated in recent twenty years, such as industrial manufacturing [62–66], network systems [67–71], biology [72–75], traffic flow, etc. [76–78]. Main challenges arise from the complicated form of the mathematical model, which may be high dimensional and/or strongly nonlinear and exhibit complex nonlinear behaviour including quasi-periodic motion, chaos, bifurcations with high codimension, various types of synchronization, etc. Besides, the time delay appearing in the system is not necessarily constant, and could be a periodic function of time and even dependent on the state variables. Special attention should be paid to these issues when tackling specific time delay systems from various applications.

1.4 Structure of the Book

This book is organised into three parts—Part I: Preliminary, Part II: Fundamental Phenomena and Methods, and Part III: Applications. Part I contains the chapter of introduction. Part II has three chapters. Chapter 2 introduces typical nonlinear phenomena related with time delay systems. Chapter 3 discusses two novel methods of quantitatively solving nonlinear systems with time delay, i.e., the perturbation-incremental scheme and the integral equation method. Chapter 4 introduces the research progress made by my group on the typical inverse problem, namely, parameter identification of time delay systems. Part III consists of six chapters, all of which provide the detailed results on the nonlinear dynamics of time delay systems arising from real applications, including vibration suppression (Chap. 5), industrial manufacturing (Chap. 6), network systems (Chap. 7), biology (Chap. 8) and traffic flow (Chaps. 9 and 10).

References

1. Hirsch MW, Smale S, Devaney RL (2013) Differential equations, dynamical systems, and an introduction to chaos. Academic Press
2. Scheck F (2018) Mechanics from newton's laws to deterministic chaos (6th ed.) Springer-Verlag
3. Guckenheimer J, Holmes P (1983) Nonlinear oscillations, dynamical systems, and bifurcations of vector fields. Springer-Verlag
4. Kuznetsov Y (1997) Elements of applied bifurcation theory (2nd ed.). Springer.
5. Wiggins S (2003) Introduction to applied nonlinear dynamical systems and chaos (2nd ed.). Springer-Verlag
6. Wang Z, Hu H (2013) Stability and bifurcation of delayed dynamic systems: from theory to application. Adv Mech 43:3–19
7. Yuan Y, Campbell SA (2004) Stability and synchronization of a ring of identical cells with delayed coupling. J Dynam Diff Equ 16:709–744
8. Cooke KL, Huang WZ (1996) On the problem of linearization for state-dependent delay differential equations. Proc Am Math Soc 124:1417–1426
9. Lani-Wayda B, Walther HO (2016) A Shilnikov phenomenon due to state-dependent delay, by means of the fixed point index. J Dynam Diff Equ 28:627–688
10. Guo S (2005) Spatio-temporal patterns of nonlinear oscillations in an excitatory ring network with delay. Nonlinearity 18:2391–2407
11. Mazenc F, Malisoff M (2017) Extensions of Razumikhin's theorem and Lyapunov-Krasovskii functional constructions for time-varying systems with delay. Automatica 78:1–13
12. Zhang XM, Han QL, Seuret A et al (2017) An improved reciprocally convex inequality and an augmented Lyapunov-Krasovskii functional for stability of linear systems with time-varying delay. Automatica 84:221–226

13. Zhang L, Orosz G (2017) Consensus and disturbance attenuation in multi-agent chains with nonlinear control and time delays. Int J Robust Nonlinear Cont 27:781–803
14. Olgac N, Sipahi R (2002) An exact method for the stability analysis of time-delayed linear time-invariant (LTI) systems. IEEE Trans Autom Control 47:793–797
15. Sipahi R, Olgac N (2005) Complete stability robustness of third-order LTI multiple time-delay systems. Automatica 41:1413–1422
16. Elmali H, Renzulli M, Olgac N (2000) Experimental comparison of delayed resonator and PD controlled vibration absorbers using electromagnetic actuators. J Dynam Syst Meas Cont 122:514–520
17. Liu ZH, Payre G (2007) Stability analysis of doubly regenerative cylindrical grinding process. J Sound Vib 301:950–962
18. Jnifene A (2007) Active vibration control of flexible structures using delayed position feedback. Syst Cont Lett 56:215–222
19. Nayfeh AH, Nayfeh NA (2012) Time-delay feedback control of lathe cutting tools. J Vib Cont 18:1106–1115
20. Dai HL, Abdelkefi A, Wang L et al (2014) Control of cross-flow-induced vibrations of square cylinders using linear and nonlinear delayed feedbacks. Nonlinear Dynam 78:907–919
21. Wu JH (1998) Symmetric functional differential equations and neural networks with memory. Trans Am Math Soc 350:4799–4838
22. Guo S, Huang L (2007) Stability of nonlinear waves in a ring of neurons with delays. J Differ Equ 236:343–374
23. Guo Y, Jiang W, Niu B (2013) Multiple scales and normal forms in a ring of delay-coupled oscillators with application to chaotic Hindmarsh-Rose neurons. Nonlinear Dyn 71:515–529
24. Zhang Q, Wei XP, Xu J (2003) An analysis of the global asymptotic stability for neural networks with variable delays. Phys Lett A 318:399–405
25. Mou SH, Gao HJ, Lam J, Qiang WY (2008) A new criterion of delay-dependent asymptotic stability for Hopfield neural networks with time delay. IEEE Trans Neural Netw 19:532–535
26. Srikant R (2004) The mathematics of internet congestion control. Birkhäuser, Boston
27. Xu WY, Cao JD, Xiao M (2014) Bifurcation analysis and control in exponential RED algorithm. Neurocomputing 129:232–245
28. Li CG, Chen GR, Liao XF, Yu JB (2004) Hopf bifurcation in an internet congestion control model. Chaos, Solitons Fractals 19:853–862
29. Xiao M, Cao JD (2007) Delayed feedback-based bifurcation control in an internet congestion model. J Math Anal Appl 332:1010–1027
30. Zhang S, Xu J (2013) Quasiperiodic motion induced by heterogeneous delays in a simplified internet congestion control model. Nonlinear Ana 14:661–670
31. Sun YX, Xu J (2015) Experiments and analysis for a controlled mechanical absorber considering delay effect. J Sound Vib 339:25–37
32. Hale J (1977) Theory of functional differential equations. Springer-Verlag
33. Guo S, Wu J (2013) Bifurcation theory of functional differential equations. Springer
34. Xu J, Pei LJ (2006) Advances in dynamics of delayed systems. Adv Mech 36:17–30
35. Yan XP (2006) Hopf bifurcation and stability for a delayed tri-neuron network model. J Comput Appl Math 196:579–595
36. Das SL, Chatterjee A (2002) Multiple scales without center manifold reductions for delay differential equations near Hopf bifurcations. Nonlinear Dyn 30:323–335
37. Nayfeh AH (2008) Order reduction of retarded nonlinear systems-the method of multiple scales versus center-manifold reduction. Nonlinear Dyn 51:483–500
38. Chai Q, Loxton R, Teo KL, Yang C (2013) Time-delay estimation for nonlinear systems with piecewise-constant input. Appl Math Comput 219:9543–9560
39. Hartung F (2013) Parameter estimation by quasilinearization in differential equations with state-dependent delays. Discrete Continuous Dynam Syst Ser B 18:1611–1631
40. Lin Q, Loxton R, Xu C, Teo KL (2015) Parameter estimation for nonlinear time-delay systems with noisy output measurements. Automatica 60:48–56

41. Abooshahab MA, Ekramian M, Ataei M, Ebrahimpour-Boroojeny A (2019) Time-delay estimation in state and output equations of nonlinear systems using optimal computational approach. J Opt The Appl 180:1036–1064
42. Zhang T, Lu ZR, Liu JK, Liu G (2021) Parameter identification of nonlinear systems with time-delay from time-domain data. Nonlinear Dyn 104:4045–4061
43. Lei Y, Xia D, Erazo K, Nagarajaiah S (2019) A novel unscented Kalman filter for recursive state-input-system identification of nonlinear systems. Mech Syst Sig Proc 127:120–135
44. He J, Qi M, Tong Z, Hua X, Chen Z (2023) An improved extended Kalman filter for parameters and loads identification without collocated measurements. Smart Struct Syst 31:131–140
45. Cui T, Ding F, Hayat T (2022) Moving data window-based partially-coupled estimation approach for modeling a dynamical system involving unmeasurable states. ISA Trans 128:437–452
46. Sun S, Xu L, Ding F (2023) Parameter estimation methods of linear continuous-time time-delay systems from multi-frequency response data. Circ Syst Sig Proc 42:3360–3384
47. Nayfeh AH, Mook DT (2008) Nonlinear oscillations. John Wiley & Sons
48. Peng X, Li DZ, Chen SN (1997) Quasi-zero stiffness vibration isolators and design for their elastic characteristics. J Vibrat Meas Diagn 17
49. Han YW, Cao QJ, Chen YS et al (2012) A novel smooth and discontinuous oscillator with strong irrational nonlinearities. Sci China Phys Mech Astron 55:1832–1843
50. Shahraeeni M, Sorokin V, Mace B et al (2022) Effect of damping nonlinearity on the dynamics and performance of a quasi-zero-stiffness vibration isolator. J Sound Vibrat 526:116822
51. Heiden UAD, Walther HO (1983) Existence of chaos in control system with delayed feedback. J Differ Eqn 47:273–295
52. Chen GR, Wang XF (2006) Chaos of dynamical systems: Theoretical methods and applications. Shanghai Jiao Tong University Press, Shanghai (in Chinese)
53. Zhang XX, Xu J (2014) Identification of time delay in nonlinear systems with delayed feedback control. J Franklin Inst 352:2987–2998
54. Olgac N, Holm-Hansen BT (1994) A novel active vibration absorption technique: delayed resonator. J Sound Vib 176:93–104
55. Alhazza KA, Majeed MA (2012) Free vibrations control of a cantilever beam using combined time delay feedback. J Vib Cont 18:609–621
56. Hamdi M, Belhaq M (2009) Self-excited vibration control for axially fast excited beam by a time delay state feedback. Chaos Solitons Fractals 41:521–532
57. Cai GP, Chen LX (2013) Some problems of delayed feedback control. Adv Mech 43:21–28 (in Chinese)
58. Cai GP, Chen LX (2010) Delayed feedback control experiments on some flexible structures. Acta Mech Sinica 6:951–965
59. Xu J, Sun YX (2015) Experimental studies on active control of a dynamic system via a time-delayed absorber. Acta Mech Sinica 31:229–247
60. Sun XT, Zhang S, Xu J (2018) Parameter design of a multi-delayed isolator with asymmetrical nonlinearity. Int J Mech Sci 138–139:398–408
61. Sun XT, Wang F, Xu J (2019) Dynamics and realization of a feedback-controlled nonlinear isolator with variable time delay. J Vib Acoust 141:021005
62. Yan Y, Xu J, Wiercigroch M (2019) Modelling of regenerative and frictional cutting dynamics. Int J Mech Sci 156:86–92
63. Yan Y, Xu J (2018) Stability and dynamics of parallel plunge grinding. Int J Adv Manuf Tech 99:881–895
64. Yao Y, Xu J, Wiercigroch M (2017) Basins of attraction of the bi-stable region of time-delayed cutting dynamics. Phys Rev E 96:032205
65. Yan Y, Xu J, Wiercigroch M (2017) Regenerative chatter in a plunge grinding process with workpiece imbalance. Int J Adv Manuf Tech 89:2845–2862
66. Yan Y, Xu J, Wiercigroch M (2016) Regenerative and frictional chatter in plunge grinding. Nonlinear Dynam 86:283–307

67. Wang SN, Zhang S, Xu J (2019) Suppression of oscillatory congestion via trunk link bandwidth and control gain in star network. Appl Math Mech (English Edition) 40:25–48
68. Zhang S, Xu J, Chung KW (2017) Desynchronization-based congestion suppression for a star-type Internet system with arbitrary dimension. Neurocomputing 266:42–55
69. Zhang S, Xu J, Chung KW (2015) On the stability and multi-stability of a TCP/RED congestion control model with state-dependent delay and discontinuous marking function. Commun Nonlinear Sci Numer Simul 22:269–284
70. Zhang S, Chung KW, Xu J (2013) Stability switch boundaries in an internet congestion control model with diverse time delays. Int J Bifurcat Chaos 23:1330016–1-24.
71. Zhang S, Xu J (2011) Oscillation control for n-dimensional congestion control model via time-varying delay. Sci China-Technol Sci 54:2044–2053
72. Ge JH, Xu J (2015) Double Hopf bifurcation in a four-neuron delayed system with inertial terms. Nonlinear Dynam 82:1969–1978
73. ZBXJ Zhen B, Xu J (2010) Fold-Hopf bifurcation analysis for a coupled FitzHugh-Nagumo neural system with time delay. Int J Bifurcat Chaos 20:3919–3914
74. Song ZG, Xu J (2013) Stability switches and double Hopf bifurcation in a two-neural network system with multiple delays. Cognitive Neurodyn 7:505–521
75. Ge JH, Xu J (2011) An efficient method for studying fold-Hopf Bifurcation in delayed neural networks. Int J Bifurcat Chaos 21:1393–1406
76. Wang CH, Xu J (2010) Effects of time delay and noise on asymptotic Stability in human quiet standing model. Math Probl Eng 2010:829484
77. Shang HL, Xu J (2009) Delayed feedbacks to control the fractal erosion of safe basins in a parametrically excited system. Chaos, Solitons & Fractals
78. Xu J, Jiang SY (2015) Delay-induced Bogdanov-Takens bifurcation and dynamical classifications in a slow-fast flexible joint system. Int J Bifurcat Chaos 25:1550121

Chapter 2
Delay Induced Nonlinear Dynamics

The 1970s, with the development of computer technology and numerical methods, researchers have gradually recognized that parameter perturbations can lead to complex dynamic behaviors in nonlinear dynamic systems. Based on the typical nonlinear examples commonly found in mechanics and physics, numerous new nonlinear phenomena, including chaos and fractals, have been discovered. The corresponding fundamental theories and methods have been proposed and established.

Time delay is ubiquitous in real systems and has become an important issue of concern for scientists in various fields such as mathematics [1, 2], mechanics [3], high-precision manufacturing engineering [4], biology [5], etc. The research results in recent years indicated that time delay can lead to multiple bifurcation sequences, resulting in the system exhibiting multi-stable activity, steady state at different frequencies, and even complex dynamic behavior [6]. In engineering technology, there are numerous nonlinear dynamic systems, among which the Van der Pol (VDP) oscillator system with a nonlinear damping term that sustains self-excited vibration and the Duffing oscillator system with a cubic nonlinear restoring force term are the most famous. Researchers have not only devoted themselves to the analysis and study of VDP systems [7] and Duffing systems [8], but also shown great interest in the nonlinear dynamic behavior of VDP-Duffing systems [9–14]. For decades, many scholars have conducted systematic research on the dynamic behavior of the VDP-Duffing system. The stability and Hopf bifurcation of a nonlinear timedelay VDP-Duffing system with cubic terms were researched by Xu et al. [15]. They demonstrated the existence of Hopf bifurcation, determined the direction of Hopf bifurcation and the stability of periodic solutions, and discussed the fluence of time delay on the Hopf bifurcation of the system. Li et al. studied the chaotic dynamic behavior of a class of VDP-Duffing oscillators and applied the direct perturbation method to construct a general solution of the system, from which the Melnikov criterion for predicting the occurrence of chaos was obtained [16]. The chaos control and synchronization problems of a class of VDP-Duffing systems were discussed by Zhang [17].

J. Xu, *Nonlinear Dynamics of Time Delay Systems*,
https://doi.org/10.1007/978-981-99-9907-1_2

Using the averaging method, the amplitude-frequency response of a class of VDP-Duffing system was investigated and their static bifurcation phenomenon through singularity theory was analyzed in Ref. [18].

In this section, based on the classic time delay nonlinear oscillator systems (including VDP system and VDP-Duffing system), the asymptotic analytical solution of periodic trajectory under the influence of time delay is theoretically obtained using the center manifold theorem and method of multiple scales. On this basis, the stability of multiple periodic solutions induced by time delay in nonlinear systems and their steady state coexistence are discussed. Thereby a computational analysis method is presented to describe the dynamic behavior of time delay systems. Specifically, a theoretical analysis method is proposed to study the strong resonance dynamics characteristics of double Hopf bifurcation in time-delayed nonlinear systems. And then the singularity of dynamic behavior induced by time delay and its dynamic classification are investigated. In addition, our research also finds that time delay can act as a "control switch" for systems. We have also discovered the period-doubling bifurcation and quasi-periodic breakdown in time delay dynamical systems as a route to chaos.

2.1 Asymptotic Analytical Method for Periodic Solution of the Delay Induced VDP-Duffing System

In this section, the time delayed VDP-Duffing system is taken as the object of research. Based on the eigenvalue distribution of its linear characteristic equation, the critical conditions of system parameters are given to produce Hopf bifurcation, demonstrating the "amplitude dead region" and multiple steady state switching under the influence of time delay. Then, using the center manifold theorem, the existence and stability of periodic solutions of the system are studied. Furthermore, based on the perturbation of the time-delayed parameter near the bifurcation point and using the averaged equations obtained by the perturbation method, an asymptotic analytical solution of periodic solutions is derived, studying the phase-locked periodic solution and phase-shifting periodic solution generated by the system. A new method for approximate analytical calculation of periodic solutions in time delay systems is proposed.

2.1.1 Hopf Bifurcation and Btability Bwitching

The mathematical model is the VDP-Duffing system with external excitation, which can be written as

$$\ddot{x} + \omega_0^2 x - (\alpha - \gamma x^2)\dot{x} + \beta x^3 = k\cos(\Omega t). \tag{2.1}$$

By adding a linear or nonlinear time delayed position feedback to (2.1), we get the following new system

$$\ddot{x} + \omega_0^2 x - (\alpha - \gamma x^2)\dot{x} + \beta x^3 = k\cos(\Omega t) + A(x_\tau - x) + B(x_\tau - x)^3, \quad (2.2)$$

where $\alpha, \gamma, k > 0$, $x_\tau = x(t-\tau)$ and τ represents a time delay. If $A, B < 0$, the position feedback is called negative feedback, and if $A, B > 0$, it is called positive feedback. At $\tau = 0$, (2.2) is simplified to (2.1).

The study of (2.2) starts with the corresponding linearized system without time delay and external forces. For $k = 0$, the linearized Eq. (2.2) can be written as

$$\ddot{x} + (\omega_0^2 + A)x - \alpha\dot{x} = Ax_\tau. \quad (2.3)$$

The characteristic of Eq. (2.3) is shown as

$$\lambda^2 - \alpha\lambda + (\omega_0^2 + A) = Ae^{(-\lambda\tau)}, \quad (2.4)$$

where ω_0 is a real positive constant.

Therefore, we can obtain that $\lambda = (\alpha \pm \sqrt{\alpha^2 - 4\omega_0^2})/2$ when $\tau = 0$. For the corresponding linearized autonomous system ($k = 0$), the stable trivial equilibrium occurs when the parameter $\alpha < 0$. For the corresponding linearized non-autonomous system ($k \neq 0$), stable periodic solutions occur when the same parameter conditions are met. So, we first discuss the stability of the trivial solution in the linearized system of (2.2) without excitation and then derived an explicit expression for the critical boundary at which the trivial solution loses its stability.

To find the explicit expression of the critical stability boundary, suppose $\lambda = a + i\omega$ is one root of the characteristic Eq. (2.4), where a and ω are real numbers. Without losing generality, let $\omega > 0$, because the root of Eq. (2.4) is in the form of a pair of conjugate complex numbers. Then substituting λ into Eq. (2.4), we can get that

$$\begin{gathered} a^2 - \alpha a - \omega^2 + (\omega_0^2 + A) - Ae^{-a\tau}\cos(\omega\tau) = 0, \\ \omega(2a - \alpha) + Ae^{-a\tau}\sin(\omega\tau) = 0. \end{gathered} \quad (2.5)$$

By setting $a = 0$, the explicit expressions at the critical stability boundary can be written as

$$(\omega_0^2 + A) - \omega^2 = A\cos(\omega\tau), \alpha\omega = A\sin(\omega\tau). \quad (2.6)$$

Eliminating τ from Eq. (2.6), we get that

$$\kappa^4\omega^4 + \kappa^2\omega^2(\kappa^2\alpha^2 - 2\kappa A - 2) + 2\kappa^2 A + 1 = 0, \quad (2.7)$$

and then the roots are

$$(\kappa\omega_{\pm})^2 = 1 + \kappa^2 A - \frac{\kappa^2\alpha^2}{2} \pm \frac{1}{2}\sqrt{(2\kappa^2 A - \kappa^2\alpha^2)^2 - (2\kappa\alpha)^2}, \tag{2.8}$$

where $\kappa = 1/\omega_0$. Since $\pm\omega$ are real numbers, the boundary of the critical values of $\kappa\alpha$, $\kappa^2 A$ can be obtained from Eq. (2.8) that

$$(2\kappa^2 A - \kappa^2\alpha^2)^2 - (2\kappa\alpha)^2 = 0.$$

The stability conditions about the time delay τ are obtained. This will be achieved by solving τ from Eq. (2.6) in terms of the other parameters. To this end, let I be the set of (α, A) such that Eq. (2.7) have real and positive roots, given by

$$I = \left\{ (\alpha, A) | 2\kappa^2 A + 1 > 0, 2 + 2\kappa^2 A - (\kappa\alpha)^2 > 0, \left(2\kappa^2 A - (\kappa\alpha)^2\right) - (2\kappa\alpha)^2 > 0 \right\}.$$

If $(\alpha, A) \in I$, the two sets of critical boundaries can be obtained for $a > 0$, $A\sin(\omega\tau) > 0$:

$$\tau_+[j, A, \alpha, \omega_0] = \begin{cases} \frac{1}{\omega_+}\left(2j\pi + \cos^{(-1)}\left(1 + \frac{\omega_0^2 - \omega_+^2}{A}\right)\right) & \text{for } A > 0, \\ \frac{1}{\omega_+}\left((2j+2)\pi - \cos^{(-1)}\left(1 + \frac{\omega_0^2 - \omega_+^2}{A}\right)\right) & \text{for } A < 0, \end{cases} \tag{2.9}$$

$$\tau_-[j, A, \alpha, \omega_0] = \begin{cases} \frac{1}{\omega_-}\left(2j\pi + \cos^{(-1)}\left(1 + \frac{\omega_0^2 - \omega_-^2}{A}\right)\right) & \text{for } A > 0, \\ \frac{1}{\omega_-}\left((2j+2)\pi - \cos^{(-1)}\left(1 + \frac{\omega_0^2 - \omega_-^2}{A}\right)\right) & \text{for } A < 0, \end{cases} \tag{2.10}$$

where $j = 0, 1, 2, \ldots$. The trivial solution of Eq. (2.3) is unstable when $a > 0$ and $\tau = 0$, while the real part of an eigenvalue becomes zero when τ reaches its first critical value $\tau_-[0, A, \alpha, \omega_0]$. If the stability of the trivial solution is changed, this critical curve should be the one with $\frac{d\alpha}{d\tau} < 0$. From Eq. (2.4), we can obtain that

$$\frac{d\lambda}{d\tau} = \frac{-A\lambda}{A\tau + e^{\lambda\tau}(2\lambda - \alpha)}, \tag{2.11}$$

and

$$\begin{aligned} \left.\frac{d\alpha}{d\tau}\right|_{a=0} &= \mathrm{Re}\left(\frac{d\lambda}{d\tau}\right)\Big|_{a=0} = G^{-1}\left[A\alpha\omega\sin(\omega\tau) - 2A\omega^2\cos(\omega\tau)\right] \\ &= G^{-1}A^2\omega^2\left[\alpha^2 - 2A - 2\omega_0^2 + 2\omega^2\right] \\ &= G^{-1}A^2\omega^2\omega_0^2\left[\kappa^2(\alpha^2 + 2\omega^2) - 2\kappa^2 A - 2\right], \end{aligned} \tag{2.12}$$

where $G = \left(2\omega\cos(\omega\tau) - \alpha\sin(\omega\tau)^2\right) + (A\tau - \alpha\cos(\omega\tau) - 2\omega\sin(\omega\tau))^2$ is real positive.

Substituting Eq. (2.8) into Eq. (2.12) yields

$$\left.\frac{d\alpha}{d\tau}\right|_{a=0} = \begin{cases} -G_-^{-1}(A\omega_-\omega_0)^2 H < 0 \text{ on } \tau_-, \\ G_+^{-1}(A\omega_-\omega_0)^2 H > 0 \text{ on } \tau_+, \end{cases} \tag{2.13}$$

where $H = \sqrt{(2\kappa^2 A - \kappa^2\alpha^2)^2 - (2\kappa\alpha)^2}$ and $G_\pm = (2\omega_\pm \cos(\omega_\pm\tau) - \alpha \sin(\omega_\pm\tau))^2 + (A\tau - \alpha\cos(\omega_\pm\tau) - 2\omega_\pm \sin(\omega_\pm\tau))^2$.

From Eq. (2.13), it can be seen that a pair of eigenvalues will cross the imaginary axis of the left (right) half-plane with τ increases to its critical value $\tau_-(\tau_+)$. So, the stable boundaries of the origin can be located by the approximate branch of τ_- and τ_+, which satisfy

$$\tau_-[j, A, \alpha, \omega_0] < \tau_+[j, A, \alpha, \omega_0] \quad \text{for } j \text{ with } j = 0, 1, 2, \ldots.$$

For example, if $\tau_-[0, A, \alpha, \omega_0] < \tau_+[0, A, \alpha, \omega_0]$ for a certain interval of A and $a > 0$, the unstable origin becomes stable as τ increased to cross τ_-. Then the origin loses its stability once again when a pair of eigenvalues pass through the imaginary half-axis from left to right. Therefore, a stable region of the origin is bounded by the curves $\tau_-[0, A, \alpha, \omega_0]$ and $\tau_+[0, A, \alpha, \omega_0]$ for the interval of A such that $\tau_+ > \tau_-$. As any vibration will disappear in this region eventually, it is called the amplitude dead region or death island.

Similarly, higher order death islands can be located by $\tau_-[j, A, \alpha, \omega_0] < \tau < \tau_+[j, A, \alpha, \omega_0]$ for $j = 1, 2, \ldots$, which will lead to the existence of multiple stable regions. Figure 2.2 shows two numerical simulations of the curves $\tau_+[j, A, \alpha, \omega_0]$ (solid lines) and $\tau_-[j, A, \alpha, \omega_0]$ (dashed lines) in the $A - \tau$ plane for the first few values of j and for $\omega_0 = 0.1.$. The two cases are (a) $\alpha = 0.2$, $A > 0$ and (b) $\alpha = 0.002$, $A < 0$, respectively. All stable areas at the origin are filled with gray.

For $A > 0$, Fig. 2.1 (a) shows that the region is the only stable region between $\tau_-[0, A, \alpha, \omega_0]$ and $\tau_+[0, A, \alpha, \omega_0]$ when τ taking some small values. In fact, along τ, the next two curves are $\tau_+[1, A, \alpha, \omega_0]$ and $\tau_+[2, A, \alpha, \omega_0]$. When τ passing through these two curves, the two pairs of eigenvalues move to the right half. When τ reached $\tau_+[1, A, \alpha, \omega_0]$, one pair of eigenvalues moves to the left half plane, and the other pair of eigenvalues remains on the right half plane. Therefore, the regional boundaries $\tau_+[1, A, \alpha, \omega_0]$ and $\tau_+[2, A, \alpha, \omega_0]$ cannot form a dead island. The higher order situations can also be analyzed in the same way.

Furthermore, it can be seen from Fig. 2.1b that there are multiple stability regions when $a = 0.002$, $A < 0$ and $j = 0, 1, 2, \ldots$. Therefore, the existence of the region not only depends on natural frequency ω_0 [19], but also on the delay feedback amplitude A. Physically speaking, these multiple death islands are very useful for controlling the stability of the system. When people consider time delay as a "switch", it provides more collectible values.

Finally, it should be noted that the case $\alpha > 0$ can also be obtained using a similar method, and this section only focuses on analyzing the periodic solutions generated by the codimension-1 Hopf bifurcation. The double Hopf bifurcation will be discussed in detail in another paper.

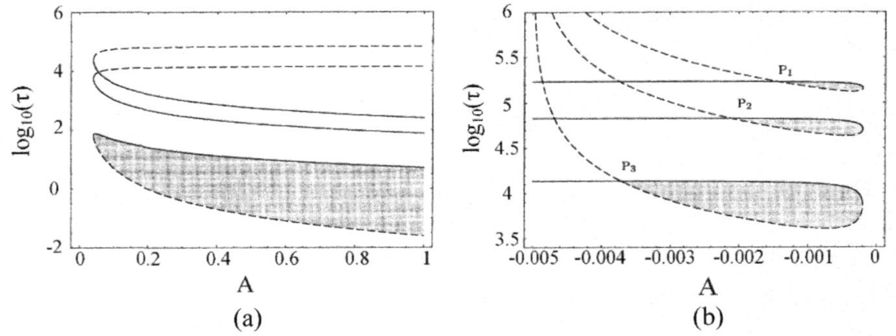

Fig. 2.1 Critical stability boundaries of the trivial solution of Eq. (2.3) in the $A - \tau$ plane for **a** $\alpha = 0.2$ and **b** $\alpha = 0.002$: the shaded portion represents stable attractors; solid line is τ_+ and dashed line is τ_-, double Hopf bifurcations at P_1, P_2 and P_3

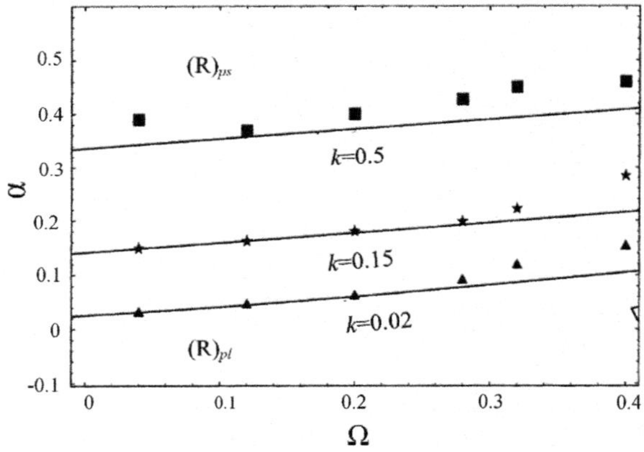

Fig. 2.2 Bifurcation diagram without time delay. Regions (R) pl and (R) ps, separated by BPSS governed by Eq. (2.48), represent the phase-locked (periodic) solutions and the phase-shifting (complicated) states, respectively. Stars, triangles and boxes symbols show the numerical calculation obtained from Eq. (2.2) for the chosen amplitudes $k = 0.02$, $k = 0.15$ and $k = 0.5$, respectively. Solid lines represent the theoretical values obtained from Eq. (2.48)

2.1.2 Reduction on Center Manifold

In this section, we will discuss the periodic solutions and their branches of Eq. (2.2). Obviously, nonlinearity, external forces, and time delay all affect the dynamics of the linearized (2.3). If all these factors are taken into account, it is very difficult to analyze this issue. Based on our motivation, we limit our analysis by setting the nonlinear coefficients and external force amplitudes to have a certain order of magnitude for the system. Finally, letting $\alpha = \alpha_c + \mu_1\varepsilon$, $A = A_c + \mu_2\varepsilon$, and rescaling $x \to \sqrt{\varepsilon}x$, $k \to \varepsilon^{3/2}k$ (which physically implies that the excitation is soft and a resonant case will be considered) in Eq. (2.2) yield

$$\ddot{x} - \alpha_c\dot{x} + A_c(x - x_\tau) + \omega_0^2 x = \varepsilon\big((\mu_1 - \gamma x^2)\dot{x} + \mu_2(x_\tau - x) \\ -\beta x^3 + k\cos(\Omega t) + B(x_\tau - x)^3\big), \tag{2.14}$$

or

$$\ddot{x} - \alpha_c\dot{x} + A_c(x - x_\tau) + \omega_0^2 x = \varepsilon f(x, x_\tau, \dot{x}, t), \tag{2.15}$$

where $0 < \varepsilon < 1$, $f(x, \dot{x}, \dot{x}_\tau, t) = (\mu_1 - \gamma x^2)\dot{x} + \mu_2(x_\tau - x) - \beta x^3 + k\cos(\Omega t) + B(x_\tau - x)^3$, α_c and A_c are the critical values satisfying Eq. (2.6). In addition, Eq. (2.15) can be rewritten as

$$\dot{u}_1 = u_2, \dot{u}_2 = -(\omega_0^2 + A)u_1 + \alpha_c u_2 + A_c u_{1\tau} + \varepsilon f(u_1, u_{1\tau}, u_2, t), \tag{2.16}$$

where $u_1 = x$, $u_{1\tau} = u_1(t - \boldsymbol{\tau})$.

We can transform Eq. (2.16) into a functional differential equation (FDE), assume that $\mathbb{C} = \mathbb{C}([-\tau, 0], \mathbb{R}^2)$, and define that $\|\phi\| = \sup_{-\tau\le\theta\le 0}|\phi(\theta)|$ and $u_t(\theta) = u(t + \theta)$, $-\tau \le \theta \le 0$. Therefore, Eq. (2.16) becomes

$$\dot{\mathbf{u}} = \mathbf{L}(0)\mathbf{u_t} + \varepsilon\mathbf{L}(\mu_1, \mu_2)\mathbf{u_t} + \varepsilon\mathbf{F}(t, u_{2t}), \tag{2.17}$$

where $\mathbf{u} = (u_1, u_2)^T, u| = u_1(t - \tau) = \int_{-\tau}^0 \delta(\theta + \tau)u_1(t + \tau)d\theta = \int_{-\tau}^0 \delta(\theta + \tau)u_{1t}d\theta$, and $\mathbf{F} : \mathbb{R}_+ \times \mathbb{C} \to \mathbb{R}^2$, which can be written as

$$\mathbf{F}(t, u_{2t}) = \begin{Bmatrix} 0 \\ k\cos\Omega t - \gamma u_{1t}^2(0)u_{2t}(0) + B[u_{1t}(-\tau) - u_{1t}(0)]^3 \end{Bmatrix}. \tag{2.18}$$

Then the linear operator of the critical case is $\mathbf{L}(0) : \mathbb{C} \to \mathbb{R}^2$, given by

$$\mathbf{L}(0)\boldsymbol{\varphi} = \int_{-\tau}^{0} [d\boldsymbol{\eta}(s)]\boldsymbol{\varphi}(s), \tag{2.19}$$

where

$$d\boldsymbol{\eta}(s) = \begin{bmatrix} 0 & \delta(s) \\ -(\omega_0^2 + A_c)\delta(s) + A_c\delta(s + \tau) & \alpha_c\delta(s) \end{bmatrix} ds. \tag{2.20}$$

Similarly,

$$\mathbf{L}(\mu_1, \mu_2)\boldsymbol{\varphi} = \int_{-\tau}^{0} [d\boldsymbol{\eta}(s, \mu_1, \mu_2)]\boldsymbol{\varphi}(s), \tag{2.21}$$

with

$$d\boldsymbol{\eta}(s, \mu_1, \mu_2) = \begin{bmatrix} 0 & 0 \\ \mu_2(\delta(s+\tau) - \delta(s)) & \mu_1\delta(s) \end{bmatrix} ds. \tag{2.22}$$

Further, for $\boldsymbol{\varphi} \in \mathbb{C}([-\tau, 0], \mathbb{R}^2)$, we define

$$\mathbf{D}(0)\boldsymbol{\varphi} = \begin{cases} \frac{d\boldsymbol{\varphi}(\theta)}{d\theta} & \text{for } \theta \in [-\tau, 0], \\ \mathbf{L}(0)\boldsymbol{\varphi} & \text{for } \theta = 0, \end{cases} \tag{2.23}$$

$$\mathbf{D}(\mu_1, \mu_2)\boldsymbol{\varphi} = \begin{cases} \mathbf{0} & \text{for } \theta \in [-\tau, 0], \\ \mathbf{L}(\mu_1, \mu_2)\boldsymbol{\varphi} & \text{for } \theta = 0, \end{cases} \tag{2.24}$$

and

$$\mathbf{Q}\varphi = \begin{cases} 0 & \text{for } \theta \in [-\tau, 0], \\ \mathbf{F}(t, \varphi) & \text{for } \theta = 0, \end{cases} \tag{2.25}$$

where

$$\mathbf{F}(t, \varphi) = \begin{Bmatrix} 0 \\ k\cos\Omega t - \gamma\varphi_1^2(0)\varphi_2(0) - \beta\varphi_1^3(0) + B[\varphi_1(-\tau) - \varphi_1(0)]^3 \end{Bmatrix}. \tag{2.26}$$

Finally, Eq. (2.17) can be rewritten as a functional differential equation (FDE):

$$\dot{\mathbf{u}}_\mathbf{t} = \mathbf{D}(0)\mathbf{u}_\mathbf{t} + \varepsilon\mathbf{D}(\mu_1, \mu_2)\mathbf{u}_\mathbf{t} + \varepsilon\mathbf{Q}\mathbf{u}_\mathbf{t}. \tag{2.27}$$

Based on the discussion in the previous section, it can be seen that the characteristic Eq. (2.4) has a single pair of pure imaginary eigenvalues $\Lambda = \pm i\omega$ for $\varepsilon = 0$ and a fixed $\tau \neq 0$. Therefore, $\mathbb{C}$ can be divided into two subspaces as $\mathbb{C} = P_\Lambda \oplus Q_\Lambda$, where P_Λ is a two-dimensional space spanned by the eigenvectors of an operator $\mathbf{D}(0)$ associated with the eigenvalues Λ, and Q_Λ is the complementary space of P_Λ.

Furthermore, we can define that

$$\mathbf{D}(0)\boldsymbol{\varphi} = \begin{cases} \frac{d\boldsymbol{\varphi}(\theta)}{d\theta} & \text{for } \theta \in [-\tau, 0], \\ \mathbf{L}(0)\boldsymbol{\varphi} & \text{for } \theta = 0, \end{cases} \tag{2.28}$$

and the bilinear form

$$\langle \boldsymbol{\psi}, \boldsymbol{\varphi} \rangle = \boldsymbol{\psi}^{-T}(0)\boldsymbol{\varphi}(0) - \int_{-\tau}^{0}\int_{0}^{\theta} \boldsymbol{\psi}^{-T}(\xi - \theta)[d\boldsymbol{\eta}(\theta)]\boldsymbol{\varphi}(\xi)d\xi, \tag{2.29}$$

where $\mathbb{C}^*$ is the dual space of $\mathbb{C}$, $\boldsymbol{\varphi} \in \mathbb{C}$ and $\boldsymbol{\psi} \in \mathbb{C}^*$. Then, $\mathbf{D}^*(0)$ and $\mathbf{D}(0)$ are adjoint operators.

Now assume that $\mathbf{q}(\theta)$ and $\mathbf{q}^*(\theta)$ are eigenvectors of $\mathbf{D}(0)$ and $\mathbf{D}^*(0)$, respectively, corresponding to the eigenvalues iω and $-$iω, that is,

$$\mathbf{D}(0)\mathbf{q}(\theta) = i\omega\mathbf{q}(\theta),\ \mathbf{D}^*(0)\mathbf{q}^*(\xi) = -i\omega\mathbf{q}^*(\xi). \tag{2.30}$$

After calculation, we can get that

$$\mathbf{q}(\theta) = \begin{Bmatrix} 1 \\ i\omega \end{Bmatrix} e^{i\omega\theta}, \tag{2.31}$$

and

$$\mathbf{q}^*(\xi) = N\begin{Bmatrix} -i[\omega + A_c \sin(\omega\tau)] \\ 1 \end{Bmatrix} e^{i\omega\xi}, \tag{2.32}$$

where $N = 1/(l - im)$, $l = \tau A_c \cos(\omega\tau) - \alpha_c$, $m = 2\omega - \tau A_c \sin(\omega\tau)$. Obviously, $\langle \mathbf{q}^*, \mathbf{q} \rangle = 1$ and $\langle \mathbf{q}^*, \overline{\mathbf{q}} \rangle = 0$. We can see from Eqs. (2.30–2.32) that the real bases for P_Λ and its dual space are displayed as $\boldsymbol{\Phi}(\xi) = (\varphi_1, \varphi_2) = \sqrt{2}(\mathrm{Re}(\mathbf{q}(\theta))\ , \mathrm{Im}(\mathbf{q}(\theta)))$ and $\boldsymbol{\Psi}(\xi) = (\psi_1, \psi_2) = \sqrt{2}(\mathrm{Re}(\mathbf{q}^*(\xi)), \mathrm{Im}(\mathbf{q}^*(\xi)))$, respectively, where the factor $\sqrt{2}$ is required for the normalization. So, the basic solution matrices can be written as

$$\boldsymbol{\Phi}(\theta) = \sqrt{2}\begin{bmatrix} \cos\omega\theta & \sin\omega\theta \\ -\omega\sin\omega\theta & \omega\cos\omega\theta \end{bmatrix}, \tag{2.33}$$

and

$$\boldsymbol{\Psi}(\omega\xi) = \frac{\sqrt{2}}{l^2 + m^2}\begin{bmatrix} \hat{c}_1\cos(\omega\xi) + \hat{c}_2\sin(\omega\xi) & \hat{c}_1\sin(\omega\xi) - \hat{c}_2\cos(\omega\xi) \\ l\cos(\omega\xi) - m\sin(\omega\xi) & m\cos(\omega\xi) + l\sin(\omega\xi) \end{bmatrix}, \tag{2.34}$$

where $\hat{c}_1 = (m\omega - l\alpha_c)$, $\hat{c}_2 = (m\alpha_c + l\omega)$. In addition, letting $\boldsymbol{v} \equiv (v_1, v_2)^T = \langle \boldsymbol{\Psi}, \mathbf{u_t} \rangle$ (It represents the local coordinate system on a two-dimensional central manifold, induced by the basis $\boldsymbol{\Psi}$), $\mathbf{u_t}$ can be decomposed into two parts according to Eqs. (2.33) and (2.34), which can be expressed as

$$\mathbf{u_t} = \mathbf{u_t^{P_\Lambda}} + \mathbf{u_t^{Q_\Lambda}} = \boldsymbol{\Phi}\langle \boldsymbol{\Psi}, \mathbf{u_t} \rangle + \mathbf{u_t^{Q_\Lambda}} = \boldsymbol{\Phi}_v + \mathbf{u_t^{Q_\Lambda}}, \tag{2.35}$$

which indicates that $\boldsymbol{\Phi}_v$ is the projection of $\mathbf{u_t}$ on the center manifold. Substituting Eq. (2.35) into Eq. (2.27) and applying the bilinear operator (2.29) with $\boldsymbol{\Psi}$ given by Eq. (2.34) to the resulting equation yield

$$\left\langle \boldsymbol{\Psi}, (\boldsymbol{\Phi}\dot{v} + \dot{\mathbf{u}}_{\mathbf{t}}^{\mathbf{Q_A}})\right\rangle = \left\langle \boldsymbol{\Psi}, [\mathbf{D}(0) + \varepsilon\mathbf{D}(\mu_1, \mu_2) + \varepsilon\mathbf{Q}]\left(\boldsymbol{\Phi}v + \mathbf{u}_{\mathbf{t}}^{\mathbf{Q_A}}\right)\right\rangle.$$

According to Eq. (2.35), we will get

$$\begin{aligned}\langle \boldsymbol{\Psi}, \boldsymbol{\Phi}\rangle\dot{\mathbf{v}} &= \langle \boldsymbol{\Psi}, \mathbf{D}(0)\boldsymbol{\Phi}\rangle\mathbf{v} + \varepsilon\langle \boldsymbol{\Psi}, \mathbf{D}(\mu_1, \mu_2)\boldsymbol{\Phi}\rangle\mathbf{v} + \varepsilon\left\langle \boldsymbol{\Psi}, \left(\boldsymbol{\Phi}v + \mathbf{u}_{\mathbf{t}}^{\mathbf{Q_A}}\right)\right\rangle \\ \Rightarrow \dot{\boldsymbol{v}} &= \begin{bmatrix} 0 & \omega \\ -\omega & 0 \end{bmatrix}\boldsymbol{v} + \varepsilon\mathbf{D}_{\boldsymbol{\varepsilon}}\boldsymbol{v} + \varepsilon\mathbf{N}_{\boldsymbol{\varepsilon}}(\boldsymbol{v}),\end{aligned} \tag{2.36}$$

where $\mathbf{D}_{\mathbf{A}v}$ is in the Jordan canonical form. $\mathbf{D}_{\boldsymbol{\varepsilon}}$ is the coefficients of the $\mathbf{O}(\varepsilon)$ and $\mathbf{N}_{\boldsymbol{\varepsilon}}(\boldsymbol{v})$ represents the nonlinear terms contributed from the original system to the center manifold, which can be written as

$$\begin{aligned}\mathbf{D}_{\boldsymbol{\varepsilon}} &= \frac{2}{l^2 + m^2}\begin{pmatrix} l\mu_2(\cos(\omega\tau) - 1) & l(\mu_1\omega - \mu_2\sin(\omega\tau)) \\ m\mu_2(\cos(\omega\tau) - 1) & m(\mu_1\omega - \mu_2\sin(\omega\tau)) \end{pmatrix}, \\ \mathbf{N}_{\boldsymbol{\varepsilon}}(\boldsymbol{v}) &= \frac{\sqrt{2}}{l^2 + m^2}\Psi\begin{Bmatrix} l \\ m \end{Bmatrix},\end{aligned} \tag{2.37}$$

where $\Psi = k\cos(\Omega t) - 2\sqrt{2}v_1^2(\gamma\omega v_2 + 2\sqrt{2}\beta v_1) + 2\sqrt{2}B((\cos(\omega\tau) - 1)v_1 - \sin(\omega\tau)v_2)^3$. According to Eq. (2.37), we should note that $\mathbf{N}_{\boldsymbol{\varepsilon}}(\boldsymbol{v})$ takes only the leading (cubic) order terms from the nonlinear function $\left\langle \boldsymbol{\Psi}, \mathbf{F}\left(t, \boldsymbol{\Phi}\boldsymbol{v} + \mathbf{u}_{\mathbf{t}}^{\mathbf{Q_A}}\right)\right\rangle$ and thus $\mathbf{N}_{\boldsymbol{\varepsilon}}(\boldsymbol{v})$ is not affected by the part of $\mathbf{u}_{\mathbf{t}}^{\mathbf{Q_A}}$.

2.1.3 *The Phase-Shifting and Phase-Locked Periodic Solutions*

When $\varepsilon = 0$, the solution of Eq. (2.36) can be written as

$$\boldsymbol{v} = (r\cos(\omega t + \theta), -r\sin(\omega t + \theta))^T, \tag{2.38}$$

where r, θ are constants and ω is the vibration frequency and it corresponds to the critical point which is defined by $\alpha = \alpha_c$ and $A = A_c$. When $\varepsilon \neq 0$ and since excitation is weak soft in the order of ε (see Eq. (2.14)), the primary resonance requires us to consider, and therefore, let

$$\Omega = \omega + \varepsilon\sigma, \tag{2.39}$$

where $\sigma = O(1)$ is a detuning parameter. We would like to find the periodic solutions of the system with small ε, which can be seen as perturbing the linear system $\dot{\mathbf{v}} = \mathbf{D}_\Lambda \mathbf{v}$. Thus we may suppose that (2.38) still expresses the solution of Eq. (2.36), but now r and θ are supposed to be functions of t rather than constants. Then substituting Eqs. (2.38) and (2.39) into Eq. (2.36) and making use of the averaging theorem [20], the variable v can be transformed into the new variables r and Θ as

$$\begin{aligned} r' &= \left(a_{11} + a_{12} r^2\right) r + b_1 \cos\Theta + b_2 \sin\Theta, \\ \Theta' &= \sigma + a_{21} + a_{22} r^2 - \frac{b_1}{r}\sin\Theta + \frac{b_2}{r}\cos\Theta, \end{aligned} \tag{2.40}$$

where $r = r(t)$, $\Theta = \Theta(t) = \sigma\varepsilon t - \theta(t)$, $(\cdot)' = d(\cdot)/d(\varepsilon t)$ and

$$a_{11} = \frac{m\omega\mu_1 + (l(\cos(\tau\omega) - 1) - m\sin(\tau\omega))\mu_2}{l^2 + m^2},$$

$$a_{12} = \frac{-(3l(3B+\beta) + m\gamma\omega + 3B(l(\cos(2\tau\omega) - 4\cos(\tau\omega)) + m(2\sin(\tau\omega) - \sin(2\tau\omega))))}{2(l^2+m^2)},$$

$$b_1 = \frac{kl}{\sqrt{2}(l^2+m^2)},\ b_2 = \frac{km}{\sqrt{2}(l^2+m^2)},$$

$$a_{21} = \frac{-l\omega\mu_1 + (m(\cos(\tau\omega) - 1) - l\sin(\tau\omega))\mu_2}{l^2 + m^2},$$

$$a_{22} = \frac{-3m(3B+\beta) + l\gamma\omega - 3B(m(\cos(2\tau\omega) - 4\cos(\tau\omega)) + l(\sin(2\tau\omega) - 2\sin(\tau\omega)))}{2(l^2+m^2)}.$$

Then according to Eq. (2.40), one can obtain that the form of the first order approximate solution as follows:

$$x(t) = u_1(t) = u_{1t}(0) = \sqrt{2}\upsilon_1 + \mathrm{O}(\varepsilon) = \sqrt{2}r\cos(\Omega t - \Theta) + \mathrm{O}(\varepsilon), \tag{2.41}$$

where r and Θ are the solutions of Eq. (2.40). It should be noted that $b_1 = 0$ and $b_2 = 0$ as $k = 0$. Equation (2.40) means that the system has steady state and nonsteady state solutions, and these solutions depend on the time delay, the frequency and the amplitude of the external forcing for $k = 0$. By setting $r = 0$ and $\Theta = 0$, one may gain the steady state solutions, resulting in periodic motions characterized by Eq. (2.41). Because the phase and amplitude of the oscillator are real constant and the frequency of the oscillator synchronizes to that of the external excitation, such a state is referred as a phase-locked. The other collective state is referred as phase shifting when a Hopf bifurcation occurs in the equilibrium point (r, Θ) of Eq. (2.40). In the next section, the formula of the boundary of the phase shifting solutions (BPSS) will be derived and we will describe the regions of driven phase locked solutions (periodic solutions).

In reality, we may gain BPSS by analyzing the stability of the equilibrium point, which is determined by analyzing the Jacobian matrix of the averaging (2.40).

The equilibria of Eq. (2.40) satisfy

$$\begin{aligned} -(a_{11}+a_{12}r^2)\,r &= b_1\cos\Theta + b_2\sin\Theta, \\ -(\sigma+a_{12})\,r - a_{22}r^3 &= -b_1\sin\Theta + b_2\cos\Theta. \end{aligned} \tag{2.42}$$

We can obtain the bifurcation equation with parameters σ, μ_1, μ_2, τ and B by eliminating Θ from Eq. (2.42) and the bifurcation equation is given by

$$E(r^2) = b_1^2 + b_2^2, \tag{2.43}$$

where

$$E(z) = (a_{12}^2 + a_{22}^2)z^3 + 2(a_{11}a_{12} + a_{22}(\sigma + a_{21}))z^2 + \big(a_{11}^2 + (\sigma + a_{21})^2\big)z. \tag{2.44}$$

The form of the Jacobian matrix at the equilibria is taken as

$$\mathbf{J}(r) = \begin{bmatrix} a_{11} + 3a_{12}r^2 & -(\sigma + a_{21})r - a_{22}r^3 \\ 3a_{22}r + \frac{\sigma + a_{21}}{r} & a_{11+a_{12}r^2} \end{bmatrix}. \tag{2.45}$$

The trace and determinant of the Jacobian matrix are, respectively,

$$tr(\mathbf{J}(r)) = 2a_{11} + 4a_{12}r^2,\ \det(\mathbf{J}(r)) = E'\big(r^2\big), \tag{2.46}$$

where E is the derivative of E with respect to z. Since the determinant of the matrix $\mathbf{J}$ represents the product changes crossing to $-a_{11}/2a_{12}$, the system undergoes a Hopf bifurcation if $\det(\mathbf{J}(r)) > 0$. We may determine BPSS for which the system undergoes a Hopf bifurcation by substituting $r_2 = -a_{11}/2a_{12}$ into Eq. (2.43), giving

$$E\left(\frac{-a_{11}}{2a_{12}}\right) = b_1^2 + b_2^2,\ E'\left(\frac{-a_{11}}{2a_{12}}\right) > 0. \tag{2.47}$$

According to Eq. (2.47), the regions of the phase-locked (periodic) solutions and the phase-shifting (complicated) solutions for the driven system with time delayed feedback can be located. And we can find the corresponding Θ component by using Eq. (2.41).

To consider the impact of time delay on BPSS, we first briefly described the situation without time delay, i.e., $\tau = 0$. Then, when τ, k or Ω changes, we study the change in the dynamical behavior of time system. We get that $A_c = 0, \alpha_c = 0, \omega = \omega_0, l = 0, m = 2\omega_0$ by setting $\tau = 0$ in Eq. (2.6) and according to (2.40), resulting in $b_1 = 0$. Then, the parameter values α, Ω and k on BPSS controlled by Eq. (2.47) satisfy the real solution of the following cubic equation:

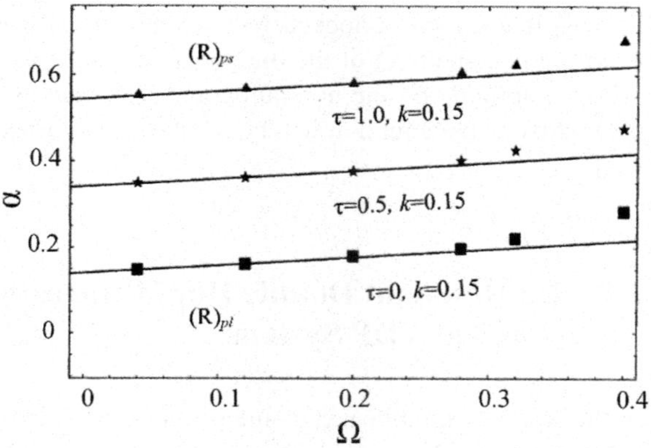

Fig. 2.3 The effect of time delay on BPSS for the driven system. Select three different values of $\tau = 0, 0.5$ and 1.0, and driven amplitude $k = 0.15$ in Eqs. (2.47) and (2.2). Other parameters are the same as those in Fig. 2.2

$$\left(\frac{1}{16\gamma} + \frac{9\beta^2}{16\gamma^3\omega_0^2}\right)\alpha^3 - \frac{3\beta(\Omega - \omega_0)}{2\gamma^2\omega_0}\alpha^2 + \frac{(\Omega - \omega_0)^2}{\gamma}\alpha - \frac{k^2}{8\omega_0^2} = 0 \text{ for } k \neq 0,$$
$$\alpha = 0 \text{ for } k = 0. \tag{2.48}$$

Equation (2.48) is consistent with the results obtained by Nayfeh and Mook [21] for an oscillating system in the absence of time delay. The results of dynamical stability analysis (Hopf bifurcation) for some special cases are plotted in Figs. 2.2 and 2.3, and the results obtained from the numerical integration of Eq. (2.2), for the non-autonomous case with or without time delay feedback. The numerical results corresponding to those special cases are presented using some symbols, such as stars, triangles, and boxes.

Firstly, Fig. 2.2 displays the distribution of the phase-locked and phase-shifting solutions derived from Eqs. (2.48) and (2.2) in the Ω, α space for three chosen amplitudes of the external forcing when $\tau = 0$. Regions (R) pl and (R) ps, separated by BPSS, represent the phase-locked (periodic) solutions and the phase- shifting (complicated) states, respectively. The numerical calculations for the selected amplitudes $k = 0.02$, $k = 0.15$ and $k = 0.5$ obtained from Eq. (2.2) are represented by triangles, stars, and boxes symbols, respectively. The theoretical values obtained from Eq. (2.48) are represented by solid lines. The other parametric values are taken as $\beta = 0.1$, $\gamma = 0.2$, $\omega_0 = 0.2$. From Fig. 2.2 and Eq. (2.48), it can be seen that there is a fundamental difference between the cases $k = 0$ (autonomous system) and $k = 0$ (non-autonomous system). For $k = 0$, the system has a stable trivial solution as $\alpha < 0$, which forms a death island. For $k = 0$, such an island is activated to become a region of phase-locked solutions. Such a region enlarges monotonically in size with increasing driven amplitude.

Next, we investigate the effect of time delay on the dynamics of the non-autonomous system. Based on the results given in Fig. 2.2, we start with $k = 0.15$ and increase τ from zero. For $\tau = 0$, the boundary curves predicted by Eq. (2.48) are plotted in Fig. 2.3 for three different values of $\tau = 0, 0.5$ and 1.0. Clearly, it can be seen that the phase-locked region grows in size with the time delay increasing from 0. Such an effect of the time delay on the system is very similar to that of the driven amplitude on the non-autonomous system as discussed above. Therefore, a delay differential equation will be treated as a non-autonomous differential equation [22].

2.2 1:3 Resonant Double Hopf Bifurcation in a Delayed Coupled VDP System

In the study of nonlinear dynamics, periodic oscillations are often related to Hopf bifurcation [23–26] and double Hopf bifurcation [27–32]. Strong resonance double Hopf bifurcation can lead to energy exchange between oscillators, which has practical applications in engineering [33–35]. Theoretically, it may be a high-dimensional bifurcation that can produce rich dynamic phenomena, making it of great theoretical research value [36]. As a matter of fact, in the power systems with time delay, even a small delay can cause significant changes of the dynamic behaviors, or even lead to complex dynamic behavior [37]. Therefore, time delay cannot be ignored in many practical systems, and more and more researchers have begun to attach importance to the study of the impact of time delay.

In this section, we mainly study the double Hopf bifurcation caused by time delay, which reflects the interaction between two modes in a system. Under the condition of internal resonance, a coupling and energy exchange mechanism is established between the linear modes of the system due to the presence of nonlinear factors. It leads to significant responses between the modes experiencing internal resonance, which are not observed in linear systems without internal resonance.

From the perspective of bifurcation theory, there are rich dynamic behaviors near the double Hopf bifurcation point. By studying the classification of dynamical behavior near the double Hopf bifurcation points in parameter space, it is possible to predict the long-term states of system evolution in different parameter regions, thereby providing theoretical guidance for controlling the system to achieve ideal states. In this section, we employ MMS (the method of multiple scales) to obtain the complex amplitude equation for 1:3 resonance. By representing the complex amplitude in a polar Cartesian hybrid form, the complex amplitude equation for 1:3 resonance is transformed into a real three-dimensional system. Through the analysis of this real system, it is possible to classify the dynamical behavior of the original system and provide a method for studying 1:3 resonance double Hopf bifurcation in time delay system. Furthermore, we validate this method by discussing 1:3 resonance double Hopf bifurcation in two VDP oscillator models with time delay coupling and classify the dynamical behavior near the resonant point.

2.2.1 Multi-scale Analysis for 1:3 Resonance Double Hopf Bifurcation

A general vector form of a dimensional delay differential equation has the following form

$$\dot{\mathbf{x}} = \mathbf{F}(\mathbf{x}, \mathbf{x}_\tau, \boldsymbol{\mu}), \tag{2.49}$$

where $\mathbf{x} \in \mathbb{R}^n$ is a state variable that dependent on parameter $\boldsymbol{\mu} \in \mathbb{R}^m$, $\mathbf{x}_\tau = \mathbf{x}(t-\tau)$ is a state variable with time delay τ. Without losing generality, we assume that $\mathbf{x} = \mathbf{0}$ is always the equilibrium point of the system for any parameter value $\boldsymbol{\mu}$. According to bifurcation theory, if a system undergoes 1:3 resonance double Hopf bifurcation, the two eigenvalues λ_1 and λ_2 of the system must satisfy three conditions, that is, $\mathrm{Re}\lambda_1 = \mathrm{Re}\lambda_2 = 0$ and $\mathrm{Im}\lambda_2 = 3\mathrm{Im}\lambda_1$. Therefore, there are three critical parameter values

$\boldsymbol{\mu}_\mathbf{c} = (\mu_{1c}, \mu_{2c}, \tau_c) \in \mathbb{R}^3$ are determined.

Assuming $\mu_1 = \mu_{1c} + \mu_{1\varepsilon}$, $\mu_2 = \mu_{2c} + \mu_{2\varepsilon}$, $\tau = \tau_c + \tau_\varepsilon$, then the point O,($\mathbf{x} = \mathbf{0}$, $\boldsymbol{\mu}_\varepsilon = (\mu_{1\varepsilon}, \mu_{2\varepsilon}, \tau_\varepsilon) = \mathbf{0}$) is a 1:3 resonance double Hopf bifurcation point. Below we will provide the conditions that 1:3 resonance double Hopf bifurcation must meet

$$\begin{aligned}\mathbf{F}_\mathbf{x}^\mathbf{0} &= \frac{\partial \mathbf{F}(\mathbf{x}, \mathbf{x}_\tau, \boldsymbol{\mu}_\varepsilon)}{\partial \mathbf{x}}|_{\mathbf{x}=\mathbf{0},\mathbf{x}_\tau=\mathbf{0},\boldsymbol{\mu}_\varepsilon=\mathbf{0}},\\ \mathbf{F}_{\mathbf{x}_\tau}^\mathbf{0} &= \frac{\partial \mathbf{F}(\mathbf{x}, \mathbf{x}_\tau, \boldsymbol{\mu}_\varepsilon)}{\partial \mathbf{x}_\tau}|_{\mathbf{x}=\mathbf{0},\mathbf{x}_\tau=\mathbf{0},\boldsymbol{\mu}_\varepsilon=\mathbf{0}}.\end{aligned} \tag{2.50}$$

(C.1) Let the Jacobian matrix $\mathbf{F}_\mathbf{x}^\mathbf{0} + \mathbf{F}_{\mathbf{x}_\tau}^\mathbf{0} e^{-\lambda\tau}$ of (2.49) has two pairs of pure imaginary eigenvalues $\lambda_{1,3} = \pm i\omega_1$ and $\lambda_{2,4} = \pm i\omega_2$, and $\omega_2 = 3\omega_1$, with all other eigenvalues $\lambda_h (h \geq 5)$ on the left side of the complex plane (with negative real parts). The right and left eigenvectors p_j and q_j corresponding to eigenvalues $\lambda_{1,3}$ and $\lambda_{2,4}$ are determined by the following equations

$$\begin{aligned}(\mathbf{F}_\mathbf{x}^\mathbf{0} + \mathbf{F}_{\mathbf{x}_\tau}^\mathbf{0} e^{-i\omega_j\tau} - i\mathbf{E}\omega_j)p_j = 0,\\ ((\mathbf{F}_\mathbf{x}^\mathbf{0} + \mathbf{F}_{\mathbf{x}_\tau}^\mathbf{0} e^{i\omega_j\tau})^T + i\mathbf{E}\omega_j)q_j = 0,\end{aligned} \tag{2.51}$$

where $\mathbf{E}$ is the identity matrix, $\mathbf{p_3} = \overline{\mathbf{p}}_\mathbf{1}$, $\mathbf{p_4} = \overline{\mathbf{p}}_\mathbf{2}$, $\mathbf{q_3} = \overline{\mathbf{q}}_\mathbf{1}$ and $\mathbf{q_4} = \overline{\mathbf{q}}_\mathbf{2}$. The right and left eigenvectors satisfy $\mathbf{q_i}^\mathbf{H}\,\mathbf{p_i} = 1$, $(i = 1, 2)$, where H represents transposed conjugation.

(C.2) The critical eigenvalues $\lambda_{1,3} = \alpha_1(\mu_{1\varepsilon}, \mu_{2\varepsilon}, \tau_\varepsilon) \pm i\omega_1(\mu_{1\varepsilon}, \mu_{2\varepsilon}, \tau_\varepsilon)$ and $\lambda_{2,4} = \alpha_2(\mu_{1\varepsilon}, \mu_{2\varepsilon}, \tau_\varepsilon) \pm i\omega_2(\mu_{1\varepsilon}, \mu_{2\varepsilon}, \tau_\varepsilon)$ satisfy the transversality conditions $\alpha_j(0, 0, 0) = 0\,(j = 1, 2)$ and $\omega_2(0, 0, 0) = 3\omega_1(0, 0, 0)$ at 1:3 resonance point O.

In the parameters space $(\mu_{1\varepsilon}, \mu_{2\varepsilon}, \tau_\varepsilon)$, $\alpha_j(\mu_{1\varepsilon}, \mu_{2\varepsilon}, \tau_\varepsilon) = 0 (j = 1, 2)$ are the critical surfaces for constraining the linear stability region of trivial solutions. A double

Hopf bifurcation occurs at the intersection of two surfaces, and the 1:3 resonance double Hopf bifurcation occurs at an intersecting line that satisfies $\omega_2 = 3\omega_1$.

Next, we will study the dynamic behavior of the system near the 1 : 3 resonance point by applying MMS. So, the solution of the differential Eq. (2.49) can be written as follows

$$\mathbf{x} = \mathbf{x}(\varepsilon, T_k, \ldots), \tag{2.52}$$

where $T_k = \varepsilon^k t$, $(k = 0, 2, \ldots)$, $(\varepsilon \ll 1)$.

Due to the correlation between 1 : 3 resonance and the influence of the cubic term, its solution does not depend on the time scale T_1. Therefore, we assume that the solution of Eq. (2.49) only depends on the time scales T_0 and T_2, which can be written as

$$\mathbf{x}(t) = \varepsilon \mathbf{x_1}(T_0, T_2) + \varepsilon^2 \mathbf{x_2}(T_0, T_2) + \varepsilon^3 \mathbf{x_3}(T_0, T_2) + \mathbf{O}(\varepsilon^4). \tag{2.53}$$

Set the three perturbation parameters as

$$\mu_{1\varepsilon} = \varepsilon^2 \hat{\mu}_{1\varepsilon}, \mu_{2\varepsilon} = \varepsilon^2 \hat{\mu}_{2\varepsilon}, \tau_\varepsilon = \varepsilon^2 \hat{\tau}_\varepsilon . \tag{2.54}$$

By Taylor expansion of the time delay term, it can be obtained that

$$\begin{aligned}
\mathbf{x}(t - \tau) &= \varepsilon \mathbf{x_1}(T_0 - \tau_c - \varepsilon^2 \hat{\tau}_\varepsilon, T_2 - \varepsilon^2 \tau_c - \varepsilon^4 \hat{\tau}_\varepsilon) \\
&\quad + \varepsilon^2 \mathbf{x_2}(T_0 - \tau_c - \varepsilon^2 \hat{\tau}_\varepsilon, T_2 - \varepsilon^2 \tau_c - \varepsilon^4 \hat{\tau}_\varepsilon) \\
&\quad + \varepsilon^3 \mathbf{x_3}(T_0 - \tau_c - \varepsilon^2 \hat{\tau}_\varepsilon, T_2 - \varepsilon^2 \tau_c - \varepsilon^4 \hat{\tau}_\varepsilon) + \cdots \\
&= \varepsilon \mathbf{x_1}(T_0 - \tau_c, T_2) + \varepsilon^2 \mathbf{x_2}(T_0 - \tau_c, T_2) \\
&\quad + \varepsilon^3 [-\hat{\tau}_\varepsilon D_0 \mathbf{x_1}(T_0 - \tau_c, T_2) - \tau_c D_2 \mathbf{x_1}(T_0 - \tau_c, T_2) \\
&\quad + \mathbf{x_3}(T_0 - \tau_c, T_2)] + \mathbf{O}(\varepsilon^4),
\end{aligned} \tag{2.55}$$

where $D_k = \partial / \partial T_k$.

Substituting Eqs. (2.53), (2.54), and (2.55) into (2.49), Taylor expands $\mathbf{F}(\mathbf{x}, \mathbf{x}_\tau, \boldsymbol{\mu})$ can be obtained. Then we make the coefficients of each power of ε at both ends of the equation equal, resulting in the following perturbation equation

$$D_0 \mathbf{x_1} - \mathbf{F_x^0} \mathbf{x_1} - \mathbf{F_{x_\tau}^0} \mathbf{x}_{1\tau} = \mathbf{0}, \tag{2.56}$$

$$D_0 \mathbf{x_2} - \mathbf{F_x^0} \mathbf{x_2} - \mathbf{F_{x_\tau}^0} \mathbf{x}_{2\tau} = \tfrac{1}{2} \mathbf{F_{xx}^0} \mathbf{x_1^2} + \mathbf{F_{xx_\tau}^0} \mathbf{x_1} \mathbf{x}_{1\tau} + \tfrac{1}{2} \mathbf{F_{x_\tau x_\tau}^0} \mathbf{x}_{1\tau}^2, \tag{2.57}$$

$$\Delta \tau_1 = -4, \tag{2.58}$$

where $\mathbf{F}^0_{\mathbf{x}\mu_\varepsilon} = \frac{\partial^2 \mathbf{F}(0,0,0)}{\partial \mathbf{x}\partial \boldsymbol{\mu}_\varepsilon}$, $\mathbf{F}^0_{\mathbf{x}_\tau \boldsymbol{\mu}_\varepsilon} = \frac{\partial^2 \mathbf{F}(0,0,0)}{\partial \mathbf{x}_\tau \partial \boldsymbol{\mu}_\varepsilon}$, and higher order partial derivatives $\mathbf{x_j}_\tau = \mathbf{x_j}(T_0 - \tau_c, T_2), j = 1, 2, 3$ can also be obtained similarly.

The solution of Eq. (2.56) has the following form:

$$\mathbf{x}_1 = \mathbf{A}_1(T_2)\mathbf{p}_1 e^{i\omega_1 T_0} + \mathbf{A}_2(T_2)\mathbf{p}_2 e^{i\omega_2 T_0} + c.c., \tag{2.59}$$

where $\mathbf{x}(t) = \mathbf{q} + d\dot{\mathbf{q}}\mathbf{A_j} (j = 1, 2)$ is the complex amplitude, $\mathbf{p_j}$ is the right eigenvector of the matrix $\mathbf{F}^0_{\mathbf{x}} + \mathbf{F}^0_{\mathbf{x}_\tau} e^{-i\omega_j \tau_c}$ corresponding to the eigenvalues ω_ji, and $c.c.$ represents the complex conjugation of the previous terms.

Substituting Eq. (2.59) into Eq. (2.57) yields

$$\begin{aligned}
D_0\mathbf{x}_2 - \mathbf{F}^0_{\mathbf{x}}\mathbf{x}_2 - \mathbf{F}^0_{\mathbf{x}_\tau}\mathbf{x}_{2\tau} &= (\frac{1}{2}\mathbf{F}^0_{\mathbf{xx}}\mathbf{p}_1^2 + \mathbf{F}^0_{\mathbf{xx}_\tau}\mathbf{p}_1^2 e^{-i\omega_1\tau_c} + \frac{1}{2}\mathbf{F}^0_{\mathbf{x}_\tau\mathbf{x}_\tau}\mathbf{p}_1^2 e^{-2i\omega_1\tau_c})\mathbf{A}_1^2 e^{2i\omega_1 T_0} \\
&+ \left(\frac{1}{2}\mathbf{F}^0_{\mathbf{xx}}\mathbf{p}_2^2 + \mathbf{F}^0_{\mathbf{xx}_\tau}\mathbf{p}_2^2 e^{-i\omega_2\tau_c} + \frac{1}{2}\mathbf{F}^0_{\mathbf{x}_\tau\mathbf{x}_\tau}\mathbf{p}_2^2 e^{-2i\omega_2\tau_c}\right)\mathbf{A}_2^2 e^{2i\omega_1 T_0} \\
&+ \left(\frac{1}{2}\mathbf{F}^0_{\mathbf{xx}}\mathbf{p}_1\,\overline{\mathbf{p}}_1 + \mathbf{F}^0_{\mathbf{xx}_\tau}\mathbf{p}_1\,\overline{\mathbf{p}}_1\,e^{i\omega_1\tau_c} + \frac{1}{2}\mathbf{F}^0_{\mathbf{x}_\tau\mathbf{x}_\tau}\mathbf{p}_1\,\overline{\mathbf{p}}_1\right)\mathbf{A}_1\overline{\mathbf{A}}_1 \\
&+ \left(\frac{1}{2}\mathbf{F}^0_{\mathbf{xx}}\mathbf{p}_2\,\overline{\mathbf{p}}_2 + \mathbf{F}^0_{\mathbf{xx}_\tau}\mathbf{p}_2\,\overline{\mathbf{p}}_2\,e^{i\omega_2\tau_c} + \frac{1}{2}\mathbf{F}^0_{\mathbf{x}_\tau\mathbf{x}_\tau}\mathbf{p}_2\,\overline{\mathbf{p}}_2\right)\mathbf{A}_2\overline{\mathbf{A}}_2 \\
&+ \left[\mathbf{F}^0_{\mathbf{xx}}\mathbf{p}_1\mathbf{p}_2 + \mathbf{F}^0_{\mathbf{xx}_\tau}(\mathbf{p}_1\mathbf{p}_2 e^{-i\omega_1\tau_c} + \mathbf{p}_1\mathbf{p}_2 e^{-i\omega_2\tau_c})\right. \\
&\left.+ \mathbf{F}^0_{\mathbf{x}_\tau\mathbf{x}_\tau}\mathbf{p}_1\mathbf{p}_2 e^{-i(\omega_1+\omega_2)\tau_c}\right]\mathbf{A}_1\mathbf{A}_2 e^{i(\omega_1+\omega_2)T_0} \\
&+ \left[\mathbf{F}^0_{\mathbf{xx}}\,\overline{\mathbf{p}}_1\,\mathbf{p}_2 + \mathbf{F}^0_{\mathbf{xx}_\tau}(\overline{\mathbf{p}}_1\,\mathbf{p}_2 e^{i\omega_1\tau_c} + \overline{\mathbf{p}}_1\,\mathbf{p}_2 e^{-i\omega_2\tau_c})\right. \\
&\left.+ \mathbf{F}^0_{\mathbf{x}_\tau\mathbf{x}_\tau}\,\overline{\mathbf{p}}_1\,\mathbf{p}_2 e^{i(\omega_1-\omega_2)\tau_c}\right]\overline{\mathbf{A}}_1\mathbf{A}_2 e^{i(\omega_2-\omega_1)T_0} + c.c..
\end{aligned} \tag{2.60}$$

The solution of Eq. (2.60) has the following form

$$\begin{aligned}
\mathbf{x}_2 = \mathbf{A}_1^2\mathbf{z}_{11}e^{2i\omega_1 T_0} + \mathbf{A}_2^2\mathbf{z}_{22}e^{2i\omega_2 T_0} + \mathbf{A}_1\overline{\mathbf{A}}_1\mathbf{z}_{1\overline{1}} \\
+\mathbf{A}_2\overline{\mathbf{A}}_2\mathbf{z}_{2\overline{2}} + \mathbf{A}_1\mathbf{A}_2\mathbf{z}_{12}e^{-i(\omega_1+\omega_2)T_0} \\
+\overline{\mathbf{A}}_1\,\mathbf{A}_2\mathbf{z}_{\overline{1}2}e^{i(\omega_2-\omega_1)T_0} + c.c.,
\end{aligned} \tag{2.61}$$

where the vectors $\mathbf{z_{rs}}$ and $\mathbf{z_{r\overline{s}}}(r, s = 1, 2) \in \mathbb{C}^n$ can be obtained by solving the following equation

$$\begin{aligned}
(2i\omega_1\mathbf{E} - \mathbf{F}^0_{\mathbf{x}} - \mathbf{F}^0_{\mathbf{x}_\tau}e^{-2i\omega_1\tau_c})\mathbf{z}_{11} &= \frac{1}{2}\mathbf{F}^0_{\mathbf{xx}}\mathbf{p}_1^2 + \mathbf{F}^0_{\mathbf{xx}_\tau}\mathbf{p}_1^2 e^{-i\omega_1\tau_c} \\
&+ \frac{1}{2}\mathbf{F}^0_{\mathbf{x}_\tau\mathbf{x}_\tau}\mathbf{p}_1^2 e^{-2i\omega_1\tau_c}, \\
(2i\omega_2\mathbf{E} - \mathbf{F}^0_{\mathbf{x}} - \mathbf{F}^0_{\mathbf{x}_\tau}e^{-2i\omega_2\tau_c})\mathbf{z}_{22} &= \frac{1}{2}\mathbf{F}^0_{\mathbf{xx}}\mathbf{p}_2^2 + \mathbf{F}^0_{\mathbf{xx}_\tau}\mathbf{p}_2^2 e^{-i\omega_2\tau_c} \\
&+ \frac{1}{2}\mathbf{F}^0_{\mathbf{x}_\tau\mathbf{x}_\tau}\mathbf{p}_2^2 e^{-2i\omega_2\tau_c},
\end{aligned}$$

$$(-\mathbf{F}_{\mathbf{x}}^{\mathbf{0}} - \mathbf{F}_{\mathbf{x}_\tau}^{\mathbf{0}})\mathbf{z}_{1\bar{1}} = \frac{1}{2}\mathbf{F}_{\mathbf{xx}}^{\mathbf{0}}\mathbf{p}_1\overline{\mathbf{p}}_1 + \mathbf{F}_{\mathbf{xx}_\tau}^{\mathbf{0}}\mathbf{p}_1\overline{\mathbf{p}}_1 e^{i\omega_1\tau_c}$$
$$+ \frac{1}{2}\mathbf{F}_{\mathbf{x}_\tau\mathbf{x}_\tau}^{\mathbf{0}}\mathbf{p}_1\overline{\mathbf{p}}_1, \quad (-\mathbf{F}_{\mathbf{x}}^{\mathbf{0}} - \mathbf{F}_{\mathbf{x}_\tau}^{\mathbf{0}})\mathbf{z}_{2\bar{2}}$$
$$= \frac{1}{2}\mathbf{F}_{\mathbf{xx}}^{\mathbf{0}}\mathbf{p}_2\overline{\mathbf{p}}_2 + \mathbf{F}_{\mathbf{xx}_\tau}^{\mathbf{0}}\mathbf{p}_2\overline{\mathbf{p}}_2 e^{i\omega_1\tau_c} + \frac{1}{2}\mathbf{F}_{\mathbf{x}_\tau\mathbf{x}_\tau}^{\mathbf{0}}\mathbf{p}_2\overline{\mathbf{p}}_2,$$
$$(i\omega_1\mathbf{E} + i\omega_2\mathbf{E} - \mathbf{F}_{\mathbf{x}}^{\mathbf{0}} - \mathbf{F}_{\mathbf{x}_\tau}^{\mathbf{0}} e^{-i(\omega_1+\omega_2)\tau_c})\mathbf{z}_{12}$$
$$= \mathbf{F}_{\mathbf{xx}}^{\mathbf{0}}\mathbf{p}_1\mathbf{p}_2 + \mathbf{F}_{\mathbf{xx}_\tau}^{\mathbf{0}}(\mathbf{p}_1\mathbf{p}_2 e^{-i\omega_1\tau_c} + \mathbf{p}_1\mathbf{p}_2 e^{-i\omega_2\tau_c})$$
$$+ \mathbf{F}_{\mathbf{x}_\tau\mathbf{x}_\tau}^{\mathbf{0}}\mathbf{p}_1\mathbf{p}_2 e^{-i(\omega_1+\omega_2)\tau_c},$$
$$(-i\omega_1\mathbf{E} + i\omega_2\mathbf{E} - \mathbf{F}_{\mathbf{x}}^{\mathbf{0}} - \mathbf{F}_{\mathbf{x}_\tau}^{\mathbf{0}} e^{i(\omega_1-\omega_2)\tau_c})\mathbf{z}_{\bar{1}2} = \mathbf{F}_{\mathbf{xx}}^{\mathbf{0}}\,\overline{\mathbf{p}}_1\,\mathbf{p}_2$$
$$+ \mathbf{F}_{\mathbf{xx}_\tau}^{\mathbf{0}}(\overline{\mathbf{p}}_1\,\mathbf{p}_2 e^{i\omega_1\tau_c} + \overline{\mathbf{p}}_1\,\mathbf{p}_2 e^{-i\omega_2\tau_c})$$
$$+ \mathbf{F}_{\mathbf{x}_\tau\mathbf{x}_\tau}^{\mathbf{0}}\,\overline{\mathbf{p}}_1\,\mathbf{p}_2 e^{i(\omega_1-\omega_2)\tau_c}.$$

By substituting Eqs. (2.59) and (2.61) into Eq. (2.58) and eliminating the secular terms, two equations for $D_2\mathbf{A}_1$ and $D_2\mathbf{A}_2$ can be obtained. Applying the left eigenvector to eliminate the coefficients before $D_2\mathbf{A}_1$, $D_2\mathbf{A}_2$ and absorb the parameters ε [38], the following complex amplitude equation can be obtained

$$\begin{aligned}\dot{\mathbf{A}}_1 &= \mathbf{C}_{1\mu}\boldsymbol{\mu}_\varepsilon\mathbf{A}_1 + \mathbf{C}_{11\bar{1}}\mathbf{A}_1^2\overline{\mathbf{A}}_1 + \mathbf{C}_{\bar{1}\bar{1}2}\overline{\mathbf{A}}_1^2\mathbf{A}_2 + \mathbf{C}_{12\bar{2}}\mathbf{A}_1\mathbf{A}_2\overline{\mathbf{A}}_2,\\ \dot{\mathbf{A}}_2 &= \mathbf{C}_{2\mu}\boldsymbol{\mu}_\varepsilon\mathbf{A}_2 + \mathbf{C}_{111}\mathbf{A}_1^3 + \mathbf{C}_{1\bar{1}2}\mathbf{A}_1\overline{\mathbf{A}}_1\mathbf{A}_2 + \mathbf{C}_{22\bar{2}}\mathbf{A}_2^2\overline{\mathbf{A}}_2,\end{aligned} \tag{2.62}$$

where the expressions for coefficients $\mathbf{C_{ijk}}$ and $\mathbf{C}_{i\mu}\boldsymbol{\mu}_\varepsilon$ are given by the following equation

$$\mathbf{C}_{1\mu}\boldsymbol{\mu}_\varepsilon = \frac{1}{\mathbf{M}_1}\mathbf{q}_1^{\mathbf{H}}(\mathbf{F}_{\mathbf{x}\mu}^{\mathbf{0}}\hat{\mu}_\varepsilon\mathbf{p}_1 + \mathbf{F}_{\mathbf{x}_\tau\boldsymbol{\mu}_\varepsilon}^{\mathbf{0}}\hat{\mu}_\varepsilon\mathbf{p}_1 e^{-i\omega_1\tau_c} - i\omega_1\hat{\tau}_\varepsilon\mathbf{F}_{\mathbf{x}_\tau}^{\mathbf{0}}\mathbf{p}_1 e^{-i\omega_1\tau_c}),$$

$$\mathbf{C}_{11\bar{1}} = \frac{1}{\mathbf{M}_1}\mathbf{q}_1^{\mathbf{H}}[\frac{1}{2}\mathbf{F}_{\mathbf{xxx}}^{\mathbf{0}}\mathbf{p}_1^2\,\overline{\mathbf{p}}_1 + \mathbf{F}_{\mathbf{xx}}^{\mathbf{0}}(\mathbf{p}_1\overline{\mathbf{z}}_{1\bar{1}} + \mathbf{z}_{11}\,\overline{\mathbf{p}}_1 + \mathbf{z}_{1\bar{1}}\mathbf{p}_1)$$
$$+ \frac{1}{2}\mathbf{F}_{\mathbf{xxx}_\tau}^{\mathbf{0}}(2\mathbf{p}_1^2\,\overline{\mathbf{p}}_1\,e^{-i\omega_1\tau_c} + \mathbf{p}_1^2\,\overline{\mathbf{p}}_1\,e^{i\omega_1\tau_c})$$
$$+ \mathbf{F}_{\mathbf{xx}_\tau}^{\mathbf{0}}(\mathbf{p}_1\,\overline{\mathbf{z}}_{1\bar{1}}\,e^{-i\omega_1\tau_c} + \mathbf{z}_{11}\,\overline{\mathbf{p}}_1\,e^{i\omega_1\tau_c} + \mathbf{z}_{1\bar{1}}\mathbf{p}_1 e^{-i\omega_1\tau_c})$$
$$+ \frac{1}{2}\mathbf{F}_{\mathbf{xx}_\tau\mathbf{x}_\tau}^{\mathbf{0}}(2\mathbf{p}_1^2\,\overline{\mathbf{p}}_1 + \mathbf{p}_1^2\,\overline{\mathbf{p}}_1\,e^{-2i\omega_1\tau_c}) + \frac{1}{2}\mathbf{F}_{\mathbf{x}_\tau\mathbf{x}_\tau\mathbf{x}_\tau}^{\mathbf{0}}\mathbf{p}_1^2\,\overline{\mathbf{p}}_1\,e^{-i\omega_1\tau_c}$$
$$+ \mathbf{F}_{\mathbf{xx}_\tau}^{\mathbf{0}}(\mathbf{p}_1\,\overline{\mathbf{z}}_{1\bar{1}} + \mathbf{z}_{11}\,\overline{\mathbf{p}}_1\,e^{-2i\omega_1\tau_c} + \mathbf{z}_{1\bar{1}}\mathbf{p}_1)$$
$$+ \mathbf{F}_{\mathbf{x}_\tau\mathbf{x}_\tau}^{\mathbf{0}}(\mathbf{p}_1\overline{\mathbf{z}}_{1\bar{1}}e^{-i\omega_1\tau_c} + \mathbf{z}_{11}\,\overline{\mathbf{p}}_1\,e^{-i\omega_1\tau_c} + \mathbf{z}_{1\bar{1}}\mathbf{p}_1 e^{-i\omega_1\tau_c})],$$

$$\mathbf{C}_{\bar{1}\bar{1}2} = \frac{1}{\mathbf{M}_1}\mathbf{q}_1^{\mathbf{H}}\Big[\frac{1}{2}\mathbf{F}_{\mathbf{xxx}}^{\mathbf{0}}\mathbf{p}_2\,\overline{\mathbf{p}}_1^2 + \mathbf{F}_{\mathbf{xx}}^{\mathbf{0}}(\overline{\mathbf{z}}_{11}\mathbf{p}_2 + \mathbf{z}_{\bar{1}2}\overline{\mathbf{p}}_1)$$
$$+ \frac{1}{2}\mathbf{F}_{\mathbf{xxx}_\tau}^{\mathbf{0}}(2\mathbf{p}_2\,\overline{\mathbf{p}}_1^2\,e^{i\omega_1\tau_c} + \mathbf{p}_2\,\overline{\mathbf{p}}_1^2\,e^{-i\omega_2\tau_c})$$

$$+ \mathbf{F}^{\mathbf{0}}_{\mathbf{xx}_\tau}(\overline{\mathbf{z}}_{11}\mathbf{p}_2 e^{-3i\omega_1\tau_c} + \mathbf{z}_{\overline{1}2}\overline{\mathbf{p}}_1 e^{i\omega_1\tau_c})$$
$$+ \frac{1}{2}\mathbf{F}^{\mathbf{0}}_{\mathbf{xx}_\tau\mathbf{x}_\tau}(2\mathbf{p}_2\overline{\mathbf{p}}_1^2 e^{-2i\omega_1\tau_c} + \mathbf{p}_2\overline{\mathbf{p}}_1^2 e^{2i\omega_1\tau_c})$$
$$+ \frac{1}{2}\mathbf{F}^{\mathbf{0}}_{\mathbf{x}_\tau\mathbf{x}_\tau\mathbf{x}_\tau}\mathbf{p}_2\overline{\mathbf{p}}_1^2 e^{-i\omega_1\tau_c} + \mathbf{F}^{\mathbf{0}}_{\mathbf{xx}_\tau}(\mathbf{p}_2\overline{\mathbf{z}}_{11}e^{2i\omega_1\tau_c} + \mathbf{z}_{\overline{1}2}\overline{\mathbf{p}}_1 e^{-2i\omega_1\tau_c})$$
$$+ \mathbf{F}^{\mathbf{0}}_{\mathbf{x}_\tau\mathbf{x}_\tau}(\mathbf{p}_2\,\overline{\mathbf{z}}_{11}\,e^{-i\omega_1\tau_c} + \mathbf{z}_{\overline{1}2}\,\overline{\mathbf{p}}_1\,e^{-i\omega_1\tau_c})],$$

$$\mathbf{C}_{12\overline{2}} = \frac{1}{\mathbf{M}_1}\mathbf{q}_1^{\mathbf{H}}[\mathbf{F}^{\mathbf{0}}_{\mathbf{xxx}}\mathbf{p}_1\mathbf{p}_2\overline{\mathbf{p}}_2$$
$$+ \mathbf{F}^{\mathbf{0}}_{\mathbf{xx}}(\overline{\mathbf{z}}_{2\overline{2}}\mathbf{p}_1 + \overline{\mathbf{z}}_{\overline{1}2}\mathbf{p}_2 + \mathbf{z}_{12}\overline{\mathbf{p}}_2 + \mathbf{z}_{2\overline{2}}\mathbf{p}_1)$$
$$+ \frac{1}{2}\mathbf{F}^{\mathbf{0}}_{\mathbf{xxx}_\tau}(2\mathbf{p}_1\mathbf{p}_2\overline{\mathbf{p}}_2 e^{-i\omega_1\tau_c}$$
$$+ 2\mathbf{p}_1\mathbf{p}_2\overline{\mathbf{p}}_2 e^{i\omega_2\tau_c} + 2\mathbf{p}_1\mathbf{p}_2\overline{\mathbf{p}}_2 e^{-i\omega_2\tau_c})$$
$$+ \mathbf{F}^{\mathbf{0}}_{\mathbf{xx}_\tau}(\overline{\mathbf{z}}_{2\overline{2}}\mathbf{p}_1 e^{-i\omega_1\tau_c}$$
$$+ \overline{\mathbf{z}}_{\overline{1}2}\mathbf{p}_2 e^{-3i\omega_1\tau_c} + \mathbf{z}_{12}\overline{\mathbf{p}}_2 e^{3i\omega_1\tau_c} + \mathbf{z}_{2\overline{2}}\mathbf{p}_1 e^{-i\omega_1\tau_c})$$
$$+ \frac{1}{2}\mathbf{F}^{\mathbf{0}}_{\mathbf{xx}_\tau\mathbf{x}_\tau}(2\mathbf{p}_1\mathbf{p}_2\overline{\mathbf{p}}_2 + 2\mathbf{p}_1\mathbf{p}_2\overline{\mathbf{p}}_2 e^{2i\omega_1\tau_c} + 2\mathbf{p}_1\mathbf{p}_2\overline{\mathbf{p}}_2 e^{-4i\omega_1\tau_c})$$
$$+ \mathbf{F}^{\mathbf{0}}_{\mathbf{x}_\tau\mathbf{x}_\tau\mathbf{x}_\tau}\mathbf{p}_1\mathbf{p}_2\,\overline{\mathbf{p}}_2\,e^{-i\omega_1\tau_c}$$
$$+ \mathbf{F}^{\mathbf{0}}_{\mathbf{xx}_\tau}(\overline{\mathbf{z}}_{2\overline{2}}\mathbf{p}_1 + \overline{\mathbf{z}}_{\overline{1}2}\mathbf{p}_2 e^{2i\omega_1\tau_c} + \mathbf{z}_{12}\overline{\mathbf{p}}_2 e^{-4i\omega_1\tau_c} + \mathbf{z}_{2\overline{2}}\mathbf{p}_1)$$
$$+ \mathbf{F}^{\mathbf{0}}_{\mathbf{x}_\tau\mathbf{x}_\tau}(\overline{\mathbf{z}}_{2\overline{2}}\mathbf{p}_1 e^{-i\omega_1\tau_c} + \overline{\mathbf{z}}_{\overline{1}2}\mathbf{p}_2 e^{-i\omega_1\tau_c} + \mathbf{z}_{12}\overline{\mathbf{p}}_2 e^{-i\omega_1\tau_c}$$
$$+ \mathbf{z}_{2\overline{2}}\mathbf{p}_1 e^{-i\omega_1\tau_c})],$$

$$\mathbf{C}_{2\mu}\boldsymbol{\mu}_{\boldsymbol{\varepsilon}} = \frac{1}{\mathbf{M}_2}\mathbf{q}_2^{\mathbf{H}}(\mathbf{F}^{\mathbf{0}}_{\mathbf{x}\mu_\varepsilon}\hat{\mu}_\varepsilon\mathbf{p}_2 + \mathbf{F}^{\mathbf{0}}_{\mathbf{x}_\tau\boldsymbol{\mu}_\varepsilon}\hat{\mu}_\varepsilon\mathbf{p}_2 e^{-i\omega_2\tau_c} - i\omega_2\hat{\tau}_\varepsilon\mathbf{F}^{\mathbf{0}}_{\mathbf{x}_\tau}\mathbf{p}_2 e^{-i\omega_2\tau_c}),$$

$$\mathbf{C}_{111} = \frac{1}{\mathbf{M}_2}\mathbf{q}_2^{\mathbf{H}}\left[\frac{1}{6}\mathbf{F}^{\mathbf{0}}_{\mathbf{xxx}}\,\mathbf{p}_1^3\right.$$
$$+ \mathbf{F}^{\mathbf{0}}_{\mathbf{xx}}\mathbf{p}_1\mathbf{z}_{11} + \frac{1}{2}\mathbf{F}^{\mathbf{0}}_{\mathbf{xxx}_\tau}\mathbf{p}_1^3 e^{-i\omega_1\tau_c}$$
$$+ \mathbf{F}^{\mathbf{0}}_{\mathbf{xx}_\tau}\mathbf{p}_1\mathbf{z}_{11}(e^{-i\omega_1\tau_c} + e^{-2i\omega_1\tau_c})$$
$$+ \frac{1}{2}\mathbf{F}^{\mathbf{0}}_{\mathbf{xx}_\tau\mathbf{x}_\tau}\mathbf{p}_1^3 e^{-2i\omega_1\tau_c} + \frac{1}{6}\mathbf{F}^{\mathbf{0}}_{\mathbf{x}_\tau\mathbf{x}_\tau\mathbf{x}_\tau}\mathbf{p}_1^3 e^{-3i\omega_1\tau_c}$$
$$\left. + \mathbf{F}^{\mathbf{0}}_{\mathbf{x}_\tau\mathbf{x}_\tau}\mathbf{p}_1\mathbf{z}_{11}e^{-3i\omega_1\tau_c}\right],$$

$$\mathbf{C}_{1\overline{1}2} = \frac{1}{\mathbf{M}_2}\mathbf{q}_2^{\mathbf{H}}[\mathbf{F}^{\mathbf{0}}_{\mathbf{xxx}}\mathbf{p}_1\mathbf{p}_2\overline{\mathbf{p}}_1 + \mathbf{F}^{\mathbf{0}}_{\mathbf{xx}}(\overline{\mathbf{z}}_{1\overline{1}}\mathbf{p}_2$$
$$+ \mathbf{z}_{12}\overline{\mathbf{p}}_1 + \mathbf{z}_{1\overline{1}}\mathbf{p}_2 + \mathbf{z}_{\overline{1}2}\mathbf{p}_1)$$
$$+ \frac{1}{2}\mathbf{F}^{\mathbf{0}}_{\mathbf{xxx}_\tau}(2\mathbf{p}_1\mathbf{p}_2\,\overline{\mathbf{p}}_1\,e^{-i\omega_2\tau_c}$$

$$
\begin{aligned}
&+ 2\mathbf{p}_1\mathbf{p}_2\overline{\mathbf{p}}_1 e^{i\omega_1\tau_c} + 2\mathbf{p}_1\mathbf{p}_2\overline{\mathbf{p}}_1 e^{-i\omega_1\tau_c})\\
&+ \mathbf{F}^0_{\mathbf{x}\mathbf{x}_\tau}(\overline{\mathbf{z}}_{1\overline{1}}\mathbf{p}_2 e^{-3i\omega_1\tau_c} + \mathbf{z}_{12}\overline{\mathbf{p}}_1 e^{i\omega_1\tau_c}\\
&+ \mathbf{z}_{1\overline{1}}\mathbf{p}_2 e^{-3i\omega_1\tau_c} + \mathbf{z}_{\overline{1}2}\mathbf{p}_1 e^{-i\omega_1\tau_c})\\
&+ \frac{1}{2}\mathbf{F}^0_{\mathbf{x}\mathbf{x}_\tau\mathbf{x}_\tau}(2\mathbf{p}_1\mathbf{p}_2\overline{\mathbf{p}}_1\\
&+ 2\mathbf{p}_1\mathbf{p}_2\,\overline{\mathbf{p}}_1\, e^{-2i\omega_1\tau_c} + 2\mathbf{p}_1\mathbf{p}_2\,\overline{\mathbf{p}}_1\, e^{-4i\omega_1\tau_c})\\
&+ \mathbf{F}^0_{\mathbf{x}_\tau\mathbf{x}_\tau\mathbf{x}_\tau}\mathbf{p}_1\mathbf{p}_2\,\overline{\mathbf{p}}_1\, e^{-3i\omega_1\tau_c} + \mathbf{F}^0_{\mathbf{x}\mathbf{x}_\tau}(\overline{\mathbf{z}}_{1\overline{1}}\,\mathbf{p}_2 + \mathbf{z}_{12}\overline{\mathbf{p}}_1 e^{-4i\omega_1\tau_c}\\
&+ \mathbf{z}_{1\overline{1}}\mathbf{p}_2 + \mathbf{z}_{\overline{1}2}\mathbf{p}_1 e^{-2i\omega_1\tau_c})\\
&+ \mathbf{F}^0_{\mathbf{x}_\tau\mathbf{x}_\tau}(\overline{\mathbf{z}}_{1\overline{1}}\,\mathbf{p}_2 e^{-3i\omega_1\tau_c} + \mathbf{z}_{12}\,\overline{\mathbf{p}}_1\, e^{-i\omega_1\tau_c} + \mathbf{z}_{1\overline{1}}\mathbf{p}_2 e^{-3i\omega_1\tau_c}\\
&+ \mathbf{z}_{\overline{1}2}\mathbf{p}_1 e^{-3i\omega_1\tau_c})],
\end{aligned}
$$

$$
\begin{aligned}
\mathbf{C}_{22\overline{2}} = \frac{1}{\mathbf{M}_2}\mathbf{q}_2^{\mathbf{H}}\Big[&\frac{1}{2}\mathbf{F}^0_{\mathbf{xxx}}\,\mathbf{p}_2^2\overline{\mathbf{p}}_2 + \mathbf{F}^0_{\mathbf{xx}}(\overline{\mathbf{z}}_{2\overline{2}}\mathbf{p}_2 + \mathbf{z}_{22}\overline{\mathbf{p}}_2 + \mathbf{z}_{2\overline{2}}\mathbf{p}_2)\\
&+ \frac{1}{2}\mathbf{F}^0_{\mathbf{xxx}_\tau}(2\mathbf{p}_2^2\,\overline{\mathbf{p}}_2\, e^{-i\omega_2\tau_c} + \mathbf{p}_2^2\,\overline{\mathbf{p}}_2\, e^{i\omega_2\tau_c})\\
&+ \mathbf{F}^0_{\mathbf{x}\mathbf{x}_\tau}(\overline{\mathbf{z}}_{2\overline{2}}\,\mathbf{p}_2 e^{-3i\omega_1\tau_c} + \mathbf{z}_{22}\,\overline{\mathbf{p}}_2\, e^{3i\omega_1\tau_c} + \mathbf{z}_{2\overline{2}}\mathbf{p}_2 e^{-3i\omega_1\tau_c})\\
&+ \mathbf{F}^0_{\mathbf{x}\mathbf{x}_\tau\mathbf{x}_\tau}(\mathbf{p}_2^2\overline{\mathbf{p}}_2 e^{-6i\omega_1\tau_c} + p_2^2\overline{p}_2)\\
&+ \frac{1}{2}\mathbf{F}^0_{\mathbf{x}_\tau\mathbf{x}_\tau\mathbf{x}_\tau}\mathbf{p}_2^2\,\overline{\mathbf{p}}_2\, e^{-3i\omega_1\tau_c}\\
&+ \mathbf{F}^0_{\mathbf{x}\mathbf{x}_\tau}(\overline{\mathbf{z}}_{2\overline{2}}\,\mathbf{p}_2 + \mathbf{z}_{22}\overline{\mathbf{p}}_2 e^{-6i\omega_1\tau_c} + \mathbf{z}_{2\overline{2}}\mathbf{p}_2)\\
&+ \mathbf{F}^0_{\mathbf{x}_\tau\mathbf{x}_\tau}(\overline{\mathbf{z}}_{2\overline{2}}\mathbf{p}_2 e^{-3i\omega_1\tau_c} + \mathbf{z}_{22}\overline{\mathbf{p}}_2 e^{-3i\omega_1\tau_c} + \mathbf{z}_{2\overline{2}}\mathbf{p}_2 e^{-3i\omega_1\tau_c})],
\end{aligned}
$$

$$
\mathbf{M}_1 = \mathbf{q}_1{}^{\mathbf{H}}(\tau_c\mathbf{F}^0_{\mathbf{x}_\tau}e^{-i\omega_1\tau_c} + \mathbf{E})\mathbf{p}_1, \qquad \mathbf{M}_2 = \mathbf{q}_2{}^{\mathbf{H}}(\tau_c\mathbf{F}^0_{\mathbf{x}_\tau}e^{-i\omega_2\tau_c} + \mathbf{E})\mathbf{p}_2.
$$

In Eq. (2.62), $\mathbf{C}_{\overline{1}\overline{1}2}\,\overline{\mathbf{A}}_1^2\,\mathbf{A}_2$ and $\mathbf{C}_{111}\mathbf{A}_1^3$ are resonance terms, which appear because the ratio of the two frequencies of the system is 1:3. Due to the existence of these two resonance terms, the complex amplitude Eq. (2.62) cannot set the two complex amplitudes in polar coordinate form as it does for nonresonance or weak resonance. Here, we set the complex amplitudes $\mathbf{A}_1$ and $\mathbf{A}_2$ in a hybrid form (polar Cartesian form) as follows

$$
\mathbf{A}_1 = \frac{1}{2}\mathbf{a}e^{i\boldsymbol{\theta}},\ \mathbf{A}_2 = \frac{1}{2}(\mathbf{u} + i\mathbf{v})e^{i3\boldsymbol{\theta}}. \tag{2.63}
$$

Substituting Eq. (2.63) into Eq. (2.62) and separating the real and imaginary parts yields the following amplitude phase equation

$$
\dot{\mathbf{a}} = \mathbf{a}R_1 + \frac{1}{4}R_{11\overline{1}}\mathbf{a}^3 + \frac{1}{4}R_{\overline{1}\overline{1}2}\mathbf{a}^2u - \frac{1}{4}I_{\overline{1}\overline{1}2}\mathbf{a}^2v + \frac{1}{4}R_{12\overline{2}}\mathbf{a}\mathbf{u}^2 + \frac{1}{4}R_{12\overline{2}}\mathbf{a}\mathbf{v}^2,
$$

$$
\begin{aligned}
\mathbf{a}\dot{\boldsymbol{\theta}} &= \mathbf{a}I_1 + \frac{1}{4}I_{11\bar{1}}\mathbf{a}^3 + \frac{1}{4}R_{\bar{1}\bar{1}2}\mathbf{a}^2\mathbf{v} + \frac{1}{4}I_{\bar{1}\bar{1}2}\mathbf{a}^2\mathbf{u} + \frac{1}{4}I_{12\bar{2}}\mathbf{a}\mathbf{v}^2 + \frac{1}{4}I_{12\bar{2}}\mathbf{a}\mathbf{u}^2,\\
\dot{\mathbf{u}} &= 3\mathbf{v}\dot{\theta} + \mathbf{u}R_2 - \mathbf{v}I_2 + \frac{1}{4}R_{111}\mathbf{a}^3 + \frac{1}{4}R_{1\bar{1}2}\mathbf{a}^2\mathbf{u} - \frac{1}{4}I_{1\bar{1}2}\mathbf{a}^2\mathbf{v} + \frac{1}{4}R_{22\bar{2}}\mathbf{u}^3\\
&\quad + \frac{1}{4}R_{22\bar{2}}\mathbf{u}\mathbf{v}^2 - \frac{1}{4}I_{22\bar{2}}\mathbf{u}^2\mathbf{v} - \frac{1}{4}I_{22\bar{2}}\mathbf{v}^3,\\
\dot{\mathbf{v}} &= -3\mathbf{u}\dot{\theta} + \mathbf{v}R_2 + \mathbf{u}I_2 + \frac{1}{4}I_{111}\mathbf{a}^3 + \frac{1}{4}R_{1\bar{1}2}\mathbf{a}^2\mathbf{v} + \frac{1}{4}I_{1\bar{1}2}\mathbf{a}^2\mathbf{u} + \frac{1}{4}I_{22\bar{2}}\mathbf{u}^3\\
&\quad + \frac{1}{4}I_{22\bar{2}}\mathbf{u}\mathbf{v}^2 + \frac{1}{4}R_{22\bar{2}}\mathbf{u}^2\mathbf{v} + \frac{1}{4}R_{22\bar{2}}\mathbf{v}^3,
\end{aligned}
\tag{2.64}
$$

where the coefficients $R_i = \mathrm{Re}(\mathbf{C_{i\mu}}\boldsymbol{\mu}_{\varepsilon})$, $I_i = \mathrm{Im}(\mathbf{C_{i\mu}}\boldsymbol{\mu}_{\varepsilon})$, $R_{ijk} = \mathrm{Re}(\mathbf{C_{ijk}})$, and $I_{ijk} = \mathrm{Im}(\mathbf{C_{ijk}})$. If $\mathbf{a} \neq \mathbf{0}$, from Eq. (2.64$_2$), we can obtain that $\dot{\boldsymbol{\theta}}$ can be expressed as a function of **a**, **u**, and **v**. By substituting $\dot{\boldsymbol{\theta}}$ into Eqs. (2.64$_3$) and (2.64$_4$), the standard normal form for 1:3 resonance can be obtained as a three-dimension system

$$
\begin{aligned}
\dot{\mathbf{a}} &= \mathbf{a}R_1 + \frac{1}{4}R_{11\bar{1}}\mathbf{a}^3 + \frac{1}{4}R_{\bar{1}\bar{1}2}\mathbf{a}^2\mathbf{u} - \frac{1}{4}I_{\bar{1}\bar{1}2}\mathbf{a}^2\mathbf{v} + \frac{1}{4}R_{12\bar{2}}\mathbf{a}\mathbf{u}^2 + \frac{1}{4}R_{12\bar{2}}\mathbf{a}\mathbf{v}^2,\\
\dot{\mathbf{u}} &= 3\mathbf{v}I_1 + \frac{3}{4}I_{11\bar{1}}\mathbf{a}^2\mathbf{v} + \frac{3}{4}R_{\bar{1}\bar{1}2}\mathbf{a}\mathbf{v}^2 + \frac{3}{4}I_{\bar{1}\bar{1}2}\mathbf{a}\mathbf{u}\mathbf{v} + \frac{3}{4}I_{12\bar{2}}\mathbf{v}^3 + \frac{3}{4}I_{12\bar{2}}\mathbf{u}^2\mathbf{v}\\
&\quad + \mathbf{u}R_2 - \mathbf{v}I_2 + \frac{1}{4}R_{111}\mathbf{a}^3 + \frac{1}{4}R_{1\bar{1}2}\mathbf{a}^2\mathbf{u} - \frac{1}{4}I_{1\bar{1}2}\mathbf{a}^2\mathbf{v} + \frac{1}{4}R_{22\bar{2}}\mathbf{u}^3\\
&\quad + \frac{1}{4}R_{22\bar{2}}\mathbf{u}\mathbf{v}^2 - \frac{1}{4}I_{22\bar{2}}\mathbf{u}^2\mathbf{v} - \frac{1}{4}I_{22\bar{2}}\mathbf{v}^3,\\
\dot{\mathbf{v}} &= -3\mathbf{u}I_1 - \frac{3}{4}I_{11\bar{1}}\mathbf{a}^2\mathbf{u} - \frac{3}{4}R_{\bar{1}\bar{1}2}\mathbf{a}\mathbf{u}\mathbf{v} - \frac{3}{4}I_{\bar{1}\bar{1}2}\mathbf{a}\mathbf{u}^2 - \frac{3}{4}I_{12\bar{2}}\mathbf{u}\mathbf{v}^2 - \frac{3}{4}I_{12\bar{2}}\mathbf{u}^3\\
&\quad + \mathbf{v}R_2 + \mathbf{u}I_2 + \frac{1}{4}I_{111}\mathbf{a}^3 + \frac{1}{4}R_{1\bar{1}2}\mathbf{a}^2\mathbf{v} + \frac{1}{4}I_{1\bar{1}2}\mathbf{a}^2\mathbf{u} + \frac{1}{4}I_{22\bar{2}}\mathbf{u}^3\\
&\quad + \frac{1}{4}I_{22\bar{2}}\mathbf{u}\mathbf{v}^2 + \frac{1}{4}R_{22\bar{2}}\mathbf{u}^2\mathbf{v} + \frac{1}{4}R_{22\bar{2}}\mathbf{v}^3.
\end{aligned}
\tag{2.65}
$$

In order to obtain Eq. (2.65), we assume that $a \neq 0$, which may cause the (2.65) to lose some of its solutions. However, by observing Eq. (2.65), we find that $a = 0$ is still a solution to Eq. (2.65). Therefore, in the process of obtaining Eq. (2.65) from Eq. (2.64), the solution of the system is not lost. If we assume $a_1 = |\mathbf{A_1}| = a$, $a_2 = |\mathbf{A_2}| = \sqrt{u^2 + v^2}$, it can be obtained from Eq. (2.65) that

$$
\begin{aligned}
\dot{\mathbf{a}} &= \mathbf{a}R_1 + \frac{1}{4}R_{11\bar{1}}\mathbf{a}^3 + \frac{1}{4}R_{\bar{1}\bar{1}2}\mathbf{a}^2 u - \frac{1}{4}I_{\bar{1}\bar{1}2}\mathbf{a}^2\mathbf{v} + \frac{1}{4}R_{12\bar{2}}\mathbf{a}\mathbf{u}^2 + \frac{1}{4}R_{12\bar{2}}\mathbf{a}\mathbf{v}^2,\\
\mathbf{a_2}\,\dot{\mathbf{a}}_2 &= \mathbf{u}\dot{\mathbf{u}} + \mathbf{v}\dot{\mathbf{v}} = R_2\mathbf{a}_2^2 + \frac{1}{4}R_{111}\mathbf{a}_1^3\mathbf{u} + \frac{1}{4}I_{111}\mathbf{a}_1^3\mathbf{v} + \frac{1}{4}R_{1\bar{1}2}\mathbf{a}_1^2\mathbf{a}_2^2 + \frac{1}{4}R_{22\bar{2}}\mathbf{a}_2^4.
\end{aligned}
\tag{2.66}
$$

Equation (2.66) has a trivial equilibrium point $(a_1, a_2) = (0, 0)$, while Eq. (2.66) has two types of nontrivial equilibrium points: a single mode equilibrium point and a bimodal equilibrium point. The equilibrium point of a single mode refers to the amplitude of one of the two modes being zero. When $a_2 = 0$, there must be $u = v = 0$, and Eq. (2.66_2) is always valid. From Eq. (2.66_1), it can be obtained that

$$a_{10} = \sqrt{-\frac{4R_1}{R_{11\bar{1}}}}.$$

So, we can obtain a single mode equilibrium point $(a_{10} = \sqrt{-4R_1/R_{11\bar{1}}}, 0)$. Similarly, Eq. (2.66_1) is identically satisfied $a_1 = 0$, while Eq. (2.66_2) leads to

$$a_{20} = \sqrt{-\frac{4R_2}{R_{22\bar{2}}}}.$$

Finally, when $a_1 \neq 0$ and $a_2 \neq 0$, the equilibrium point of the system is a bimodal equilibrium point. Due to the existence of resonance terms $R_{\bar{1}\bar{1}2}\mathbf{a^2u}$ and $R_{111}\mathbf{a_1^3u}$ in Eq. (2.66), it is difficult to solve the bimodal equilibrium point (a_{21}, a_{22}) according to Eq. (2.66). However, we can provide the existence region of the equilibrium point (a_{21}, a_{22}) and classify the dynamic behavior of the system in the parameter plane according to Eq. (2.66). It can be obtained from Eqs. (2.66_1) and 2.66_2).

$$\begin{aligned}
\frac{1}{4}a_1 u = &\frac{1}{4a_1^2(R_{\bar{1}\bar{1}2}I_{111} + R_{111}I_{\bar{1}\bar{1}2})}(-R_{11\bar{1}}I_{111}a_1^4 - 4R_1I_{111}a_1^2 \\
&-R_{12\bar{2}}I_{111}a_1^2a_2^2 - R_{1\bar{1}2}I_{\bar{1}\bar{1}2}a_1^2a_2^2) - 4R_2I_{\bar{1}\bar{1}2}a_2^2 - R_{22\bar{2}}I_{\bar{1}\bar{1}2}a_2^4), \\
\frac{1}{4}a_1 v = &\frac{1}{4a_1^2(R_{\bar{1}\bar{1}2}I_{111} + R_{111}I_{\bar{1}\bar{1}2})}(R_{11\bar{1}}R_{111}a_1^4 + 4R_1R_{111}a_1^2 \\
&+R_{12\bar{2}}R_{111}a_1^2a_2^2 - R_{1\bar{1}2}R_{\bar{1}\bar{1}2}a_1^2a_2^2) - 4R_2R_{\bar{1}\bar{1}2}a_2^2 - R_{22\bar{2}}R_{\bar{1}\bar{1}2}a_2^4).
\end{aligned}$$

By applying the relationship $u^2 + v^2 = a_2^2$ between u and v, the following relationship can be obtained:

$$\left(\frac{1}{4}a_1 u\right)^2 + \left(\frac{1}{4}a_1 v\right)^2 = \frac{1}{16}a_1^2a_2^2. \tag{2.67}$$

By substituting $a_1 u$ and $a_1 v$ into Eq. (2.67), the existence region of bimodal solution and the relationship between a_1 and a_2 can be obtained.

2.2.2 1:3 Resonance Analysis of Time Delay Coupled VDP Oscillator System

We study the dynamic behavior near the 1:3 resonance double Hopf bifurcation of two VDP oscillators with time delay coupling. Coupled VDP oscillators are used to describe laser dynamics and have a wide range of application backgrounds. The time delay caused by the propagation of signals during the coupling process of two oscillators is referred to as the time delay, and its dynamic equation is as follows [39]

$$\begin{aligned} \ddot{y}_1 + \omega_{10}^2 y_1 + \beta\,\dot{y}_1 + \gamma y_1^2\,\dot{y}_1 &= \alpha\,\dot{y}_2(t-\tau), \\ \ddot{y}_2 + \omega_{20}^2 y_2 + \beta\,\dot{y}_2 + \gamma y_2^2\,\dot{y}_2 &= \alpha\,\dot{y}_1(t-\tau). \end{aligned} \tag{2.68}$$

Let $x_1 = y_1$, $x_2 = \dot{y}_1$, $x_3 = y_2$ and $x_4 = \dot{y}_2$, we will write Eq. (2.68) as a first order system as follows

$$\begin{aligned} \dot{x}_1 &= x_2, \\ \dot{x}_2 &= -\omega_{10}^2 x_1 - \beta x_2 + \alpha x_4(t-\tau) - \gamma x_1^2 x_2, \\ \dot{x}_3 &= x_4, \\ \dot{x}_4 &= -\omega_{20}^2 x_3 - \beta x_4 + \alpha x_2(t-\tau) - \gamma x_3^2 x_4. \end{aligned} \tag{2.69}$$

Obviously, the origin $O(0, 0, 0, 0)$ is an equilibrium point of (2.69), and the linearized equation of (2.69) at $O(0, 0, 0, 0)$ is

$$\begin{aligned} \dot{x}_1 &= x_2, \\ \dot{x}_2 &= -\omega_{10}^2 x_1 - \beta x_2 + \alpha x_4(t-\tau), \\ \dot{x}_3 &= x_4, \\ \dot{x}_4 &= -\omega_{20}^2 x_3 - \beta x_4 + \alpha x_2(t-\tau). \end{aligned} \tag{2.70}$$

The characteristic equation corresponding to Eq. (2.70) is

$$(\lambda^2 + \beta\lambda + \omega_{10}^2)(\lambda^2 + \beta\lambda + \omega_{20}^2) = (\alpha\lambda e^{-\lambda\tau})^2. \tag{2.71}$$

Because we want to discuss the dynamic behavior of (2.68) when 1:3 resonance double Hopf bifurcation occurs, we assume that (2.68) has two pairs of pure imaginary roots, $\lambda_{1,3} = \pm i\omega$ and $\lambda_{2,4} = \pm 3i\omega$ $(\omega > 0)$. Firstly, substituting $\lambda = i\omega$ into the characteristic equation yields

$$\begin{aligned} &(\omega_{10}^2 - \omega^2)(\omega_{20}^2 - \omega^2) - \beta^2\omega^2 + i\beta\omega(\omega_{10}^2 + \omega_{20}^2 - 2\omega^2) \\ &= -\alpha^2\omega^2 \cos 2\omega\tau + \alpha^2\omega^2 \sin 2\omega\tau. \end{aligned} \tag{2.72}$$

Separating the real and imaginary parts, we will get

$$\begin{aligned}(\omega_{10}^2-\omega^2)(\omega_{20}^2-\omega^2)-\beta^2\omega^2&=-\alpha^2\omega^2cos2\omega\tau,\\ \beta\omega(\omega_{10}^2+\omega_{20}^2-2\omega^2)&=\alpha^2\omega^2sin2\omega\tau.\end{aligned} \tag{2.73}$$

Similarly, substituting $\lambda = 3i\omega$ into Eq. (2.71) and separating the real and imaginary parts, it can be obtained that

$$\begin{aligned}(\omega_{10}^2-9\omega^2)(\omega_{20}^2-9\omega^2)-9\beta^2\omega^2&=-9\alpha^2\omega^2cos6\omega\tau,\\ 3\beta\omega(\omega_{10}^2+\omega_{20}^2-18\omega^2)&=9\alpha^2\omega^2sin6\omega\tau.\end{aligned} \tag{2.74}$$

From Eq. (2.73) and $\sin^2\omega\tau+\cos^2\omega\tau=1$, it can be concluded that

$$[(\omega_{10}^2-\omega^2)(\omega_{20}^2-\omega^2)-\beta^2\omega^2]^2+[\beta\omega(\omega_{10}^2+\omega_{20}^2-2\omega^2)]^2=(\alpha^2\omega^2)^2. \tag{2.75}$$

From Eq. (2.74) and $\sin^2 3\omega\tau+\cos^2 3\omega\tau=1$, it can be concluded that

$$[(\omega_{10}^2-9\omega^2)(\omega_{20}^2-9\omega^2)-9\beta^2\omega^2]^2+[3\beta\omega(\omega_{10}^2+\omega_{20}^2-18\omega^2)]^2=(9\alpha^2\omega^2)^2. \tag{2.76}$$

Noticing that $\cos 3\omega\tau=(\cos\omega\tau)^3-3\cos\omega\tau(\sin\omega\tau)^2$, so we will get

$$\begin{aligned}\frac{9\beta^2\omega^2-(\omega_{10}^2-9\omega^2)(\omega_{20}^2-9\omega^2)}{9\alpha^2\omega^2}&=\left(\frac{\beta^2\omega^2-(\omega_{10}^2-\omega^2)(\omega_{20}^2-\omega^2)}{\alpha^2\omega^2}\right)^3\\ -3\left(\frac{\beta^2\omega^2-(\omega_{10}^2-\omega^2)(\omega_{20}^2-\omega^2)}{\alpha^2\omega^2}\right)&\left(\frac{\beta\omega(\omega_{10}^2+\omega_{20}^2-2\omega^2)}{\alpha^2\omega^2}\right)^2.\end{aligned} \tag{2.77}$$

Fixing $\omega_{10} = 1.1$ and $\omega_{20} = 2.9$ and solving Eqs. (2.75), (2.76), and (2.77), $\omega = 1.03118$, $\beta = 0$, $\alpha = 1.00664$ can be obtained. Substituting these values into Eqs. (2.75) and (2.76) yields

$$\begin{aligned}\sin 2\omega\tau&=0,\ \cos 2\omega\tau=-1,\\ \sin 6\omega\tau&=0,\ \cos 6\omega\tau=-1.\end{aligned} \tag{2.78}$$

Therefore, it can be obtained that

$$2\omega\tau=\pi+2j\pi,\ j=0,1,2,\ldots, \tag{2.79}$$

that is,

$$\tau=\frac{\pi+2j\pi}{2\omega},\ j=0,1,2,\ldots. \tag{2.80}$$

Then, at the point $(\beta_c, \alpha_c, \tau_c) = (0, 1.00664, 1.5233)$, a 1:3 resonance double Hopf bifurcation may occur. In order to further verify the eigenvalue conditions, we provided some of the maximum eigenvalues of (2.69) through numerical methods, as

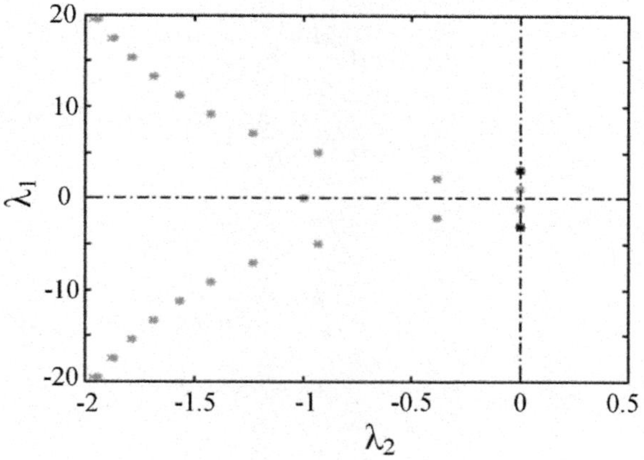

Fig. 2.4 Eigenvalues distribution of (2.69) at $(\beta_c, \alpha_c, \tau_c) = (0, 1.00664, 1.5233)$ when $\omega_{10} = 1.1$, $\omega_{20} = 2.9$

shown in Fig. 2.4. It can be seen from Fig.2.4 that (2.69) has two pairs of pure imaginary roots and all other eigenvalues $\lambda_h (h \geq 5)$ are distributed in the left half plane of the complex plane, which has a negative real part at the bifurcation point. Therefore, a 1:3 resonance double Hopf bifurcation occurs at $(\beta_c, \alpha_c, \tau_c) = (0, 1.00664, 1.5233)$ with the frequencies $\omega_1 = 1.03118$ and $\omega_2 = 3\omega_1$.

To analyze the dynamic behavior of (2.68) near the 1:3 resonance double Hopf bifurcation point, let $\beta = \beta_c + \beta_\varepsilon$, $\alpha = \alpha_c + \alpha_\varepsilon$,$\tau = \tau_c + \tau_\varepsilon$, where β_ε,α_ε and τ_ε are small perturbation parameters. Equation (2.68) can be rewritten as

$$\begin{aligned}
\dot{x}_1 &= x_2, \\
\dot{x}_2 &= -\omega_{10}^2 x_1 - \beta_\varepsilon x_2 + \alpha_c x_4(t-\tau) + \alpha_\varepsilon x_4(t-\tau) - \gamma x_1^2 x_2, \\
\dot{x}_3 &= x_4, \\
\dot{x}_4 &= -\omega_{20}^2 x_3 - \beta_\varepsilon x_4 + \alpha_c x_2(t-\tau) + \alpha_\varepsilon x_2(t-\tau) - \gamma x_3^2 x_4.
\end{aligned}$$

Then, it can be concluded that

$$\mathbf{F_x^0} = \begin{pmatrix} 0 & 1 & 0 & 0 \\ -\omega_{10}^2 & 0 & 0 & 0 \\ 0 & 0 & 0 & 1 \\ 0 & 0 & -\omega_{20}^2 & 0 \end{pmatrix}, \quad \mathbf{F_{x_\tau}^0} = \begin{pmatrix} 0 & 0 & 0 & 0 \\ 0 & 0 & 0 & \alpha_c \\ 0 & 0 & 0 & 0 \\ 0 & \alpha_c & 0 & 0 \end{pmatrix}.$$

The left and right eigenvectors corresponding to the eigenvalues $\lambda_1 = i\omega$ and $\lambda_2 = 3i\omega$ are

$$\begin{aligned}
\mathbf{p_1} &= \left(1, i\omega_1, \frac{\omega_{10}^2 - \omega_1^2}{\alpha_c \omega_1}, \frac{i(\omega_{10}^2 - \omega_1^2)}{\alpha_c}\right), \\
\mathbf{p_2} &= \left(1, i\omega_2, \frac{\omega_2^2 - \omega_{10}^2}{\alpha_c \omega_2}, \frac{i(\omega_2^2 - \omega_{10}^2)}{\alpha_c}\right),
\end{aligned}$$

$$\mathbf{q_1} = e_1(-\frac{i\omega_{10}^2}{\omega_1}, 1, -\frac{i(\alpha_c^2 - \omega_1^2 + \omega_{10}^2)}{\alpha_c}, \frac{\omega_{10}^2 - \omega_1^2}{\alpha_c \omega_1}),$$
$$\mathbf{q_2} = e_2(-\frac{i\omega_{10}^2}{\omega_2}, 1, \frac{i(\alpha_c^2 - \omega_2^2 + \omega_{10}^2)}{\alpha_c}, \frac{\omega_2^2 - \omega_{10}^2}{\alpha_c \omega_1}),$$

where

$$e_1 = \frac{i\alpha_c^2 \omega_1}{2[\alpha_c^2 \omega_{10}^2 + (\omega_{10}^2 - \omega_1^2)^2]},$$
$$e_2 = \frac{i\alpha_c^2 \omega_2}{2[\alpha_c^2 \omega_{10}^2 + (\omega_{10}^2 - \omega_2^2)^2]}.$$

Therefore, if we set $\beta_\varepsilon = \varepsilon^2\, \hat{\beta}_\varepsilon$, $\alpha_\varepsilon = \varepsilon^2\, \hat{\alpha}_\varepsilon$ and $\tau_\varepsilon = \varepsilon^2\, \hat{\tau}_\varepsilon$ and apply the method derived in the previous section, we can obtain the corresponding coefficients $C_{j\mu}\mu_\varepsilon$ and C_{ijk}. Substituting $C_{j\mu}\mu_\varepsilon$ and C_{ijk} into Eq. (2.65) yields a three-dimension real bifurcation equation:

$$\begin{aligned}\dot{\mathbf{a}} &= 0.022061\mathbf{a}\hat{\alpha}_\varepsilon - 0.425539\mathbf{a}\hat{\beta}_\varepsilon - 0.122382\mathbf{a}\hat{\tau}_\varepsilon \\ &- 0.0173907\mathbf{a^3} - 0.0175154\mathbf{a^2}u - 0.0397698\mathbf{au^2} \\ &+ 0.00327745\mathbf{a^2v} - 0.0397698\mathbf{av^2},\end{aligned}$$

$$\begin{aligned}\dot{\mathbf{u}} &= 0.156259\, \mathbf{u}\hat{\alpha}_\varepsilon - 0.632057\, \mathbf{v}\hat{\alpha}_\varepsilon - 0.425485\, \mathbf{u}\hat{\beta}_\varepsilon \\ &- 0.477725\, \mathbf{v}\hat{\beta}_\varepsilon + 0.866844\, \mathbf{u}\hat{\tau}_\varepsilon + 0.417904\, \mathbf{v}\hat{\tau}_\varepsilon \\ &- 0.00072552\mathbf{a^3} - 0.00494202\mathbf{a^2u} - 0.114359\mathbf{u^3} \\ &- 0.0125365\mathbf{a^2v} - 0.00983234\mathbf{auv} - 0.0865207\mathbf{u^2v} \\ &- 0.0525461\mathbf{av^2} - 0.114359\mathbf{uv^2} - 0.0865207\mathbf{v^3},\end{aligned}$$

$$\begin{aligned}\dot{\mathbf{v}} &= 0.632057\mathbf{u}\hat{\alpha}_\varepsilon + 0.156259\mathbf{v}\hat{\alpha}_\varepsilon + 0.477725\mathbf{u}\hat{\beta}_\varepsilon \\ &- 0.425485\mathbf{v}\hat{\beta}_\varepsilon - 0.417904\mathbf{u}\hat{\tau}_\varepsilon + 0.866844\mathbf{v}\hat{\tau}_\varepsilon \\ &+ 0.000407271\mathbf{a^3} + 0.0125365\mathbf{a^2u} + 0.00983234\mathbf{au^2} \\ &+ 0.0865207\mathbf{u^3} - 0.00494202\mathbf{a^2v} + 0.0525461\mathbf{auv} \\ &- 0.114359\mathbf{u^2v} + 0.0865207\mathbf{uv^2} - 0.114359\mathbf{v^3}.\end{aligned}$$

Next, we will conduct a dynamic analysis of (2.68) near the bifurcation point (0, 1.00664, 1.5233). For (2.68), our main concern is the impact of time delay τ and coupling strength α on the dynamic behavior of the system. We study the influence of parameter $\hat{\tau}_\varepsilon$, $\hat{\alpha}_\varepsilon$ changes on the dynamic behavior of the system in the parameter plane $(\hat{\tau}_\varepsilon, \hat{\alpha}_\varepsilon)$. Figure 2.5 shows the classification and phase diagram of the dynamic behavior of the system near the 1:3 resonance double Hopf bifurcation point. The

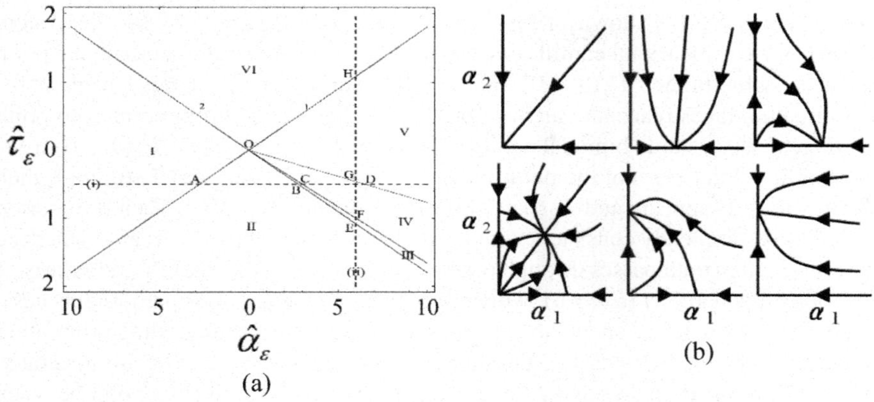

Fig. 2.5 Two parameter cross-section and phase diagram of 1:3 resonance double Hopf bifurcation: **a** cross-section at $\hat{\beta}_\varepsilon = 0$; **b** phase diagram within region (I)–(VI), (i) and (ii) are the two paths of bifurcation diagrams

Hopf bifurcation corresponding to the state variable x of the original system is a static bifurcation of the amplitude variable (a_1, a_2), whose critical boundaries are marked as D_1 and D_2 in Fig. 2.5a. By analyzing the equilibria and their stability of the amplitude equation, we can divide the plane $(\hat{\tau}_\varepsilon, \hat{\alpha}_\varepsilon)$ into six regions. Within each region, the number and stability of equilibrium points are different, and their topological structure is shown in Fig. 2.5(I)–(VI).

As shown in Fig. 2.5, the system has a stable trivial equilibrium point $E_0(0, 0)$ in region (I), which is the so-called amplitude death region. Two oscillators that undergo periodic vibrations suppress each other's vibrations due to their mutual coupling. As $(\hat{\tau}_\varepsilon, \hat{\alpha}_\varepsilon)$ changes, when the parameter enters region (II), the trivial equilibrium point becomes unstable, and a stable single mode equilibrium point $E_1(a_{10}, 0)$ appears, which corresponds to the periodic solution of the original system frequency ω_1. In region (III), the equilibria $E_0(0, 0)$ and $E_1(a_{10}, 0)$ still exist, and a new unstable single mode equilibrium point $E_0(0, a_{20})$ is generated. In region (IV), a stable bimodal solution $E_3(a_{12}, a_{22})$ appears, the single modal solution $E_1(a_{10}, 0)$ becomes unstable, while the equilibrium points $E_0(0, 0)$ and $E_2(0, a_{20})$ remain unstable. When the parameters change to region (V), the bimodal solution $E_3(a_{12}, a_{22})$ disappears, and the equilibrium point $E_0(0, a_{20})$ is stable. In region (VI), the equilibrium point $E_1(a_{10}, 0)$ disappears and equilibrium point $E_2(0, a_{20})$ maintains its stability.

2.2.3 *Numerical Simulation*

In order to observe the effects of parameters τ and α on the bifurcation behavior of the system more clearly, we selected two paths in Fig. 2.5a. Along these two paths, the bifurcation diagrams of two amplitudes a_1 and a_2 with respect to parameters $\hat{\tau}_\varepsilon$

and $\hat{\alpha}_\varepsilon$ are given (as shown in Fig. 2.6). Along path (i) ($\hat{\tau}_\varepsilon = -0.5$), the system starts from a stable trivial equilibrium point and a bifurcation takes place at the point A and a stable branch $E_1(a_{10}, 0)$ arises. When the parameter changes to point B, a bifurcation takes place and an unstable branch $E_2(0, a_{20})$. From point C to point D, there exists a stable bimodal equilibrium point $E_3(a_{12}, a_{22})$ and $E_1(a_{10}, 0)$ loses its stability. After passing the point D, $E_3(a_{12}, a_{22})$ collapses with $E_2(0, a_{20})$, then $E_3(a_{12}, a_{22})$ disappears and $E_2(0, a_{20})$ becomes stable (Fig. 2.6i). Such bifurcation behavior is completely consistent with the dynamic behavior of each region when the first bifurcation path passes through regions (I), (II), (III), (IV), and (V). Similarly, if we follow the path (ii) ($\hat{\alpha}_\varepsilon = 6$), which is opposite to the previous path. In this case, we fix the parameter $\hat{\alpha}_\varepsilon$ and change the parameter $\hat{\tau}_\varepsilon$. The corresponding bifurcation diagram is shown in Fig. 2.6ii. Equilibria $E_1(a_{10}, 0)$ and $E_2(0, a_{20})$ are generated by bifurcation at points H and E, and there is a bimodal equilibrium point between points F and G. These results are consistent with previous analytical predictions.

To verify the correctness of the previous theoretical predictions, numerical simulations of the original system in the parameter plane $\alpha - \tau$ are given. Corresponding to the six regions in Fig. 2.5, the $\alpha - \tau$ plane is also divided into six regions. In

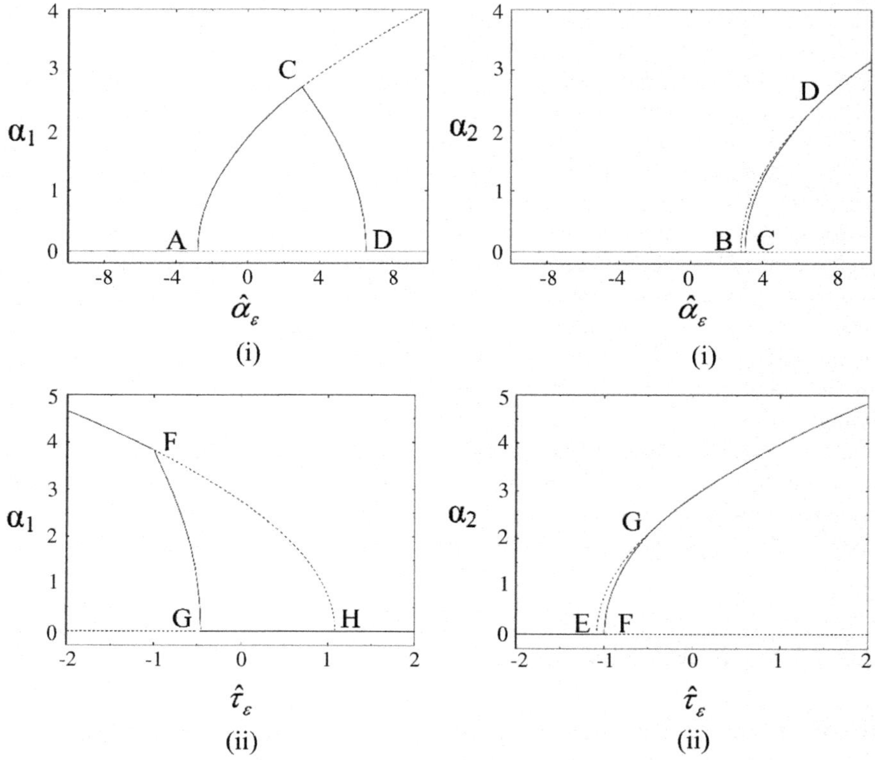

Fig. 2.6 Bifurcation diagram along paths (i) and (ii)

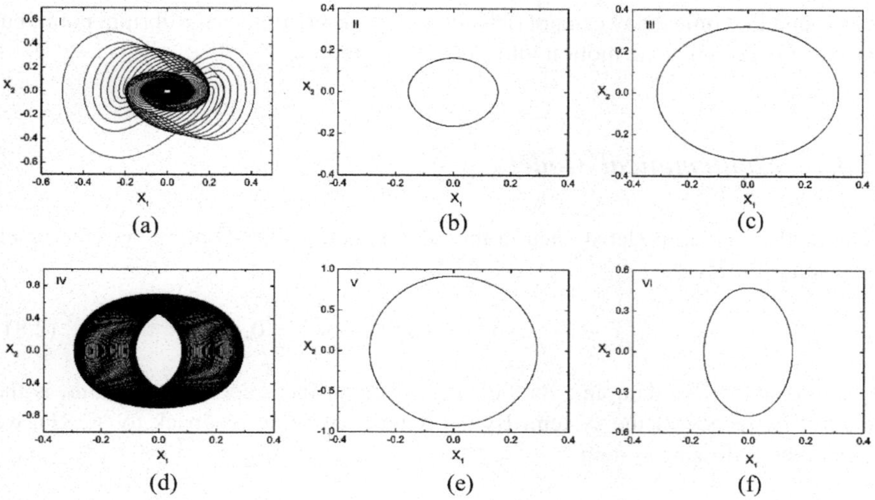

Fig. 2.7 Cross section and phase diagram at $\beta = 0$. Figures **a**–**f** show the phase diagram within region (I)–(VI)

region (I), there is a stable trivial equilibrium point. In regions (II) and (III), there is a stable periodic solution with a frequency of ω_1, which corresponds to the single mode equilibrium point $E_1(a_{10}, 0)$ in the amplitude system. In regions (V) and (VI), there is a stable periodic solution with frequency ω_2, which corresponds to the single mode equilibrium point $E_2(0, a_{20})$ in the amplitude system. In region (IV), there exists a stable quasi-periodic solution. Obviously, the numerical simulation results are given in Fig. 2.7, which are completely consistent with the theoretical results in Fig. 2.5.

2.3 Multistability and Attraction Basin of VDP-Duffing System with Delayed Feedback

There is a common phenomenon of multiple attractors coexisting in time delay nonlinear dynamical systems, where each attractor has its independent domain of attraction in the phase plane. Therefore, the basin of attraction problem naturally arises [40]. In this section, we introduce linear time delay displacement feedback control into the VDP-Duffing system. By using the averaging method, we obtain the first order approximate analytical form of the periodic solution and determine its stability. We find that time delay can trigger the coexistence of multiple steady state phenomena in the system. The basin of attraction for the multi-steady state motion of the system caused by time delay was divided through numerical simulation. Research has found that time delay can not only alter the stability of the equilibrium point, but also transform unstable motion into steady state motion.

2.3.1 Mathematical Model

The mathematical model studied in this section is the VDP-Duffing system, which is given as follows

$$\ddot{x} - (\alpha - \gamma x^2)\dot{x} + \omega_0^2 x + \beta x^3 = 0, \tag{2.81}$$

where α and γ are damping coefficients, β is a stiffness coefficient and ω_0 is the natural frequency of the system. By applying time delay feedback to (2.81), we obtain the following system

$$\ddot{x} - (\alpha - \gamma x^2)\dot{x} + \omega_0^2 x + \beta x^3 = A(x_\tau - x), \tag{2.82}$$

where $x_\tau = x(t - \tau)$ and τ is time delay, and A is the feedback gain coefficient. When $A < 0$, it is a negative displacement feedback; when $A > 0$, it is positive displacement feedback. When $\tau = 0$, (2.82) degenerates into (2.81).

2.3.2 Approximated Analytical Solution of Periodic Solution

The characteristic equation of the linear system corresponding to (2.82) is

$$\lambda^2 - \alpha\lambda + (\omega_0^2 + A) - Ae^{-\lambda\tau} = 0. \tag{2.83}$$

Let $\lambda = a \pm \omega i$, where a, ω are real numbers. Substituting $a = 0$ into Eq. (2.83), we obtain the necessary condition for the occurrence of Hopf bifurcation in (2.82)

$$\begin{cases} -\omega^2 + \omega_0^2 + A - A\cos\omega\tau_0 = 0, \\ A\sin\omega\tau_0 = \alpha\omega. \end{cases} \tag{2.84}$$

Assuming I is a set of (α, A):

$$\begin{aligned} I = \{(\alpha, A) | 2A + \omega_0^2 > 0, 2\omega_0^2 + 2A - \alpha^2 \\ > 0, (2A - 2\alpha\omega_0 - \alpha^2)(2A + 2\alpha\omega_0 - \alpha^2) > 0\}. \end{aligned} \tag{2.85}$$

When $(\alpha, A) \in I$, Eq. (2.83) has two unequal positive roots $\omega_+ > \omega_- > 0$. According to the Hopf bifurcation theorem, the critical values of τ for the occurrence of Hopf bifurcation in (2.82) are

$$\tau_{\pm}(j) = \begin{cases} \frac{1}{\omega_{\pm}}[(2j+2)\pi - \cos^{-1}(1 + \frac{\omega_0^2 - \omega_{\pm}^2}{A})], & A > 0, \alpha < 0, \\ \frac{1}{\omega_{\pm}}[2j\pi + \cos^{-1}(1 + \frac{\omega_0^2 - \omega_{\pm}^2}{A})], & A < 0, \alpha < 0, \end{cases} \tag{2.86}$$

$$\tau_{\pm}(j) = \begin{cases} \frac{1}{\omega_{\pm}}[2j\pi + \cos^{-1}(1 + \frac{\omega_0^2 - \omega_{\pm}^2}{A})], & for\ A > 0, \alpha > 0, \\ \frac{1}{\omega_{\pm}}[(2j+2)\pi - \cos^{-1}(1 + \frac{\omega_0^2 - \omega_{\pm}^2}{A})], & for\ A < 0, \alpha > 0. \end{cases} \tag{2.87}$$

The equilibrium point of (2.82) is stable at $\alpha < 0$ and $\tau = 0$, and the equilibrium point of (2.82) remains stable until τ increases from 0 to the first critical value. For the cases of $\alpha < 0$ and $A > 0$, when τ reaches the critical value τ_+, the characteristic Eq. (2.83) has a pair of pure imaginary eigenvalues, and Hopf bifurcation will occur in the system. The critical values of time delay τ for equilibrium stability switching are τ_- and τ_+, where $\tau_-(j) > \tau_+(j)(j = 0, 1, 2\ldots)$. When $\tau_-(j) < \tau < \tau_+(j+1)$ or $0 < \tau < \tau_+(0)$, the equilibrium point of (2.82) is stable. Hopf bifurcation occurs at the equilibrium point at the critical values $\tau_+(j)$ and $\tau_-(j)$.

We assume that (2.82) is weakly nonlinear and the delay gain coefficient is small enough. Might as well set $\alpha = \varepsilon\alpha_1$, $\beta = \varepsilon\beta_1$, $A = \varepsilon A_1$, $\alpha_1 = O(1)$, $\beta_1 = O(1)$, $\gamma_1 = O(1)$, $A_1 = O(1)$ and $0 < \varepsilon << 1$. Then (2.82) becomes

$$\ddot{x} + \omega^2 x = \varepsilon[(\sigma - A_1)x + \alpha_1\dot{x} - \gamma_1 x^2\dot{x} - \beta_1 x^3 + A_1 x_\tau]. \tag{2.88}$$

To study the dynamic behavior of the system near the Hopf bifurcation point by using MMS, we can obtain the average equation of the periodic solution as

$$\begin{cases} \dot{a}_1 = D_0 a_1 + \varepsilon D_1 a_1 = \dfrac{\omega(4\alpha a_1 - \gamma a_1^3) - 4Aa_1 \sin\omega\tau}{8\omega}, \\ \dot{b}_1 = D_0 b_1 + \varepsilon D_1 b_1 = \dfrac{4A(1 - \cos\omega\tau) + 3\beta a_1^2 - 4(\omega^2 - \omega_0^2)}{8\omega}. \end{cases} \tag{2.89}$$

To obtain the expression of the periodic solution, we set $\dot{a}_1 = 0$, $\dot{b}_1 = 0$. Then we can acquire the amplitude and frequency of the steady state solution of the system, i.e.

$$\begin{cases} 4(\omega\alpha - A\sin\omega\tau) - \omega\gamma a_1^2 = 0, \\ 4A(1 - \cos\omega\tau) - 4(\omega^2 - \omega_0^2) + 3\beta a_1^2 = 0. \end{cases} \tag{2.90}$$

If $\gamma = 0$, the relationship between frequency ω and time delay τ can be found in the first sub-equation of Eq. (2.90); if $\gamma \neq 0$, removed a_1 from Eq. (2.90), then the relationship between frequency ω and time delay τ is

$$3\beta(\omega\alpha - A\sin\omega\tau) + \omega\gamma(A + \omega_0^2 - \omega^2) = A\omega\gamma\cos\omega\tau. \tag{2.91}$$

By removing the time delay τ from Eq. (2.90), the first order approximate relationship between amplitude a_1 and time delay τ as follows

$$(4\omega\alpha - \omega\gamma a_1^2)^2 + (4A - 4\omega^2 + 4\omega_0^2 + 3\beta a_1^2)^2 = 16A^2. \tag{2.92}$$

Obviously, Eq. (2.91) is a transcendental equation about ω, with infinitely many solutions. The multiplicity of solutions is a characteristic of time delayed equations. When other system parameters are given, the number of solutions of ω in Eq. (2.91) depends on the time delay τ. But not every root of ω in Eq. (2.91) corresponds to a periodic solution, only those that can make a_1 in Eq. (2.92) have positive real roots can correspond to periodic solutions.

To analyze the stability of these periodic solutions, Eq. (2.89) is transformed into a rectangular coordinate form

$$\begin{aligned}\dot{x} = &\frac{\omega\alpha - A\sin\omega\tau}{2\omega}x - (\omega + \frac{A(1-\cos\omega\tau) - \omega^2 + \omega_0^2}{2\omega})y \\ &- \frac{\omega\gamma x + 3\beta y}{8\omega}(x^2 + y^2) + O(\varepsilon^2),\end{aligned} \tag{2.93}$$

$$\begin{aligned}\dot{y} = &(\omega + \frac{A(1-\cos\omega\tau) - \omega^2 + \omega_0^2}{2\omega})x + \frac{\omega\alpha - A\sin\omega\tau}{2\omega}y \\ &+ \frac{3\beta x - \omega\gamma y}{8\omega}(x^2 + y^2) + O(\varepsilon^2),\end{aligned} \tag{2.94}$$

where $y = a_1\sin\theta + O(\varepsilon^2)$.

According to Eq. (2.91), the frequency ω is a function of the amplitude a_1. By combining Eq. (2.91), we can immediately obtain the Frechet derivatives of Eqs. (2.93) and (2.94) regarding periodic solutions. The corresponding characteristic equation is

$$\lambda^2 - \lambda K(\omega, a_1, \tau) + \omega^2 + F(\omega, a_1, \tau) = 0, \tag{2.95}$$

where

$$\begin{aligned}&K(\omega, a_1, \tau) \\ &= \frac{a_1^2(2\omega^3\gamma - 3\beta A\sin\omega\tau - A\omega^2\gamma\tau\sin\omega\tau + 3\beta A\omega\tau\cos\omega\tau)\cos\omega\tau}{4\omega(A\sin^2\omega\tau - 2\omega^2\cos\omega\tau)},\end{aligned}$$

$$\begin{aligned}&F(\omega, a_1, \tau) \\ &= \frac{a_1^2(-6\omega^2\beta\cos\omega\tau + (3\beta A + A\omega^2\gamma\tau)\sin^2\omega\tau - 3\beta A\omega\tau\cos\omega\tau\sin\omega\tau)}{4(A\sin^2\omega\tau - 2\omega^2\cos\omega\tau)},\end{aligned}$$

with ω, a_1 and τ satisfy Eq. (2.90).

Then its characteristic value is

$$\lambda = K(\omega, a_1, \tau) \pm i\sqrt{\omega^2 - \frac{K^2(\omega, a_1, \tau) - 4F(\omega, a_1, \tau)}{4}}. \tag{2.96}$$

2.3.3 *Basin of Attraction for Periodic Solutions of Multiple Steady States*

Next, we provide the distribution of the attractive domain for the multi-steady state solutions of the system. For this purpose, it is assumed that the initial conditions of the system satisfy

$$\begin{cases} x(t) = 0.0, & -\tau \le t < 0, \\ \dot{x}(t) = 0.0, & -\tau \le t < 0. \end{cases} \tag{2.97}$$

System parameters are set to

$$\alpha = -0.04, \beta = 0.3, \gamma = 0, \omega_0 = 1, A = 0.3. \tag{2.98}$$

From Sect. 2.3.2, we obtained the first order approximate analytical form of the periodic solution using the averaging method and multiple periodic solutions generated based on multi-level Hopf bifurcation under time delay. The amplitude of the periodic solution is given, as shown in Fig. 2.8. To verify the correctness of the theoretical analysis, we use the fourth order Runge–Kutta method to simulate (2.82) for stable periodic solutions, and obtain the amplitude value of the periodic solution represented by “x”. From Fig. 2.8, we can see that when time delay increases from 0 to the value corresponding to the first saddle-node bifurcation point, there is no change in the dynamic behavior of (2.82), and the equilibrium point remains stable. When time delay τ exceeds the value corresponding to the first saddle-node bifurcation point, (2.82) will add a stable periodic solution and an unstable periodic solution, that is, the system will have a stable equilibrium point coexisting with the periodic solution. When time delay τ increases to the subcritical Hopf bifurcation point τ_{11}, the unstable periodic solution degenerates into an equilibrium point, and the equilibrium point undergoes a stability transition from stable to unstable. As τ further increases, it increases to the value corresponding to the saddle-node bifurcation point of the second solution brunch. (2.82) will add a stable periodic solution and an unstable periodic solution, that is, the system will experience the coexistence of two steady state solutions. As τ increases to the critical value τ_{10}, the stable periodic solution on the first solution branch degenerates into an equilibrium point, which once again undergoes a stability switch from unstable to stable.

In fact, in Fig. 2.8, the critical values $\tau_{j1}(\tau_+(j))$ and $\tau_{i0}(\tau_-(i))$ at the Hopf bifurcation point are obtained from Eqs. (2.86) and (2.87). At the saddle-node bifurcation point of each solution brunch, Eq. (2.83) satisfies the critical condition $\text{Re}(\lambda) = 0$, namely

$$2\omega^3\gamma - 3\beta A \sin\omega\tau - A\omega^2\gamma\tau \sin\omega\tau + 3\beta A\omega\tau \cos\omega\tau = 0. \tag{2.99}$$

Equation (2.99) holds if and only if ω has multiple roots in Eq. (2.91). In addition, due to $\tau_{41} < \tau_{30}$ in Fig. 2.8, the equilibrium point no longer undergoes stability switching as τ changes and the intersection of the third and fourth solution branches. It is worth noting that complex motions such as quasi-periodic motions often occur at the intersection of stable and unstable solution branches.

Figure 2.9 shows the attraction basin for the multi-steady state solution of (2.82) with different time delays. From Fig. 2.9a–d, it can be seen that when $\tau = 0.8$, the system only exists with one stable periodic solution and one stable equilibrium point. As the time delay increases, the system undergoes multi-level Hopf bifurcation, resulting in the steady state coexistence of multiple different periodic solutions. We found that when $\tau = 4.5$, there is a coexistence of five periodic attractors in (2.82). In addition, the comparison between the boundary of the attraction basin and the amplitude of the unstable periodic solution is shown in Fig. 2.8. From Fig. 2.8, it can be seen that the analytical results and numerical solutions are quite consistent both qualitatively and quantitatively.

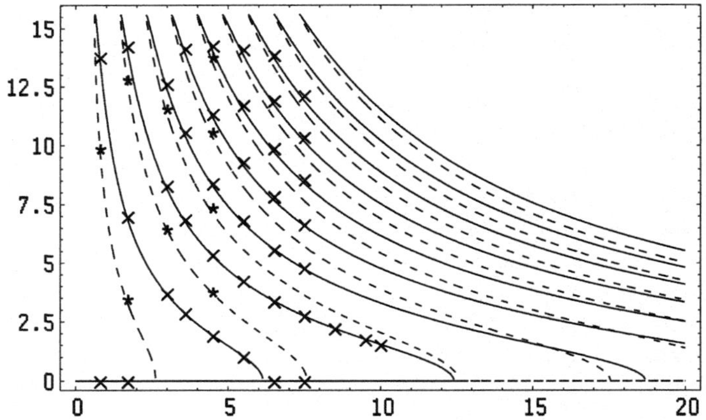

Fig. 2.8 The variation curve of the amplitude of the periodic solution at any time lag under the parameter values of Eq. (2.98). The solid line represents a stable periodic solution, while the dashed line represents an unstable solution "×" Represents numerical results, "*" represents the boundary of the basin of attraction, and "•" represents the saddle-node bifurcation point

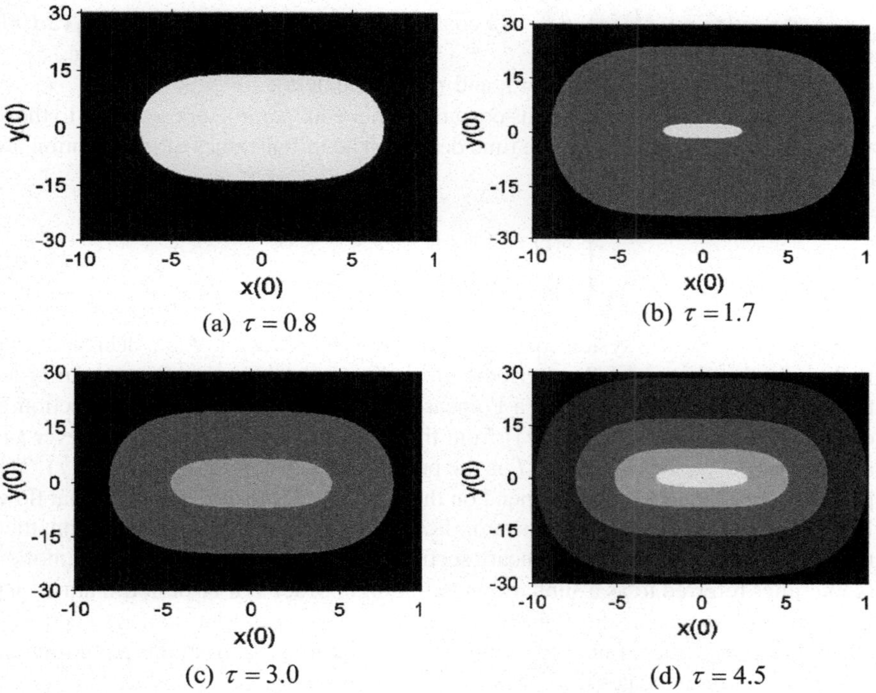

Fig. 2.9 Attraction basin with different time delays. **a** $\tau = 0.8$; **b** $\tau = 1.7$; **c** $\tau = 3.0$; **d** $\tau = 4.5$. The black area represents the attractive domain of large amplitude periodic solutions; the dark gray area represents the attractive domain of the second largest amplitude periodic solution; the gray area represents the attractive domain of the third largest amplitude periodic solution; the light gray area represents the attractive domain of the fourth largest amplitude periodic solution; the white area represents the attractive domain of equilibria

2.4 Quasi-Periodic Solutions and Chaos in Time Delay Systems

At the end of this section, to further elucidate the complex nonlinear dynamic behavior induced by time delay, we further take the delayed VDP-Duffing system in the first part as the research object. Based on Poincare cross-section technology, phase portrait, and bifurcation diagram, we explore the quasi-periodic bifurcation and its chaos route caused by time delay. In this section, we found that the system can achieve chaotic motion through two routes: period-doubling bifurcation and quasi-periodic bifurcation under the influence of time delay. We also demonstrated the "control switch" effect of time delay, which means that time delay can not only lead to complex motion of the system, but also turn complex motion into simple motion.

The system model studied in this section is

$$\ddot{x} + \omega_0^2 x - (\alpha - \gamma x^2)\dot{x} + \beta x^3 = k\cos(\Omega t) + A(x_\tau - x) + B(x_\tau - x)^3, \quad (2.100)$$

where $\alpha, \gamma, k > 0$, $x_\tau = x(t-\tau)$, and τ is a time delay.

In all the numerical simulations considered here, all parameters are fixed to those values for Fig. 2.10 except for the time delay τ. The initial values of the equation are determined by

$$\begin{cases} x(t) = 1.0 \text{ for } -\tau < t \leq 0, \\ \dot{x}(t) = 0.0 \text{ for } t = 0. \end{cases} \quad (2.101)$$

When $\tau > \tau_c$, the system has only one periodic attractor. As τ decreases, we will study the change of this periodic attractor across the point P at τ_c. In order to determine the route to chaos, a Poincaré diagram is used. A Poincaré section is defined as a projection $(x(t), \dot{x}(t))$ from the three dimensional space whenever t is a multiple of $T = 2\pi/\Omega$, where T is the period of the excitation $(x(t), \dot{x}(t), t)$. The points in the Poincaré section depend on the behavior of the three-dimensional flow. If the final motion of the system is periodic with the frequency of the excitation, then there is only one point in the Poincaré section. For a period-n ($n = 2, 3, \ldots$) motion (sometimes referred to as a subharmonic motion of order $1/n$ or period-n attractor), n points will appear in the Poincaré section. For non-periodic motions such as a chaotic response, the number of points becomes infinite. An irregular pattern in the Poincaré section indicates the existence of a strange attractor.

On the one hand, using the Poincaré cross-section technique defined above, a detailed bifurcation diagram as a function of delay is shown in Fig. 2.11a. Figure 2.11b is an enlarged part of Fig. 2.11a for $\tau \in [0.145, 0.153]$. A scenario of dynamics in the original (2.2) is shown in Fig. 2.11a and b with the time delay decreasing: phase-locked solution $\Rightarrow$ Hopf bifurcation $\Rightarrow$ periodic doubling bifurcation $\Rightarrow$ chaos $\Rightarrow$ quasi-periodic solution. The Poincaré section of solutions at several sampled values of the time delay is shown in Fig. 2.12, and it reveals the change of solutions and bifurcations to chaos as the time delay τ decreases from 0.2 to 0.

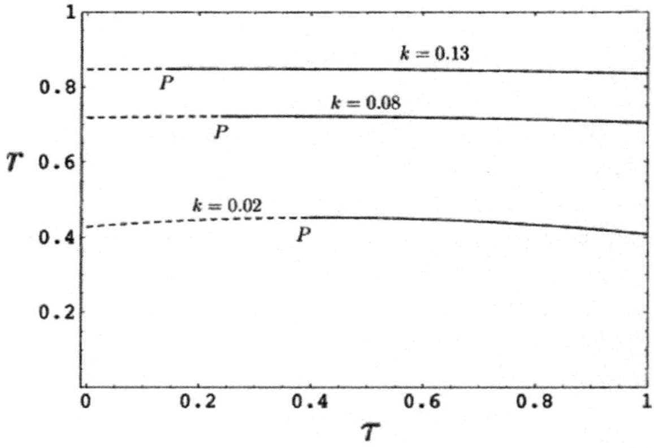

Fig. 2.10 Amplitude-response curves of periodic solution as a function of time delay for finite excited amplitude and $\sigma = 0$. Solid lines represent stable periodic solutions and dashed lines unstable ones

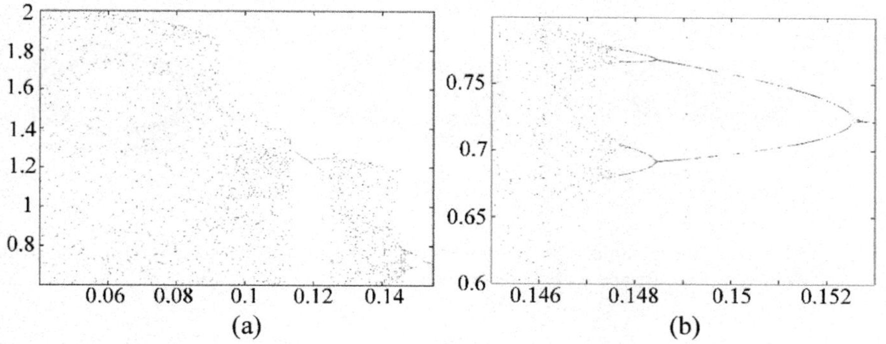

Fig. 2.11 Bifurcation diagram derived from Eq. (2.2) by Poincare section for **a** $\tau \in [0.04, 0.155]$ and **b** enlarged part for $\tau \in [0.145, 0.153]$, where $x = x(nT)$, $T = 2\pi/\Omega$, $n = 1, 2, 3, \ldots$

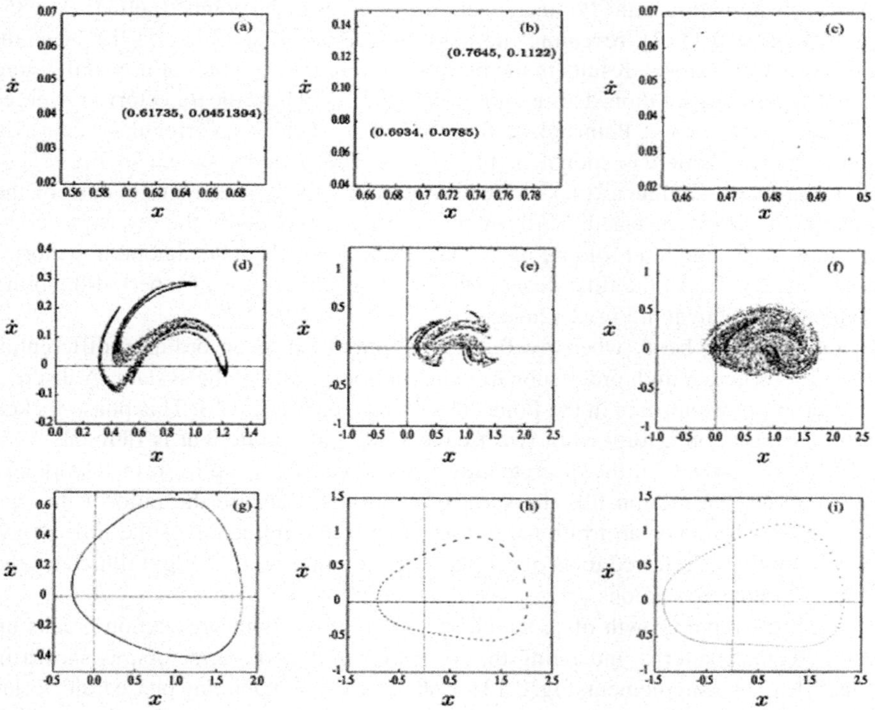

Fig. 2.12 Route to chaos for the motion of (2.2) with time delay τ varying: **a** $\tau = 0.2$ (period-1 attractor); **b** $\tau = 0.149$ (period-2 attractor); **c** $\tau = 0.148$ (period-4 attractor); **d** $\tau = 0.135$ (chaotic attractor); **e** $\tau = 0.1$(chaotic attractor); **f** $\tau = 0.09308$ (chaotic attractor); **g** $\tau = 0.093075$ (quasi-periodic attractor); **h** $\tau = 0.05$ (subharmonic attractor with high order); **i** $\tau = 0.0$ (subharmonic attractor with high order)

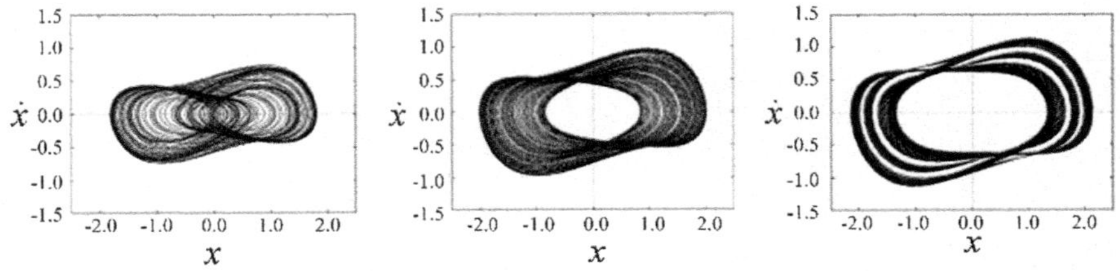

Fig. 2.13 Phase plane of trajectories corresponding to Fig. 2.12i, h and g for Eq. (2.2) with **a** $\tau = 0.093075$, **b** $\tau = 0.05$ and **c** $\tau = 0$, respectively

The analytical prediction described in Fig. 2.10 is confirmed in Fig. 2.12a. Since there is only one point (0.61735, 0.0451394) appearing in the Poincaré section, a stable period-1 motion occurs in the system at $\tau = 0.2$. As time delay τ decreases to 0.148, the trajectory in the Poincaré section converges to the points (0.6934, 0.0785) and (0.7645, 0.1122), revealing the periodic-2 motion (see Fig. 2.12b). A slight decrease in time delay results in the period-2 solution losing its stability and giving rise to a period-4 solution. This is shown in Fig. 2.12c where the trajectory converges to four points in the Poincaré section. When $\tau = 0.135$, an irregular pattern in the Poincaré section is shown in Fig. 2.12d. The patterns shown in Fig. 2.12d and e are very similar to two of Ueda's attractors [41]. The fractal nature of the patterns is clearly revealed. This complex structure is actually the consequence of a simple stretching and folding of the ensemble of steady state chaotic trajectories (see Fig. 2.12f). As the time delay decreases, a route to chaos by period-doubling sequences is shown in Fig. 2.12a–e.

On the other hand, when $\tau = 0$, (2.2) is simplified as an ordinary differential equation, where a high order subharmonic motion occurs in the system. A discrete cluster of points appears in the Poincaré section (see Fig. 2.12i). This phase-locked motion moves on a torus with two frequencies, and its quotient is rational. As τ increases to 0.093075, the discrete cluster of points closes up to form a loop, and at this time, the motion fills the surface of a torus. Because the ratio of the two frequencies becomes an irrational number due to the influence of the time delay, such a motion is referred to as quasi-periodic motion. Figure 2.12g–i illustrates the selected Poincaré sections.

With the rapid growth of τ, the closed curve in the Poincaré section breaks up into irregular patterns, indicating the formation of chaotic attractors, as shown in Fig. 2.12f. As a supplement, Fig. 2.13a–c shows the corresponding phase trajectories described in Fig. 2.12i–g, respectively. Even for small τ, the time delay does change the phase trajectory of the motion qualitatively as shown in these figures.

We have seen two routes in which the system under consideration may become chaotic. The former one is a period-doubling sequence after a Hopf bifurcation as τ decreases from τ_c. The latter one is phase-locked, quasi-periodic motions followed by torus breaking as τ increases from zero. In a delayed system, the simultaneous occurrence of these two routes leading to chaos is a very interesting phenomenon.

References

1. Song ZG, Zhen B, Hu DP (2020) Multiple bifurcations and coexistence in an inertial two-neuron system with multiple delays. Cogn Neurodynam 14(3):359–374
2. Song ZG, Wang CH, Zhen B (2016) Codimension-two bifurcation and multistability coexistence in an inertial two-neuron system with multiple delays. Nonlinear Dyn 85(4):2099–2113
3. Wang F, Sun XT, Meng H, Xu J (2021) Time delayed feedback control design and its application for vibration absorption. IEEE T Ind Electron 68(9):8593–8602
4. Sun XT, Wang F, Xu J (2019) Dynamics and realization of a feedback-controlled nonlinear isolator with variable time delay. J Vib Acoust 141(2):021005
5. Guo L, Song ZG, Xu J (2014) Complex dynamics in the Leslie-Gower type of the food chain system with multiple delays. Commun Nonlinear Sci 19(8):2850–2865
6. Song ZG, Xu J (2023) Multiple switching and bifurcations of in-phase and anti-phase periodic orbits to chaos coexistence in a delayed half-center CPG oscillator. Nonlinear Dyn 111(17):16569–16584
7. Song ZG, Huang XJ, Xu J (2022) Spatiotemporal pattern of periodic rhythrms in delayed Van der Pol oscillators for the CPG-based locomotion of snake-like robot. Nonlinear Dyn 110:3377–3393
8. Salas AH, El-Tantawy SA (2020) On the approximate solutions to a damped harmonic oscillator with higher-order nonlinearities and its application to plasma physics: semi-analytical solution and moving boundary method. Eur Phys J Plus 135(10):833
9. Wang WY, Xu J (2011) Multiple scales analysis for double Hopf bifurcationwith 1:3 resonance. Nonlinear Dyn 66(1–2):39–51
10. Ding YT, Jiang WH, Yu P (2013) Double Hopf bifurcation in delayed van der Pol-Duffing equation. Int J Bifurcat Chaos 23(1):1350014
11. Ge JH, Xu J (2019) An analytical method for studying double Hopf bifurcations induced by two delays in nonlinear differential systems. Sci China Tech Sci 63(4):597–602
12. Xu J, Chung KW (2003) Effects of time delayed position feedback on a van der Pol-Duffing oscillator. Physica D 180:17–39
13. Zhao ZH, Yang SP (2015) Application of van der Pol-Duffing oscillator in weak signal detection. Comput Electr Eng 41:1–8
14. Peng HH, Xu XM, Yang BC, Yin LZ (2016) Implication of two-coupled differential van der Pol Duffing oscillator in weak signal detection. J Phys Soc Jpn 85(4):044005
15. Xu J, Lu QS, Wang C (2000) Stability and bifurcations in a Van der Pol-Duffing time-delay system. Acta Mech Sin 32:112–116 (in Chinese)
16. Li F, Fang JS (2009) Chaos and control in a van der pol-duffing system. J Hunan City Univ 18:29–32 (in Chinese)
17. Zhang LY, Jian N, Li Y, Peng JK (2007) Chaos control for the periodically excited Van der Pol-Duffing. J Wenzhou Univ Nat Sci 28:11–14 (in Chinese)
18. Xu L, Lu M, Cao Q (2002) Nonlinear dynamical bifurcation analysis of Van der Pol-Duffing equation. Chinese J Appl Mech 19:130–133 (in Chinese)
19. Reddy DVR, Sen A, Johnston GL (1999) Time delay effects on coupled limit cycle oscillators at Hopf bifurcation. Phys D 129:15–34
20. Guckenheimer J, Holmes P (1993) Nonlinear oscillations, dynamical systems, and bifurcations of vector fields. Springer, New York
21. Nayfeh AH, Mook DT (1979) Nonlinear oscillations. Wiley, New York
22. Uçr A (2003) On the chaotic behaviour of a prototype delayed dynamical system. Chaos Soliton Fract 16(2):187–194
23. Moiola J L, Chen G (1997) Hopf bifurcation analysis: a frequency domain approach. Singapore, Word Scientific
24. Song ZG, Xu J (2009) Bursting near Bautin bifurcation in a neural network with delay coupling. Int J Neural Syst 19(5):359–373
25. Li L, Xu J (2018) Bifurcation analysis and spatiotemporal patterns in unidirectionally delay-coupled vibratory gyroscopes. Int J Bifurcat Chaos 28(2):1850029

26. Xu J (1999) Hopf bifurcation of time-delay lienard equations. Int J Bifurcat Chaos 9(5):939–951
27. Pei LJ, Xu J (2005) Nonresonant double Hopf bifurcation in delayed stuart-landau system. J Vibrat Eng 18(1):24–29 (in Chinese)
28. Ma SQ, Lu QS, Feng ZS (2008) Double Hopf bifurcation for Van der Pol-Duffing oscillator with parametric delay feedback control. J Math Anal Appl 338:993–1007
29. Ma SQ, Lu QS, Hogan SJ (2007) Double Hopf bifurcation for Stuart-Landau system with nonlinear delay feedback and delay-dependent parameters. Adv Complex Syst 10(4):423–448
30. Rand RH, Keith WL (1986) Normal forms and center manifold calculations on MACSYMA. Kluwer Academic Press, Boston
31. Yu P (2001) Symbolic computation of normal forms for resonant double Hopf bifurcations using multiple time scales. J Sound Vib 247(4):615–632
32. Chedjou JC, Kyamakya K, Moussa I, Kuchenbecker HP, Mathis W (2006) Behavior of a self-sustained electromechanical transducer and routes to chaos. Trans ASME 128(3):282–293
33. Boivin N, Pierre C, Shaw SW (1995) Nonlinear modal analysis of structural systems featuring internal resonance. J Sound Vib 1982(2):336–3341
34. Shaw SW, Holmes PJ (1983) A periodically forced piecewise linear oscillator. J Sound Vib 90(1):129–155
35. Shaw SW, Pierre C (1994) Normal modes of vibration for nonlinear continuous systems. J Sound Vib 169(3):319–347
36. Wang WY, Xu J (2011) Multiple scales analysis for double Hopf bifurcation with 1:3 resonance. Nonlinear Dyn 66:39–51
37. Xu J, Pei Y (2003) Delay-induced bifurcations in a nonautonomous system with delayed velocity feedbacks. Int J Bssifurcat Chaos 14(8):2777–2798
38. Luongo A, Paolone A, Egidio AD (2003) Multiple timescales analysis for 1:2 and 1:3 resonant Hopf bifurcations. Nonlinear Dyn 34(3–4):269–291
39. Wirkus S, Rand R (2002) The dynamics of two coupled van der Pol oscillators with delay coupling. Nonlinear Dyn 30(3):205–221
40. Shang HL, Xu J (2005) Multiple periodic solutions in delayed Duffing equation. J Taiyuan Univ Techn 36(6):749–751 (in Chinese)
41. Ueda Y, Akamatsu N (1981) Chaotically transitional phenomena in the forced negative resistance oscillator. IEEE Trans Circuits Syst 28(3):217–223

Chapter 3
Perturbation-Incremental Scheme and Integral Equation Method for Solving Time Delay Systems

This chapter presents two new methods for analyzing dynamics of nonlinear systems with time delay. The first method is called the perturbation-incremental scheme (PIS). The perturbation step and incremental step, which are the two key components of the PIS, allow us to track the bifurcated solution arising from Hopf or double Hopf bifurcation with high accuracy, without the tedious calculations needed for center manifold reduction or normal form computation. In doing so, the PIS avoides the shortcomings of the traditional incremental harmonic balance method in addition to inheriting the advantages of the method of multiple scales.

The second method is the integral equation method. Similar like the PIS, the integral equation method transforms the problem of solving differential equations to the problem of solving algebraic equations. Based on the solutions obtained by this method, the Floquet theorem can be used for analyzing the stability of the solution. Comprisons with numerical results show that the integral equation method exhibits remarkable efficiency when handling delay differential equations.

3.1 Perturbation-Incremental Scheme for Studying Hopf Bi-Furcation in Delay Differential Systems

3.1.1 Introduction

A class of delayed models can be represented as follows:

$$\dot{\mathbf{Z}}(t) = \mathbf{C}\mathbf{Z}(t) + \mathbf{D}\mathbf{Z}(t-\tau) + \varepsilon\mathbf{F}(\mathbf{Z}(t), \mathbf{Z}(t-\tau)), \tag{3.1}$$

J. Xu, *Nonlinear Dynamics of Time Delay Systems*,
https://doi.org/10.1007/978-981-99-9907-1_3

where $\mathbf{Z}(t) \in \mathbb{R}^n$, $\mathbf{C}$ and $\mathbf{D}$ are $n \times n$ constant real matrix, $\mathbf{F}(\cdot)$ represents a nonlinear function in its variables with $\mathbf{F}(\mathbf{0}, \mathbf{0}) = \mathbf{0}$, with ε being an arbitrary magnitude parameter, τ denoting the time delay, and n a positive integer.

It has been demonstrated that, even for modest time delays, time delays in a variety of systems affect dynamics in both quantitative and qualitative ways [1, 2]. Such situation has led to the development and expansion of a large number of theories of delayed differential equations (DDEs), among which the Hopf bifurcation generated by delay might be the simplest and most fundamental bifurcation in stability analysis.

Obtaining analytical form for the periodic solutions arising from the Hopf bifurcation is the main goal of the quantitative research of such bifurcation. It can, however, frequently be a laborious procedure to reduce a given DDE to a finite-dimensional system. The method of multiple scales was used by Das and Chatterjee [3] to get analytical solutions for bifurcation parameters close to the Hopf bifurcation point. The numerical solutions and the analytical approximations derived from Das' version are in good agreement for values near the Hopf bifurcation point. This approach, however, becomes invalid for values that are very far from the bifurcation point. In these situations, periodic or even double periodic solutions can be analytically expressed using the IHB approach [4]. Finding an initial value for an iteration is a prerequisite for the IHB approach, and this can be a difficult task in general. Casal and Freedmann [5] introduced a Poincaré-Lindstedt method for bifurcation problems in DDEs. The approach works well for bifurcation parameters that are somewhat close to the Hopf bifurcation point, but becomes invalid for those that are much farther away.

In this section, we address these issues by concentrating on Eq. (3.1) and introducing a method for studying delay-induced Hopf bifurcations and periodic solutions: the perturbation-incremental scheme (PIS). The primary benefit of the PIS is that it avoids the complex computations that are frequently involved in center manifold reduction. The incremental iteration starts with an initial guess based on the results of the perturbation step. Periodic solutions for bifurcation parameters far from the bifurcation point can be precisely found to any required level of precision by employing the incremental step.

3.1.2 Perturbation-Incremental Scheme

3.1.2.1 Linear Analysis ($\varepsilon = 0$)

To assess the stability of the trivial solution when $\tau \neq 0$, Eq. (3.1) is linearized around $\mathrm{Z} = 0$, resulting in the following characteristic equation:

$$\det\left(\lambda \mathrm{I} - \mathrm{C} - \mathrm{D}e^{-\lambda\tau}\right) = 0, \tag{3.2}$$

where $\mathbf{I}$ represents the identity matrix, and $\tau > 0$.

When either zero or purely imaginary pairs are roots of Eq. (3.2), the trivial equilibrium's stability may change. In the former case, the number of equilibrium points varies as the bifurcation parameters vary, corresponding to a static bifurcation of equilibrium points. The latter is related to a Hopf bifurcation, for which the dynamic behaviour of the system shifts from a stable, static state to a periodic motion, or the other way around. In particular, Eq. (3.1) experiences a simple Hopf bifurcation if all eigenvalues—aside from the purely imaginary ones—have negative real parts and the critical root shows nonzero velocity when crossing the imaginary axis.

In this section, we choose τ as the bifurcation parameter and assume τ_c is the critical value for a simple Hopf bifurcation, i.e., Eq. (3.2) has a simple purely imaginary pair $\lambda_0 = \pm i\omega \neq 0$ at $\tau = \tau_c$, and the crossing-speed of the roots is nonzero. Meanwhile, none of other characteristic roots are zero or purely imaginary pairs.

3.1.2.2 Perturbation Method at a Critical Value (Small $\varepsilon\tau_\varepsilon$)

The first stage in the PIS is to approximate a small-amplitude periodic solution at a Hopf bifurcation point. Perturbing the parameter, namely, letting $\tau = \tau_c + \varepsilon\tau_\varepsilon$, results in

$$\dot{Z}(t) = CZ(t) + DZ(t-\tau_c) + \tilde{F}(Z(t), Z(t-\tau_c), Z(t-\tau_c-\varepsilon\tau_\varepsilon), \varepsilon), \quad (3.3)$$

where

$$\begin{aligned}\tilde{F}(Z(t), Z(t-\tau_c), Z(t-\tau_c-\varepsilon\tau_\varepsilon), \varepsilon) = {} & D[Z(t-\tau_c-\varepsilon\tau_\varepsilon) - Z(t-\tau_c)] \\ & +\varepsilon F(Z(t), Z(t-\tau_c-\varepsilon\tau_\varepsilon)).\end{aligned} \quad (3.4)$$

There may be computational difficulties when attempting to solve the periodic solution of Eq. (3.3) directly. This led to the development of the adjoint equation, which in some circumstances can assist in lowering the computational complexity of the differential equations. We define the following inner product:

$$\langle \mathbf{U}, \mathbf{V} \rangle = \int_0^b \mathbf{V}^{\mathrm{T}}(t)\mathbf{U}(t)\mathrm{d}t, \quad (3.5)$$

where $(\cdot)^T$ denotes transpose and b represents a constant.

The next step is to find the adjoint equation that correspond to the linearized equation in Eq. (3.3). Equation (3.3) can be rewritten in the operator form for $\varepsilon = 0$ as

$$\mathbf{L}(\mathbf{Z}) = 0, \quad (3.6)$$

where $\mathbf{L}(\mathbf{Z}) = \dot{Z}(t) - CZ(t) - DZ(t-\tau_c)$. The operator $\mathbf{L}^*$ is said to be an adjoint operator corresponding to $\mathbf{L}$, and given $\mathbf{L}$, $\mathbf{L}^*$ will satisfy

$$\langle \mathbf{L}(\mathbf{u}), \mathbf{v}\rangle = \langle \mathbf{u}, \mathbf{L}^*(\mathbf{v})\rangle. \tag{3.7}$$

Using Eq. (3.5) and noting $b = 2\pi/\omega$ for $\varepsilon = 0$ and ω is the frequency of the periodic solution derived from the Hopf bifurcation at $\tau = \tau_c$, it can be obtained that

$$\begin{aligned}\langle \mathbf{L}(\mathbf{Z}), \mathbf{W}\rangle &= \int_0^{2\pi/\omega} \mathbf{W}^{\mathrm{T}}(t)\big(\dot{Z}(t) - CZ(t) - DZ(t-\tau_c)\big)\mathrm{d}t \\ &= \int_0^{2\pi/\omega} \big(-\dot{\mathbf{W}}(t) - \mathbf{C}^{\mathrm{T}}\mathbf{W}(t) - \mathbf{D}^{\mathrm{T}}\mathbf{W}(t+\tau_c)\big)^{\mathrm{T}}\mathbf{Z}(t)\mathrm{d}t \\ &+ \left(\mathbf{W}\left(\frac{2\pi}{\omega}\right) - \mathbf{W}(0)\right)^{\mathrm{T}}\mathbf{Z}(t) \\ &- \int_{\tau_c}^{0}\left(\mathbf{W}\left(t + \frac{2\pi}{\omega} + \tau_c\right) - \mathbf{W}(t+\tau_c)\right)^{\mathrm{T}}\mathbf{DZ}(\mathrm{t})\mathrm{d}t.\end{aligned} \tag{3.8}$$

If $\mathbf{W}(t) = \mathbf{W}(t + 2\pi/\omega)$, one has

$$\langle \mathbf{L}(\mathbf{Z}), \mathbf{W}\rangle = \langle \mathbf{Z}, \mathbf{L}^*(\mathbf{W})\rangle, \tag{3.9}$$

where $\mathbf{L}^*(\mathbf{W}) = -\dot{\mathbf{W}}(t) - \mathbf{C}^T\mathbf{W}(t) - \mathbf{D}^T\mathbf{W}(t+\tau_c)$. Consequently, the following form can be used to define the adjoint equation that corresponds to the linearized equation of Eq. (3.3)

$$\dot{\mathbf{W}}(t) = -\mathbf{C}^{\mathrm{T}}\mathbf{W}(t) - \mathbf{D}^{\mathrm{T}}\mathbf{W}(t+\tau_c), \tag{3.10}$$

where $\mathbf{W}(t) = \mathbf{W}(t + 2\pi/\omega)$. Suppose that the periodic solution of Eq. (3.10) with period $2\pi/\omega$ is in the following form

$$\mathbf{W}(\mathrm{t}) = \mathbf{p}\cos(\phi) + \mathbf{q}\sin(\phi), \tag{3.11}$$

where $\phi = \omega t$, so that

$$\mathbf{W}(t+\tau_c) = \mathbf{p}\cos(\phi + \omega\tau_c) + \mathbf{q}\sin(\phi + \omega\tau_c), \tag{3.12}$$

where $\mathbf{p} = (p_1, p_2, ..., p_n)^T$, $\mathbf{q} = (q_1, q_2, ..., q_n)^T \in \mathbf{R}^n$, τ_c and ω are determined by (3.2). Substituting Eqs. (3.11) and (3.12) into Eq. (3.10) and using the harmonic balance, one may have

$$\mathbf{M}^{\mathrm{T}}\mathbf{p} = \mathbf{N}^{\mathrm{T}}\mathbf{q}, \tag{3.13}$$

and

$$\mathbf{M}^{\mathrm{T}}\mathbf{q} = -\mathbf{N}^{\mathrm{T}}\mathbf{p}, \tag{3.14}$$

where $\mathbf{M} = \omega\mathbf{I} + \mathbf{D}\sin(\omega\tau_c)$, $\mathbf{N} = \mathbf{C} + \mathbf{D}\cos(\omega\tau_c)$, which are real and imaginary parts of the characteristic matrix $(\lambda I - C - De^{-\lambda\tau})$ respectively at $\tau = \tau_c$, where Eq. (3.2) will have a simple purely imaginary root. Therefore, there are only $(2n - 2)$ independent equations to determine $p_1, p_2, ..., p_n$ and $q_1, q_2, ..., q_n$ in Eqs. (3.13) and (3.14). If p_1 and q_1 are chosen to be independent, then p_i and q_i $(i = 2, ..., n)$ can be determined by Eqs. (3.13) and (3.14) in terms of p_1 and q_1. Similarly, for $\varepsilon = 0$, the periodic solution of Eq. (3.3) with period $2\pi/\omega$ can be obtained by

$$\mathbf{Z}(\mathrm{t}) = \mathbf{a}\cos(\phi) + \mathbf{b}\sin(\phi), \tag{3.15}$$

where $\mathbf{a} = (a_1, a_2, ..., a_n)^T$, $\mathbf{b} = (b_1, b_2, ..., b_n)^T \in \mathbf{R}^n$, a_i and b_i (i = 2,…,n) are functions of a_1 and b_1, which can be expressed by

$$\mathbf{Mb} = \mathbf{Na}, \tag{3.16}$$

and

$$-\mathbf{Ma} = \mathbf{Nb}. \tag{3.17}$$

Equation (3.15) in a polar coordinate can be expressed as

$$\mathbf{Z}(\mathrm{t}) = \mathbf{r}\cos(\phi + \theta), \tag{3.18}$$

where $r = (r_1, r_2, ..., r_n)^T$ and r_i $(i = 2, ..., n)$ are functions of r_1. Therefore, for a small ε, the solution of Eq. (3.3) may be considered as a perturbation of (3.18), and can be given by

$$\mathbf{Z}(\mathrm{t}) = \mathbf{r}(\varepsilon)\cos((\omega + \sigma(\varepsilon))t + \theta), \tag{3.19}$$

where $\mathbf{r}(0) = \mathbf{r}$, $\sigma(0) = 0$, $r_i(\varepsilon) = r_i(r_1(\varepsilon))$ $(i = 2, .., n)$. To obtain $r_1(\varepsilon)$ and $\sigma(\varepsilon)$, both sides of (3.3) are multiplied by $\mathbf{W}^T(t)$, and integrated with respect to t from zero to $2\pi/\omega$, so we have

$$\begin{aligned} &\int_0^{2\pi/\omega} \mathbf{W}^{\mathrm{T}}(t)\dot{\mathbf{Z}}(t)\mathrm{d}t \\ &= \int_0^{2\pi/\omega} \mathbf{W}^{\mathrm{T}}(t)\begin{bmatrix} \mathbf{CZ}(t) + \mathbf{DZ}(t - \tau_c) \\ + \tilde{F}(Z(t), Z(t - t_c), Z(t - \tau_c - \varepsilon\tau_\varepsilon), \varepsilon) \end{bmatrix}\mathrm{d}t, \end{aligned} \tag{3.20}$$

which yields [6]

$$\begin{aligned}&\int_{-\tau_c}^{0}\mathbf{W}^{\mathrm{T}}(t+\tau_c)\mathbf{D}[\mathbf{Z}(t)-\mathbf{Z}(\mathrm{t}+2\pi/\omega)]\mathrm{d}t-\mathbf{W}^{\mathrm{T}}(0)[\mathbf{Z}(2\pi/\omega)-\mathbf{Z}(0)]\\&+\int_{0}^{2\pi/\omega}\mathbf{W}^{\mathrm{T}}(t)\tilde{F}(\mathbf{Z}(t),\mathbf{Z}(t-\tau_c),\mathbf{Z}(t-\tau_c-\varepsilon\tau_\varepsilon),\varepsilon)\mathrm{d}t=0.\end{aligned}\tag{3.21}$$

From the above analysis, the conclusion can be obtained that if $\mathbf{Z}(t)$ in Eq. (3.19) is a periodic solution induced from the trivial equilibrium by the Hopf bifurcation at τ_c. Then it satisfies Eq. (3.21). Thus, substituting Eqs. (3.11) and (3.18) into Eq. (3.21) and noting the independence of p_1 and q_1, a set of algebraic equations can be obtained, which can be used in the determination of $r_1(\varepsilon)$ and $\sigma(\varepsilon)$.

It is worth noting that Eq. (3.21) comprises explicit algebraic equations in $r_1(\varepsilon)$ and $\sigma(\varepsilon)$. The roots of Eq. (3.21) cannot be expressed in a closed form due to the presence of transcendental functions. To obtain analytical forms for $r_1(\varepsilon)$ and $\sigma(\varepsilon)$, the transcendental functions in Eq. (3.21) are expanded using Taylor's series in ε, with high-order terms in power ε being neglected. Consequently, an approximated formula for $r_1(\varepsilon)$ and $\sigma(\varepsilon)$ is derived. Accordingly, an approximate analytical expression for the periodic solution resulting from a Hopf bifurcation is provided in Eq. (3.19). Equation (3.19) shows that the value of ε has a significant impact on how accurate this analytical expression is. When the time delay value is pretty close to the Hopf bifurcation point (i.e., $\varepsilon\tau_\varepsilon$ is very small), the approximation (3.19) can accurately describe the periodic solutions of Eq. (3.3). However, this approximation becomes quantitatively invalid when the time delay value is far from the Hopf bifurcation point. As will be covered in the following, the incremental step can be used to modify the approximation expression in order to get closer to the precise solution.

3.1.2.3 Parameter Incremental Method (Large $\varepsilon\tau_\varepsilon$)

From the previous subsection, the periodic solution can be approximated as

$$\mathbf{Z}(t)=\mathbf{r}(\varepsilon)\cos\varphi,\quad \dot{\mathbf{Z}}(t)=-\mathbf{r}(\varepsilon)\frac{\mathrm{d}\varphi}{\mathrm{d}t}\sin\varphi,\tag{3.22}$$

where $\varphi=\omega+\sigma(\varepsilon))t+\theta$ and $\frac{d\varphi}{dt}=\omega+\sigma(\varepsilon)$ for a small $\varepsilon\tau_\varepsilon$. Such approximated expression needs to be improved to approach required accuracy for a large $\varepsilon\tau_\varepsilon$. To assure that the solution is periodic in φ, then $\frac{d\varphi}{dt}$ should also be periodic in φ. Therefore, a time transformation is introduced as [7]

$$\frac{\mathrm{d}\varphi}{\mathrm{d}t}=\Phi(\varphi),\quad \Phi(\varphi+2\pi)=\Phi(\varphi),\tag{3.23}$$

which has been used by many authors (see [8, 9]). Furthermore, $\Phi(\phi)$ can be approximately expanded in a truncated Fourier's series about ϕ as

$$\Phi(\varphi) = \sum_{j=0}^{m} \left(p_j \cos j\varphi + q_j \sin j\varphi\right). \tag{3.24}$$

In the φ domain, Eq. (3.1) is rewritten as

$$\Phi Z' = CZ + DZ_\tau + \varepsilon F(Z, Z_\tau), \tag{3.25}$$

where prime denotes differentiation with respect to φ and $\tau = \tau_c + \varepsilon\tau_\varepsilon$.

To rescale the time delay τ into the φ domain, we introduce φ_1 as the new time corresponding to $t - \tau$. It follows from (3.23) that

$$\mathrm{d}t = \frac{\mathrm{d}\varphi}{\Phi(\varphi)} = \frac{\mathrm{d}\varphi_1}{\Phi(\varphi_1)} \Rightarrow \Phi(\varphi)\frac{\mathrm{d}\varphi_1}{\mathrm{d}\varphi} = \Phi(\varphi_1), \tag{3.26}$$

which implies that $\varphi_1 - \varphi$ is a periodic function in φ with period 2π. Similarly, we can also expand $\varphi_1 - \varphi$ in a truncated Fourier's series about φ as

$$\varphi_1 = \varphi + \sum_{j=0}^{m} \left(r_j \cos j\varphi + s_j \sin j\varphi\right). \tag{3.27}$$

The integration constant of (3.26) provides information about the delay τ. Since φ_1 is the new time corresponding to $t - \tau$, it follows from (3.26) that

$$\int_{t-\tau}^{t} \mathrm{d}t_1 = \int_{\varphi_1}^{\varphi} \frac{\mathrm{d}\theta}{\Phi(\theta)} \Rightarrow \tau = \int_{\varphi_1}^{\varphi} \frac{\mathrm{d}\theta}{\Phi(\theta)}. \tag{3.28}$$

If Eq. (3.25) possesses a periodic solution at $\tau = \tau_0 = \tau_c + \epsilon\tau_\epsilon$ by the perturbation step, then it can be assumed that the expression is given as

$$\mathbf{Z}(t) = \sum_{j=0}^{m} \left(\mathbf{a}_j \cos j\varphi + \mathbf{b}_j \sin j\varphi\right), \tag{3.29}$$

is close to a periodic solution at $\tau = \tau_0 + \Delta\tau$, where $\mathbf{a}_j, \mathbf{b}_j \in \mathbf{R}^n$.

Initially, for $\Delta\tau = 0$, one can obtain that

$$\mathbf{a}_j = \begin{cases} r(\varepsilon) & j = 1, \\ 0 & j \neq 1, \end{cases} \quad \mathbf{b}_j = 0 \;\; for\, any\; j. \tag{3.30}$$

As a result, we have

$$\Phi(\varphi) = \sum_{j=0}^{m} (p_j \cos j\varphi + q_j \sin j\varphi), \tag{3.31}$$

where

$$\mathrm{p}_0 = \omega + \sigma(\varepsilon), \quad \mathrm{p}_\mathrm{j} = 0, \quad \mathrm{q}_0 = \mathrm{q}_\mathrm{j} = 0 \quad for\, all \;\; \mathrm{j} > 0. \tag{3.32}$$

Afterwards, an increment of τ from τ_0 to $\tau_0 + \Delta\tau$ corresponds to changes of the following quantities

$$\begin{aligned} &\mathbf{Z} \to \mathbf{Z} + \Delta\mathbf{Z}, \mathbf{Z}_\tau \to \mathbf{Z}_\tau + \Delta\mathbf{Z}_\tau, \Phi \to \Phi + \Delta\Phi \\ &\text{and } \varphi_1 \to \varphi_1 + \Delta\varphi_1. \end{aligned} \tag{3.33}$$

Substituting Eq. (3.33) into Eqs. (3.25) and (3.26), and expanding to Taylor series around an initial guess, the linearized incremental equations by ignoring all the nonlinear terms of small increments can be expressed by the following form

$$\begin{aligned} &\mathbf{Z}'\Delta\Phi(\varphi) + \Phi(\varphi)\Delta\mathbf{Z}' \\ &- C\Delta\mathbf{Z} - D\Delta\mathbf{Z}_\tau - \varepsilon\left(\frac{\partial F(\mathbf{Z}, \mathbf{Z}_\tau)}{\partial \mathbf{Z}}|_0 \Delta Z - \frac{\partial F(\mathbf{Z}, \mathbf{Z}_\tau)}{\partial \mathbf{Z}_\tau}|_0 \Delta\mathbf{Z}_\tau\right) \\ &= C\mathbf{Z} + D\mathbf{Z}_\tau + \varepsilon F(\mathbf{Z}, \mathbf{Z}_\tau) - \Phi(\varphi)\mathbf{Z}', \end{aligned} \tag{3.34}$$

$$\varphi_1'\Delta\Phi(\varphi) + \Phi(\varphi)\Delta\varphi_1' - \Delta\Phi(\varphi_1) - \Phi'(\varphi_1)\Delta\varphi_1 = \Phi(\varphi_1) - \Phi(\varphi)\varphi_1', \tag{3.35}$$

where the subscript 0 represents the evaluation of the relevant quantities corresponding to the initial solution. From (3.29), the terms ΔZ and $\Delta Z'$ can be expressed as follow respectively,

$$\begin{aligned} \Delta\mathbf{Z} &= \sum_{j=0}^{m} (\Delta a_j \cos j\varphi + \Delta b_j \sin j\varphi), \\ \Delta\mathbf{Z}' &= \sum_{j=1}^{m} j(\Delta b_j \cos j\varphi - \Delta a_j \sin j\varphi). \end{aligned} \tag{3.36}$$

Since Φ and $\varphi_1 - \varphi$ are both periodic functions in φ with period 2π, we have

$$\begin{aligned} \Phi(\varphi) &= \sum_{j=0}^{m} (p_j \cos j\varphi + q_j \sin j\varphi), \\ \Delta\Phi(\varphi) &= \sum_{j=0}^{m} (\Delta p_j \cos j\varphi + \Delta q_j \sin j\varphi), \\ \Delta\Phi'(\varphi) &= \sum_{j=1}^{m} j(\Delta q_j \cos j\varphi - \Delta p_j \sin j\varphi), \end{aligned} \tag{3.37}$$

and

$$\begin{aligned}\varphi_1 &= \varphi + \sum_{j=0}^{m}\left(r_j \cos j\varphi + s_j \sin j\varphi\right),\\ \Delta\varphi_1 &= \sum_{j=0}^{m}\left(\Delta r_j \cos j\varphi + \Delta s_j \sin j\varphi\right),\\ \Delta\varphi_1' &= \sum_{j=1}^{m} j\left(\Delta s_j \cos j\varphi - \Delta r_j \sin j\varphi\right).\end{aligned} \tag{3.38}$$

Similarly, for $\Delta\tau = 0$, the initial guess of φ_1 can be chosen as

$$r_0 = -(\omega + \sigma(\varepsilon))\tau_c,\ r_j = 0\,(j \neq 0),\ s_j = 0\ for\ any\ j. \tag{3.39}$$

For the delay term $\mathbf{Z}_\tau$, we have

$$\begin{aligned}Z_\tau &= \sum_{j=0}^{m}\left(\mathbf{a}_j \cos j\varphi_1 + \mathbf{b}_j \sin j\varphi_1\right),\\ \Delta Z_\tau &= \sum_{j=0}^{m}\left(\Delta\mathbf{a}_j \cos j\varphi_1 + \Delta\mathbf{b}_j \sin j\varphi_1\right) + \tfrac{\partial Z_\tau}{\partial\varphi_1}\Delta\varphi_1.\end{aligned} \tag{3.40}$$

For a small increment of τ to $\tau + \Delta\tau$, the linearized incremental equation of (3.28) is given by

$$\int_{\varphi_1}^{\varphi} \frac{\Delta\Phi(\theta)}{\Phi^2(\theta)} d\theta + \frac{\Delta\varphi_1}{\Phi(\varphi_1)} = \int_{\varphi_1}^{\varphi} \frac{d\theta}{\Phi(\theta)} - \tau - \Delta\tau, \tag{3.41}$$

which means that for $\varphi = 0$,

$$\int_{\xi}^{0} \frac{\Delta\Phi(\theta)}{\Phi^2(\theta)} d\theta + \frac{\Delta\varphi_1(0)}{\Phi(\alpha)} = \int_{\xi}^{0} \frac{d\theta}{\Phi(\theta)} - \tau - \Delta\tau, \tag{3.42}$$

where $\xi = \varphi_1(0)$. Then The harmonic balance method can be applied to (3.34), (3.35) and (3.42). Rewriting the linearized equation (3.34) in terms of the increments $\Delta\mathbf{a}_j$, $\Delta\mathbf{b}_j$, Δp_j, Δq_j, Δr_j and Δs_j, then we have

$$\begin{aligned}\sum_{j=0}^{m}[\ &\Psi_{1,j}\Delta\mathbf{a}_j+\Psi_{2,j}\Delta\mathbf{b}_j+\Psi_{3,j}\Delta p_j+\Psi_{4,j}\Delta q_j\\ &+\Psi_{5,j}\Delta r_j+\Psi_{6,j}\Delta s_j]=\Lambda_1,\end{aligned}\tag{3.43}$$

where

$$\begin{aligned}\Psi_{1,j}&=-j\Phi(\varphi)\sin j\varphi-\mathbf{C}\cos j\varphi-\mathbf{D}\cos j\varphi_1\\ &-\epsilon\left(\frac{\partial\mathbf{F}}{\partial\mathbf{Z}}\bigg|_0\cos j\varphi+\frac{\partial\mathbf{F}}{\partial\mathbf{Z}_\tau}\bigg|_0\cos j\varphi_1\right),\\ \Psi_{2,j}&=j\Phi(\varphi)\cos j\varphi-\mathbf{C}\sin j\varphi-\mathbf{D}\mathrm{sinj}\varphi_1\\ &-\varepsilon\left(\frac{\partial\mathbf{F}}{\partial\mathbf{Z}}\bigg|_0\mathrm{sinj}\varphi+\frac{\partial\mathbf{F}}{\partial\mathbf{Z}_\tau}\bigg|_0\sin j\varphi_1\right),\\ \Psi_{3,j}&=\mathbf{Z}'\cos j\varphi,\\ \Psi_{4,j}&=\mathbf{Z}'\sin j\varphi,\\ \Psi_{5,j}&=-\mathbf{D}\frac{\partial\mathbf{Z}_\tau}{\partial\varphi_1}\cos j\varphi-\frac{\partial\mathbf{F}}{\partial\mathbf{Z}_\tau}\bigg|_0\frac{\partial\mathbf{Z}_\tau}{\partial\varphi_1}\cos j\varphi,\\ \Psi_{6,j}&=-\mathbf{D}\frac{\partial\mathbf{Z}_\tau}{\partial\varphi_1}\sin j\varphi-\frac{\partial\mathbf{F}}{\partial\mathbf{Z}_\tau}\bigg|_0\frac{\partial\mathbf{Z}_\tau}{\partial\varphi_1}\sin j\varphi,\\ \Lambda_1&=\mathbf{CZ}+\mathbf{DZ}_\tau+\varepsilon F(\mathbf{Z},\mathbf{Z}_\tau)-\Phi(\varphi)\mathbf{Z}.\end{aligned}\tag{3.44}$$

Similarly, from (3.35) and (3.42), it can be obtained respectively that

$$\sum_{j=0}^{m}\left[\Psi_{7,j}\Delta p_j+\Psi_{8,j}\Delta q_j+\Psi_{9,j}\Delta r_j+\Psi_{10,j}\Delta s_j\right]=\Lambda_2,\tag{3.45}$$

and

$$\sum_{j=0}^{m}\left[\Psi_{11,j}\Delta p_j+\Psi_{12,j}\Delta q_j+\Psi_{13,j}\Delta r_j\right]=\Lambda_3,\tag{3.46}$$

where

$$
\begin{aligned}
\Psi_{7,j} &= \varphi_1' \cos j\varphi - \cos j\varphi_1, \\
\Psi_{8,j} &= \varphi_1' \sin j\varphi - \text{sinj}\varphi_1, \\
\Psi_{9,j} &= -j\Phi(\varphi)\sin j\varphi - \Phi'(\varphi_1)\cos j\varphi, \\
\Psi_{10,j} &= j\Phi(\varphi)\cos j\varphi - \Phi'(\varphi_1)\sin j\varphi, \\
\Psi_{11,j} &= \int\limits_{\xi}^{0} \frac{\cos j\theta}{\Phi^2(\theta)} d\theta, \\
\Psi_{12,j} &= \int\limits_{\xi}^{0} \frac{\sin j\theta}{\Phi^2(\theta)} d\theta, \\
\Psi_{13,j} &= \frac{1}{\Phi(\xi)}, \\
\Lambda_2 &= \Phi(\varphi_1) - \Phi(\varphi)\varphi_1', \\
\Lambda_3 &= \int\limits_{\xi}^{0} \frac{d\theta}{\Phi(\theta)} - \tau - \Delta\tau.
\end{aligned}
\tag{3.47}
$$

$\Psi_{i,j}(1 \le i \le 13, 1 \le j \le m)$ and $\Lambda_k(1 \le k \le 3)$ are periodic functions in φ, and therefore can be expressed in Fourier series whose coefficients can easily be obtained by Fast Fourier Transform. Let a_{ij}, $b_{ij} \in \mathbf{R}$ $(1 \le i \le n, 0 \le j \le m)$ be the ith element in $\mathbf{a}_j$ and $\mathbf{b}_j$, respectively. Compare the coefficients of harmonic terms of (3.43), (3.45) and (3.46), so that a system of linear equations is obtained with unknowns Δa_{ij}, Δb_{ij}, Δp_j, Δq_j, Δr_j and Δs_j in the form of

$$
\begin{aligned}
&\sum_{i=1}^{n}\sum_{j=0}^{m}\left(A_{k,ij}\Delta a_{ij} + B_{k,ij}\Delta b_{ij}\right) \\
&+\sum_{j=0}^{m}\left(P_{k,j}\Delta p_j + Q_{k,j}\Delta q_j + R_{k,j}\Delta r_j + S_{k,j}\Delta s_j\right) = T_k,
\end{aligned}
\tag{3.48}
$$

where T_k are residue terms. The values of a_j, b_j, p_j, q_j, r_j and s_j are updated by adding the original values and the corresponding incremental values. The iteration process continues until $T_k \to 0$ for all k.

The stability of a periodic solution can be studied by the Floquet method [10, 11]. Let $\boldsymbol{\zeta} \in \mathbb{R}^n$ be a small perturbation to a periodic solution of Eq. (3.1). Then we have

$$
\frac{d\zeta}{d\varphi} = \frac{1}{\Phi}[\mathbf{A}(\varphi, \varphi_1)\zeta + \mathbf{B}(\varphi, \varphi_1)\zeta_\tau] + O\left(\zeta^2, \zeta_\tau^2\right), \tag{3.49}
$$

where $\mathbf{A}(\varphi, \varphi_1) = \mathbf{C} + \varepsilon\frac{\partial \mathbf{F}(\mathbf{Z},\mathbf{Z}_\tau)}{\partial \mathbf{Z}}$ and $\mathbf{B}(\varphi, \varphi_1) = \mathbf{D} + \varepsilon\frac{\partial \mathbf{F}(\mathbf{Z},\mathbf{Z}_\tau)}{\partial \mathbf{Z}_\tau}$. The entities of $\mathbf{A}$ and $\mathbf{B}$ are all periodic functions of φ with period 2π, and can be determined by the incremental procedure. The time delay interval $I_1 = [-\tau, 0]$ corresponds to

$I_2 = [\alpha, 0]$ in the φ domain. Discrete points in I_2 are selected to compute the Floquet multipliers.

From the incremental procedure, the Fourier coefficients of φ_1 in (3.38) can be obtained. Assume that $\varphi = \beta$ when $\varphi_1 = 0$ and let $I_3 = [0, \beta]$. For each $\varphi \in I_3$, there exists a unique $\varphi_1 \in I_2$. A mesh whose size is $h = \frac{\beta}{N-1}$ and discrete points $\varphi^{(i)} = ih(0 \le i \le N-1)$ in I_3 are chosen, which correspond to $\varphi_1^{(i)} = \varphi_1\left(\varphi^{(i)}\right)$ in I_2. Let $\zeta\left(\varphi_1^{(i)}\right)$ be the $(i+1)$ th unit vector in $\mathbb{R}^n$. By integrating Eq. (3.49), the monodromy matrix M is obtained as

$$M = \left[\zeta\left(\varphi_1^{(0)} + 2\pi\right), \zeta\left(\varphi_1^{(1)} + 2\pi\right), \ldots, \zeta\left(\varphi_1^{(N-1)} + 2\pi\right)\right]. \tag{3.50}$$

We use the eigenvalues of M to evaluate the periodic solution's stability. It is known that one of the eigenvalues has to be unity and the periodic solution is stable if all the other eigenvalues are located inside the unit circle.

3.1.3 Synchronization Solution in a Network of Three Identical Neurons

The final example considers a network of three identical electronic (artificial) neurons connected by their closest neighborhoods. According to [12], the Hopfield's model with delay governs the dynamics of this system as follows

$$\begin{aligned} \dot{x}_1 &= -x_1 + \alpha f\left(x_{1,\tau}\right) + \beta\left[f\left(x_{2,\tau}\right) + f\left(x_{3,\tau}\right)\right], \\ \dot{x}_2 &= -x_2 + \alpha f\left(x_{2,\tau}\right) + \beta\left[f\left(x_{1,\tau}\right) + f\left(x_{3,\tau}\right)\right], \\ \dot{x}_3 &= -x_3 + \alpha f\left(x_{3,\tau}\right) + \beta\left[f\left(x_{1,\tau}\right) + f\left(x_{2,\tau}\right)\right], \end{aligned} \tag{3.51}$$

where $x_{i,\tau} = x_i(t-\tau)$ $(i = 1, 2, 3)$, $f(x) = \tanh(x) = \frac{e^x - e^{-x}}{e^x + e^{-x}}$, α and β measure, respectively, the coupled strength of self-connection and neighborhood-interaction, and τ is the time delay due to the finite switching speed of amplifiers. Equation (3.51) has an approximated synchronization solution that Wu et al. [12] have obtained using the center manifold reduction. This solution bifurcates from the trivial solution by a Hopf bifurcation at $\tau = \tau_c = \frac{\pi - \cos^{-1}(1/|\alpha+2\beta|)}{\sqrt{(\alpha+2\beta)^2 - 1}}$ when $(\alpha, \beta) \in D$ and $\beta < 0$, where $D = \{(\alpha, \beta) : \alpha - \beta < -1, \alpha + 2\beta < -1\}$. This synchronization solution is obtained here using the PIS, and can then be compared to that obtained from the numerical simulation and the center manifold reduction. To this end, Eq. (3.51) is rewritten the following form

$$\mathbf{Z}(t)=\begin{Bmatrix} x_1(t)\\ x_2(t)\\ x_3(t)\end{Bmatrix}\in\mathbf{R}^3,$$

$$\mathbf{C}=\begin{bmatrix}-1&0&0\\0&-1&0\\0&0&-1\end{bmatrix}\mathbf{D}=\begin{bmatrix}\alpha&\beta&\beta\\ \beta&\alpha&\beta\\ \beta&\beta&\alpha\end{bmatrix},$$

$$\mathbf{F}(\mathbf{Z}_{\tau_c},\mathbf{Z}_\tau)=\mathbf{D}\big[\mathbf{Z}_\tau-\mathbf{Z}_{\tau_c}\big]+\mathbf{F}(\mathbf{Z}_\tau),$$

$$\mathbf{F}(\mathbf{Z}_\tau)=\begin{pmatrix}\alpha\big(f(x_{1,\tau 1})-x_{1,\tau}\big)+\beta\big(f(x_{2,\tau})-x_{2,\tau}\big)+\beta\big(f(x_{3,\tau})-x_{3,\tau}\big)\\ \beta\big(f(x_{1,\tau})-x_{1,\tau}\big)+\alpha\big(f(x_{2,\tau})-x_{2,\tau}\big)+\beta\big(f(x_{3,\tau})-x_{3,\tau}\big)\\ \beta\big(f(x_{1,\tau})-x_{1,\tau c}\big)+\beta\big(f(x_{2,\tau})-x_{2,\tau}\big)+\alpha\big(f(x_{3,\tau})-x_{3,\tau}\big)\end{pmatrix}. \tag{3.52}$$

where $\tau=\tau_c+\varepsilon^2\tau_\varepsilon$. If $\mathbf{W}(t)$ is assumed to be a periodic solution of Eq. (3.10), then it can be represented in the form of

$$W(t)=\begin{pmatrix}p_1\cos(\omega t)+q_1\sin(\omega t)\\ p_2\cos(\omega t)+q_2\sin(\omega t)\\ p_3\cos(\omega t)+q_3\sin(\omega t)\end{pmatrix}, \tag{3.53}$$

where $\omega=\sqrt{(\alpha+2\beta)^2-1}$ for $\beta<0$. Substituting (3.53) into (3.13) and (3.14) yields that $p_1=p_2=p_3=p$ and $q_1=q_2=q_3=q$. Therefore, the synchronization solution of Eq. (3.51) can be rewritten as

$$Z(t)=\begin{pmatrix}\varepsilon r\cos\big((\omega+\varepsilon^2\sigma)t+\theta\big)\\ \varepsilon r\cos\big((\omega+\varepsilon^2\sigma)t+\theta\big)\\ \varepsilon r\cos\big((\omega+\varepsilon^2\sigma)t+\theta\big)\end{pmatrix}. \tag{3.54}$$

Considering Eqs. (3.53) and (3.54) and noting the independence of p and q, one can obtain that

$$\varepsilon r=2\sqrt{\frac{A_1}{B_1}(\tau-\tau_c)},\ \varepsilon^2\sigma=-\frac{C_1}{D_1}(\tau-\tau_c), \tag{3.55}$$

where

$$\begin{aligned}A_1&=2\big((\alpha+2\beta)^2-1\big)|\alpha+2\beta|\sqrt{|\alpha+2\beta|^2-1}\\&\quad+(\alpha+2\beta)\sqrt{(\alpha+2\beta)^2-1}\big(|\alpha+2\beta|^2-1\big),\end{aligned} \tag{3.56}$$

$$\begin{aligned}B_1&=2\sqrt{(\alpha+2\beta)^2-1}|\alpha+2\beta|+(\alpha+2\beta)\Big(\sqrt{|\alpha+2\beta|^2-1}\\&\quad-|\alpha+2\beta|^2\big(\pi-\sec^{-1}(|\alpha+2\beta|)\big)\Big),\end{aligned} \tag{3.57}$$

$$C_1 = (\alpha + 2\beta)\big((\alpha + 2\beta)^2 - 1\big)|\alpha + 2\beta|^2, \tag{3.58}$$

$$D_1 = -B_1. \tag{3.59}$$

To compare the perturbation solution of Eq. (3.55) with that from the center manifold reduction [12] and the numerical simulation, the parameters are set as $\alpha = -2$, $\beta = -0.5$, which results in

$$\varepsilon r = 2.12604\sqrt{\tau - \tau_c},\ \varepsilon^2\sigma = -3.59566(\tau - \tau_c), \tag{3.60}$$

where $\tau_c = \frac{\pi - \cos^{-1}(1/|\alpha+2\beta|)}{\sqrt{(\alpha+2\alpha)^2-1}} = 0.675511$.

The bifurcation curves that are generated using various methods from the Hopf bifurcation at $\tau = \tau_c$ are shown in Fig. 3.1. Significantly, compared to the center manifold reduction solution [12], the perturbation solution derived from the PIS's perturbation step shows improved accuracy. The difference between the approximate solutions (from the perturbation and the center manifold reduction) and the numerical solution gets increasingly noticeable as the time delay increases. This is especially clear from Fig. 3.2. For the value of τ close to $\tau_c = 0.67551$, for example $\tau = 0.7$, the perturbation solution exhibits very high accuracy, whereas the center manifold reduction solution [12] does not, as depicted in Fig. 3.2. However, the perturbation solution becomes quantitatively invalid for $\tau = 5$. The PIS's incremental step is executed to modify this solution. For Eq. (3.51), the initial guess is chosen for the incremental step of the PIS as follows

$$\begin{aligned} x_1(\varphi) &= x_2(\varphi) = x_3(\varphi) = \sum_{j=0}^{m}\big(a_j \cos j\varphi + b_j \sin j\varphi\big), \\ \Phi(\varphi) &= p_0 + \sum_{j=1}^{m}\big(p_j \cos j\varphi + q_j \sin j\varphi\big), \\ \varphi_1 &= \varphi + r_0 + \sum_{j=1}^{m}\big(r_j \cos j\varphi + s_j \sin j\varphi\big), \end{aligned} \tag{3.61}$$

where $a_0 = 0, a_1 = \varepsilon r, a_j = 0\,(j > 1), b_j = 0\,(j > 0)$, $p_0 = \omega + \varepsilon^2\sigma$, $p_j = q_j = 0$ $(j > 0)$, $r_0 = -\big(\omega + \varepsilon^2\sigma\big)\tau_c$, $r_j = s_j = 0\ (j > 0)$ and εr and $\varepsilon^2\sigma$ are determined by (3.55). Next, we use the PIS's incremental step—which is described in Sect. 3.1.2 to get the closed-form PIS solution for any value of τ. The PIS solution for $\tau = 5$ is represented by the red line in Fig. 3.2. Furthermore, Fig. 3.1 shows the thick line, which represents the continuation of the bifurcated periodic solution obtained by PIS, as the time delay varies. Figures 3.1 and 3.2 clear show that the PIS solutions can be computed to attain any required level of accuracy.

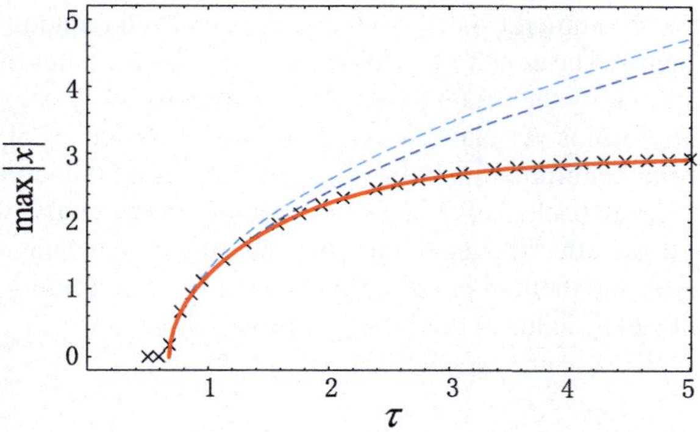

Fig. 3.1 Synchronization solution derived from Hopf bifurcation in $Max(x)$ versus τ for Eq. (3.51) when $\alpha = -2$ and $\beta = -0.5$, where thin dashed line denotes perturbation solution (3.54) with (3.55)–(3.59), thick solid line center manifold reduction solution from [12], thick solid line the PIS solution and crossing symbol the numerical simulation

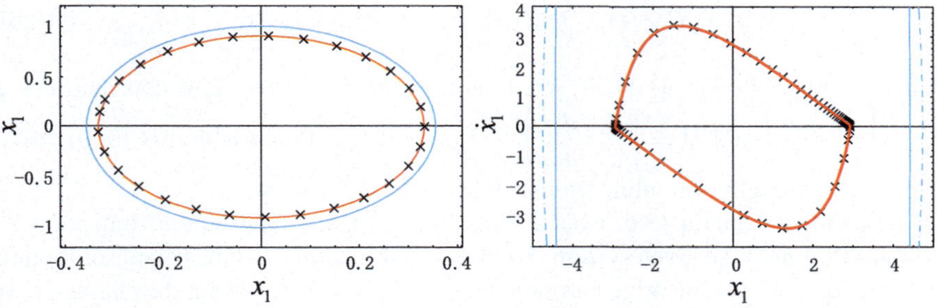

Fig. 3.2 A comparison between the perturbation solution (3.55) (thin), CMR solution (thick red) [12], PIS solution (thick red) and the numerical simulation (crossing symbol) in phase plane for Eq. (3.51), where $\alpha = -2$, $\beta = -0.5$, (a) $\tau = 0.7$, (b) $\tau = 5$

3.2 Perturbation-Incremental Scheme for Studying Weak Resonant Double Hopf Bifurcation of Nonlinear Systems with Time Delay

The double Hopf bifurcation may occur in delayed feedback control systems with different time delays and feedback gains, as demonstrated by [1, 13]. The fundamental mechanism hasn't been completely understood, though. Campbell et al. [14] and Reddy, Sen and Johnston [15] have both observed such phenomenon. For certain categories of retarded functional differential equations (RFDEs), Buono and Bélair [16] studied the universal unfolding and normal form of a vector field at non-resonant double Hopf bifurcation points using the techniques presented by Faria and Magalhães [17]. Their study placed limitations on the possible flows for specific singularities on a center manifold. Because DDEs and RFDEs are infinite-dimensional, many researchers recommend starting with center manifold reduction when investigating delay systems. The center manifold reduction, however, has disadvantages of its own. First, computing the center manifold reduction for a codimension-2 bifurcation using normal form is very laborious. Furthermore, for values of the bifurcation parameter far from the bifurcation point, the center manifold reduction is considered as invalid. These drawbacks serve as the motivation of our study, which proposes a straightforward yet effective technique that inherits the advantages of the center manifold reduction and method of multiple scales, thereby providing insight into the mechanics of the delay-induced double Hopf bifurcation.

3.2.1 Weak Resonant Double Hopf Bifurcation

Two first-order DDEs with nonlinearities in the general form and linear delayed feedback are taken into consideration as follows

$$\dot{Z}(t) = C\,Z(t) + D\,Z(t-\tau) + \varepsilon\,F(\,Z(t)\,,\,Z(t-\tau)\,), \tag{3.62}$$

where $Z(t) = \{x(t), y(t)\}^T \in R^2$, C and D are 2×2 real constant matrices as $C = \begin{bmatrix} c_{11} & c_{12} \\ c_{21} & c_{22} \end{bmatrix}$ and $D = \begin{bmatrix} d_{11} & d_{12} \\ d_{21} & d_{22} \end{bmatrix}$, F is a nonlinear function with $F(0, 0) = 0$, ε represents strength of nonlinearities, and τ is the time delay.

It is evident from Eq. (3.62) that $Z = 0$ always constitutes an equilibrium point or a trivial solution of the given system. To ascertain the stability of the trivial solution for $\tau \neq 0$, Eq. (3.62) is linearized around the trivial solution to obtain the characteristic equation as follows

$$\det\bigl(\lambda\,I - C - D\,e^{-\lambda\,\tau}\bigr) = 0, \tag{3.63}$$

where I is the identity matrix. The characteristic Eq. (3.63) can be reformulated as

$$\lambda^2 - \lambda\bigl(c_1 + d_1\,e^{-\lambda\,\tau}\bigr) + c_d\,e^{-\lambda\,\tau} + c_2 + \det(D)\,e^{-2\lambda\tau} = 0, \tag{3.64}$$

where

$$\begin{aligned} c_1 &= c_{11} + c_{22}, \\ d_1 &= d_{11} + d_{22}, \\ c_2 &= c_{11}\,c_{22} - c_{12}\,c_{21}, \\ c_d &== c_{22}\,d_{11} - c_{21}\,d_{12} - c_{12}\,d_{21} + c_{11}\,d_{22}. \end{aligned} \tag{3.65}$$

When the system poccesses two pairs of purely imaginary eigenvalues at a critical time delay, the dynamics become highly complicated. Assume that $c_d + c_2 + \det(D) \neq 0$. Thus, $\lambda = 0$ is not a solution to the characteristic equation (3.64). The exact expressions for the critical boundaries of stability switch in the following two cases are easy to determine:

1. $\det(D) = 0, c_d + c_2 \neq 0;$
2. $\det(D) \neq 0, c_{11} = c_{22}, c_{12} = -c_{21}, d_{11} = d_{22}, d_{12} = d_{21} = 0, c_{11} > 0, c_{12} > 0, d_{11} \neq 0.$

For the first case, substituting $\lambda = a + i\omega$ into (3.64) and setting the real and imaginary parts to zero yield

$$\begin{aligned} a^2 - \omega^2 - a\,c_1 + c_2 - e^{-a\,\tau}\,\omega\sin(\tau\,\omega)\,d_1 + e^{-a\,\tau}\cos(\tau\,\omega)(\,c_d - a\,d_1) = 0,\\ 2\,a\,\omega - \omega\,c_1 - e^{-a\,\tau}\,\omega\cos(\tau\,\omega)\,d_1 + e^{-a\,\tau}\sin(\tau\,\omega)(\,a\,d_1 - c_d) = 0. \end{aligned} \tag{3.66}$$

If a = 0, the equations to determine the critical boundaries for stability switch can be calculated as follows

$$\begin{aligned} -\omega^2 + c_2 + c_d\cos(\tau\,\omega) - \omega\,d_1\sin(\tau\,\omega) = 0,\\ -\omega\,c_1 - c_d\sin(\tau\,\omega) - \omega\,d_1\cos(\tau\,\omega) = 0. \end{aligned} \tag{3.67}$$

Eliminating τ from Eq. (3.66), it can be obtained that

$$\omega_{\pm} = \frac{\sqrt{d_1^2 - c_1^2 + 2\,c_2 \pm \sqrt{\left(d_1^2 - c_1^2 + 2\,c_2\right)^2 - 4\left(c_2^2 - c_d^2\right)}}}{\sqrt{2}} \tag{3.68}$$

with the following conditions being satisfied

$$\begin{aligned} c_2^2 - c_d^2 > 0,\\ \left(d_1^2 - c_1^2 + 2\,c_2\right)^2 > 4\left(c_2^2 - c_d^2\right). \end{aligned} \tag{3.69}$$

Then, two families of surfaces, denoted by τ_- and τ_+ in terms of c_d and d_1 corresponding to ω_- and ω_+, respectively, can be derived from Eq. (3.66) and expressed by

$$\begin{aligned} \cos(\,\omega_-\tau_-) = \frac{\omega_-^2\,c_d - c_2\,c_d - \omega_-^2\,c_1\,d_1}{c_d^2 + \omega_-^2\,d_1^2},\\ \cos(\,\omega_+\tau_+) = \frac{\omega_+^2\,c_d - c_2\,c_d - \omega_+^2\,c_1\,d_1}{c_d^2 + \omega_+^2\,d_1^2}. \end{aligned} \tag{3.70}$$

It should be noted that $\omega_- < \omega_+$. Therefore, a possible double Hopf bifurcation point arises when two sets of surfaces intersect each other where

$$\tau_- = \tau_+. \tag{3.71}$$

Equation (3.71) not only determines the linearized system around the trivial equilibrium but also establishes a connection between ω_- and ω_+. If

$$\omega_- : \omega_+ = k_1 : k_2, \tag{3.72}$$

then a possible double Hopf bifurcation point appears with frequencies in the ratio $k_1 : k_2$. If $k_1, k_2 \in Z^+, k_1 < k_2, k \neq 1$, and $k_2 \neq 1$, then such a point is called the $k_1 : k_2$ weak resonant double Hopf bifurcation point. Equations (3.71) and (3.72) provide the necessary conditions for the occurrence of such bifurcation. Equation (3.72) implies

$$d_1^2 = c_1^2 - 2\,c_2 + \frac{k_1^2 + k_2^2}{k_1\,k_2}\sqrt{c_2^2 - c_d^2}. \tag{3.73}$$

If Eq. (3.69) are satisfied, the frequencies in the simple expressions can be obtained by substituting Eq. (3.73) into Eq. (3.68) as follows

$$\omega_- = \sqrt{\frac{k_1}{k_2}\sqrt{c_2^2 - c_d^2}}, \quad \omega_+ = \sqrt{\frac{k_2}{k_1}\sqrt{c_2^2 - c_d^2}}. \tag{3.74}$$

The other parameters can be determined based on Eq. (3.71) or the following equality

$$\begin{aligned}
&\arccos\left(\begin{array}{l} -(c_2\,c_d\,k_2) + \sqrt{c_2^2 - c_d^2}\,(c_d - c_1\,d_1)\,k_1/c_d^2\,k_2 \\ +\sqrt{c_2^2 - c_d^2}\,d_1^2\,k_1 \end{array}\right) \\
&= k_1/k_2 \arccos\left(\begin{array}{l} -(c_2\,c_d\,k_1) + \sqrt{c_2^2 - c_d^2}\,(c_d - c_1\,d_1)\,k_2/c_d^2\,k_1 \\ +\sqrt{c_2^2 - c_d^2}\,d_1^2\,k_2 \end{array}\right),
\end{aligned} \tag{3.75}$$

where d_1 is given in Eq. (3.73). The corresponding value of the time delay at the double Hopf bifurcation point is determined by

$$\begin{aligned}
\tau_c &= \tau_- = \tau_+ \\
&= \sqrt{\frac{k_1}{k_2\sqrt{c_2^2 - c_d^2}}} \arccos\left(\frac{-(c_2\,c_d\,k_1) + \sqrt{c_2^2 - c_d^2}(c_d - c_1\,d_1)k_2}{c_d^2\,k_1 + \sqrt{c_2^2 - c_d^2}\,d_1^2\,k_2}\right).
\end{aligned} \tag{3.76}$$

For the second case, the characteristic Eq. (3.63) becomes

$$\lambda = c_{11} \pm i c_{12} + d_{11} e^{-\lambda\tau}. \tag{3.77}$$

By substituting $\lambda = a + i\omega$ into Eq. (3.77), we have

$$\begin{aligned}\omega = \omega_{\pm} = c_{12} \pm \sqrt{d_{11}^2\, e^{-2a\tau} - (a - c_{11})^2},\\ a = c_{11} - \omega - c_{21}/\tan(\omega\tau),\end{aligned} \tag{3.78}$$

where only one set of curves is considered by choosing $\omega = c_{12} \pm \sqrt{\cdot}$. Since the eigenvalues are always in the form of complex conjugate pairs, another set of resulting curves is implicit in the above. It follows from Eq. (3.78) that ω is real only when $d_{11}^2 \geq (a - c_{11})^2\, e^{2a\tau}$. In order to obtain the critical boundary, let $a = 0$ which yields

$$\begin{aligned}\omega_- = c_{12} - \sqrt{d_{11}^2 - c_{11}^2},\\ \omega_+ = c_{12} + \sqrt{d_{11}^2 - c_{11}^2},\end{aligned} \tag{3.79}$$

and

$$\begin{aligned}\tau_-[j] = 1/\omega_-\big(2j\pi - cos^{-1}(-c_{11}/d_{11})\big),\\ \tau_+[j] = 1/\omega_+\big(2j\pi + cos^{-1}(-c_{11}/d_{11})\big),\end{aligned} \tag{3.80}$$

where $j = 1, 2, \ldots$. We can obtain the necessary conditions for the $k_1 : k_2$ occurrence of the resonant double Hopf bifurcation by setting $\tau_-[j] = \tau_+[j]$ and $\frac{\omega_-}{\omega_+} = \frac{k_1}{k_2}$. This yields

$$(d_{11})_c = -\frac{c_{11}}{\cos(2j\pi(k_2 - k_1)/(k_1 + k_2))},\ \tau_c = \frac{4jk_1\pi}{(k_1 + k_2)\omega_-}. \tag{3.81}$$

Note that the parameters cannot be solved in a closed form from Eq. (3.76) because of the trigonometric function. We are able to get the values numerically, though. With frequencies in the ratio of $k_1 : k_2$, these parameters are referred to as the crucial values at the resonant double Hopf bifurcation point. Thus, for given physical parameters and in Eq. (3.62), we can obtain that

$$\varepsilon\tau_{\int} = \tau - \tau_c,\ \varepsilon\, D_\varepsilon = D - D_c. \tag{3.82}$$

Such that Eq. (3.62) can be rewritten as

$$\dot{Z} = C\,Z + D_c\,Z_{\tau_c} + \tilde{F}\big(Z, Z_{\tau_c}, Z_{\tau_c+\varepsilon\tau_\varepsilon}, \varepsilon\big), \tag{3.83}$$

where $Z = Z(t)$, $Z_\tau = Z(t - \tau)$, τ_c is given by Eq. (3.76), and

$$\tilde{F}(\cdot) = D_c\big[Z_{\tau_c+\varepsilon\tau_\varepsilon} - Z_{\tau_c}\big] + \varepsilon\big[D_\varepsilon Z_{\tau_c+\varepsilon\tau_\varepsilon} + F\big(Z, Z_{\tau_c+\varepsilon\tau_\varepsilon}\big)\big]. \tag{3.84}$$

It is clear that $\tilde{F} = 0$ for $\varepsilon = 0$ and the resonant double Hopf bifurcation may occur in the system for $\varepsilon = 0$.

3.2.2 Perturbation-Incremental Scheme

3.2.2.1 Perturbation Step of PIS

As demonstrated in the preceding subsection, we have identified a double Hopf bifurcation with weak resonance of $k_1 : k_2$ in Eq. (3.62) occurring at $\tau = \tau_c$ and $D = D_c$. When τ and D are treated as bifurcation parameters, the point (D_c, τ_c) marks a double Hopf bifurcation point with weak resonance. In this subsection, We want to extract the analytical form of harmonic solutions arising from the weak resonant double Hopf bifurcation in Eq. (3.62) or (3.83) when τ and D are around τ_c and D_c, respectively.

For $\epsilon = 0$, it can be seen from Eqs. (3.82) and (3.84) that $\tau = \tau_c$, $D = D_c$, and $\tilde{F}(\cdot) = 0$. Thus, the solution of Eq. (3.83) may be supposed as the following form

$$Z_0(t) = \sum_{i=1}^{2}\left[\begin{Bmatrix} a_{k_i} \\ c_{k_i} \end{Bmatrix}\cos(k_i\,\phi) + \begin{Bmatrix} b_{k_i} \\ d_{k_i} \end{Bmatrix}\sin(k_i\,\phi)\right], \tag{3.85}$$

which results in

$$Z_0(t-\tau_c) = \sum_{i=1}^{2}\left[\begin{Bmatrix} a_{k_i} \\ c_{k_i} \end{Bmatrix}\cos(k_i\,\phi - k_i\omega\tau_c) + \begin{Bmatrix} b_{k_i} \\ d_{k_i} \end{Bmatrix}\sin(k_i\,\phi - k_i\omega\tau_c)\right], \tag{3.86}$$

where τ_c is given by Eq. (3.76), $\phi = \omega t$, and $\omega = \frac{\omega_-}{k_1} = \frac{\omega_+}{k_2}$ is determined by Eq. (3.74). Substituting Eqs. (3.85) and (3.86) into Eq. (3.83) for $\varepsilon = 0$ and utilizing the harmonic balance, it can be obtained that

$$M_{k_i}\begin{Bmatrix} b_{k_i} \\ d_{k_i} \end{Bmatrix} = N_{k_i}\begin{Bmatrix} a_{k_i} \\ c_{k_i} \end{Bmatrix}, \tag{3.87}$$

and

$$-M_{k_i}\begin{Bmatrix} a_{k_i} \\ c_{k_i} \end{Bmatrix} = N_{k_i}\begin{Bmatrix} b_{k_i} \\ d_{k_i} \end{Bmatrix}, \tag{3.88}$$

where $M_{k_i} = k_i\omega I + D_c\sin(k_i\omega\tau)$, $N_{k_i} = C + D_c\cos(k_i\omega\tau)$, and I is the 2×2 identity matrix. Equations (3.87) and (3.88) are actually identical. Let $\begin{Bmatrix} \tilde{a}_{k_i} \\ \tilde{b}_{k_i} \end{Bmatrix} =$

$\frac{N_{k_i}}{\det(N_{k_i})}\begin{Bmatrix} a_{k_i} \\ b_{k_i} \end{Bmatrix}$ and then note that $\det(N_{k_i}) = \det(M_{k_i})$. It follows from (3.87) that the harmonic solution of Eq. (3.83) for $\varepsilon = 0$ is given by

$$Z_0(t) = \sum_{i=1}^{2}\left[\tilde{N}_{k_i}\cos(k_i\,\phi) + \tilde{M}_{k_i}\sin(k_i\,\phi)\right]\begin{Bmatrix} \tilde{a}_{k_i} \\ \tilde{b}_{k_i} \end{Bmatrix}, \tag{3.89}$$

with $\tilde{N}_{k_i} = N_{k_i}^{-1}\det(N_{k_i})$ and $\tilde{M}_{k_i} = M_{k_i}^{-1}\det(M_{k_i})$ for $i = 1, 2$.

The harmonic solution of Eq. (3.83) can be considered to be a perturbation to that of Eq. (3.89), given by

$$Z(t) = \sum_{i=1}^{2}\left[\tilde{N}_{k_i}\cos\left(k_i\,\omega t + \int \sigma_i t\right) + \tilde{M}_{k_i}\sin\left(k_i\,\omega t + \int \sigma_i t\right)\right]\begin{Bmatrix} a_{k_i}(\int) \\ b_{k_i}(\int) \end{Bmatrix}, \tag{3.90}$$

where $a_{k_i}(0) = \tilde{a}_{k_i}$, $b_{k_i}(0) = \tilde{b}_{k_i}$, and σ_1 and σ_2 are detuning parameters. The following theorem provides a novel approach to determine $\begin{Bmatrix} a_{k_i}(\varepsilon) \\ b_{k_i}(\varepsilon) \end{Bmatrix}$ in Eq. (3.90).

Theorem [18] *If* $W(t)$ *is a periodic solution of the equation*

$$\dot{W}(t) = -C^T\,W(t) - D_c^T\,W(t+\tau_c), \tag{3.91}$$

and $W(t) = W(t + 2\pi/\omega)$, *then*

$$\int_{-\tau_c}^{0}\left[D_c^T\,W(t+\tau_c)\right]^T[Z(t) - Z(t+2\pi/\omega)]dt$$

$$-\,[W(0)]^T\,[Z(2\pi/\omega) - Z(0)] + \int_0^{2\pi/\omega}[W(t)]^T\,\tilde{F}\left(Z, Z_{\tau_c}, Z_{\tau_c+\int\tau_\int}, \int\right)dt = 0. \tag{3.92}$$

The expression of $W(t)$ must be obtained in order to apply the theorem to determine $\begin{Bmatrix} a_{k_i}(\varepsilon) \\ b_{k_i}(\varepsilon) \end{Bmatrix}$. It is evident that the periodic solution of Eq. (3.91) can be expressed as

$$W(t) = \sum_{i=1}^{2}\left[\left(-\tilde{N}_{k_i}\right)^T\cos(k_i\,\phi) + \left(\tilde{M}_{k_i}\right)^T\sin(k_i\,\phi)\right]\begin{Bmatrix} p_{k_i} \\ q_{k_i} \end{Bmatrix}, \tag{3.93}$$

where p_{k_i} and q_{k_i} are independent constants. Substituting Eqs. (3.90) and (3.93) into Eq. (3.92), and noticing the independence of p_{k_i} and q_{k_i} we obtain four algebraic equations in $a_{k_i}(\varepsilon)$, $b_{k_i}(\varepsilon)$, σ_1, and σ_2. For $\varepsilon \neq 0$, $a_{k_i}(\varepsilon)$ and $b_{k_i}(\varepsilon)$ are dependent. Therefore, the algebraic equations are changed to the form of polar coordinate by setting

$$
\begin{aligned}
a_{k_i}(\varepsilon) &= \frac{-r_{k_i}(\varepsilon)(c_{12}\sin(\theta_i) + d_{12}\sin(k_i\omega\tau_c + \theta_i))}{c_{12}(k_i\omega + d_{22}\sin(k_i\omega\tau_c)) + d_{12}(k_i\omega\cos(k_i\omega\tau_c) - c_{22}\sin(k_i\omega\tau_c))}, \\
b_{k_i}(\varepsilon) &= \frac{-r_{k_i}(\varepsilon)(c_{22}\sin(\theta_i) + d_{22}\sin(k_i\omega\tau_c + \theta_i) + k_i\omega\cos(\theta_i))}{c_{12}(k_i\omega + d_{22}\sin(k_i\omega\tau_c)) + d_{12}(k_i\omega\cos(k_i\omega\tau_c) - c_{22}\sin(k_i\omega\tau_c))},
\end{aligned}
\tag{3.94}
$$

where $i = 1, 2$ and $\left(r_{k_1}, r_{k_2}, \theta_1, \theta_2\right)$ is a polar coordinate system. Togther with the trigonometric identity $\cos^2(\theta_i) + \sin^2(\theta_i) = 1$ $(i = 1, 2)$, one can solve $r_{k_1}(\varepsilon)$, $r_{k_2}(\varepsilon)$, σ_1, and σ_2. Thus, when the time delay and feedback gain are pretty close to the double Hopf bifurcation point (i.e., $\varepsilon\tau_c$ and ϵD_ϵ are very small), the solution in $O(\varepsilon)$ can be approximated as follows

$$
Z(t) = \begin{Bmatrix} r_{k_1}\cos((k_1\omega + \varepsilon\sigma_1)t + \theta_1) + r_{k_2}\cos((k_2\omega + \varepsilon\sigma_2)t + \theta_2) \\ (\cdot) \end{Bmatrix}. \tag{3.95}
$$

Furthermore, once θ_1 and θ_2 are identified from the initial conditions, $a_{k_i}(\varepsilon)$ and $b_{k_i}(\varepsilon)$ in Eq. (3.90) can be obtained from Eq. (3.94), which are denoted by $a^*_{k_i}(\varepsilon)$ and $b^*_{k_i}(\varepsilon)$. Therefore, Eq. (3.90) can be rewritten as

$$
Z(t) = \sum_{i=1}^{2}\left[\tilde{N}_{k_i}\cos(k_i\,\omega t + \varepsilon\sigma_i t) + \tilde{M}_{k_i}\sin\left(k_i\,\omega t + \int \sigma_i t\right)\right]\begin{Bmatrix} a^*_{k_i}(\varepsilon) \\ b^*_{k_i}(\varepsilon) \end{Bmatrix}. \tag{3.96}
$$

Thus far, we have been able to get an approximate solution using the PIS method's perturbation stage.

3.2.2.2 Incremental Step of the PIS

In order to analyze the effects of time delay, we elaborate on the incremental step of the PIS method in this subsection. By employing the transformation introduced in the previous part, we can rewrite Eq. (3.62) or (3.83) as follows:

$$
\Phi Z' = C\,Z + D\,Z_\tau + \varepsilon\,F(Z\,, Z_\tau\,), \tag{3.97}
$$

where the prime notation indicates the derivative with respect to ϕ, $D = D_c + \varepsilon D_\varepsilon$, and $\tau = \tau_c + \varepsilon\tau_\varepsilon$. The subsequent steps of utilizing the PIS method resembles those of the Hopf bifurcation case and will not be repeated here. Further details can be found in [18].

3.2.3 *Weak Resonant Double Hopf Bifurcation in the Van Der Pol–Duffing System with Delayed Feedback*

The following example focuses on the Van der Pol-Duffing oscillator with linear delayed position feedback, which is governed by

$$\ddot{u} - \left(\alpha - \varepsilon\gamma\, u^2\right)\dot{u} + \omega_0^2\, u + \epsilon\beta\, u^3 = A(u_\tau - u), \tag{3.98}$$

where α, γ, and β are positive constants, $A < 0$, τ is the time delay, ε is the small perturbation parameter, and $u_\tau = u(t-\tau)$. The Eq. (3.98) can be expressed in the form of Eq. (3.62), where

$$\begin{aligned} c_{11} = 0, c_{12} = 1, c_{21} = -\omega_0^2 - A, c_{22} = \alpha, \\ d_{11} = 0, d_{12} = 0, d_{21} = A, d_{22} = 0, \end{aligned} \tag{3.99}$$

which imply

$$c_1 = \alpha, c_2 = A + \omega_0^2, d_1 = 0, c_d = -A. \tag{3.100}$$

Substituting Eq. (3.99) into Eqs. (3.68)–(3.70), we have

$$\begin{aligned} 2\,A + \omega_0^2 > 0, \\ \left(-2\,A + \alpha^2\right)^2 - 4\,\alpha^2\,\omega_0^2 > 0, \end{aligned} \tag{3.101}$$

then

$$\omega_\pm = \sqrt{A - \frac{\alpha^2}{2} + \omega_0^2 \pm \sqrt{\left(A - \frac{\alpha^2}{2}\right)^2 - \alpha^2\,\omega_0^2}}, \tag{3.102}$$

$$\cos(\omega_\pm\,\tau) = 1 + \frac{\omega_0^2 - \omega_\pm^2}{A},\ \sin(\omega_\pm\,\tau) = \frac{\alpha}{A}\,\omega_\pm. \tag{3.103}$$

For $A < 0$ and $\alpha \geq 0$, it follows from Eq. (3.103) that

$$\tau_\pm[j] = \frac{1}{\omega_\pm}\left[2j\pi - \arccos\left(1 + \frac{\omega_0^2 - \omega_\pm^2}{A}\right)\right], \tag{3.104}$$

where $j = 1, 2, 3, \ldots$, and $\omega_\pm$ are obtained by Eq. (3.102). The essential criteria for the existence of resonant double Hopf bifurcaxtion points with frequencies in the ratio of $k_1 : k_2$ in terms of the critical values A_c and τ_c can be provided by utilizing Eqs. (3.73) and (3.76).

$$\alpha^2 - 2\left(A_c + \omega_0^2\right) + \frac{(k_1^2 + k_2^2)\omega_0}{k_1\,k_2}\sqrt{2A_c + \omega_0^2} = 0, \tag{3.105}$$

$$\tau_c = \tau_+ = \tau_-, \tag{3.106}$$

and the corresponding frequencies are

$$\omega_- = \sqrt{\frac{k_1}{k_2}\omega_0\sqrt{2A_c+\omega_0^2}},\ \omega_+ = \sqrt{\frac{k_2}{k_1}\omega_0\sqrt{2A_c+\omega_0^2}}, \tag{3.107}$$

where ω_0 is a constant.

For $\alpha \neq 0$, A_c cannot be analytically solved from Eq. (3.106), but numerical solutions can be easily obtained for a fixed ω_0. The diagrams for the case with $\alpha = 0.17468$ are plotted in Fig. 3.3. The region in green indicates stable trivial solutions of Eq. (3.98), i.e., amplitude death regions. In order to acquire the solutions that result from this type of double Hopf bifurcation, we let $A = A_c + \varepsilon A_\varepsilon$ and $\tau = \tau_c + \varepsilon\tau_\varepsilon$ for a given α, where $\varepsilon A_\varepsilon$ and $\varepsilon\tau_\varepsilon$ are small perturbations. Therefore, Eq. (3.98) now reads

$$\dot{Z}(t) = C\,Z(t) + D_c\,Z(t-\tau_c) + \tilde{F}(Z(t), Z(t-\tau_c), Z(t-\tau_c-\epsilon\tau_\epsilon), \epsilon), \tag{3.108}$$

where

$$\begin{gathered} Z(t) = \begin{Bmatrix} u(t) \\ v(t) \end{Bmatrix},\ C = \begin{bmatrix} 0 & 1 \\ -(A_c+\omega_0^2) & \alpha \end{bmatrix},\ D_c = \begin{bmatrix} 0 & 0 \\ A_c & 0 \end{bmatrix}, \\ \tilde{F} = \begin{Bmatrix} 0 \\ A_c\big(u_{\tau_c+\epsilon\tau_\epsilon} - u_{\tau_c}\big) + \epsilon\big(A_\epsilon\big(u_{\tau_c+\epsilon\tau_\epsilon} - u\big) - u^2(\beta u + \gamma v)\big) \end{Bmatrix}, \end{gathered} \tag{3.109}$$

with $u_\tau = u(t-\tau)$.

As an illustration, we study the case as depicted in Fig. 3.3 with $\omega_0 = 1$, $\alpha = 0.17468$, in which a 3:5 weak resonant double Hopf bifurcation arises within Eq. (3.110) at $A_c = -0.336338$ and $\tau_c = 5.88976$ (cf. Lemma 8.15 in [12]). It

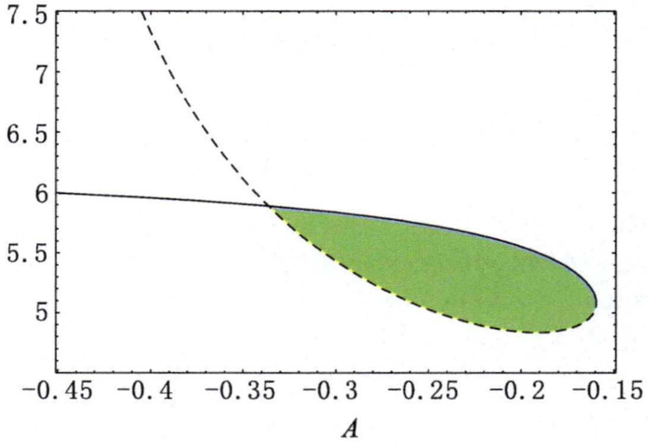

Fig. 3.3 High-order resonant double Hopf bifurcation diagram with frequencies in the ratio $\omega_- : \omega_+ = k_1 : k_2$ for $\alpha = 0.17468$ and $\omega_0 = 1.0$, where the solid line represents τ_+, the dashed line τ_-, and the region in green the amplitude death

follows from Eqs. (3.93), (3.94), and (3.95) that

$$W(t) = \begin{Bmatrix} w_1(t) \\ w_2(t) \end{Bmatrix}, \tag{3.110}$$

with

$$\begin{aligned} w_1(t) &= -0.17468 p_{k_1} \cos(0.585895t) \\ &- 0.17468 p_{k_2} \cos(0.976492t) \\ &- 0.343274 q_{k_1} \cos(0.585895t) - 0.953538 q_{k_2} \cos(0.976492t) \\ &+ 0.585895 p_{k_1} \sin(0.585895t) - 0.102345 q_{k_1} \sin(0.585895t) \\ &+ 0.976492 p_{k_2} \sin(0.976492t) - 0.170574 q_{k_2} \sin(0.976492t), \\ w_2(t) &= p_{k_1} \cos(0.585895t) + p_{k_2} \cos(0.976492t) \\ &+ 0.585895 q_{k_1} \sin(0.585895t) + 0.976492 q_{k_2} \sin(0.976492t), \end{aligned} \tag{3.111}$$

$$\begin{aligned} a_{k_1} &= -1.70679 r_{k_1} \sin(\theta_1), \\ b_{k_1} &= -r_{k_1} \cos(\theta_1) - 0.298142 r_{k_1} \sin(\theta_1), \\ a_{k_2} &= -1.02407 r_{k_2} \sin(\theta_2), \\ b_{k_2} &= -r_{k_2} \cos(\theta_2) - 0.178885 r_{k_2} \sin(\theta_2), \end{aligned} \tag{3.112}$$

and

$$\begin{aligned} u(t) &= r_{k_1} \cos(\theta_1 + t(0.585895 + \varepsilon\sigma_1)) \\ &+ r_{k_2} \cos(\theta_2 + t(0.976492 + \varepsilon\sigma_2)), \\ v(t) &= -0.585896 r_{k_1} \sin(\theta_1 + t(0.585895 + \varepsilon\sigma_1)) \\ &- 0.976492 r_{k_2} \sin(\theta_2 + t(0.976492 + \varepsilon\sigma_2)). \end{aligned} \tag{3.113}$$

Substituting Eqs. (3.109), (3.110), (3.111), and (3.113) into Eq. (3.92), noticing that p_1, p_2, q_1, and q_2 are independent, and using the trigonometric identity $\cos^2(\theta_i) + \sin^2(\theta_i) = 1$ $(i = 1, 2)$, yield

$$\begin{aligned} & r_{k_1}\varepsilon\left(A_\varepsilon + 0.384107 r_{k_1}^2 \beta + 0.768215 r_{k_2}^2 \beta - 0.291413\sigma_1 + 0.0307097\tau_\varepsilon\right) = 0, \\ & r_{k_1}\varepsilon\left(A_\varepsilon - 0.481362 r_{k_1}^2 \gamma - 0.962723 r_{k_2}^2 \gamma - 5.62729\sigma_1 - 0.616891\tau_\varepsilon\right) = 0, \\ & r_{k_2}\varepsilon\left(A_\varepsilon + 10.8584 r_{k_1}^2 \beta + 5.42919 r_{k_2}^2 \beta - 6.865\sigma_2 + 1.20574\tau_\varepsilon\right) = 0, \\ & r_{k_2}\varepsilon\left(A_\varepsilon - 0.962726 r_{k_1}^2 \gamma - 0.481363 r_{k_2}^2 \gamma + 3.71089\sigma_2 + 0.558141\tau_\varepsilon\right) = 0. \end{aligned} \tag{3.114}$$

It is evident from Eq. (3.113) that r_{k_1}, r_{k_2}, σ_1, and σ_2 play a crucial role in manifesting the motion features of Eq. (3.108) when the double Hopf point (A_c, τ_c) is perturbed by A_ε and τ_ε for given values of ε, β, and γ. As a result, it is required to categorize the solutions of algebraic equation (3.114) in the neighborhood of the double Hopf point (A_c, τ_c).

It is crucial to distinguish between "simple" and "difficult" double Hopf cases (cf. Sect. 8.6 in [12] for further insights). In the simple case, regarding the truncated cubic amplitude Eq. (3.114), it does not allow for periodic solutions and its bifurcation diagram will not be altered by adding any fourth- and fifth-order terms. On the other hand, one has to take into account the fifth-order term to predict all possible neighboring dynamics in a theoretical manner. This bifurcation indicates the presence of three-dimensional invariant tori in the complete four-dimensional system on the center manifold. Both scenarios are apparent when $\beta = 0$ and $\gamma \neq 0$ for the simple case, and when $\beta \neq 0$, $\gamma = 0$ for the difficult case, as depicted in Fig. 3.4.

Let us study the difficult case as an example. With $\gamma = 0$, by solving Eq. (3.114) we can obtain that $\left(r_{k_1}, r_{k_2}\right) = \left(r_{k_10}, r_{k_20}\right) = (0, 0)$, which is always a root and that there exist at maximum three other roots, given by:

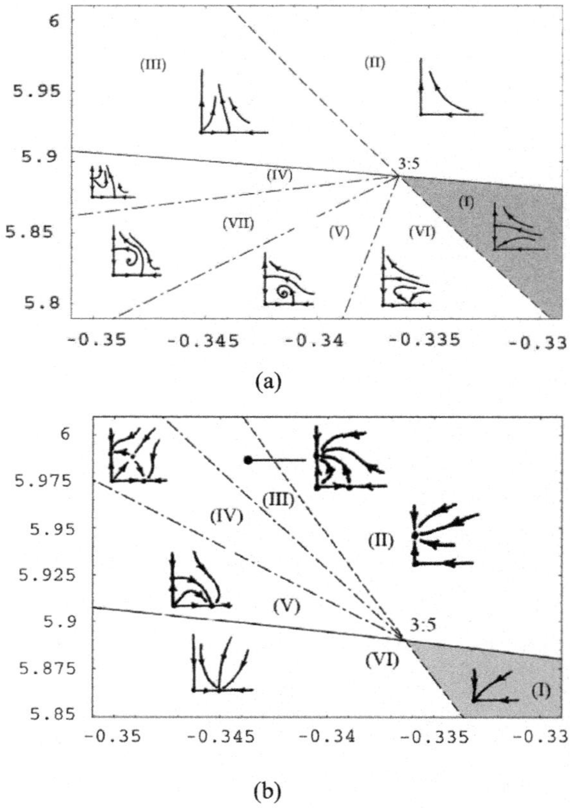

Fig. 3.4 Classification and bifurcation sets of the solution for the system (3.108) resulting from 3:5 resonant double Hopf bifurcation, where solid lines, dashed lines, and dot-dashed lines represent boundaries and amplitude death region is in grey for **a** the simple case: $\gamma = 1$, $\beta = 0$ and **b** the difficult case: $\gamma = 0$, $\beta = 4$

$$
\begin{aligned}
&\left(r_{k_1}, r_{k_2}\right) = \left(r_{k_1 1}, r_{k_2 0}\right)\\
&= \left(1.57118\sqrt{-A_\varepsilon - 0.0660777\tau_\varepsilon/\beta}, 0\right) \quad \text{for } A_\varepsilon + 0.0660777\tau_\varepsilon/\beta < 0,\\
&\left(r_{k_1}, r_{k_2}\right) = \left(r_{k_1 0}, r_{k_2 1}\right)\\
&= \left(0, 0.724522\sqrt{-A_\varepsilon - 0.785373\tau_\varepsilon/\beta}\right) \text{ for } A_\varepsilon + 0.785373\tau_\varepsilon/\beta < 0,\\
&\left(r_{k_1}, r_{k_2}\right) = \left(r_{12}, r_{22}\right)\\
&= \left(0.68769\sqrt{A_\varepsilon - 0.46195\tau_\varepsilon/\beta}, 1.21275\sqrt{-A_\varepsilon - 0.019497\tau_\varepsilon/\beta}\right)\\
&\quad \text{for } A_\varepsilon - 0.46195\tau_\varepsilon/\beta > 0,\\
&A_\varepsilon + 0.019497\tau_\varepsilon/\beta < 0 \sigma_1 = 0.177705(A_\varepsilon - 0.616891\tau_\varepsilon),\\
&\sigma_2 = -0.269477(A_\varepsilon + 0.558141\tau_\varepsilon).
\end{aligned}
\tag{3.115}
$$

By utilizing Eq. (3.114), the stability of the solution of Eq. (3.113), determined by Eq. (3.115), can be readily studied. As a result, centered at (A_c, τ_c), the parameter plane (A, τ) is divided into seven regions (I)–(VII), bounded by Eq. (3.115), as depicted in Fig. 3.4. This classification is very similar to the one presented by Guckenheimer and Holmes (cf. Fig. 7.5 in [9]).

In Fig. 3.4, both a stable trivial solution $(0, 0)$ and an unstable periodic solution $\left(0, r_{k_2 1}\right)$ in region (I) are observed, which correspond to an amplitude death region. As (A, τ) moves to region (II), the trivial solution becomes unstable, and no local solution is found. In region (III), there exist two unstable solutions $(0, 0)$ and $\left(r_{k_1 1}, 0\right)$. As for region (IV), three unstable solutions are found, namely, $(0, 0)$, $\left(r_{k_1 1}, 0\right)$, and $\left(0, r_{k_2 1}\right)$. The stable nontrivial solutions of Eq. (3.114) exist in areas (V) and (VI), given by $\left(r_{k_1 2}, r_{k_2 2}\right)$ and $\left(r_{k_1 1}, 0\right)$, respectively. Additionally, there are other unstable solutions in region (V) at $(0, 0)$, $\left(r_{k_1 1}, 0\right)$, and $\left(0, r_{k_2 1}\right)$, and in region (VI) at $(0, 0)$ and $\left(0, r_{k_2 1}\right)$. It is noteworthy that the stable solution in region (V) exhibits quasi-periodic behaviour since σ_1/σ_2 is not a rational number, as indicated by Eq. (3.113). Specifically, the Hopf bifurcation of the nontrivial equilibrium at $\left(r_{k_1}, r_{k_2}\right) = \left(r_{12}, r_{22}\right)$ appears in the cubic amplitude system (3.114). This results in the boundary between region (VII) and region (VIII), as well as three-dimensional torus in the four-dimensional system on the center manifold in region (VIII). Consequently, in order to investigate the cycle generated by the Hopf bifurcation, it becomes imperative to explore the truncated fifth-order amplitude system. The boundary between regions (VIII) and (V) can be found as follows [12]

$$
\tau = -15.377 - 134.333A - 211.40A^2, \; A \le A_c, \tag{3.116}
$$

which indicates coexistence of the cycle with three saddles at $\left(r_{k_1 0}, r_{k_2 0}\right)$, $\left(r_{k_1 1}, r_{k_2 0}\right)$, and $\left(r_{k_1 0}, r_{k_2 1}\right)$. Hence, the cycle disappears via a heteroclinic bifurcation when the parameter shifts from regions (VIII) to (V), as illustrated by Fig. 3.4b.

For the challenging cases depicted in Fig. 3.4b, we will now evaluate the accuracy and validity of the perturbation step—the first stage of the PIS—using numerical simulation. We consider two cases: for the first one the delay and gain values are close to the double Hopf bifurcation point at (A_c, τ_c), and for the second the delay and gain values are far away from this point. We use the Runge–Kutta scheme to obtain the numerical results, with $\beta = 4, \gamma = 0$, and the other parameters are fixed as in Fig. 3.3. We keep the gain fixed for both cases, as our key objective is to investigate the impact of time delay on the system.

Figure 3.5 presents a comparison between the approximated solution (3.113) and the numerical simulation from Eq. (3.98) for different values of (A, τ): (a) $A = -0.3348$, $\tau = 5.852$; (b) $A = -0.33$, $\tau = 5.7$; (c) $A = -0.31$, $\tau = 5.4$; and (d) $A = -0.33$, $\tau = 5.2$. In Fig. 3.5a and b where the values of (A, τ) are close to (A_c, τ_c), the analytical prediction agrees well with the numerical results. Meanwhile, for Fig. 3.5c, d, it can be inferred that the theoretical prediction fails. This indicates that the periodic solutions obtained by this method are accurate even for large ε, as long as (A, τ) is close to the double Hopf bifurcation point at (A_c, τ_c). However, as (A, τ) drifts away from (A_c, τ_c), the accuracy decreases, as shown in Fig. 3.5b) and d. This trend can be clearly observed from Fig. 3.5a, b, where τ decreases from τ_c while A remains fixed.

A similar conclusion can be obtained for (A, τ) in region (V). The time history plot of the quasi-periodic solution of Eq. (3.98) is provided in Fig. 3.7 with small and large ϵ. Even for quasi-periodic motions, Step 1 of PIS provides highly accurate analytical expressions near the double Hopf bifurcation point, as can be seen from Fig. 3.7, where there is good agreement of analytical prediction with the numerical simulation. However, when (A, τ) deviates from the bifurcation point, the method will be ineffective, as shown in Fig. 3.8.

Figures 3.5, 3.6, 3.7 and 3.8 show that, when the parameters are close to the weak resonant double Hopf bifurcation point, the first stage of the PIS generates a highly accurate analytical expression in addition to a reasonably precise qualitative prediction for both periodic and quasi-periodic motions. Nevertheless, as the bifurcation parameters deviate from the bifurcation point, the analytical expression's quantitative correctness is not guaranteed. When this occurs, the approximate expression now acts as a rough approximation for the PIS's second step, which keeps track of periodic solutions for bifurcation parameters that are far away from (A_c, τ_c).

To this end, we select $\varepsilon = 1$ and $(\tau, A) = (5.85, -0.3365)$ which is close to (A_c, τ_c) as the starting point. According to Fig. 3.4b, the stable periodic solution located in region (VI) is given by

$$Z(t) = \left\{ \begin{array}{c} r_{k_1 1} \cos((0.585895 + \varepsilon\sigma_1)t + \theta_1) \\ -0.585896\, r_{k_1 1} \sin((0.585895 + \varepsilon\sigma_1)t + \theta_1) \end{array} \right\}, \tag{3.117}$$

where $r_{k_1 1}$ and σ_1 are determined by (3.115), $\tau_\varepsilon = \tau - \tau_c$, $A_\varepsilon = A - A_c$, and θ_1 is determined by initial values. Therefore, $\theta_1 = 0$ by setting $v(0) = 0$.The solution of

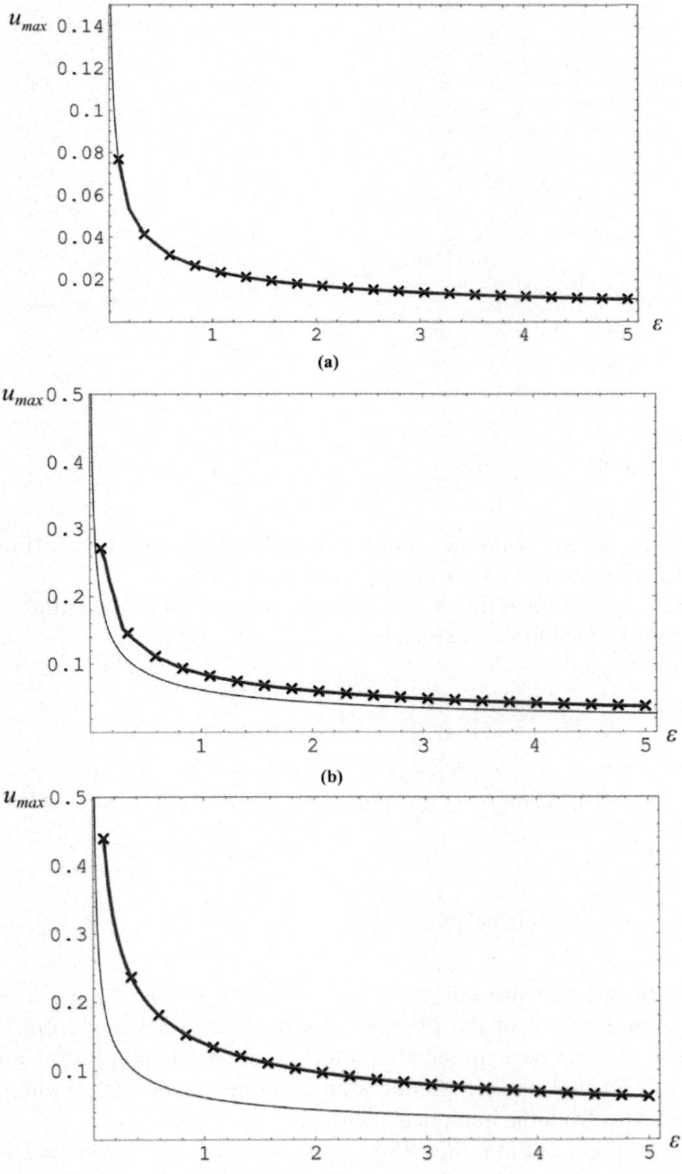

Fig. 3.5 A comparison among the approximate solution (3.113) (solid), the solution from step two of the PIS (thick blue), and the numerical simulation (crossing symbols) in Max(u) versus ε for the periodic solution of system (3.98) with (A, τ) located in region (VI) in Fig. 3.4 **a** $A = -0.3348$, $\tau = 5.852$, **b** $A = -0.33$, $\tau = 5.7$, **c** $A = -0.31$, $\tau = 5.4$, and **d** $A = -0.33$, $\tau = 5.2$

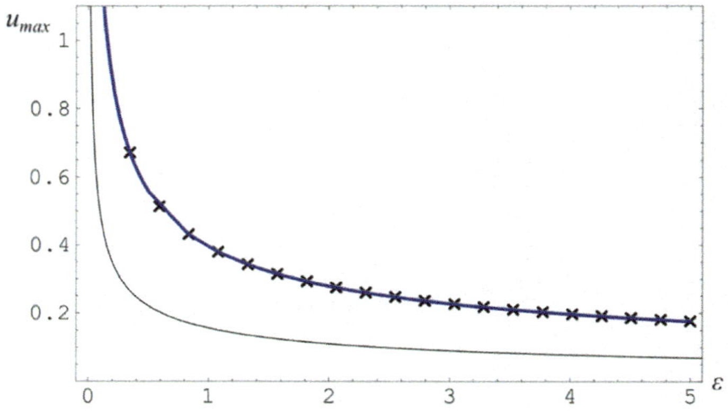

(d)

Fig. 3.5 (continued)

Eq. (3.117) agrees well with the numerical one illustrated in Fig. 3.9. However, this conformity disappears at $\tau = 3$, as displayed in Fig. 3.9.

If Eq. (3.117) is used as the initial estimate for Step 2, then the initial coefficients in the incremental solution provided by

$$\begin{gathered} a_0 = 0,\ a_1 = \begin{Bmatrix} r_{k_1 1} \\ 0 \end{Bmatrix},\ b_1 = \begin{Bmatrix} 0 \\ -0.585895\, r_{k_1 1} \end{Bmatrix}, \\ a_j = b_j = 0\ for\ j = 2, \ldots, m, \\ p_0 = 0.195298 + \tfrac{\sigma_1}{3},\ p_j = q_j = 0\ for\ j = 1, \ldots, m, \end{gathered} \tag{3.118}$$

and

$$r_0 = -p_0\,\tau = -5.85\left(0.195298 + \frac{\sigma_1}{3}\right),\ r_j = s_j = 0\ for\ j = 1, \ldots, m. \tag{3.119}$$

With the incremental procedure from $\tau = 5.85$ to $\tau = 5.85 + n\Delta\tau = 3$ ($n \in Z^+$, $|\Delta\tau| \ll 1$) and Step 2 of the PIS, Fig. 3.9 illustrates that a converged periodic solution at $\tau = 3$ can be acquired after a few iterations, demonstrating that the PIS method's incremental process produces an accurate solution even when the intial estimate deviates from the numerical solution.

Finally, we point out that the PIS is applicable not only for the weak resonant double Hopf bifurcation, but also for the strong resonant one. The procedure of extending the PIS to that case will not be discussed here and the readers are refered to [19] for details.

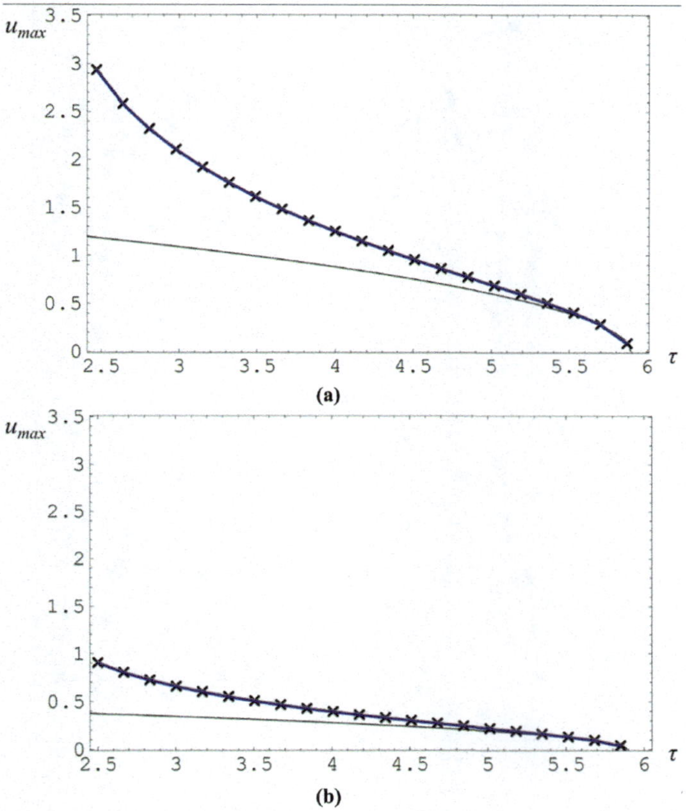

Fig. 3.6 A comparison between the approximated solution (3.113) (solid), the solution from step two of the PIS (thick blue), and the numerical simulation (crossing symbols) in Max(u) versus τ for the periodic solution of the system (3.98) when (A, τ) is located in region (VI) of Fig. 3.4, where **a** $\varepsilon = 0.1$ and **b** $\varepsilon = 1.0$ and $A = -0.3365$

3.3 Integral Equation Method for Nonlinear Systems with Time Delay

In this section, we employ a novel analytical approach, called integral equation method to study primary resonance and internal resonance in non-autonomous system with delays. G. Schimidt initially proposed this approach [20], and his findings can be extended to include the application involving differential equations with delays. After an in-depth comparison between the results obtained from numerical solutions, it is clear that the method of integral equation displays exceptional precision. Utilizing Floquet's theorem in combination with the Nyquist criterion and the Routh-Hurwitz criterion, we evaluate the stability of periodic motions. Additionally, we show the

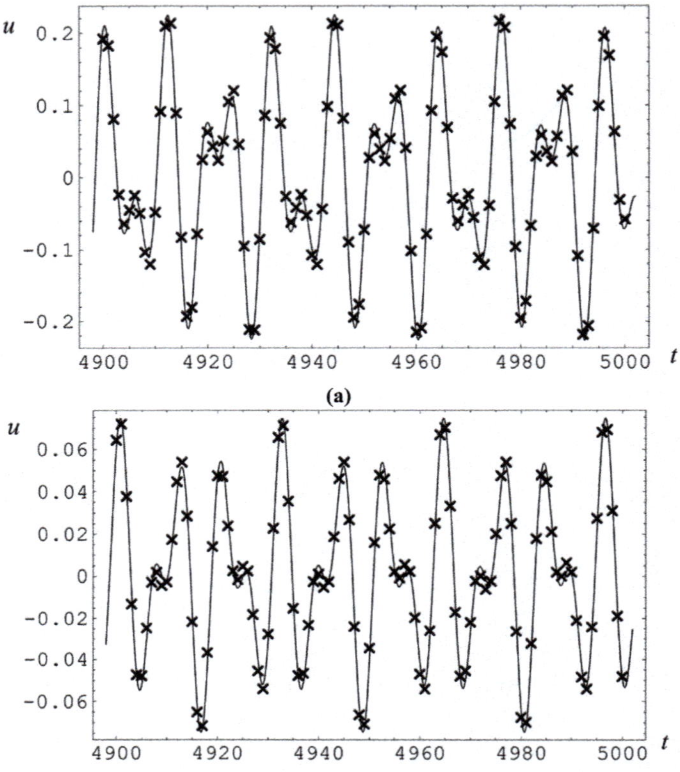

(b)

Fig. 3.7 A comparison between the approximated solution given in (3.113) (solid) and the numerical simulation (crossing symbols) in time history for the quasi-periodic solution of system (3.98), where **a** $\varepsilon = 0.1$, **b** $\epsilon = 1$, and $A = -0.34$, $\tau = 5.85$ are chosen in region (V) of Fig. 3.4 and close to (A_c, τ_c)

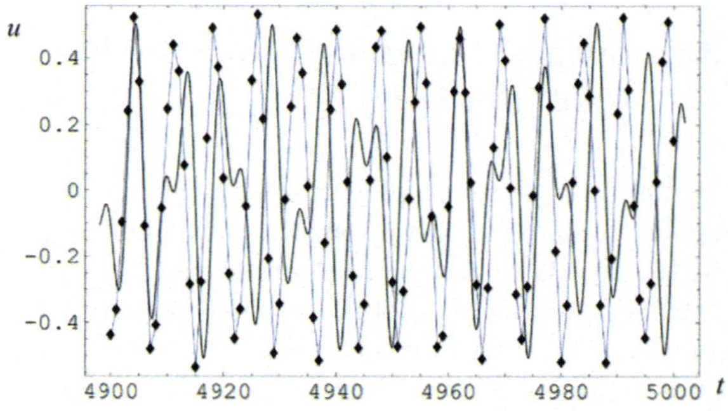

Fig. 3.8 A comparison between the approximate solution (3.113) (solid) and the numerical simulation (line with diamond symbols) in time history for the quasi-periodic solution of system (3.98) when (A, τ) is in region (V) but far away from (A_c, τ_c), where $\varepsilon = 1$, $A = -0.5$, $\tau = 4$

Fig. 3.9 A comparison between the approximate solution (3.113) (solid), solution (thick blue) from step two of PIS and the numerical simulation (crossing symbols) in phase plane for the periodic solution of system (3.98), where (**a**) $\tau = 5.85$, (**b**) $\tau = 3$, $\varepsilon = 1$, and $A = -0.3365$ throughout.

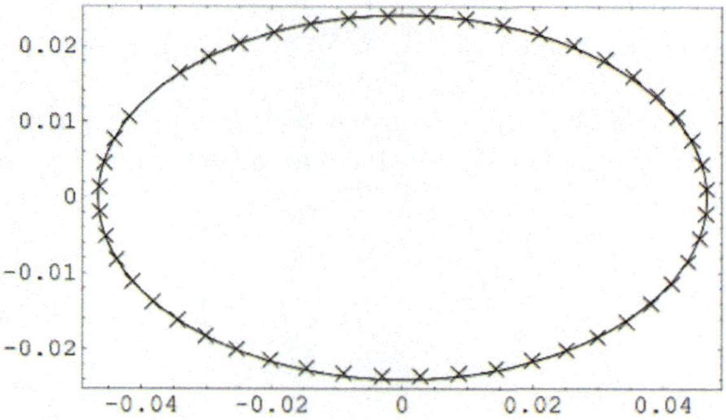

(a)

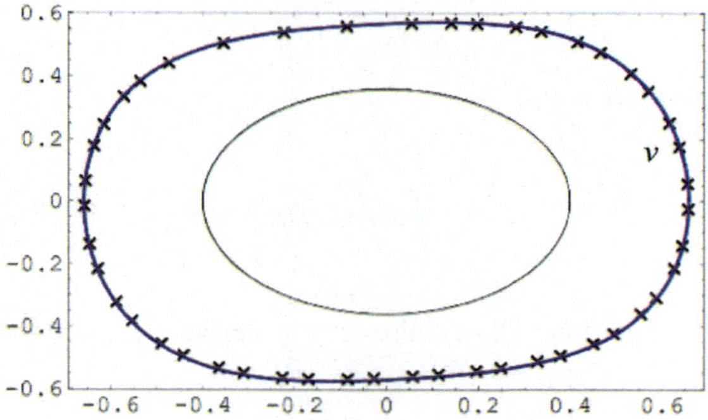

(b)

performance of the proposed method in the accuracy of prediction by conducting numerical simulations for comparisons.

3.3.1 The Integral Equation Method

The equations that govern vibrations with time delay assume the following form:

$$\begin{aligned} \ddot{x}_i + \lambda_i x_i &= \phi_i\big[x_i, \dot{x}_i, x_{i\tau}, \dot{x}_{i\tau}, x_j, \dot{x}_j, x_{j\tau}, \dot{x}_{j\tau}; t\big] \\ &= \phi_i[t]. \end{aligned} \tag{3.120}$$

Theorem 1 [21] *Each periodic solution of Eq.* (3.120) *is a solution to the integro-differential equations (refer to* [20])

$$x_i(t) = \int_0^{2\pi} G_i[t, \sigma]\phi_i[\sigma]d\sigma + \delta_{\lambda_i}^{n_i^2}(r_i \cos n_i t + s_i \sin n_i t), \tag{3.121}$$

where $G_i[t,\sigma] = \frac{1}{\pi}\left(\frac{1}{2\lambda_i} + \sum_{j=1}^{\infty} \frac{\cos j(t-\sigma)}{\vartheta\lambda_i - j^2}\right)$ *is the corresponding Generalized Green's function. And for* $\lambda_i = n_i^2$ *where* n_i *is an integer, the following periodicity equations can be used to calculate the parameters* r_i *and* s_i *based on existing studies (refer to* [20]*):*

$$\begin{aligned} r_i &= \frac{1}{\pi}\int_0^{2\pi} x_i(t) cos n_i t \,\mathrm{d}t, \\ s_i &= \frac{1}{\pi}\int_0^{2\pi} x_i(t) sin n_i t \,\mathrm{d}t, \end{aligned} \tag{3.122}$$

or equivalently

$$\int_0^{2\pi} \phi_i[t] cos n_i t \,\mathrm{d}t = \int_0^{2\pi} \phi_i[t] sin n_i t \,\mathrm{d}t = 0. \tag{3.123}$$

Equation (3.121) allows us to derive the successive approximations $x_{ik}(t),\ k = 1,\ 2,\ 3, \cdots$, which is:

$$x_{ik}(t) = \int_0^{2\pi} G_i[t,\sigma]\phi_{i,k-1}[\sigma]\,\mathrm{d}\sigma + \delta_{\lambda_i}^{n_i^2}(r_i \cos n_i t + s_i \sin n_i t), \tag{3.124}$$

where

$$\begin{aligned} &\phi_{i0}[t] = \phi_i\big[0, 0, 0, 0, ; \varepsilon_p, t\big], \\ &\phi_{ik}[t] = \phi_i[x_{ik}(t), \dot{x}_{jk}(t), x_{ik}(t-\tau), \dot{x}_{ik}(t-\tau), \\ &\quad x_{jk}(t), \dot{x}_{jk}(t), x_{jk}(t-\tau), \dot{x}_{jk}(t-\tau); \varepsilon_p, t], \\ &k = 1,\ 2,\ 3,\ \cdots. \end{aligned} \tag{3.125}$$

3.3.2 The Primary Resonance and 1:1 Internal Resonance of a Two-Degrees-of-Freedom Model

The subsequent analysis focuses on a mathematical model that characterizes the behaviour of a vibration absorber incorporating feedback with time delay (refer to [22]). Dynamic equations of the system are given by:

$$
\begin{aligned}
\ddot{y}_1 + \omega_1^2(y_1 - y_2) =& \xi_1(\dot{y}_2 - \dot{y}_1) - \alpha(y_1 - y_2)^3 - g(y_{1\tau} - y_1),\\
\ddot{y}_2 + \omega_2^2 y_2 =& - \xi_2\dot{y}_2 + \mu\omega_1^2(y_1 - y_2) + F\cos(\Omega t) + \xi_1\mu(\dot{y}_1 - \dot{y}_2)\\
&+ \mu\alpha(y_1 - y_2)^3 + \mu g(y_{1\tau} - y_1).
\end{aligned} \tag{3.126}
$$

3.3.2.1 The Integral Equation Method

Firstly, two detuning parameters $\sigma_1,\ \sigma_2$ are introduced in the study of the primary resonance and 1:1 internal resonance, for which $\Omega \approx \omega_2,\ \omega_1 \approx \omega_2$:

$$
\omega_1 = \omega_2 + \sigma_1, \qquad \Omega = \omega_2 + \sigma_2. \tag{3.127}
$$

Then the first order approximated solution is acquired through application of the integral equation method:

$$
y_{11} = r_1\cos t + s_1\sin t, \tag{3.128}
$$

$$
y_{21} = r_2\cos t + s_2\sin t. \tag{3.129}
$$

And the second approximation is given by:

$$
\begin{aligned}
y_{12} =& \int_0^{2\pi} \frac{1}{\pi}\left(\frac{1}{2} - \cos(t-\delta) + \sum_{j=2}^{5}\frac{\cos(j(t-\delta))}{1-j^2}\right)\\
&\frac{1}{\Omega^2}(-(2\Omega(\sigma_1-\sigma_2) + (\sigma_1-\sigma_2)^2)y_{11}(\delta) + \omega_1^2 y_{21}(\delta)\\
&+ \xi_1\Omega(\dot{y}_{21}(\delta) - \dot{y}_{11}(\delta)) - \alpha(y_{11}(\delta) - y_{21}(\delta))^3\\
&- g(y_{11}(\delta-\tau) - y_{11}(\delta)))d\delta + r_1\cos t + s_1\sin t\\
=& (\frac{r_1^3\alpha}{32\Omega^2} - \frac{3r_1^2r_2\alpha}{32\Omega^2} + \frac{3r_1r_2^2\alpha}{32\Omega^2} - \frac{r_2^3\alpha}{32\Omega^2} - \frac{3r_1s_1^2\alpha}{32\Omega^2}\\
&+ \frac{3r_2s_1^2\alpha}{32\Omega^2} + \frac{3r_1s_1s_2\alpha}{16\Omega^2} - \frac{3r_2s_1s_2\alpha}{16\Omega^2} - \frac{3r_1s_2^2\alpha}{32\Omega^2}\\
&+ \frac{3r_2s_2^2\alpha}{32\Omega^2})\cos(3t) + (\frac{3r_1^2s_1\alpha}{32\Omega^2} - \frac{3r_1r_2s_1\alpha}{16\Omega^2} + \frac{3r_2^2s_1\alpha}{32\Omega^2}\\
&- \frac{s_1^3\alpha}{32\Omega^2} - \frac{3r_1^2s_2\alpha}{32\Omega^2} + \frac{3r_1r_2s_2\alpha}{16\Omega^2} - \frac{3r_2^2s_2\alpha}{32\Omega^2} + \frac{3s_1^2s_2\alpha}{32\Omega^2}\\
&- \frac{3s_1s_2^2\alpha}{32\Omega^2} + \frac{s_2^3\alpha}{32\Omega^2})\sin(3t) + (s_1 - \frac{gs_1}{\Omega^2} + \frac{3r_1^2s_1\alpha}{4\Omega^2}\\
&- \frac{3r_1r_2s_1\alpha}{2\Omega^2} + \frac{3r_2^2s_1\alpha}{4\Omega^2} - \frac{3r_1^2s_2\alpha}{4\Omega^2} + \frac{3s_1^3\alpha}{4\Omega^2} + \frac{3r_1r_2s_2\alpha}{2\Omega^2}
\end{aligned}
$$

$$
\begin{aligned}
&-\frac{3r_2^2 s_2\alpha}{4\Omega^2}-\frac{9s_1^2 s_2\alpha}{4\Omega^2}+\frac{9s_1 s_2^2\alpha}{4\Omega^2}-\frac{3s_2^3\alpha}{4\Omega^2}+\frac{s_1\sigma_1^2}{\Omega^2}\\
&-\frac{2s_1\sigma_1\sigma_2}{\Omega^2}+\frac{s_1\sigma_2^2}{\Omega^2}-\frac{r_1\xi_1}{\Omega}+\frac{r_2\xi_1}{\Omega}+\frac{2s_1\sigma_1}{\Omega}-\frac{2s_1\sigma_2}{\Omega}\\
&-\frac{s_2\omega_1^2}{\Omega^2}+\frac{gs_1\cos\tau}{\Omega^2}+\frac{gr_1\sin\tau}{\Omega^2})\sin t+(r_1-\frac{gr_1}{\Omega^2}\\
&+\frac{3r_1^3\alpha}{4\Omega^2}-\frac{9r_1^2 r_2\alpha}{4\Omega^2}+\frac{9r_1 r_2^2\alpha}{4\Omega^2}-\frac{3r_2^3\alpha}{4\Omega^2}+\frac{3r_1 s_1^2\alpha}{4\Omega^2}\\
&-\frac{3r_2 s_1^2\alpha}{4\Omega^2}-\frac{3r_1 s_1 s_2\alpha}{2\Omega^2}+\frac{3r_2 s_1 s_2\alpha}{2\Omega^2}+\frac{3r_1 s_2^2\alpha}{4\Omega^2}-\frac{3r_2 s_2^2\alpha}{4\Omega^2}\\
&+\frac{r_1\sigma_1^2}{\Omega^2}-\frac{2r_1\sigma_1\sigma_2}{\Omega^2}+\frac{r_1\sigma_2^2}{\Omega^2}+\frac{s_1\xi_1}{\Omega}-\frac{s_2\xi_1}{\Omega}+\frac{2r_1\sigma_1}{\Omega}\\
&-\frac{2r_1\sigma_2}{\Omega}-\frac{r_2\omega_1^2}{\Omega^2}+\frac{gr_1\cos\tau}{\Omega^2}-\frac{gs_1\sin\tau}{\Omega^2})\cos t,
\end{aligned}
\tag{3.130}
$$

$$
\begin{aligned}
y_{22}=&\int_0^{2\pi}\frac{1}{\pi}\left(\frac{1}{2}-\cos(t-\delta)+\sum_{j=2}^{5}\frac{\cos(j(t-\delta))}{1-j^2}\right)\\
&\frac{1}{\Omega^2}((2\Omega\sigma_2-\sigma_2^2)y_{21}(\delta)-\xi_2\Omega\dot{y}_{21}(\delta)+F\cos(\delta)\\
&+\mu\omega^2(y_{11}(\delta)-y_{21}(\delta))+\xi_1\mu\Omega(\dot{y}_{11}(\delta)-\dot{y}_{21}(\delta))\\
&+\mu\alpha(y_{11}(\delta)-y_{21}(\delta))^3+\mu g(y_{11}(\delta-\tau)\\
&-y_{11}(\delta)))d\delta+r_2\cos t+s_2\sin t\\
=&(-\frac{r_1^3\alpha\mu}{32\Omega^2}+\frac{3r_1^2 r_2\alpha\mu}{32\Omega^2}-\frac{3r_1 r_2^2\alpha\mu}{32\Omega^2}+\frac{r_2^3\alpha\mu}{32\Omega^2}+\frac{3r_1 s_1^2\alpha\mu}{32\Omega^2}\\
&-\frac{3r_2 s_1^2\alpha\mu}{32\Omega^2}-\frac{3r_1 s_1 s_2\alpha\mu}{16\Omega^2}+\frac{3r_2 s_1 s_2\alpha\mu}{16\Omega^2}+\frac{3r_1 s_2^2\alpha\mu}{32\Omega^2}\\
&-\frac{3r_2 s_2^2\alpha\mu}{32\Omega^2})\cos(3t)+(-\frac{3r_1^2 s_1\alpha\mu}{32\Omega^2}+\frac{3r_1 r_2 s_1\alpha\mu}{16\Omega^2}\\
&-\frac{3r_2^2 s_1\alpha\mu}{32\Omega^2}+\frac{s_1^3\alpha\mu}{32\Omega^2}-\frac{3r_1 r_2 s_2\alpha\mu}{16\Omega^2}+\frac{3r_1^2 s_2\alpha\mu}{32\Omega^2}+\frac{3r_2^2 s_2\alpha\mu}{32\Omega^2}\\
&-\frac{3s_1^2 s_2\alpha\mu}{32\Omega^2}+\frac{3s_1 s_2^2\alpha\mu}{32\Omega^2}-\frac{s_2^3\alpha\mu}{32\Omega^2})\sin(3t)+\sin t(s_2\\
&+\frac{gs_1\mu}{\Omega^2}+\frac{3r_1 r_2 s_1\alpha\mu}{2\Omega^2}-\frac{3r_1^2 s_1\alpha\mu}{4\Omega^2}-\frac{3r_2^2 s_1\alpha\mu}{4\Omega^2}-\frac{3s_1^3\alpha\mu}{4\Omega^2}\\
&+\frac{3r_1^2 s_2\alpha\mu}{4\Omega^2}-\frac{3r_1 r_2 s_2\alpha\mu}{2\Omega^2}+\frac{3r_2^2 s_2\alpha\mu}{4\Omega^2}+\frac{9s_1^2 s_2\alpha\mu}{4\Omega^2}\\
&-\frac{9s_1 s_2^2\alpha\mu}{4\Omega^2}+\frac{3s_2^3\alpha\mu}{4\Omega^2}+\frac{s_2\sigma_2^2}{\Omega^2}+\frac{r_1\mu\xi_1}{\Omega}-\frac{r_2\mu\xi_1}{\Omega}\\
&-\frac{r_2\xi_2}{\Omega}-\frac{2s_2\sigma_2}{\Omega}-\frac{s_1\mu\omega_1^2}{\Omega^2}+\frac{s_2\mu\omega_1^2}{\Omega^2}-\frac{gs_1\mu\cos\tau}{\Omega^2}
\end{aligned}
$$

$$
\begin{aligned}
&- \frac{g r_1 \mu \sin\tau}{\Omega^2}) + \cos t\,(r_2 - \frac{F}{\Omega^2} + \frac{g r_1 \mu}{\Omega^2} \\
&- \frac{3 r_1^3 \alpha\mu}{4\Omega^2} + \frac{9 r_1^2 r_2 \alpha\mu}{4\Omega^2} - \frac{9 r_1 r_2^2 \alpha\mu}{4\Omega^2} + \frac{3 r_2^3 \alpha\mu}{4\Omega^2} - \frac{3 r_1 s_1^2 \alpha\mu}{4\Omega^2} \\
&+ \frac{3 r_2 s_1^2 \alpha\mu}{4\Omega^2} + \frac{3 r_1 s_1 s_2 \alpha\mu}{2\Omega^2} - \frac{3 r_2 s_1 s_2 \alpha\mu}{2\Omega^2} - \frac{3 r_1 s_2^2 \alpha\mu}{4\Omega^2} \\
&+ \frac{3 r_2 s_2^2 \alpha\mu}{4\Omega^2} + \frac{r_2 \sigma_2^2}{\Omega^2} - \frac{s_1 \mu \xi_1}{\Omega} + \frac{s_2 \mu \xi_1}{\Omega} + \frac{s_2 \xi_2}{\Omega} - \frac{2 r_2 \sigma_2}{\Omega} \\
&- \frac{r_1 \mu \omega_1^2}{\Omega^2} + \frac{r_2 \mu \omega_1^2}{\Omega^2} - \frac{g r_1 \mu \cos\tau}{\Omega^2} + \frac{g s_1 \mu \sin\tau}{\Omega^2}).
\end{aligned}
\tag{3.131}
$$

The periodicity Eq. (3.122) now becomes

$$
\begin{aligned}
&\frac{3 r_1^3 \alpha}{4\Omega^2} - \frac{g r_1}{\Omega^2} - \frac{9 r_1^2 r_2 \alpha}{4\Omega^2} + \frac{9 r_1 r_2^2 \alpha}{4\Omega^2} - \frac{3 r_2^3 \alpha}{4\Omega^2} + \frac{3 r_1 s_1^2 \alpha}{4\Omega^2} \\
&- \frac{3 r_2 s_1^2 \alpha}{4\Omega^2} - \frac{3 r_1 s_1 s_2 \alpha}{2\Omega^2} + \frac{3 r_2 s_1 s_2 \alpha}{2\Omega^2} + \frac{3 r_1 s_2^2 \alpha}{4\Omega^2} - \frac{3 r_2 s_2^2 \alpha}{4\Omega^2} \\
&+ \frac{r_1 \sigma_1^2}{\Omega^2} - \frac{2 r_1 \sigma_1 \sigma_2}{\Omega^2} + \frac{r_1 \sigma_2^2}{\Omega^2} + \frac{s_1 \xi_1}{\Omega} - \frac{s_2 \xi_1}{\Omega} + \frac{2 r_1 \sigma_1}{\Omega} - \frac{2 r_1 \sigma_2}{\Omega} \\
&- \frac{r_2 \omega_1^2}{\Omega^2} + \frac{g r_1 \cos(\Omega\tau)}{\Omega^2} - \frac{g s_1 \sin(\Omega\tau)}{\Omega^2} = 0 \quad ,
\end{aligned}
\tag{3.132}
$$

$$
\begin{aligned}
&\frac{3 r_1^2 s_1 \alpha}{4\Omega^2} - \frac{g s_1}{\Omega^2} - \frac{3 r_1 r_2 s_1 \alpha}{2\Omega^2} + \frac{3 r_2^2 s_1 \alpha}{4\Omega^2} + \frac{3 s_1^3 \alpha}{4\Omega^2} - \frac{3 r_1^2 s_2 \alpha}{4\Omega^2} \\
&+ \frac{3 r_1 r_2 s_2 \alpha}{2\Omega^2} - \frac{3 r_2^2 s_2 \alpha}{4\Omega^2} - \frac{9 s_1^2 s_2 \alpha}{4\Omega^2} + \frac{9 s_1 s_2^2 \alpha}{4\Omega^2} - \frac{3 s_2^3 \alpha}{4\Omega^2} \\
&+ \frac{s_1 \sigma_1^2}{\Omega^2} - \frac{2 s_1 \sigma_1 \sigma_2}{\Omega^2} + \frac{s_1 \sigma_2^2}{\Omega^2} - \frac{r_1 \xi_1}{\Omega} + \frac{r_2 \xi_1}{\Omega} + \frac{2 s_1 \sigma_1}{\Omega} \\
&- \frac{2 s_1 \sigma_2}{\Omega} - \frac{s_2 \omega_1^2}{\Omega^2} + \frac{g s_1 \cos(\Omega\tau)}{\Omega^2} + \frac{g r_1 \sin(\Omega\tau)}{\Omega^2} = 0,
\end{aligned}
\tag{3.133}
$$

$$
\begin{aligned}
&\frac{g r_1 \mu}{\Omega^2} - \frac{F}{\Omega^2} - \frac{3 r_1^3 \alpha\mu}{4\Omega^2} + \frac{9 r_1^2 r_2 \alpha\mu}{4\Omega^2} - \frac{9 r_1 r_2^2 \alpha\mu}{4\Omega^2} \\
&+ \frac{3 r_2^3 \alpha\mu}{4\Omega^2} - \frac{3 r_1 s_1^2 \alpha\mu}{4\Omega^2} + \frac{3 r_2 s_1^2 \alpha\mu}{4\Omega^2} + \frac{3(r_1 - r_2) s_1 s_2 \alpha\mu}{2\Omega^2} \\
&- \frac{3 r_1 s_2^2 \alpha\mu}{4\Omega^2} + \frac{3 r_2 s_2^2 \alpha\mu}{4\Omega^2} + \frac{r_2 \sigma_2^2}{\Omega^2} - \frac{s_1 \mu \xi_1}{\Omega} + \frac{s_2 \mu \xi_1}{\Omega} \\
&+ \frac{s_2 \xi_2}{\Omega} - \frac{2 r_2 \sigma_2}{\Omega} - \frac{r_1 \mu \omega_1^2}{\Omega^2} + \frac{r_2 \mu \omega_1^2}{\Omega^2} - \frac{g r_1 \mu \cos(\Omega\tau)}{\Omega^2} \\
&+ \frac{g s_1 \mu \sin(\Omega\tau)}{\Omega^2} = 0,
\end{aligned}
\tag{3.134}
$$

$$\frac{gs_1\mu}{\Omega^2}-\frac{3r_1^2s_1\alpha\mu}{4\Omega^2}+\frac{3r_1r_2s_1\alpha\mu}{2\Omega^2}-\frac{3r_2^2s_1\alpha\mu}{4\Omega^2}-\frac{3s_1^3\alpha\mu}{4\Omega^2}$$
$$+\frac{3r_1^2s_2\alpha\mu}{4\Omega^2}-\frac{3r_1r_2s_2\alpha\mu}{2\Omega^2}+\frac{3r_2^2s_2\alpha\mu}{4\Omega^2}+\frac{9s_1^2s_2\alpha\mu}{4\Omega^2}$$
$$-\frac{9s_1s_2^2\alpha\mu}{4\Omega^2}+\frac{3s_2^3\alpha\mu}{4\Omega^2}+\frac{s_2\sigma 2^2}{\Omega^2}+\frac{r_1\mu\xi_1}{\Omega}-\frac{r_2\mu\xi_1}{\Omega}-\frac{r_2\xi_2}{\Omega}$$
$$-\frac{2s_2\sigma_2}{\Omega}-\frac{(s_1-s_2)\mu\omega_1^2}{\Omega^2}-\frac{gs_1\mu\cos(\Omega\tau)}{\Omega^2}-\frac{gr_1\mu\sin(\Omega\tau)}{\Omega^2}=0. \tag{3.135}$$

3.3.2.2 Stability Analysis of Eq. (3.126)

We employ the linear variational equation to analyze the stability of the vibrations of Eq. (3.127),

$$\begin{aligned}\ddot{z}_1+z_1&=\frac{1}{\Omega^2}(-(2\Omega(\sigma_1-\sigma_2)+(\sigma_1-\sigma_2)^2)z_1+\omega_1^2z_2\\&+\xi_1\Omega(\dot{z}_2-\dot{z}_1)-3\alpha(y_1-y_2)^2(z_1-z_2)-g(z_{1\tau}-z_1)),\\ \ddot{z}_2+z_2&=\frac{1}{\Omega^2}((2\Omega\sigma_2-\sigma_2^2)z_2-\xi_2\Omega\dot{z}_2+\mu\omega_1^2(z_1-z_2)\\&+\xi_1\mu\Omega(\dot{z}_1-\dot{z}_2)+3\mu\alpha(y_1-y_2)^2(z_1-z_2)+\mu g(z_{1\tau}-z_1)),\end{aligned} \tag{3.136}$$

where $y_1=r_1\cos t+s_1\sin t,\quad y_2=r_2\cos t+s_2\sin t.$.

The Floquet theorem allows us to express the solution of Eq. (3.136) as:

$$z_i(t)=e^{\rho t}Z_i(t),\qquad i=1,\ 2. \tag{3.137}$$

Equation (3.136) now reads:

$$\begin{aligned}\ddot{Z}_1+Z_1&=(a_{z11}+c_{z11}\cos(2t)+s_{z11}\sin(2t)-\rho^2\\&-\frac{\xi_1\rho}{\Omega})Z_1-\frac{\exp(-\rho\tau)g}{\Omega^2}Z_{1\tau}+(a_{z12}+\frac{\xi_1\rho}{\Omega}\\&+c_{z12}\cos(2t)+s_{z12}\sin(2t))Z_2+(-2\rho\\&-\frac{\xi_1}{\Omega})\dot{Z}_1+\frac{\xi_1}{\Omega}\dot{Z}_2=\Phi_{z1}(t),\end{aligned} \tag{3.138}$$

$$\begin{aligned}\ddot{Z}_2+Z_2&=\left(a_{z21}+\frac{\mu\xi_1}{\Omega}\rho+c_{z21}\cos(2t)+s_{z21}\sin(2t)\right)Z_1\\&+\frac{\exp(-\rho\tau)g\mu}{\Omega^2}Z_{1\tau}+(a_{z22}-\rho^2-\frac{\mu\xi_1\rho}{\Omega}-\frac{\xi_2\rho}{\Omega}\\&+c_{z22}\cos(2t)+s_{z22}\sin(2t))Z_2+\frac{\mu\xi_1}{\Omega}\dot{Z}_1\end{aligned}$$

$$+\left(-2\rho-\frac{\mu\xi_1}{\Omega}-\frac{\xi_2}{\Omega}\right)\dot{Z}_2=\Phi_{z2}(t), \tag{3.139}$$

where the coeffcients $c_{zij}, s_{zij}, a_{zij}, i, j = 1, 2$ are given in the Appendix in the end of this chapter.

$F = 0.1, \alpha = 0.2$, and (a, b) $g = -0.1$, (c, d) $g = -0.2$.

The first approximation of Eq. (3.138) is:

$$\begin{aligned} Z_1(t) &= p_1\cos t + q_1\sin t,\\ Z_2(t) &= p_2\cos t + q_2\sin t. \end{aligned} \tag{3.140}$$

The periodicity Eq. (3.123) shows

$$\begin{aligned} &q_2\left(\frac{s_{z12}}{2}+\frac{\xi_1}{\Omega}\right)+p_2\left(a_{z12}+\frac{c_{z12}}{2}+\frac{\xi_1}{\Omega}\rho\right)\\ &+p_1(a_{z11}+\frac{c_{z11}}{2}-\rho^2-\frac{\xi_1}{\Omega}\rho-\frac{g\exp(-\rho\tau)}{\Omega^2}\cos\tau)\\ &+q_1(\frac{s_{z11}}{2}-2\rho-\frac{\xi_1}{\Omega}+\frac{\exp(-\rho\tau)g\sin\tau}{\Omega^2})=0, \end{aligned} \tag{3.141}$$

$$\begin{aligned} &p_2\left(\frac{s_{z12}}{2}-\frac{\xi_1}{\Omega}\right)+q_2\left(a_{z12}-\frac{c_{z12}}{2}+\frac{\xi_1}{\Omega}\rho\right)\\ &+q_1(a_{z11}-\frac{c_{z11}}{2}-\rho^2-\frac{\xi_1}{\Omega}\rho-\frac{g\exp(-\rho\tau)}{\Omega^2}\cos\tau)\\ &+p_1(\frac{s_{z11}}{2}+2\rho+\frac{\xi_1}{\Omega}-\frac{\exp(-\rho\tau)g\sin\tau}{\Omega^2})=0, \end{aligned} \tag{3.142}$$

$$\begin{aligned} &q_2\left(\frac{s_{z22}}{2}-2\rho-\frac{\mu\xi_1}{\Omega}-\frac{\xi_2}{\Omega}\right)+p_2(a_{z22}+\frac{c_{z22}}{2}-\rho^2-\frac{\mu\xi_1\rho}{\Omega}-\frac{\xi_2\rho}{\Omega})\\ &+p_1(a_{z21}+\frac{c_{z21}}{2}+\frac{\mu\xi_1\rho}{\Omega}+\frac{exp(-\rho\tau)g\mu cos\tau}{\Omega^2})\\ &+q_1(\frac{s_{z21}}{2}+\frac{\mu\xi_1}{\Omega}-\frac{exp(-\rho\tau)g\mu sin\tau}{\Omega^2})=0, \end{aligned} \tag{3.143}$$

$$\begin{aligned} &p_2\left(\frac{s_{z22}}{2}+2\rho+\frac{\mu\xi_1}{\Omega}+\frac{\xi_2}{\Omega}\right)+q_2(a_{z22}-\frac{c_{z22}}{2}-\rho^2-\frac{\mu\xi_1\rho}{\Omega}-\frac{\xi_2\rho}{\Omega})\\ &+q_1(a_{z21}-\frac{c_{z21}}{2}+\frac{\mu\xi_1\rho}{\Omega}+\frac{exp(-\rho\tau)g\mu cos\tau}{\Omega^2})\\ &+p_1(\frac{exp(-\rho\tau)g\mu sin\tau}{\Omega^2}+\frac{s_{z21}}{2}-\frac{\mu\xi_1}{\Omega})=0, \end{aligned} \tag{3.144}$$

Hence, by assuming that the values of p_1, q_1, p_2, q_2 are not at the same time equal to zero, we can derive the following formula from Eqs. (3.141)–(3.143):

$$
\begin{aligned}
p_z(\rho) = {} & a_{zc} + a_{z1}\rho + a_{z2}\rho^2 + a_{z3}\rho^3 + a_{z4}\rho^4 \\
& + a_{z5}\rho^5 + a_{z6}\rho^6 + a_{z7}\rho 7 + \rho 8 + exp(-\rho\Omega\tau)(b_{zc} \\
& + b_{z1}\rho + b_{z2}\rho^2 + b_{z3}\rho^3 + b_{z4}\rho^4 + b_{z5}\rho^5 + b_{z6}\rho^6) \\
& + exp(-2\rho\Omega\tau)(c_{zc} + c_{z1}\rho + c_{z2}\rho^2 + c_{z3}\rho^3 + c_{z4}\rho^4).
\end{aligned}
\tag{3.145}
$$

In the next, we consider the following two case.

Case (i) Time delay equals zero ($\tau = 0$):
Equation (3.145) indicates that:

$$
\begin{aligned}
p_z(\rho) = {} & a_{zc} + b_{zc} + c_{zc} + (a_{z1} + b_{z1} + c_{z1})\rho \\
& + (a_{z2} + b_{z2} + c_{z2})\rho^2 + (a_{z3} + b_{z3} + c_{z3})\rho^3 \\
& + (a_{z4} + b_{z4} + c_{z4})\rho^4 + (a_{z5} + b_{z5})\rho^5 + (a_{z6} + b_{z6})\rho^6 \\
& + a_{z7}\rho^7 + \rho^8.
\end{aligned}
\tag{3.146}
$$

To examine whether the vibration is stable or not, the Routh-Hurwitz criterion can be applied in this particular scenario.

Case (ii) Time delay does not equal to zero ($\tau \neq 0$):
Whether the system is stable or not can be examined by applying the Nyquist criterion. For the sake of illustration, we look into the stability for $\tau = 2$ and $\tau = 4.5$ when $g = -0.2$. Figure 3.10 indicates that the periodic solution for $\tau = 2$, $g = -0.2$ is stable, while the periodic solution for $\tau = 4.5$, $g = -0.2$ is unstable. And in Fig. 3.11, the validity of the stability analysis has been confirmed.

Figure 3.12a, b provide a comparision of analytical solutions and numerical solutions, through amplitude-delay response curves for $g = -0.1$, whereas Figs. 3.12c, d show the comparison for $g = -0.2$. Our findings reveal that the stability of amplitude-delay response curves persist regardless of the value of time delay for $g = -0.1$. When $g = -0.2$, the periodic solutions within interval [3.8, 6] are unstable and will develop to quasi-periodic motion, as shown in Fig. 3.11.

Comparison of the analytical and numerical results in terms frequency response curves of are displayed in Fig. 3.13 where for (a) and (b) $\tau = 0.2$, (c) and (d) $\tau = 10$. These figures clearly show the effectiveness and reliability of the proposed integral equation method in solving nonlinear systems with time delay.

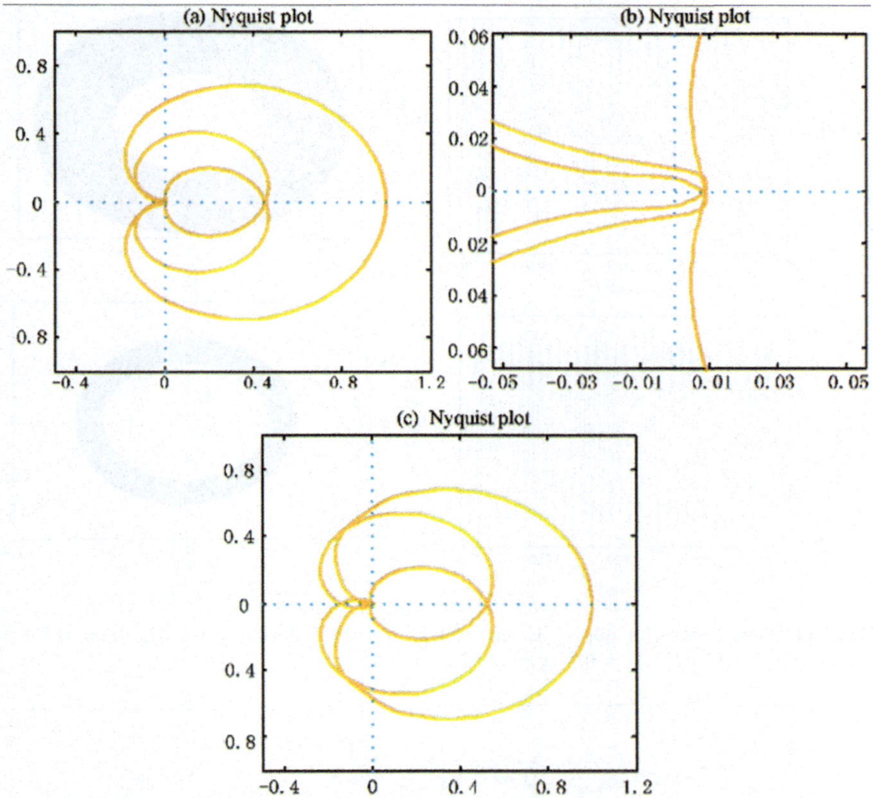

Fig. 3.10 **a** The Nyquist plot for $p(i\omega)/(1+i\omega)^4$ using parameters $\omega_1 = 1$, $\omega_2 = 1.2$, $\sigma_1 = -0.2$, $\sigma_2 = 0$, $\mu = 0.1$, $\xi_1 = 0.1$, $\xi_2 = 0.1$, $F = 0.1, \alpha = 0.2$ and $\tau = 2$. **b** Local magnification of (**a**), indicating that the periodic solution is stable when $\tau = 2$, $g = -0.2$; likewise **c** shows the instability of periodic motion when $\tau = 4.5$, $g = -0.2$.

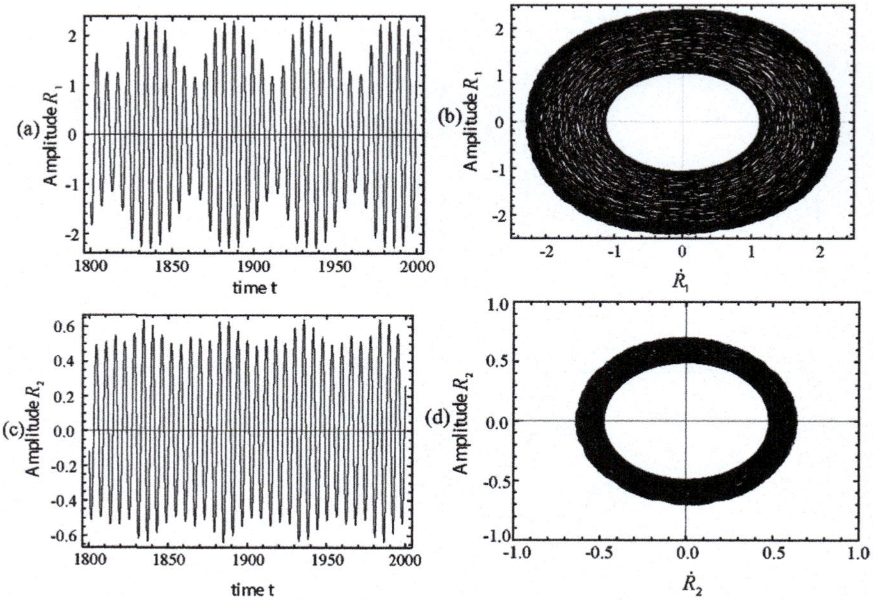

Fig. 3.11 Time history for ($\omega_1 = 1$, $\omega_2 = 1.2$, $\sigma_1 = -0.2$, $\sigma_2 = -0.02$, $\mu = 0.1\xi_1 = 0.1$, $\xi_2 = 0.1$, $F = 0.1, \alpha = 0.2$, $g = -0.2$, $\tau = 4.5$)

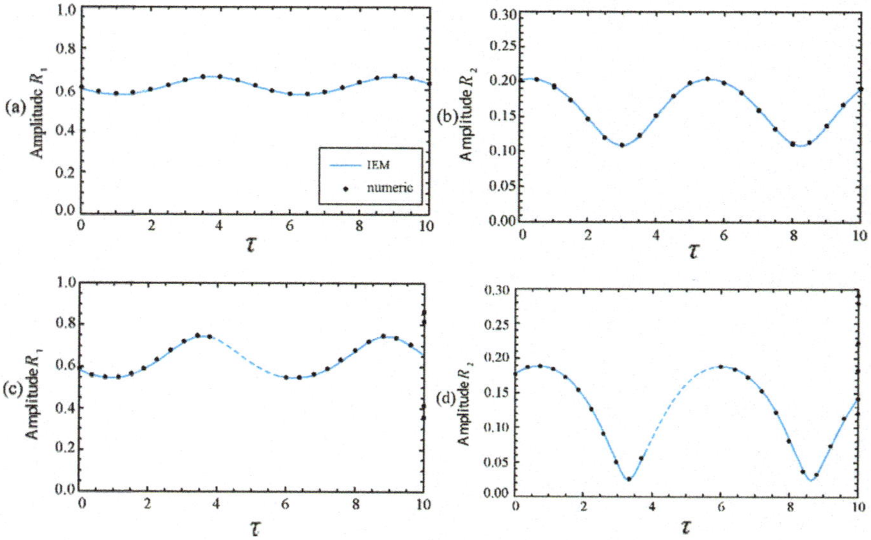

Fig. 3.12 The analytical solutions obtained by integral equation method (blue curve), which is compared with numerical results (dot) for $\omega_1 = 1$, $\omega_2 = 1.2$, $\sigma_1 = -0.2$, $\sigma_2 = 0$, $\mu = 0.1$, $\xi_1 = 0.1$, $\xi_2 = 0.1$,

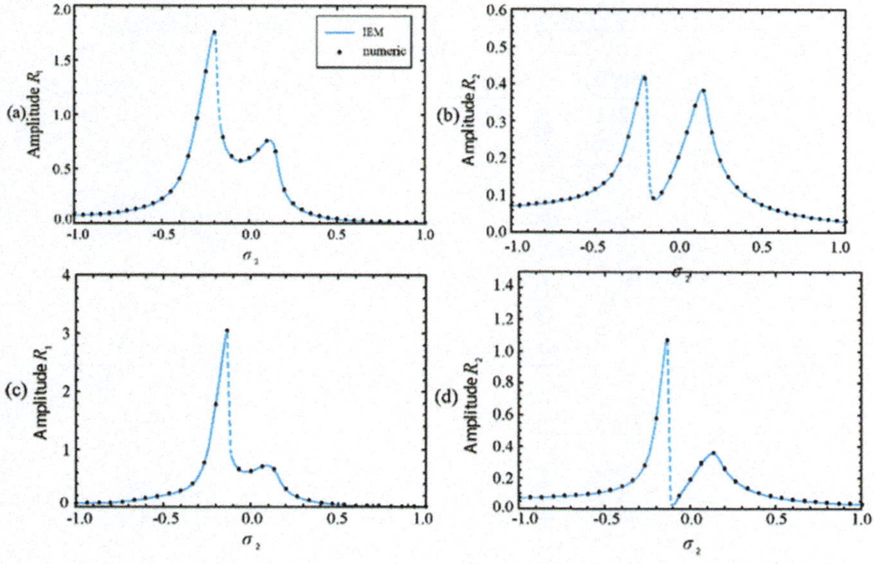

Fig. 3.13 Sorry for the mistake about the figure. In fact, Fig. 3.11 and its caption is exactly correct, but wrong picture and caption has been used in Fig. 3.13. The correct picture for Fig. 3.13 is provided in the attachment, which can be used to replace the current picture. In the meantime, the caption of Fig. 3.13 has been corrected. ($\omega_1 = 1,\ \omega_2 = 1.2,\ \sigma_1 = -0.2,\ \mu = 0.1,\ \xi_1 = 0.1,\ \xi_2 = 0.1,\ F = 0.1,\ \alpha = 0.2,\ g = -0.1$)(($a$) ($b$) *where* $\tau = 0.2$, (c) (d) *where* $\tau = 10$)

Appendix

$$c_{z11} = -\frac{3r_1^2\alpha}{2\Omega^2} + \frac{3r_1r_2\alpha}{\Omega^2} - \frac{3r_2^2\alpha}{2\Omega^2} + \frac{3s_1^2\alpha}{2\Omega^2} - \frac{3s_1s_2\alpha}{\Omega^2} + \frac{3s_2^2\alpha}{2\Omega^2}$$

$$s_{z11} = -\frac{3r_1s_1\alpha}{\Omega^2} + \frac{3r_2s_1\alpha}{\Omega^2} + \frac{3r_1s_2\alpha}{\Omega^2} - \frac{3r_2s_2\alpha}{\Omega^2}$$

$$c_{z12} = \frac{3r_1^2\alpha}{2\Omega^2} - \frac{3r_1r_2\alpha}{\Omega^2} + \frac{3r_2^2\alpha}{2\Omega^2} - \frac{3s_1^2\alpha}{2\Omega^2} + \frac{3s_1s_2\alpha}{\Omega^2} - \frac{3s_2^2\alpha}{2\Omega^2}$$

$$s_{z12} = \frac{3r_1s_1\alpha}{\Omega^2} - \frac{3r_2s_1\alpha}{\Omega^2} - \frac{3r_1s_2\alpha}{\Omega^2} + \frac{3r_2s_2\alpha}{\Omega^2}$$

$$a_{z11} = \frac{g}{\Omega^2} - \frac{3r_1^2\alpha}{2\Omega^2} + \frac{3r_2r_2\alpha}{\Omega^2} - \frac{3r_2^2\alpha}{2\Omega^2} - \frac{3s_1^2\alpha}{2\Omega^2} + \frac{3s_1s_2\alpha}{\Omega^2} - \frac{3s_2^2\alpha}{2\Omega^2} - \frac{(\sigma_1 - \sigma_2)^2}{\Omega^2} - \frac{2(\sigma_1 - \sigma_2)}{\Omega}$$

$$a_{z12} = \frac{3r_1^2\alpha}{2\Omega^2} - \frac{3r_1r_2\alpha}{\Omega^2} + \frac{3r_2^2\alpha}{2\Omega^2} + \frac{3s_1^2\alpha}{2\Omega^2} - \frac{3s_1s_2\alpha}{\Omega^2} + \frac{3s_2^2\alpha}{2\Omega^2} + \frac{\omega_1^2}{\Omega^2}$$

$$c_{z21} = \frac{3r_1^2\alpha\mu}{2\Omega^2} - \frac{3r_1r_2\alpha\mu}{\Omega^2} + \frac{3r_2^2\alpha\mu}{2\Omega^2} - \frac{3s_1^2\alpha\mu}{2\Omega^2} + \frac{3s_1s_2\alpha\mu}{\Omega^2} - \frac{3s_2^2\alpha\mu}{2\Omega^2}$$

$$s_{z21} = \frac{3r_1s_1\alpha\mu}{\Omega^2} - \frac{3r_2s_1\alpha\mu}{\Omega^2} - \frac{3r_1s_2\alpha\mu}{\Omega^2} + \frac{3r_2s_2\alpha\mu}{\Omega^2}$$

$$a_{z21} = -\frac{g\mu}{\Omega^2} + \frac{3r_1^2\alpha\mu}{2\Omega^2} - \frac{3r_1r_2\alpha\mu}{\Omega^2} + \frac{3r_2^2\alpha\mu}{2\Omega^2} + \frac{3s_1^2\alpha\mu}{2\Omega^2} - \frac{3s_1s_2\alpha\mu}{\Omega^2} + \frac{3s_2^2\alpha\mu}{2\Omega^2} + \frac{\mu\omega_1^2}{\Omega^2}$$

$$c_{z22} = -\frac{3r_1^2\alpha\mu}{2\Omega^2} + \frac{3r_1r_2\alpha\mu}{\Omega^2} - \frac{3r_2^2\alpha\mu}{2\Omega^2} + \frac{3s_1^2\alpha\mu}{2\Omega^2} - \frac{3s_1s_2\alpha\mu}{\Omega^2} + \frac{3s_2^2\alpha\mu}{2\Omega^2}$$

$$s_{z22} = -\frac{3r_1s_1\alpha\mu}{\Omega^2} + \frac{3r_2s_1\alpha\mu}{\Omega^2} + \frac{3r_1s_2\alpha\mu}{\Omega^2} - \frac{3r_2s_2\alpha\mu}{\Omega^2}$$

$$a_{z22} = -\frac{3r_1^2\alpha\mu}{2\Omega^2} + \frac{3r_1r_2\alpha\mu}{\Omega^2} - \frac{3r_2^2\alpha\mu}{2\Omega^2} - \frac{3s_1^2\alpha\mu}{2\Omega^2} + \frac{3s_1s_2\alpha\mu}{\Omega^2} - \frac{3s_2^2\alpha\mu}{2\Omega^2} - \frac{\sigma_2^2}{\Omega^2} + \frac{2\sigma_2}{\Omega} - \frac{\mu\omega_1^2}{\Omega^2}$$

References

1. Xu J, Chung KW (2003) Effects of time delayed position feedback on a van der pol-duffing oscillator. Physica D 180(1–2):17–39
2. Fridman E (2002) Effects of small delays on stability of singularly perturbed systems. Automatica 38(5):897–902
3. Das SL, Chatterjee A (2002) Multiple scales without center manifold reductions for delay differential equations near hopf bifurcation. Nonlinear Dyn 30(2):323–335
4. Maccari (2001) The response of a parametrically excited van der pol oscillator to a time delay state feedback. Nonlinear Dyn 26(2):105–119
5. Casal A, Freedman M (1980) Poincaré-lindstedt approach to bifurcation problems for differential-delay equations. IEEE Trans Automat Cont AC–25(5):967–973
6. Xu J, Chung KW (2009) A perturbation-incremental scheme for studying Hopf bifurcation in delayed differential systems. Sci China Ser E: Technol Sci 52(3):698–708
7. Chan HSY, Chung KW, Xu Z (1996) A perturbation-incremental method for strongly non-linear oscillators. Int J Nonlin Mech 31(1):59–72
8. Li L, Ye HL (2003) Energy method for computing periodic solutions of strongly nonlinear autonomous systems with multi-degree-of-freedom. Nonlinear Dyn 31(1):23–47
9. Chen SH, Yan XM, Cheung YK (1998) Periodic solutions of strongly non-linear oscillations by the elliptic perturbation method. J Sound Vib 212(5):771–780
10. Hale JK, Luncl SV (1993) Introduction to functional differential equations. Springer-Verlag, New York
11. Diekmann OD et al (1995) Delay equations. Springer-Verlag, New York
12. Wu J, Faria T, Huang YS (1999) Synchronization and stable phase-locking in a network of neurons with memory. Math Comput Modell 30(1):117–138
13. Yu P, Yuan Y, Xu J (2002) Study of double hopf bifurcation and chaos for oscillator with time delay feedback. Commun Nonlinear Sci Numer Simul 7:69–91
14. Campbell SA, Bélair J, Ohira T, Milton J (1995) Limit cycles, tori, and complex dynamics in a second-order differential equations with delayed negative feedback. J Dynam Different Equat 7:213–236
15. Reddy DVR, Sen A, Johnston GL (2000) Dynamics of a limit cycle oscillator under time delayed linear and nonlinear feedbacks. Phys D 144:335–357

16. Buono PL, Bélair J (2003) Restrictions and unfolding of double hopf bifurcation in functional differential equations. J Differential Equations 189:234–266
17. Faria T, Magalh LT (1995) Realisation of ordinary differential equations by retarded functional differential equations in neighborhoods of equilibrium points. Proc Roy Soc Edinburgh Sect A 125:759–776
18. Xu J, Chung KW, Chan CL (2007) An efficient method for studying weak resonant double hopf bifurcation in nonlinear systems with delayed feedback. SIAM J Appl Dynam Syst 6(1):29–60
19. Xu J, Chung KW, Zhao YY (2009) Double hopf bifurcation with strong resonances in delayed systems with nonlinerities. Math Probl Eng vol, Art no. 759363
20. Schmidt G, Tondl A (1986) Nonlinear vibrations. Cambridge University Press, Cambridge
21. Chen YL, Xu J (2013) Applications of the integral equation method to delay differential equations. Nonlinear Dyn 73(4):2241–2260
22. Zhao YY, Xu J (2008) Mechanism analysis of delayed non-linear vibration absorber. Chinese J f Theoret Appl Mech 40(1):98–106

Chapter 4
Inverse Problem of Systems with Time Delay

As mentioned in the preface, time delay identification in the time domain [1–5] may be unstable when the initial estimation is improperly set. In contrast, the representation of time delay in the frequency domain is algebraic and periodic, indicating that one may use the frequency domain form to construct time delay identification. Besides, conventional filters [6–9] cannot distinguish coupled noise from nonlinear response components, *e.g.*, the so-called "streaming" [10], indicating that newly developed signal processing strategies should be applied.

Focusing on the above two challenging issues, the organization of this section is arranged as follows. Section 4.1 first illustrates the periodicity of time delay in the frequency domain. Sections 4.2 and 4.3 provide some preliminary principles of using such periodicity to identify time delay. At last, Sect. 4.5. introduces the digital-flow technique to realize accurate time delay tuning and verifies the proposed identification methods by some experiments.

4.1 Characteristics of Time Delay in the Frequency Domain

For the convenience of understanding how to identify the time delay, this section first gives a systematic insight into its characteristics. Without loss of generality, a dimensionless linear dynamic system with single time delay can be given as

$$\ddot{x}(t) + 0.1\dot{x}(t) + x(t) + 0.2x(t-\tau) = f(t), \tag{4.1}$$

where τ represents the time delay and $f(t)$ is the external excitation. Considering that solving this equation in the time domain is a process of functional analysis, which is complex and frustrating, it would be simpler to inspect the delay-induced characteristics from the frequency domain.

J. Xu, *Nonlinear Dynamics of Time Delay Systems*,
https://doi.org/10.1007/978-981-99-9907-1_4

When the initial state is negligible, it is not difficult to apply the integration-by-parts to derive that

$$\mathscr{F}[\dot{x}(t)] = \int_{-\infty}^{\infty} \dot{x}(t)e^{-j\omega t}\mathrm{d}t = \int_{-\infty}^{\infty} e^{-j\omega t}\mathrm{d}x(t) = x(t)e^{-j\omega t}\Big|_{-\infty}^{\infty} - \int_{-\infty}^{\infty} x(t)\mathrm{d}e^{-j\omega t}$$
$$= j\omega \int_{-\infty}^{\infty} x(t)e^{-j\omega t}\mathrm{d}t = j\omega\mathscr{F}[x(t)],$$

$$\mathscr{F}[\ddot{x}(t)] = j\omega\mathscr{F}[\dot{x}(t)] = -\omega^2\mathscr{F}[x(t)],$$

and

$$\mathscr{F}[x(t-\tau)] = \int_{-\infty}^{\infty} x(t-\tau)e^{-j\omega t}\mathrm{d}t = e^{-j\omega\tau}\int_{-\infty}^{\infty} x(t-\tau)e^{-j\omega(t-\tau)}\mathrm{d}(t-\tau)$$
$$= e^{-j\omega\tau}\mathscr{F}[x(t)],$$

where $\mathscr{F}[\cdot]$ represents Fourier transform, j is the imaginary number, and ω denotes the circular frequency. For the convenience of reading, this section uses lower-case letters as state variables in the time domain, say x or $x(t)$, and corresponding capital letters as state variables in the frequency domain, say X or $X(j\omega)$.

By applying the Fourier transform to Eq. (4.1) and considering the above properties, one obtains

$$H(j\omega) = \frac{X(j\omega)}{F(j\omega)} = \frac{1}{-\omega^2 + 0.1j\omega + 1 + 0.2e^{-j\omega\tau}}, \tag{4.2}$$

where $H(j\omega)$ is called the frequency response function, $X(j\omega)$ and $F(j\omega)$ are spectrums of $x(t)$ and $f(t)$ through Fourier transform.

Figure 4.1 shows the amplitudes of $H(j\omega)$ to different time delays, where one can find that (a) the resonant frequency varies with the time delay τ, (b) extra resonant frequencies emerge when the time delay τ is large, and (c) the emerged resonant frequencies tend to distribute uniformly when the time delay τ is sufficiently large. These features inspire the idea that one may identify time delay and other system parameters from the frequency domain. Specifically, when the time delay is sufficiently large, the interval between the emerged resonant frequencies (see Fig. 4.1c) is approximately $2\pi/\tau$.

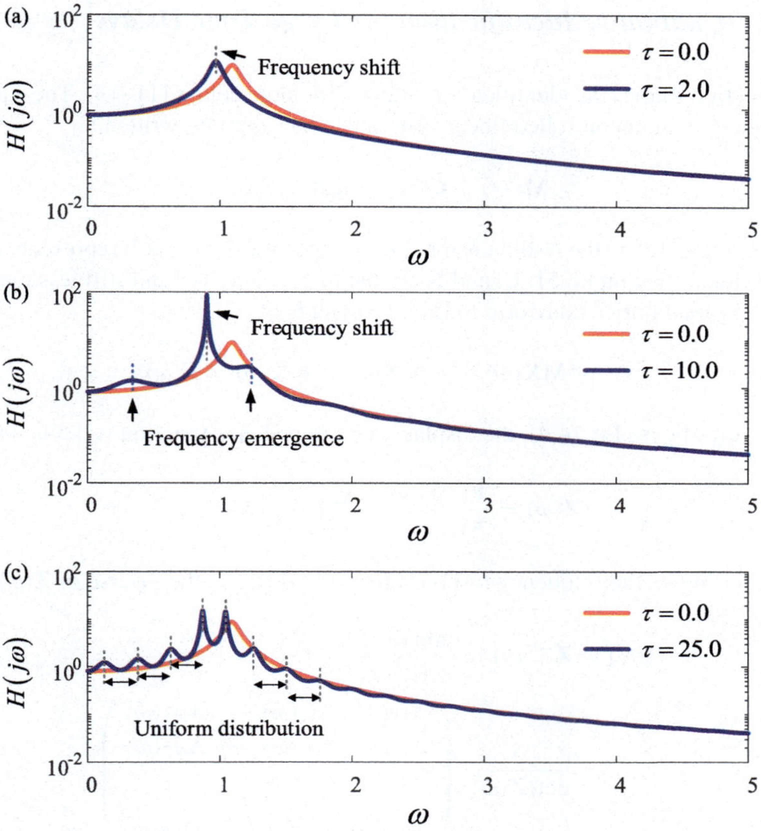

Fig. 4.1 Frequency response characteristics to different time delays: **a** $\tau = 2.0$, **b** $\tau = 10.0$, and **c** $\tau = 25.0$

4.2 Parameter Identification for Linear Time Delay Systems

Section 4.1 verified that if a system has time delay, one can identify its parameters from the frequency domain. Figure 4.1 clearly demonstrates that the frequency response with a large time delay is qualitatively distinct from that with a small delay. Therefore, the algorithm construction for the delay-related identification should be designed by considering the delay magnitude.

4.2.1 Qualitative Identification for Large Time Delays

This section studies the identification strategy for large delays [11–13]. The equation of motion of an uncontrolled linear system can be generally written as

$$\mathbf{M}\ddot{\mathbf{x}}(t) + \mathbf{C}\dot{\mathbf{x}}(t) + \mathbf{K}\mathbf{x}(t) = \mathbf{f}(t), \tag{4.3}$$

where $\mathbf{x}(t) \in \mathbb{R}^N$ is the N-dimensional state response, $\mathbf{f}(t) \in \mathbb{R}^N$ represents the N-dimensional state input, $\mathbf{M}$, $\mathbf{C}$, and $\mathbf{K}$ are the mass, damping, and stiffness matrices. Applying the Fourier transform to Eq. (4.3) yields

$$-\omega^2\mathbf{M}\mathbf{X}(\omega) + j\omega\mathbf{C}\mathbf{X}(\omega) + \mathbf{K}\mathbf{X}(\omega) = \mathbf{F}(\omega). \tag{4.4}$$

By referring to Eq. (4.4), the displacement impedance function is defined as

$$\mathbf{Z}(\omega) = \frac{\mathbf{F}(\omega)}{\mathbf{X}(\omega)} = -\omega^2\mathbf{M} + j\omega\mathbf{C} + \mathbf{K}. \tag{4.5}$$

Accordingly, the frequency response function (FRF) is the inverse of $\mathbf{Z}(\omega)$, i.e.,

$$\begin{aligned}\mathbf{H}(\omega) = \mathbf{Z}^{-1}(\omega) &= \frac{\mathrm{adj}(\mathbf{Z}(\omega))}{\det(\mathbf{Z}(\omega))}\\ &= \frac{1}{\det(\mathbf{Z}(\omega))}\begin{pmatrix} A_{11}(\omega) & A_{21}(\omega) & \cdots & A_{N1}(\omega)\\ A_{12}(\omega) & A_{22}(\omega) & \cdots & A_{N2}(\omega)\\ \vdots & \vdots & \ddots & \vdots\\ A_{1N}(\omega) & A_{2N}(\omega) & \cdots & A_{NN}(\omega)\end{pmatrix},\end{aligned} \tag{4.6}$$

where $\mathrm{adj}(\mathbf{Z}(\omega))$ and $\det(\mathbf{Z}(\omega))$ represent the adjoint matrix and determinant of $\mathbf{Z}(\omega)$, $A_{pq}(\omega)$ is the algebra cofactor of $\mathbf{Z}(\omega)$. Specifically, both $\det(\mathbf{Z}(\omega))$ and $A_{pq}(\omega)$ are polynomial functions of ω. For the convenience of writing in the following derivations, $\det(\mathbf{Z}(\omega))$ is rewritten as $\Phi(\omega)$. Hence, any element, say $H_{pq}(\omega)$, of $\mathbf{H}(\omega)$ can be expressed as

$$H_{pq}(\omega) = \frac{A_{qp}(\omega)}{\det(\mathbf{Z}(\omega))} = \frac{A_{qp}(\omega)}{\Phi(\omega)},\ p, q = 1, 2, \ldots, N. \tag{4.7}$$

In the case of displacement time-delay feedback, the equation of motion can be given as

$$\mathbf{M}\ddot{\mathbf{x}}(t) + \mathbf{C}\dot{\mathbf{x}}(t) + \mathbf{K}\mathbf{x}(t) = \mathbf{f}(t) + \mathbf{G}\mathbf{x}(t-\tau), \tag{4.8}$$

where $\mathbf{G}$ denotes the displacement feedback gain. Applying the Fourier transform to Eq. (4.8) yields

$$-\omega^2\mathbf{MX}(\omega)+j\omega\mathbf{CX}(\omega)+\mathbf{KX}(\omega)=\mathbf{F}(\omega)+e^{-j\omega\tau}\mathbf{GX}(\omega). \quad (4.9)$$

The corresponding displacement impedance function matrix can be expressed as

$$\mathbf{Z}^D(\omega,\tau)=-\omega^2\mathbf{M}+j\omega\mathbf{C}+\mathbf{K}-\mathbf{G}e^{-j\omega\tau}. \quad (4.10)$$

Set point r as the reference point and all the measuring points as the feedback points. The feedback gain matrix $\mathbf{G}$ has only one column of nonzero elements:

$$\mathbf{G}=\begin{pmatrix} 0 & \cdots & 0 & g_{1r} & 0 & \cdots & 0 \\ 0 & \cdots & 0 & g_{2r} & 0 & \cdots & 0 \\ \vdots & \ddots & \vdots & \vdots & \vdots & \ddots & \vdots \\ 0 & \cdots & 0 & g_{Nr} & 0 & \cdots & 0 \end{pmatrix}. \quad (4.11)$$

Then $\mathbf{Z}^{\mathrm{D}}(\omega,\tau)$ can be expressed in detail as

$$\mathbf{Z}^{\mathrm{D}}(\omega,\tau)=$$

$$\begin{pmatrix} \begin{matrix} k_{11}-\omega^2 m_1 \\ +j\omega c_{11} \end{matrix} & k_{12}+j\omega c_{12} & \cdots & \begin{matrix} k_{1r}+j\omega c_{1r} \\ -g_{1r}e^{-j\omega\tau} \end{matrix} & \cdots & k_{1N}+j\omega c_{1N} \\ k_{21}+j\omega c_{21} & \begin{matrix} k_{22}-\omega^2 m_2 \\ +j\omega c_{22} \end{matrix} & \cdots & \begin{matrix} k_{21}+j\omega c_{2r} \\ -g_{2r}e^{-j\omega\tau} \end{matrix} & \cdots & k_{2N}+j\omega c_{2N} \\ \vdots & \vdots & \ddots & \vdots & \vdots & \vdots \\ k_{r1}+j\omega c_{r1} & k_{r2}+j\omega c_{r1} & \cdots & \begin{matrix} k_{rr}-\omega^2 m_r+j\omega c_{rr} \\ -g_{rr}e^{-j\omega\tau} \end{matrix} & \cdots & k_{rN}+j\omega c_{rN} \\ \vdots & \vdots & \vdots & \vdots & \ddots & \vdots \\ k_{N1}+j\omega c_{N1} & k_{N1}+j\omega c_{N2} & \cdots & \begin{matrix} k_{Nr}+j\omega c_{Nr} \\ -g_{Nr}e^{-j\omega\tau} \end{matrix} & \cdots & \begin{matrix} k_{NN}-\omega^2 m_N \\ +j\omega c_{NN} \end{matrix} \end{pmatrix}. \quad (4.12)$$

The displacement FRF matrix $\boldsymbol{H}^{\mathrm{D}}(\omega,\tau)$ of the delayed system can be written as

$$\mathbf{H}^D(\omega,\tau)=\frac{1}{\det(\mathbf{Z}^D(\omega,\tau))}\begin{pmatrix} A^D_{11}(\omega,\tau) & A^D_{21}(\omega,\tau) & \cdots & A^D_{r1}(\omega,\tau) & \cdots & A^D_{N1}(\omega,\tau) \\ A^D_{12}(\omega,\tau) & A^D_{22}(\omega,\tau) & \cdots & A^D_{r2}(\omega,\tau) & \cdots & A^D_{N2}(\omega,\tau) \\ \vdots & \vdots & \ddots & \vdots & \vdots & \vdots \\ A^D_{1r}(\omega) & A^D_{2r}(\omega) & \cdots & A^D_{rr}(\omega) & \cdots & A^D_{Nr}(\omega) \\ \vdots & \vdots & \vdots & \vdots & \ddots & \vdots \\ A^D_{1N}(\omega,\tau) & A^D_{2N}(\omega,\tau) & \cdots & A^D_{rN}(\omega,\tau) & \cdots & A^D_{NN}(\omega,\tau) \end{pmatrix}, \quad (4.13)$$

where $\det\left(\mathbf{Z}^D(\omega,\tau)\right)$ represents the determinant of $\mathbf{Z}^D(\omega,\tau)$ and $A_{pq}^D(\omega,\tau)$ is the algebraic cofactor of $\mathbf{Z}^D(\omega,\tau)$.

As can be seen in Eq. (4.13), all the algebraic cofactors of $\mathbf{Z}^D(\omega,\tau)$ are functions of ω. It should be noted that $A_{1r}^D(\omega)$, $A_{2r}^D(\omega)$, ..., $A_{Nr}^D(\omega)$ do not contain $e^{-j\omega\tau}$. This means elements in the rth row of $\mathbf{H}^D(\omega,\tau)$ do not include $e^{-j\omega\tau}$ in their numerators. In other words, numerators of the elements in the rth row of $\mathbf{H}^D(\omega,\tau)$ are the same as those in the rth row of $\mathbf{H}(\omega)$. The expressions can be written as follows:

$$H_{r\beta}^D(\omega) = \frac{A_{\beta r}^D(\omega)}{\det\left(\mathbf{Z}^D(\omega,\tau)\right)} = \frac{A_{\beta r}(\omega)}{\det\left(\mathbf{Z}^D(\omega,\tau)\right)},\ \beta = 1, 2, \ldots, N. \tag{4.14}$$

According to Eqs. (4.5) and (4.10), we can obtain the relationship between $\det\left(\mathbf{Z}^{\mathrm{D}}(\omega,\tau)\right)$ and $\det(\mathbf{Z}(\omega))$:

$$\det\left(\mathbf{Z}^{\mathrm{D}}(\omega,\tau)\right) = \det(\mathbf{Z}(\omega)) - \sum_{s=1}^{N} g_{sr} A_{sr}(\omega) e^{-j\omega\tau}. \tag{4.15}$$

Since $\det(\mathbf{Z}(\omega))$ has been denoted as $\Phi(\omega)$, Eq. (4.15) can be rewritten as

$$\det\left(\mathbf{Z}^{\mathrm{D}}(\omega,\tau)\right) = \Phi(\omega) - \sum_{s=1}^{N} g_{sr} A_{sr}(\omega) e^{-j\omega\tau}. \tag{4.16}$$

Assume that point i is the input point. Then, elements in the ith column of displacement FRFs measured in the experiment can be expressed as

$$\left\{H_{1i}^{\mathrm{D}}(\omega,\tau), H_{2i}^{\mathrm{D}}(\omega,\tau), \ldots H_{ri}^{\mathrm{D}}(\omega,\tau), \ldots, H_{Ni}^{\mathrm{D}}(\omega,\tau)\right\}^{\mathrm{T}}. \tag{4.17}$$

According to Eq. (4.13), the expression in Eq. (4.17) can be rewritten as

$$\frac{\left\{A_{i1}^{\mathrm{D}}(\omega,\tau), A_{i2}^{\mathrm{D}}(\omega,\tau), \ldots, A_{ir}^{\mathrm{D}}(\omega,\tau), \ldots, A_{iN}^{\mathrm{D}}(\omega,\tau)\right\}^{\mathrm{T}}}{\det\left(\mathbf{Z}^{\mathrm{D}}(\omega,\tau)\right)}. \tag{4.18}$$

Equation (4.13) shows that all FRFs have the same denominator, which inevitably contains a time delay τ. However, if the denominator and numerator of FRFs contain τ, it isn't easy to extract τ. To construct an algorithm to identify τ, an FRF whose numerator does not contain τ must be found.

According to Eqs. (4.14) and (4.18), $H_{ri}^{\mathrm{D}}(\omega,\tau)$ can be measured in experiments. Moreover, it does not contain τ in its numerator. Therefore, $H_{ri}^{\mathrm{D}}(\omega,\tau)$ is the right function to use to identify τ:

$$H_{ri}^{\mathrm{D}}(\omega,\tau) = \frac{A_{ir}^{\mathrm{D}}(\omega)}{\det\left(\mathbf{Z}^{\mathrm{D}}(\omega,\tau)\right)} = \frac{A_{ir}(\omega)}{\det\left(\mathbf{Z}^{\mathrm{D}}(\omega,\tau)\right)}. \tag{4.19}$$

Substituting Eq. (4.16) into Eq. (4.19), we obtain

$$H_{ri}^{\mathrm{D}}(\omega,\tau) = \frac{A_{ir}(\omega)}{\Phi(\omega) - \Sigma_{s=1}^{N} g_{sr} A_{sr}(\omega) e^{-j\omega\tau}}. \tag{4.20}$$

We define the pseudo-displacement impedance functions at the reference point r of the uncontrolled and delayed system as

$$\begin{aligned}\tilde{Z}_{ri}(\omega) &= \frac{1}{H_{ri}(\omega)},\\ \tilde{Z}_{ri}^{\mathrm{D}}(\omega,\tau) &= \frac{1}{H_{ri}^{\mathrm{D}}(\omega,\tau)}.\end{aligned} \tag{4.21}$$

Obviously, the pseudo-displacement impedance function is the reciprocal of the FRF. It should be noted that the pseudo-displacement impedance functions $\tilde{Z}_{ri}(\omega)$ and $\tilde{Z}_{ri}^{\mathrm{D}}(\omega,\tau)$ are functions, while the displacement impedance function matrices $\mathbf{Z}(\omega)$ and $\mathbf{Z}^{\mathrm{D}}(\omega)$ are function matrices. Use " ~ " to show the difference between pseudo-displacement impedance functions and displacement impedance functions.

Substituting Eqs. (4.7) and (4.20) into Eq. (4.21), we obtain

$$\tilde{Z}_{ri}(\omega) = \frac{\Phi(\omega)}{A_{ir}(\omega)} \tag{4.22}$$

and

$$\tilde{Z}_{ri}^{\mathrm{D}}(\omega,\tau) = \frac{\Phi(\omega)}{A_{ir}(\omega)} - \frac{\Sigma_{s=1}^{N} g_{sr} e^{-j\omega\tau} A_{sr}(\omega)}{A_{ir}(\omega)}, \tag{4.23}$$

It can be seen that $\tilde{Z}_{ri}^{\mathrm{D}}(\omega,\tau)$ consists of two parts. One is $\tilde{Z}_{ri}(\omega)$, the pseudo-displacement impedance function at point i of the uncontrolled system. The other part is an amplitude-changing periodic function.

The difference between the two previous pseudo-displacement impedance functions can be expressed as

$$\tilde{Z}_{ri}^{\mathrm{D}}(\omega,\tau) - \tilde{Z}_{ri}(\omega) = -\frac{\Sigma_{s=1}^{N} g_{sr} A_{sr}(\omega)}{A_{ir}(\omega)} e^{-j\omega\tau}. \tag{4.24}$$

The amplitude of the difference function, namely, $\Sigma_{s=1}^{N} g_{sr} A_{sr}(\omega) / A_{ir}(\omega)$, is related to all the parameters except the time delay τ, while the "frequency" of the difference function is τ.

Therefore, the real part of the pseudo-displacement impedance function of a delayed system can be treated as a superposition of the real part of the pseudo-displacement impedance function of an uncontrolled system and a cosine curve, while the imaginary part of the pseudo-displacement impedance function of the delayed system can be treated as a superposition of the imaginary part of the pseudo-displacement impedance function of the uncontrolled system and a sine curve. This means that if the difference between the real parts of the two pseudo-displacement impedance functions is obtained, τ can be identified. Similarly, τ can also be identified from the difference between the imaginary parts of the two pseudo-displacement impedance functions.

If the delayed system has only one feedback point f, Eq. (4.24) can be simplified to

$$\tilde{Z}_{ri}^{\mathrm{D}}(\omega,\tau) - \tilde{Z}_{ri}(\omega) = -\frac{g_{fr}A_{fr}(\omega)}{A_{ir}(\omega)}e^{-j\omega\tau}. \tag{4.25}$$

Note that $A_{fr}(\omega)/A_{ir}(\omega)$ in Eq. (4.25) is a rational function about ω, indicating that it has limited fluctuations in the frequency domain and thus is a slowly varying process. In contrast, the fluctuation induced by time delay is more intensive, especially when the delay is significant. In other words, the fluctuations induced by the rational cofactors and time delay have distinct "frequencies" in the frequency domain. Therefore, one may apply the Fourier transform to Eq. (4.25) to detect the delay.

Example 4.1 This section first introduces a single-degree-of-freedom linear system, say Eq. (4.1), to illustrate the qualitative identification. In such cases, the pseudo-impedance functions of the systems without and with time delay are identical to their true impedance functions, i.e.,

$$\tilde{Z}_{11}(\omega) = \frac{\Phi(\omega)}{A_{11}(\omega)} = -\omega^2 + 0.1j\omega + 1 = Z_{11}(\omega) \tag{4.26}$$

and

$$\tilde{Z}_{11}^{\mathrm{D}}(\omega,\tau) = -\omega^2 + 0.1j\omega + 1 + 0.2e^{-j\omega\tau} = Z_{11}^{\mathrm{D}}(\omega,\tau). \tag{4.27}$$

Substituting Eqs. (4.26) and (4.27) into Eq. (4.25) yields

$$\tilde{Z}_{11}^{\mathrm{D}}(\omega,\tau) - \tilde{Z}_{11}(\omega) = 0.2e^{-j\omega\tau}. \tag{4.28}$$

The real part of Eq. (4.28) is

$$\Re\left(\tilde{Z}_{11}^{\mathrm{D}}(\omega,\tau) - \tilde{Z}_{11}(\omega)\right) = 0.2\cos(\omega\tau), \tag{4.29}$$

which is a harmonic function. Apparently, one can identify the time delay τ by applying the Fourier transform to Eq. (4.29).

Example 4.2 Suppose that there is a multiple-degree-of-freedom linear system as Fig. 4.2 shows. The masses, damping coefficients, and stiffnesses of all elements are identical. Let $m = 1$, $c = 0.1$, and $k = 10$. In addition, the reference point is four, and the feedback points are 1 and 3. The feedback gains are 5 and -4, respectively.

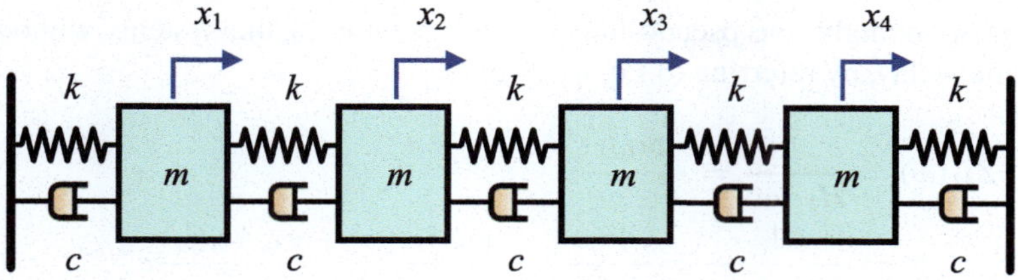

Fig. 4.2 Schematic view of a multiple-degree-of-freedom linear system

By referring to Eq. (4.3), the mass, damping, and stiffness matrices are respectively given as

$$\mathbf{M} = \begin{pmatrix} 1 & 0 & 0 & 0 \\ 0 & 1 & 0 & 0 \\ 0 & 0 & 1 & 0 \\ 0 & 0 & 0 & 1 \end{pmatrix}, \tag{4.30}$$

$$\mathbf{C} = \begin{pmatrix} 0.2 & -0.1 & 0 & 0 \\ -0.1 & 0.2 & -0.1 & 0 \\ 0 & -0.1 & 0.2 & -0.1 \\ 0 & 0 & -0.1 & 0.2 \end{pmatrix}, \tag{4.31}$$

and

$$\mathbf{K} = \begin{pmatrix} 20 & -10 & 0 & 0 \\ -10 & 20 & -10 & 0 \\ 0 & -10 & 20 & -10 \\ 0 & 0 & -10 & 20 \end{pmatrix}. \tag{4.32}$$

By referring to Eq. (4.8), the feedback gain matrix is

$$\mathbf{G} = \begin{pmatrix} 0 & 0 & 0 & 5 \\ 0 & 0 & 0 & 0 \\ 0 & 0 & 0 & -4 \\ 0 & 0 & 0 & 0 \end{pmatrix}. \tag{4.33}$$

According to Eq. (4.20), if one wants to identify the time delay τ, one should choose transfer functions such as $H_{4i}^{\mathrm{D}}(\omega, \tau)$, where $i = 1$, 2, 3, or 4. Without loss of generality, one may choose $i = 1$. Hence, the detailed form of $H_{41}^{\mathrm{D}}(\omega, \tau)$ can be specified as

$$H_{41}^{\mathrm{D}}(\omega, \tau) = \frac{A_{14}(\omega)}{\Phi(\omega) - (g_{14}A_{14}(\omega) + g_{34}A_{34}(\omega))e^{-j\omega\tau}}. \tag{4.34}$$

Correspondingly, the pseudo-impedance functions of the systems without and with time delay, by referring to Eq. (4.21), are

$$\tilde{Z}_{41}(\omega) = \frac{1}{H_{41}(\omega)} = \frac{\Phi(\omega)}{A_{14}(\omega)},$$
$$\tilde{Z}_{41}^{\mathrm{D}}(\omega,\tau) = \frac{1}{H_{41}^{\mathrm{D}}(\omega,\tau)} = \frac{\Phi(\omega)}{A_{14}(\omega)} - \frac{g_{14}A_{14}(\omega) + g_{34}A_{34}(\omega)}{A_{14}(\omega)}e^{-j\omega\tau}. \quad (4.35)$$

Substituting Eq. (4.35) into Eq. (4.24) yields

$$\tilde{Z}_{41}^{\mathrm{D}}(\omega,\tau) - \tilde{Z}_{41}(\omega) = -\frac{g_{14}A_{14}(\omega) + g_{34}A_{34}(\omega)}{A_{14}(\omega)}e^{-j\omega\tau}. \quad (4.36)$$

Note that Eq. (4.33) indicates $g_{14} = 5$ and $g_{34} = -4$. Besides, Eqs. (4.30), (4.31), and (4.32) imply that

$$A_{14}(\omega) = (0.1j\omega + 10)^3 \quad (4.37)$$

and

$$A_{34}(\omega) = 0.1j\omega^5 + 10.04\omega^4 - 8.003j\omega^3 - 400.9\omega^2 + 90j\omega + 3000. \quad (4.38)$$

As can be seen, both $A_{14}(\omega)$ and $A_{34}(\omega)$ are low-order polynomials, indicating that $(g_{14}A_{14}(\omega) + g_{34}A_{34}(\omega))/A_{14}(\omega)$ is also a low-order rational term.

By setting different time delays, the transfer functions $H_{41}(\omega)$ and $H_{41}^{\mathrm{D}}(\omega,\tau)$ are shown in Fig. 4.3. One can find that, similar to the feature shown in Fig. 4.1, the density of delay-induced fluctuations gets thicker when the time delay increases. Such consistency verifies that the fluctuation variation is an intrinsic characteristic to the time-delayed systems, whether the system is s-DOF or m-DOF. However, it is difficult to directly estimate the delay from transfer function fluctuations because the original transfer curves distort them. To this end, one should apply Eq. (4.36) further to weaken the influence of the original transfer feature.

By substituting $H_{41}(\omega)$ and $H_{41}^{\mathrm{D}}(\omega,\tau)$ into Eq. (4.35) to get $\tilde{Z}_{41}(\omega)$ and $\tilde{Z}_{41}^{\mathrm{D}}(\omega,\tau)$, and then substituting them into Eq. (4.36), one can finally obtain the pseudo-impedance deviation. Furthermore, one can qualitatively estimate the time delay by applying the Fourier transform to the real part of the obtained impedance deviation. As can be seen from Fig. 4.4, the spectrum peaks approximately indicate the value of time delay. More specifically, the estimation accuracy increases with the growth of time delay. Such a feature proves that the delay can be detected and estimated from the frequency domain.

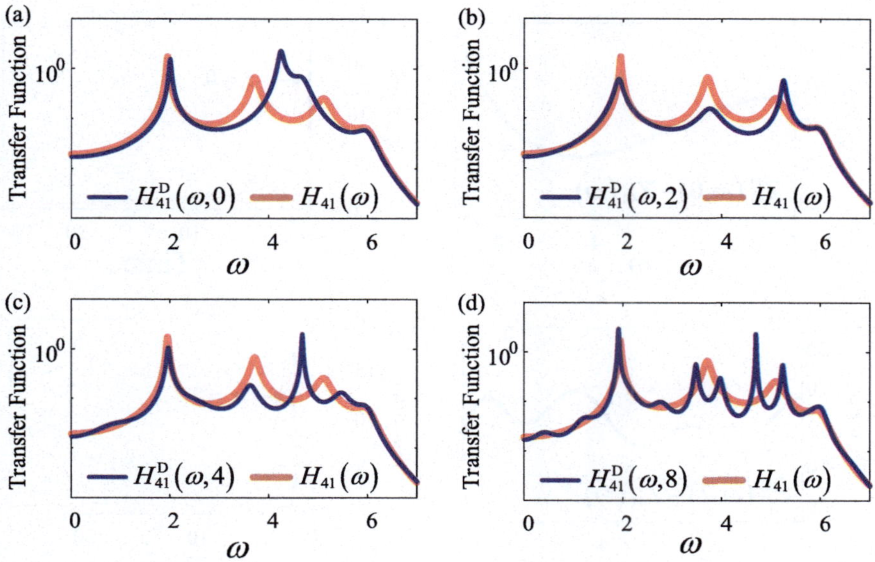

Fig. 4.3 Transfer functions with and without time delays: **a** $\tau = 0$, **b** $\tau = 2$, **c** $\tau = 4$, and **d** $\tau = 8$

4.2.2 Quantitative Identification for Practical Applications

Section 4.2.1 illustrates a qualitative strategy for delay identification. This strategy is effective when the time delay is significant. However, it cannot help us identify small time delays or other parameters, such as feedback gain g. To fix this problem, a refined approach should be proposed.

Compared with Eq. (4.8), a more general form of the multiple time-delay coupled linear system [14] can be represented as

$$\mathbf{M}\ddot{\mathbf{x}}(t) + \mathbf{C}\dot{\mathbf{x}}(t) + \mathbf{K}\mathbf{x}(t) = \sum_{l=1}^{L} \mathbf{G}_l \mathbf{x}(t - \tau_l) + \mathbf{f}(t), \tag{4.39}$$

where τ_l is the time delay and $\mathbf{G}_l$ represents the displacement feedback gain. This section further considers that the system is partially observable and excitable. Hence, the displacement $\mathbf{x}(t)$ can be divided into two contributions as

$$\mathbf{x}(t) = \mathbf{U}_o \mathbf{x}_o(t) + \mathbf{U}_{uo} \mathbf{x}_{uo}(t), \tag{4.40}$$

where $\mathbf{x}_o(t) \in \mathbb{R}^{N_o}$ is the observable state that $N_o \leq N$, $\mathbf{x}_{uo}(t) \in \mathbb{R}^{N-N_o}$ denotes the unobservable state, $\mathbf{U}_o \in \mathbb{R}^{N \times N_o}$ and $\mathbf{U}_{uo} \in \mathbb{R}^{N \times (N-N_o)}$ are elementary matrices that span $\mathbf{x}_o(t)$ and $\mathbf{x}_{uo}(t)$ into the full state space. For example, suppose that $\mathbf{x}(t) \in \mathbb{R}^3$ and the observable points are x_2 and x_3, then

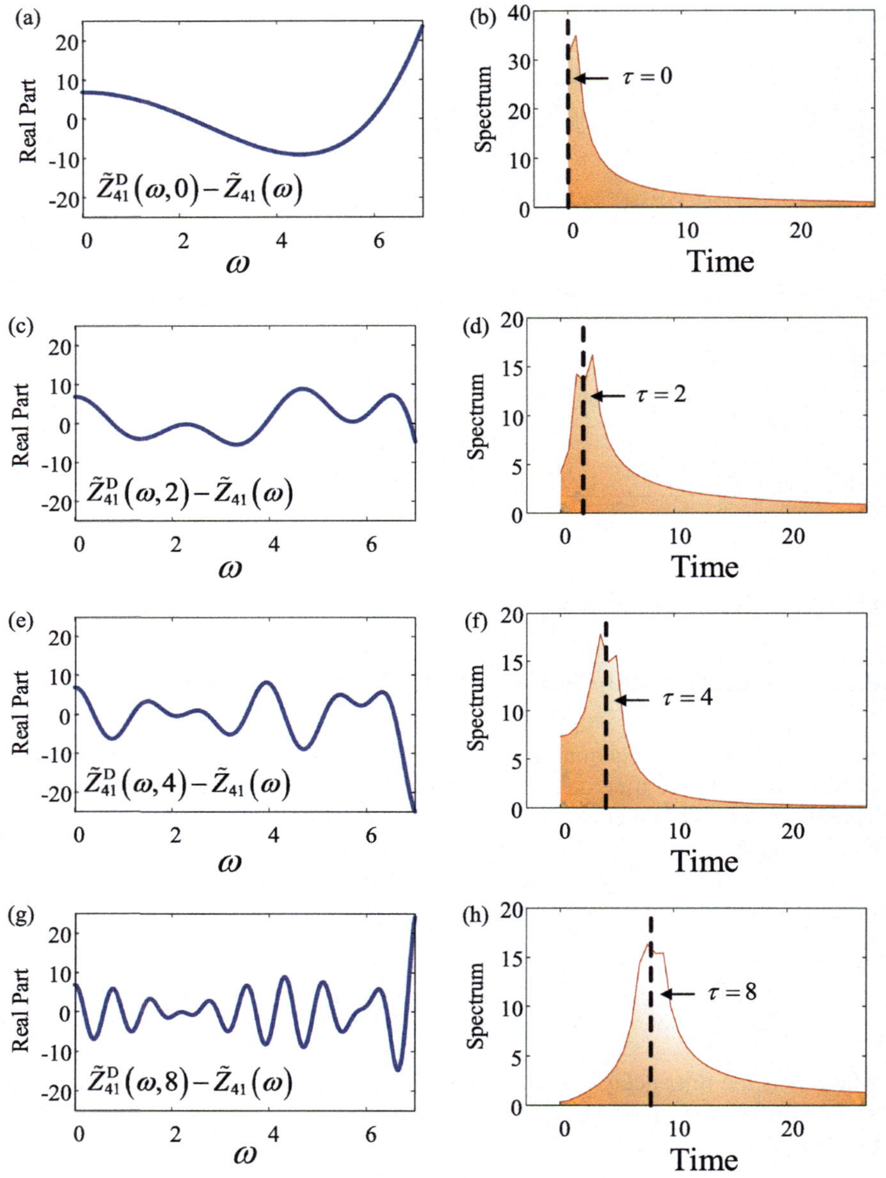

Fig. 4.4 Pseudo-impedance deviation and corresponding spectrums: **a**, **b** $\tau = 0$, **c**, **d** $\tau = 2$, **e**, **f** $\tau = 4$, and **g**, **h** $\tau = 8$

$$\mathbf{x}_o = \begin{pmatrix} x_2 \\ x_3 \end{pmatrix}, \mathbf{x}_{uo} = x_1, \mathbf{U}_o = \begin{pmatrix} 0\,0 \\ 1\,0 \\ 0\,1 \end{pmatrix}, \text{ and } \mathbf{U}_{uo} = \begin{pmatrix} 1 \\ 0 \\ 0 \end{pmatrix}.$$

Similarly, the state input can be written as

$$\mathbf{f}(t) = \mathbf{V}_e\mathbf{f}_e(t) + \mathbf{V}_{ue}\mathbf{f}_{ue}(t), \tag{4.41}$$

where $\mathbf{f}_e \in \mathbb{R}^{N_e}$ is the excitable state input that $N_e \leq N$, $\mathbf{f}_{ue} \in \mathbb{R}^{N-N_e}$ is the unexcitable state input which implies that $\mathbf{f}_{ue}(t) \equiv \mathbf{0}$ because the excitation cannot be applied onto these unexcitable points, and $\mathbf{V}_e \in \mathbb{R}^{N\times N_e}$ is the elementary matrix. Substituting Eqs. (4.40) and (4.41) into Eq. (4.39) and applying the Fourier transform, one can get

$$\begin{aligned}&\left(-\omega^2\mathbf{M} + j\omega\mathbf{C} + \mathbf{K}\right)(\mathbf{U}_o\mathbf{X}_o(\omega) + \mathbf{U}_{uo}\mathbf{X}_{uo}(\omega))\\ &= \left(\sum_{l=1}^{L} e^{-j\omega\tau_l}\mathbf{G}_l\right)\mathbf{U}_o\mathbf{X}_o(\omega) + \mathbf{V}_e\mathbf{F}_e(\omega) + \left(\sum_{l=1}^{L} e^{-j\omega\tau_l}\mathbf{G}_l\right)\mathbf{U}_{uo}\mathbf{X}_{uo}(\omega),\end{aligned} \tag{4.42}$$

where $\mathbf{X}_o(\omega)$, $\mathbf{X}_{uo}(\omega)$ and $\mathbf{F}_e(\omega)$ are the frequency forms of $\mathbf{x}_o(t)$, $\mathbf{x}_{uo}(t)$, and $\mathbf{f}_e(t)$ through Fourier transform, respectively.

Theorem 4.1 *If the coupling signal of the delayed active control is acquired from the observable state variable and is acted on the excitable nodes, time delay τ_l and delay coupling matrix $\mathbf{G}_l$ would be identifiable.*

Proof Suppose that the delay parameters are unidentifiable, which means under the same input $\mathbf{F}_e(\omega)$ and the same observation $\mathbf{X}_o(\omega)$, there are other parameters like $\tilde{\tau}_r$ and $\tilde{\mathbf{G}}_r$ satisfy Eq. (4.42), i.e.,

$$\begin{aligned}&\left(-\omega^2\mathbf{M} + j\omega\mathbf{C} + \mathbf{K}\right)(\mathbf{U}_o\mathbf{X}_o(\omega) + \mathbf{U}_{uo}\mathbf{X}_{uo}(\omega))\\ &= \left(\sum_{r=1}^{R} e^{-j\omega\tilde{\tau}_r}\tilde{\mathbf{G}}_r\right)\mathbf{U}_o\mathbf{X}_o(\omega) + \mathbf{V}_e\mathbf{F}_e(\omega) + \left(\sum_{r=1}^{R} e^{-j\omega\tilde{\tau}_r}\tilde{\mathbf{G}}_r\right)\mathbf{U}_{uo}\mathbf{X}_{uo}(\omega).\end{aligned} \tag{4.43}$$

Considering that the coupling signal of the delayed active control is acquired from the observable state variable and is acted on the excitable nodes, the delay coupling matrices $\mathbf{G}_l$ and $\tilde{\mathbf{G}}_r$ can be represented by

$$\mathbf{G}_l = \mathbf{V}_e\mathbf{\Gamma}_l\mathbf{U}_o^T \tag{4.44}$$

and

$$\tilde{\mathbf{G}}_r = \mathbf{V}_e\tilde{\mathbf{\Gamma}}_r\mathbf{U}_o^T, \tag{4.45}$$

where $\mathbf{\Gamma}_l \in \mathbb{R}^{N_e \times N_o}$ and $\tilde{\Gamma}_r \in \mathbb{R}^{N_e \times N_o}$ are compacted coupling matrices. Substituting Eqs. (4.44) and (4.45) into Eqs. (4.42) and (4.43), respectively, and taking into account that $\mathbf{U}_o^T \mathbf{U}_o = \mathbf{E}_{N_o}$ and $\mathbf{U}_o^T \mathbf{U}_{uo} = \mathbf{0}$, one can get

$$\mathbf{X}_o(\omega) - \mathbf{H}(\omega)\mathbf{F}_e(\omega) = \mathbf{H}(\omega)\left(\sum_{l=1}^{L} e^{-j\omega\tau_l}\mathbf{\Gamma}_l\right)\mathbf{X}_o(\omega) \tag{4.46}$$

and

$$\mathbf{X}_o(\omega) - \mathbf{H}(\omega)\mathbf{F}_e(\omega) = \mathbf{H}(\omega)\left(\sum_{r=1}^{R} e^{-j\omega\tilde{\tau}_r}\tilde{\Gamma}_r\right)\mathbf{X}_o(\omega), \tag{4.47}$$

where $\mathbf{H}(\omega) := \mathbf{U}_o^T\left(-\omega^2\mathbf{M} + j\omega\mathbf{C} + \mathbf{K}\right)^{-1}\mathbf{V}_e \in \mathbb{R}^{N_o \times N_e}$ is the transfer matrix between the system's observable state variable and excitable state input when the delayed control is not activated. Subtracting Eq. (4.46) by Eq. (4.47) yields

$$\mathbf{H}(\omega)\left(\sum_{l=1}^{L} e^{-j\omega\tau_l}\mathbf{\Gamma}_l - \sum_{r=1}^{R} e^{-j\omega\tilde{\tau}_r}\tilde{\Gamma}_r\right)\mathbf{X}_o(\omega) = \mathbf{0}. \tag{4.48}$$

Because $\mathbf{H}(\omega)$ and $\mathbf{X}_o(\omega)$ are not always zero when the frequency ω varies, Eq. (4.48) implies that

$$\sum_{l=1}^{L} e^{-j\omega\tau_l}\mathbf{\Gamma}_l - \sum_{r=1}^{R} e^{-j\omega\tilde{\tau}_r}\tilde{\Gamma}_r = \mathbf{0}. \tag{4.49}$$

Furthermore, filtering Eq. (4.49) on time delay τ_l yields

$$\begin{aligned}
&\lim_{n\to\infty}\left(\frac{\tau_l}{2\pi n}\int_0^{\frac{2\pi n}{\tau_l}}\left(\sum_{l=1}^{L} e^{-j\omega\tau_l}\mathbf{\Gamma}_l - \sum_{r=1}^{R} e^{-j\omega\tilde{\tau}_r}\tilde{\Gamma}_r\right)e^{j\omega\tau_l}\mathrm{d}\omega\right) \\
&= \mathbf{\Gamma}_l - \sum_{r=1}^{R}\delta(\tau_l - \tilde{\tau}_r)\tilde{\Gamma}_r = \mathbf{0},
\end{aligned} \tag{4.50}$$

where $\delta(\tau_l - \tilde{\tau}_r)$ is the Dirac function holds $\delta(\tau_l - \tilde{\tau}_r) = 1$ for $\tau_l = \tilde{\tau}_r$ and $\delta(\tau_l - \tilde{\tau}_r) = 0$ for the others. Thus, Eq. (4.50) indicates that $\tau_l = \tilde{\tau}_r$ and $\mathbf{\Gamma}_l = \tilde{\Gamma}_r$, i.e. $\mathbf{G}_l = \tilde{\mathbf{G}}_r$. It is contrary to the supposed condition. Thus, the delay parameters are identifiable. ■

The identifiability of the partially observable and excitable system is proved. The delay parameters can be identified from Eq. (4.46). In the identification, the observable state variable $\mathbf{X}_o(\omega)$ and the excitable state input $\mathbf{F}_e(\omega)$ are measured by acceleration and force sensors. The transfer matrix $\mathbf{H}(\omega)$ is calculated by transfer function estimators under the condition that the active delayed control is shut down. However, the delay parameters cannot be solved by using the inversing method directly since $\mathbf{H}(\omega)$ and $\mathbf{X}_o(\omega)$ are not square matrices. To fix this problem, the iterative procedure is adopted, i.e., the Newton–Raphson method.

Let

$$\tau_l = \tau_l^0 + \Delta\tau_l^0 \tag{4.51}$$

and

$$\mathbf{\Gamma}_l = \mathbf{\Gamma}_l^0 + \Delta\mathbf{\Gamma}_l^0, \tag{4.52}$$

where τ_l^0 and $\mathbf{\Gamma}_l^0$ are initial parameters, $\Delta\tau_l^0$ and $\Delta\mathbf{\Gamma}_l^0$ are increments. Substituting Eq. (4.14) into Eq. (4.8) and eliminating higher order terms yield the linearized increment equation, i.e.

$$\mathbf{R}^0(\omega) = \mathbf{H}(\omega)\left(\sum_{l=1}^{L}\left(e^{-j\omega\tau_l^0}\Delta\mathbf{\Gamma}_l^0 - j\omega e^{-j\omega\tau_l^0}\mathbf{\Gamma}_l^0\Delta\tau_l^0\right)\right)\mathbf{X}_o(\omega), \tag{4.53}$$

where

$$\mathbf{R}^0(\omega) := \mathbf{X}_o(\omega) - \mathbf{H}(\omega)\mathbf{F}_e(\omega) - \mathbf{H}(\omega)\left(\sum_{l=1}^{L} e^{-j\omega\tau_l^0}\mathbf{\Gamma}_l^0\right)\mathbf{X}_o(\omega) \tag{4.54}$$

is called the redundant term. Equation (4.53) is a matrix equation about $\Delta\mathbf{\Gamma}_l^0$. It must be vectorized before releasing the increments' solution. The vectorized form of Eq. (4.53) is given as

$$\mathbf{R}^0(\omega) = \mathbf{X}_o^T(\omega) \otimes \mathbf{H}(\omega)\left(\mathbf{Q}_\Gamma^0(\omega)\ \mathbf{Q}_\tau^0(\omega)\right)\begin{pmatrix}\Delta\mathbf{\Gamma}_{vec}^0 \\ \Delta\boldsymbol{\tau}^0\end{pmatrix}, \tag{4.55}$$

where

$$\mathbf{Q}_K^0(\omega) = \left(e^{-j\omega\tau_1^0}\mathbf{E}_{N_eN_o}\ e^{-j\omega\tau_2^0}\mathbf{E}_{N_eN_o}\ \cdots\ e^{-j\omega\tau_L^0}\mathbf{E}_{N_eN_o}\right), \tag{4.56}$$

$$\mathbf{Q}_\tau^0(\omega) = -j\omega\left(e^{-j\omega\tau_1^0}\mathscr{V}(\mathbf{\Gamma}_1^0)\ e^{-j\omega\tau_2^0}\mathscr{V}(\mathbf{\Gamma}_2^0)\ \cdots\ e^{-j\omega\tau_L^0}\mathscr{V}(\mathbf{\Gamma}_L^0)\right), \tag{4.57}$$

$$\Delta\mathbf{\Gamma}_{vec}^0 = \begin{pmatrix}\mathscr{V}(\Delta\mathbf{\Gamma}_1^0) \\ \mathscr{V}(\Delta\mathbf{\Gamma}_2^0) \\ \vdots \\ \mathscr{V}(\Delta\mathbf{\Gamma}_L^0)\end{pmatrix}, \text{ and } \Delta\boldsymbol{\tau}^0 = \begin{pmatrix}\Delta\tau_1^0 \\ \Delta\tau_2^0 \\ \vdots \\ \Delta\tau_L^0\end{pmatrix}.$$

In Eqs. (4.55) and (4.57), $\otimes$ means the Kronecker product of two matrices, and $\mathscr{V}(\cdot)$ means the vectorization of a matrix.

In the experiment, the acquired frequency response data is discrete. Suppose that the frequency sample is ω_k and $k = 1, 2, \ldots, K$, Eq. (4.53) could be specified into

$$\boldsymbol{\Psi}^0 = \boldsymbol{\Phi}^0 \begin{pmatrix} \Delta\boldsymbol{\Gamma}^0_{vec} \\ \Delta\boldsymbol{\tau}^0 \end{pmatrix}, \tag{4.58}$$

where

$$\boldsymbol{\Psi}^0 = \begin{pmatrix} \mathbf{R}^0(\omega_1) \\ \mathbf{R}^0(\omega_2) \\ \vdots \\ \mathbf{R}^0(\omega_K) \end{pmatrix}, \tag{4.59}$$

$$\boldsymbol{\Phi}^0 = \begin{pmatrix} \mathbf{X}_o^T(\omega_1) \otimes \mathbf{H}(\omega_1)\big(\mathbf{Q}^0_\Gamma(\omega_1)\ \mathbf{Q}^0_\tau(\omega_1)\big) \\ \mathbf{X}_o^T(\omega_2) \otimes \mathbf{H}(\omega_2)\big(\mathbf{Q}^0_\Gamma(\omega_2)\ \mathbf{Q}^0_\tau(\omega_2)\big) \\ \vdots \\ \mathbf{X}_o^T(\omega_K) \otimes \mathbf{H}(\omega_K)\big(\mathbf{Q}^0_\Gamma(\omega_K)\ \mathbf{Q}^0_\tau(\omega_K)\big) \end{pmatrix}. \tag{4.60}$$

Hence, the iteration equation can be given from Eq. (4.58) as

$$\begin{pmatrix} \boldsymbol{\Gamma}^{i+1}_{vec} \\ \boldsymbol{\tau}^{i+1} \end{pmatrix} = \begin{pmatrix} \boldsymbol{\Gamma}^{i}_{vec} \\ \boldsymbol{\tau}^{i} \end{pmatrix} + \begin{pmatrix} \Delta\boldsymbol{\Gamma}^{i}_{vec} \\ \Delta\boldsymbol{\tau}^{i} \end{pmatrix} = \begin{pmatrix} \boldsymbol{\Gamma}^{i}_{vec} \\ \boldsymbol{\tau}^{i} \end{pmatrix} + \left(\left(\boldsymbol{\Phi}^i\right)^T \boldsymbol{\Phi}^i\right)^{-1} \left(\boldsymbol{\Phi}^i\right)^T \boldsymbol{\Psi}^i, \tag{4.61}$$

where $\boldsymbol{\Phi}^i$ and $\boldsymbol{\Psi}^i$ are obtained by substituting $\boldsymbol{\Gamma}^i_{vec}$ and $\boldsymbol{\tau}^i$ into Eqs. (4.59) and (4.60).

Example 4.3 Let's continue with the system given in Example 4.2. Suppose only the state on the reference point, i.e., point 4, is observable, and the actuation points, i.e., points 1 and 3, are excitable. In this case, the observable and unobservable states and corresponding elementary matrices $\mathbf{U}_o$ and $\mathbf{U}_{uo}$, by referring to Eq. (4.40), should be

$$\mathbf{x}_o = x_4,\ \mathbf{x}_{uo} = \begin{pmatrix} x_1 \\ x_2 \\ x_3 \end{pmatrix},\ \mathbf{U}_o = \begin{pmatrix} 0 \\ 0 \\ 0 \\ 1 \end{pmatrix},\ \text{and}\ \mathbf{U}_{uo} = \begin{pmatrix} 1\ 0\ 0 \\ 0\ 1\ 0 \\ 0\ 0\ 1 \\ 0\ 0\ 0 \end{pmatrix}.$$

Furthermore, the excitable and unexcitable states and corresponding elementary matrices $\mathbf{V}_e$ and $\mathbf{V}_{ue}$ should be

$$\mathbf{f}_e = \begin{pmatrix} f_1 \\ f_3 \end{pmatrix}, \mathbf{f}_{ue} = \begin{pmatrix} f_2 \\ f_4 \end{pmatrix}, \mathbf{V}_e = \begin{pmatrix} 1\,0 \\ 0\,0 \\ 0\,1 \\ 0\,0 \end{pmatrix}, \text{ and } \mathbf{V}_{ue} = \begin{pmatrix} 0\,0 \\ 1\,0 \\ 0\,0 \\ 0\,1 \end{pmatrix}.$$

By substituting Eq. (4.33) into Eq. (4.44), one can find that the compacted coupling matrix $\mathbf{\Gamma}$ is

$$\mathbf{\Gamma} = \mathbf{V}_e^T \mathbf{G} \mathbf{U}_o = \begin{pmatrix} 1\,0\,0\,0 \\ 0\,0\,1\,0 \end{pmatrix} \begin{pmatrix} 0\,0\,0 & 5 \\ 0\,0\,0 & 0 \\ 0\,0\,0 & -4 \\ 0\,0\,0 & 0 \end{pmatrix} \begin{pmatrix} 0 \\ 0 \\ 0 \\ 1 \end{pmatrix} = \begin{pmatrix} 5 \\ -4 \end{pmatrix}. \tag{4.62}$$

The transfer matrix between the system's observable state variable and excitable state input when the delayed control is not activated should be

$$\mathbf{H}(\omega) = \mathbf{U}_o^T \left(-\omega^2 \mathbf{M} + j\omega \mathbf{C} + \mathbf{K}\right)^{-1} \mathbf{V}_e = \left(H_{41}(\omega)\ H_{43}(\omega) \right). \tag{4.63}$$

Substituting Eqs. (4.62) and (4.63) into Eq. (4.46) yields

$$\begin{aligned} &X_4(\omega) - H_{41}(\omega) F_1(\omega) - H_{43}(\omega) F_3(\omega) \\ &= (g_{14} H_{41}(\omega) + g_{34} H_{43}(\omega)) X_4(\omega) e^{-j\omega\tau}. \end{aligned} \tag{4.64}$$

Because $X_4(\omega)$, $F_1(\omega)$, and $F_3(\omega)$ are measurable variables. Meanwhile, $H_{41}(\omega)$ and $H_{43}(\omega)$ can also be conveniently obtained when delayed control is not activated. One can certainly identify the feedback gains, i.e., g_{14} and g_{34}, and the time delay τ from Eq. (4.64). Such a conclusion verifies the correctness of Theorem 4.1.

Substituting them into Eqs. (4.56) and (4.57) yields

$$\mathbf{Q}_K^0(\omega) = \begin{pmatrix} e^{-j\omega\tau^0} & 0 \\ 0 & e^{-j\omega\tau^0} \end{pmatrix} \tag{4.65}$$

and

$$\mathbf{Q}_\tau^0(\omega) = \begin{pmatrix} -j\omega g_{14}^0 e^{-j\omega\tau^0} \\ -j\omega g_{34}^0 e^{-j\omega\tau^0} \end{pmatrix}. \tag{4.66}$$

Substituting Eqs. (4.65) and (4.66) into Eq. (4.55) further yields

$$\begin{aligned} &X_4(\omega) - H_{41}(\omega) F_1(\omega) - H_{43}(\omega) F_3(\omega) - \left(g_{14}^0 H_{41}(\omega) + g_{34}^0 H_{43}(\omega)\right) X_4(\omega) e^{-j\omega\tau^0} \\ &= X_4(\omega) \otimes \left(H_{41}(\omega)\ H_{43}(\omega) \right) \begin{pmatrix} e^{-j\omega\tau^0} & 0 & -j\omega g_{14}^0 e^{-j\omega\tau^0} \\ 0 & e^{-j\omega\tau^0} & -j\omega g_{34}^0 e^{-j\omega\tau^0} \end{pmatrix} \begin{pmatrix} \Delta g_{14}^0 \\ \Delta g_{34}^0 \\ \Delta\tau^0 \end{pmatrix} \end{aligned}$$

$$= \big(X_4(\omega)H_{41}(\omega)\ X_4(\omega)H_{43}(\omega) \big) \begin{pmatrix} e^{-j\omega\tau^0} & 0 & -j\omega g_{14}^0 e^{-j\omega\tau^0} \\ 0 & e^{-j\omega\tau^0} & -j\omega g_{34}^0 e^{-j\omega\tau^0} \end{pmatrix} \begin{pmatrix} \Delta g_{14}^0 \\ \Delta g_{34}^0 \\ \Delta\tau^0 \end{pmatrix}$$

$$= \big(X_4(\omega)H_{41}(\omega)e^{-j\omega\tau^0}\ X_4(\omega)H_{43}(\omega)e^{-j\omega\tau^0}\ \phi_3^0(\omega) \big) \begin{pmatrix} \Delta g_{14}^0 \\ \Delta g_{34}^0 \\ \Delta\tau^0 \end{pmatrix}, \tag{4.67}$$

where $\phi_3^0(\omega) = -j\omega e^{-j\omega\tau^0} X_4(\omega)\big(g_{14}^0 H_{41}(\omega) + g_{34}^0 H_{43}(\omega)\big)$. By comparing it with Eq. (4.58), one can find that

$$\Psi^0 = \begin{pmatrix} X_4(\omega_1) - H_{41}(\omega_1)F_1(\omega_1) - H_{43}(\omega_1)F_3(\omega_1) \\ \quad -\big(g_{14}^0 H_{41}(\omega_1) + g_{34}^0 H_{43}(\omega_1)\big)X_4(\omega_1)e^{-j\omega_1\tau^0} \\ X_4(\omega_2) - H_{41}(\omega_2)F_1(\omega_2) - H_{43}(\omega_2)F_3(\omega_2) \\ \quad -\big(g_{14}^0 H_{41}(\omega_2) + g_{34}^0 H_{43}(\omega_2)\big)X_4(\omega_2)e^{-j\omega_2\tau^0} \\ \vdots \end{pmatrix} \tag{4.68}$$

and

$$\mathbf{\Phi}^0 = \begin{pmatrix} X_4(\omega_1)H_{41}(\omega_1)e^{-j\omega\tau^0}, X_4(\omega_1)H_{43}(\omega_1)e^{-j\omega_1\tau^0}, \phi_3^0(\omega_1) \\ X_4(\omega_2)H_{41}(\omega_2)e^{-j\omega_2\tau^0}, X_4(\omega_2)H_{43}(\omega_2)e^{-j\omega_2\tau^0}, \phi_3^0(\omega_2) \\ \vdots \end{pmatrix}. \tag{4.69}$$

Therefore, the iteration equation, by referring to Eq. (4.61), can be written as

$$\begin{pmatrix} g_{14}^{i+1} \\ g_{34}^{i+1} \\ \tau^{i+1} \end{pmatrix} = \begin{pmatrix} g_{14}^{i} \\ g_{34}^{i} \\ \tau^{i} \end{pmatrix} + \left(\left(\mathbf{\Phi}^i\right)^T \mathbf{\Phi}^i \right)^{-1} \left(\mathbf{\Phi}^i\right)^T \mathbf{\Psi}^i, \tag{4.70}$$

where $\mathbf{\Phi}^i$ and $\mathbf{\Psi}^i$ are those substituting g_{14}^0, g_{34}^0, and τ^0 by g_{14}^i, g_{34}^i, and τ^i.

Suppose that the initial estimations of g_{14} and g_{34} are 3 and -3. The initial guess of the time delay is set as $\tau^0 = 0.9\tau - 0.1$. In addition, the excitation $f_1(t)$ and $f_3(t)$ are assumed to be Dirac delta functions that make $F_1(\omega) = F_2(\omega) = 1$. The parameters identified by Eq. (4.70) are shown in Fig. 4.5, where one can find that the quantitative identification can accurately identify the time delay and corresponding feedback gains even when the delay is zero.

Example 4.4 The truss structure is usually used in an environment that requires a large span but is lightweight, *e.g.*, the solar panel skeleton or the antenna holder on the spacecraft. The payload is often mounted at the tip of the structure. To stabilize the payload, active control is introduced into the structure. Figure 4.6 shows a simple truss structure with tip mass. With this kind of mechanism, the monitoring of all nodes

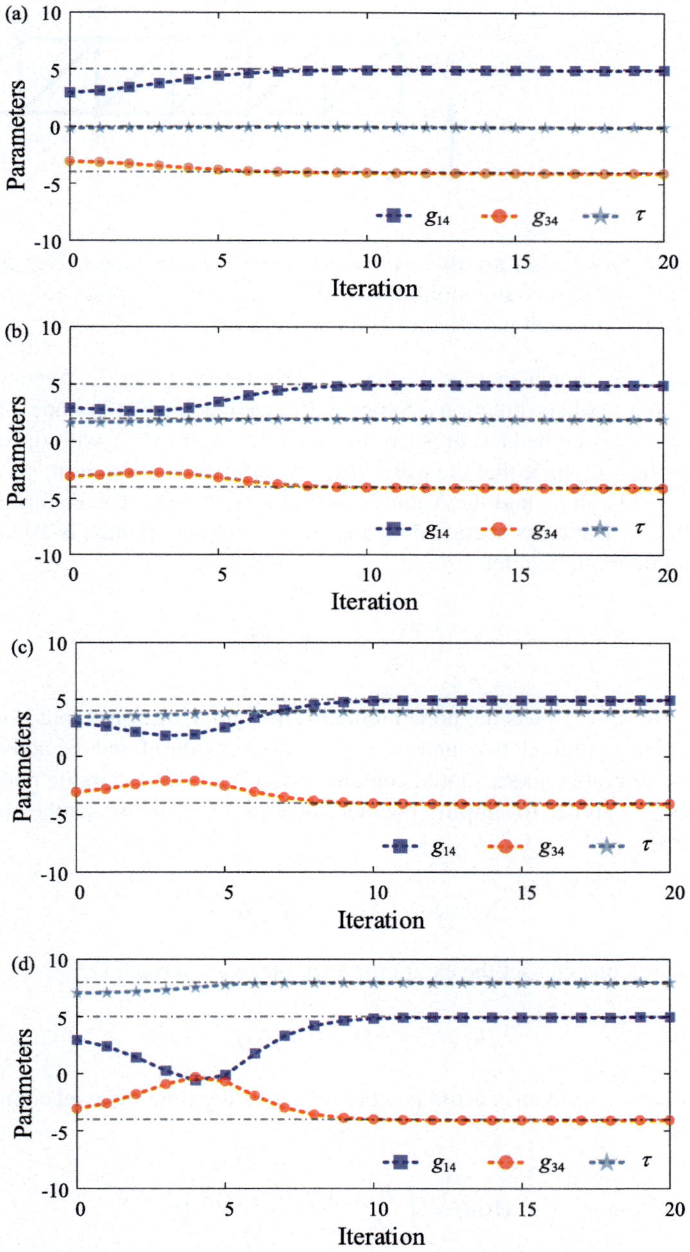

Fig. 4.5 Identification processes of g_{14}, g_{34}, and τ: **a**, **b** τ = 0, **c**, **d** τ = 2, **e**, **f** τ = 4, and **g**, **h** τ = 8

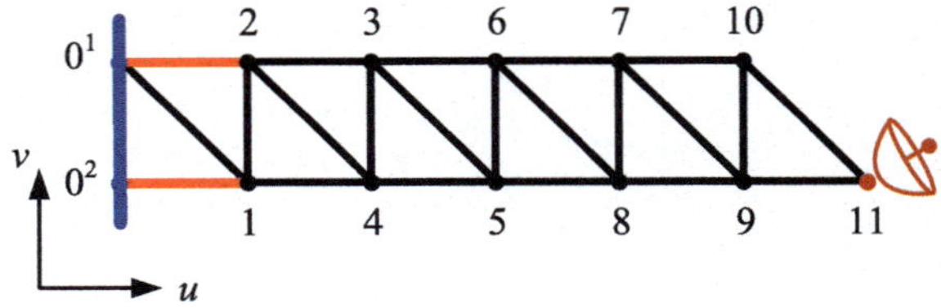

Fig. 4.6 Schematic view of a truss-structured antenna

is impossible. Otherwise, it will cost too much space or resources from the carrier. Thus, laying out sensors and monitors as few as possible is a practical problem for dynamic monitoring and parameter identification.

The structure shown in Fig. 4.6 has 11 nodes, indicating 22 freedoms in the horizontal and vertical directions. Suppose the feedback point is node 11, and the delayed control is embedded in active elements 0^1–2 and 0^2–1 with different time delays. Besides, suppose that the truss structure is made of aluminum, i.e., the mass density is 2700 kg/m^3, and the Young's modulus is 69 GPa. The distance between nodes is 0.8 m. The cross-sectional area is 5 cm^2, and the tip mass is 10 kg.

In dynamic modeling, let

$$\begin{pmatrix} x_{2i-1} \\ x_{2i} \end{pmatrix} = \begin{pmatrix} u_i \\ v_i \end{pmatrix}, \tag{4.71}$$

where $i = 1, 2, \ldots, 11$ is the node number. Theorem 4.1 implies that if one wants to identify delay parameters connecting the feedback node 11 and the active nodes 1 and 2, the observable nodes should contain node 11, and the excitable nodes should contain nodes 1 and 2. To simplify the identification, one can just set the observation variable as u_{11} and v_{11}, i.e.,

$$\mathbf{X}_o(\omega) = \left(X_{21}(\omega) \; X_{22}(\omega) \right)^T. \tag{4.72}$$

Meanwhile, one can set the excitation variable as u_1 and u_2, i.e.,

$$\mathbf{F}_e(\omega) = \left(F_1(\omega) \; F_3(\omega) \right)^T. \tag{4.73}$$

Thus, the transfer matrix estimated before activating time-delayed control can be specified as

$$\mathbf{H}(\omega) = \begin{pmatrix} H_{21,1}(\omega) & H_{21,3}(\omega) \\ H_{22,1}(\omega) & H_{22,3}(\omega) \end{pmatrix}, \tag{4.74}$$

which is much easier to achieve than the entire transfer function with 22 freedoms. By setting proper damping coefficients, the frequency response curves of Eq. (4.74) are shown in Fig. 4.7.

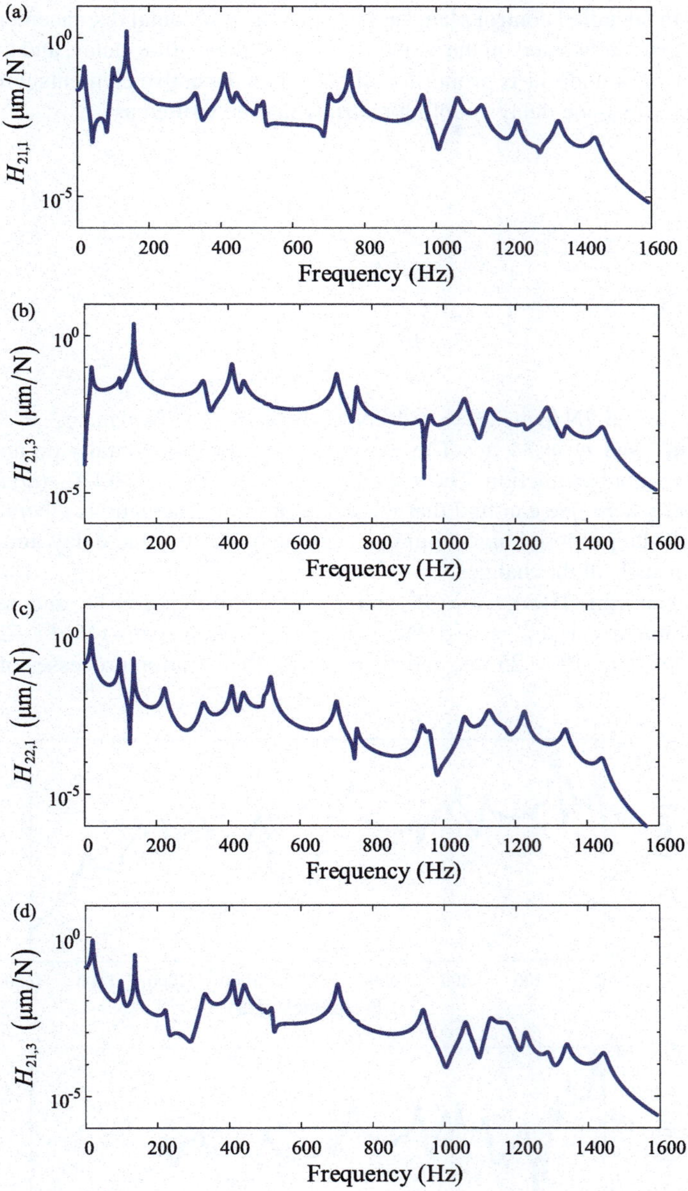

Fig. 4.7 Transfer functions before activating time-delayed control: **a** $H_{21,1}$, **b** $H_{21,3}$, **c** $H_{22,1}$, and **d** $H_{22,3}$

As for the detailed control plan, suppose that the horizontal displacement of node 11 is proportionally acted on the active elements with one time delay, and the vertical displacement of node 11 is proportionally acted on the active elements with another time delay. Thus, the delay coupling matrix could be written as

$$\boldsymbol{\Gamma}_1 = \begin{pmatrix} g_1 & 0 \\ g_2 & 0 \end{pmatrix} \tag{4.75}$$

and

$$\boldsymbol{\Gamma}_2 = \begin{pmatrix} 0 & g_3 \\ 0 & g_4 \end{pmatrix}. \tag{4.76}$$

Let $g_1 = -3.7\text{N}/\mu\text{m}$, $g_2 = -4.2\text{N}/\mu\text{m}$, $g_3 = -3.3\text{N}/\mu\text{m}$, $g_4 = 3.9\text{N}/\mu\text{m}$, $\tau_1 = 27$ ms, and $\tau_2 = 34$ ms. The structure is excited horizontally on node 1 by a Dirac delta impulse function. The response spectrums $X_{21}(\omega)$ and $X_{22}(\omega)$ are shown in Fig. 4.8, where one can find that the delay-induced fluctuations emerge slightly. Let's see if the proposed algorithm can accurately identify the delay and feedback gains from such slight changes.

By substituting $\mathbf{H}(\omega)$, $X_{21}(\omega)$, and $X_{22}(\omega)$ into Eq. (4.61), and setting the initial parameters as $g_1^0 = -3.2\text{N}/\mu\text{m}$, $g_2^0 = -3.6\text{N}/\mu\text{m}$, $g_3^0 = -2.7\text{N}/\mu\text{m}$, $g_4^0 = -4.4\text{N}/\mu\text{m}$, $\tau_1^0 = 25$ ms, and $\tau_2^0 = 39$ ms, the iteration processes of different

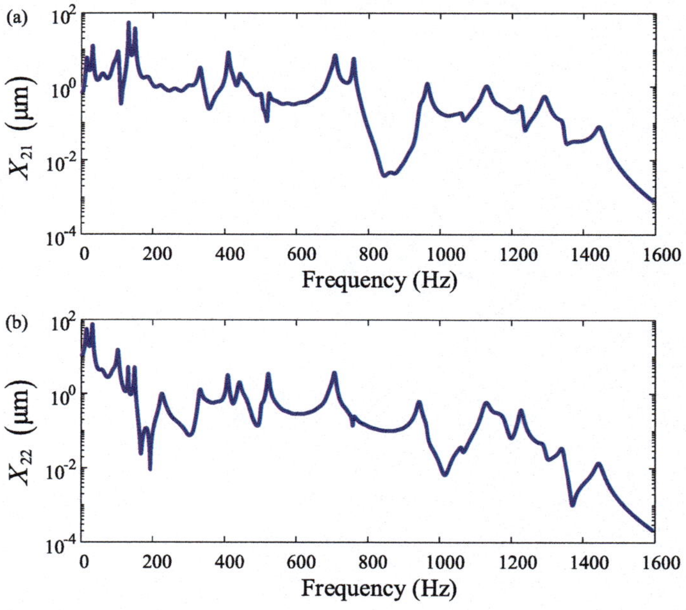

Fig. 4.8 Response spectrums after activating time-delayed control: **a** X_{21} and **b** X_{22}

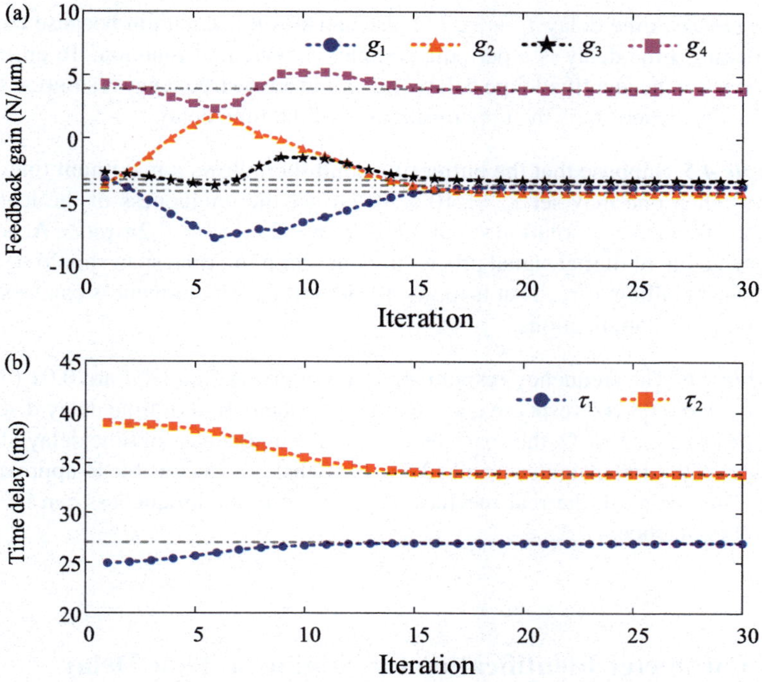

Fig. 4.9 Identification process of different parameters: **a** feedback gains and **b** time delays

parameters are shown in Fig. 4.9. One can find that the iteration converges to the set parameters very fast, indicating the effectiveness of the proposed algorithm.

4.2.3 Uniqueness in Delay Identification and Its Solution

Theorem 4.2 *indicates that the identified parameters will be unique. However, the conclusion is proved using the continuous frequency function. In the experiment, the frequency function is discrete, which may cause a unique problem in the identification procedure.*

Suppose that the frequency sample is uniform, i.e.

$$\omega_s = s\Omega, \tag{4.77}$$

where Ω is the frequency resolution. Substituting it into Eq. (4.46) yields

$$\mathbf{X}_o(n\Omega) - \mathbf{H}(n\Omega)\mathbf{F}_e(n\Omega) = \mathbf{H}(n\Omega)\left(\sum_{l=1}^{L} e^{-jn\Omega\tau_l}\mathbf{\Gamma}_l\right)\mathbf{X}_o(n\Omega). \tag{4.78}$$

Apparently, time delay $\tau_l + 2\pi / \Omega$ also satisfies this equation because the term that contains time delay is a periodic complex exponential function. To ensure the uniqueness of the identified time delay, one should let the frequency resolution satisfy $\Omega < 2\pi / \tau_u$, where τ_u is the upper limitation of the time delay.

Example 4.5 Suppose that the buffer of a controller allows a maximum time delay of 10 s. Then one may set $\tau_u = 10$ s. To ensure the uniqueness of the identified delay, the frequency resolution should satisfy $\Omega < 2\pi / \tau_u = 0.2\pi$ rad/s. According to the principle of digital signal processing, the sampling time span should be larger than 10 s, equalling to τ_u. That is to say, the identification uniqueness can be simply achieved in real applications.

Example 4.6 The frequency resolutions in Examples 4.3 and 4.4 are 0.01 (dimensionless) and 0.08 Hz, respectively. According to the period of time delay discussed in Eq. (4.78), i.e., $2\pi / \Omega$, the spans that ensure the uniqueness of time delay identification are 628 (dimensionless) and 12.5 s. Note that it is impossible to approve such a large time delay in the real mechanism. Therefore, the uniqueness can be easily realized in practice.

4.3 Parameter Identification for Nonlinear Time Delay Systems

Section 4.2 provides a general and explicit form representation of the linear time-delay systems, i.e., Eq. (4.39). When the system becomes nonlinear, it is impossible to give an explicit representation because of various nonlinearities. For the convenience of algorithm construction, this section unifies the representation of the nonlinear time-delay system as

$$\dot{\mathbf{x}}(t) = \mathbf{g}(\mathbf{x}(t), \mathbf{x}(t - \tau), \mathbf{f}(t); \boldsymbol{\mu}), \tag{4.79}$$

where $\mathbf{x}(t) \in \mathbb{R}^M$ is the M-dimensional state response, $\mathbf{g}(\cdot) \in \mathbb{R}^M$ denotes the M-dimensional governing function, τ represents the time delay, $\mathbf{f}(t) \in \mathbb{R}^N$ is the N-dimensional external excitation, and $\boldsymbol{\mu} \in \mathbb{R}^P$ is the P-dimensional unknown parameter. For a system with multiple delays, Eq. (4.79) can be easily modified to

$$\dot{\mathbf{x}}(t) = \mathbf{g}(\mathbf{x}(t), \mathbf{x}(t - \tau_1), \mathbf{x}(t - \tau_2), \ldots, \mathbf{x}(t - \tau_L), \mathbf{f}(t); \boldsymbol{\mu}). \tag{4.80}$$

4.3.1 Algorithm Construction in Frequency Domain: Quasi-Linear Method

Suppose that the external excitation $\mathbf{f}(t)$ is sinusoidal, i.e.,

$$\mathbf{f}(t) = \mathbf{F}_s \sin(\omega t) + \mathbf{F}_c \cos(\omega t), \tag{4.81}$$

where $\mathbf{F}_s$ and $\mathbf{F}_c$ are harmonic coefficients of sine and cosine terms. In addition, when the system given by Eq. (4.79) is weakly nonlinear and globally stable, the steady-state response can be represented as

$$\mathbf{x}(t) = \mathbf{X}_s \sin(\omega t), \tag{4.82}$$

where $\mathbf{X}_s$ is the sinusoidal coefficient. Here, the harmonic coefficient $\mathbf{X}_c$ is omitted because the sinusoidal response given by Eq. (4.82) can be achieved by tuning the excitation harmonic coefficients $\mathbf{F}_s$ and $\mathbf{F}_c$. Besides, the algorithm construction will be more compact based on these representations.

Substituting Eqs. (4.81) and (4.82) into Eq. (4.79) yields

$$\begin{aligned}&\omega \mathbf{X}_s \cos(\omega t)\\&= \mathbf{g}(\mathbf{X}_s \sin(\omega t), \mathbf{X}_s \sin(\omega(t-\tau)), \mathbf{F}_s \sin(\omega t) + \mathbf{F}_c \cos(\omega t); \boldsymbol{\mu}).\end{aligned} \tag{4.83}$$

Apparently, the governing function is also periodic because the terms in it are all periodic. Therefore, one can get a coefficient balance equation [15] that

$$\omega \mathbf{X}_s = \mathbf{G}_c(\mathbf{X}_s, \mathbf{F}_s, \mathbf{F}_c; \tau, \boldsymbol{\mu}), \tag{4.84}$$

where

$$\begin{aligned}&\mathbf{G}_c(\mathbf{X}_s, \mathbf{F}_s, \mathbf{F}_c; \boldsymbol{\mu})\\&= \frac{1}{\pi}\int_0^{2\pi} \mathbf{g}(\mathbf{X}_s \sin(\varphi), \mathbf{X}_s \sin(\varphi - \omega\tau), \mathbf{F}_s \sin(\varphi) + \mathbf{F}_c \cos(\varphi); \boldsymbol{\mu})\cos(\varphi)\mathrm{d}\varphi\end{aligned} \tag{4.85}$$

and $\varphi = \omega t$.

Because Eq. (4.84) is nonlinear to the parameters τ and $\boldsymbol{\mu}$, one cannot get an explicit identification form from it. To this end, the Newton–Raphson method is further employed. Let

$$\tau = \tau^0 + \Delta\tau^0 \tag{4.86}$$

and

$$\boldsymbol{\mu} = \boldsymbol{\mu}^0 + \Delta\boldsymbol{\mu}^0, \tag{4.87}$$

where τ^0 and $\boldsymbol{\mu}^0$ are the initial guesses. $\Delta\tau^0$ and $\Delta\boldsymbol{\mu}^0$ are their increments. Substituting Eqs. (4.86) and (4.87) into Eq. (4.84) and eliminating higher-order terms yield

$$\omega \mathbf{X}_s = \mathbf{G}_c\left(\mathbf{X}_s, \mathbf{F}_s, \mathbf{F}_c; \tau^0, \boldsymbol{\mu}^0\right) + \frac{\partial \mathbf{G}_c}{\partial \tau}\Delta\tau^0 + \frac{\partial \mathbf{G}_c}{\partial \boldsymbol{\mu}^T}\Delta\boldsymbol{\mu}^0. \tag{4.88}$$

By referring to Eqs. (4.79) and (4.85), the gradient terms in Eq. (4.88) are equivalent to

$$\begin{aligned}\frac{\partial \mathbf{G}_c}{\partial \tau} &= -\frac{1}{\pi}\int_0^{2\pi} \left.\frac{\partial \mathbf{g}}{\partial \mathbf{x}^T\left(t-\tau^0\right)}\right|_{\mathbf{x}(t)=\mathbf{X}_s\sin(\varphi)} \dot{\mathbf{x}}\left(t-\tau^0\right)\cos(\varphi)\mathrm{d}t \\ &= -\frac{\omega}{\pi}\int_0^{2\pi} \left.\frac{\partial \mathbf{g}}{\partial \mathbf{x}^T\left(t-\tau^0\right)}\right|_{\mathbf{x}(t)=\mathbf{X}_s\sin(\varphi)} \mathbf{X}_s\cos\left(\varphi-\omega\tau^0\right)\cos(\varphi)\mathrm{d}t\end{aligned} \tag{4.89}$$

and

$$\frac{\partial \mathbf{G}_c}{\partial \boldsymbol{\mu}^T} = \frac{1}{\pi}\int_0^{2\pi} \left.\frac{\partial \mathbf{g}}{\partial \boldsymbol{\mu}^T}\right|_{\mathbf{x}(t)=\mathbf{X}_s\sin(\varphi)} \cos(\varphi)\mathrm{d}t. \tag{4.90}$$

In the experiment, the acquired frequency response data is discrete. Suppose that the frequency sample is ω_q and $q = 1, 2, \ldots, Q$, Eq. (4.88) could be augmented to

$$\boldsymbol{\Psi}^0 = \boldsymbol{\Phi}^0\begin{pmatrix}\Delta\boldsymbol{\mu}^0 \\ \Delta\tau^0\end{pmatrix}, \tag{4.91}$$

where

$$\boldsymbol{\Psi}^0 = \begin{pmatrix} \omega_1\mathbf{X}_{s,1} - \mathbf{G}_c\left(\mathbf{X}_{s,1}, \mathbf{F}_{s,1}, \mathbf{F}_{c,1}; \tau^0, \boldsymbol{\mu}^0\right) \\ \omega_2\mathbf{X}_{s,2} - \mathbf{G}_c\left(\mathbf{X}_{s,2}, \mathbf{F}_{s,2}, \mathbf{F}_{c,2}; \tau^0, \boldsymbol{\mu}^0\right) \\ \vdots \\ \omega_Q\mathbf{X}_{s,Q} - \mathbf{G}_c\left(\mathbf{X}_{s,Q}, \mathbf{F}_{s,Q}, \mathbf{F}_{c,Q}; \tau^0, \boldsymbol{\mu}^0\right) \end{pmatrix} \in \mathbb{R}^{MQ} \tag{4.92}$$

and

$$\boldsymbol{\Phi}^0 = \begin{pmatrix} \frac{\partial \mathbf{G}_c\left(\mathbf{X}_{s,1},\mathbf{F}_{s,1},\mathbf{F}_{c,1};\tau^0,\boldsymbol{\mu}^0\right)}{\partial \boldsymbol{\mu}^T} & \frac{\partial \mathbf{G}_c\left(\mathbf{X}_{s,1},\mathbf{F}_{s,1},\mathbf{F}_{c,1};\tau^0,\boldsymbol{\mu}^0\right)}{\partial \tau} \\ \frac{\partial \mathbf{G}_c\left(\mathbf{X}_{s,2},\mathbf{F}_{s,2},\mathbf{F}_{c,2};\tau^0,\boldsymbol{\mu}^0\right)}{\partial \boldsymbol{\mu}^T} & \frac{\partial \mathbf{G}_c\left(\mathbf{X}_{s,2},\mathbf{F}_{s,2},\mathbf{F}_{c,2};\tau^0,\boldsymbol{\mu}^0\right)}{\partial \tau} \\ \vdots & \vdots \\ \frac{\partial \mathbf{G}_c\left(\mathbf{X}_{s,Q},\mathbf{F}_{s,Q},\mathbf{F}_{c,Q};\tau^0,\boldsymbol{\mu}^0\right)}{\partial \boldsymbol{\mu}^T} & \frac{\partial \mathbf{G}_c\left(\mathbf{X}_{s,Q},\mathbf{F}_{s,Q},\mathbf{F}_{c,Q};\tau^0,\boldsymbol{\mu}^0\right)}{\partial \tau} \end{pmatrix}. \\ \in \mathbb{R}^{MQ\times(P+1)} \tag{4.93}$$

Hence, the iteration equation can be given from Eq. (4.91) as

$$\begin{pmatrix} \boldsymbol{\mu}^{j+1} \\ \tau^{j+1} \end{pmatrix} = \begin{pmatrix} \boldsymbol{\mu}^{j} \\ \tau^{j} \end{pmatrix} + \begin{pmatrix} \Delta\boldsymbol{\mu}^{j} \\ \Delta\tau^{j} \end{pmatrix} = \begin{pmatrix} \boldsymbol{\mu}^{j} \\ \tau^{j} \end{pmatrix} + \left(\left(\boldsymbol{\Phi}^{j}\right)^{T} \boldsymbol{\Phi}^{j} \right)^{-1} \left(\boldsymbol{\Phi}^{j}\right)^{T} \boldsymbol{\Psi}^{j}, \tag{4.94}$$

where $\boldsymbol{\Phi}^j$ and $\boldsymbol{\Psi}^j$ are obtained by substituting $\boldsymbol{\mu}^j$ and τ^j into Eqs. (4.92) and (4.93). Equation (4.94) also implies the identifiability that $\boldsymbol{\Phi}^j$ should be full column rank, and the necessary condition satisfies $MQ \geq P + 1$. For example, if the system is one dimensional, i.e., $M = 1$, and there are 4 unknown structural parameters, i.e., $P = 4$, then one should collect at least 5 frequency samples ($Q \geq (P + 1)\big/ M = 5$) to promise a successful parameter identification.

Example 4.7 Suppose that an s-DOF system with time delay is governed by

$$\ddot{y}(t) + c\dot{y}(t) + k_1 y(t) + k_3 y^3(t) = f(t) + g y(t - \tau). \tag{4.95}$$

By letting $\mathbf{x}(t) = \left(y(t)\ \dot{y}(t) \right)^T$, the state-space equation of Eq. (4.95) can be written as

$$\begin{pmatrix} \dot{y} \\ \ddot{y} \end{pmatrix} = \begin{pmatrix} \dot{y} \\ -c\dot{y} - k_1 y - k_3 y^3 + f + g y(t - \tau) \end{pmatrix}. \tag{4.96}$$

According to Eqs. (4.81) and (4.82), the external excitation $f(t)$ and the steady-state response $y(t)$ can be given as

$$f(t) = F_s \sin(\omega t) + F_c \cos(\omega t) \tag{4.97}$$

and

$$y(t) = Y_s \sin(\omega t). \tag{4.98}$$

Substituting Eqs. (4.97) and (4.98) into Eq. (4.96) yields

$$\begin{pmatrix} F_s + \omega^2 Y_s \\ F_c \end{pmatrix} = \begin{pmatrix} k_1 Y_s + \frac{3}{4} k_3 Y_s^3 - g Y_s \cos(\omega\tau) \\ c\omega Y_s + g Y_s \sin(\omega\tau) \end{pmatrix}. \tag{4.99}$$

Suppose that the unknown parameters are c, k_1, k_3, g, and τ, and their initial guesses are c^0, k_1^0, k_3^0, g^0, and τ^0, the incremental function of Eq. (4.99) becomes

$$\begin{pmatrix} F_s + \omega^2 Y_s \\ F_c \end{pmatrix} = \begin{pmatrix} k_1^0 Y_s + \frac{3}{4}k_3^0 Y_s^3 - g^0 Y_s \cos(\omega\tau^0) \\ c^0\omega Y_s + g^0 Y_s \sin(\omega\tau^0) \end{pmatrix}$$

$$+ \begin{pmatrix} 0 & Y_s & \frac{3}{4}Y_s^3 & Y_s\cos(\omega\tau^0) & \omega g^0 Y_s \sin(\omega\tau^0) \\ \omega Y_s & 0 & 0 & Y_s \sin(\omega\tau^0) & \omega g^0 Y_s \cos(\omega\tau^0) \end{pmatrix} \begin{pmatrix} \Delta c^0 \\ \Delta k_1^0 \\ \Delta k_3^0 \\ \Delta g^0 \\ \Delta \tau^0 \end{pmatrix}. \tag{4.100}$$

By referring to Eq. (4.91), the augmented equation with different frequencies can be further written as

$$\mathbf{\Psi}^0 = \mathbf{\Phi}^0 \begin{pmatrix} \Delta c^0 \\ \Delta k_1^0 \\ \Delta k_3^0 \\ \Delta g^0 \\ \Delta \tau^0 \end{pmatrix}, \tag{4.101}$$

where

$$\mathbf{\Psi}^0 = \begin{pmatrix} F_{s,1} + \omega_1^2 Y_{s,1} - k_1^0 Y_{s,1} - \frac{3}{4}k_3^0 Y_{s,1}^3 + g^0 Y_{s,1}\cos(\omega_1\tau^0) \\ F_{c,1} - c^0\omega_1 Y_{s,1} - g^0 Y_{s,1}\sin(\omega_1\tau^0) \\ \vdots \\ F_{s,Q} + \omega_1^2 Y_{s,Q} - k_1^0 Y_{s,Q} - \frac{3}{4}k_3^0 Y_{s,Q}^3 + g^0 Y_{s,Q}\cos(\omega_Q\tau^0) \\ F_{c,Q} - c^0\omega_Q Y_{s,Q} - g^0 Y_{s,Q}\sin(\omega_Q\tau^0) \end{pmatrix} \tag{4.102}$$

and

$$\mathbf{\Phi}^0 = \begin{pmatrix} 0 & Y_{s,1} & \frac{3}{4}Y_{s,1}^3 & Y_{s,1}\cos(\omega_1\tau^0) & \omega_1 g^0 Y_{s,1}\sin(\omega_1\tau^0) \\ \omega_1 Y_{s,1} & 0 & 0 & Y_{s,1}\sin(\omega_1\tau^0) & \omega_1 g^0 Y_{s,1}\cos(\omega_1\tau^0) \\ & & \vdots & & \\ 0 & Y_{s,Q} & \frac{3}{4}Y_{s,Q}^3 & Y_{s,Q}\cos(\omega_Q\tau^0) & \omega_Q g^0 Y_{s,Q}\sin(\omega_Q\tau^0) \\ \omega_Q Y_{s,Q} & 0 & 0 & Y_{s,Q}\sin(\omega_Q\tau^0) & \omega_Q g^0 Y_{s,Q}\cos(\omega_Q\tau^0) \end{pmatrix} \tag{4.103}$$

Hence, the iteration equation can be given as

$$\begin{pmatrix} c^{j+1} \\ k_1^{j+1} \\ k_3^{j+1} \\ g^{j+1} \\ \tau^{j+1} \end{pmatrix} = \begin{pmatrix} c^j \\ k_1^j \\ k_3^j \\ g^j \\ \tau^j \end{pmatrix} + \left((\mathbf{\Phi}^j)^T \mathbf{\Phi}^j\right)^{-1} (\mathbf{\Phi}^j)^T \mathbf{\Psi}^j, \tag{4.104}$$

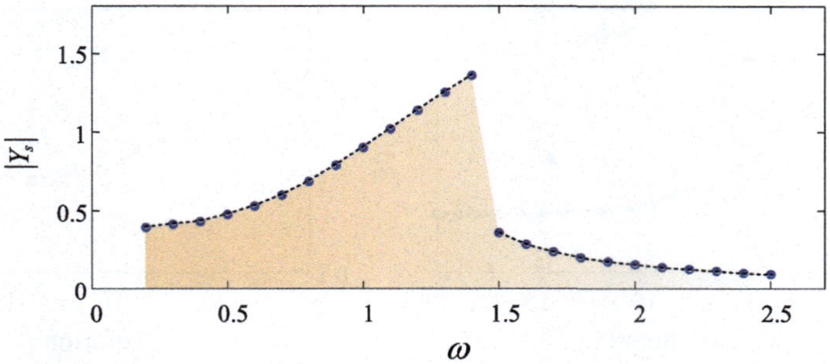

Fig. 4.10 Response amplitudes at the constant excitation amplitude of 0.5

where $\mathbf{\Phi}^j$ and $\mathbf{\Phi}^j$ are obtained by substituting c^j, k_1^j, k_3^j, g^j, and τ^j into Eqs. (4.103) and (4.102).

In the numerical simulation, the system parameters are set as $c = 0.2$, $k_1 = 1$, $k_3 = 1$, $g = 0.2$, and $\tau = 2$. The excitation amplitude, i.e., $\sqrt{F_s^2 + F_c^2}$, is set as a constant of 0.5. In addition, the excitation frequency ω_k is set as $\omega_k = 0.1+0.1k$ and $k = 1, 2, \ldots, 24$. The amplitude of the steady-state responses, i.e., Y_s, are extracted and displayed in Fig. 4.10, where one can find that the example system shows a strong nonlinearity.

By setting the initial guess as $c^0 = 0.25$, $k_1^0 = 1.25$, $k_3^0 = 1.25$, $g^0 = 0.25$, and $\tau^0 = 1.50$, and substituting the excitation information ω_q, $F_{s,q}$, and $F_{c,q}$ together with the response information $Y_{s,q}$ into Eq. (4.104), one can finally identify the unknown system parameters. The identification processes shown in Fig. 4.11 indicate that the quasi-linear identification method is effective.

Example 4.8 Let's continue with the example system given by Eq. (4.95). The parameters of the linear part, i.e., c, k_1, g, and τ. First, the external excitation remains the same as Example 4.7 sets. When the cubic nonlinearity, i.e., k_3, increases, the identified results are listed in Table 4.1. Apparently, when the cubic nonlinearity gets stronger, the identification error significantly grows up. For example, when $k_3 = 4$, the identification error of c becomes 20.5%, indicating that the identification is invalid. In addition, Table 4.2. shows the identified results corresponding to different excitation amplitudes, where one can find that the identification accuracy also deteriorates with the increase of excitation amplitude. The reason is that the system's nonlinear response gets stronger when the cubic nonlinearity or the excitation amplitude increases. Consequently, the quasi-linear procedure, which assumes a sinusoidal response, cannot efficiently represent the nonlinear response. To this end, a modified identification procedure that considers higher-order harmonic components should be proposed.

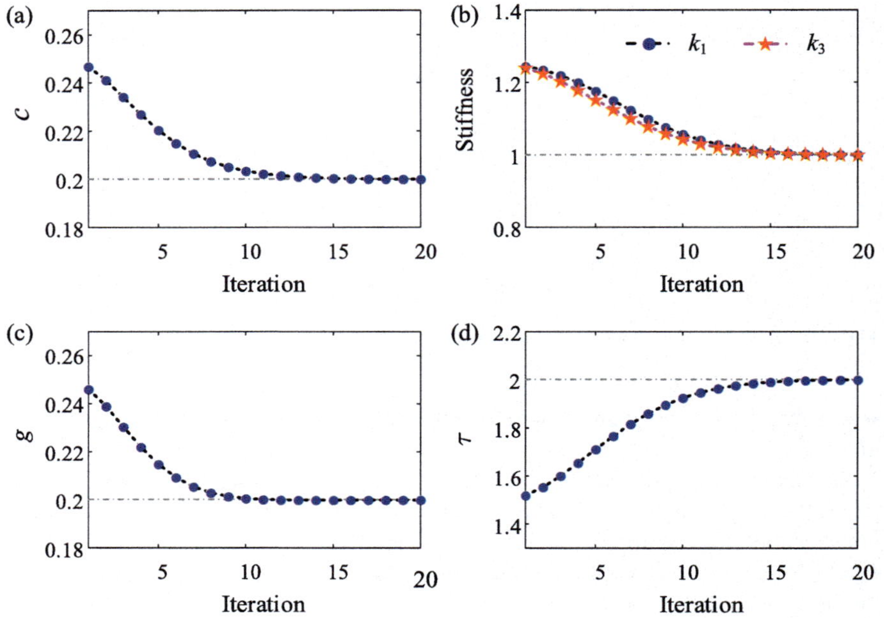

Fig. 4.11 Identification process of different parameters: **a** damping coefficients, **b** stiffness parameters, **c** feedback gain, and **d** time delay

Table 4.1 Identification results corresponding to different cubic nonlinearities

Parameters	c	k_1	k_3	g	τ
True	0.200	1.000	–	0.200	2.000
$k_3 = 1.000$	0.202	0.999	1.037	0.200	1.998
$k_3 = 2.000$	0.210	1.010	2.067	0.203	1.968
$k_3 = 3.000$	0.230	0.986	3.157	0.213	1.875
$k_3 = 4.000$	0.241	0.974	4.244	0.220	1.822

Table 4.2 Identification results corresponding to different excitation amplitudes

Parameters	c	k_1	k_3	g	τ
True	0.200	1.000	1.000	0.200	2.000
$F = 0.500$	0.202	0.999	1.037	0.200	1.998
$F = 1.000$	0.241	0.974	1.061	0.220	1.822
$F = 1.500$	0.184	1.166	0.948	0.142	2.055
$F = 2.000$	0.196	1.205	0.934	0.109	1.846

4.3.2 Algorithm Construction in Frequency Domain: Harmonic Balance Method

Since higher-order harmonic components are considered, there is no need to assume that the excitation is still sinusoidal. Without loss of generality, the excitation is in the form of

$$\mathbf{f}(t) = \sum_{i=1}^{I} \left(\mathbf{F}_{s,i} \sin(i\omega t) + \mathbf{F}_{c,i} \cos(i\omega t) \right), \tag{4.105}$$

where ω is the fundamental frequency, $\mathbf{F}_{s,i}$ and $\mathbf{F}_{c,i}$ are sine and cosine coefficients corresponding to the ith order of harmonics. Similarly, the steady-state response is assumed as

$$\mathbf{x}(t) = \mathbf{X}_0 + \sum_{i=1}^{I} \left(\mathbf{X}_{s,i} \sin(i\omega t) + \mathbf{X}_{c,i} \cos(i\omega t) \right). \tag{4.106}$$

Substituting Eq. (4.106) into Eq. (4.79) yields

$$\sum_{i=1}^{I} i\omega \left(\mathbf{X}_{s,i} \cos(i\omega t) - \mathbf{X}_{c,i} \sin(i\omega t) \right) = \mathbf{g}(\mathbf{x}(t), \mathbf{x}(t-\tau), \mathbf{f}(t); \boldsymbol{\mu}). \tag{4.107}$$

Apparently, the governing function is also periodic because the terms in it are all periodic. Therefore, one can get a coefficient balance equation [16] corresponding to different harmonic orders that

$$\begin{cases} -j\omega \mathbf{X}_{c,i} = \mathbf{G}_{s,i}\left(\mathbf{X}_{s,i}, \mathbf{X}_{c,i}, \mathbf{F}_{s,i}, \mathbf{F}_{c,i}; \tau, \boldsymbol{\mu}\right) \\ j\omega \mathbf{X}_{s,i} = \mathbf{G}_{c,i}\left(\mathbf{X}_{s,i}, \mathbf{X}_{c,i}, \mathbf{F}_{s,i}, \mathbf{F}_{c,i}; \tau, \boldsymbol{\mu}\right) \end{cases}, \tag{4.108}$$

where

$$\begin{cases} \mathbf{G}_{s,i}\left(\mathbf{X}_{s,i}, \mathbf{X}_{c,i}, \mathbf{F}_{s,i}, \mathbf{F}_{c,i}; \tau, \boldsymbol{\mu}\right) = \frac{1}{\pi} \int_0^{2\pi} \mathbf{g}(\mathbf{x}(t), \mathbf{x}(t-\tau), \mathbf{f}(t); \boldsymbol{\mu}) \sin(i\varphi) \mathrm{d}\varphi \\ \mathbf{G}_{c,i}\left(\mathbf{X}_{s,i}, \mathbf{X}_{c,i}, \mathbf{F}_{s,i}, \mathbf{F}_{c,i}; \tau, \boldsymbol{\mu}\right) = \frac{1}{\pi} \int_0^{2\pi} \mathbf{g}(\mathbf{x}(t), \mathbf{x}(t-\tau), \mathbf{f}(t); \boldsymbol{\mu}) \cos(i\varphi) \mathrm{d}\varphi \end{cases} \tag{4.109}$$

and $\varphi = \omega t$.

Again, Eq. (4.109) is nonlinear to the parameters τ and $\boldsymbol{\mu}$, the Newton–Raphson method should be employed to solve this equation. The incremental representations of τ and $\boldsymbol{\mu}$ have been defined in Eqs. (4.86) and (4.87). Substituting them into Eq. (4.108) yields

$$\begin{cases} -i\omega \mathbf{X}_{c,i} = \mathbf{G}_{s,i}\left(\mathbf{X}_{s,i}, \mathbf{X}_{c,i}, \mathbf{F}_{s,i}, \mathbf{F}_{c,i}; \tau^0, \boldsymbol{\mu}^0\right) + \frac{\partial \mathbf{G}_{s,i}}{\partial \tau}\Delta\tau^0 + \frac{\partial \mathbf{G}_{s,i}}{\partial \boldsymbol{\mu}^T}\Delta\boldsymbol{\mu}^0 \\ i\omega \mathbf{X}_{s,i} = \mathbf{G}_{c,i}\left(\mathbf{X}_{s,i}, \mathbf{X}_{c,i}, \mathbf{F}_{s,i}, \mathbf{F}_{c,i}; \tau^0, \boldsymbol{\mu}^0\right) + \frac{\partial \mathbf{G}_{c,i}}{\partial \tau}\Delta\tau^0 + \frac{\partial \mathbf{G}_{c,i}}{\partial \boldsymbol{\mu}^T}\Delta\boldsymbol{\mu}^0 \end{cases}, \tag{4.110}$$

where $i = 1, 2, \ldots, I$,

$$\begin{cases} \frac{\partial \mathbf{G}_{s,i}}{\partial \tau} = -\frac{1}{\pi}\int\limits_0^{2\pi} \frac{\partial \mathbf{g}}{\partial \mathbf{x}^T(t-\tau^0)}\dot{\mathbf{x}}\left(t-\tau^0\right)\sin(i\varphi)\mathrm{d}t \\ \frac{\partial \mathbf{G}_{c,i}}{\partial \tau} = -\frac{1}{\pi}\int\limits_0^{2\pi} \frac{\partial \mathbf{g}}{\partial \mathbf{x}^T(t-\tau^0)}\dot{\mathbf{x}}\left(t-\tau^0\right)\cos(i\varphi)\mathrm{d}t \end{cases}, \tag{4.111}$$

and

$$\begin{cases} \frac{\partial \mathbf{G}_{s,i}}{\partial \boldsymbol{\mu}^T} = \frac{1}{\pi}\int\limits_0^{2\pi} \frac{\partial \mathbf{g}}{\partial \boldsymbol{\mu}^T}\sin(i\varphi)\mathrm{d}t \\ \frac{\partial \mathbf{G}_{c,i}}{\partial \boldsymbol{\mu}^T} = \frac{1}{\pi}\int\limits_0^{2\pi} \frac{\partial \mathbf{g}}{\partial \boldsymbol{\mu}^T}\cos(i\varphi)\mathrm{d}t \end{cases}. \tag{4.112}$$

By letting

$$\begin{cases} \mathbf{R}^0_{s,i}(\omega) = -i\omega \mathbf{X}_{c,i} - \mathbf{G}_{s,i}\left(\mathbf{X}_{s,i}, \mathbf{X}_{c,i}, \mathbf{F}_{s,i}, \mathbf{F}_{c,i}; \tau^0, \boldsymbol{\mu}^0\right) \\ \mathbf{R}^0_{c,i}(\omega) = i\omega \mathbf{X}_{s,i} - \mathbf{G}_{c,i}\left(\mathbf{X}_{s,i}, \mathbf{X}_{c,i}, \mathbf{F}_{s,i}, \mathbf{F}_{c,i}; \tau^0, \boldsymbol{\mu}^0\right) \end{cases}, \tag{4.113}$$

Equation (4.110) can be rewritten in the vector form as

$$\begin{pmatrix} \mathbf{R}^0_{s,1}(\omega) \\ \mathbf{R}^0_{c,1}(\omega) \\ \vdots \\ \mathbf{R}^0_{s,I}(\omega) \\ \mathbf{R}^0_{c,I}(\omega) \end{pmatrix} = \begin{pmatrix} \frac{\partial \mathbf{G}_{s,1}}{\partial \tau} & \frac{\partial \mathbf{G}_{s,1}}{\partial \boldsymbol{\mu}^T} \\ \frac{\partial \mathbf{G}_{c,1}}{\partial \tau} & \frac{\partial \mathbf{G}_{c,1}}{\partial \boldsymbol{\mu}^T} \\ \vdots & \vdots \\ \frac{\partial \mathbf{G}_{s,I}}{\partial \tau} & \frac{\partial \mathbf{G}_{s,I}}{\partial \boldsymbol{\mu}^T} \\ \frac{\partial \mathbf{G}_{c,I}}{\partial \tau} & \frac{\partial \mathbf{G}_{c,I}}{\partial \boldsymbol{\mu}^T} \end{pmatrix} \begin{pmatrix} \Delta\tau^0 \\ \Delta\boldsymbol{\mu}^0 \end{pmatrix}, \tag{4.114}$$

or in the compact form as

$$\mathbf{R}^0(\omega) = \boldsymbol{\Theta}^0(\omega)\begin{pmatrix} \Delta\tau^0 \\ \Delta\boldsymbol{\mu}^0 \end{pmatrix}, \tag{4.115}$$

where $\mathbf{R}^0(\omega) \in \mathbb{R}^{2MI}$ and $\boldsymbol{\Theta}^0(\omega) \in \mathbb{R}^{2MI\times(P+1)}$.

In the experiment, the acquired frequency response data is discrete. Suppose that the frequency sample is ω_q and $q = 1, 2, \ldots, Q$, Eq. (4.88) could be augmented to

$$\begin{pmatrix} \mathbf{R}^0(\omega_1) \\ \vdots \\ \mathbf{R}^0\left(\omega_Q\right) \end{pmatrix} = \begin{pmatrix} \boldsymbol{\Theta}^0(\omega_1) \\ \vdots \\ \boldsymbol{\Theta}^0\left(\omega_Q\right) \end{pmatrix}\begin{pmatrix} \Delta\tau^0 \\ \Delta\boldsymbol{\mu}^0 \end{pmatrix}. \tag{4.116}$$

If one uses $\boldsymbol{\Psi}^0$ and $\boldsymbol{\Phi}^0$ to denote the left-hand side and the coefficient matrix of the right-hand side of Eq. (4.116), then the iterative identification equation can be represented in the same form as Eq. (4.94) shows, i.e.,

$$\begin{pmatrix} \boldsymbol{\mu}^{k+1} \\ \tau^{k+1} \end{pmatrix} = \begin{pmatrix} \boldsymbol{\mu}^{k} \\ \tau^{k} \end{pmatrix} + \begin{pmatrix} \Delta\boldsymbol{\mu}^{k} \\ \Delta\tau^{k} \end{pmatrix} = \begin{pmatrix} \boldsymbol{\mu}^{k} \\ \tau^{k} \end{pmatrix} + \left(\left(\boldsymbol{\Phi}^k\right)^T \boldsymbol{\Phi}^k \right)^{-1} \left(\boldsymbol{\Phi}^k\right)^T \boldsymbol{\Psi}^k.$$

Specifically, $\boldsymbol{\Psi}^k \in \mathbb{R}^{2MQI}$ and $\boldsymbol{\Phi}^k \in \mathbb{R}^{2MQI \times (P+1)}$. Hence, the identifiability requires that $2MQI \geq P + 1$. It seems that if one wants to promise the identifiability, he only needs to collect $1/(2I)$ of the samples compared with that required by the quasi-linear identification method. However, taking the fact that the higher order responses are often weak and thus have a limited signal-to-noise ratio, the recommended number of samples should satisfy $Q \geq (P+1)/(2M)$.

Example 4.9 Let's continue with the system discussed in Example 4.8 to check whether the harmonic balance identification method can deal with strong nonlinearity. For the convenience of reading, we recall the state space model given by Eq. (4.96), i.e.,

$$\begin{pmatrix} \dot{y} \\ \ddot{y} \end{pmatrix} = \begin{pmatrix} \dot{y} \\ -c\dot{y} - k_1 y - k_3 y^3 + f + g y(t-\tau) \end{pmatrix}.$$

It is easy to find that the governing function corresponding to this system is

$$\mathbf{g}(\mathbf{x}(t), \mathbf{x}(t-\tau), \mathbf{f}(t); \boldsymbol{\mu}) = \begin{pmatrix} \dot{y} \\ -c\dot{y} - k_1 y - k_3 y^3 + f + g y(t-\tau) \end{pmatrix}, \quad (4.117)$$

where

$$\mathbf{x}(t) = \begin{pmatrix} y \\ \dot{y} \end{pmatrix}, \mathbf{f}(t) = \begin{pmatrix} 0 \\ f \end{pmatrix}, \text{ and } \boldsymbol{\mu} = \begin{pmatrix} c & k_1 & k_3 & g \end{pmatrix}^T.$$

Therefore, the detailed forms of the integration terms in Eqs. (4.111) and (4.112) are

$$\frac{\partial \mathbf{g}}{\partial \mathbf{x}^T\left(t-\tau^0\right)} \dot{\mathbf{x}}\left(t-\tau^0\right) = \begin{pmatrix} 0 \\ g\dot{y}(t-\tau) \end{pmatrix} \quad (4.118)$$

and

$$\frac{\partial \mathbf{g}}{\partial \boldsymbol{\mu}^T} = \begin{pmatrix} 0 & 0 & 0 & 0 \\ -\dot{y} & -y & -y^3 & y\left(t-\tau^0\right) \end{pmatrix}. \quad (4.119)$$

Hence, the harmonic terms $\mathbf{G}_{s,i}$, $\mathbf{G}_{c,i}$, $\partial\mathbf{G}_{s,i}/\partial\tau$, $\partial\mathbf{G}_{c,i}/\partial\tau$, $\partial\mathbf{G}_{s,i}/\partial\boldsymbol{\mu}^T$, and $\partial\mathbf{G}_{c,i}/\partial\boldsymbol{\mu}^T$ can be calculated by implementing Eqs. (4.109), (4.111), and (4.112).

By substituting these terms into Eq. (4.116), the time delay and system parameters can be identified.

In the process of parameter identification, all the initial settings are identical to those used in Example 4.8. However, the harmonic order in the harmonic balance identification procedure is set as $J = 5$. According to the cases discussed in Tables 4.1 and 4.2, the identification results using the harmonic balance method are listed in Tables 4.3 and 4.4, respectively. One can find that the harmonic balance method that considers higher-order responses successfully fixed the problem shown in Example 4.8.

Example 4.10 The comparison of the identification results in Example 4.8 and Example 4.9 clearly demonstrates that the harmonic balance identification procedure is more competent for the mission of strong nonlinearity identification. This example will investigate whether the harmonic balance identification is noise-resistant. The parameters are also set as $c = 0.2$, $k_1 = 1$, $k_3 = 1$, $g = 0.2$, and $\tau = 2$. The excitation amplitude is $F = 1$. Here, we add some Gaussian white noise into the clean response. Figure 4.12 shows some responses with different signal-to-noise ratios (SNR), defined in Eq. (4.120), at the excitation frequency of 0.5.

$$\mathrm{SNR} = 10\log_{10}\left(\frac{P_{\mathrm{signal}}}{P_{\mathrm{noise}}}\right), \tag{4.120}$$

where P_{signal} is the power of the clean signal and P_{noise} represents the power of noise.

Because the signal used for parameter identification is stochastic, the identification performance should also be evaluated in the stochastic approach. To this end, the

Table 4.3 Identification results with cubic nonlinearities based on the harmonic balance method

Parameters	c	k_1	k_3	g	τ
True	0.200	1.000	–	0.200	2.000
$k_3 = 1.000$	0.200	1.000	1.000	0.200	2.000
$k_3 = 2.000$	0.200	1.000	2.000	0.200	2.000
$k_3 = 3.000$	0.200	1.000	3.001	0.200	1.999
$k_3 = 4.000$	0.200	1.000	4.001	0.200	1.999

Table 4.4 Identification results with different excitation amplitudes based on the harmonic balance method

Parameters	c	k_1	k_3	g	τ
True	0.200	1.000	1.000	0.200	2.000
$F = 0.500$	0.200	1.000	1.000	0.200	2.000
$F = 1.000$	0.200	1.000	1.000	0.200	1.999
$F = 1.500$	0.200	1.003	1.000	0.201	2.004
$F = 2.000$	0.200	1.006	0.999	0.201	2.003

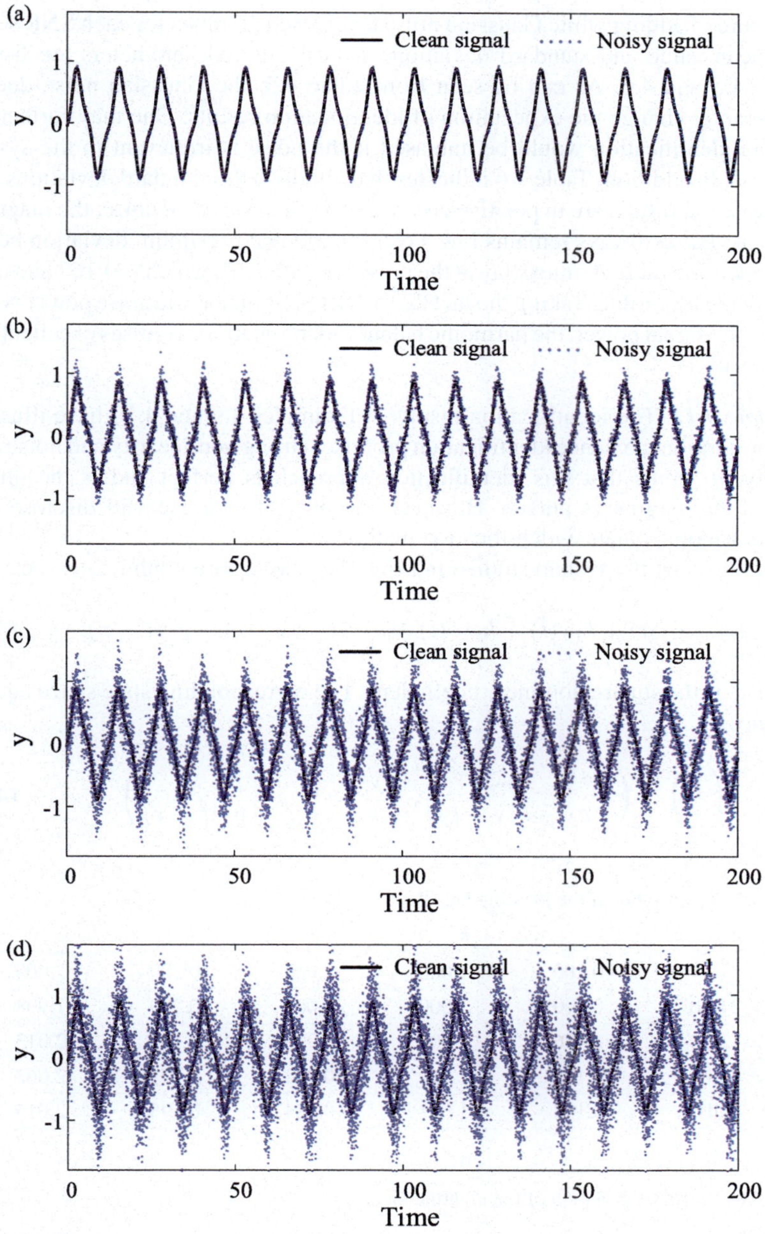

Fig. 4.12 Steady-state response with different levels of noise: **a** SNR = 24 dB, **b** SNR = 12 dB, **c** SNR = 6 dB, and **d** SNR = 3 dB

procedure of adding white Gaussian noise is repeated 25 times for each SNR setting. The mean value and standard deviation of the identified parameters are listed in Tables 4.5 and 4.6. As can be seen from Table 4.5, the Gaussian noise does not significantly change the expectation of identification results. One may further infer that the identification would be unbiased if the noise is irrelevant to the system's response. In addition, Table 4.6 indicates that although the standard deviations of the identification results are in positive correlation with the level of noise, the magnitude of the deviation always remains low. For example, the maximum deviation belongs to the identification of time delay at the noise level of 3 dB, which is 0.162 and is only 8.1% of the true value. Taking the fact that a 3 dB SNR means the noise power is about 50% of the signal power, the harmonic balance identification promises excellent noise resistance.

Example 4.11 The identification results in Examples 4.9 and 4.10 have illustrated that the harmonic balance identification promises strong nonlinearity and noise applicability. It seems that this identification approach is perfect and is the ultimate method that engineers pursue. However, the present example will disclose a bug in the harmonic balance identification method.

Suppose that the system studied in Eq. (4.95) has square nonlinearity, i.e.,

$$\ddot{y}(t) + c\dot{y}(t) + k_1 y(t) + k_2 y^2(t) + k_3 y^3(t) = f(t) + gy(t - \tau), \tag{4.121}$$

where k_2 is the square nonlinear coefficient. The corresponding state space equation becomes

$$\begin{pmatrix} \dot{y} \\ \ddot{y} \end{pmatrix} = \begin{pmatrix} \dot{y} \\ -c\dot{y} - k_1 y - k_2 y^2 - k_3 y^3 + f + gy(t - \tau) \end{pmatrix}. \tag{4.122}$$

Table 4.5 Mean value of the identified results

Parameters	c	k_1	k_3	g	τ
True	0.200	1.000	1.000	0.200	2.000
SNR = 24 dB	0.200	1.000	1.001	0.200	1.998
SNR = 12 dB	0.196	1.004	0.996	0.197	2.013
SNR = 6 dB	0.188	1.012	0.992	0.191	2.063
SNR = 3 dB	0.197	1.011	0.984	0.195	2.015

Table 4.6 Standard deviation of the identified results

Parameters	c	k_1	k_3	g	τ
SNR = 24 dB	0.003	0.004	0.003	0.002	0.012
SNR = 12 dB	0.010	0.014	0.013	0.008	0.040
SNR = 6 dB	0.023	0.028	0.024	0.017	0.117
SNR = 3 dB	0.031	0.033	0.027	0.023	0.162

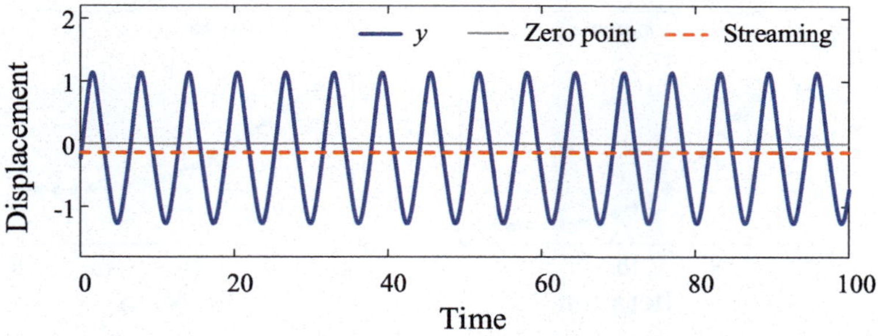

Fig. 4.13 Steady-state response when the excitation amplitude and frequency are both 1

In the numerical simulation, the system parameters are set as $c = 0.2$, $k_1 = 1$, $k_2 = 0.5$, $k_3 = 1$, $g = 0.2$, and $\tau = 2$. When the excitation amplitude is 1, and the excitation frequency is also 1, the steady-state response is shown in Fig. 4.13. One can find from this figure that the response center is no longer zero, which is called streaming in nonlinear dynamics.

In real applications, accelerometers and Doppler laser vibrometers are the most commonly used sensors for vibration acquisition. These kinds of sensors only measure the vibration's acceleration or velocity. Therefore, they cannot record the streaming component in the displacement response. Besides, most signal acquisition devices activate high-pass filtering by default to avoid signal drifting induced by thermal or charge accumulation. The streaming component in the displacement will also be omitted even if displacement sensors are employed.

Hence, let's study the influence of missing streaming on parameter identification accuracy. To simulate the case of streaming missing, the steady-state responses are all centralized by mean value compensation. The initial guesses are set as $c^0 = 0.25$, $k_1^0 = 1.25$, $k_2^0 = 0.65$, $k_3^0 = 1.25$, $g^0 = 0.25$, and $\tau^0 = 1.50$. The identified parameters are shown in Fig. 4.14, where one can find that they are significantly biased. It should also be noted that such failure is common to other identification approaches if the streaming is unknown. To solve this problem, the next subsection will introduce active noise-resistant identification.

4.4 Algorithm Modification for Noise-Correction Identification

Section 4.3 illustrates that the identification results will be significantly biased if the streaming components in the response are missing. Note that the streaming is the zero-order harmonics induced by the system's nonlinearity. One may use the nonlinear characteristic of the system to recover the missing streaming. However, the unknown system parameters, which should be identified from the missing streaming and other response components, in turn, incommode the streaming recovery. In other

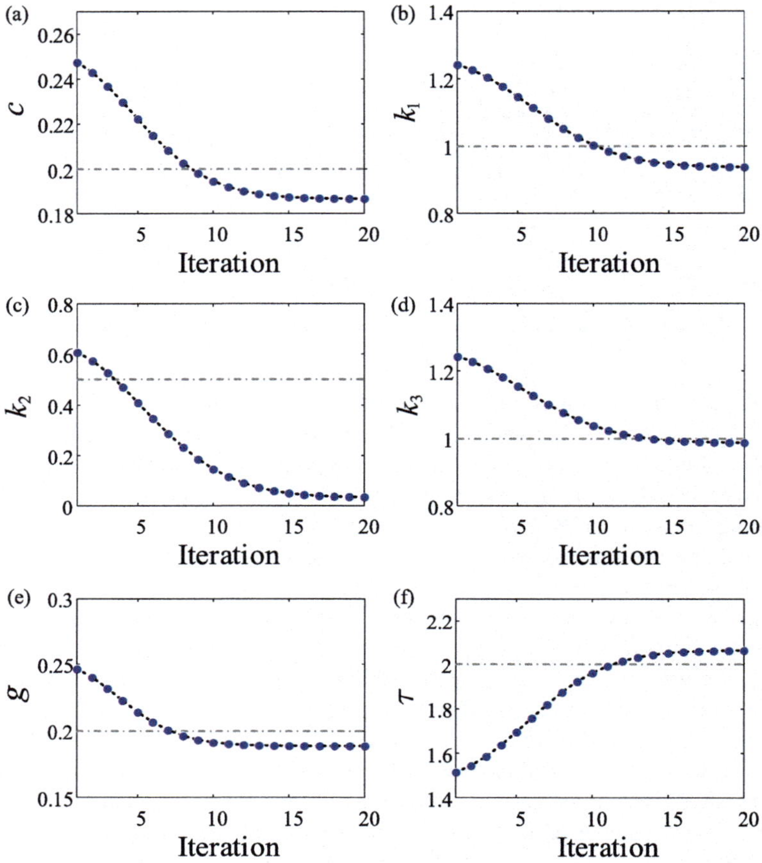

Fig. 4.14 Identification process at the case of streaming missing: **a** damping coefficients, **b** linear stiffness, **c** square stiffness, **d** cubic stiffness, **e** feedback gain, and **f** time delay

words, the streaming recovery and the parameter identification are coupled. To solve this problem, this section will give a detailed algorithm construction and rigorous convergence proving.

4.4.1 Problem Representation with Noise Correction

Section 4.3 provides the basic principle for parameter identification of time-delayed nonlinear systems. Because the adaptive noise correction will introduce more parameters, the problem of identification should now be represented in a more compact form. To this end, this section denotes the estimations of τ and μ as $\tilde{\tau}$ and $\tilde{\mu}$. Hence, the problem of the aforementioned parameter identification can be represented as

$$\min_{\tilde{\tau},\tilde{\boldsymbol{\mu}}} \|\dot{\mathbf{x}}(t) - \mathbf{g}(\mathbf{x}(t), \mathbf{x}(t-\tilde{\tau}), \mathbf{f}(t); \tilde{\boldsymbol{\mu}})\|. \tag{4.123}$$

In addition, the measured response that needs noise correction is denoted as $\tilde{\mathbf{x}}(t;\omega_q)$, where q means the qth excitation frequency. Suppose that the noise correction is $\delta\tilde{\mathbf{x}}(t;\omega_q)$, which makes

$$\mathbf{x}(t;\omega_q) \approx \tilde{\mathbf{x}}(t;\omega_q) + \delta\tilde{\mathbf{x}}(t;\omega_q), \tag{4.124}$$

then the noise-correction-inspired parameter identification [17–20] can be modified to

$$\begin{aligned}&\min_{\tilde{\tau},\tilde{\boldsymbol{\mu}},\delta\tilde{\mathbf{x}}(t;\omega_q)} \Big\|\dot{\tilde{\mathbf{x}}}(t;\omega_q) + \delta\dot{\tilde{\mathbf{x}}}(t;\omega_q)\\ &- \mathbf{g}\big(\tilde{\mathbf{x}}(t;\omega_q) + \delta\tilde{\mathbf{x}}(t;\omega_q), \tilde{\mathbf{x}}(t-\tilde{\tau};\omega_q) + \delta\tilde{\mathbf{x}}(t-\tilde{\tau};\omega_q), \mathbf{f}(t); \tilde{\boldsymbol{\mu}}\big)\Big\|.\end{aligned} \tag{4.125}$$

According to Eq. (4.106), the measured response $\tilde{\mathbf{x}}(t;\omega_k)$ may also be represented in Fourier series as

$$\begin{aligned}&\tilde{\mathbf{x}}(t;\omega_q)\\ &= \big(\tilde{\mathbf{X}}_0(\omega_q)\ \tilde{\mathbf{X}}_{s,1}(\omega_q)\ \tilde{\mathbf{X}}_{c,1}(\omega_q)\ \cdots\ \tilde{\mathbf{X}}_{s,I}(\omega_q)\ \tilde{\mathbf{X}}_{c,I}(\omega_q)\big)\varphi(t;\omega_q),\end{aligned} \tag{4.126}$$

where

$$\varphi(t;\omega_q) = \big(1\ \sin(\omega_q t)\ \cos(\omega_q t)\ \cdots\ \sin(I\omega_q t)\ \cos(I\omega_q t)\big)^T \in \mathbb{R}^{2I+1}.$$

For the convenience of calculation, Eq. (4.126) is further written in the vectorized form as

$$\tilde{\mathbf{x}}(t;\omega_q) = \boldsymbol{\Phi}(t;\omega_q)\tilde{\mathbf{X}}(\omega_q), \tag{4.127}$$

where

$$\boldsymbol{\Phi}(t;\omega_q) = \varphi^T(t;\omega_q) \otimes \mathbf{E}_M \in \mathbb{R}^{M\times(2I+1)M},$$

$$\tilde{\mathbf{X}}(\omega_q) = \begin{pmatrix} \tilde{\mathbf{X}}_0(\omega_q) \\ \tilde{\mathbf{X}}_{s,1}(\omega_q) \\ \tilde{\mathbf{X}}_{c,1}(\omega_q) \\ \vdots \\ \tilde{\mathbf{X}}_{s,I}(\omega_q) \\ \tilde{\mathbf{X}}_{c,I}(\omega_q) \end{pmatrix} \in \mathbb{R}^{(2I+1)M},$$

$\mathbf{E}_M$ is the $M \times M$ unit matrix, the superscription T means the transpose, and $\otimes$ denotes the Kronecker product.

By considering that the primary harmonic component is dominant in $\tilde{\mathbf{x}}(t;q)$ and has the highest signal-to-noise ratio (SNR), we can use it as the reference to correct the other components. That is to say, the coefficients of the primary harmonic component should be fixed, and the others should be treated as correctable variables. Hence, the noise correction can be represented as

$$\begin{aligned} &\delta\tilde{\mathbf{x}}(t;\omega_q) \\ &= \left(\delta\tilde{\mathbf{X}}_0(\omega_q)\ \delta\tilde{\mathbf{X}}_{s,2}(\omega_q)\ \delta\tilde{\mathbf{X}}_{c,2}(\omega_q)\ \cdots\ \delta\tilde{\mathbf{X}}_{s,I}(\omega_q)\ \delta\tilde{\mathbf{X}}_{c,I}(\omega_q)\right)\boldsymbol{\psi}(t;\omega_q), \end{aligned} \tag{4.128}$$

where

$$\boldsymbol{\psi}(t;\omega_q) = \left(1\ \sin(2\omega_q t)\ \cos(2\omega_q t)\ \cdots\ \sin(I\omega_q t)\ \cos(I\omega_q t)\right)^T \in \mathbb{R}^{2I-1}$$

is a subspace of $\varphi(t;\omega_q)$. The relationship between $\boldsymbol{\psi}(t;\omega_q)$ and $\varphi(t;\omega_q)$ can be expressed as

$$\boldsymbol{\psi}(t;\omega_q) = \mathbf{v}\varphi(t;\omega_q), \tag{4.129}$$

where

$$\mathbf{v} = \left(\begin{array}{ccc|ccc} 1 & 0 & 0 & 0 & \cdots & 0 \\ \hline 0 & 0 & 0 & 1 & & \\ \vdots & \vdots & \vdots & & \ddots & \\ 0 & 0 & 0 & & & 1 \end{array}\right) \in \mathbb{R}^{(2I-1)\times(2I+1)}$$

is called the reduction transformation. Accordingly, the vectorized form of Eq. (4.128) can be written as

$$\delta\tilde{\mathbf{x}}(t;\omega_q) = \boldsymbol{\Phi}(t;\omega_q)\mathbf{V}\delta\tilde{\mathbf{X}}(\omega_q), \tag{4.130}$$

where

$$\mathbf{V} = \mathbf{v}^T \otimes \mathbf{E}_M \in \mathbb{R}^{(2I+1)M\times(2I-1)M}$$

and

$$\delta\tilde{\mathbf{X}}(\omega_q) = \begin{pmatrix} \delta\tilde{\mathbf{X}}_0(\omega_q) \\ \delta\tilde{\mathbf{X}}_{s,2}(\omega_q) \\ \delta\tilde{\mathbf{X}}_{c,2}(\omega_q) \\ \vdots \\ \delta\tilde{\mathbf{X}}_{s,I}(\omega_q) \\ \delta\tilde{\mathbf{X}}_{c,I}(\omega_q) \end{pmatrix} \in \mathbb{R}^{(2I-1)M}.$$

By substituting Eq. (4.130) into Eq. (4.125), the problem of noise-correction-based parameter identification can hence be represented as

$$\min_{\tilde{\tau},\tilde{\mu},\delta\tilde{\mathbf{X}}(\omega_q)} \Big\| \dot{\tilde{\mathbf{x}}}(t;\omega_q) + \dot{\Phi}(t;\omega_q)\mathbf{V}\delta\tilde{\mathbf{X}}(\omega_q)$$
$$- \mathbf{g}(\tilde{\mathbf{x}}(t;\omega_q) + \mathbf{\Phi}(t;\omega_q)\mathbf{V}\delta\tilde{\mathbf{X}}(\omega_q), \tilde{\mathbf{x}}(t-\tilde{\tau};\omega_q)$$
$$+ \mathbf{\Phi}(t-\tilde{\tau};\omega_q)\mathbf{V}\delta\tilde{\mathbf{X}}(\omega_q), \mathbf{f}(t;\omega_q); \tilde{\boldsymbol{\mu}})||. \tag{4.131}$$

4.4.2 Local Linearization and Regularization

Equation (4.131) is strongly nonlinear with respect to $\tilde{\tau}$, $\tilde{\boldsymbol{\mu}}$, and $\delta\tilde{\mathbf{X}}(\omega_q)$. It is impossible to give an analytic solution for this problem. By assuming that $\delta\tilde{\tau}$ and $\delta\tilde{\boldsymbol{\mu}}$ are local corrections of $\tilde{\tau}$ and $\tilde{\boldsymbol{\mu}}$, Eq. (4.131) can be linearized to

$$\min_{\tilde{\tau},\tilde{\boldsymbol{\mu}},\delta\tilde{\mathbf{X}}(\omega_q)} \Big\| \tilde{\boldsymbol{\gamma}}(t;\omega_q) + \tilde{\mathbf{h}}_\tau(t;\omega_q)\dot{\tilde{\mathbf{x}}}(t-\tilde{\tau})\delta\tilde{\tau} - \tilde{\mathbf{h}}_\mu(t;\omega_q)\delta\tilde{\boldsymbol{\mu}}$$
$$+ \left(\dot{\Phi}(t;\omega_q) - \tilde{\mathbf{h}}_x(t;\omega_q)\mathbf{\Phi}(t;\omega_q) - \tilde{\mathbf{h}}_\tau(t;\omega_q)\mathbf{\Phi}(t-\tilde{\tau};\omega_q)\right)\mathbf{V}\delta\tilde{\mathbf{X}}(\omega_q)\Big\|, \tag{4.132}$$

where

$$\tilde{\mathbf{h}}_x(t;\omega_q) = \frac{\partial \mathbf{g}}{\partial \tilde{\mathbf{x}}^T(t;\omega_q)},\ \tilde{\mathbf{h}}_\tau(t;\omega_q) = \frac{\partial \mathbf{g}}{\partial \tilde{\mathbf{x}}^T(t-\tilde{\tau};\omega_q)},\ \text{and}\ \tilde{\mathbf{h}}_\mu(t;\omega_q) = \frac{\partial \mathbf{g}}{\partial\, \tilde{\boldsymbol{\mu}}^T}$$

are Jacobian matrices, and

$$\tilde{\boldsymbol{\gamma}}(t;\omega_q) = \dot{\tilde{\mathbf{x}}}(t;\omega_q) - \mathbf{g}(t, \tilde{\mathbf{x}}(t;\omega_q), \tilde{\mathbf{x}}(t-\tilde{\tau};\omega_q), \mathbf{f}(t;\omega_q); \tilde{\boldsymbol{\mu}})$$

is called the residue vector.

Furthermore, it can be proven that $\tilde{\mathbf{h}}_\tau(t;\omega_q)$, $\tilde{\mathbf{h}}_\mu(t;\omega_q)$, $\tilde{\mathbf{h}}_x(t;\omega_q)$, and $\tilde{\boldsymbol{\gamma}}(t;\omega_q)$ are also periodic. They can be represented in the harmonic basis as

$$\begin{cases} \tilde{\mathbf{h}}_x(t;\omega_q) = \mathbf{\Phi}(t;\omega_q)\tilde{\mathbf{H}}_x(\omega_q) \\ \tilde{\mathbf{h}}_\tau(t;\omega_q) = \mathbf{\Phi}(t;\omega_q)\tilde{\mathbf{H}}_\tau(\omega_q) \\ \tilde{\mathbf{h}}_\mu(t;\omega_q) = \mathbf{\Phi}(t;\omega_q)\tilde{\mathbf{H}}_\mu(\omega_q) \end{cases} \tag{4.133}$$

and

$$\tilde{\boldsymbol{\gamma}}(t;\omega_q) = \mathbf{\Phi}(t;\omega_q)\tilde{\mathbf{\Gamma}}(\omega_q), \tag{4.134}$$

where

$$\tilde{\mathbf{H}}_x(\omega_q) = \left(\tilde{\mathbf{H}}_x^1(\omega_q)\ \tilde{\mathbf{H}}_x^2(\omega_q)\ \cdots\ \tilde{\mathbf{H}}_x^M(\omega_q)\right) \in \mathbb{R}^{(2I+1)M\times M},$$

$$\tilde{\mathbf{H}}_\tau(\omega_q) = \left(\tilde{\mathbf{H}}_\tau^1(\omega_q)\ \tilde{\mathbf{H}}_\tau^2(\omega_q)\ \cdots\ \tilde{\mathbf{H}}_\tau^M(\omega_q)\right) \in \mathbb{R}^{(2I+1)M\times M},$$

$$\tilde{\mathbf{H}}_\mu(\omega_q) = \left(\tilde{\mathbf{H}}_\mu^1(\omega_q)\ \tilde{\mathbf{H}}_\mu^2(\omega_q)\ \cdots\ \tilde{\mathbf{H}}_\mu^P(\omega_q)\right) \in \mathbb{R}^{(2I+1)M\times P}.$$

and

$$\tilde{\mathbf{\Gamma}}(\omega_q) = \begin{pmatrix} \tilde{\mathbf{\Gamma}}_0(\omega_q) \\ \tilde{\mathbf{\Gamma}}_{s,1}(\omega_q) \\ \tilde{\mathbf{\Gamma}}_{c,1}(\omega_q) \\ \vdots \\ \tilde{\mathbf{\Gamma}}_{s,I}(\omega_q) \\ \tilde{\mathbf{\Gamma}}_{c,I}(\omega_q) \end{pmatrix} \in \mathbb{R}^{(2I+1)M}.$$

$\tilde{\mathbf{\Gamma}}_{s,i}(\omega_q)$ and $\tilde{\mathbf{\Gamma}}_{c,i}(\omega_q)$ are the ith sine and cosine coefficients corresponding to $\tilde{\boldsymbol{\gamma}}(t;\omega_q)$, i.e.,

$$\begin{cases} \tilde{\mathbf{\Gamma}}_{s,i}(\omega_q) = \frac{1}{\pi}\int_0^{2\pi} \tilde{\boldsymbol{\gamma}}(t;\omega_q)\sin(i\varphi)\mathrm{d}\varphi \\ \tilde{\mathbf{\Gamma}}_{c,i}(\omega_q) = \frac{1}{\pi}\int_0^{2\pi} \tilde{\boldsymbol{\gamma}}(t;\omega_q)\cos(i\varphi)\mathrm{d}\varphi \end{cases}.$$

The detailed expressions of $\tilde{\mathbf{H}}_x^m(\omega_q)$ are

$$\tilde{\mathbf{H}}_x^m(\omega_q) = \begin{pmatrix} \tilde{\mathbf{H}}_{x,0}^m(\omega_q) \\ \tilde{\mathbf{H}}_{x,s,1}^m(\omega_q) \\ \tilde{\mathbf{H}}_{x,c,1}^m(\omega_q) \\ \vdots \\ \tilde{\mathbf{H}}_{x,s,I}^m(\omega_q) \\ \tilde{\mathbf{H}}_{x,c,I}^m(\omega_q) \end{pmatrix} \in \mathbb{R}^{(2I+1)M},$$

where $\tilde{\mathbf{H}}_{x,s,i}^m(\omega_q)$ and $\tilde{\mathbf{H}}_{x,c,i}^m(\omega_q)$ are the ith order of sine and cosine coefficients of the mth column of $\tilde{\mathbf{h}}_x(t;\omega_q)$, i.e.,

$$\begin{cases} \tilde{\mathbf{H}}_{x,s,i}^{m}(\omega_q) = \frac{1}{\pi}\int\limits_0^{2\pi} \tilde{\mathbf{h}}_x^m(t;\omega_q)\sin(i\varphi)\mathrm{d}\varphi \\ \tilde{\mathbf{H}}_{x,c,i}^{m}(\omega_q) = \frac{1}{\pi}\int\limits_0^{2\pi} \tilde{\mathbf{h}}_x^m(t;\omega_q)\cos(i\varphi)\mathrm{d}\varphi \end{cases}.$$

In addition, the detailed expressions of $\tilde{\mathbf{H}}_{\tau}^{m}(\omega_q)$ and $\tilde{\mathbf{H}}_{\mu}^{m}(\omega_q)$ are similarly constructed.

Substituting Eqs. (4.133) and (4.134) into Eq. (4.132) yields

$$\begin{aligned} \min_{\tilde{\tau},\tilde{\mu},\delta\tilde{\mathbf{X}}(\omega_q)} & \Big\| \mathbf{\Phi}(t;\omega_q)\tilde{\mathbf{\Gamma}}(\omega_q) + \mathbf{\Phi}(t;\omega_q)\tilde{\mathbf{H}}_{\tau}(\omega_q)\mathbf{\Phi}(t-\tilde{\tau};\omega_q)\tilde{\mathbf{X}}(\omega_q)\delta\tilde{\tau} \\ & - \mathbf{\Phi}(t;\omega_q)\tilde{\mathbf{H}}_{\mu}(\omega_q)\delta\tilde{\mu} + \Big\{\dot{\Phi}(t;\omega_q) - \mathbf{\Phi}(t;\omega_q)\tilde{\mathbf{H}}_{x}(\omega_q)\mathbf{\Phi}(t;\omega_q) \\ & - \mathbf{\Phi}(t;\omega_q)\tilde{\mathbf{H}}_{\tau}(\omega_q)\mathbf{\Phi}(t-\tilde{\tau};\omega_q)\Big\}\mathbf{V}\delta\tilde{\mathbf{X}}(\omega_q)\Big\|. \end{aligned} \tag{4.135}$$

As can be seen, Eq. (4.135) contains derivative, delay, and high-order terms of the basis function $\mathbf{\Phi}(t;\omega_q)$. The following rules can be applied to regularize this expression.

Rule 1: Regularization of $\dot{\Phi}(t;\omega_q)$. The time derivative of $\mathbf{\Phi}(t;\omega_q)$ can be regularized to

$$\dot{\Phi}(t;\omega_q) = \mathbf{\Phi}(t;\omega_q)\mathbf{\Lambda}(\omega_q). \tag{4.136}$$

Proof The time derivative of $\mathbf{\Phi}(t;\omega_q)$ is given by

$$\begin{aligned} \dot{\Phi}(t;\omega_q) &= \dot{\varphi}^{T}(t;\omega_q)\otimes\mathbf{E}_M = \big(\boldsymbol{\lambda}(\omega_q)\varphi(t;\omega_q)\big)^{T}\otimes\mathbf{E}_M \\ &= \big(\varphi^{T}(t;\omega_q)\otimes\mathbf{E}_M\big)\big(\boldsymbol{\lambda}^{T}(\omega_q)\otimes\mathbf{E}_M\big) = \mathbf{\Phi}(t;\omega_q)\big(\boldsymbol{\lambda}^{T}(\omega_q)\otimes\mathbf{E}_M\big), \end{aligned} \tag{4.137}$$

where

$$\boldsymbol{\lambda}(\omega) = \left(\begin{array}{c|c|c|c} 0 & & & \\ \hline & \begin{matrix} 0 & \omega \\ -\omega & 0 \end{matrix} & & \\ \hline & & \ddots & \\ \hline & & & \begin{matrix} 0 & I\omega \\ -I\omega & 0 \end{matrix} \end{array}\right) \in \mathbb{R}^{(2I+1)\times(2I+1)}.$$

By comparing Eq. (4.137) with Eq. (4.136), it is not difficult to find that

$$\mathbf{\Lambda}(\omega_q) = \boldsymbol{\lambda}^{T}(\omega_q)\otimes\mathbf{E}_M \in \mathbb{R}^{(2I+1)M\times(2I+1)M} \tag{4.138}$$

Rule 2: Regularization of $\boldsymbol{\Phi}(t-\tilde{\tau};\omega_q)$. The delayed $\boldsymbol{\Phi}(t;\omega_q)$ can be regularized to

$$\boldsymbol{\Phi}(t-\tilde{\tau};\omega_q)=\boldsymbol{\Phi}(t;\omega_q)\boldsymbol{\Theta}(\omega_q). \tag{4.139}$$

Proof The time delay of $\boldsymbol{\Phi}_q(t)$ is given by

$$\begin{aligned}\boldsymbol{\Phi}(t-\tilde{\tau};\omega_q)&=\varphi^T(t-\tilde{\tau};\omega_q)\otimes\mathbf{E}_M=\left(\boldsymbol{\theta}(\omega_q)\varphi(t;\omega_q)\right)^T\otimes\mathbf{E}_M\\&=\left(\varphi^T(t;\omega_q)\otimes\mathbf{E}_M\right)\left(\boldsymbol{\theta}^T(\omega_q)\otimes\mathbf{E}_M\right)\\&=\boldsymbol{\Phi}(t;\omega_q)\left(\boldsymbol{\theta}^T(\omega_q)\otimes\mathbf{E}_M\right),\end{aligned} \tag{4.140}$$

where

$$\boldsymbol{\theta}(\omega)=\left(\begin{array}{c|c|c|c}1&&&\\\hline&\begin{matrix}\cos(\omega\tilde{\tau})&-\sin(\omega\tilde{\tau})\\\sin(\omega\tilde{\tau})&\cos(\omega\tilde{\tau})\end{matrix}&&\\\hline&&\ddots&\\\hline&&&\begin{matrix}\cos(I\omega\tilde{\tau})&-\sin(I\omega\tilde{\tau})\\\sin(I\omega\tilde{\tau})&\cos(I\omega\tilde{\tau})\end{matrix}\end{array}\right)\in\mathbb{R}^{(2I+1)\times(2I+1)}.$$

By comparing Eq. (4.140) with Eq. (4.139), it is convenient to find that

$$\boldsymbol{\Theta}(\omega_q)=\boldsymbol{\theta}^T(\omega_q)\otimes\mathbf{E}_M\in\mathbb{R}^{(2I+1)M\times(2I+1)M}. \tag{4.141}$$

Rule 3: Regularization of $\boldsymbol{\Phi}(t;\omega_q)\tilde{\mathbf{H}}_x(\omega_q)\boldsymbol{\Phi}(t;\omega_q)$. This high-order term can be regularized to

$$\boldsymbol{\Phi}(t;\omega_q)\tilde{\mathbf{H}}_x(\omega_q)\boldsymbol{\Phi}(t;\omega_q)=\boldsymbol{\Phi}(t;\omega_q)\tilde{\mathbf{G}}_x(\omega_q). \tag{4.142}$$

Proof According to the definition of $\boldsymbol{\Phi}(t;\omega_q)$, $\boldsymbol{\Phi}(t;\omega_q)\tilde{\mathbf{H}}_x(\omega_q)\boldsymbol{\Phi}(t;\omega_q)$ can be expressed as

$$\begin{aligned}&\boldsymbol{\Phi}(t;\omega_q)\tilde{\mathbf{H}}_x(\omega_q)\boldsymbol{\Phi}(t;\omega_q)\\&=\varphi^T(t;\omega_q)\otimes\tilde{\mathbf{h}}_x(t;\omega_q)\\&=\Big(\boldsymbol{\Phi}(t;\omega_q)\tilde{\mathbf{H}}_x(\omega_q),\sin(\omega_qt)\boldsymbol{\Phi}(t;\omega_q)\tilde{\mathbf{H}}_x(\omega_q),\cos(\omega_qt)\boldsymbol{\Phi}(t;\omega_q)\tilde{\mathbf{H}}_x(\omega_q)\\&\cdots\sin(I\omega_qt)\boldsymbol{\Phi}(t;\omega_q)\tilde{\mathbf{H}}_x(\omega_q),\cos(I\omega_qt)\boldsymbol{\Phi}(t;\omega_q)\tilde{\mathbf{H}}_x(\omega_q)\Big).\end{aligned} \tag{4.143}$$

It is not difficult to find that

$$\sin(i\omega_qt)\varphi(t;\omega_q)=\mathbf{u}_{i,s}\varphi(t;\omega_q), \tag{4.144}$$

where the construction of $\mathbf{u}_{i,s}\in\mathbb{R}^{(2I+1)\times(2I+1)}$ follows Table 4.7. Similarly,

$$\cos(i\omega_q t)\varphi(t;\,\omega_q) = \mathbf{u}_{i,c}\varphi(t;\,\omega_q), \tag{4.145}$$

where the construction of $\mathbf{u}_{i,c} \in \mathbb{R}^{(2I+1)\times(2I+1)}$ follows Table 4.8.

$j = 1, 2, \ldots, I$. *If the column index exceeds* $[1, 2I+1]$, *the entry assignment should be ignored.*

$j = 1, 2, \ldots, I$. *If the column index exceeds* $[1, 2I+1]$, *the entry assignment should be ignored.*

Hence, $\sin(i\omega_q t)\mathbf{\Phi}(t;\,\omega_q)$ and $\cos(i\omega_q t)\mathbf{\Phi}(t;\,\omega_q)$ can be represented as

$$\begin{aligned}\sin(i\omega_q t)\mathbf{\Phi}(t;\,\omega_q) &= \left(\sin(i\omega_q t)\varphi(t;\,\omega_q)\right)^T \otimes \mathbf{E}_M = \left(\mathbf{u}_{i,s}\varphi(t;\,\omega_q)\right)^T \otimes \mathbf{E}_M \\ &= \left(\varphi^T(t;\,\omega_q) \otimes \mathbf{E}_M\right)\left(\mathbf{u}_{i,s}^T \otimes \mathbf{E}_M\right) = \mathbf{\Phi}(t;\,\omega_q)\mathbf{U}_{i,s}\end{aligned} \tag{4.146}$$

and

$$\begin{aligned}\cos(i\omega_q t)\mathbf{\Phi}(t;\,\omega_q) &= \left(\cos(i\omega_q t)\varphi(t;\,\omega_q)\right)^T \otimes \mathbf{E}_M = \left(\mathbf{u}_{i,c}\varphi(t;\,\omega_q)\right)^T \otimes \mathbf{E}_M \\ &= \left(\varphi^T(t;\,\omega_q) \otimes \mathbf{E}_M\right)\left(\mathbf{u}_{i,c}^T \otimes \mathbf{E}_M\right) = \mathbf{\Phi}(t;\,\omega_q)\mathbf{U}_{i,c},\end{aligned} \tag{4.147}$$

where $\mathbf{U}_{i,s} = \mathbf{u}_{i,s}^T \otimes \mathbf{E}_M \in \mathbb{R}^{(2I+1)M\times(2I+1)M}$ and $\mathbf{U}_{i,c} = \mathbf{u}_{i,c}^T \otimes \mathbf{E}_M \in \mathbb{R}^{(2I+1)M\times(2I+1)M}$.

Substituting Eqs. (4.146) and (4.147) into Eq. (4.143) yields

$$\begin{aligned}&\mathbf{\Phi}(t;\,\omega_q)\tilde{\mathbf{H}}_x(\omega_q)\mathbf{\Phi}(t;\,\omega_q) \\ &= \mathbf{\Phi}(t;\,\omega_q)\left(\tilde{\mathbf{H}}_x(\omega_q), \mathbf{U}_{1,s}\tilde{\mathbf{H}}_x(\omega_q), \mathbf{U}_{1,c}\tilde{\mathbf{H}}_x(\omega_q), \ldots, \mathbf{U}_{I,s}\tilde{\mathbf{H}}_x(\omega_q), \mathbf{U}_{I,c}\tilde{\mathbf{H}}_x(\omega_q)\right).\end{aligned} \tag{4.148}$$

By comparing Eq. (4.148) with Eq. (4.142), it is easy to find that

$$\begin{aligned}\tilde{\mathbf{G}}_x(\omega_q) = &\left(\tilde{\mathbf{H}}_x(\omega_q)\ \mathbf{U}_{1,s}\tilde{\mathbf{H}}_x(\omega_q)\ \mathbf{U}_{1,c}\tilde{\mathbf{H}}_x(\omega_q)\ \cdots\ \mathbf{U}_{I,s}\tilde{\mathbf{H}}_x(\omega_q)\ \mathbf{U}_{I,c}\tilde{\mathbf{H}}_x(\omega_q)\right) \\ &\in \mathbb{R}^{(2I+1)M\times(2I+1)M}.\end{aligned}$$

Table 4.7 The assignment of entries in $\mathbf{u}_{i,s}$

Row	1	$2j$	$2j$	$2j+1$	$2j+1$
Column	$2i$	$2\|i-j\|+1$	$2(i+j)+1$	$2\|i-j\|$	$2(i+j)$
Assignment	1	$1/2$	$-1/2$	$\mathrm{sgn}(i-j)/2$	$1/2$

Table 4.8 The assignment of entries in $\mathbf{u}_{i,c}$

Row	1	$2j$	$2j$	$2j+1$	$2j+1$
Column	$2i+1$	$2\|i-j\|$	$2(i+j)$	$2\|i-j\|+1$	$2(i+j)+1$
Assignment	1	$-\mathrm{sgn}(i-j)/2$	$1/2$	$1/2$	$1/2$

Rule 4: Regularization of $\boldsymbol{\Phi}(t;\omega_q)\tilde{\mathbf{H}}_\tau(\omega_q)\boldsymbol{\Phi}(t;\omega_q)$. This high-order term can be regularized to

$$\boldsymbol{\Phi}(t;\omega_q)\tilde{\mathbf{H}}_\tau(\omega_q)\boldsymbol{\Phi}(t;\omega_q) = \boldsymbol{\Phi}(t;\omega_q)\tilde{\mathbf{G}}_\tau(\omega_q). \tag{4.149}$$

Proof Similar to the proof of Rule 3, it is not difficult to infer that

$$\tilde{\mathbf{G}}_\tau(\omega) = \left(\tilde{\mathbf{H}}_\tau(\omega)\ \mathbf{U}_{1,s}\tilde{\mathbf{H}}_\tau(\omega)\ \mathbf{U}_{1,c}\tilde{\mathbf{H}}_\tau(\omega)\ \cdots\ \mathbf{U}_{I,s}\tilde{\mathbf{H}}_\tau(\omega)\ \mathbf{U}_{I,c}\tilde{\mathbf{H}}_\tau(\omega)\right) \in \mathbb{R}^{(2I+1)M\times(2I+1)M}.$$

By applying the above four rules to Eq. (4.135), the objective function becomes

$$\min_{\tilde{\tau},\tilde{\boldsymbol{\mu}},\delta\tilde{\mathbf{X}}(\omega_q)} \left\| \boldsymbol{\Phi}(t;\omega_q)\left(\tilde{\boldsymbol{\Gamma}}(\omega_q) + \tilde{\mathbf{G}}_\tau(\omega_q)\boldsymbol{\Lambda}(\omega_q)\boldsymbol{\Theta}(\omega_q)\tilde{\mathbf{X}}(\omega_q)\delta\tilde{\tau} - \tilde{\mathbf{H}}_\mu(\omega_q)\delta\tilde{\boldsymbol{\mu}}\right)\right.$$
$$\left. + \boldsymbol{\Phi}(t;\omega_q)\left(\boldsymbol{\Lambda}(\omega_q) - \tilde{\mathbf{G}}_x(\omega_q) - \tilde{\mathbf{G}}_\tau(\omega_q)\boldsymbol{\Theta}(\omega_q)\right)\mathbf{V}\delta\tilde{\mathbf{X}}(\omega_q)\right\|. \tag{4.150}$$

Note that $\boldsymbol{\Phi}(t;\omega_q)$ is the basis function, Eq. (4.150) implies that the objective function given by Eq. (4.135) can be regularized to

$$\min_{\tilde{\tau},\tilde{\boldsymbol{\mu}},\delta\tilde{\mathbf{X}}(\omega_q)} \left\| \tilde{\boldsymbol{\Gamma}}(\omega_q) + \tilde{\mathbf{G}}_\tau(\omega_q)\boldsymbol{\Lambda}(\omega_q)\boldsymbol{\Theta}(\omega_q)\tilde{\mathbf{X}}(\omega_q)\delta\tilde{\tau} - \tilde{\mathbf{H}}_\mu(\omega_q)\delta\tilde{\boldsymbol{\mu}}\right.$$
$$\left. + \left(\boldsymbol{\Lambda}(\omega_q) - \tilde{\mathbf{G}}_x(\omega_q) - \tilde{\mathbf{G}}_\tau(\omega_q)\boldsymbol{\Theta}(\omega_q)\right)\mathbf{V}\delta\tilde{\mathbf{X}}(\omega_q)\right\|. \tag{4.151}$$

By comparing Eq. (4.151) with Eq. (4.135), it can be found that the objective function in the time domain has been transformed into the frequency domain through harmonic basis. Because of the orthogonality of the harmonic functions, this transformation also ensures a good property of resistance for broadband noise, *e.g.*, white noise.

4.4.3 Algorithm Construction

Since a standard objective function through local linearization and regularization has been obtained, it is convenient to give the identification process from Eq. (4.151). Note that the dimension of the error function shown in Eq. (4.151) is $M(2I+1)$ and the number of undetermined parameters is $1+P+M(2I-1)$. When $P > 2M-1$, the parameters will be underdetermined. To fix this problem, we consider multiple response samples with distinct primary frequencies, e.g. $\{\omega_1, \omega_2, \ldots, \omega_Q\}$, where $Q > (1+P)/(2M)$. Hence, Eq. (4.151) can be augmented to

$$\min_{\tilde{\tau},\tilde{\boldsymbol{\mu}},\delta\tilde{\mathbf{X}}(\omega_1),\ldots,\delta\tilde{\mathbf{X}}(\omega_Q)} \|\mathbf{A}\delta\mathbf{Y} + \mathbf{B}\|, \tag{4.152}$$

where

$$\mathbf{A} = \left(\mathbf{A}_l \ \mathbf{A}_r\right) \in \mathbb{R}^{(2I+1)MQ \times (1+P+(2I-1)MQ)}$$

$$\mathbf{A}_l = \begin{pmatrix} \tilde{\mathbf{G}}_\tau(\omega_1)\mathbf{\Lambda}(\omega_1)\mathbf{\Theta}(\omega_1)\tilde{\mathbf{X}}(\omega_1) & -\tilde{\mathbf{H}}_\mu(\omega_1) \\ \vdots & \vdots \\ \tilde{\mathbf{G}}_\tau(\omega_Q)\mathbf{\Lambda}(\omega_Q)\mathbf{\Theta}(\omega_Q)\tilde{\mathbf{X}}(\omega_Q) & -\tilde{\mathbf{H}}_\mu(\omega_Q) \end{pmatrix}$$

$$\mathbf{A}_r = \begin{pmatrix} \left(\mathbf{\Lambda}(\omega_1) - \tilde{\mathbf{G}}_x(\omega_1) - \tilde{\mathbf{G}}_\tau(\omega_1)\mathbf{\Theta}(\omega_1)\right)\mathbf{V} & & \\ & \ddots & \\ & & \left(\mathbf{\Lambda}(\omega_Q) - \tilde{\mathbf{G}}_x(\omega_Q) - \tilde{\mathbf{G}}_\tau(\omega_Q)\mathbf{\Theta}(\omega_Q)\right)\mathbf{V} \end{pmatrix}$$

$$\mathbf{B} = \begin{pmatrix} \tilde{\mathbf{\Gamma}}(\omega_1) \\ \vdots \\ \tilde{\mathbf{\Gamma}}(\omega_Q) \end{pmatrix} \in \mathbb{R}^{(2I+1)MQ}, \text{ and } \delta\mathbf{Y} = \begin{pmatrix} \delta\tilde{\tau} \\ \delta\tilde{\boldsymbol{\mu}} \\ \hline \delta\tilde{\mathbf{X}}(\omega_1) \\ \vdots \\ \delta\tilde{\mathbf{X}}(\omega_Q) \end{pmatrix} \in \mathbb{R}^{1+P+(2I-1)MQ}.$$

The optimal solution of $\delta\mathbf{Y}$ can be given as

$$\delta\mathbf{Y} = -\left(\mathbf{A}^T\mathbf{A}\right)^{-1}\mathbf{A}^T\mathbf{B}. \tag{4.153}$$

Then, one can use the calculated corrections to update $\tilde{\tau}$, $\tilde{\boldsymbol{\mu}}$, and $\tilde{\mathbf{x}}\left(t; \omega_q\right)$. By repeating the calculation of corrections and the updating of parameters until $\|\delta\mathbf{Y}\|$ deceases to a small level, say ϵ, we then consider that the noise-correction-based parameter identification finishes.

4.4.4 Convergence Analysis

The above subsection gave the scheme of how to identify unknown parameters with adaptive noise correction. The convergence of this scheme should be verified before its application to real examples. To this end, some preliminary statements should be stated prior.

Lemma 4.1 If the columns of $\tilde{\mathbf{h}}_x\left(t; \omega_q\right)$, $\tilde{\mathbf{h}}_\tau\left(t; \omega_q\right)\dot{\tilde{\mathbf{x}}}\left(t - \tilde{\tau}; \omega_q\right)$, and $\tilde{\mathbf{h}}_\mu\left(t; \omega_q\right)$ are linearly independent, then the columns of the corresponding coefficient matrices

$$\tilde{\mathbf{G}}_\tau\left(\omega_q\right)\mathbf{\Lambda}\left(\omega_q\right)\mathbf{\Theta}\left(\omega_q\right)\tilde{\mathbf{X}}\left(\omega_q\right), -\tilde{\mathbf{H}}_\mu\left(\omega_q\right),$$

and

$$\left(\mathbf{\Lambda}(\omega_q) - \tilde{\mathbf{G}}_x(\omega_q) - \tilde{\mathbf{G}}_\tau(\omega_q)\mathbf{\Theta}(\omega_q)\right)\mathbf{V}$$

are linearly independent.

Proof By referring to the definition of noise correction, the error function can be written as

$$\begin{aligned}\boldsymbol{\varepsilon}(t;\omega_q) &= \tilde{\boldsymbol{\gamma}}(t;\omega_q) + \tilde{\mathbf{h}}_\tau(t;\omega_q)\dot{\tilde{\mathbf{x}}}(t-\tilde{\tau};\omega_q)\delta\tilde{\tau} - \tilde{\mathbf{h}}_\mu(t;\omega_q)\delta\tilde{\boldsymbol{\mu}} \\ &+ \left\{\dot{\boldsymbol{\psi}}^T(t;\omega_q)\otimes\mathbf{E}_M - \boldsymbol{\psi}^T(t;\omega_q)\otimes\tilde{\mathbf{h}}_x(t;\omega_q)\right. \\ &\left. - \boldsymbol{\psi}^T(t-\tilde{\tau};\omega_q)\otimes\tilde{\mathbf{h}}_\tau(t;\omega_q)\right\}\delta\tilde{\mathbf{X}}(\omega_q),\end{aligned} \tag{4.154}$$

where the detailed form of $\boldsymbol{\psi}^T(t;\omega_q)\otimes\tilde{\mathbf{h}}_x(t;\omega_q)$ is

$$\begin{aligned}\boldsymbol{\psi}^T(t;\omega_q)\otimes\tilde{\mathbf{h}}_x(t;\omega_q) = (&\tilde{\mathbf{h}}_x(t;\omega_q)\ \sin(2\omega_q t)\tilde{\mathbf{h}}_x(t;\omega_q)\ \cos(2\omega_q t)\tilde{\mathbf{h}}_x(t;\omega_q) \\ &\cdots\ \sin(I\omega_q t)\tilde{\mathbf{h}}_x(t;\omega_q)\ \cos(I\omega_q t)\tilde{\mathbf{h}}_x(t;\omega_q)).\end{aligned}$$

Since the harmonic functions are independent variables and the columns of $\tilde{\mathbf{h}}_x(t;\omega_q)$ are linearly independent, the columns of $\boldsymbol{\psi}^T(t;\omega_q)\otimes\tilde{\mathbf{h}}_x(t;\omega_q)$ are linearly independent. Furthermore, considering the independence of $\tilde{\mathbf{h}}_x(t;\omega_q)$, $\tilde{\mathbf{h}}_\tau(t;\omega_q)\dot{\tilde{\mathbf{x}}}(t-\tilde{\tau};\omega_q)$, and $\tilde{\mathbf{h}}_\mu(t;\omega_q)$, we can infer that the columns of

$$\tilde{\mathbf{h}}_\tau(t;\omega_q)\dot{\tilde{\mathbf{x}}}(t-\tilde{\tau};\omega_q), \tilde{\mathbf{h}}_\mu(t;\omega_q).$$

and

$$\dot{\boldsymbol{\psi}}^T(t;\omega_q)\otimes\mathbf{E}_M - \boldsymbol{\psi}^T(t;\omega_q)\otimes\tilde{\mathbf{h}}_x(t;\omega_q) - \boldsymbol{\psi}^T(t-\tilde{\tau};\omega_q)\otimes\tilde{\mathbf{h}}_\tau(t;\omega_q),$$

as shown in Eq. (4.1.154), are linearly independent.

The Fourier series expansion, denoted as $\mathscr{F}(\bullet)$, is a linear transformation.

$$\tilde{\mathbf{G}}_\tau(\omega_q)\mathbf{\Lambda}(\omega_q)\mathbf{\Theta}(\omega_q)\tilde{\mathbf{X}}(\omega_q), -\tilde{\mathbf{H}}_\mu(\omega_q),$$

and

$$\left(\mathbf{\Lambda}(\omega_q) - \tilde{\mathbf{G}}_x(\omega_q) - \tilde{\mathbf{G}}_\tau(\omega_q)\mathbf{\Theta}(\omega_q)\right)\mathbf{V}$$

are the coefficient matrices of

$$\tilde{\mathbf{h}}_\tau(t;\omega_q)\dot{\mathbf{x}}(t-\tilde{\tau};\omega_q),\ -\tilde{\mathbf{h}}_\mu(t;\omega_q),$$

and

$$\dot{\boldsymbol{\psi}}^T(t;\omega_q)\otimes\mathbf{E}_M-\boldsymbol{\psi}^T(t;\omega_q)\otimes\tilde{\mathbf{h}}_x(t;\omega_q)-\boldsymbol{\psi}^T(t-\tilde{\tau};\omega_q)\otimes\tilde{\mathbf{h}}_\tau(t;\omega_q)$$

in harmonic basis. If the columns of $\tilde{\mathbf{G}}_\tau(\omega_q)\boldsymbol{\Lambda}(\omega_q)\boldsymbol{\Theta}(\omega_q)\tilde{\mathbf{X}}(\omega_q)$, $-\tilde{\mathbf{H}}_\mu(\omega_q)$, and $\left(\boldsymbol{\Lambda}(\omega_q)-\tilde{\mathbf{G}}_x(\omega_q)-\tilde{\mathbf{G}}_\tau(\omega_q)\boldsymbol{\Theta}(\omega_q)\right)\mathbf{V}$ are linearly dependent, then there must be a nonzero constant vector $\mathbf{u}$ satisfying

$$\Big(\tilde{\mathbf{h}}_\tau(t;\omega_q)\dot{\mathbf{x}}(t-\tilde{\tau};\omega_q)\ -\tilde{\mathbf{h}}_\mu(t;\omega_q)$$
$$\dot{\boldsymbol{\psi}}^T(t;\omega_q)\otimes\mathbf{E}_M-\boldsymbol{\psi}^T(t;\omega_q)\otimes\tilde{\mathbf{h}}_x(t;\omega_q)-\boldsymbol{\psi}^T(t-\tilde{\tau};\omega_q)\otimes\tilde{\mathbf{h}}_\tau(t;\omega_q)\Big)\mathbf{u}=\mathbf{0}, \tag{4.155}$$

which is contrary to the condition of linear independence. Therefore, the columns of

$$\tilde{\mathbf{G}}_\tau(\omega_q)\boldsymbol{\Lambda}(\omega_q)\boldsymbol{\Theta}(\omega_q)\tilde{\mathbf{X}}(\omega_q),\ -\tilde{\mathbf{H}}_\mu(\omega_q),$$

and

$$\left(\boldsymbol{\Lambda}(\omega_q)-\tilde{\mathbf{G}}_x(\omega_q)-\tilde{\mathbf{G}}_\tau(\omega_q)\boldsymbol{\Theta}(\omega_q)\right)\mathbf{V}$$

are linearly independent. ■

Corollary *The set of true parameters is a fixed point of the iteration given by ANCPI.*

Proof Substituting the true parameters into the residue function yields $\tilde{\boldsymbol{\gamma}}(t;\omega)=\mathbf{0}$. In this case, the solution that minimizes the error, given by Eq. (4.154), should satisfy

$$\begin{aligned}\mathbf{0}=&\,\tilde{\mathbf{h}}_\tau(t;\omega_q)\dot{\mathbf{x}}(t-\tilde{\tau};\omega_q)\delta\tilde{\tau}-\tilde{\mathbf{h}}_\mu(t;\omega_q)\delta\tilde{\boldsymbol{\mu}}\\&+\Big\{\dot{\boldsymbol{\psi}}^T(t;\omega_q)\otimes\mathbf{E}_M-\boldsymbol{\psi}^T(t;\omega_q)\otimes\tilde{\mathbf{h}}_x(t;\omega_q)\\&-\boldsymbol{\psi}^T(t-\tilde{\tau};\omega_q)\otimes\tilde{\mathbf{h}}_\tau(t;\omega_q)\Big\}\delta\tilde{\mathbf{X}}(\omega_q).\end{aligned} \tag{4.156}$$

Since the coefficient matrices are linearly independent, the solutions of $\delta\tilde{\tau}$, $\delta\tilde{\boldsymbol{\mu}}$, and $\delta\tilde{\mathbf{X}}(\omega_q)$ must be zero. This implies that the set of true parameters is the fixed point. ■

Additionally, since the columns of coefficient matrices

$$\tilde{\mathbf{G}}_\tau(\omega_q)\boldsymbol{\Lambda}(\omega_q)\boldsymbol{\Theta}(\omega_q)\tilde{\mathbf{X}}(\omega_q),\ -\tilde{\mathbf{H}}_\mu(\omega_q),$$

and

$$\left(\mathbf{\Lambda}\left(\omega_q\right) - \tilde{\mathbf{G}}_x\left(\omega_q\right) - \tilde{\mathbf{G}}_\tau\left(\omega_q\right)\mathbf{\Theta}\left(\omega_q\right)\right)\mathbf{V}$$

are linearly independent, we can properly select a frequency sequence $\{\omega_1, \omega_2, \ldots, \omega_Q\}$, where $Q > (1 + P)/(2M)$, so that the augmented matrix $\mathbf{A}$, as shown in Eq. (4.152), is of full column rank.

Lemma 4.2 *If the coefficient matrix* $\mathbf{A}$ *is of full column rank, then its Gram matrix* $\mathbf{A}^T\mathbf{A}$ *is invertible.*

Proof Suppose that the Gram matrix $\mathbf{A}^T\mathbf{A}$ is not invertible, which implies that $\mathbf{A}^T\mathbf{A}$ is not full rank, then there must be a nonzero vector $\mathbf{u}$ satisfying

$$\mathbf{A}^T\mathbf{A}\mathbf{u} = \mathbf{0}. \tag{4.157}$$

Left multiplying $\mathbf{u}^T$ to Eq. (4.157) yields

$$\mathbf{u}^T\mathbf{A}^T\mathbf{A}\mathbf{u} = (\mathbf{A}\mathbf{u})^T(\mathbf{A}\mathbf{u}) = \|\mathbf{A}\mathbf{u}\|^2 = \mathbf{0}, \tag{4.158}$$

which means the matrix $\mathbf{A}$ is not full-column rank. It is contrary to the assumption in this lemma. Therefore, the Gram matrix $\mathbf{A}^T\mathbf{A}$ is invertible. ■

By using the above lemmas, it is easy to give the theorem of convergence for the constructed algorithm.

Theorem 4.3 *If* $\tilde{\mathbf{h}}_x\left(t; \omega_q\right)$, $\tilde{\mathbf{h}}_\tau\left(t; \omega_q\right)\dot{\mathbf{x}}\left(t - \tilde{\tau}; \omega_q\right)$, *and* $\tilde{\mathbf{h}}_\mu\left(t; \omega_q\right)$ *are bounded and the columns of these matrices are linearly independent, then the identification is convergent.*

Proof The objective function given by Eq. (4.152) is a locally linearized form. It is in the same order with $\|\delta\mathbf{Y}\|^2$. Because of the condition of boundedness, there must be a constant α that makes

$$\left\|\mathbf{A}^T\mathbf{A}\delta\mathbf{Y} + \mathbf{A}^T\mathbf{B}\right\| \le \alpha\|\delta\mathbf{Y}\|^2. \tag{4.159}$$

In addition, from Lemmas 4.1 and 4.2, we know that the Gram matrix $\mathbf{A}^T\mathbf{A}$ is invertible. Thus, there must be a constant β, which makes

$$\left\|\left(\mathbf{A}^T\mathbf{A}\right)^{-1}\right\| \le \beta. \tag{4.160}$$

Suppose that the fixed point is $\mathbf{Y}_*$ and the estimated parameter in the kth iteration is $\mathbf{Y}_k$, then the norm of $\mathbf{Y}_1 - \mathbf{Y}_*$ can be obtained as

$$\begin{aligned}\|\mathbf{Y}_1 - \mathbf{Y}_*\| &= \|\mathbf{Y}_0 + \delta\mathbf{Y}_0 - \mathbf{Y}_*\| = \left\|\mathbf{Y}_0 - \left(\mathbf{A}_0^T\mathbf{A}_0\right)^{-1}\mathbf{A}_0^T\mathbf{B}_0 - \mathbf{Y}_*\right\| \\ &= \left\|\left(\mathbf{A}_0^T\mathbf{A}_0\right)^{-1}\left((\mathbf{A}_0^T\mathbf{A}_0)(\mathbf{Y}_0 - \mathbf{Y}_*) - \mathbf{A}_0^T\mathbf{B}_0\right)\right\|. \end{aligned} \tag{4.161}$$

Furthermore, substituting Eqs. (4.159) and (4.160) into Eq. (4.161) yields

$$\begin{aligned}\|\mathbf{Y}_1 - \mathbf{Y}_*\| &= \left\|\left(\mathbf{A}_0^T\mathbf{A}_0\right)^{-1}\left((\mathbf{A}_0^T\mathbf{A}_0)(\mathbf{Y}_0 - \mathbf{Y}_*) - \mathbf{A}_0^T\mathbf{B}_0\right)\right\| \\ &\le \left\|\left(\mathbf{A}_0^T\mathbf{A}_0\right)^{-1}\right\| \left\|(\mathbf{A}_0^T\mathbf{A}_0)(\mathbf{Y}_0 - \mathbf{Y}_*) - \mathbf{A}_0^T\mathbf{B}_0\right\| \\ &\le \alpha\beta\|\mathbf{Y}_0 - \mathbf{Y}_*\|^2. \end{aligned} \tag{4.162}$$

If the initial guess $\mathbf{Y}_0$ satisfies $\|\mathbf{Y}_0 - \mathbf{Y}_*\| \le d/\alpha\beta$, where $d < 1$, then Eq. (4.162) implies $\|\mathbf{Y}_1 - \mathbf{Y}_*\| \le d\|\mathbf{Y}_0 - \mathbf{Y}_*\|$. Applying the induction rule, we further have

$$\|\mathbf{Y}_k - \mathbf{Y}_*\| \le d^k\|\mathbf{Y}_0 - \mathbf{Y}_*\|. \tag{4.163}$$

Since $d < 1$, Eq. (4.163) implies that the iteration is a contraction mapping and $\mathbf{Y}_k$ will converge to the fixed point $\mathbf{Y}_*$. Because the set of true parameters is a fixed point, which has been proved in the Corollary, the identification algorithm will finally give results in true ones. ■

Example 4.12 Example 4.11 has demonstrated that when the streaming of the nonlinear response cannot be recorded, the identification based on the harmonic balance would be significantly biased. This example will show whether the modified identification could fix this problem.

To apply the noise-correction-based identification procedure, one should first three Jacobian matrices of the governing function, i.e., $\tilde{\mathbf{h}}_\tau(t;\omega_q)$, $\tilde{\mathbf{h}}_\mu(t;\omega_q)$, and $\tilde{\mathbf{h}}_x(t;\omega_q)$. Note that the governing function corresponding to Eq. (4.122) is

$$\mathbf{g}(\mathbf{x}(t), \mathbf{x}(t-\tau), \mathbf{f}(t); \boldsymbol{\mu}) = \begin{pmatrix} \dot{y} \\ -c\dot{y} - k_1 y - k_2 y^2 - k_3 y^3 + f + gy(t-\tau) \end{pmatrix}. \tag{4.164}$$

Substituting it into the definitions of $\tilde{\mathbf{h}}_\tau(t;\omega_q)$, $\tilde{\mathbf{h}}_\mu(t;\omega_q)$, and $\tilde{\mathbf{h}}_x(t;\omega_q)$, which are given in Eq. (4.132), yields

$$\tilde{\mathbf{h}}_x(t;\omega_q) = \frac{\partial\mathbf{g}}{\partial\tilde{\mathbf{x}}^T(t;\omega_q)} = -\begin{pmatrix} 0 & 0 \\ \tilde{k}_1 + 2\tilde{k}_2\tilde{y} + 3\tilde{k}_3\tilde{y}^2 & c \end{pmatrix},$$

$$\tilde{\mathbf{h}}_\tau(t;\omega_q) = \frac{\partial\mathbf{g}}{\partial\tilde{\mathbf{x}}^T(t-\tilde{\tau};\omega_q)} = \begin{pmatrix} 0 & 0 \\ g & 0 \end{pmatrix},$$

and

$$\tilde{\mathbf{h}}_{\mu}\left(t;\omega_q\right)=\frac{\partial \mathbf{g}}{\partial\,\tilde{\boldsymbol{\mu}}^T}=\begin{pmatrix}0 & 0 & 0 & 0 & 0\\ -\dot{\tilde{y}} & -\tilde{y} & -\tilde{y}^2 & -\tilde{y}^3 & \tilde{y}(t-\tilde{\tau})\end{pmatrix}.$$

By further substituting them into Eq. (4.133), one can collect the harmonic coefficient matrices, i.e., $\tilde{\mathbf{H}}_x\left(\omega_q\right)$, $\tilde{\mathbf{H}}_\tau\left(\omega_q\right)$, and $\tilde{\mathbf{H}}_\mu\left(\omega_q\right)$.

Note that once the harmonic order has been determined, the regularization matrices $\boldsymbol{\Lambda}\left(\omega_q\right)$, $\boldsymbol{\Theta}\left(\omega_q\right)$, $\mathbf{U}_{i,s}$, and $\mathbf{U}_{i,c}$ can be constructed by referring to Eqs. (4.138) and (4.141), Tables 4.7, and 4.8, respectively. By substituting the harmonic coefficient matrices, i.e., $\tilde{\mathbf{H}}_x\left(\omega_q\right)$, $\tilde{\mathbf{H}}_\tau\left(\omega_q\right)$, and $\tilde{\mathbf{H}}_\mu\left(\omega_q\right)$, and the regularization matrices $\boldsymbol{\Lambda}\left(\omega_q\right)$, $\boldsymbol{\Theta}\left(\omega_q\right)$, $\mathbf{U}_{i,s}$, and $\mathbf{U}_{i,c}$ into Eq. (4.153), the unknown parameters, together with the missing streaming, can be identified. Figure 4.15 shows the identification process of the structural parameters and time delay, where one can find by comparing them with those without noise correction that the modified identification algorithm can accurately identify the unknown parameters.

Example 4.13 Note that the proposed identification algorithm is designed for multiple DOF nonlinear systems with multiple time delays. However, the examples mentioned above did not demonstrate such practicability. To this end, here we take the following system for example.

$$\dot{\mathbf{x}}(t)=\mathbf{g}(\mathbf{x}(t),\mathbf{x}(t-\tau_1),\mathbf{x}(t-\tau_2),\mathbf{f}(t);\boldsymbol{\mu}),\tag{4.165}$$

where

$$\mathbf{x}=\left(x_1\ x_2\ x_3\ x_4\ x_5\ x_6\ x_7\ x_8\right)^T,$$

$$\mathbf{g}(\mathbf{x}(t),\mathbf{x}(t-\tau_1),\mathbf{x}(t-\tau_2),\mathbf{f}(t);\boldsymbol{\mu})=\begin{pmatrix} x_5\\ x_6\\ x_7\\ x_8\\ -k_1(x_1-x_3)-c_1(x_5-x_7)-k_4(x_1(t-\tau_1)-x_3(t-\tau_1))\\ -k_2(x_2-x_4)-c_2(x_6-x_8)-k_5(x_2(t-\tau_2)-x_4(t-\tau_2))\\ g_7\\ g_8\end{pmatrix},$$

$$\begin{aligned}g_7=&\,k_1(x_1-x_3)+c_{s1}(x_5-x_7)+k_4(x_1(t-\tau_1)-x_3(t-\tau_1))\\ &-k_3\left(1-\frac{4}{\sqrt{(x_3-x_4)^2+16}}\right)(x_3-x_4)-c_3\frac{(x_3-x_4)^2}{(x_3-x_4)^2+16}(x_7-x_8)\end{aligned}$$

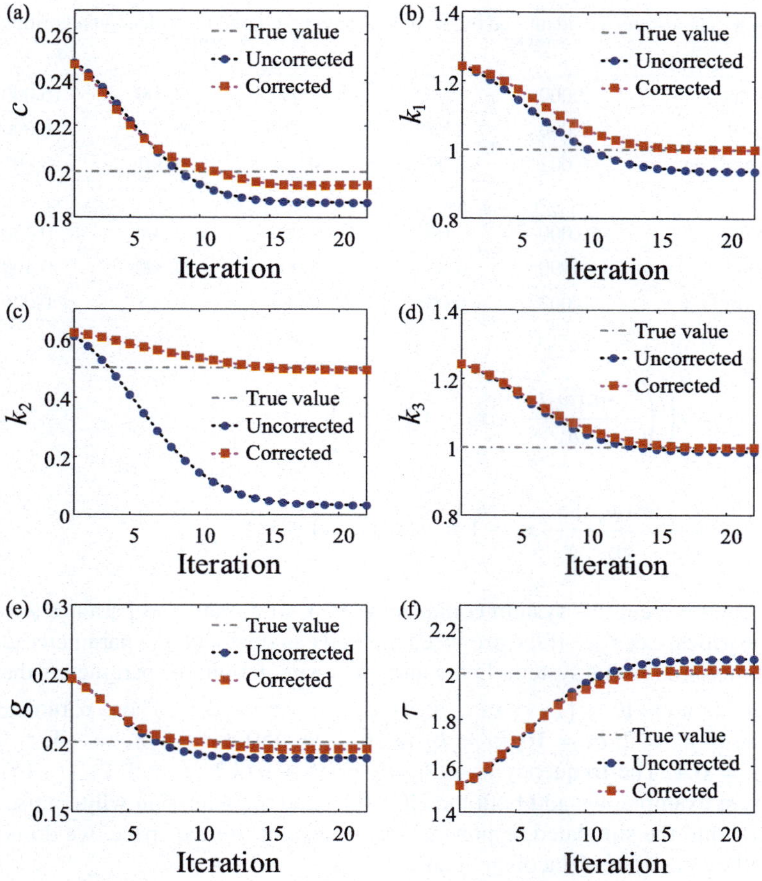

Fig. 4.15 Identification process with streaming correction: **a** damping coefficients, **b** linear stiffness, **c** square stiffness, **d** cubic stiffness, **e** feedback gain, and **f** time delay

$$+40\left(\left(\frac{\sin(\omega_q t)}{\omega_q}-x_3+1\right)^2-1\right)+\left(\frac{\sin(\omega_q t)}{\omega_q}-x_3+1\right)^{3/2}(\cos(\omega_q t)-x_7),$$

and

$$g_8 = k_2(x_2-x_4)+c_2(x_6-x_8)+k_5(x_2(t-\tau_2)-x_4(t-\tau_2))$$
$$+k_3\left(1-\frac{4}{\sqrt{(x_3-x_4)^2+16}}\right)(x_3-x_4)+c_3\frac{(x_3-x_4)^2}{(x_3-x_4)^2+16}(x_7-x_8)$$

Table 4.9 The parameters identified by the noise-correction based identification algorithm

	c_1	c_2	c_3	k_1	k_2
True value	2.000	2.000	10.000	16.000	16.000
Identified	2.000	2.000	10.013	16.000	16.000
Rel. Error (%)	0.002	0.001	0.051	0.001	0.001
	k_3	k_4	k_5	τ_1	τ_2
True value	450.000	4.000	– 4.000	0.300	0.400
Identified	449.990	4.000	– 4.000	0.300	0.400
Rel. Error (%)	0.002	0.001	0.003	0.001	0.000

$$+40\left(\left(-\frac{\sin(\omega_q t)}{\omega_q} - x_4 + 1\right)^2 - 1\right)$$

$$-\left(-\frac{\sin(\omega_q t)}{\omega_q} - x_4 + 1\right)^{3/2} \left(\cos(\omega_q t) + x_8\right).$$

As can be seen, this system contains rational, irrational, and parametric excited nonlinearities. Let's see if the proposed method can deal with the parameter identification mission for this system. In the numerical simulation, the parameters that need identification are $\boldsymbol{\mu} = \left(c_1\ c_2\ c_3\ k_1\ k_2\ k_3\ k_4\ k_5\right)^T$, τ_1, and τ_2. These parameters are set as $c_1 = c_2 = 2$, $c_3 = 10$, $k_1 = k_2 = 16$, $k_3 = 450$, $k_4 = 4$, $k_5 = -4$, $\tau_1 = 0.3$, and $\tau_2 = 0.4$. The frequency is set as $\omega_q = (5+q)/2$ $(q = 1, 2, \ldots, 15)$. Like the above example, we add both the DC offset and the Gaussian white noise (SNR = 6 dB) into the simulated response. For the sake of limited space, we do not give detailed expressions of Jacobian matrices.

In the identification process, we set the state dimension M as 8, the excitation dimension N as 1, the parameter dimension P as 8, the harmonic order I as 5, and the number of samples Q as 15. The initial parameters are set as $\widetilde{\boldsymbol{\mu}} = \left(2.6\ 2.6\ 13\ 20.8\ 20.8\ 585\ 5.2\ {-5.2}\right)^T$, $\tau_1 = 0.39$, and $\tau_2 = 0.52$, which are 30% deviated from the true values. By loading the response samples, the parameters identified by the proposed identification method are given in Table 4.9. As can be seen, the identified results are very accurate.

4.5 Experiment Realization and Validations

The problem of parameter identification for nonlinear systems with time delay has been comprehensively studied in the above sections, including qualitative feature analysis, identification algorithm construction, uniqueness and convergence analysis, and abundant numerical simulations. However, the algorithms' practicability for real applications has not been verified.

Note that the highlight of the proposed algorithms is accurate time delay identification. Therefore, the experiment example can be considered valid unless the mentioned highlight can be directly verified. To this end, the most challenging issue is how to realize precise time delay tuning.

4.5.1 Realization of the Time-Delayed Control via YASKAWA Hardware

The conventional technique usually uses analog low-pass filters, which produce the so-called group delay, to realize the effect of time delay. However, such group delay is phase-dependent and often causes frequency dispersion. Figure 4.16 shows the input of a half-sine pulse with a duration of 0.01 s and the output signal through low pass filters with different cutoff frequencies. Note that the main lobe width of the half-sine pulse with a duration of 0.01 s is 150 Hz. One can find from Fig. 4.16 that when the cutoff frequency is higher than the main lobe width of the input signal, the output will not be distorted, indicating that the low pass filter is an ideal time delayer. In addition, the lower the cutoff frequency is, the larger time delay the low pass filter will induce, indicating that one can tune the time delay by changing the cutoff frequency. Hence, some limitations for real applications emerge. First, the range of time delay is limited because the cutoff frequency cannot be set too low to avoid frequency dispersion. Second, the time delay cannot be accurately tuned because changing the cutoff frequency is an indirect parameter tuning method. To fix this problem, a novel and reliable time delay realization platform should be constructed.

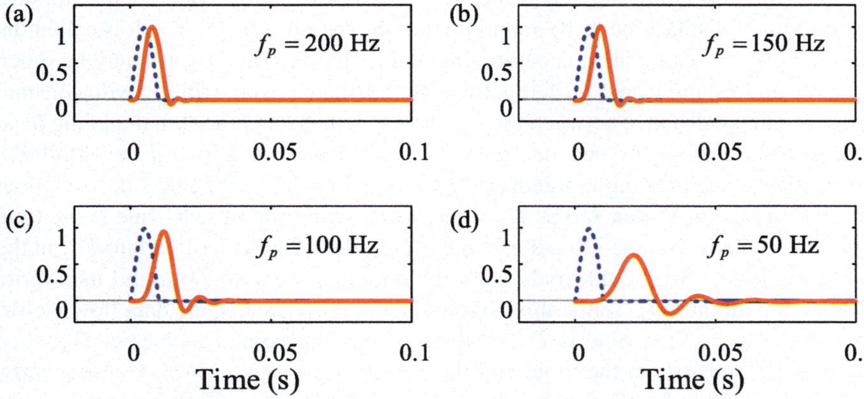

Fig. 4.16 The effect of time delay induced by low pass filtering: **a** cutoff frequency is 200 Hz, **b** cutoff frequency is 150 Hz, **c** cutoff frequency is 100 Hz, and **d** cutoff frequency is 50 Hz, where blue lines are input signal and solid red lines are outputs

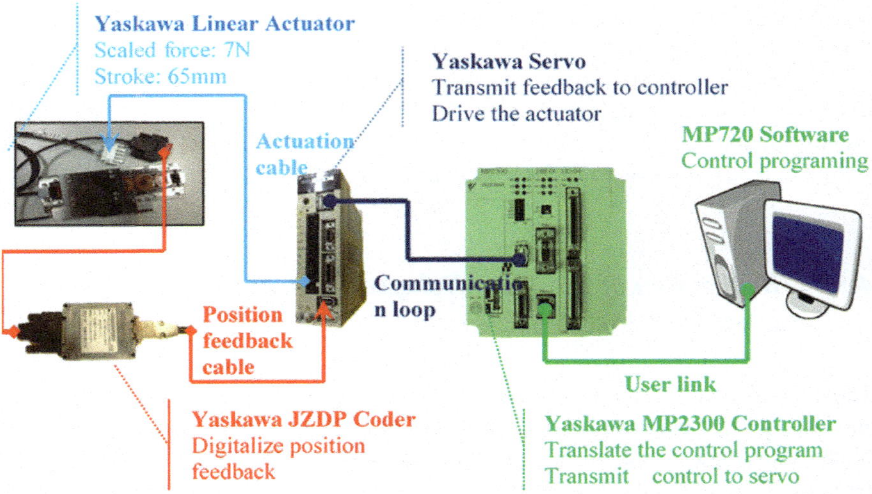

Fig. 4.17 Control loop of the time delay realization platform

This study chooses YASKAWA control hardware to construct the time delay realization platform [15]. Figure 4.17 illustrates the whole architecture of the control loop, which consists of a grating ruler as the displacement sensor, a linear motor as the actuator, and a controller. The linear motor (YASKAWA SGTMM03-065AH20A) has a rated thrust of 7 N with a linear stroke of 65 mm. The embedded grating ruler has a displacement resolution of 0.5 μm, which promises high-accuracy sensing at long stroke. The controller (YASKAWA MP2300), which can be programmed through the MPE720 software on a PC, supports a scan time as low as 1 ms, indicating that the control frequency can be set as high as 1000 Hz.

The MP2300 controller has a memory of 1024 kB to provide a table to save the measured displacement. By manipulating the memory table, which is called the digital flow, one can realize accurate time delay tuning. The process may be much easier to understand from Fig. 4.18a. Imagine that there is a tape rolling with constant speed v and the distance between the signal writing head and the signal reading head is l, then the time-delay between the input and the output is $\tau = l/v$. Similarly, we virtualize such rolling in the digital memory. To achieve this, let's define a float array with length N, denoted as $\mathbf{D}$. In the initial scan time (a scan time is the time cost of completing one fetch-and-execute cycle), the signal is written into D_1. In the next scan time, one should firstly copy the data in $D_{1:N-1}$ to $D_{2:N}$ and then write the new signal into D_1. While this copy-write operation keeps, the data flow yields. Suppose that the scan time is T_s and the data reading pointer locates at D_N, then the time-delay between the input and the output is $\tau = (N-1)T_s$. Consequently, the time delay can be physically tuned by changing N. If multiple time-delays are needed, multiple data reading pointers shall be applied.

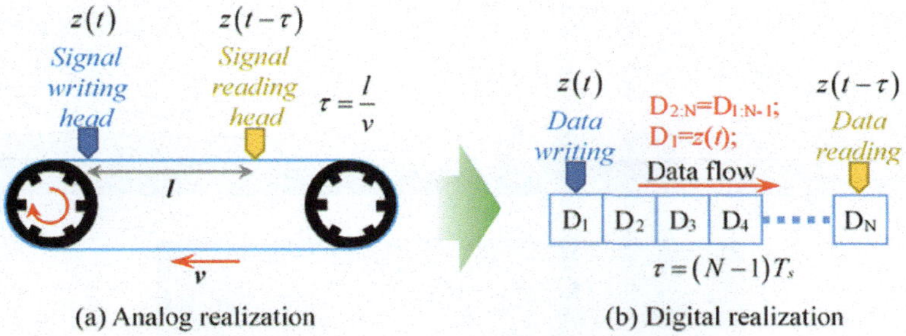

Fig. 4.18 Mechanism of time delay realization

Since the mechanism of time delay realization has been explained, the corresponding programming in MPE720 software can be conveniently introduced. The start interface of the MPE720 is shown in Fig. 4.19, where one can set the scan time for the controller by clicking the "Communications Setting", say, 1 ms as shown in Fig. 4.20, at first. Then, one may apply a linear table with a certain length, say, 200, to save the measured data. Note that the length of the default data format, headed as "MW" in the MPE720 programs, occupies two bytes. However, the signal used for feedback control is the floating number with double precision, which is headed as "ML" and occupies four bytes. Therefore, the linear table with 200 elements can only save 100 floating numbers. According to the digital flow shown in Fig. 4.20, the maximum time delay that this table can realize is 100 ms.

To move the data saved in the linear table, one may apply the "COPYW" command. As the example shown in Fig. 4.21, the "COPYW" command forward shifts the table, which has 200 elements, by two elements at each scan time. Because the head point of the table is "MW20000", the new data array will start from "MW20002" after the 2-element forward shifting. After forward shifting, the new displacement data that is read from the grating ruler is written to "MW20000". Then, the operation of forward shifting and new data saving repeat at each scan time so that the data float is achieved.

Suppose that the time-delayed feedback control is $f_c = gx(t - \tau)$, where f_c is the actuation force that the linear motor should generate, x means the displacement of the linear motor that is measured by the grating ruler, g is the tunable feedback gain, and τ is the tunable time delay. As is shown in Line-0004 in Fig. 4.22, the actuation force is saved in "DF00014" and the feedback gain is saved in "MF00300". "DF00002" and "DF00004" are rescaling constants. If one wants to tune the time delay as 10 ms, then the displacement data should be read from "ML20020". By setting "MB010001" in Line 0006 as 1, the actuation will be enabled. The subsequent "STORE" command first saves an int "24" to "OW8108" to enable force control mode, and then saves the data from "DF00014" to "OL810C" to make the actuator generate the actuation force as indicated in "DF00014".

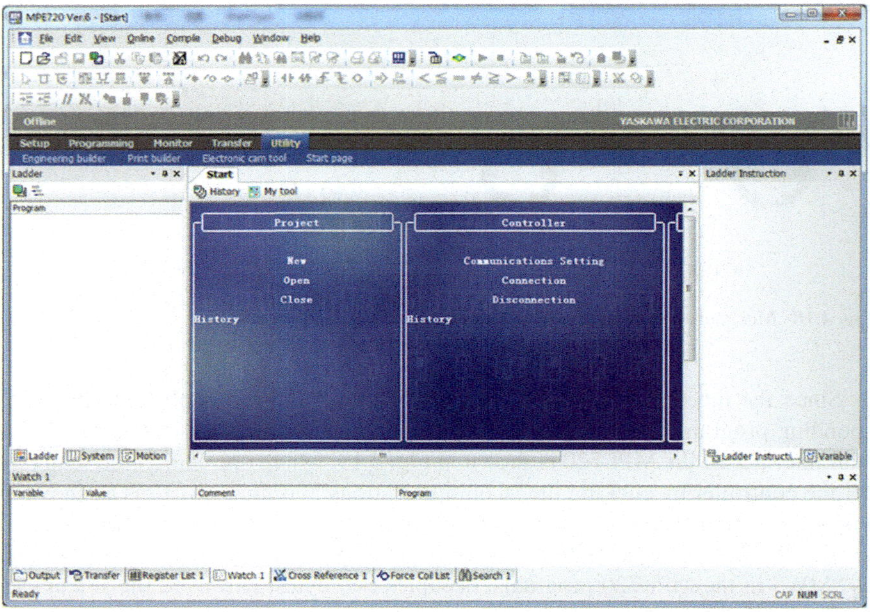

Fig. 4.19 The start interface of the MPE720 software

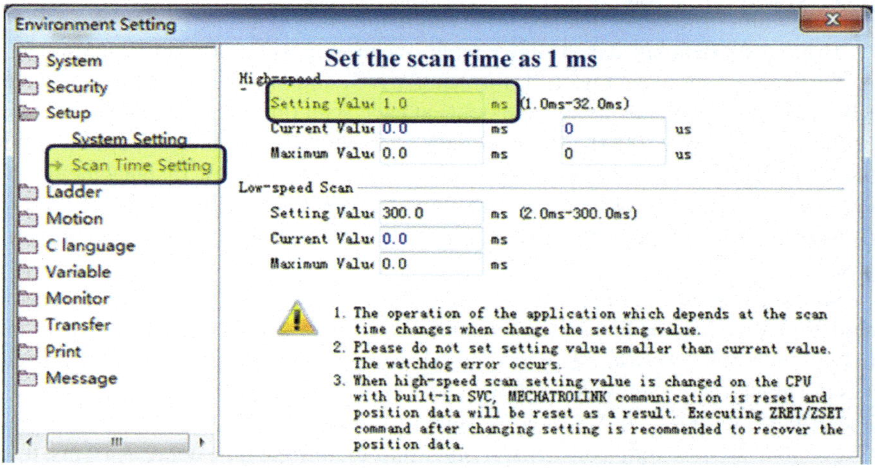

Fig. 4.20 Set the scan time for the MP2300 controller

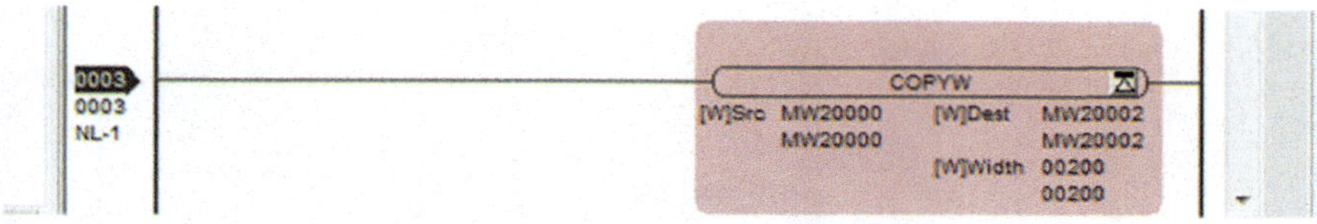

Fig. 4.21 Programs for the data flow realization

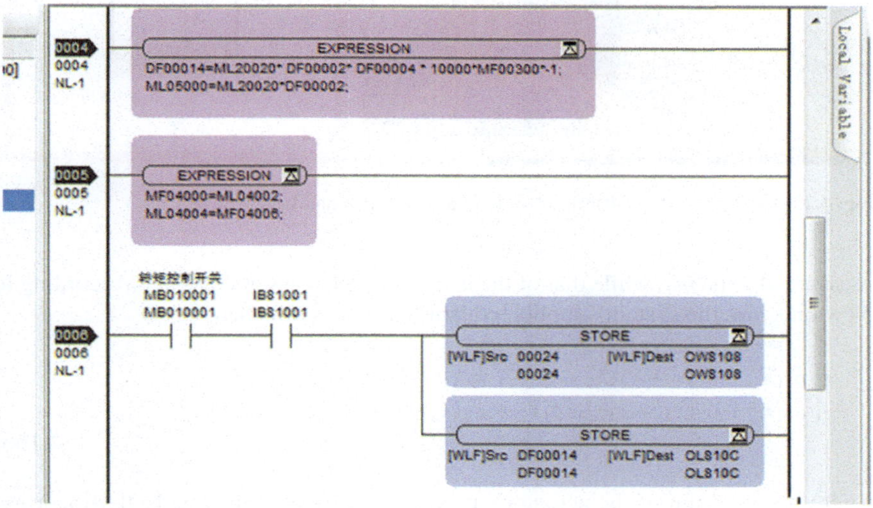

Fig. 4.22 Programs for the time-delayed feedback control

4.5.2 Identification Experiment of a Linear Time Delay System

As shown in Fig. 4.23, the whole dynamic system [15, 21] is encapsulated in the yellow assemble frame. The primary part, hung on the ceiling by sheet springs, consists of a block payload and a payload holder. Its vibration is assumed to be induced by a force excitation. The auxiliary part, namely the dynamic absorber, consists of a linear actuator and an actuator adaptor. As can be seen, only the actuator's mover is connected to the primary part by sheet springs, whereas its stator is attached to the aluminum frame so that the primary part is free from unnecessary payload. Considering that the principle of dynamic absorber design is shifting the absorber's resonant frequency, which generally prefers stiffness tuning, the linear actuator adopts the proportional position feedback of x_2 with time delay.

In system modeling, let the stiffness and damping coefficients of the upper layer sheet springs be k_1 and c_1, and that of the lower layer sheet springs be k_2 and c_2. The weak friction resistance between the mover and stator is generalized as a viscous damping combined with elastic stiffness, i.e. c_3 and k_3. The mass of the primary part

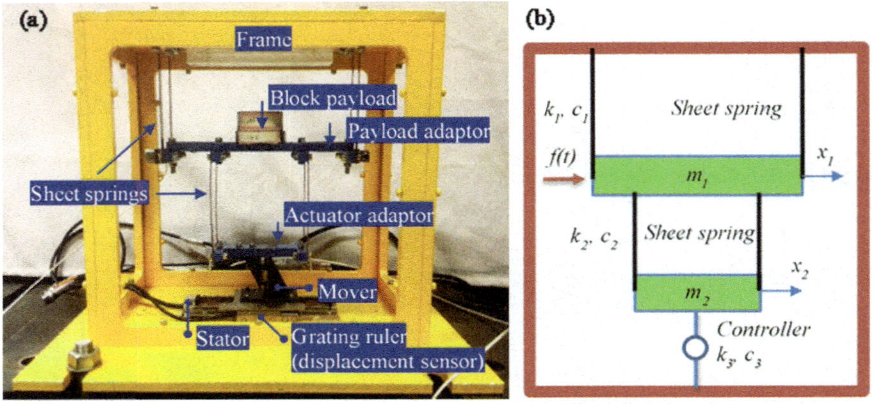

Fig. 4.23 Active vibration absorbing system: **a** real device and **b** schematic view

is assumed to be m_1, while that of the auxiliary part is assumed as m_2. According to Newton's law, the system's motion equation can be written as

$$\begin{cases} m_1\ddot{x}_1 + c_1\dot{x}_1 + c_2(\dot{x}_1 - \dot{x}_2) + k_1x_1 + k_2(x_1 - x_2) = f(t) \\ m_2\ddot{x}_2 + c_2(\dot{x}_2 - \dot{x}_1) + c_3\dot{x}_2 + k_2(x_2 - x_1) + k_3x_2 = gkx_2(t - \tau_{in} - \tau) \end{cases}, \tag{4.166}$$

where g is the gain of the actuator's power amplifier, k is the feedback proportion given in the controller, τ_{in} is the inherent time delay in the control loop, and τ is the artificial delay given in the controller. Applying Fourier transforming on Eq. (4.166), one can get the system's frequency response function (FRF), which is given as

$$\begin{pmatrix} Z_{11} & -j\omega c_2 - k_2 \\ -j\omega c_2 - k_2 & Z_{22} \end{pmatrix}\begin{pmatrix} X_1(j\omega) \\ X_2(j\omega) \end{pmatrix} = \begin{pmatrix} F(j\omega) \\ 0 \end{pmatrix}, \tag{4.167}$$

where

$$Z_{11} = -m_1\omega^2 + j\omega(c_1 + c_2) + k_1 + k_2,$$
$$Z_{22} = -m_2\omega^2 + j\omega(c_2 + c_3) + k_2 + k_3 - gke^{-j\omega(\tau_{in}+\tau)},$$

j is the pure imaginary number; ω is the radial frequency; $X_1(j\omega)$, $X_2(j\omega)$ and $F(j\omega)$ are Fourier transforms of $x_1(t)$, $x_2(t)$ and $f(t)$, respectively.

Let $-\omega^2X_1(j\omega)/F(j\omega)$ be $H_1(j\omega)$, and $-\omega^2X_2(j\omega)/F(j\omega)$ be $H_2(j\omega)$, Eq. (4.167) can be further deduced to

$$H_1(j\omega) = \\ -\omega^2 \frac{-\beta_2\omega^2 + \beta_0 + j\beta_1\omega - gke^{-j\omega(\tau_{in}+\tau)}}{a_4\omega^4 - a_2\omega^2 + a_0 + j(-a_3\omega^3 + a_1\omega) - gke^{-j\omega(\tau_{in}+\tau)}(-\alpha_2\omega^2 + j\alpha_1\omega + \alpha_0)}, \tag{4.168}$$

and

$$H_2(j\omega) = \\ -\omega^2 \frac{-\alpha_2\omega^2 + j\alpha_1\omega + \alpha_0}{a_4\omega^4 - a_2\omega^2 + a_0 + j(-a_3\omega^3 + a_1\omega) - gke^{-j\omega(\tau_{in}+\tau)}(-\alpha_2\omega^2 + j\alpha_1\omega + \alpha_0)}, \tag{4.169}$$

where

$$\begin{aligned}
a_4 &= m_1 m_2, \\
a_3 &= m_1 c_2 + m_1 c_3 + m_2 c_1 + m_2 c_2, \\
a_2 &= c_1 c_2 + c_1 c_3 + c_2 c_3 + m_1 k_2 + m_1 k_3 + m_2 k_1 + m_2 k_2, \\
a_1 &= c_1 k_2 + c_1 k_3 + c_2 k_1 + c_2 k_3 + c_3 k_1 + c_3 k_2, \\
a_0 &= k_1 k_2 + k_1 k_3 + k_2 k_3, \\
\alpha_2 &= m_1, \alpha_1 = c_1 + c_2, \alpha_0 = k_1 + k_2, \\
\beta_2 &= m_2, \beta_1 = c_2 + c_3, \beta_0 = k_2 + k_3.
\end{aligned}$$

In the experiment, the artificial time delay τ is set as 0. The feedback gain g is chosen from − 400 to 400 N/m with an interval of 50 N/m. The system is driven by burst random excitation, and the excitation force is recorded by a piezoelectric force sensor. The responses of x_1 and x_2 are recorded by two accelerometers. The system's acceleration FRFs, namely $H_1(j\omega)$ and $H_2(j\omega)$, corresponding to some feedback gains are shown in Fig. 4.24.

By fitting the FRF representations, given by Eqs. (4.168) and (4.169), to the measured FRF samples, the unknown parameters, i.e., m_1, m_2, c_1, c_2, c_3, k_1, k_2, k_3, k and τ_{in}, corresponding to different feedback gain settings can be identified. Figure 4.25 shows the identified structural parameters, where one can find that the identified parameters stay in good coincidence whatever the feedback gain varies. Furthermore, Fig. 4.26 shows the identified control parameters, where one can find from Fig. 4.26a that the identified feedback gain is proportional to the feedback gain settings, and one can find from Fig. 4.26b that the inherent time delay τ_{in} is stable to different feedback gains. Hence, one may use the mean value of the identified structural and control parameters as the calibrated results, i.e., $m_1 = 0.803$ kg, $m_2 = 0.383$ kg, $c_1 = 2.524$ Ns/m, $c_2 = -0.473$ Ns/m, $c_3 = 8.793$ Ns/m, $k_1 = 2271.228$ N/m, $k_2 = 1469.887$ N/m, $k_3 = 185.645$ N/m, $k = 1.155$, and $\tau_{in} = 7.892$ ms.

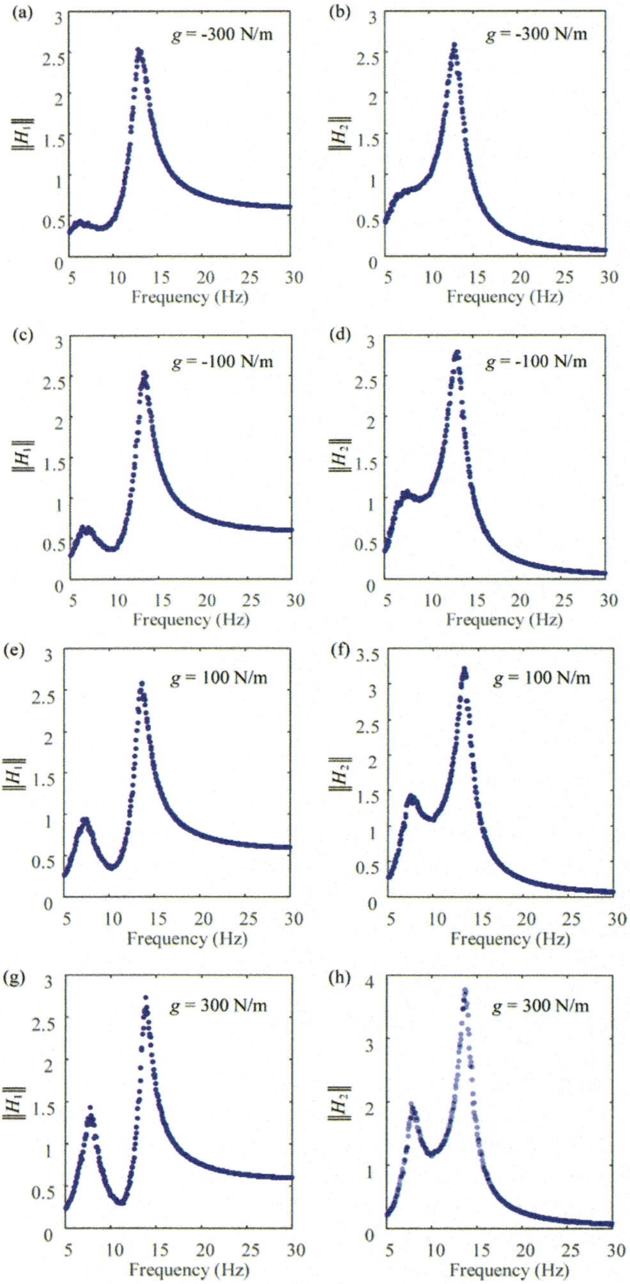

Fig. 4.24 The measured frequency response functions: **a** H_1 for $g = -$ 300 N/m, **b** H_2 for $g = -$ 300 N/m, **c** H_1 for $g = -$ 100 N/m, **d** H_2 for $g = -$ 100 N/m, **e** H_1 for $g = 100$ N/m, **f** H_2 for $g = 100$ N/m, **g** H_1 for $g = 300$ N/m, **h** H_2 for $g = 300$ N/m

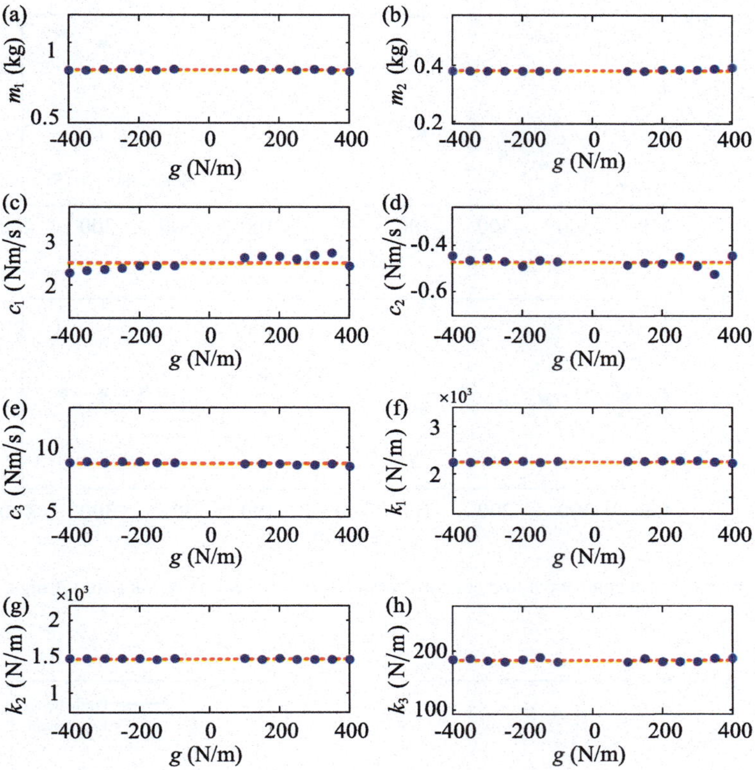

Fig. 4.25 Identified structural parameters corresponding to different feedback gain settings: **a** m_1, **b** m_2, **c** c_1, **d** c_2, **e** c_3, **f** k_1, **g** k_2, and **h** k_3

In these parameters, one may wonder that the calibrated value of c_2 is minus, which is in contrary with common sense because it should be an energy-consuming unit. A reasonable explanation is that the friction effect in the actuator is treated as viscous damping combined with elastic stiffness. Thus, it needs another parameter to balance its energy consumption. Furthermore, c_2 is too tiny to affect the entire system's stability. Another reason we insist on accepting it is mainly because the identified c_2 almost remains the same no matter how the feedback gain is changed.

Nevertheless, it is still too early to conclude the accuracy and robustness of the proposed identification algorithm. A good algorithm should not only well represent the measured data but also efficiently predict the system's response under other conditions. To this end, we reset the feedback gain, i.e., g, as 200 N/m, and the artificial time delay, i.e., τ, as 200 ms. By using the identified parameters and given artificial time delay, the system's frequency responses are simulated and shown in Fig. 4.27. As can be seen, the predicted curves considering the identified internal delay fit the measured transfer ratio curves precisely, indicating the practicability of the identification algorithm.

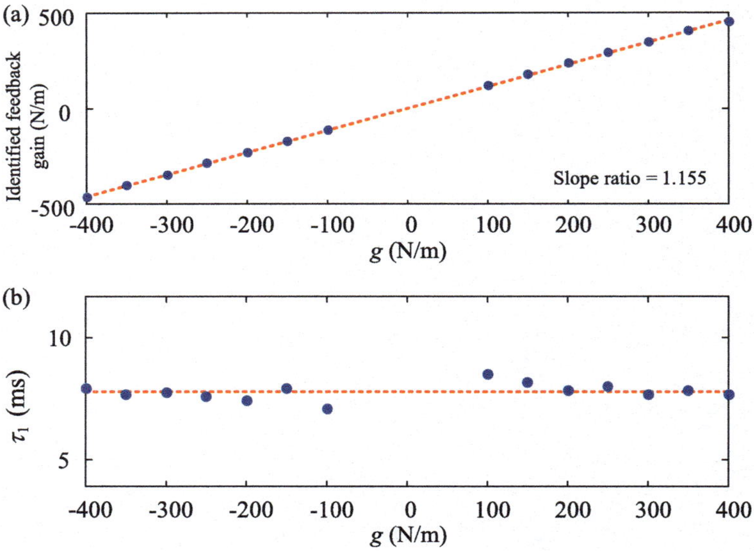

Fig. 4.26 Identified control parameters corresponding to different feedback gain settings: **a** k and **b** τ_{in}

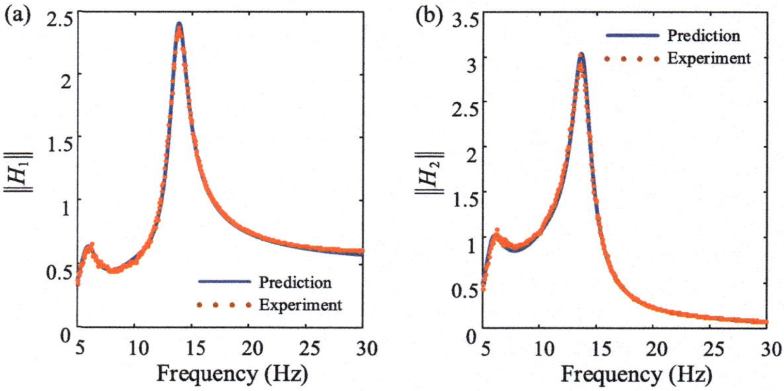

Fig. 4.27 Comparison of the predicated and experimental FRFs: **a** H_1 and **b** H_2

4.5.3 Identification Experiment of a Nonlinear Time Delay Nonlinear System

As is shown in Fig. 4.28, the nonlinear system with two degrees of freedom [17, 22] is hung on the ceiling of a rigid aluminum framework. The sheet springs produce linear restore force, whereas the Nd-Fe-B magnet pairs produce symmetric nonlinear force.

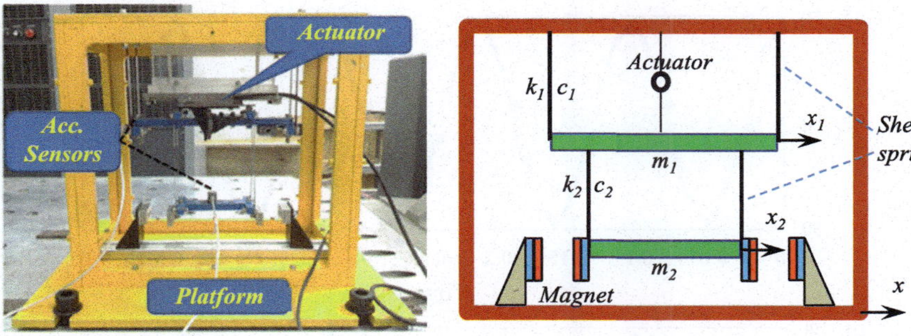

Fig. 4.28 Experiment device and schematic view of the 2-DOF nonlinear system

The controller acts time-delayed force, which is proportional to the relative displacement between the framework and the primary system. The mass of the primary system is m_1, which is calibrated as 0.698 kg. The mass of the secondary system is m_2, which is calibrated as 0.236 kg. The stiffness and damping of the primary sheet spring are k_1 and c_1, respectively. The stiffness and damping of the secondary sheet spring are k_2 and c_2, respectively. This nonlinear device is fixed onto a vibrator's horizontal platform, which generates displacement excitation, i.e. x.

By letting $y_1 = x_1 - x$ and $y_2 = x_2 - x$, the motion equation of the nonlinear system can be represented as

$$\begin{cases} m_1\ddot{y}_1 + c_1\dot{y}_1 + c_2(\dot{y}_1 - \dot{y}_2) + k_1y_1 + k_2(y_1 - y_2) + gy_2(t-\tau) + m_1\ddot{x} = 0 \\ m_2\ddot{y}_2 + c_2(\dot{y}_2 - \dot{y}_1) + k_2(y_2 - y_1) + \varphi_m(y_2) + m_2\ddot{x} = 0 \end{cases}, \tag{4.170}$$

where $\varphi_m(y_2)$ is the restoring force generated by the magnet pairs. When the magnet cells are assembled symmetrically, the resultant nonlinear force can be approximately written as

$$\varphi_m(y_2) = k_3y_2 + k_4y_2^3. \tag{4.171}$$

In this system, the parameters that need identification are c_1, c_2, k_1, k_2, k_3, k_4, g, and τ.

In the experiment, three acceleration sensors are mounted on the framework, the primary system, and the secondary system, to measure the excitation acceleration $\ddot{x}$ and the two response accelerations, i.e., $\ddot{x}_1$ and $\ddot{x}_2$. The vibrator generates sine waves, where the amplitude of the framework's displacement is 2 mm, and the excitation frequency is set from 8 to 20 Hz with an interval of 0.25 Hz. The coupling coefficient g is carefully set as − 150 N/m, which ensures all-delay stability to the system, and the time delay is set as 120 ms. Considering that the feedback proportion, i.e., k introduced in Eq. (4.166), and the calibrated inherent delay, i.e., τ_{in}, are 1.155 and

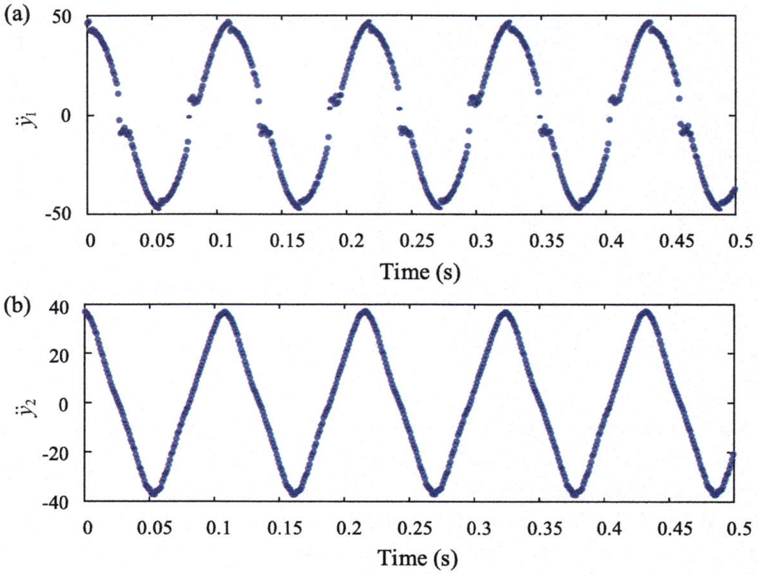

Fig. 4.29 Acceleration of steady-state responses at an excitation frequency of 9.25 Hz: **a** $\ddot{y}_1$ and **b** $\ddot{y}_2$

7.892 ms, respectively, the actual coupling coefficient and time delay in the control loop are – 173 N/m and 128 ms.

Figure 4.29 illustrates the measured responses at an excitation frequency of 9.25 Hz, where one can find that the responses show significant nonlinearity and are blurred by noise. Therefore, one should employ the method proposed in Sect. 4.4 to identify the unknown parameters.

To apply the noise-correction-based parameter identification, Eq. (4.170) should be rewritten as the state space form as Eq. (4.79) shows. In detail,

$$\mathbf{g}(\mathbf{x}(t), \mathbf{x}(t-\tau), \mathbf{f}(t); \boldsymbol{\mu})$$
$$= \begin{pmatrix} x_3 \\ x_4 \\ -\frac{c_1}{m_1}x_3 - \frac{c_2}{m_1}(x_3 - x_4) - \frac{k_1}{m_1}x_1 - \frac{k_2}{m_1}(x_1 - x_2) - \frac{g}{m_1}x_2(t-\tau) - \ddot{x} \\ -\frac{c_2}{m_2}(x_4 - x_3) - \frac{k_2}{m_2}(x_2 - x_1) - \frac{k_3}{m_2}x_2 - \frac{k_4}{m_2}x_2^3 - \ddot{x} \end{pmatrix}, \tag{4.172}$$

where $m_1 = 0.698$ kg and $m_2 = 0.236$ kg. The corresponding Jacobian matrices are

$$\tilde{\mathbf{h}}_x\left(t;\omega_q\right)=\frac{\partial\mathbf{g}}{\partial\tilde{\mathbf{x}}^T\left(t;\omega_q\right)}=\begin{pmatrix}0 & 0 & 1 & 0\\ 0 & 0 & 0 & 1\\ -\frac{\tilde{k}_1+\tilde{k}_2}{m_1} & \frac{\tilde{k}_2}{m_1} & -\frac{\tilde{c}_1+\tilde{c}_2}{m_1} & \frac{\tilde{c}_2}{m_1}\\ \frac{\tilde{k}_2}{m_2} & -\frac{\tilde{k}_2+\tilde{k}_3}{m_2}-3\frac{\tilde{k}_4}{m_2}\tilde{x}_2^2 & \frac{\tilde{c}_2}{m_2} & -\frac{\tilde{c}_2}{m_2}\end{pmatrix},$$

$$\tilde{\mathbf{h}}_\tau\left(t;\omega_q\right)=\frac{\partial\mathbf{g}}{\partial\tilde{\mathbf{x}}^T\left(t-\tilde{\tau};\omega_q\right)}=\begin{pmatrix}0 & 0 & 0\,0\\ 0 & 0 & 0\,0\\ 0 & -\frac{\tilde{g}}{m_1} & 0\,0\\ 0 & 0 & 0\,0\end{pmatrix},$$

and

$$\tilde{\mathbf{h}}_\mu\left(t;\omega_q\right)=\frac{\partial\mathbf{g}}{\partial\,\tilde{\boldsymbol{\mu}}^T}$$

$$=\begin{pmatrix}0 & 0 & 0 & 0 & 0 & 0 & 0\\ 0 & 0 & 0 & 0 & 0 & 0 & 0\\ -\frac{\tilde{x}_3}{m_1} & -\frac{\tilde{x}_3-\tilde{x}_4}{m_1} & -\frac{\tilde{x}_1}{m_1} & -\frac{\tilde{x}_1-\tilde{x}_2}{m_1} & 0 & 0 & -\frac{\tilde{x}_2(t-\tilde{\tau})}{m_1}\\ 0 & -\frac{\tilde{x}_4-\tilde{x}_3}{m_2} & 0 & -\frac{\tilde{x}_2-\tilde{x}_1}{m_2} & -\frac{\tilde{x}_2}{m_2} & -\frac{\tilde{x}_2^3}{m_2} & 0\end{pmatrix}.$$

By further substituting them into Eq. (4.133), one can collect the harmonic coefficient matrices, i.e., $\tilde{\mathbf{H}}_x\left(\omega_q\right)$, $\tilde{\mathbf{H}}_\tau\left(\omega_q\right)$, and $\tilde{\mathbf{H}}_\mu\left(\omega_q\right)$. Note that once the harmonic order has been determined, the regularization matrices $\boldsymbol{\Lambda}\left(\omega_q\right)$, $\boldsymbol{\Theta}\left(\omega_q\right)$, $\mathbf{U}_{i,s}$, and $\mathbf{U}_{i,c}$ can be constructed by referring to Eqs. (4.138) and (4.141), Table 4.7, and Table 4.8, respectively. By substituting the harmonic coefficient matrices, i.e., $\tilde{\mathbf{H}}_x\left(\omega_q\right)$, $\tilde{\mathbf{H}}_\tau\left(\omega_q\right)$, and $\tilde{\mathbf{H}}_\mu\left(\omega_q\right)$, and the regularization matrices $\boldsymbol{\Lambda}\left(\omega_q\right)$, $\boldsymbol{\Theta}\left(\omega_q\right)$, $\mathbf{U}_{i,s}$, and $\mathbf{U}_{i,c}$ into Eq. (4.153), the unknown parameters, together with the missing streaming, can be identified.

Figure 4.30 shows the identification process of the structural parameters and time delay, where one can find that all parameters converge to constants. Specifically, the identified parameters are

$$\begin{aligned}&c_1=6.183\ \text{Ns/m},\ c_2=5.016\times10^{-2}\ \text{Ns/m},\ k_1=2.220\times10^3\ \text{N/m},\\&k_2=5.513\times10^2\ \text{N/m},\ k_3=8.221\times10^2\ \text{N/m},\ k_4=2.535\times10^6\ \text{N/m}^3,\\&g=-1.674\times10^2\ \text{N/m},\ \tau=128.5\ \text{ms}.\end{aligned}$$

Note that the true value of feedback gain and time delay are – 173 N/m and 128 ms. It is easy to calculate that the relative errors of them are 3.2 and 0.5%, respectively. Hence, one may conclude that the identified results are reliable.

To further confirm the efficiency of the identification of the time delay and nonlinear parameters for the experimental results, as a typical performance, we substitute the identified parameters into Eq. (4.170) to perform the numerical simulation. With the excitation frequency in $f_{s.}=2\pi\omega_s$ varying from 8 to 20 Hz,

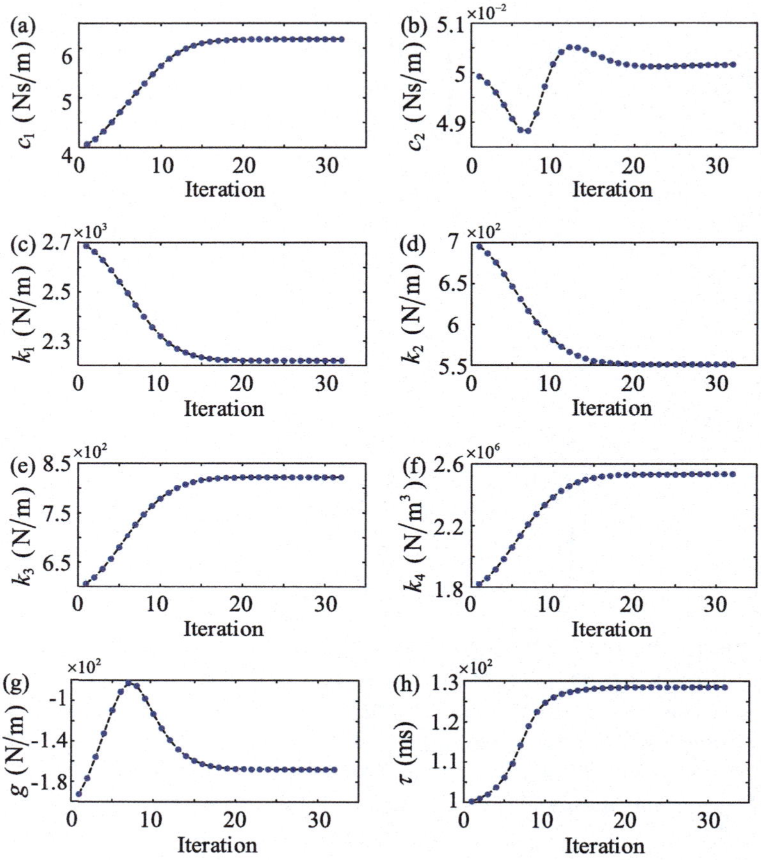

Fig. 4.30 Identification process with noise correction: **a** c_1, **b** c_2, **c** k_1, **d** k_2, **e** k_3, **f** k_4, **g** g, and **h** τ

one may numerically obtain the frequency response. Figure 4.31 shows that the frequency–response curves of the numerical simulation from Eq. (4.170) are consistent with the original measurements. Therefore, it is sufficient to conclude that the noise-correction-based parameter identification, which is proposed in Sect. 4.4, is competent for real applications.

4.6 Conclusions

This section studied the inverse problem of systems with time delay from the viewpoint of parameter identification. The characteristics of the time-delayed system in the frequency domain and methods of using the characteristics to identify unknown

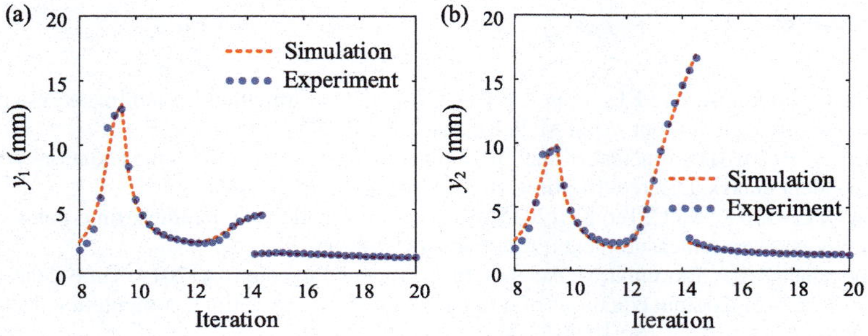

Fig. 4.31 Comparisons between the simulated and experimentally measured responses: **a** y_1 and **b** y_2

parameters were comprehensively discussed. Some conclusions may be drawn as follows.

In the frequency domain, the state with time delay is a sinusoidal function of frequency. When the time delay is large, the sinusoidal function will show complete periods over the limited frequency range. Therefore, one may employ the Fourier transform to the extracted frequency response functions to detect and approximately estimate the delay. However, if one wants to accurately identify time delay and other unknown parameters, the refined identification algorithm, which was proposed in Sect. 4.2 should be employed. This algorithm also figured out that when the frequency interval is sufficiently small, at least smaller than the reciprocal of the upper bound of time delay, the uniqueness of time delay identification will be promised.

When the system is nonlinear, the linear or quasi-linear identification methods will no longer be applicable. An efficient approach for parameter identification in the frequency domain is the harmonic balance method. Besides, when the system's response couples with noise, say, streaming missing, the identification algorithm based on the harmonic balance principle can be modified by correcting part of the harmonic coefficients adaptively. Note that streaming missing is common for motion sensing via accelerometers or laser vibrometers. It is recommended to use the noise-correction-based identification algorithm if system nonlinearity is detected.

In real applications, the time delay can be accurately realized by the operation of digital flow in any kind of digital signal processor. The YASKAWA hardware supports a minimum scan time of 1 ms. Therefore, the resolution of time delay tuning on this platform is 1 ms. If other devices with smaller scan time, i.e., higher processing speed, can be applied, a refined resolution of time delay can be achieved. It should be noted that because of signal transferring and communication, inherent time delay always exists. One should calibrate the inherent time delay before performing active control.

References

1. Chai Q, Loxton R, Teo KL, Yang C (2013) Time-delay estimation for nonlinear systems with piecewise-constant input. Appl Math Comput 219(17):9543–9560
2. Hartung F (2013) Parameter estimation by quasilinearization in differential equations with state dependent delays. Discrete Contin Dyn Syst Ser B 18:1611–1631
3. Lin Q, Loxton R, Xu C, Teo KL (2015) Parameter estimation for nonlinear time-delay systems with noisy output measurements. Automatica 60:48–56
4. Abooshahab MA, Ekramian M, Ataei M, Ebrahimpour-Boroojeny A (2019) Time-delay estimation in state and output equations of nonlinear systems using optimal computational approach. J Optim Theory Appl 180:1036–1064
5. Zhang T, Lu ZR, Liu JK, Liu G (2021) Parameter identification of nonlinear systems with time-delay from time-domain data. Nonlinear Dyn 104:4045–4061
6. Lei Y, Xia D, Erazo K, Nagarajaiah S (2019) A novel unscented Kalman filter for recursive state-input-system identification of nonlinear systems. Mech Syst Signal Process 127:120–135
7. He J, Qi M, Tong Z, Hua X, Chen Z (2023) An improved extended Kalman filter for parameters and loads identification without collocated measurements. Smart Struct Syst 31(2):131–140
8. Cui T, Ding F, Hayat T (2022) Moving data window-based partially-coupled estimation approach for modeling a dynamical system involving unmeasurable states. ISA Trans 128:437–452
9. Sun S, Xu L, Ding F (2023) Parameter estimation methods of linear continuous-time time-delay systems from multi-frequency response data. Circuit Syst Signal Process 42(6):3360–3384
10. Nayfeh AH, Mook DT (2008) Nonlinear oscillations. Wiley
11. Sun Y, Song H, Xu J (2013) Time-delay identification for linear controlled systems. Theor Appl Mech Lett 3(6):063010
12. Sun Y, Jin M, Song H, Xu J (2016) Time-delay identification for vibration systems with multiple feedback. Acta Mech Sin 32:1138–1148
13. Jin M, Sun Y, Song H, Xu J (2017) Experiment-based identification of time delays in linear systems. Acta Mech Sin 33:429–439
14. Zhang X, Xu J (2016) Time delay identifiability and estimation for the delayed linear system with incomplete measurement. J Sound Vib 361:330–340
15. Zhang X, Xu J, Ji J (2018) Modelling and tuning for a time-delayed vibration absorber with friction. J Sound Vib 424:137–157
16. Zhang X, Xu J (2015) Identification of time delay in nonlinear systems with delayed feedback control. J Franklin Inst 352(8):2987–2998
17. Zhang X, Xu J, Feng Z (2017) Nonlinear equivalent model and its identification for a delayed absorber with magnetic action using distorted measurement. Nonlinear Dyn 88:937–954
18. Zhang X, Ji J, Fu J, Xu J (2019) Denoising identification for nonlinear systems with distorted streaming. Int J Non-Linear Mech 117:103224
19. Zhang X, Ji J, Xu J (2019) Parameter identification of time-delayed nonlinear systems: An integrated method with adaptive noise correction. J Franklin Inst 356(11):5858–5880
20. Wang S, Diao B, Zhang X, Xu J, Chen L (2022) Adaptive signal-correction-based identification for friction perception of the vibration-driven limbless robot. Nonlinear Dyn 108(4):3817–3837
21. Zhang X, Xu J, Huang Y (2015) Experiment on parameter identification of a time delayed vibration absorber. IFAC-PapersOnLine 48(12):57–62
22. Zhang X, Xu J, Zhang S, Huang Y (2015) Experiment on Parameter Identification of a Time Delay Coupled Nonlinear System. IFAC-PapersOnLine 48(11):694–699

Chapter 5
Time-Delayed Control of Vibration

Active control can effectively inhibit vibration responses in aerospace and vehicle engineering applications [1]. Notably, the control loop's time lag cannot be overlooked as it significantly affects the system's stability and vibration characteristics, which can alter fluidly with the adjustment of variables like time delay and control strength [2, 3]. The control strategy's simplicity allows for the enhancement of its intelligent performance by adjusting the time delay without affecting energy consumption. Therefore, it is worthwhile to consider an artificially adjustable time-delay control and establish the correlation between the control parameters and their effect. The oscillation characteristics of the vibration isolator across different frequency ranges serve as a standard for assessing the efficacy of the time-delayed control and classifying control parameters. The association between the vibration isolation effect and the control parameters classification range is thereby deduced. Active control of nonlinear vibration isolation is a meaningful engineering task that can inhibit vibration by interrupting the vibration energy propagation process or by adding an energy-dissipation body to consume vibration energy. Hence, the introduction of time-delay control in suppressing low-frequency, high-amplitude vibration is referred to as time-delay vibration elimination technology.

Our previous researches confirmed the existence of time lag in controlled loop through parameter identification. This shows that both the theoretical modeling and practical application of the delayed feedback control mechanism are plausible. Olgac and Holm-Hansen [4] introduced a time-delay vibration absorber into a primary system. They discovered that the time delay could manipulate the anti-resonance point in order to stifle vibration. Hamdi and Belhaq [5] investigated the suppression effect of time-delay displacement feedback on the self-induced vibration of a beam that is simply supported and subjected to high-frequency axial excitation. The team of Professor Cai Guoping employs time-delayed feedback control in managing continuous structures. The experimental results revealed that delay in control could enhance the effectiveness of vibration suppression. However, many issues require further exploration under time-delay feedback control. These include the design of

J. Xu, *Nonlinear Dynamics of Time Delay Systems*,
https://doi.org/10.1007/978-981-99-9907-1_5

the control rate, the resilience of time-delay control, the application of time-delay control in nonlinear structures, the experimental methodology of time-delay control, and equipment preparation [6, 7]. Xu and Sun [8, 9] developed and utilized the linear-time-delay vibration absorber to subdue the vibration of a linear spring oscillator. This was based on the concept of a time-delay vibration absorber and was accomplished using a motion controller, servo motor, and elastic coupling element. Subsequent to this, Sun Xiuting and associates [10, 11] constructed a nonlinear vibration isolation prototype utilizing time-delay control. After conducting both qualitative and quantitative analysis, they confirmed the influence of time-delay on the stability, frequency, and other dynamic qualities of nonlinear systems. They also provided the most favorable value of time-delay across different frequency ranges, highlighting the practical value of time-delay control in the area of nonlinear vibration isolation. The impact of time delay on vibration suppression has been observed in active control. It was noted that even for linear time-delay feedback or coupling terms, varying design requirements can be proposed based on differing operational circumstances. The mechanism of vibration suppression brought about by the optimal time-delay parameters derived from this method tends to differ. As such, vibration suppression triggered by a time delay is termed as a time-delay vibration elimination technique in this chapter.

5.1 Effect of Time Delay on Vibration Isolation

As Fig. 5.1 depicted, time-delayed feedback control is incorporated into a nonlinear vibration system subject to the external displacement [12], denoted as $z(t)$.

The model in Fig. 5.1 displays a vibration system with a mass, denoted as M, connected to a base through a non-linear spring with a linear damper. The stiffness of the nonlinear isolation structure is hypothesized to adhere to a nonlinear function expressed as $f = k_l(\cdot) + k_n(\cdot)^3$, executable by the Quasi-Zero-Stiffness (QZS) structure [4–8]. A damping impact is also accounted for with the characteristic $f = c{\cdot}d(\cdot)/dt$. The complete motion of mass M is shown by $x(t)$, the base stimulation is indicated as $z(t)$, and the relative motion distinguishing the base and mass is $x(t) - z(t)$. The control signal $u(x, \dot{x}, x_\tau, \dot{x}_\tau)$ adheres to the linear function of the full motion and speed of the mass. Consequently, the platform dynamic equation can be

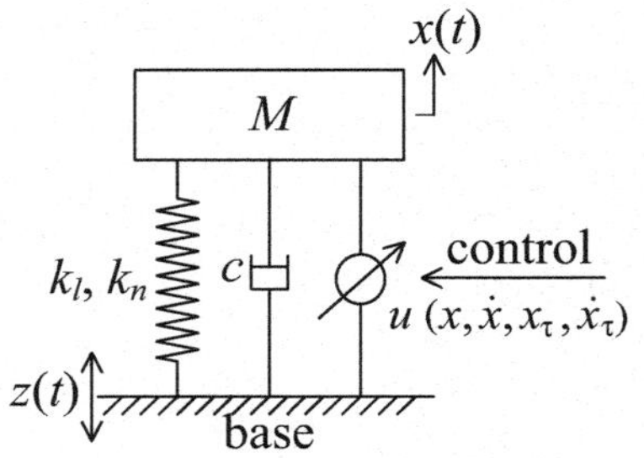

Fig. 5.1 Design of a nonlinear isolation system utilizing time-delayed feedback control

derived

$$M\ddot{x} + k_l(x - z) + k_n(x - z)^3 + c(\dot{x} - \dot{z}) = u(x, \dot{x}, x_\tau, \dot{x}_\tau). \tag{5.1}$$

To emphasize the impact of the active management time-delay, the control signal $u(x, \dot{x}, x_\tau, \dot{x}_\tau)$ is designated as $u(x, \dot{x}, x_\tau, \dot{x}_\tau) = p_1(x_\tau - x) + p_2(\dot{x}_\tau - \dot{x})$ wherein p_1 and p_2 are symbolize control strength, the system operates devoid of control when the time delay is zero. Dimensionless transfer is then ushered in

$$\omega_0 = \sqrt{\frac{k_l}{M}}, \tilde{t} = \omega_0 t, \gamma = \frac{k_n}{k_l}, \xi_1 = \frac{c}{2\sqrt{Mk_l}}, \tau_k = \omega_0 \tau, g_1 = \frac{p_1}{k_l}, g_2 = \frac{p_2}{\sqrt{Mk_l}}. \tag{5.2}$$

Then, the dynamic equation is as

$$x'' + (x - z) + \gamma(x - z)^3 + 2\xi_1(x' - z') = g_1(x_{\tau_k} - x) + g_2(x'_{\tau_k} - x'), \tag{5.3}$$

where $(\cdot)'' = \frac{d^2(\cdot)}{d\tilde{t}^2}$, $(\cdot)' = \frac{d(\cdot)}{d\tilde{t}}$. Subsequent sections discuss the impacts of time delay and nonlinearity on the system's stability and response. It is necessary to determine the most suitable time delay value for distinct forms of external excitation.

5.1.1 Stability Criteria

The equilibrium of the vibration system, as per the linearization system, is zero and should maintain stability. The stability of this equilibrium for $z(\tilde{t}) = 0$ can be ascertained through a characteristic equation

$$\lambda^2 + 1 + 2\xi_1\lambda - g_1(e^{-\lambda\tau_k} - 1) - g_2\lambda(e^{-\lambda\tau_k} - 1) = 0. \tag{5.4}$$

The eigenvalue λ is $\lambda = \alpha + i\beta$ where α is the real-part and β is the imaginary part. If all real components of the eigenvalues are negative, the zero equilibrium remains stable. However, altering parameter values can bifurcate the zero equilibrium if a single eigenvalue has a positive real-part. This modification may prevent the system's free vibration from being dampened. In the instance of a Hopf bifurcation, the real component of the eigenvalues is null, shifting from negative to positive as the time delay increases. To determine the critical parameters of a Hopf bifurcation, eigenvalues are designated as $\lambda = i\Omega_c$, where Ω_c indicates the vibration frequency prompted by the Hopf bifurcation. Upon replacing $\lambda = i\Omega_c$ in the characteristic Eq. (5.4) yields

$$-\Omega_c^2 + 1 + 2\xi_1 i\Omega_c - g_1(e^{-i\Omega_c\tau_{kc}} - 1) - g_2 i\Omega_c(e^{-i\Omega_c\tau_{kc}} - 1) = 0, \tag{5.5}$$

where τ_{kc} denotes the critical time delay value regarding Hopf bifurcation. By distinguishing between the real and imaginary components of (5.5), the following can be deduced

$$\begin{cases} -\Omega_c^2 + 1 - g_1(\cos\Omega_c\tau_{kc} - 1) - g_2\Omega_c \sin\Omega_c\tau_{kc} = 0, \\ 2\xi_1\Omega_c + g_1 \sin\Omega_c\tau_{kc} - g_2\Omega_c(\cos\Omega_c\tau_{kc} - 1) = 0. \end{cases} \tag{5.5a}$$

Solving $\cos\Omega_c\tau_{kc}$ and $\sin\Omega_c\tau_{kc}$ from Eq. (5.5a), it has

$$\begin{cases} \sin\Omega_c\tau_{kc} = \frac{g_2\Omega_c - 2g_1\xi_1\Omega_c - g_2\Omega_c^3}{g_1^2 + g_2^2\Omega_c^2}, \\ \cos\Omega_c\tau_{kc} = \frac{g_2\Omega_c(g_2\Omega_c + 2\xi_1\Omega_c) + g_1(g_1 - \Omega_c^2 + 1)}{g_1^2 + g_2^2\Omega_c^2}. \end{cases} \tag{5.5b}$$

For $\cos^2\Omega_c\tau_{kc} + \sin^2\Omega_c\tau_{kc} = 1$, the following equation could be derived

$$\begin{aligned} &\left(g_2\Omega_c - 2g_1\xi_1\Omega_c - g_2\Omega_c^3\right)^2 \\ &\quad + \left[g_2\Omega_c(g_2\Omega_c + 2\xi_1\Omega_c) + g_1\left(g_1 - \Omega_c^2 + 1\right)\right]^2 = \left(g_1^2 + g_2^2\Omega_c^2\right)^2. \end{aligned} \tag{5.5c}$$

The Hopf bifurcation frequency Ω_c can be determined from Eq. (5.5c) at varying control strengths. By inserting Ω_c into Eq. (5.5b), it is possible to ascertain the critical time delay value for Hopf bifurcation. Figure 5.2 illustrates the critical curves of the time delay τ_{kc} on the parameter plane.

Figure 5.2 illustrates the critical Hopf bifurcation curves for varying τ_k, g_1, and g_2. In Region I, where the control parameters are steady, the highest real part of the eigenvalues is negative. Conversely, in Region II, where the control parameters are stable, there is one real part eigenvalue that is positive. This suggests that the 0 equilibrium of the vibrational system is stable in Region I. It is noticeable that as the equivalent damping of the system is enhanced, leading to a more rapid dissipation of free vibration with a larger g_2, the span of Region I correspondingly expands with the escalation of g_2.

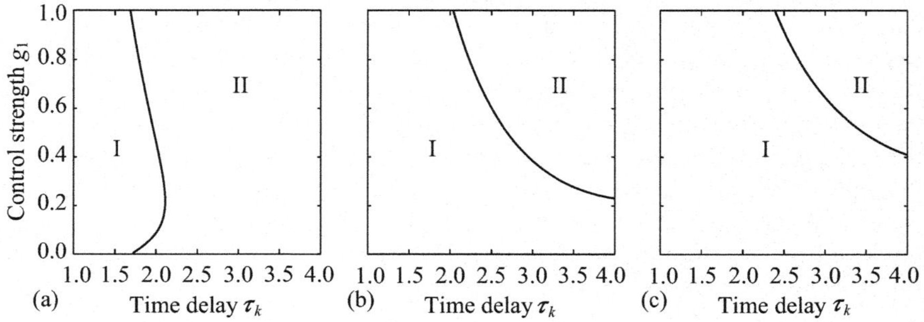

Fig. 5.2 Illustrates the Hopf bifurcation critical curves on the parameters plane (τ_k, g_1), while varying the value of g_2. **a** $g_2 = -0.15$; **b** $g_2 = 0.0$; **c** $g_2 = 0.15$

By identifying the scenario where the eigenvalues real-part equals zero, one can ascertain the occurrence of Hopf bifurcation for the equilibrium point at zero. Additionally, in Region I, the vibration attenuation time depends on the real part of the eigenvalues, while the resonance frequency depends on the eigenvalues' imaginary part.

5.1.2 Real and Imaginary Parts for Different Time Delays

By substituting the $\lambda = \alpha + i\beta$ into the characteristic equation, one has

$$(\alpha + i\beta)^2 + 1 + 2\xi_1(\alpha + i\beta) - g_1\left[e^{-(\alpha+i\beta)\tau_k} - 1\right] - g_2(\alpha + i\beta)\left[e^{-(\alpha+i\beta)\tau_k} - 1\right] = 0. \tag{5.6}$$

The calculation process involves determining the eigenvalues' real and imaginary parts for varied time delays. In Fig. 5.3, the highest value of the real parts is displayed, while in Fig. 5.4, the associated imaginary part is showcased. This analysis considers different time delays and control strengths, specifically when $\xi_1 = 0.1$.

From Fig. 5.3, as the time delay increases, the real part of eigenvalues initially decreases, then increases until it passes through zero. When the eigenvalues have a negative maximum real part, the larger equivalent damping property results in a quicker vibration dissipation. Hence, the optimal time delay value for suppressing impact response relates to eigenvalues having the minimal real part.

On the other hand, Fig. 5.4 shows that for $g_2 = -0.15$ and $g_2 = 0.0$, the resonant frequency of the system, represented by the imaginary part, first increases and then decreases as the time delay increases. In the instance of $g_2 = 0.15$, the resonant frequency initially decreases, reaching a minimum value, then sharply rises, and finally decreases. Due to the need for a greater natural frequency in a vibrating system to avoid resonance during periodic excitation in engineering practice, the

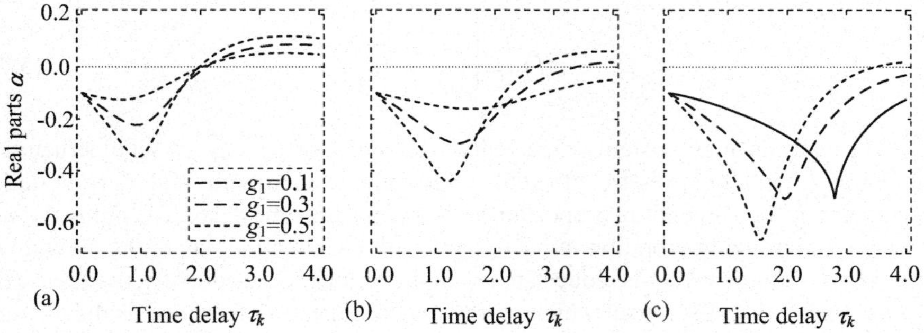

Fig. 5.3 Illustration of the variations in the real parts α of eigenvalues with different- time-delays and control-strengths. **a** $g_2 = -0.15$; **b** $g_2 = 0.0$; **c** $g_2 = 0.15$

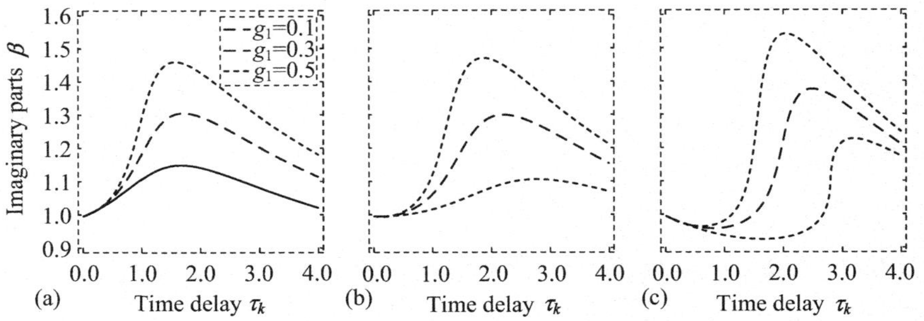

Fig. 5.4 The imaginary value parts β of eigenvalues for different-time-delays and control strengths. **a** $g_2 = -0.15$; **b** $g_2 = 0.0$; **c** $g_2 = 0.15$

periodic excitation optimum time delay is the point of maximum natural frequency value. However, if the vibrating system is an isolator that requires a lower natural frequency, the control strength g_2 should be positive to reduce the natural frequency and broaden the effective isolation range.

The effectiveness of an isolation system can be heightened when there is a lower natural frequency and quicker vibration energy dissipation. This is demonstrated through the smallest values of the real and imaginary parts depicted in Figs. 5.3 and 5.4, where the optimal time delay was identified. Moreover, the effectiveness of suppressing vibrations can also be improved for varying external stimuli. This is based on the different real and imaginary components of eigenvalues found at different time lags and control strengths.

5.1.3 Optimal Control for Time-Delayed Control

The negative real part of eigenvalues decreases while the decay speed of the system response increases. When $z(\tilde{t})$ is an impact excitation as

$$z(\tilde{t}) = \begin{cases} 0.1 \; \tilde{t} < 0 \\ 0.0 \; \tilde{t} \geq 0 \end{cases}. \tag{5.7}$$

The analysis of the system's eigenvalues revealed that, for a given set of structure-parameters and control-strengths, the eigenvalues maximum real part initially declines and then ascends in proportion to the increasing time delay. The optimal time delay in minimizing vibration impact aligns with the situation where the maximum real part α of eigenvalues is reduced to its smallest value. Consequently, in the case of impact excitation, the rate at which vibrations diminish is determined as the parameter to be optimized. Figure 5.5 indicates the preferred time delays for impacted excitation at different control strengths, for instance, when $\xi_1 = 0.1$.

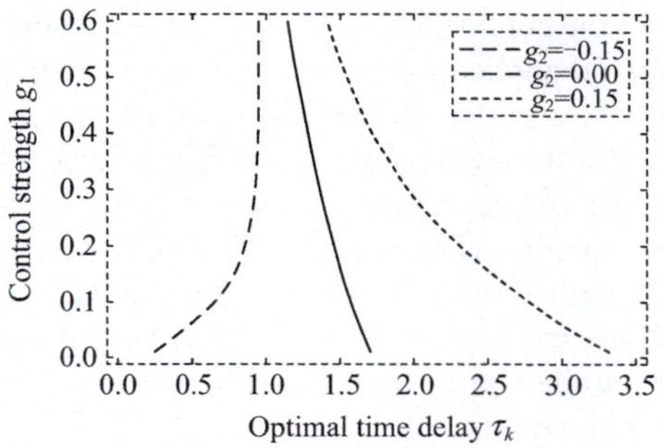

Fig. 5.5 The ideal time delay for impacted excitation for different-control-strengths

After analyzing the equivalent damping property reflected by the eigenvalue real part α, the optimal time delay values for dissipating impact vibration are determined from Fig. 5.5. Figure 5.6 illustrates the comparisons of the responses between the optimal time delay and various other values with $\xi_1 = 0.1$.

Referring to Fig. 5.6, one can observe that the vibration resulting from impact excitation can be subdued faster when the time delay is set at the optimum value, as demonstrated in Fig. 5.5. As such, when the controlled parameters are established within Region I, as depicted in Fig. 5.4, the optimum time delay can be determined for modifying the control strength, in line with the suggested optimization standard. With

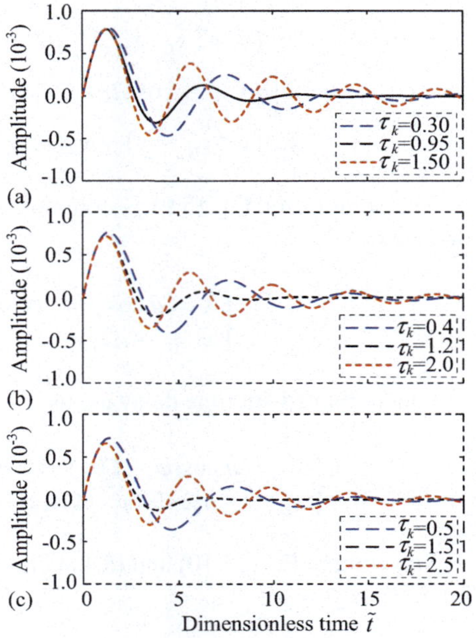

Fig. 5.6 Responses comparison for impact excitation among the optimum-time-delay and some other values for $g_1 = 0.5$ and **a** $g_2 = -0.15$; **b** $g_2 = 0.0$; **c** $g_2 = 0.15$

the optimal time delay, the control is beneficial in hastening the dissipation of vibration. This has widespread applications in many areas, including vehicle suspension and the protection of instruments, among others.

The ideal time delay values for impact excitation have been previously determined. However, in most engineering scenarios, this vibration excitation can be represented as periodic excitation encompassing multiple frequencies. In situations where excitation frequency falls within the effective isolation-frequency-band, the vibration originating from the base can be efficiently contained and regulated. Therefore, this section will examine and analyze the impact of time delay on the amplitude-frequency attribute in harmonic excitation.

By using the Averaging Method, the solutions for the time-delayed control nonlinear isolator for different parameters are determined. In situations where the excitation is denoted as $z(t) = z_0\cos\omega t$, after going through a dimensionless transformation, one can articulate the vibration equation as such

$$\begin{aligned} &x'' + \left(x - z_0 \cos \Omega\tilde{t}\right) + \gamma\left(x - z_0 \cos \Omega\tilde{t}\right)^3 + 2\xi_1\left(x' + z_0\Omega \sin \Omega\tilde{t}\right) \\ &= g_1\left(x_{\tau_k} - x\right) + g_2\left(x'_{\tau_k} - x'\right), \end{aligned} \tag{5.8}$$

where Ω is the dimensionless frequency and can be calculated as $\Omega = \frac{\omega}{\omega_0}$. Introducing a tuning parameter σ, the excited frequency Ω must obey the condition that $\Omega^2 = 1 + \varepsilon\sigma$. Therefore, in Eq. (5.7), the dimensionless natural frequency can be described as $1 = \Omega^2 - \varepsilon\sigma$. Taking into account that the damping, nonlinearity, and active control terms are minor perturbations to the vibration, Eq. (5.7) can be expressed as

$$\begin{aligned} x'' + \Omega^2 x &= \varepsilon f\left(x, x', x_{\tau_k}, x'_{\tau_k}, \tilde{t}\right) \\ &= \varepsilon\left[\sigma x + z_0 \cos \Omega\tilde{t} - 2\xi_1\left(x' + z_0\Omega \sin \Omega\tilde{t}\right) - \gamma\left(x - z_0 \cos \Omega\tilde{t}\right)^3\right] \\ &\quad + \varepsilon\left[g_1\left(x_{\tau_k} - x\right) + g_2\left(x'_{\tau_k} - x'\right)\right]. \end{aligned} \tag{5.9}$$

The solution of Eq. (5.9) is commonly recognized as a periodic vibration, as depicted

$$\begin{cases} x = a \cos \varphi = a \cos\left(\Omega\tilde{t} + \theta\right) \\ x' = -a\Omega \sin \varphi = -a\Omega \sin\left(\Omega\tilde{t} + \theta\right) \end{cases}, \tag{5.10}$$

and the motions with time delay are as

$$\begin{cases} x_{\tau_k} = a \cos(\varphi - \Omega\tau_k) = a \cos\left(\Omega\tilde{t} + \theta - \Omega\tau_k\right) \\ x'_{\tau_k} = -a\Omega \sin(\varphi - \Omega\tau_k) = -a\Omega \sin\left(\Omega\tilde{t} + \theta - \Omega\tau_k\right) \end{cases}. \tag{5.11}$$

Substituting Eqs. (5.10) and (5.11) into Eq. (5.9), the normal form expression of Eq. (5.9) is as

$$
\begin{aligned}
\frac{da}{d\tilde{t}} &= -\frac{\varepsilon}{2\pi}\int_0^{2\pi} f(a\cos\varphi, -a\Omega\sin\varphi, a\cos(\varphi-\Omega\tau_k),\\
&\quad -a\Omega\sin(\varphi-\Omega\tau_k), \tilde{t}) \sin\varphi d\varphi\\
&= -\frac{\varepsilon}{2\pi}\int_0^{2\pi} \{\sigma a\cos\varphi + z_0\cos(\varphi-\theta) - 2\xi_1(-a\Omega\sin\varphi + z_0\Omega\sin(\varphi-\theta))\\
&\quad - \gamma[a\cos\varphi - z_0\cos(\varphi-\theta)]^3 + g_1 a[\cos(\varphi-\Omega\tau_k) - \cos\varphi]\\
&\quad + g_2 a\Omega[-\sin(\varphi-\Omega\tau_k) - \cos\varphi]\}\sin\varphi d\varphi,
\end{aligned} \tag{5.12}
$$

and

$$
\begin{aligned}
\frac{d\theta}{d\tilde{t}} &= -\frac{\varepsilon}{2\pi a}\int_0^{2\pi} f(a\cos\varphi, -a\Omega\sin\varphi, a\cos(\varphi-\Omega\tau_k),\\
&\quad -a\Omega\sin(\varphi-\Omega\tau_k), \tilde{t}) \cos\varphi d\varphi\\
&= -\frac{\varepsilon}{2\pi a}\int_0^{2\pi} \{\sigma a\cos\varphi + z_0\cos(\varphi-\theta)\\
&\quad - 2\xi_1(-a\Omega\sin\varphi + z_0\Omega\sin(\varphi-\theta))\\
&\quad - \gamma[a\cos\varphi - z_0\cos(\varphi-\theta)]^3 + g_1 a[\cos(\varphi-\Omega\tau_k) - \cos\varphi]\\
&\quad + g_2 a\Omega[-\sin(\varphi-\Omega\tau_k) - \cos\varphi]\}\cos\varphi d\varphi.
\end{aligned} \tag{5.13}
$$

From Eqs. (5.12) and (5.13), it can be got that

$$
\begin{aligned}
\frac{da}{d\tilde{t}} &= -\frac{\varepsilon}{2}\left[\Phi(a,\Omega) + \frac{z_0\left(4+3\gamma z_0^2\right)}{4}\sin\theta - 2\xi_1\Omega z_0\cos\theta\right]\\
&= -\frac{\varepsilon}{2}\left[(g_2\Omega + 2\xi_1\Omega - g_2\Omega\cos\Omega\tau_k + g_1\sin\Omega\tau_k)a\right.\\
&\quad \left.+\frac{z_0\left(4+3\gamma z_0^2\right)}{4}\sin\theta - 2\xi_1\Omega z_0\cos\theta\right],
\end{aligned} \tag{5.14}
$$

$$
\begin{aligned}
\frac{d\theta}{d\tilde{t}} &= \frac{\varepsilon}{2a}\left[\Psi(a,\Omega) - \frac{z_0\left(4+3\gamma z_0^2\right)}{4}\cos\theta - 2\xi_1\Omega z_0\sin\theta\right]\\
&= \frac{\varepsilon}{2a}\left[(g_1 - \sigma - g_1\cos\Omega\tau_k - g_2\Omega\sin\Omega\tau_k)a + \frac{3}{4}\gamma a^3\right.\\
&\quad \left.-\frac{z_0\left(4+3\gamma z_0^2\right)}{4}\cos\theta - 2\xi_1\Omega z_0\sin\theta\right].
\end{aligned} \tag{5.15}
$$

where

$$
\begin{cases}
\Phi(a,\Omega) = (g_2\Omega + 2\xi_1\Omega - g_2\Omega\cos\Omega\tau_k + g_1\sin\Omega\tau_k)a,\\
\Psi(a,\Omega) = (g_1 - \sigma - g_1\cos\Omega\tau_k - g_2\Omega\sin\Omega\tau_k)a + \frac{3}{4}\gamma a^3.
\end{cases} \tag{5.16}
$$

The terms of $\cos\theta$ and $\sin\theta$ can be resolved given the conditions $da/d\tilde{t} = 0$ and $d\theta/d\tilde{t} = 0$. From the condition $\sin^2\theta + \cos^2\theta = 1$, the relationship between frequency Ω and response amplitude a can be obtained

$$W(a, \Omega) = d_3 a^6 + d_2 a^4 + d_1 a^2 + d_0 = 0. \tag{5.17}$$

We can ascertain the nonlinear isolator amplitude-frequency curves for various parameters using Eq. (5.17). The system's nonlinearity creates a multi-steady state band in the amplitude-frequency plane. The curve with the lowest amplitude in this multi-steady state band is referred to as the amplitude death because the amplitude value is nearly zero. Consequently, the time delay, which results in the largest region of amplitude death, can be regarded as the optimal value for vibration isolation when faced with consistent external excitation.

The solution found for $da/d\tilde{t} = 0$ and $d\theta/d\tilde{t} = 0$ in Eqs. (5.14)–(5.15) is denoted as (a_0, θ_0). We can assess its stability by causing a slight disturbance to the standard form of Eqs. (5.14)–(5.15). By defining the variation parameters $\chi_1 = a - a_0$ and $\chi_2 = \theta - \theta_0$, the first order approach equations at (a_0, θ_0) are

$$2\chi_1' = -\varepsilon\left[\left(\frac{\partial\Phi}{\partial a}\right)\bigg|_{(a_0,\theta_0)}\chi_1 + \left(\frac{z_0(4+3\gamma z_0^2)}{4}\cos\theta_0 + 2\xi\Omega z_0\sin\theta_0\right)\chi_2\right], \tag{5.18}$$

$$\begin{aligned}2\chi_2' = \varepsilon\Bigg\{&\frac{1}{a_0}\Bigg[\left(\frac{\partial\Psi}{\partial a}\right)\bigg|_{(a_0,\theta_0)} \\ &-\frac{1}{a_0^2}\left(\Psi|_{(a_0,\theta_0)} - \frac{z_0(4+3\gamma z_0^2)}{4}\cos\theta_0 - 2\xi\Omega z_0\sin\theta_0\right)\Bigg]\chi_1 \\ &+\left[\frac{z_0(4+3\gamma z_0^2)}{4}\sin\theta_0 - 2\xi\Omega z_0\cos\theta_0\right]\chi_2\Bigg\}\end{aligned} \tag{5.19}$$

Rearranging Eqs. (5.18)–(5.19), the equation above can be written as

$$\begin{aligned}&2\chi_1' + \varepsilon\left(\frac{\partial\Phi}{\partial a}\right)\bigg|_{(a_0,\theta_0)}\chi_1 + \varepsilon\,\Psi|_{(a_0,\theta_0)}\chi_2 = 0, \\ &2\chi_2' - \frac{\varepsilon}{a_0}\left(\frac{\partial\Psi}{\partial a}\right)\bigg|_{(a_0,\theta_0)}\chi_1 + \frac{\varepsilon}{a_0}\,\Phi|_{(a_0,\theta_0)}\chi_1 = 0.\end{aligned} \tag{5.20}$$

The characteristic equation of Eq. (5.20) is

$$\lambda^2 + \frac{\varepsilon}{2a_0}\left[\frac{\partial(a\Phi)}{\partial a}\right]\bigg|_{(a_0,\theta_0)}\lambda + \frac{\varepsilon^2}{8a_0}\left(\frac{\partial W}{\partial a}\right)\bigg|_{(a_0,\theta_0)} = 0. \tag{5.21}$$

If it satisfies the condition that

$$\left.\frac{\partial W(a,\Omega)}{\partial a}\right|_{(a_0,\theta_0)} < 0, \tag{5.22}$$

then the solution (a_0, θ_0) is stable. Otherwise, the solution is unstable. The displacement transmissibility is characterized by the proportion of absolute vibration amplitude a to the excitation amplitude z_0 as

$$T_d = \left|\frac{a}{z_0}\right|. \tag{5.23}$$

Equation (5.23) is substituted into the equation for relating amplitude and frequency, i.e., Eq. (5.21). Given the critical condition (5.22), displacement transmissibility T_d can be used to represent the two conditions. Equations (5.21) and (5.22) refer to specific conditions that affect displacement transmissibility T_d:

$$\widetilde{W}(T_d,\Omega) = z_0^5 d_3 T_d^6 + z_0^3 d_2 T_d^4 + z_0 d_1 T_d^2 + \frac{d_0}{z_0} = 0, \tag{5.24}$$

and

$$\left.\frac{\partial \widetilde{W}(T_d,\Omega)}{\partial T_d}\right|_{\left(\frac{a_0}{z_0},\theta_0\right)} = \left.\frac{\partial \widetilde{W}(T_d,\Omega)}{\partial T_d}\right|_{(T_d,\theta_0)} = 6z_0^5 d_3 T_d^5 + 4z_0^3 d_2 T_d^3 + 2z_0 d_1 T_d < 0. \tag{5.25}$$

Equation (5.24) can be used to get the displacement transmissibility curves T_d, based on the preceding analysis. The determination of unstable and stable regions on the plane (Ω, T_d) relies on condition (5.25), while the critical boundary can be identified by determining the derivative of $\left.\frac{\partial \tilde{W}(T_d,\Omega)}{\partial T_d}\right|_{(T_d,\theta_0)} = 0$. By simultaneously solving Eq. (5.24) and $\left.\frac{\partial \tilde{W}(T_d,\Omega)}{\partial T_d}\right|_{(T_d,\theta_0)} = 0$, the domain of the sentence is mathematics, and the background is the study of multi-steady states and their determination. The determination of the multi-steady states band (Ω_1, Ω_2) can be achieved by the simultaneous solution of Eq. (5.24) and the condition $\left.\frac{\partial \tilde{W}(T_d,\Omega)}{\partial T_d}\right|_{(T_d,\theta_0)} = 0$.

Figure 5.7 illustrates the impact of nonlinearity on the system, bifurcation, and stability by displaying the unstable region, displacement transmissibility curves, and multi-steady states band of the system. This is demonstrated for various nonlinear coefficients (γ), while as $\xi_1 = 0.1$, $\tau_k = 0$ and $z_0 = 0.05$.

Figure 5.7 depicts the vibration stability displacement transmissibility curves and the critical curves as the isolator undergoes a periodic excitation without time-delayed active control. These curves are shown in parts (a), (c), and (e) of the figure and are influenced by a nonlinear coefficient, γ, which increases over time. When the displacement transmissibility is calculated via Eq. (5.24) in the stable zone as defined in Eq. (5.25), there is a frequency range for multi-steady states. This frequency range enlarges as γ increases, as demonstrated in parts (b), (d), and (f) of Fig. 5.7. When

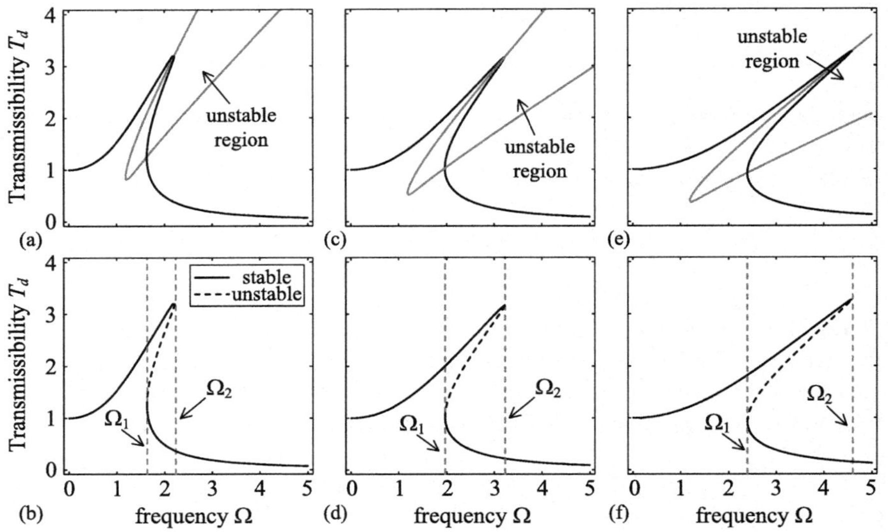

Fig. 5.7 The unstable region, displacement transmissibility curves, and multi-steady state band of the system for different nonlinear coefficients γ without time-delayed control. **a** and **b** $\gamma = 200$; **c** and **d** $\gamma = 500$; **e** and **f** $\gamma = 1000$

the isolator embodies a strong nonlinearity, its effective isolation frequency range starts from Ω_1 as the lower curves in the multi-steady state band are less than 1, a fact that has been validated. Figure 5.7 shows that a higher γ coefficient leads to a higher Ω_1 and broader multi-steady states band, thereby adversely affecting the isolation performance. Therefore, the role of time-delayed control in expanding the effective isolation range necessitates further study.

When the nonlinearity of the nonlinear isolator is increased, the multi-steady states band expands while the effective isolation frequency range contracts. This means that the impact of time delay on improving isolation effectiveness needs to be taken into account. In Fig. 5.8, the parametrical plane (Ω, τ_k) illustrates the multi-steady states band obtained using Eqs. (5.24)–(5.25) for different control strengths as $\xi_1 = 0.1$, $\gamma = 500$ and $z_0 = 0.05$.

Figure 5.8 illustrates how time delay impacts the nonlinear isolator vibration properties. The active time-delayed control could expand the effective isolation frequency range and remove the multi-steady states phenomenon as a result of an increase in time delay τ_k, which simultaneously reduces the multi-steady states band (Ω_1, Ω_2) and the value of Ω_1. Stronger control strengths g_1 and g_2 also effectively reduce the multi-steady states band and resonant peak. However, a longer time delay leads to a greater equivalent damping property based on eigenvalue analysis and consequently increases the displacement transmissibility T_d in the high-frequency band, as seen in Figure 5.8b, d, and f. Therefore, the optimal time delay must take into account the displacement transmissibility and the effective isolation frequency range in the high-frequency band. The optimal time delay value is either the intersection of the

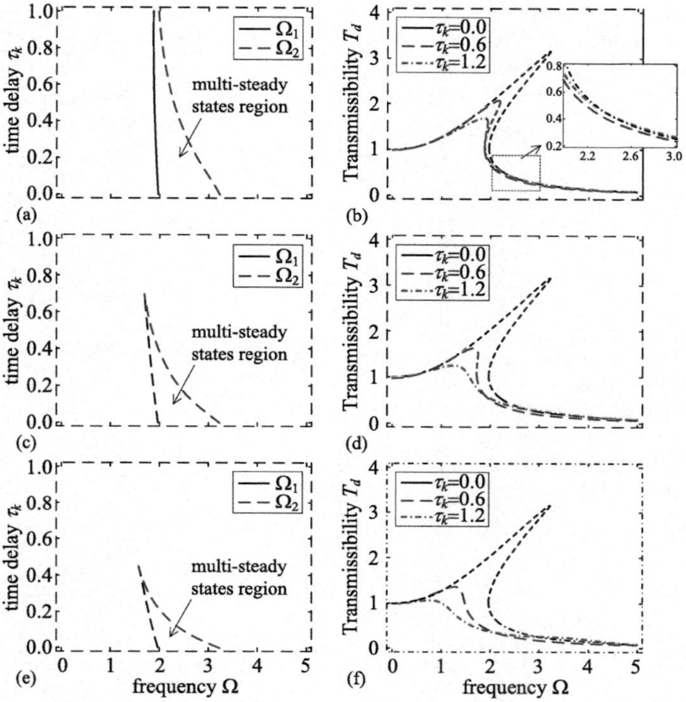

Fig. 5.8 Multi-steady states band (Ω_1, Ω_2) and displacement transmissibility curves T_d for different control strengths and time delay. **a** and **b** are for $g_1 = g_2 = 0.2$; **c** and **d** are for $g_1 = g_2 = 0.5$; **e** and **f** are for $g_1 = g_2 = 1.0$

Ω_1 and Ω_2 curves or the smallest Ω_1 value if there is no intersection. Settling time delay at the intersection of Ω_1 and Ω_2 can prevent the multi-steady state frequency band and provide the system with the biggest effective isolation frequency band. Though choosing a longer time delay can prevent the multi-steady states band, it would increase the displacement transmissibility in the high-frequency band due to its effect on the equivalent damping property. The optimal time delay is determined based on the control mechanism that regulates the system's equivalent stiffness and damping properties via linear time-delayed control. With $\xi_1 = 0.1$, $\gamma = 500$ and $z_0 = 0.05$, Fig. 5.9 showcases the optimal time delay values for maximizing the effective isolation band and minimizing the multi-steady states band.

Figure 5.9 displays the ideal time delay for the isolator under periodic excitation. When the time delay and control strengths are set based on the curves in Fig. 5.9, the vibration simultaneously exhibits the narrowest range of multi-steady states band and the least displacement transmissibility in the high-frequency band. With values of $\xi_1 = 0.1$, $\gamma = 500$ and $z_0 = 0.05$, Fig. 5.10 illustrates the comparison of transmissibility curves between the optimal time delay, $\tau_k = 0$ and $\tau_k = 1.5$.

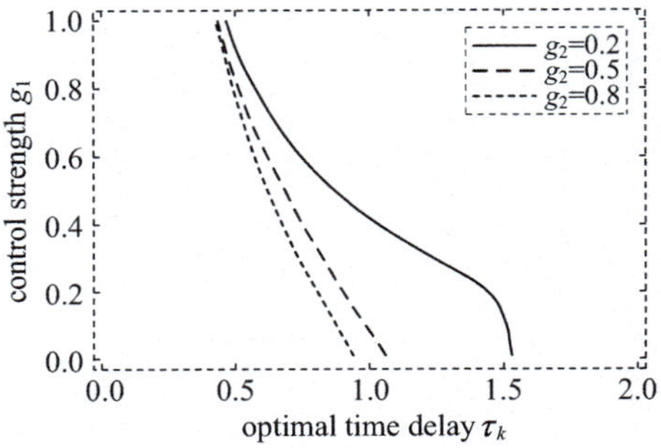

Fig. 5.9 The optimal values of time delay for different control strengths

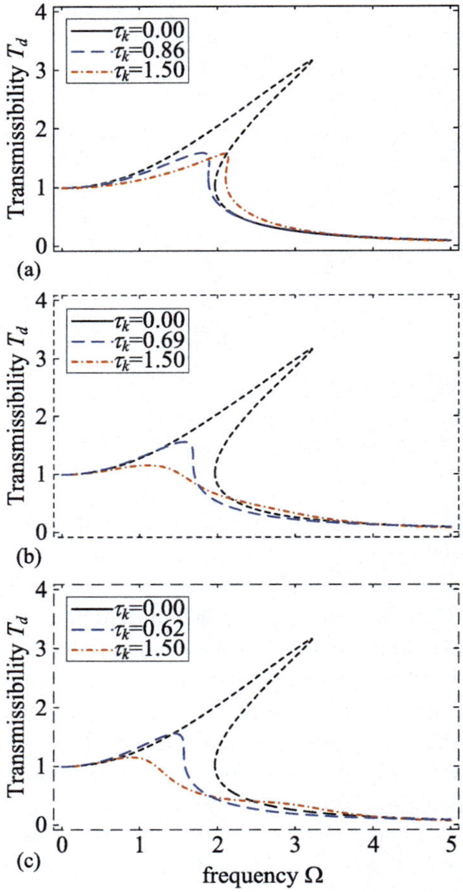

Fig. 5.10 The transmissibility curves comparison among optimum time delay, $\tau_k = 0$ and $\tau_k = 1.5$ for $g_1 = 0.5$. **a** $g_2 = 0.2$; **b** $g_2 = 0.5$; **c** $g_2 = 0.8$

The optimal time delay calculations, as illustrated in Fig. 5.9, show that when $g_1 = 0.5$, for $g_2 = 0.2$, 0.5, and 0.8, optimal time delay, $\tau_k = 0.857$, 0.695, and 0.617, respectively. Figure 5.10 presents the optimal time delay and additional values comparison of transmissibility curves. The findings illustrate that a larger effective isolation range and a reduced multi-steady states band are achievable without increasing the transmissibility in the high-frequency band for an optimal time delay. Conclusively, the comparative analyses confirm that the isolation effectiveness of the optimal time delay surpasses that of other time delay values.

5.2 Time-Delayed Vibration Isolator

5.2.1 Asymmetrical Isolation System

The mechanical model of a nonlinear vibration isolation system with a multi-time-delayed actuator [13] is shown in Fig. 5.11a. The mass of the platform is M_1, and the mass of the isolation object is M_2. The base excitation is denoted as $z(t)$, and the absolute response is denoted as $x(t)$, resulting in the relative vibration motion $\hat{x}(t) = x(t) - z(t)$. To ensure the stability of springs and the potential for multi-directional isolation, two pre-compressed SLSs are used, as illustrated in Fig. 5.11b, to achieve the desired nonlinear stiffness property $K(\hat{x})$. Figure 5.11b shows the configuration with 2-layer SLSs. The springs in the SLSs have a stiffness of k_h and an original length of l_0, while the vertical spring has a stiffness of k_v. We consider the signal to include both displacement and velocity as

$$u(\mu, \tau) = -g_1\left[\dot{\hat{x}}(t - \tau_1 - \tau_2) - \dot{\hat{x}}(t - \tau_1)\right] - g_2\left[\hat{x}(t - \tau_1 - \tau_2) - \hat{x}(t - \tau_1)\right], \tag{5.26}$$

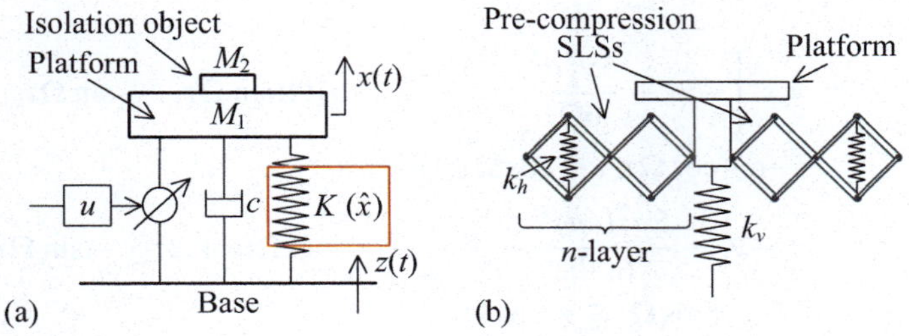

Fig. 5.11 **a** Structural diagram of an isolation system for different load M_2 with active feedback device; **b** The realization structure for $K(\hat{x})$

where τ_1 represents the natural time delay in signal transmission and processing, while τ_2 is seen as a modifiable parameter. The variables g_1 and g_2 stand for the strengths of velocity feedback and displacement feedback, respectively. As per Eq. (5.1), the signal generated by an active actuator relies on the variation between the feedback signal with an adjustable time delay and the one with an inherent time lag. Consequently, we can analyze how the adjustable time delay influences the vibration performances and comprehend its mechanism.

The dimensionless dynamical equation is written as

$$\begin{aligned}&\hat{x}'' + \omega_1^2\hat{x} + \gamma_1\hat{x} + \gamma_2\hat{x}^2 + \gamma_3\hat{x}^3 + \zeta\hat{x}' + \tilde{g}_1\left(\hat{x}'_{\tau_{k3}} - \hat{x}'_{\tau_{k1}}\right)\\ &+ \tilde{g}_2\left(\hat{x}_{\tau_{k3}} - \hat{x}_{\tau_{k1}}\right) = \Omega^2\tilde{z}_0\cos\Omega\tilde{t},\end{aligned} \tag{5.27}$$

where $(\cdot)'' = \frac{\mathrm{d}^2(\cdot)}{\mathrm{d}\tilde{t}^2}$, $(\cdot)' = \frac{\mathrm{d}(\cdot)}{\mathrm{d}\tilde{t}}$. This investigation considers various vibration properties, such as stability conditions, resonance frequency, and bifurcations. Its objective is to offer design principles and criteria for selecting appropriate time delays for different types of nonlinearity.

When the base excitation involves an impact motion, the vibration must be damped quickly while maintaining a smoothness to guarantee the isolation object effectively. By setting $\tilde{z}_0 = 0$, the vibration dissipation could be described as

$$\begin{aligned}a(\tilde{t}) &= A_0 e^{\left[-\frac{\xi}{2} - \frac{\alpha_1(\cos\Omega\tau_{k3} - \cos\Omega\tau_{k1})}{2} + \frac{\alpha_2(\sin\Omega\tau_{k3} - \sin\Omega\tau_{k1})}{2\Omega}\right]T_2}\\ &= A_0 e^{\left[-\frac{\varepsilon^2\xi}{2} - \frac{\varepsilon^2\alpha_1(\cos\Omega\tau_{k3} - \cos\Omega\tau_{k1})}{2} + \frac{\varepsilon^2\alpha_2(\sin\Omega\tau_{k3} - \sin\Omega\tau_{k1})}{2\Omega}\right]\tilde{t}}\\ &= A_0 e^{\left[-\frac{\zeta}{2} - \frac{\tilde{g}_1(\cos\Omega\tau_{k3} - \cos\Omega\tau_{k1})}{2} + \frac{\tilde{g}_2(\sin\Omega\tau_{k3} - \sin\Omega\tau_{k1})}{2\Omega}\right]\tilde{t}},\end{aligned} \tag{5.28}$$

where A_0 represents the initial displacement, Eq. (5.28) presents the amplitude envelope curve that can be utilized to assess the decay rate of free vibration. The symbol Ω within the amplitude expression (5.28) denotes the transient frequency of free vibration. Based on the given conditions $\frac{d\varphi}{dT_2} = 0$ and $\varepsilon^2\sigma = \omega_1^2 - \Omega^2$, the following equations apply:

$$\begin{aligned}\omega_1^2 - \Omega^2 &= \varepsilon^2\left[-\chi_1 + \frac{5\gamma_2^2 a^2}{6\Omega^2} - \frac{3\gamma_3 a^2}{4} - \alpha_1\Omega(\sin\Omega\tau_{k3} - \sin\Omega\tau_{k1})\right.\\ &\quad \left.-\alpha_2(\cos\Omega\tau_{k3} - \cos\Omega\tau_{k1})\right]\\ &= -\gamma_1 + \frac{5\gamma_2^2(\varepsilon a)^2}{6\Omega^2} - \frac{3\gamma_3(\varepsilon a)^2}{4} - \tilde{g}_1\Omega(\sin\Omega\tau_{k3} - \sin\Omega\tau_{k1})\\ &\quad - \tilde{g}_2(\cos\Omega\tau_{k3} - \cos\Omega\tau_{k1}).\end{aligned} \tag{5.29}$$

Hence, one can calculate the amplitude of free vibration for impact excitation and determine the dynamic stability of the trivial equilibrium using Eqs. (5.28)–(5.29). In Fig. 5.12, the dynamic stability region of the trivial equilibrium on the time-delay parameter plane is illustrated along with the corresponding amplitude envelope of

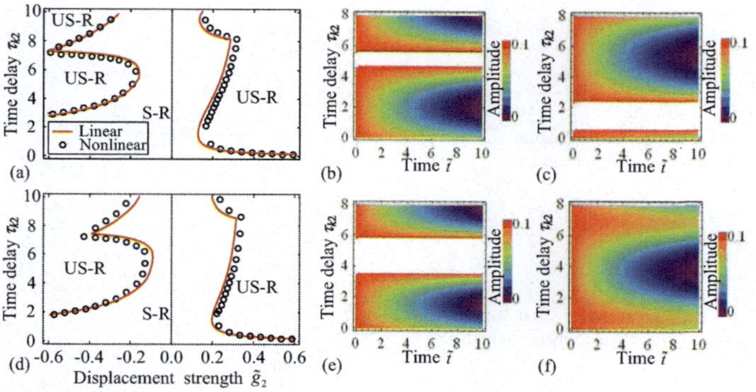

Fig. 5.12 Critical curves for dynamic stability of the zero equilibrium by linear analysis of eigenvalues and nonlinear analysis based on Eqs. (5.28)–(5.29), and the free-vibration amplitude envelope for different delayed parameters. **a** Critical curves for $\tau_{k1} = 0$; **b** Amplitude envelope for $\tilde{g}_2 = -0.2$ and $\tau_{k1} = 0$; **c** Amplitude envelope for $\tilde{g}_2 = 0.2$ and $\tau_{k1} = 0$; **d** Critical curves for $\tau_{k1} = 0.5$; **e** Amplitude envelope for $\tilde{g}_2 = -0.2$ and $\tau_{k1} = 0.5$; **f** Amplitude envelope for $\tilde{g}_2 = 0.2$ and $\tau_{k1} = 0.5$

free vibration. The plane of delayed parameters is divided into the Dynamical Stable Region (D-S-R) and the Dynamical Unstable Region (D-US-R).

Figure 5.12 demonstrates the trivial equilibrium's dynamic stability obtained through Eqs. (5.28)–(5.29) differs from the result obtained through eigenvalues analysis for the linearized system. Additionally, it shows the impact of time delays on the dissipation of free vibration for different strengths $\tilde{g}_2$. In Fig. 5.12b, c and e, f, the darker color indicates a lower amplitude envelope, while the white region represents dynamical instability. Hence, it can be observed that the decay rate of vibration initially decreases and then increases for negative feedback gain $\tilde{g}_2$, while it initially increases for positive $\tilde{g}_2$. The appropriate value of time delay τ_{k2} can be advantageous for suppressing vibration within a short period. Consequently, based on the stability region of parameters for the trivial equilibrium and the expressions of the amplitude envelope of free vibration, a criterion for parameter design is proposed to efficiently suppress free vibration under impact excitation.

$$I_1(\tau_{k_2}) = \min\left\{\frac{d\Phi}{da}\right\}. \tag{5.30}$$

Under such criterion, one could choose the appropriate time delay τ_{k2} for the most rapid dissipation of free vibration. Figure 5.13 shows the optimal values of time delay τ_{k2} and the comparison of free vibration between the optimal time delay and other values for different feedback gain $\tilde{g}_2$ and the inherent time delay τ_{k1}.

When using this criterion, individuals can select the suitable time delay τ_{k2} to dissipate free vibration quickly. Figure 5.13 illustrates the best values of τ_{k2} and compares the free vibration using the optimal time delay against various values of feedback gain $\tilde{g}_2$ and the inherent time delay τ_{k1}.

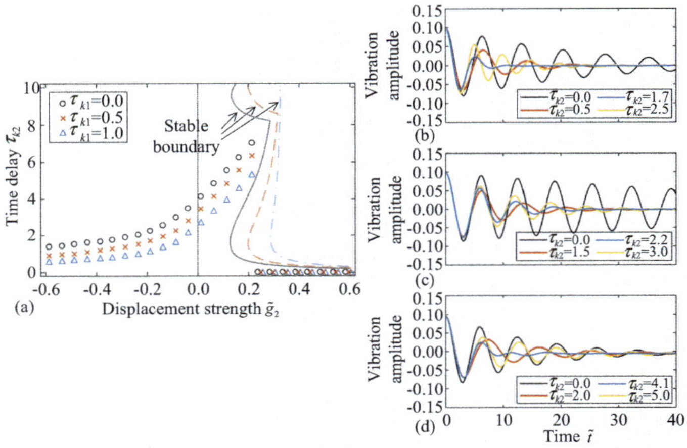

Fig. 5.13 **a** Optimum values of time delay τ_{k2} for different $\tilde{g}_2$ for impact excitation. **b** Free vibration for $\tau_{k1} = 0.0$, $\tilde{g}_2 = -0.4$; **c** Free vibration for $\tau_{k1} = 0.5$, $\tilde{g}_2 = -0.1$; **d** Free vibration for $\tau_{k1} = 1.0$, $\tilde{g}_2 = 0.1$

The delayed parameters should be fixed in the D-S-R to avoid bifurcation. When the optimal τ_{k2} obtained by Eq. (5.25) is in the D-US-R, we consider $\tau_{k2} = 0$ is the most appropriate value. When $\tau_{k1} = 0.0$, $\tilde{g}_2 = -0.4$ (the case in Fig. 5.13b), the optimal time delay is $\tau_{k2} = 1.69$; when $\tau_{k1} = 0.5$, $\tilde{g}_2 = -0.1$ (Fig. 5.13c), the optimal time delay is $\tau_{k2} = 2.204$; when $\tau_{k1} = 1.0$, $\tilde{g}_2 = 0.1$ (Fig. 5.13d), the optimal time delay is $\tau_{k2} = 4.082$. Figure 5.13b–d compares the vibration response for the optimal time delay and other values. It can be observed that the vibration decays rapidly when the optimal time delay is used. Evidently, the time history of the vibration confirms the effectiveness of the proposed structural parameter criterion for impact excitation.

The stiffness coefficient function is the main factor determining the types of the amplitude-frequency curve. It is known that the linear stiffness is dependent on the pre-deformation of SLSs and the mass of platform M_1, and the nonlinear stiffness coefficientsγ_2 and γ_3 are determined by the assembly process and the masses of M_1 and M_2. Since nonlinear stiffness coefficients γ_2 and $\gamma_3$represent the asymmetry and nonlinearity, the masses of M_1 and M_2 are actually the main changeable parameters. Figure 5.14 illustrates the divisional chart on the plane of (M_1, M_2) for different nonlinear stiffness properties. The relevant amplitude-frequency curves, multi-steady states frequency band, and the beginning frequency Ω_e of the effective isolation band are shown in Fig. 5.14.

The stiffness coefficient function primarily determines the types of the amplitude-frequency curve. Linear stiffness depends on the pre-deformation of SLSs and the mass of platform M_1, while the nonlinear stiffness coefficients γ_2 and γ_3 are determined by the assembly process and the masses of M_1 and M_2. Therefore, the masses of M_1 and M_2 are the main adjustable parameters as they influence the asymmetry and nonlinearity represented by the nonlinear stiffness coefficients γ_2 and γ_3. In

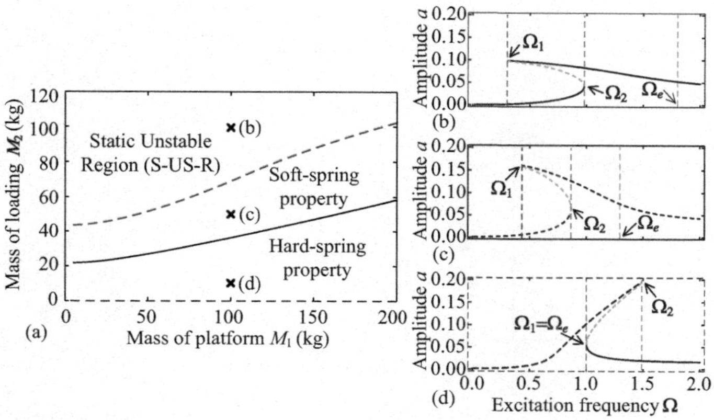

Fig. 5.14 **a** The divisional chart on the plane of (M_1, M_2) for different stiffness properties, and the amplitude-frequency curves of relative motion for selecting the masses in **b** hard-spring property region as $M_1 = 100$ kg and $M_2 = 10$ kg, **c** soft-spring property region as $M_1 = 100$ kg and $M_2 = 50$ kg, **d** static Unstable Region (S-US-R) as $M_1 = 100$ kg and $M_2 = 100$ kg

Fig. 5.14, there is a divisional chart on the (M_1, M_2) plane that illustrates different nonlinear stiffness properties. It also shows the corresponding amplitude-frequency curves, the multi-steady state frequency band, and the beginning frequency Ω_e of the effective isolation band.

As shown in Fig. 5.14, various types of nonlinearity may appear if different masses of platform M_1 and isolation object M_2 are used, which consequently leads to distinct amplitude-frequency curves.

The frequency band where the multi-steady states occur starts at Ω_1, which is also the starting frequency for effective isolation in cases of hard-spring nonlinearity. If the excitation frequency falls within the range of $[\Omega_1, \Omega_2]$, the amplitude of vibration in the isolator can transition from the lower branch to a higher branch. Thus, for systems exhibiting the hard-spring property, the guidelines for parameter design are established based on the defined bands of effective isolation frequency $[\Omega_1, \infty)$ and multi-steady states $[\Omega_1, \Omega_2]$

$$\begin{cases} I_2(\tau_{k2}) = \min\{\Omega_1\}, \\ I_3(\tau_{k2}) = \min\{|\Omega_2 - \Omega_1|\}. \end{cases} \tag{5.31}$$

After applying the criteria described in Eq. (5.31), utilizing structural nonlinearity can effectively prevent the occurrence of the multi-steady states phenomenon in the context of hard-spring behaviour. The mass of the isolated object M_2, assumed to be 10 kg, is significantly smaller than the mass of isolation object M_1 of 100 kg. The dimensionless parameters are designated as $\omega_1 = 1$, $\gamma_1 = -0.3609$,$\gamma_2 = -4.4054$, $\gamma_3 = 88.027$, $\xi = 0.095$. Figure 5.15 illustrates the most favorable amount of delay in time for various gains $\tilde{g}_2$ related to displacement feedback, as well as the comparison

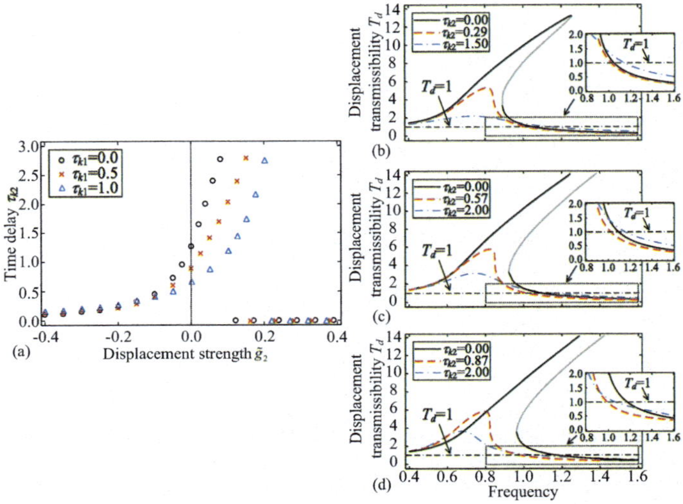

Fig. 5.15 **a** Optimal values of adjustable time delay τ_{k2} for different displacement feedback gain $\tilde{g}_2$ with increasing inherent time delayτ_{k1}. **b** amplitude-frequency curve for $\tau_{k1} = 0.0$, $\tilde{g}_2 = -0.2$; **c** amplitude-frequency curve for $\tau_{k1} = 0.5$, $\tilde{g}_2 = -0.05$; **d** amplitude-frequency curve for $\tau_{k1} = 1.0$, $\tilde{g}_2 = 0.05$

of amplitude-frequency curves between the said optimal values and alternative time delay values.

The optimal values of adjustable time delay τ_{k2} are shown in Fig. 5.15 according to the design criterions described in Eq. (5.31). In Fig. 5.15a, when $\tau_{k1} = 0.0$, $\tilde{g}_2 = -0.2$ as depicted in Fig. 5.15b, the optimal value of the time delay τ_{k2} is 0.288. Given that $\tau_{k1} = 0.5$, $\tilde{g}_2 = -0.05$, and referring to the situation illustrated in Fig. 5.15c, it can be determined that the value of τ_{k2} is 0.567. As shown in Fig. 5.15d, the value of τ_{k2} is 0.889 when $\tau_{k1} = 1.0$, $\tilde{g}_2 = 0.05$. Based on the information provided in Fig. 5.15, it can be observed that the amplitude-frequency curve depicted by the Red Dashed Lines corresponds to the optimal time delay. The lowest frequency occurs at the jumping-off point, which leads to the largest effective isolation frequency band. The optimal time delay minimizes the range of multiple stable states to zero. Despite the fact that the resonance peak for the optimal time delay is greater than the case with a larger τ_{k2}, in the effective frequency band, the optimal value of τ_{k2} results in the smallest displacement transmissibility. Since the frequency falls within the range of $[0, \Omega_1]$, T_d, the transmissibility of displacement, is always greater than 1, which indicates that the working environment of isolation object gets worsen by the isolation system. However, the nonlinear property of the system remains unchanged while the frequency band for effective isolation can be extended through the implementation of time-delayed feedback. Thus, in reality, instead of resonance peak, the displacement transmissibility characteristics in high frequency band should be focused on. According to Fig. 5.15b–d, utilizing multiple time-delayed feedback can be served to following three goals: enlarging the frequency band of effective isolation, suppressing the multi-state phenomenon and improving the isolation effectiveness.

On the contrary, when the mass of object M_2 is increased and eventually enters the region with soft-spring properties, as shown in Fig. 5.15a, the asymmetry grows more pronounced and adversely affects the high-static and low-dynamic (HSLD) characteristics. In the case of soft-spring property, considering that the effective isolation frequency band can be expressed as $\left[\sqrt{2}\Omega_0,\ \infty\right)$, where Ω_0 represents the resonance frequency, the criterion of parameter design for maximal frequency band could be written as

$$I_4(\tau_{k2}) = \min\{\Omega_0\}. \tag{5.32}$$

Given the parameter values $M_2 = 50$ kg, $\omega_1 = 1$, $\gamma_1 = 0.476$, $\gamma_2 = -16.52$, $\gamma_3 = 114.55$, $\xi = 0.0816$, it can be concluded that the isolator exhibits a soft-spring characteristic. Figure 5.16 illustrates the time delay value τ_{k2} for different τ_{k1} and $\tilde{g}_2$ as well as the corresponding amplitude-frequency curves associated with the optimal value and alternative time delays.

As shown in Fig. 5.16, when the optimal value of time delay τ_{k2} is fixed under different displacement gain $\tilde{g}_2$, the largest effective isolation frequency band can be obtained. By examining the amplitude-frequency curves, differences can be observed between the optimal time delay value and alternative values. The delayed feedback

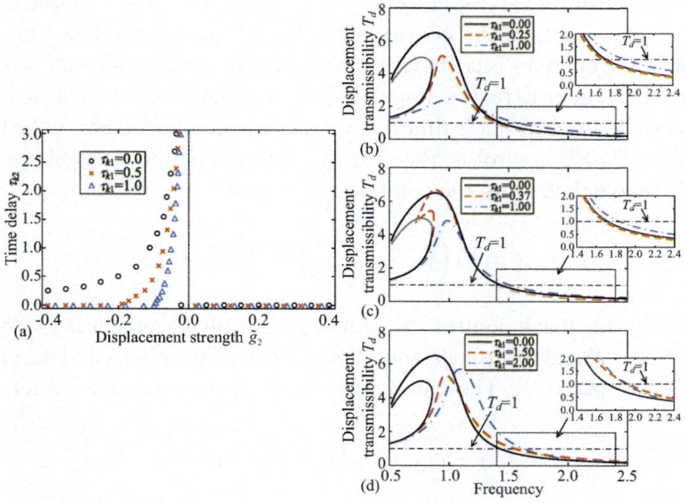

Fig. 5.16 Optimal value of time delay τ_{k2} for soft-spring property. **a** The optimal time delay; **b** Amplitude-frequency curve for $\tau_{k1} = 0.0$ and $\tilde{g}_2 = -0.4$; **c** Amplitude-frequency curve for $\tau_{k1} = 0.5$ and $\tilde{g}_2 = -0.05$; **d** Amplitude-frequency curve for $\tau_{k1} = 1.0$ and $\tilde{g}_2 = 0.05$

system effectively reduces the resonance frequency and lowers the displacement transmissibility. It is important to notice that when considering the third case with $\tau_{k1} = 1.0$ and $\tilde{g}_2 = 0.05$, the optimal value of τ_{k2} can be determined as zero based on Fig. 5.16a. Therefore, the Black line depicted in Fig. 5.16d exhibits the optimal degree of isolation effectiveness, demonstrating a significantly broader effective isolation band in comparison to the remaining two cases.

Therefore, by analyzing the nonlinear vibration properties, for different structural nonlinearities, the parameter design criterions can be obtained under different types of excitation, while the adjustable time delay τ_{k2} can be optimized to the most appropriate values. By implementing delayed feedback, one can improve various aspects of vibration performance, such as the ability to isolate specific frequency ranges effectively and the overall dampening effect. Simultaneously, without altering the structural nonlinearity, the isolation effectiveness in the high frequency band is maximized while suppressing resonance peaks.

5.2.2 *Adjustable Time Delay*

As Fig. 5.17a, we consider a nonlinear-stiffness and damping-properties isolation system controlled by TD-FB signals [14]. For the excitation $z(t)$ from base, the response of the isolation platform M, providing an isolation environment, is $y(t)$. Setting the relative motion as $\hat{y} = y - z$, the stiffness K($\mathbf{u}$) and damping properties C($\mathbf{u}$) are expressed respectively, where $\mathbf{u} = \left\{\hat{y}, \dot{\hat{y}}\right\}^T$. The control loop is as Fig. 5.17b. The TD-FB control force is assumed as $f(\mathbf{u}, \mathbf{u}_{\boldsymbol{\tau}})$ where $\boldsymbol{\tau}$ is the multiple time delays as $\boldsymbol{\tau} = \{\tau_1, \tau_2, \ldots\}^T$. In the control loop, the inherent time delay τ_1 cannot be ignored and time delay τ_2 is artificially introduced to improve the nonlinear isolation effectiveness. In order to realize improvement of isolation effectiveness in different frequency bands, the adjustable time delay τ_2 is considered variable on the frequency. The function TD-FB control is assumed as a linear function to highlight the effect of variable time delay. Thus, the control force f is as

$$f(\mathbf{u}, \boldsymbol{\tau}) = f_1\left(\dot{\hat{y}}_{\tau_1} - \dot{\hat{y}}_{\tau_1+\tau_2}\right) + f_2\left(\hat{y}_{\tau_1} - \hat{y}_{\tau_1+\tau_2}\right), \tag{5.33}$$

where velocity feedback control strength is f_1 and displacement is f_2. We assume that the stiffness K and damping C must have positive linear parts and the coefficients are k_l and c_l, respectively. Then, from Fig. 5.17a, the dynamical equation forms as

$$M\ddot{\hat{y}} + k_l\hat{y} + K_n\left(\hat{y}\right) + c_l\dot{\hat{y}} + C_n\left(\hat{y}, \dot{\hat{y}}\right) = f(\mathbf{u}, \boldsymbol{\tau}) - M\ddot{z}(t). \tag{5.34}$$

Figure 5.18 shows the stability regions of equilibrium on the parametrical plane (τ_{k2}, g_2) for different values of g_1 and τ_{k1}.

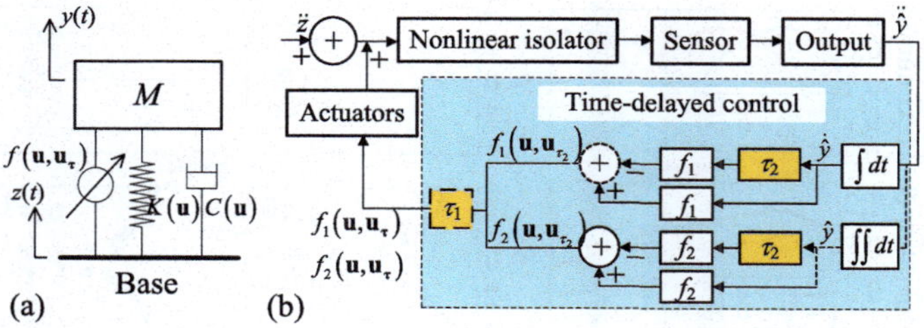

Fig. 5.17 **a** The time-delayed N-VI mechanical model; **b** Feedback control loop with multiple time-delays

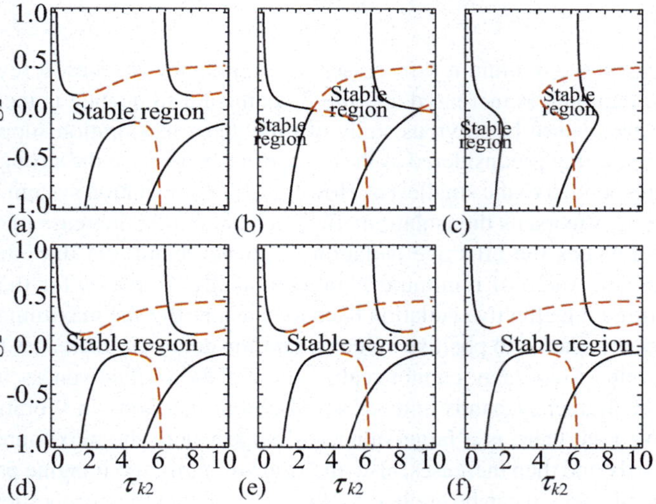

Fig. 5.18 The stability boundary of equilibrium on parametrical plane (τ_{k2}, g_2) as $\tau_{k1} = 0.1$ for **a** $g_1 = -0.05$, **b** $g_1 = 0.05$, **c** $g_1 = 0.1$; the boundary as $g_1 = 0.05$ for **d** $\tau_{k1} = 0.2$, **e** $\tau_{k1} = 0.4$ and **f** $\tau_{k1} = 0.6$

In Fig. 5.18, the Solid lines correspond the critical condition when one real part changes from negative to positive and the Dashed lines correspond the critical condition for it changes from positive to negative. From the comparison of stable regions between negative and positive g_1 in Fig. 5.18a–c, the stable range is larger for negative g_1 than the positive one. On the other hand, the increase of τ_{k1} slightly affects the stable region as Fig. 5.18d–f. The results shown in Fig. 5.18 reveal the limitation of control parameters for dynamical stability.

Figure 5.19 shows the variations of α and β for increasing adjustable time delay τ_{k2} with increasing control gains g_1.

The value of g_2 significantly influences the values of maximum α and minimum β. For negative g_2, as time delay τ_{k2} increases, the maximum α reduces firstly and

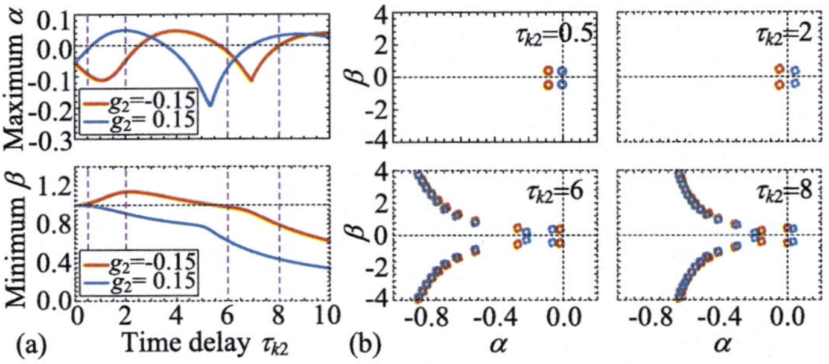

Fig. 5.19 **a** The maximum real part α and minimum imagine part β; **b** All real and imagine parts for different τ_{k2}

then increases; the minimum β increases firstly and then reduces. It reveals that the equivalent damping is increased and the fundamental frequency is increased when τ_{k2} varies from zero. In pervious study of time delay in vibration suppression, this range of time delay is considered as the appropriate values for the strongest damping effect corresponded to the smallest α. However, for the isolation system, the stronger damping effect worsens the isolation effectiveness and the increase of fundamental frequency reduces the effective isolation frequency band. On the other hand, for positive g_2, the value of minimum β monotonically decreases for the increase of τ_{k2}, benefiting the effective isolation band expansion. But the maximum α increases quickly from negative to positive value when time delay τ_{k2} increases from zero.

Small-value time-delay cannot realize the isolation effectiveness improvement in different frequency bands and resolve the contradictions in vibration isolation system. We note there is a large-value τ_{k2} range where the maximum real part α reduces firstly and then increases, also, the value of minimum imagine part β reduces monotonically, which can be applied for the different damping properties in different frequency bands and expansion of effective isolation band realization. As shown in Fig. 5.19b, for large-value τ_{k2}, multiple imagine parts are gathered following the fundamental one. Thus, for the utilization of large range time-delay, the bandgap phenomenon would occur. Therefore, the large-value time delay control mechanism on vibration isolation compounds the change of equivalent stiffness, damping and time-delayed-induced multiple bandgap.

It has known that the resonance frequencies number and values are dependent on the imagine part β and the equivalent damping effect is dependent on the real part α. As the analysis and results shown in Fig. 5.3, for g_1 is fixed less zero, as Fig. 5.19a, b, when the value of g_2 is small and close to zero, the equilibrium is global stable; when g_1 is fixed positive and g_2 is increased, for different τ_{k2}, there occurs a range for stability for large time delay; for further increasing g_2, the equilibrium is stable only for small time delay. Thus, based on the idea that time delay τ_{k2} is variable and dependent on frequency to meet the different requirements of isolation effectiveness, the functions of the variable time delay τ_{k2} for different control gains g_2 are shown in Fig. 5.20.

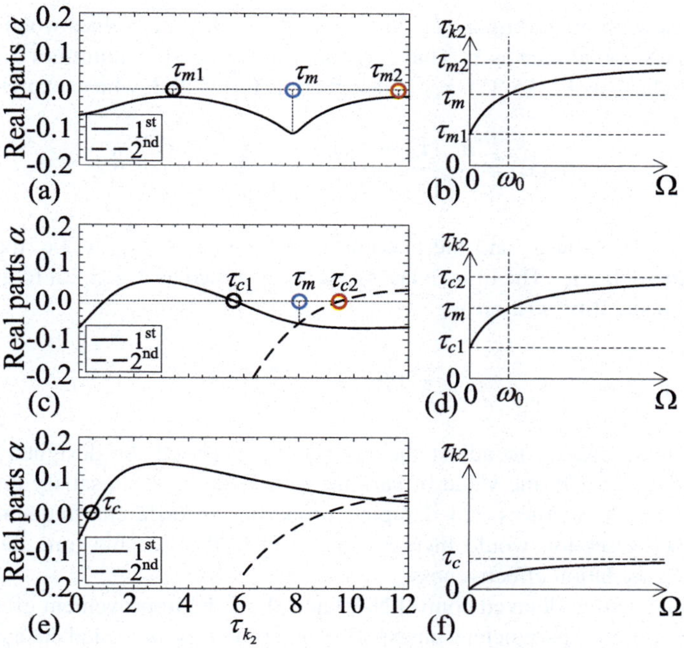

Fig. 5.20 The two largest real parts of eigenvalues for different τ_{k2} as **a** $g_2 = 0.05$, **c** $g_2 = 0.15$ and **e** $g_2 = 0.3$; the designed function of adjustable time delay τ_{k2} for different Ω for **b** $g_2 = 0.05$, **d** $g_2 = 0.15$ and **f** $g_2 = 0.3$

As shown in Fig. 5.20, for different values of g_2, there are three situations for the designed functions of variable time delay. For the case as Fig. 5.20a, the values of τ_{k2} where the real parts reaching the maximum value are defined as τ_{m1} and τ_{m2}, and the one corresponds to minimum real part defined as τ_m. When the real part α is negative and its absolute value is increasing, the equivalent damping effect is enhanced. As the response frequency is close to the natural frequency ω_0, time delay τ_{k2} should be fixed close to τ_m for negative and minimum α to reduce the resonance peak. On the other hand, when the frequency is increased and much larger than natural frequency ω_0, the adjustable time delay τ_{k2} should be close to τ_{m2} as the equivalent damping effect is reduced in order to induce nicer isolation effectiveness in high frequency band. Then, the function of variable time delay τ_{k2} is designed as Fig. 5.20b, which realizes the obvious damping effect for resonance frequency while less effect for high frequency band. Therefore, according to Fig. 5.20b, the function $\tau_{k2}(\Omega)$ on (Ω, τ_{k2}) should satisfies the points (0, 0), (0, τ_{m1}) and (∞,τ_{m2}). We write the expression of τ_{k2} as

$$\tau_{k_2}(\Omega) = \frac{\tau_{m_2}(\tau_{m_1} - \tau_m)\Omega - \tau_{m_1}(\tau_{m_2} - \tau_m)\omega_0}{(\tau_{m_1} - \tau_m)\Omega - (\tau_{m_2} - \tau_m)\omega_0}. \tag{5.35}$$

For the case with increasing g_2, as Fig. 5.20c, the critical τ_{k2} values when α crosses the zero axes are defined as τ_{c1} and τ_{c2}, and also the one for minimum α is defined as τ_m. Similarly as Eq. (5.35), τ_{k2} is designed as Fig. 5.20d, whose expression is as

$$\tau_{k_2}(\Omega) = \frac{\tau_{c_2}\left(\tau_{c_1} - \tau_m\right)\Omega - \tau_{c_1}\left(\tau_{c_2} - \tau_m\right)\omega_0}{\left(\tau_{c_1} - \tau_m\right)\Omega - \left(\tau_{c_2} - \tau_m\right)\omega_0}. \tag{5.36}$$

For the case of large g_2 and the maximum real part is as Fig. 5.20e, there is one critical time-delay τ_c. The time delay $\tau_{k2}(\Omega)$ is designed as Fig. 5.20f to grantee the equilibrium stability, written as

$$\tau_{k_2}(\Omega) = \frac{\tau_c \Omega}{\Omega + C}. \tag{5.37}$$

where C is a constant to adjust the $\tau_{k2}(\Omega)$ slope. Based the designed functions $\tau_{k2}(\Omega)$ by Eq. (5.10), the value of variable time delay τ_{k2} can satisfy the requirement of variable stiffness and damping properties in different frequency bands. The following section would theoretically analyze the variable time delay effect on vibration isolation effectiveness.

To realize a time-delayed control N-VI and show the improvement effectiveness by designed control parameters, an experimental prototype is carried out as Fig. 5.21.

As shown in Fig. 5.21, the nonlinearity of the isolation system is realized based on the concept of Quasi-Zero Stiffness (QZS). Positive and negative elastic components called Quadrilateral-Linkage Structures (QLSs) and Elastic Steel Slices (ESSs) are utilized. With the parallel elastic components, the mechanical model of the isolation system is abstracted as a nonlinear stiffness and damping properties isolation system.

Firstly, the structural parameters and inherent time delay τ_{k1} are identified. The relevant dynamical experiment is carried out at the resonance frequency band and the τ_{k2} is set as zero. Figure 5.22a is the displacement transmissibility at resonance frequency band, obtained by the time series as Fig. 5.22b.

According to the theoretical results of Eqs. (5.22)–(5.23), we utilize the LSM to obtain the structural parameters and inherent time delay τ_{k1}, whose values are listed in Table 5.1.

Based on the identified parameters, we fix three values of control strength f_2 to correspond the three cases of the designed time delay. Figure 5.23 shows the displacement transmissibility T_d for different control parameters and its comparison by theoretical results.

Figure 5.23 not only reveals that the experimental results verify the theoretical analysis of transmissibility on frequency band, but also shows the effect of time delay on isolation effectiveness. In the measurement process, the dimensionless value g_2 values are fixed as $g_2 = 0.1$, $g_2 = 0.2$ and $g_2 = 0.3$, corresponding to the three cases for the variable time delay τ_{k2}. From the experiment results of displacement transmissibility as Fig. 5.23a, the values of T_d for the three cases are all smaller than the case without control ($g_2 = 0/\tau_{k2} = 0$) in the effective isolation frequency band. Also, the experiment results in Fig. 5.23a demonstrate that the effective isolation

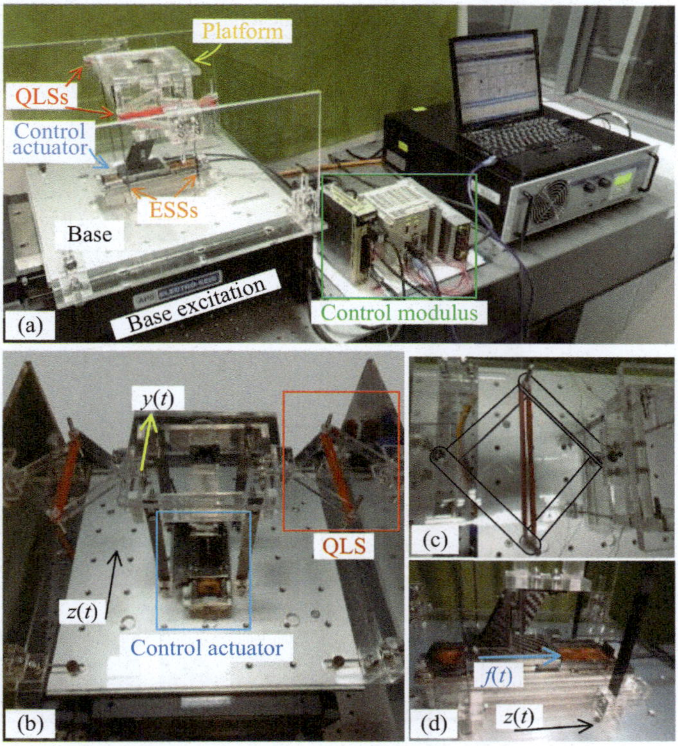

Fig. 5.21 The experimental prototype, elastic components and control modulus. **a** The experimental devices; **b** the isolation platform with time-delayed control; **c** the construction of QLS; **d** the relation of timed-delayed actuator

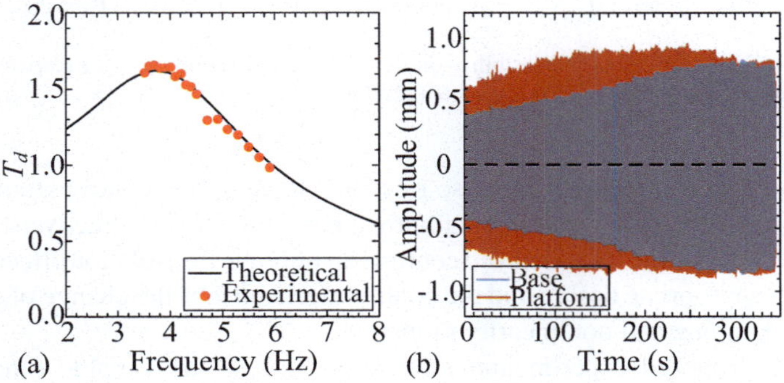

Fig. 5.22 The displacement transmissibility at resonance frequency band for identification of damping and inherent time delay

Table 5.1 Structural and control parameters by identification

Symbol	ω_0	γ	ξ_1	ξ_2	τ_{k1}	g_1
Value	34.48	0.8509	0.2743	0.018	0.032	0.05

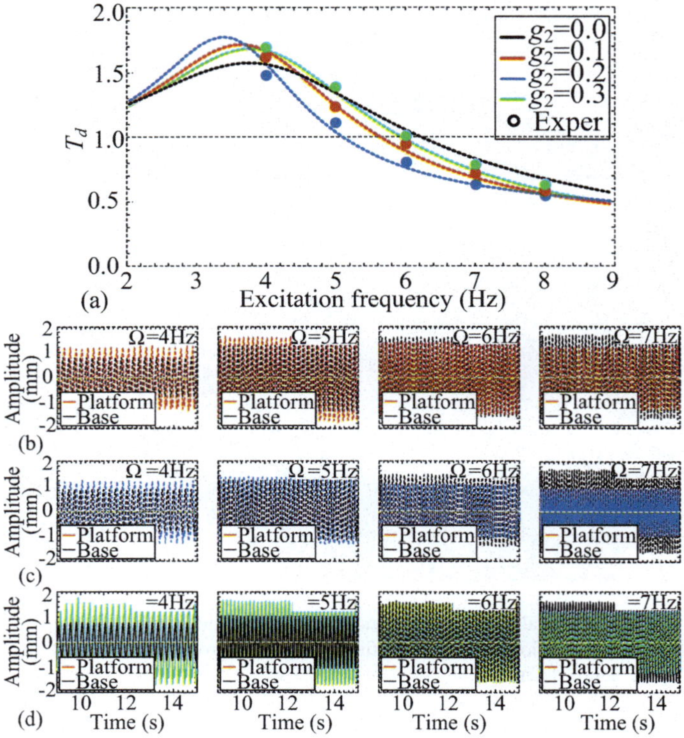

Fig. 5.23 **a** Displacement transmissibility on frequency band; Time series of responses between the platform and base for **b** $g_2 = 0.1$; **c** $g_2 = 0.2$; **d** $g_2 = 0.3$

frequency band is largest for the case $g_2 = 0.2$. Because the control strength for $g_2 = 0.2$ is stronger than $g_2 = 0.1$, the improvement of isolation effectiveness is more significant. Therefore, time-delayed control can improve the isolation effectiveness of N-VI with appropriate design and the control mechanism is the change of equivalent stiffness, damping and nonlinearity properties.

For the proposed experimental prototype, because the variable time delay is inducing multiple local resonances phenomenon, the isolation high frequency band effectiveness is effectively improved. In the following study, the time delay on the bandwidth of effective isolation frequency effect would be discussed by both theoretical and experimental results. The beginning frequency for effective isolation frequency band is defined as

Fig. 5.24 The beginning frequency Ω_e for effective isolation for different values of control strength g_2 and structural parameter γ

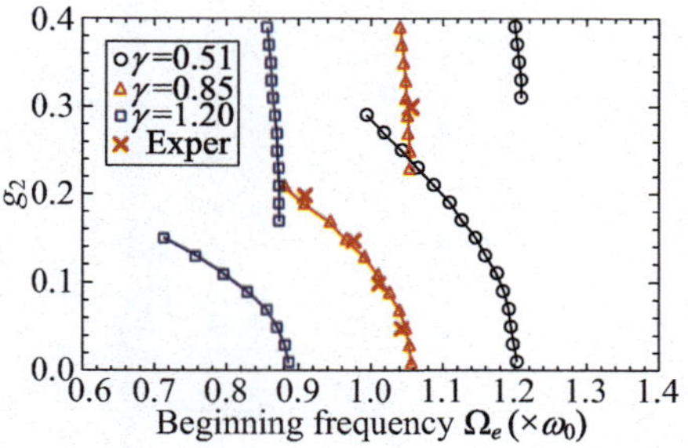

$$\Omega_e = \{\Omega_e | T_d(\Omega_e) = 1\}\backslash\{0\}. \tag{5.38}$$

According to Eq. (5.38) and experimental results, Fig. 5.24 shows the γ and g_2 on the Ω_e for effective isolation frequency band effect.

In the experiment, the values of stiffness k_1 and k_2 are identified as $k_1 = 101.4$ and $k_2 = 174.2$ respectively, and the parameter γ in experiment is as $\gamma = 2k_2/4k_1 \approx 0.85$. Due to the QZS structure as nonlinear component, Ω_e reduces with theγ increase. The utilization of time-delayed control can effectively extend the effective isolation band wider about 20% compared to dimensionless (normalized) frequency. In addition, as the value of γ increases, although the beginning frequency Ω_e reduces from a lower level resulted from stronger negative stiffness, the range of g_2 is smaller for expand the isolation band. For instance, as g_2 increases from zero to 0.2, for $\gamma = 0.51$, the beginning frequency is reduced from $1.2\omega_0$ to $1\omega_0$, even lower than the case as $\gamma = 0.85$ and $g_2 = 0.3$.

Therefore, the time-delayed control can realize effective isolation in a wide frequency band, which solves the problem as extending the isolation band to ultra-low frequency band without changing the structure construction and parameters. The above analysis shows the significance of time delay in vibration isolation and provides the guidance for appropriate control parameters in applications.

5.3 Experimental Investigation on the Time-Delayed Vibration Absorber

5.3.1 Time-Delayed Vibration Absorber

In this section, the experimental system of the proposed TDVA (time-delayed vibration absorber) coupled on a continuous primary structure is built [15]. As illustrated in Fig. 5.25, the system is composed of a primary beam and a TDVA with adjustable stiffness. The primary system is made up of a lumped mass m_1, a continuous beam and pre-stressed spring with stiffness k_{1p}. The vibration absorber consists of a lumped

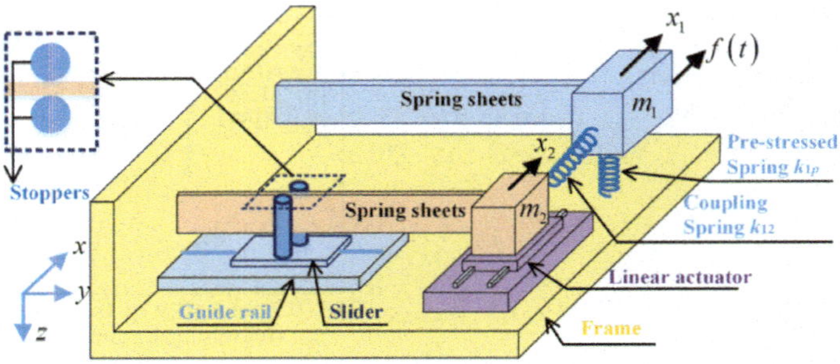

Fig. 5.25 With the help of TDVA, the proposed system's experimental setup is depicted in a schematic diagram

mass m_2, a continuous beam and a motion limiter made up of a guide rail, a slider and two stoppers.

A linear gear is linked to absorption system that applies control forces with time delays necessary for executing prescribed algorithm of control. And linear spring couples the primary system and damping device together. The primary system and absorber have lumped masses that shift in the x direction. The linear gear exerts positive control powers of the mass m_2 at the specified x direction. Beams are oriented along y axis. Within primary system, pre-tensioned coil is compressed along z axis, imparting a nonlinear characteristic. In the TDVA structure, the motion stopper can be adjusted to alter the spring sheet's length, thereby allowing for the adjustment of TDVA's equivalent stiffness. Consequently, the TDVA is appropriate for a range of excitation frequencies.

The delayed-coupled nonlinear system can be described by a set of dynamic equations

$$M_1\ddot{x}_1 + c_1\dot{x}_1 + k_1x_1 + k_{12}(x_1 - x_2) + c_{12}(\dot{x}_1 - \dot{x}_2) + G(x_1) = f_0\cos\omega t, \quad (5.39)$$

$$M_2\ddot{x}_2 + c_2\dot{x}_2 + k_2x_2 + k_{12}(x_2 - x_1) + c_{12}(\dot{x}_2 - \dot{x}_1) + h\left(x_2 - x_{2\tau_k}\right) = 0, \quad (5.40)$$

where $f(t)$ is resonant force stimulation as $f_0\cos\omega t$; Control provided by linear gears is represented by the term $h\left(x_2 - x_{2\tau_k}\right)$, where $x_{2\tau_k} = x_2(t - \tau_k)$. In Eq. (5.39), the nonlinear restoring force $G(x_1)$ is caused by spring which is prestressed.

The boundaries of stability for the delay coupled system could be derived for

$$\begin{aligned}
&\omega_c^4 - i\omega_c^3(\xi_1 + \xi_{12} + \mu\xi_{12} + \xi_2)\\
&\quad - \omega_c^2\left(1 + a_{12} - a_l + g - e^{-i\omega_c\tau}g + a_{12}\mu + \mu\xi_1\xi_{12} + \xi_1\xi_2 + \xi_{12}\xi_2 + \omega_2^2\right)\\
&\quad - i\omega_c\left(\left[\left(e^{-i\omega_c\tau} - 1\right)g - \omega_2^2\right](\xi_1 + \xi_{12}) - a_{12}(\mu\xi_1 + \xi_2) + (a_l - 1)(\mu\xi_{12} + \xi_2)\right)\\
&\quad + g(1 + a_{12} - a_l) - e^{-i\omega_c\tau}g(1 + a_{12} - a_l) + a_{12}\mu(1 - a_l)\\
&\quad + (1 + a_{12} - a_l)\omega_2^2 = 0. \qquad (5.41)
\end{aligned}$$

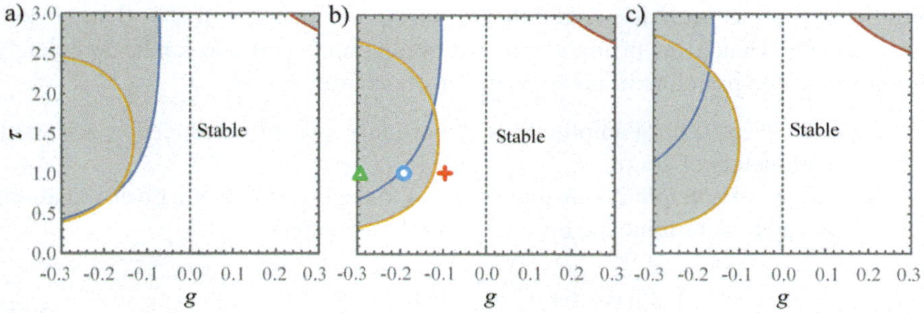

Fig. 5.26 The stability chart displays the control parameters g and τ for **a** $\omega_2 = 0.50$; **b** $\omega_2 = 0.65$; **c** $\omega_2 = 0.80$

Real part and imaginary part of Eq. (5.41) could be obtained, by taking into account $\cos^2(\omega_c\tau) + \sin^2(\omega_c\tau) = 1$, it is possible to calculate the critical stable boundaries, which is shown in Fig. 5.26.

Stabilities of the zero equilibrium in white regions (referred to as "Stable") shown in Fig. 5.26 is characterized by all eigenvalues having negative real parts. The grey regions indicate instability of the zero equilibrium on account of the presence of positive-real-part eigenvalues. It can be observed from Fig. 5.26a–c that system's stable boundaries change based on different structural parameters of the absorber.

The parameter design principle based on anti-resonance frequency is considered to achieve a low-grade response amplitude of primary systems when subjected to harmonic external excitation forces. Given the flexibility to adjust parameters of control, importance of desired excitation periodicity are appropriate control parameters in order to tune the anti-resonance frequency. As a result, the parameters of control's design practice is determined.

Principle 1. The anti-resonant frequency of the primary system is equal to the excitation frequency.

The following equations confirm that the anti-resonance frequency adheres to ***Principle 1***

$$\frac{\partial a}{\partial \Omega}\Big|_{\Omega=\Omega_a} = 0, \tag{5.42}$$

$$\frac{\partial^2 a}{\partial \Omega^2}\Big|_{\Omega=\Omega_a} > 0. \tag{5.43}$$

From Eqs. (5.42) and (5.43), when excitation amplitude and parameters of control remain constant, system's behaviours is solely determined by the frequency of the excitation. When nonlinear factor $\alpha = 0$, the proposed system converges to systems which is linear. Responses of primary system can be completely removed of optimal parameters of control in a linear primary system coupled with a TDVA. For situations with nonlinearity, it is not possible for both a and b to be zero simultaneously.

Fortunately, it has been discovered that acquiring the outcome of $a = 0$ and $b \neq 0$ is achievable. Hence, the primary system's oscillation magnitude can be completely minimized. The principle of parameter project is outlined here.

Principle 2. The response amplitude of the primary system is reduced to zero with optimal parameters of control.

According to ***Principle 2***, substitute $a = 0$ into the Eq. (5.42), the given equations provide a way to determine the optimal control parameters

$$\begin{cases} \mu\xi_{12}\Omega + \xi_2\Omega + g\sin\Omega\tau = 0, \\ g + \mu a_{12} - \Omega^2 + \omega_2^2 - g\cos\Omega\tau = 0. \end{cases} \tag{5.44}$$

Based on Eq. (5.44), the optimal parameters of control are expressed as a function of the excitation frequency in

$$g_{opt} = -\frac{\left(\mu a_{12} - \Omega^2 + \omega_2^2\right)^2 + \Omega^2(\mu\xi_{12}+\xi_2)^2}{2\left(\mu a_{12} - \Omega^2 + \omega_2^2\right)}, \tag{5.45}$$

$$\tau_{opt} = \begin{cases} \left[-\arcsin\left(\frac{\mu\xi_{12}+\xi_2}{g_{opt}}\Omega\right) + 2n\pi\right]/\Omega, \\ \left[\arcsin\left(\frac{\mu\xi_{12}+\xi_2}{g_{opt}}\Omega\right) + (2n-1)\pi\right]/\Omega, \end{cases} \quad n = 1, 2, 3, \ldots. \tag{5.46}$$

The response amplitude of the primary system could be completely suppressed when parameters of control satisfy Eqs. (5.45) and (5.46), for different excitation frequencies.

For the case $\omega_2 = 0.5$ and $\Omega = 0.9$. The optimal values for the parameters of control are $g = -0.054931$ and $\tau = 1.9502$ by ***Principle 2***. Figure 5.27 displays the response of the primary system when equipped with optimum control parameters for $\Omega = 0.9$. In Fig. 5.27, In the case of a passive absorber, when the frequency approaches 0.9, the primary system exhibits a point of anti-resonance, where the primary system experiences the lowest level of response amplitude. When ***Principle 2*** is used to select the optimal control parameters, it is possible to completely eliminate responses of primary system to the excitation frequency $\Omega = 0.9$.

By utilizing ***Principle 2***, the optimum control parameters can suppress the oscillation amplitude of the primary system entirely. Through implementing control with time delays, response amplitudes of the primary systems could be completely suppressed when subjected to varying excitation frequencies. This result shows the powerful effectiveness of the control parameter design principle, suitable for the improvement of vibration absorption performances for nonlinear system subjected to the excitation within a range of frequency.

The experimental setup is displayed of Fig. 5.28. The layout of the experiment is organized according to Fig. 5.28a. The harmonic voltage output from the signal generator gets transmitted to signal amplifiers of oscillation exciters through a connection. Voltage is then increased to reach necessary level to power the vibrational exciter. The primary system is subjected to harmonic driving forces of specified amplitude

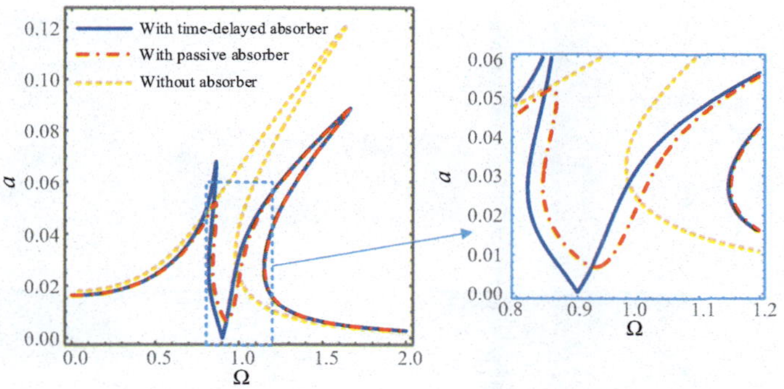

Fig. 5.27 The response of the primary system with optimum control settings for $\Omega = 0.9$ compared to scenarios without an absorber and with passive control

and frequency, generated by the vibration exciter in the subsequent step. The force sensor accurately measures the excitation forces. The proposed system is shown in detail by Fig. 5.28b. The fundamental setup includes a beam connected to a concentrated mass positioned at the beam's tip. The composition of the vibration absorber includes a flat spiral, a concentrated mass located at the beam's extremity, and a linear motor. The stoppers attached to sliders in Fig. 5.28c allow for adjustment of sheet spring's effective length in the absorber. Shox could achieve easily adjustable stiffness by forcing the stepping motor to remove sliders back and forth. The linear actuator shown in Fig. 5.28d produces delayed forces for regulating the absorber.

Analysis of the frequency responses of the primary system, both with and without control of time delays, at an anti-resonance frequency of 2.8 Hz is displayed in Fig. 5.29. Graphs display frequency data obtained through the Averaging Method, while the dots represent the actual experimental results. From Fig. 5.29, presence of a time-delayed control leads to a suppression in the oscillation amplitude of primary system near frequencies at which anti-resonance occurs. Additionally, the frequency response curves presented in Fig. 5.29 demonstrate a strong correlation with the experimental results for 2.8 Hz. The findings indicate that the frequency at which anti-resonance occurs is approximately 2.85 Hz. The primary system exhibits a response amplitude of approximately 3 mm with control and 4.5 mm without control (Fig. 5.30). The implementation of control with time delays has been found to lead to a considerable decrease of over 30% in response amplitude of the primary system when subjected to excitation at its resonant frequency (Figs 5.31 and 5.32).

Hence, the results indicate the suitability of the suggested variable stiffness TDVA for damping vibrations caused by varying driving frequencies. The anti-resonance frequency of response remains the same as driving frequency for various excitation frequencies, provided that the factors are selected ground on the suggested tenets. By utilizing the suggested design principles for parameters of control, the efficiency of oscillation absorption could be enhanced. When selecting parameters of control according to the suggested project tenets, primary system's response amplitude at

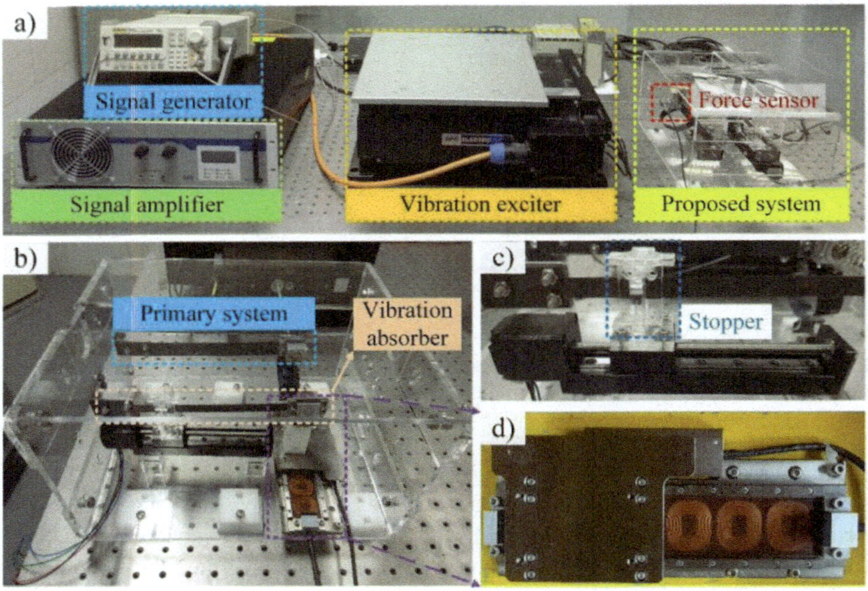

Fig. 5.28 The photograph showcases the arrangement of the experimental setup; **a** the proposed system and overall arrangement; **b** the systems with delayed coupling; **c** the absorber contains a slider, guide rail, and stopper; **d** linear actuator

the excitation frequency experiences a reduction of more than 30% (Figs. 5.29, 5.30, 5.31 and 5.32).

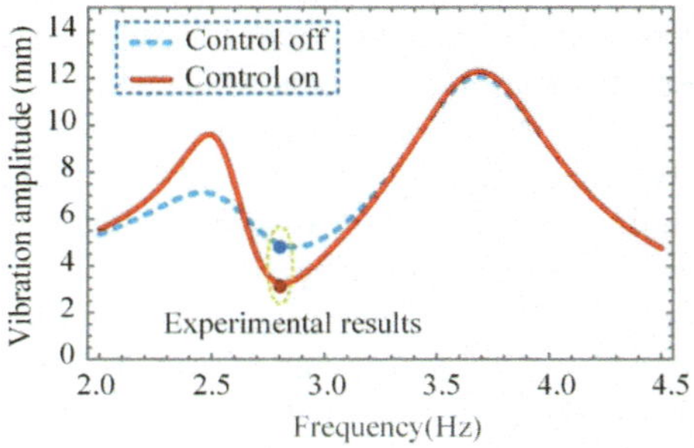

Fig. 5.29 The frequency responses of the primary system are observed both with and without control of time delays, using parameters of control intended for 2.8 Hz

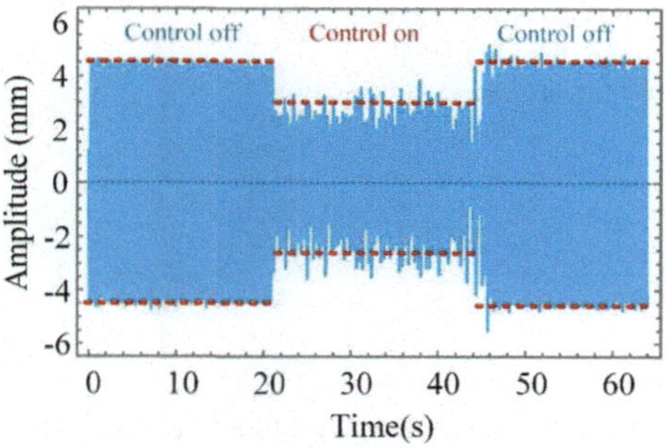

Fig. 5.30 The primary system has a time record for excitation 2.8 Hz

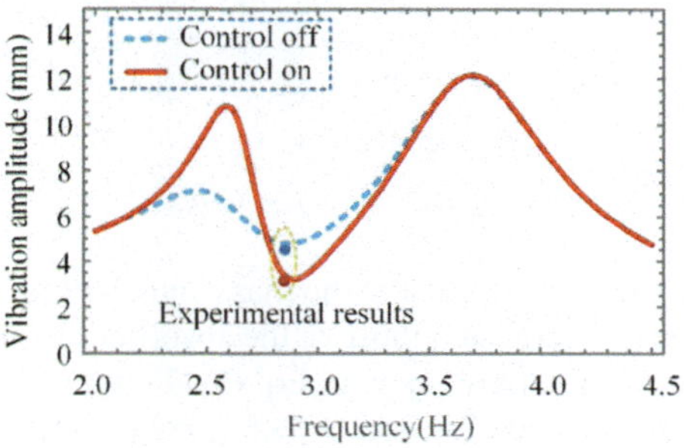

Fig. 5.31 The frequency responses of the primary system are analyzed with and without control of time delays, using control parameters specifically projected for the value of 2.85 Hz

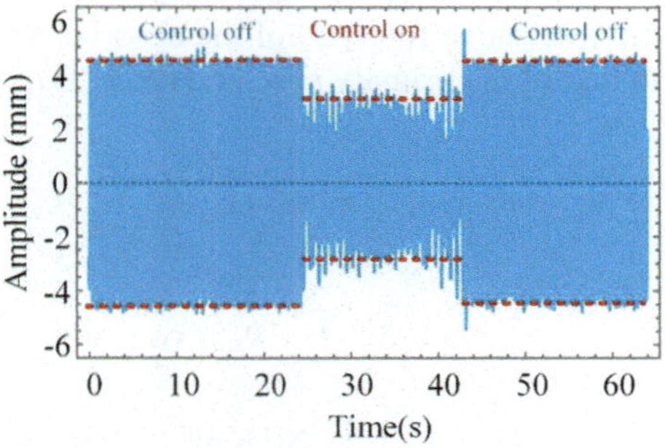

Fig. 5.32 The primary system has a time record for activation 2.85 Hz

5.3.2 *Time-Delayed Absorber for Nonlinear System*

The mechanical model in Fig. 5.33 depicts inclusion of both nonlinear terms G_2 of NLTVA and delay-coupled term F_c to expand the equal-peak approaches for the delay-coupled nonlinear system [16, 17].

In Fig. 5.33, the primary system experiences external excitation with absolute motion x_1, and the auxiliary system undergoes absolute motion denoted as x_2. The restorative forces by the primary and auxiliary systems are expressed as smooth functions $G_1(x_1)$ and $G_2(x_1 - x_2)$, respectively. Assuming symmetry of the restoring

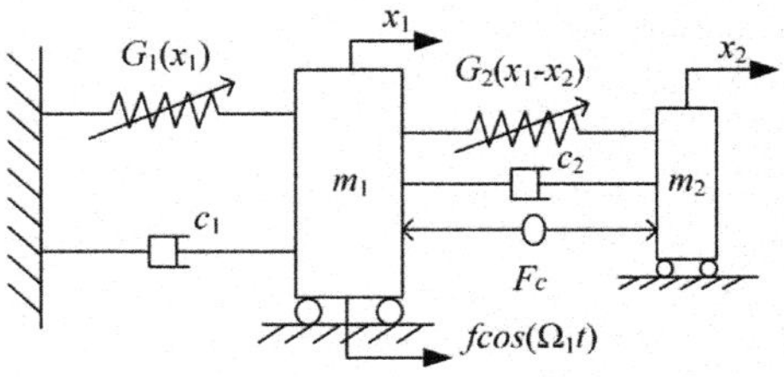

Fig. 5.33 Mechanical model of the delay-coupled nonlinear system

force in the primary system at equilibrium, the function $G_1(x_1)$ satisfies the conditions $G_1(x_1) = -G_1(-x_1)$ and $G_1(0) = 0$. Dynamic equations for the delay-coupled nonlinear system can be expressed as

$$\begin{aligned} m_1\ddot{x}_1 + c_1\dot{x}_1 + G_1(x_1) + c_2(\dot{x}_1 - \dot{x}_2) + G_2(x_1 - x_2) + F_c(\mathbf{x}(t), \mathbf{x}(t - \tau_1)) \\ = f\cos(\Omega_1 t), \\ m_2\ddot{x}_2 - c_2(\dot{x}_1 - \dot{x}_2) - G_2(x_1 - x_2) - F_c(\mathbf{x}(t), \mathbf{x}(t - \tau_1)) = 0, \end{aligned} \tag{5.47}$$

where m_1, c_1 denote the mass and damping coefficients of the primary system, while m_2, c_2 represent those of the absorber. f signifies the force amplitude, while Ω_1 represents the frequency. In Eq. (5.47), $\mathbf{x}(t) = \{x_1(t), \dot{x}_1(t), x_2(t), \dot{x}_2(t)\}^T \in \mathbf{R}^4$ denotes the state vector, and $F_c(\mathbf{x}(t), \mathbf{x}(t - \tau_1))$ represents the delay-coupled force with a time delay τ_1. Preserving generality, the restorative force $G_1(x)$ can be expanded using a Taylor series. According to the generalized nonlinear equal-peak method, the restorative force G_2 of the absorber should exhibit an identical functional form to that of the primary system. Hence, Eq. (5.47) can be expressed as

$$\begin{aligned} m_1\ddot{x}_1 + c_1\dot{x}_1 + k_1x_1 + \sum_{i=3,5,7}^{\infty} k_{1ni}x_1^i + c_2(\dot{x}_1 - \dot{x}_2) + k_2(x_1 - x_2) \\ + \sum_{i=3,5,7}^{\infty} k_{2ni}(x_1 - x_2)^i \\ + F_c(\mathbf{x}(t), \mathbf{x}(t - \tau_1)) = f\cos(\Omega_1 t), \\ m_2\ddot{x}_2 - c_2(\dot{x}_1 - \dot{x}_2) - k_2(x_1 - x_2) - \sum_{i=3,5,7}^{\infty} k_{2ni}(x_1 - x_2)^i \\ - F_c(\mathbf{x}(t), \mathbf{x}(t - \tau_1)) = 0, \end{aligned} \tag{5.48}$$

where $k_1 = \left.\frac{\mathrm{d}G_1}{\mathrm{d}x_1}\right|_{x_1=0}$, $k_{1ni} = \left.\frac{1}{i!}\frac{\mathrm{d}^iG_1}{\mathrm{d}x_1^i}\right|_{x_1=0}$, $k_2 = \left.\frac{\mathrm{d}G_2}{\mathrm{d}r}\right|_{r=0}$, $k_{2ni} = \left.\frac{1}{i!}\frac{\mathrm{d}^iG_2}{\mathrm{d}r^i}\right|_{r=0}$, $r = x_1 - x_2$ represent the linear stiffness coefficient, the ith-order nonlinear stiffness coefficient, and the relative displacement of the primary system and the absorber, respectively.

In alignment with the optimization objective for nonlinear systems, a corresponding optimization goal is posited in this research. The objective of Time-Delayed Variable Amplitude (TDVA) within the equal-peak method in this study is to ascertain the optimal values for the time delay τ and control gain g to

$$\begin{aligned}\text{Min}\|a(\mathbf{p},\Omega,g,\tau)\|_\infty &\to \text{Min}\left\{\text{Max}\left[a(\mathbf{p},\overline{\Omega},g,\tau),a\left(\mathbf{p},\widehat{\Omega},g,\tau\right)\right]\right\}\\ &\to a\left(\mathbf{p},\overline{\Omega},g,\tau\right)=a\left(\mathbf{p},\widehat{\Omega},g,\tau\right),\end{aligned} \tag{5.49}$$

where $\mathbf{p}=\{\mu,\zeta_1,\zeta_2,\lambda,\alpha_i\}$, $i=3,5,7,\ldots$ denotes the structural parameters of the system. In practical scenarios, the actual structural parameters $\mathbf{p}$ may deviate from their optimal passive counterparts $\mathbf{p_0}$, as determined by Eq. (5.49) owing to factors such as the presence of damping, nonlinearity, limitations in supporting capability, manufacturing errors, and assembly errors. Furthermore, despite the potential emergence of complex dynamical behaviours such as quasi-periodic motions and chaos due to nonlinearities and time delay, this study focuses solely on the fundamental harmonic steady-state response.

Analyzing the linear stability of the delay-coupled system is essential. The zero equilibrium is stable if and only if the real parts of all eigenvalues are negative,, a condition expressible as $\{(g,\tau)|\alpha(\mathbf{p},g,\tau)<0\}$. Furthermore, the actuator applies an active delay feedback force, implying that g and τ should be constrained by the actuator's performance (hardware restrictions). In this study, the hardware constraints are denoted as $r_{hr}=\{(g,\tau)|0\le g\le 2, 0\le\tau\le 2\}$. Therefore, the assumed ranges for g and τ are assumed as

$$r_1=\{(g,\tau)|\alpha(\mathbf{p},g,\tau)<0\}\cap r_{hr}. \tag{5.50}$$

To achieve the equal-peak property, the amplitudes of the first resonance peak $\overline{a}$ and the second peak $\hat{a}$ should be identical. Therefore, the criterion for the equal-peak method can be formulated as

$$\begin{cases} f_i\left(\overline{A}_1,\overline{A}_2,\overline{B}_1,\overline{B}_2,\overline{\Omega},g,\tau,\mathbf{p}\right)=0,\\ f_i\left(\widehat{A}_1,\widehat{A}_2,\widehat{B}_1,\widehat{B}_2,\widehat{\Omega},g,\tau,\mathbf{p}\right)=0,\\ \dfrac{d\overline{a}}{d\overline{\Omega}}=0,\\ \dfrac{d\hat{a}}{d\widehat{\Omega}}=0,\\ \overline{a}^2-\hat{a}^2=0,\end{cases} \tag{5.51}$$

where $i=1,\ldots,4$. $\overline{(\cdot)}$ and $\widehat{(\cdot)}$ denote the coefficients at the first and second resonance frequencies, respectively. The condition for applying the equal-peak method can be articulated as

$$r_2=\left\{(g,\tau)|\overline{a}\left(g,\tau,\overline{\Omega}\right)=\hat{a}\left(g,\tau,\widehat{\Omega}\right)\right\}\cap r_1. \tag{5.52}$$

Given the distinction between the equal-peak case and the peak-minimum case through the critical condition, an analysis on the conditions for the occurrence of merging is needed. The merging of the DRC (detached resonance curve) with the main frequency response curve at Q_4 is linked to the fold bifurcation, defined as the "simple bifurcation," corresponding to the second resonance frequency $\widehat{\Omega}$. This can be elucidated through the analysis on the eigenvalues for the response at $\widehat{\Omega}$. By substituting the values of g, τ and $\widehat{A}_1$, $\widehat{A}_2$, $\widehat{B}_1$, $\widehat{B}_2$, $\widehat{\Omega}$ of $Q_1 \sim Q_5$ from Eq. (5.52) into the Jacobian matrix J of Eq. (5.12), we have

$$J = J(A_1, A_2, B_1, B_2, \Omega, g, \tau, \mathbf{p}). \tag{5.53}$$

To summarize, the difference between the equal-peak case and the peak-minimum case is identified through an analysis of the eigenvalues at the merging point. Thus, the ranges for g and τ in the equal-peak case can be derived as

$$r_3 = \left\{(g, \tau)|\mathrm{Max}\left[\mathrm{Real}\left[eig\left(J\left(g, \tau, \hat{\Omega}\right)\right)\right]\right] < 0\right\} \cap r_2. \tag{5.54}$$

As illustrated in Fig. 5.34, varying control gains g and time delay τ result in distinct peak values. Therefore, there exist optimal values for g and τ in achieving the equal-peak scenario with the minimum resonance peak, which can be determined as

$$r_4 = \underbrace{\mathrm{Min}}_{a}\left\{(g, \tau)|\mathrm{Max}\left[a(\mathbf{p}, \Omega, g, \tau)|_{(g,\tau)\in r_3}\right]\right\}. \tag{5.55}$$

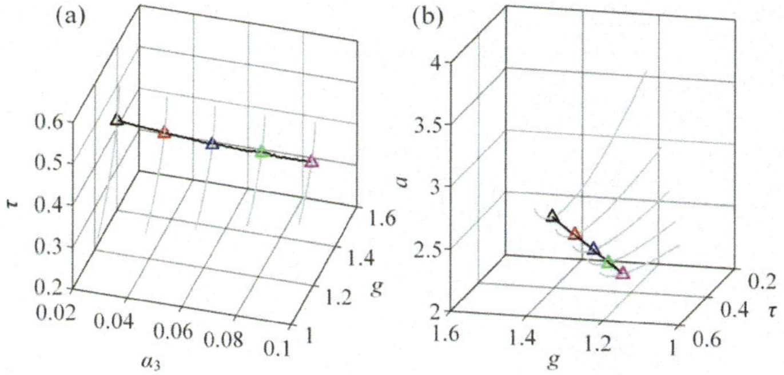

Fig. 5.34 The optimal g and τ for the minimum resonance peak are obtained for various values of α_3 using Eq. (5.52) (depicted by the black line): $\alpha_3 = 0.02$ (black triangle), $\alpha_3 = 0.04$ (red triangle), $\alpha_3 = 0.06$ (blue triangle), $\alpha_3 = 0.08$ (green triangle), $\alpha_3 = 0.1$ (magenta triangle). **a** illustrates the relationship between α_3, g and τ, **b** presents the relationship among the minimum peak amplitude a, g and τ. The grey lines indicate the ranges of g and τ for the equal-peak scenario obtained from Eq. (5.50)

Figure 5.34 illustrates the optimal control gain g and time delay τ as determined by Eq. (5.52) for the equal-peak scenario with the minimum resonance peak a. Figure 5.34 suggests that α_3 and a exhibit an almost linear relationship with g and τ. Additionally, the minimum peak amplitude a decreases with an increase in nonlinear stiffness α_3. For various values of α_3, achieving frequency response curves characterized by minimum resonance peaks is possible by means of determining g and τ through Eq. (5.52).

The structural parameters for the primary system and various types of attached absorbers are provided in Table 5.2. The optimal values for g_1 and τ_1 in the presence of TDVA for achieving the minimum resonance peak are determined using Eq. (5.52).

In Fig. 5.35, the response curves of a Duffing primary system equipped with LTVA, NLTVA, and TDVA (Present Study) are compared across variable external excitation amplitudes ranging from 0.02N to 0.2N. Figure 5.35a illustrates a slight detuning of LTVA for small amplitude excitation $f = 0.02$N, as LTVA is valid for undamped linear primary systems. For $f = 0.16$N in Fig. 5.35b, the equal-peak property for LTVA is no longer achieved due to the significant difference in two resonance frequencies. Conversely, for NLTVA, the property is approximately realized. For $f = 0.2$N in Fig. 5.35c, a DRC merges with the second resonance peak of NLTVA, and f reaches the critical value for merging. Therefore, the application of the equal-peak method to NLTVA is rendered ineffective. Additionally, in all examined instances of TDVA, the equal-peak property can be precisely achieved. In conclusion, TDVA proves applicable to higher amplitude excitations and results in a lower resonance peak compared to both LTVA and NLTVA.

The structural parameters of the primary system, along with various types of attached absorbers, are provided in Table 5.3. The optimal values for g_1 and τ_1 in the context of TDVA for achieving the minimum resonance peak are determined using Eq. (5.52).

Figure 5.36 illustrates primary system's frequency responses with third, fifth, and seventh-order nonlinearities attached by various absorbers for $f = 0.04$ N, 0.09 N and 0.18 N, respectively. In Fig. 5.36a–c, for $f = 0.04$ N, noticeable detuning occurs for LTVA and NLTVA with fifth or seventh-order nonlinearity. For $f = 0.09$ N in

Table 5.2 Structural parameters of the duffing primary system with attached LTVA, NLTVA, and TDVA

	Primary system	LTVA	NLTVA	TDVA (Present Study)
Mass [kg]	$m_1 = 1$	$m_2 = 0.05$	$m_2 = 0.05$	$m_2 = 0.05$
Linear stiffness [N/m]	$k_1 = 1$	$k_2 = 0.0454$	$k_2 = 0.0454$	$k_2 = 0.1125$
Linear damping [Ns/m]	$c_1 = 0.002$	$c_2 = 0.0128$	$c_2 = 0.0128$	$c_2 = 0.0003$
Nonlinear stiffness [N/m^3]	$k_{1n3} = 1$		$k_{2n3} = 0.0042$	

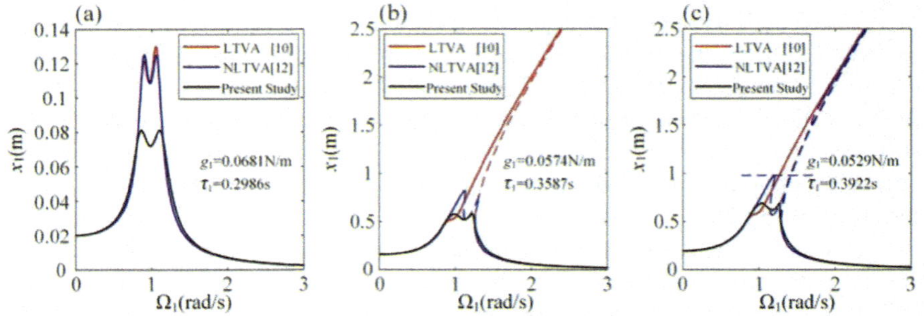

Fig. 5.35 Frequency response curves of a duffing primary system equipped with LTVA, NLTVA, and TDVA (Present Study) under variable external excitation amplitudes: **a** $f = 0.02$N, **b** $f = 0.16$N, **c** $f = 0.2$N. Dashed lines represent unstable responses

Table 5.3 Structural parameters of the primary system with multiple nonlinearities and attached LTVA, NLTVA, and TDVA

	Primary system	LTVA	NLTVA	TDVA (Present study)
Mass [kg]	$m_1 = 1$	$m_2 = 0.05$	$m_2 = 0.05$	$m_2 = 0.05$
Linear stiffness [N/m]	$k_1 = 1$	$k_2 = 0.0454$	$k_2 = 0.0454$	$k_2 = 0.1125$
Linear damping [Ns/m]	$c_1 = 0.002$	$c_2 = 0.0128$	$c_2 = 0.0128$	$c_2 = 0.0003$
Nonlinear stiffness ([N/m^3], [N/m^5], [N/m^7])	$k_{1n3} = 1$ $k_{1n5} = 1$ $k_{1n7} = 1$		$k_{2n3} = 0.0042$ $k_{2n5} = 3.96 \times 10^{-4}$ $k_{2n7} = 3.59 \times 10^{-5}$	

Fig. 5.36d–f, the equal-peak property for LTVA and NLTVA with single nonlinearity is no longer achieved. Conversely, for the complete NLTVA with all nonlinearities, the property is approximately realized due to the additive property. As the excitation amplitude f increases to 0.18N in Fig. 5.36g–i, a DRC merges with the second resonance peak of NLTVA, making the equal-peak method for NLTVA inapplicable. In all studied TDVA instances, the equal-peak property can be achieved. In summary, Fig. 5.36 demonstrates that TDVA is applicable to a primary system with damping or multiple nonlinearities under large amplitude excitation, and its performance surpasses LTVA and NLTVA.

5.3.3 *Time-Delayed Absorber for M-DOF System*

The mechanical representation of the M-DOF (multi-degrees of freedom) nonlinear primary system with multiple time-delayed vibration absorbers [18] is depicted in

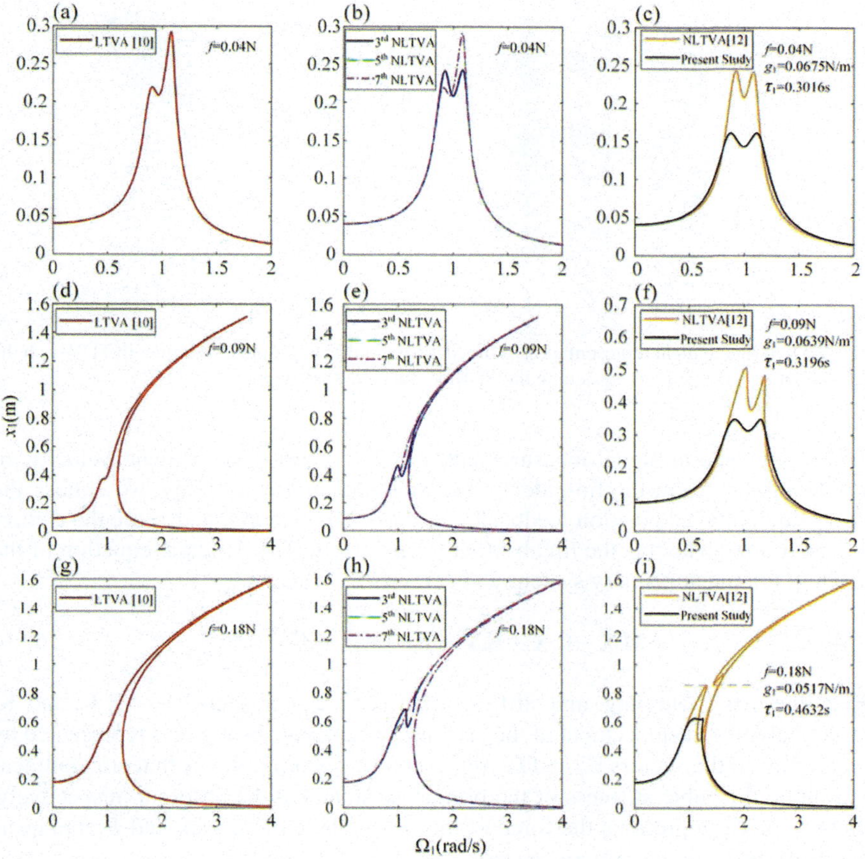

Fig. 5.36 Frequency response curves of a primary system with 3rd, 5th and 7th order nonlinearities under variable excitation amplitudes, attached by different absorbers: $f = 0.04$N attached by **a** LTVA, **b** NLTVA with single nonlinearity of 3rd, 5th and 7th order; **c** TDVA (Present Study) and the complete NLTVA; $f = 0.09$N attached by **d** LTVA, **e** NLTVA with single nonlinearity of 3rd, 5th and 7th order; **f** TDVA and the complete NLTVA; $f = 0.18$N attached by **g** LTVA, **h** NLTVA with single nonlinearity of 3rd, 5th and 7th order; **i** TDVA and the complete NLTVA

Fig. 5.37a. A harmonic excitation $f\cos(\Omega t)$ is applied to the first oscillator of the primary system, where f and Ω represent the amplitude and frequency of the excitation, respectively. The ith oscillator of the primary system and the corresponding absorber are presented Fig. 5.37b, c, respectively.

As shown in Fig. 5.37b, $x_{p,i}$ is the displacement of ith oscillator in the primary system; $m_{p,i}$, $c_{p,i}$, $k_{p,i}$ and $k_{p,i,nl3}$ are its mass, viscous damping, linear and cubic stiffness coefficients, respectively. In Fig. 5.37c, $x_{a,i}$ is the displacement of ith time-delayed absorber; $m_{a,i}$, $c_{a,i}$, $k_{a,i}$ and $k_{a,i,nl3}$ are its mass, viscous damping, linear and cubic stiffness coefficients, respectively. Considering the type of sensor and

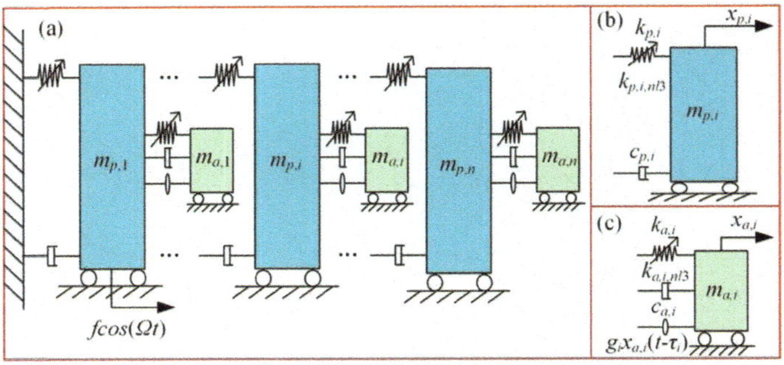

Fig. 5.37 **a** Mechanical representation of the M-DOF nonlinear primary system with TDVAs, **b** the ith oscillator in the primary system, **c** the ith time-delayed Absorber

the convenience of signal measurement, one can select displacement, velocity, or acceleration to form the time delay feedback signal. In this study, we choose the linear time-delayed position feedback $g_i x_{a,i}(t - \tau_i)$, where g_i and τ_i are the control gain and time delay for the ith absorber, respectively. The dynamic equation of the M-DOF nonlinear primary system with TDVAs is written as

$$\mathbf{M}\ddot{\mathbf{X}} + \mathbf{C}\dot{\mathbf{X}} + \mathbf{K}\mathbf{X} + \mathbf{F}_n + \mathbf{G}_n + \mathbf{G}_\tau = \mathbf{F}. \tag{5.56}$$

Here, the mass, damping, and stiffness matrices are represented by **M**, **C**, and **K**, repectively. The displacement of the primary system and absorbers is represented by vector **X**, and the vectors $\mathbf{F}_n$ and $\mathbf{G}_n$ represent the restoring forces that are nonlinear caused by the cubic stiffness of the primary system and the absorber, respectively. The vector $\mathbf{G}_\tau$ represents the force of the time-delayed feedback and **F** applies as the stimulation on the primary system.

The mathematical formulation of optimization can be written as

$$\begin{aligned} \mathrm{Min}\|a_1(\mathbf{b}, \Omega)\|_\infty &\rightarrow \mathrm{Min}\{\mathrm{Max}[a_1(\mathbf{b}, \Omega_{2i-1}), a_1(\mathbf{b}, \Omega_{2i})]\} \\ &\rightarrow a_1(\mathbf{b}, \Omega_{2i-1}) = a_1(\mathbf{b}, \Omega_{2i}), i = 1, 2, \end{aligned} \tag{5.57}$$

where $a_1(\mathbf{b}, \Omega)$ represents the FRC (frequency response curve) of $m_{p,1}$; Ω_{2i-1}, Ω_{2i} are the two resonant frequencies closse to the ith mode of the FRC.

The real parts of the eigenvalues determing the stability of the system. Therefore, the time-delayed parameters should satisfy the condition $\{(g_i, \tau_i)|\mathrm{Max}[\alpha(\mathbf{p}, g_i, \tau_i)] < 0\}$. Based on previous experimental studies on TDVAs, it is assumed that there are some restrictions on the time-delayed parameters, given by $\{(g_i, \tau_i)|0 \le g_i \le 2, 0 \le \tau_i \le 2\}$. Therefore, the stability condition is stated as

$$r_1 = \{(g_i, \tau_i)|\mathrm{Max}[\alpha(\mathbf{p}, g_i, \tau_i)] < 0\} \cap \{(g_i, \tau_i)|0 \le g_i \le 2, 0 \le \tau_i \le 2\}, i = 1, 2. \tag{5.58}$$

The sufficient condition can be written in a concise form

$$r_2 = \{(g_i, \tau_i)|a_1(\mathbf{p}, g_i, \tau_i, \Omega_{2i-1}) = a_1(\mathbf{p}, g_i, \tau_i, \Omega_{2i})\} \cap r_1, \quad i = 1, 2. \quad (5.59)$$

If there are two equal resonance peaks near the *i*th mode, Eq. (5.59) will be satisfied. However, taking the first order derivative shows that the calculated second peak points $a_1(\Omega_{2i})$ may also be a local minimum of FRC within the region with strong nonlinearity. Therefore, the equal-peak property can not be completely determined by Eq. (5.59). Namely, Eq. (5.59) serves as a necessary condition for the generalized equal-peak principle. When f = 0.01 N and the nonlinearity is not significant, the condition (5.59) is satisfied by time-delayed parameters and the corresponding FRCsof the primary system are illustrated in Fig. 5.38.

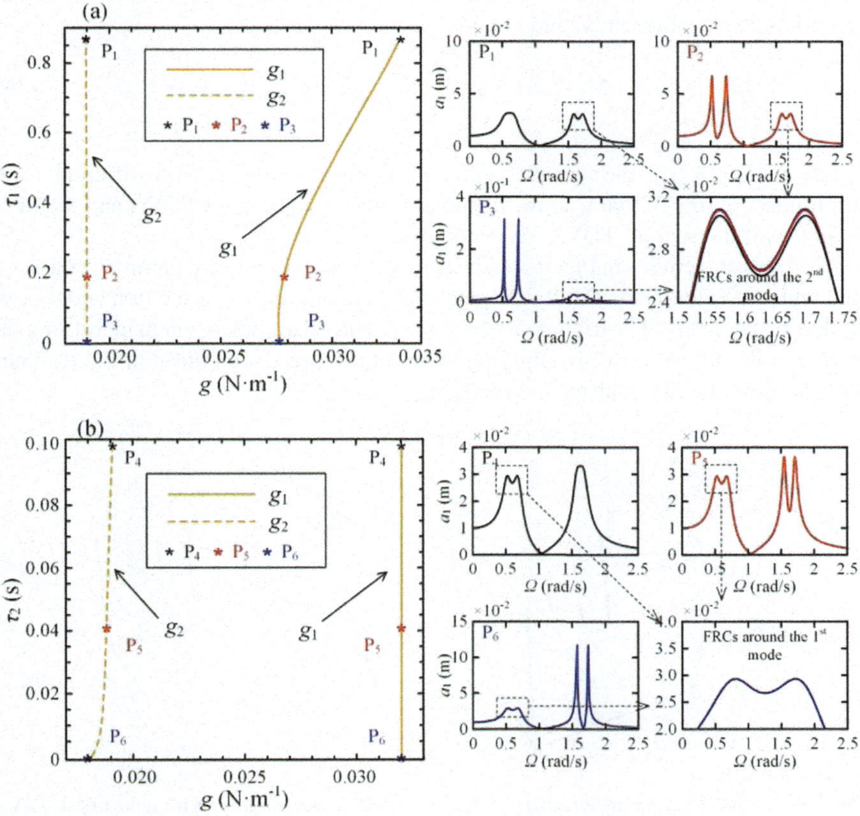

Fig. 5.38 The time-delayed parameters acquired through the necessary condition for $f = 0.01$ N and the corresponding FRCs at various points along the curves of time-delayed parameters for **a** the first mode ($\tau_2 = 0.0729$ s), **b** the second mode ($\tau_1 = 0.740$ s)

The parameters acquired through the essential requirement r_2 and the FRCs of the primary system for the first and second modes are shown in Fig. 5.38a, b. It is evident that the two resonance peaks are identical near the first or second mode. Increasing the magnitude of τ_1 (τ_2) results in reduced peaks near the first (second) mode, suggesting that manipulating time delay can control the resonance peaks and adjust the damping characteristics of t-he TDVAs. Besides, g_2 in Fig. 5.38a and g_1 in Fig. 5.38b come very close to form vertical lines, suggesting that the TDVA designed for a specific mode almost has no impact on the resonances around other modes.

Considering that different time-delayed parameters have the potential to exhibit the equal-peak property with various peak values, we propose a condition to find the optimum time-delayed parameters that result in the minimum peak value. This condition is referred to as the minimum peak condition, as it seeks to minimize the peaks. The obtained optimal parameters exhibit the ability to realize the equal-peak characteristic, consequently benefiting the control of peaks in both the first and second modes concurrently, and are denoted as

$$r_4 = \underbrace{\text{Min}}_{a_1}\left\{ (g_i, \tau_i) | \text{Max}\left[a_1(\mathbf{p}, g_i, \tau_i, \Omega_{2i-1}) |_{(g_i, \tau_i) \in r_3} \right] \right\}, \quad i = 1, 2. \tag{5.60}$$

Using Eq. (5.60), the optimal parameters are determined with different force amplitudes. Figure 5.39a, b show the ideal g_1 and τ_1 of the first TDVA and the ideal g_2 and τ_2 of the second TDVA, respectively.

The lines and circles in Fig. 5.39a, b represent the time-delayed parameters associated with the equal-peak, minimum-peak, and critical cases near the first and second modes, respectively. The triangles are the ideal delay parameters when increasing the force amplitude, obtained by applying the minimum peak condition of the resonant peaks around the first and second modes.

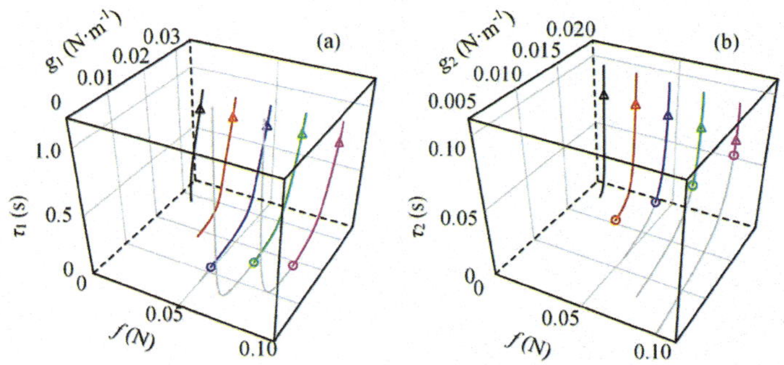

Fig. 5.39 The optimal parameters that satisfy the minimum peak conditions for **a** the first TDVA of the first mode when the force amplitude f is increased, **b** the second TDVA for the second mode when increasing the force amplitude f. The colored lines, gray lines, and circles in **a** and **b** correspond to the time-delayed parameters associated with equal-peak, peak-minimum, and critical scenarios near the first and second modes, respectively where $f = 0.01$ N (black), $f = 0.03$ N (red), $f = 0.05$ N (blue), $f = 0.07$ N (green), $f = 0.09$ N (magenta)

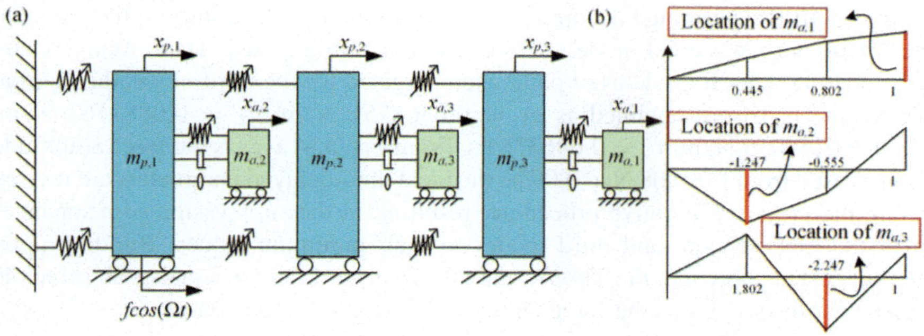

Fig. 5.40 **a** The mechanical representation of the 3-DOF primary system with three absorbers. **b** The coordination of the absorbers determined by the linear mode configuration of the primary system

The second scenario depicts the effect of vibration control by deploying three TDVAs to a three-DOF (degrees of freedom) nonlinear primary structure, as illustrated in Fig. 5.40a.

In Fig. 5.40a, assuming that the mass and cubic stiffness coefficients of the primary structure are unitary. Set the damping coefficient of the primary system to 0 so as to investigate the control effect of TDVA on large amplitude vibration. Those absorbers in the primary system constitute only 10% of the overall mass. The assembly position of the absorber depends on the linear mode of the primary system shown in Fig. 5.40b. Assemble the ith (i = 1, 2, 3) absorber to reduce the peak near the ith (i = 1, 2, 3) mode. The mass, stiffness and damping coefficients of LTVAs are the same as those of NLTTVAs. The adjustment procedure of TDVAs' time-delayed parameters at $f =$ 0.09 N for attaining the equal-peak property near the first, second, and third mode are shown in Fig. 5.42a–c, respectively. The initial time-delayed parameters are chosen as follows: $g_1^0 = 0.0369696$ N/m, $g_2^0 = 0.0885828$ N/m, $g_3^0 = 0.016029$ N/m, $\tau_1^0 = 1.32243$ s, $\tau_2^0 = 0.230316$ s, and $\tau_3^0 = 0.0458753$ s, which are the ideal set of parameters for $f = 0.08$ N. By incrementally iterating the procedure, the optimal time delays and control gains are adjusted as the force amplitude f increases from 0.08 N to 0.09 N.

In Fig. 5.41, FP, OP, MP, and LP are abbreviations used to represent specific points on the lines of time-delayed parameters. The vibrations of the first, second, and third modes are denoted for simplicity as I, II, and III, respectively. As shown in Fig. 5.41a, in order to adjust the delay parameter of TDVA in the first mode, τ_2^0 and τ_3^0 are fixed, and the time-delayed parameters of the first TDVA undergo iterative adjustments, leading to value updates. The findings illustrate that when the time-delayed parameters are at point OP-I, the resonance peaks around the first mode are equal and minimized, where the optimal value $\tau_1 = 1.32774$ s is obtained based on the minimum peak condition described in Eq. (5.59). Next, with the values of τ_1 and τ_3^0 fixed, the time-delayed parameters for the second TDVA are revised to target the second mode. As illustrated in Fig. 5.41b, for the optimal value $\tau_2 = 0.233074$ s at point OP-II determined according to the minimum peak condition, the resonant peaks around the second mode are both equal and minimized. Then, fixing τ_1 and τ_2, updating other time-delayed parameters targeted for the third mode, the optimal values at OP-III are obtained as $g_1 = 0.0368751$ N/m, $g_2 = 0.0881162$ N/m, $g_3 = 0.0158015$ N/m, $\tau_3 = 0.0463765$ s. From Fig. 5.41a–c, as the force amplitude is increased from f = 0.08 N to 0.09 N, the ideal time-delayed parameters are revised using the proposed iterative procedure, resulting in the suppression of resonances near the first, second, and third modes to their minimum values. Building upon the suggested approach, the TDVAs can effectively exhibit the equal-peak principle across various modes as the force amplitude progressively increases.

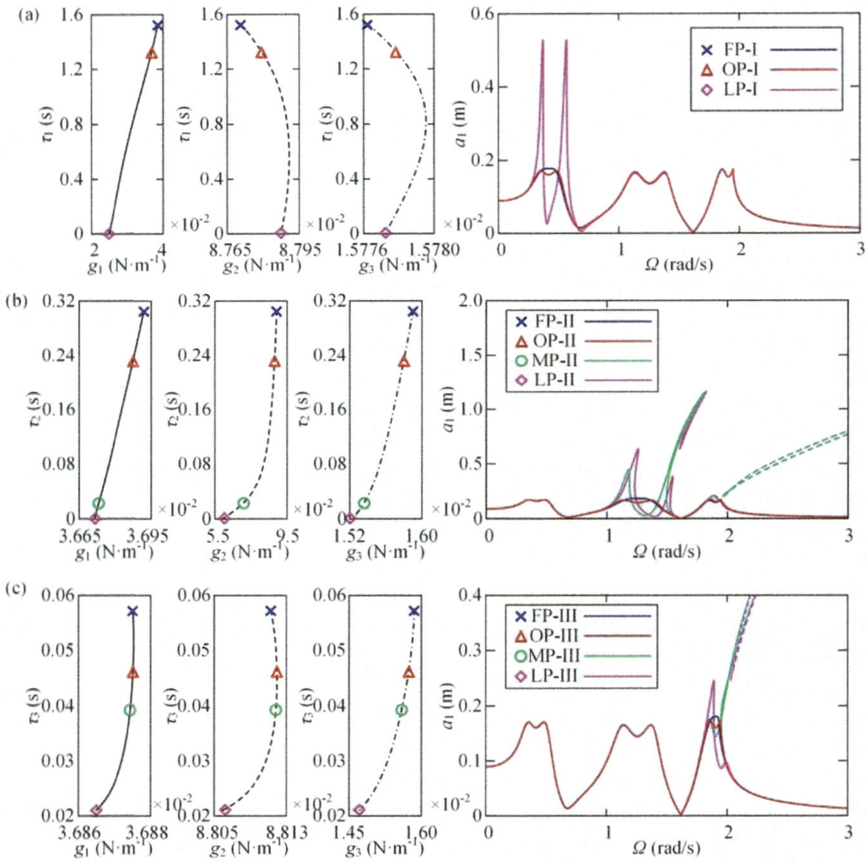

Fig. 5.41 The vibration control effects of the TDVAs at $f = 0.09$ N observed for **a** the first mode, **b** the second mode, and **c** the third mode. FP, OP, MP, and LP indicate the first point, optimal point, merging point, and last point on the curves of time-delayed parameters, respectively. Additionally, I, II, and III correspond to the first, second, and third modes, respectively

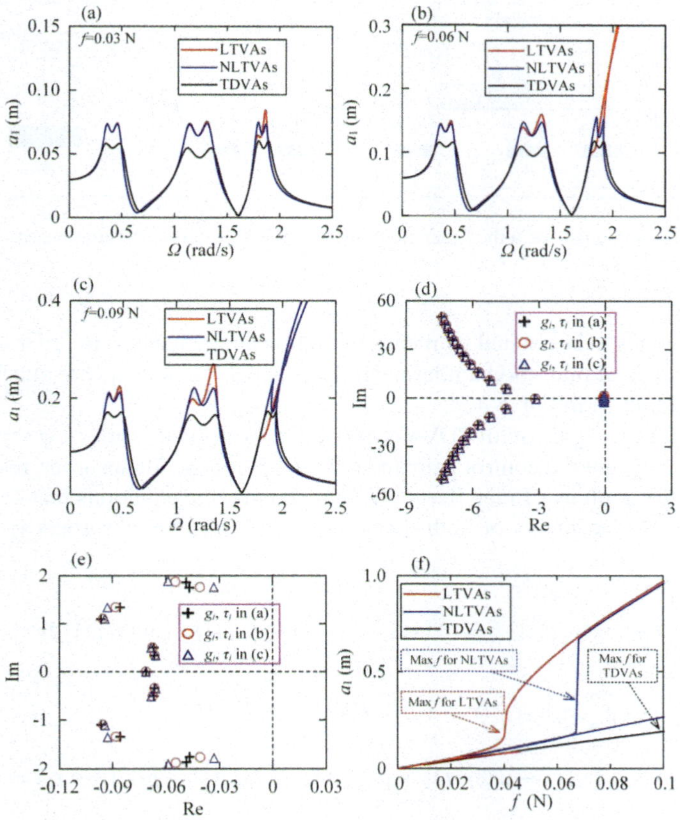

Fig. 5.42 The FRCs of LTVAs, NLTVAs and TDVAs for **a** $f = 0.03$ N, **b** 0.06 N, and **c** 0.09 N. **d** The eigenvalues of the equilibrium point obtained by Eq. (5.6); **e** the enlargement of **d**; **f** the peak values near the third mode for different amplitudes of the force

When increasing the amplitudes of the force, a similar procedure is applied as shown in Fig. 5.41. The effects of vibration control under LTVA, NLTVA, and TDVA are shown in Fig. 5.42.

5.3.4 Time-Delayed Absorber for Continuous Structure

The primary system under consideration is an Euler–Bernoulli beam exhibiting geometrical nonlinearity [19], illustrated in Fig. 5.43a. This form of geometrical nonlinearity is assumed to follow the strain–displacement relationship of the von Kármán type. In scenarios involving a lightly-damped nonlinear beam subjected to excitations with a broad frequency range and significant amplitude, it is common for

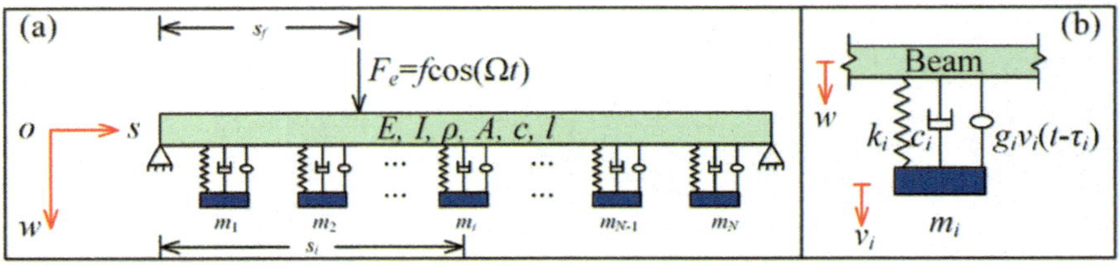

Fig. 5.43 The schematic of **a** the Euler-Bernoulli beam with geometrical nonlinearity attached by TDVAs, **b** the *i*th TDVA

multiple modes to be excited by means of nonlinear resonances. In order to mitigate the occurrence of multimodal nonlinear resonances, N absorbers are attached to the beam, as depicted in Fig. 5.43b.

The feedback signal of ith TDVA is utilized in its original form $g_i v_i(t-\tau_i)$, where g_i and τ_i represent the control gain and time delay for the ith absorber, respectively. Based on the nonlinear Euler-Bernoulli beam theory, the equations that describe the transverse displacements of both the beam $w(s,t)$ and the absorbers $v_i(t)$ can be written as

$$\begin{aligned}&\rho A\ddot{w}(s,t)+c\dot{w}(s,t)+EIw''''(s,t)-\frac{EA}{2l}w''(s,t)\int_0^l\left(w'(s,t)\right)^2\mathrm{d}s\\&\quad+\sum_{i=1}^{N}m_i\ddot{v}_i(t)\delta(s-s_i)=F_e\delta\left(s-s_f\right),\\&m_i\ddot{v}_i(t)+k_i[v_i(t)-w(s_i,t)]+c_i[\dot{v}_i(t)-\dot{w}(s_i,t)]-g_iv_i(t-\tau_i)=0,\end{aligned}\tag{5.61}$$

where the nonlinear term $-\frac{EA}{2l}w''\int_0^l\left(w'\right)^2\mathrm{d}s$ is induced by the von Kármán type nonlinear strain–displacement relation. Dirac delta functions $\delta(s-s_i)$ and $\delta\left(s-s_f\right)$ represent the concentrated force applied on the beam by the ith TDVA and the external excitation. The derivative with respect to time is denoted by the over dot, and the derivative with respect to spatial coordinate is denoted by the prime.

The dimensionless truncated equations are obtained by implementing the Galerkin truncation method and utilizing the dimensionless approach, given by

$$\begin{aligned}&\ddot{x}_p+2\zeta_p\dot{x}_p+\lambda_p^2x_p+f_{nl,p}+\sum_{i=1}^{N}\mu_i\ddot{y}_i(t)\phi_p(s_i)=f_p,\ p=1,2,\ldots,P,\\&\ddot{y}_i+\beta_i^2\left[y_i-\sum_{p=1}^{P}\phi_p(s_i)x_p\right]+2\gamma_i\beta_i\left[\dot{y}_i-\sum_{p=1}^{P}\phi_p(s_i)\dot{x}_p\right]\\&\quad-\lambda_i^2g_iy_i(t-\tau_i)=0,\ i=1,2,\ldots,N.\end{aligned}\tag{5.62}$$

To overcome the limitations associated with passive LTVAs and to mitigate the nonlinear resonances, the effective strategy should involve the incorporation of multiple TDVAs. The main objective is to optimize the multimodal equal-peak principle, aiming to effectively attenuate resonance peaks occurring around specific modes at the location s_c of the beam. This goal can be expressed mathematically as follows:

$$\begin{aligned}&\text{Find } \mathbf{p}_\tau=\{g_i,\tau_i\} \text{ that satisfies}\\ &\mathrm{Min}\|a(s_c,\mathbf{p},\mathbf{p}_\tau,\Omega)\|_\infty \to \mathrm{Min}\big\{\mathrm{Max}\big[a(s_c,\mathbf{p},\mathbf{p}_\tau,\Omega_{2i-1}),a(s_c,\mathbf{p},\mathbf{p}_\tau,\Omega_{2i})\big]\big\}\\ &\quad\to a(s_c,\mathbf{p},\mathbf{p}_\tau,\Omega_{2i-1})=a(s_c,\mathbf{p},\mathbf{p}_\tau,\Omega_{2i}),\ i\in\mathbf{\Psi}.\end{aligned} \tag{5.63}$$

In Eq. (5.63), $a(s_c,\mathbf{p},\mathbf{p}_\tau,\Omega)$ is the FRC of the beam at the location s_c, $\mathbf{p}=\{\beta_i,\gamma_i,f\}, i\in\mathbf{\Psi}$ is the vector that contains the structural parameters of the TDVAs and force amplitudes, $\mathbf{p}_\tau=\{g_i,\tau_i\}$, $i\in\mathbf{\Psi}$ is the vector that contains control gains and time delays of the TDVAs, $\mathbf{\Psi}$ are the set of modes around which the resonance peaks are to be suppressed, Ω_{2i-1} and Ω_{2i} are the resonance frequencies around the ith mode of the FRC.

The sufficient condition can be formulated as follows

$$r_{2,a}=\{\mathbf{p}_\tau|a(s_c,\mathbf{p},\mathbf{p}_\tau,\Omega_{2i-1})=a(s_c,\mathbf{p},\mathbf{p}_\tau,\Omega_{2i})\}\cap r_1,\ i\in\mathbf{\Psi}. \tag{5.64}$$

After applying (5.64) to the ith mode, one figures that the obtained time-delayed parameters can equally tune the two extremum. Figures 5.45 and 5.46 illustrate the time-delayed parameters obtained from (5.64) and corresponding FRCs for the first and second mode for $f=200$ kN, respectively (Fig. 5.44).

The extrema equal criterion that characterizes the equal-peak case can be written as

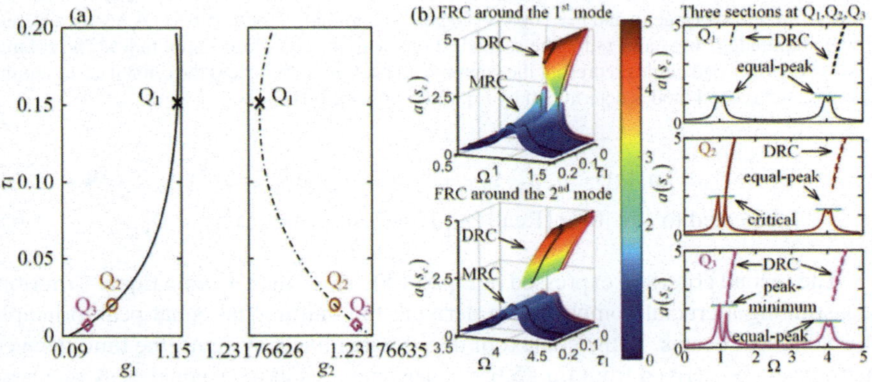

Fig. 5.44 **a** The time-delayed parameters $\mathbf{p}_\tau$ according to Eq. (5.64) for the first mode at $f=200$ kN, $\tau_2=0.0079636$, **b** 3D diagram of FRCs for three sections at points Q_1-Q_3, marked in **a**

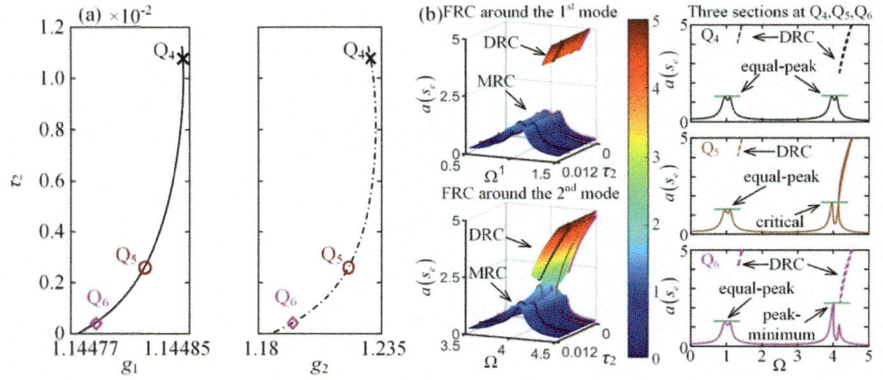

Fig. 5.45 **a** The time-delayed parameters $\mathbf{p}_\tau$ for the second mode at $f = 200$ kN, $\tau_1 = 0.151554$ s, **b** 3D diagram of FRCs for three sections at points Q_4–Q_6, marked in **a**

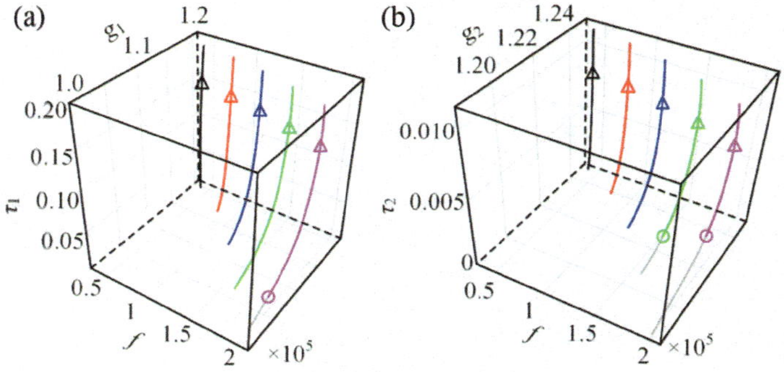

Fig. 5.46 The optimal time-delayed parameters (triangles) for the minimum resonance peaks with various force amplitudes according to Eq. (5.65): $f = 40$ kN (black), $f = 80$ kN (red), $f = 120$ kN (blue), $f = 160$ kN (green), and $f = 200$ kN (magenta). **a** g_1 and τ_1 for the first TDVA with different force amplitude f, **b** g_2 and τ_2 for the second TDVA with different force amplitudes. The colored lines, gray lines and circles represent the equal-peak, peak-minimum, and the critical cases around the first and second modes determined by Eq. (5.66), respectively

$$
\begin{aligned}
& r_2 = r_{2,a} = \{\mathbf{p}_\tau | a(s_c, \mathbf{p}, \mathbf{p}_\tau, \Omega_{2i-1}) = a(s_c, \mathbf{p}, \mathbf{p}_\tau, \Omega_{2i})\} \cap r_1, \ i \in \boldsymbol{\Psi}. \\
& \text{subjected to } \left\{ \mathbf{p}_\tau | \mathrm{Max}\left\{ \mathrm{Real}\left[\mathrm{eig}\left[\mathbf{J}(a(s_c, \mathbf{p}, \mathbf{p}_\tau, \Omega_{2i})) \right] \right] \right\} < 0 \right\}. \qquad (5.65)
\end{aligned}
$$

The second criterion, expressed in Eq. (5.65), called the Extrema Equal Criterion, constituting a crucial component in deriving the multimodal equal-peak principle for nonlinear beams. In the context of degenerated linear systems, the time-delayed parameters $\mathbf{p}_\tau$ derived from Eq. (5.65) consistently result in an equal-peak scenario. This arises from the fact that the DRCs are exclusively induced by the nonlinearity of the systems. However, in nonlinear systems, the first equation in Eq. (5.65) is insufficient as a result of merging of DRCs and MRCs (main resonance curves). To address this limitation, the second equation of Eq. (5.65) is introduced as a supplementary condition. Its purpose is to distinguish the equal-peak case specifically for nonlinear systems. Consequently, the Extremes Equal Criterion, as defined by Eq. (5.65), ensures that the derived time-delayed parameters consistently yield an equal-peak outcome, even in the presence of nonlinearity.

The minimum peak criterion identifies the best time-delayed parameters to minimize the peaks around the ith mode, which is described as

$$r_3 = \left\{ \mathbf{p}_\tau | \mathrm{Min}\left[a(s_c, \mathbf{p}, \mathbf{p}_\tau, \Omega_{2i-1}) = a(s_c, \mathbf{p}, \mathbf{p}_\tau, \Omega_{2i}) \right], \mathbf{p}_\tau \in r_2 \right\}, \quad i \in \boldsymbol{\Psi}. \quad (5.66)$$

Equation (5.66) serves as the third criterion constituing the multimodal equal-peak principle, refining the selection of optimal time-delayed parameters identified by the extremes equal criterion given in Eq. (5.65). Utilizing the minimum peak criterion, Eq. (5.66) determines the optimal time-delayed parameters for the first and second TDVAs across different amplitude of the force. The corresponding results are illustrated in Fig. 5.46a and b.

Implementing the suggested multimodal equal-peak principle along with the three proposed criteria yields the optimal time-delayed parameters for the TDVAs. Figure 5.47 illustrates the FRCs of the nonlinear beam, equipped with TDVAs under the time-delayed parameters, while retaining the first five modes ($P = 5$).

Figure 5.47a–c provide a comparative analysis of the control effect for multimodal vibration under LTVAs and TDVAs for a nonlinear beam. In Fig. 5.47a, LTVAs for $f = 50$ kN exhibits a slight divergence in resonance peaks for the first and second modes. When the force amplitude increases to $f = 100$ kN as in Fig. 5.47b and $f = 150$ kN in Fig. 5.47c, the resonance peaks around the second and first modes are effectively detuned, respectively. This observation implies that passive LTVAs become impractical for nonlinear multimodal vibration control as force amplitudes increase. Conversely, the TDVAs, as depicted in Fig. 5.47a–c, are optimized based on the proposed principle, employing three criteria for different force amplitudes. The optimal time-delayed parameters enable TDVAs to concurrently suppress multimodal nonlinear resonances of the nonlinear beam, and ensuring equal and minimized resonance values. Consequently, TDVAs, together with the proposed multimodal equal-peak principle, prove to be effective tools for suppressing multimodal resonances in nonlinear beams.

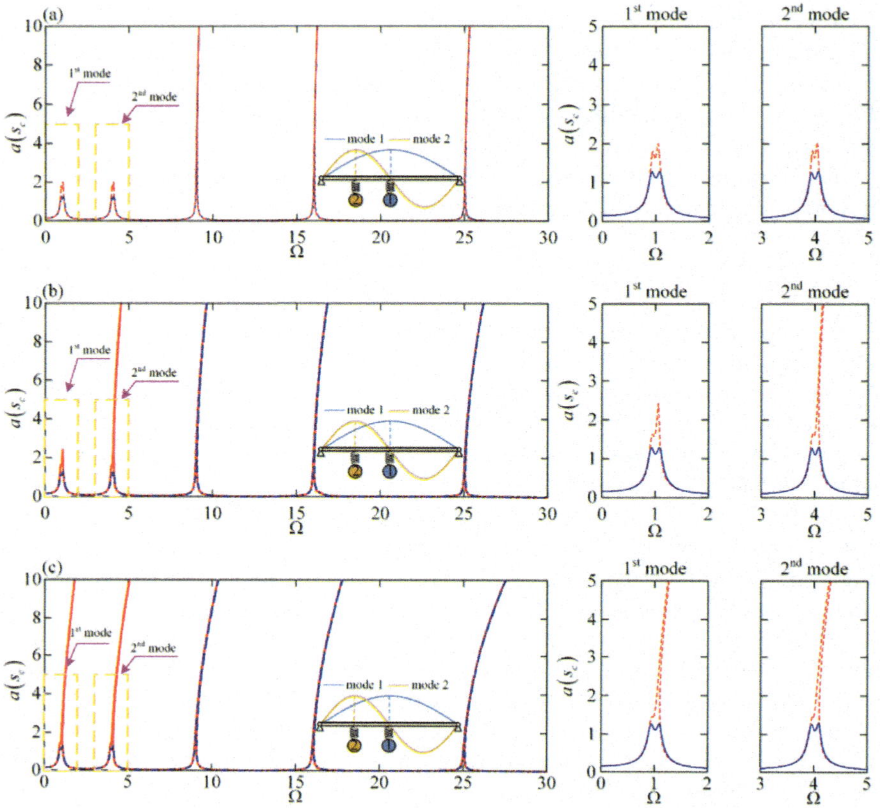

Fig. 5.47 The FRCs for the nonlinear beam attached by the LTVAs (red dashed lines) and TDVAs (blue lines) for **a** $f = 50$ kN, **b** $f = 100$ kN, and **c** $f = 150$ kN. The locations of the two absorbers are $s_1 = l/2$, $s_2 = l/4$, respectively

References

1. Xu J, Pei LJ (2006) Advances in dynamics for delayed systems. Advan Mech 36:17–30
2. Heiden UAD, Walther HO (1983) Existence of chaos in control system with delayed feedback. J Differ Equ 47:273–295
3. Chen GR, Wang XF (2006) Chaos of dynamical systems: theoretical methods and applications. Shanghai Jiao Tong University Press, Shanghai
4. Olgac N, Holm-Hansen BT (1994) A novel active vibration absorption technique: delayed resonator. J Sound Vib 176:93–104
5. Hamdi M, Belhaq M (2009) Self-excited vibration control for axially fast excited beam by a time delay state feedback. Chaos, Solitons Fractals 41:521–532
6. Cai GP, Chen LX (2010) Delayed feedback control experiments on some flexible structures. Acta Mech Sin 6:951–965
7. Cai GP, Chen LX (2013) Some problems of delayed feedback control. Advan Mech 43:21–28
8. Xu J, Sun XT (2015) A. A multi-directional vibration isolator based on Quasi-Zero-Stiffness structure and time-delayed active control. Int J Mech Sci 100:126–135

9. Xu J, Sun YX (2015) B. Experimental studies on active control of a dynamic system via a time-delayed absorber. Acta Mech Sin 31:229–247
10. Sun XT, Zhang S, Xu J (2018) C. Parameter design of a multi-delayed isolator with asymmetrical nonlinearity. Int J Mech Sci 138–139:398–408
11. Sun XT, Wang F, Xu J (2019) B. Dynamics and realization of a feedback-controlled nonlinear isolator with variable time delay. J Vibr Acoust 141:021005
12. Sun XT, Xu J, Fu JS (2017) The effect and design of time delay in feedback control for a nonlinear isolation system. Mech Syst Signal Proc 87(Part A):206–217. https://doi.org/10.1016/j.ymssp.2016.10.022. ISSN 0888-3270
13. Sun XT, Zhang S, Xu J (2018) Parameter design of a multi-delayed isolator with asymmetrical nonlinearity. Int J Mech Sci 398–408
14. Sun XT, Feng W, Xu J (2018) Dynamics and realization of a feedback-controlled nonlinear isolator with variable time delay. J Vib Acoust 141(2)
15. Wang F, Sun XT, Meng H, et al (2020) Time-delayed feedback control design and its application for vibration absorption. IEEE Trans Ind Electron (99):1–1
16. Meng H, Sun XT, Xu J et al (2020) The generalization of equal-peak method for delay-coupled nonlinear system. Physica D 403(4):132340
17. Meng H, Sun XT, Xu J, et al (2021) Establishment of the equal-peak principle for a multiple-DOF nonlinear system with multiple time-delayed vibration absorbers. Nonlinear Dyn (5)
18. Wang F, Sun XT, Meng H, et al (2022) Tunable broadband low-frequency band gap of multiple-layer metastructure induced by time-delayed vibration absorbers. Nonlinear Dyn
19. Meng H, Sun XT, Xu J, et al (2021) Multimodal vibration suppression of nonlinear Euler–Bernoulli beam by multiple time-delayed vibration absorbers. Meccanica 1–21

Chapter 6
Effects of Time Delay on Manufacturing

Manufacturing effectiveness has the potential for significant growth, yet its progress is impeded by the dynamics of the manufacturing processes. Specifically, the material removal rate (MRR) in machining processes experiences a reduction to avoid chatter. Consequently, the anticipation and management of chatter have emerged as prominent subjects over the last two decades. Cutting instabilities have frictional, regenerative, mode-coupling, and thermo-mechanical types, where the primary (frictional) and secondary (regenerative) are the most significant ones [1]. These groups have extensively investigated self-excited chatter within turning [2] and other manufacturing processes.

While the regenerative theory has proven useful in predicting cutting stability, it has become increasingly evident that this predictive model falls short when dealing with varying cutting speeds. Specifically, there has been a gradual recognition that this prediction lacks adequacy when applied to different cutting velocities. In particular, the boundaries of stability are shifted in an upward direction, offering improved stability at lower cutting speeds. To date, two alternative explanations have been put forth to elucidate this phenomenon. In scenarios involving lower cutting velocities, an additional factor contributing to cutting instability has been identified: the Stribeck effect, which gives rise to frictional (primary) chatter [3]. The root of this frictional instability can be traced back to the interactions between the flow of the chip and the rake surface of the tool. As the workpiece material traverses the primary shear zone, it forms a chip that flows along the tool's rake surface. This flow generates heat and causes the chip to adhere to the tooltip before reaching the secondary shear zone. As a result of the alternating sticking and sliding phases that the chip undergoes along the rake surface, a complex frictional force with a velocity-dependent nature is experienced by the chip on the tool [4].

A similar challenge arises in grinding processes. In the context of grinding, the grinding wheel continually interacts with the workpiece, resulting in the generation of not only the cutting force but also plowing and frictional forces [5]. Notably, the frictional force takes on a notably significant role, at times surpassing the cutting

J. Xu, *Nonlinear Dynamics of Time Delay Systems*,
https://doi.org/10.1007/978-981-99-9907-1_6

force itself. Consequently, owing to the interplay of tangential plowing and the influential frictional force, a departure from the considerations applicable in turning or milling is observed. Mannan et al. discussed that when investigating grinding chatter, it becomes imperative to account for torsional vibration [6]. Addressing this, Hesterman and Stone [7] proposed an enhanced kinematic model designed for the grinding chatter simulation. Their work confirmed that the torsional motion distinctly influences the stability of plunge grinding, setting it apart from the dynamics observed in other machining processes.

In addition to the inherent self-excited vibration, it is imperative to also consider rotor dynamics in spindles with eccentricity. To elucidate this, let's take a closer look at turning dynamics analysis. Dassanayake et al. undertook a study on vibration stemming from workpiece imbalance. The rotating vibration of a spindle with three disks is observed by means of simulation [8]. Similarly, Huang et al. delved into the dynamics of turning, specifically examining the impact of unbalanced aerostatic bearing spindles, noting variations in vibration amplitude corresponding to different spindle speeds [9]. Moving beyond turning, Inazaki and Yonetsu conducted both theoretical and experimental investigations into the consequences of wheel imbalance and eccentricity on the amplitude of workpiece undulations in surface grinding processes [10]. Considering the interplay of self-excited regenerative chatter and forced vibration, Badger et al. took on the task of investigating the mass eccentricity of grinding wheels and its influence on grinding force and workpiece waviness, providing insights into a comprehensive understanding of the phenomenon [11].

Extensive experimental examinations in the realm of transverse grinding were undertaken by Shimizu [12], yielding noteworthy findings regarding the amplitude of chatter. Notably, a pronounced fluctuation in chatter amplitude was unveiled. To elaborate further, it was determined that the chatter amplitude reaches its peak when the grinding wheel is close to the center of the workpiece, demonstrating a subsequent gradual decline as the wheel distances itself from the center. Additionally, the spectral patterns exhibited by the chatter vibration underwent substantial shifts as the grinding wheel moved along the workpiece. This signifies that the lateral movement of the grinding wheel emerges as the foremost determinant influencing the characteristics of chatter. This influence extends to parameters such as amplitude and frequency, underscoring the pivotal role played by the transverse motion of the grinding wheel in shaping the nature of chatter in this context.

In instances where cutting tools experience chatter-induced detachment from the workpiece surface, the presence of non-smooth behavior poses challenges to numerical continuation algorithms. Consequently, direct numerical integration is commonly adopted for studying global bifurcation. Additionally, through local perturbation analysis, it has been revealed that periodic chatter emerging at stability boundaries often exhibits subcritical characteristics. This leads to the introduction of substantial-amplitude chatter into regions that are linearly stable, coexisting with stationary cutting processes [13]. This coexistence of multiple stability states is regarded as precarious due to the uncertainties it introduces into the long-term dynamics of cutting processes [9]. However, despite the risks associated with it, the unsafe zones (UZs) situated near stability boundaries hold the potential for achieving highly efficient cutting owing to their elevated material removal rate. Accordingly, there is a motivation to explore these zones, even in the face of potential chatter-related issues. To this end, the establishment of a clearly defined "Basin of Attraction" (BoA) for Delay Differential Equations (DDEs) becomes pivotal. However, as of now, a widely accepted standard for such a BoA remains absent. When time delays are present in systems incorporating feedback control, certain studies have employed strategies like switching off time-delayed control at the outset, thereby eliminating the need for functional Initial Conditions (ICs). In this approach, there is no distinction between Ordinary Differential Equations (ODEs) and DDEs regarding the BoA definition. Alternatively, Various elementary functions can be used to approximate functional ICs. The simplest approximation requires the use of a constant function, as demonstrated by Ji [14], who explored BoA for coexisting periodic attractors in delayed nonlinear oscillators. An improved variant of this approximation involves linear polynomials, such as those used by Wang et al. [15] in their study of delayed Duffing oscillators. In this method, the coefficients of the polynomial help build the construction of the BoA. This approximate method was further developed by Zheng and Sun [16], who used second-order polynomials for BoA analysis of time delay Duffing systems. Furthermore, the Fourier series proves to be a valuable candidate for approximating functional ICs, especially when the delay system behaves oscillating behavior [17]. It is important to note that all of the above studies, regardless of the basic functions employed, limited the use of no more than three basis functions to characterize BoA for DDEs.

To sum up, modeling and analysis of cutting dynamics given complex tool-workpiece interactions and other effects are still very challenging. Moreover, the nonlinear cutting dynamics is still very hard to analyze due to the regenerative time delay and multiple stabilities. This section is going to discuss some related topics.

6.1 Modeling of Cutting Dynamics by Delay Differential Equation

6.1.1 Cutting Dynamics with Single Time Delay—Turning

6.1.1.1 Modeling of Turning Dynamics

A one degree-of-freedom (DOF) model of the turning process is displayed in Fig. 6.1a, where the workpiece has a radius of R [m]. The workpiece rotates at a rotary speed N, which results in a cutting speed of $V_c = \frac{R\pi N}{30}$. The chip velocity V_{ch}, yielding normal and tangential contact forces, $F_{\gamma\text{n}}$ and F_γ. Alternatively, the resultant force, F_{r} [N], can be decomposed into X and Y directions as well for cutting and feeding forces, F_{c} and F_{f}. At the same time, the workpiece moves along the tool flank face for a process damping force F_{d} in Y direction. The overall model is [18]:

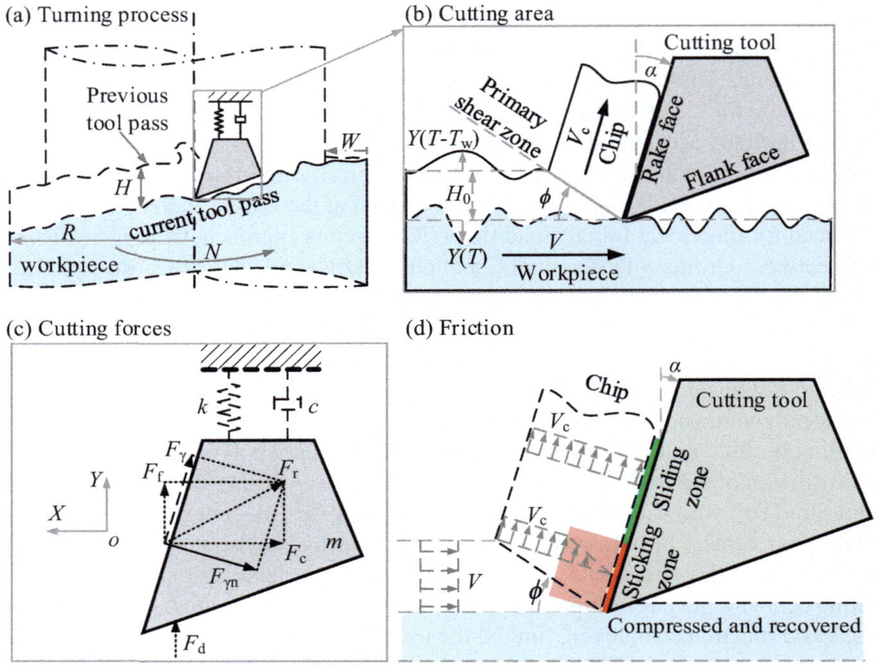

Fig. 6.1 Model of cutting process

$$m\ddot{Y}(t) + c\dot{Y}(t) + kY(t) = F_{\mathrm{f}} + F_{\mathrm{d}}, \tag{6.1}$$

where m, c, and k are the mass, damping, and stiffness of the tool.

The feeding force is:

$$F_{\mathrm{f}} = F_{\gamma}\cos(\gamma) - F_{\gamma\mathrm{n}}\sin(\gamma), \tag{6.2}$$

where F_{γ} and $F_{\gamma\mathrm{n}}$ are the cutting forces in tangential and normal directions. The normal force is proportional to the cutting depth and instantaneous chip thickness $H(t)$ as follows:

$$F_{\gamma\mathrm{n}} = Ka_{\mathrm{p}}H(t), \tag{6.3}$$

where K is force coefficient, and a_{p} is radial cutting depth. According to regenerative theory, the nominal chip thickness is perturbed by previous and current tool displacements, $Y(t - t_{\mathrm{w}})$ and $Y(t)$, as follows:

$$H(t) = H_{\mathrm{D}} - Y(t) + Y(t - t_{\mathrm{w}}), \tag{6.4}$$

where $t_{\mathrm{w}} = \frac{60}{N}$ is the rotational period of the workpiece.

The frictional force is proportional to the normal force as follows:

$$\mu = \frac{F_\gamma}{F_{\gamma n}}, \tag{6.5}$$

where μ is velocity-soften equivalent frictional coefficient [19] and can be approximated by:

$$\mu = \text{sign}(V_\gamma)\left(\mu_d + (\mu_s - \mu_d)\exp\left(-\frac{|V_\gamma|}{V_s}\right)\right), \tag{6.6}$$

where V_γ is the tangential frictional velocity of the chip in relation to the tool, V_s is Stribeck velocity [4], and μ_s and μ_d are the static and dynamic coefficients. Here, the frictional velocity is:

$$V_\gamma = V_{ch} - \dot{Y}(t)\cos(\gamma), \tag{6.7}$$

where the chip velocity is:

$$V_{ch} = V\frac{\sin(\phi)}{\cos(\phi - \gamma)}, \tag{6.8}$$

with ϕ representing the shear angle in Fig. 6.1d.

The tooltip exerts pressure on the uncut workpiece material, generating a damping force, denoted as F_d, on the flank face [20]. The so-called process damping force can be expressed by:

$$F_d = C_y a_p \frac{\dot{Y}(t)}{V_c}, \tag{6.9}$$

where C_y is process damping coefficient.

6.1.1.2 Linear Analysis

By introducing the following dimensionless parameters:

$$\xi = \frac{c}{\sqrt{mk}},\ w = a_p\frac{K}{k},\ v_s = \frac{30V_S\cos(\gamma - \phi)}{\pi R\sin(\phi)}\sqrt{\frac{m}{k}},$$
$$v = \frac{H_D}{V_s}\sqrt{\frac{k}{m}},\ n = N\sqrt{\frac{m}{k}},\ c_y = \frac{30c_y}{\pi RK},\ \tau_w = t_w\sqrt{\frac{k}{m}}, \tag{6.10}$$

and variables:

$$\tau = t\sqrt{\frac{k}{m}},\ y_1(\tau) = \frac{Y(t)}{H_{\mathrm{D}}},\ y_2(\tau) = y_1'(\tau) = \frac{\dot{Y}(t)}{H_{\mathrm{D}}}\sqrt{\frac{m}{k}},$$
$$h(\tau) = \frac{H(t)}{H_{\mathrm{D}}} = 1 - y_1(\tau) + y_1(\tau - \tau_{\mathrm{w}}), \tag{6.11}$$

$$v_\gamma(\tau) = \frac{v_\gamma}{H_{\mathrm{D}}}\sqrt{\frac{m}{k}} = \frac{n}{v_{\mathrm{S}}} - v\cos(\gamma)y_2(\tau),$$

one obtains the following dimensionless equation:

$$\begin{aligned} y_1'(\tau) &= y_2(\tau), \\ y_2'(\tau) &= -y_1(\tau) - \xi y_2(\tau) + w(\mu\cos(\gamma) - \sin(\gamma))h(\tau) - wc_y\frac{y_2(\tau)}{n}, \end{aligned} \tag{6.12}$$

with:

$$\mu = \mathrm{sign}(v_{\mathrm{f}})\big(\mu_{\mathrm{d}} + (\mu_{\mathrm{s}} - \mu_{\mathrm{d}})\exp(-|v_\gamma|)\big). \tag{6.13}$$

The equilibrium is $(y_1(\tau), y_2(\tau)) \equiv (y_{10}, 0)$, where:

$$y_{10} = w\left(\mu_{\mathrm{d}} + (\mu_{\mathrm{s}} - \mu_{\mathrm{d}})\exp\left(\frac{n}{v_{\mathrm{s}}}\right)\right)\cos(\gamma) - w\sin(\gamma). \tag{6.14}$$

The local stability of $(y_{10}, 0)$ is determined by the linear part as follows:

$$\mathbf{y}'(\boldsymbol{\tau}) = \mathbf{A}\mathbf{y}(\boldsymbol{\tau}) + \mathbf{D}\mathbf{y}(\boldsymbol{\tau} - \boldsymbol{\tau}_{\mathrm{w}}). \tag{6.15}$$

The vector and matrices are:

$$\mathbf{y}(\boldsymbol{\tau}) = \begin{pmatrix} \mathbf{y}_1(\boldsymbol{\tau}) \\ \mathbf{y}_2(\boldsymbol{\tau}) \end{pmatrix}, \mathbf{A} = \begin{pmatrix} 0 & -1 \\ 1 + wa & \xi + wb \end{pmatrix}, \mathbf{D} = \begin{pmatrix} 0 & 0 \\ -wa & 0 \end{pmatrix}, \tag{6.16}$$

where:

$$a = \left(\mu_{\mathrm{d}} + (\mu_{\mathrm{s}} - \mu_{\mathrm{d}})\exp\left(-\frac{n}{v_{\mathrm{s}}}\right)\right)\cos(\gamma) - \sin(\gamma), \tag{6.17}$$

$$b = \frac{c_y}{n} + (\mu_{\mathrm{d}} - \mu_{\mathrm{s}})v\cos^2(\gamma)\exp\left(-\frac{n}{v_{\mathrm{s}}}\right). \tag{6.18}$$

The corresponding eigenvalue problem is:

$$\det(\lambda\mathbf{I} - \mathbf{A} - \mathbf{D}\exp(-\lambda\tau_{\mathrm{w}})) = 0, \tag{6.19}$$

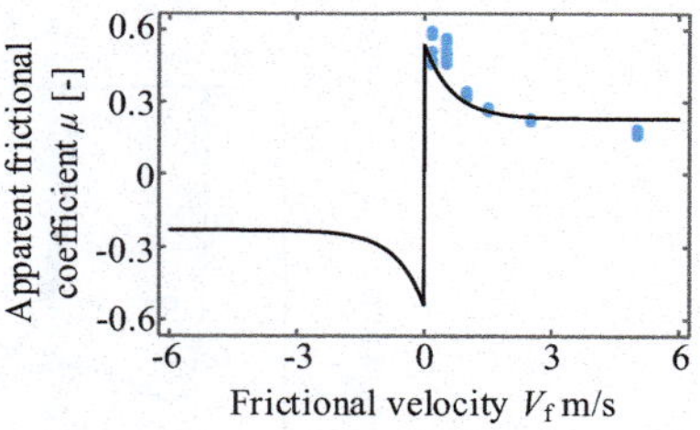

Fig. 6.2 Velocity softening equivalent friction coefficient fitting curve [4]

where det(·) is the determinant, and λ is the eigenvalues. Stability is critical for a pair of imaginary eigenvalues, $\lambda = \pm i\omega$:

$$\det(i\omega\mathbf{I} - \mathbf{A} - \mathbf{D}\exp(-i\omega\tau_w)) = 0. \tag{6.20}$$

Next, the following parameters are used for the numerical calculation of eigenvalues:

$$\begin{aligned} &m = 0.561 \text{ kg}, c = 145 \text{ Ns/m}, k = 6.48 \times 10^6 \text{ N/m}, \\ &K = 6.02 \times 10^9 \text{ N/m}^2, C_y = 6.11 \times 10^5 \text{ N/m}, \\ &R = 0.0175 \text{ m}, H_D = 0.0005 \text{ m}, \gamma = 0 \text{ deg}, \phi = 45 \text{ deg}. \end{aligned} \tag{6.21}$$

We use the data measured by Claudin et al. [4], yielding the following coefficients:

$$V_s = 0.65 \text{ m/s}, \mu_d = 0.23, \mu_s = 0.54, \tag{6.22}$$

and

$$\xi = 0.07605, v_s = 0.10436, v = 2.61434, c_y = 0.05541. \tag{6.23}$$

By using Eq. (6.20) along with the provided parameter values, the stability boundaries (dashed black) in Fig. 6.5 are determined. The newly introduced model performs admirably in forecasting cutting stability, particularly in the low-velocity region, surpassing Eynian's model [21] (Figs. 6.2 and 6.3).

6.1.2 *Cutting Dynamics with Two Time Delays—Grinding*

6.1.2.1 Modeling of the Grinding Process

The workpiece in Fig. 6.4 has one end ($S = 0$) clamped, and the other end ($S = L$) simply supported. At $S = P$, a lumped mass disc is rubbed by a wheel. The rotational speeds are Ω_g and Ω_w. The wheel advances toward the workpiece at a rate of f per workpiece revolution. The workpiece deforms in horizontal, vertical,

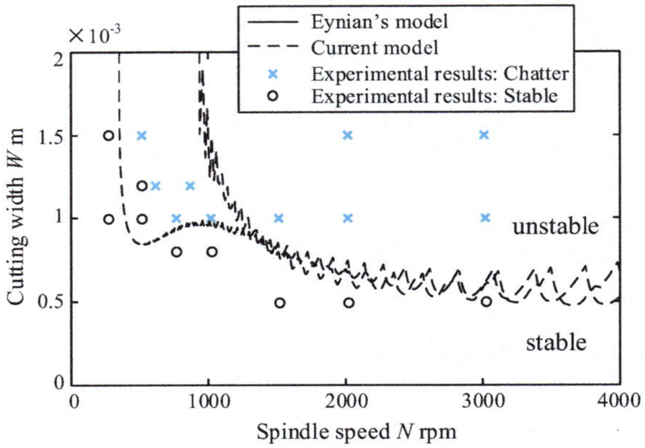

Fig. 6.3 Comparison of the stability boundaries [21]

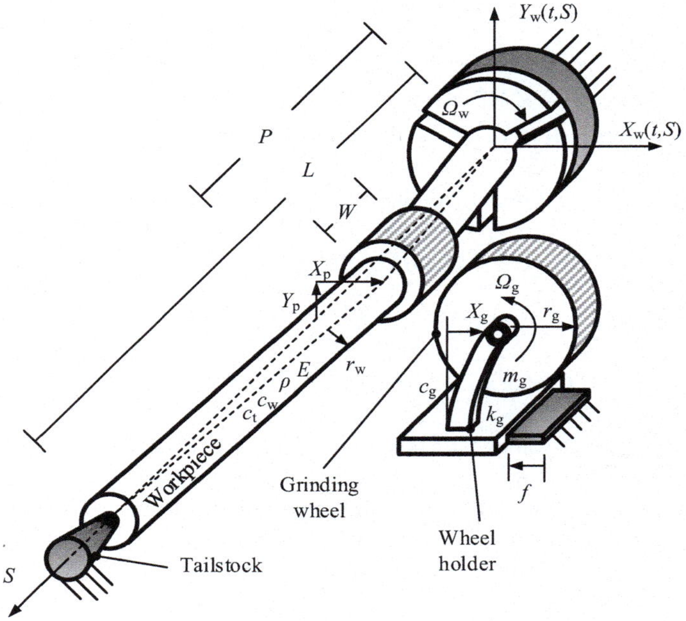

Fig. 6.4 Schematic of the plunge grinding process

and torsional directions, and the wheel bends horizontally [22]. The disc angle is $\theta_{\mathrm{p}}(t) = \phi_{\mathrm{p}}(t) - \phi(t, 0)$. The contact angle is:

$$\gamma = \tan^{-1}\left(\frac{a}{b}\right) = \tan^{-1}\left(\frac{Y_{\mathrm{W}}(t, P)}{R_{\mathrm{g}} + R_{\mathrm{p}} - f - X_{\mathrm{W}}(t, P) + X_{\mathrm{g}}(t)}\right). \tag{6.24}$$

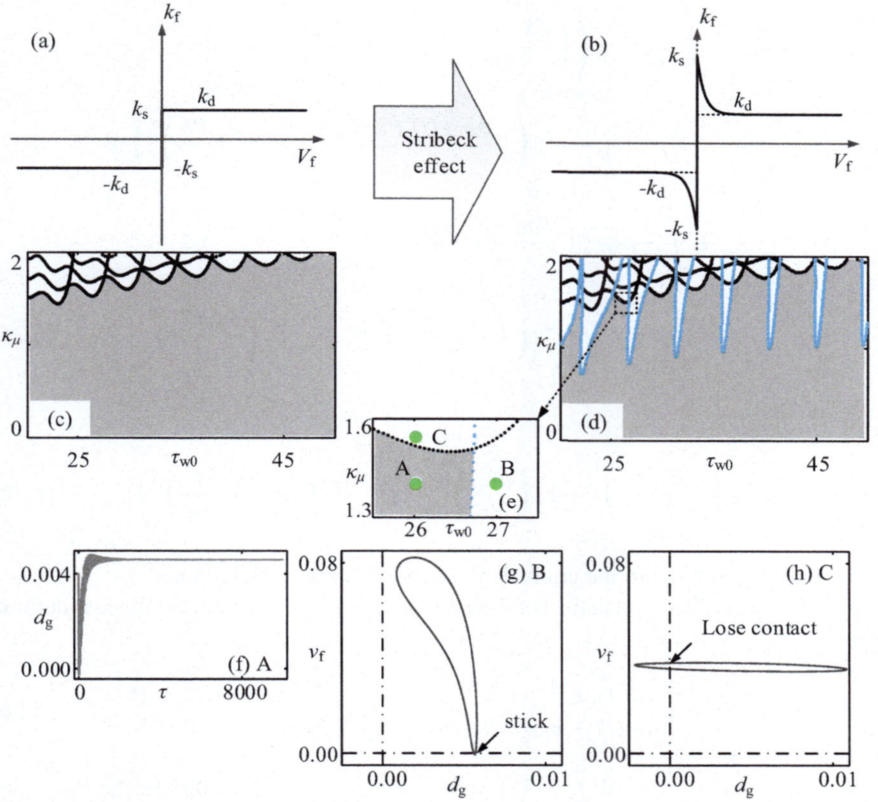

Fig. 6.5 Regarding both the **a** Coulomb and Coulomb Stribeck models, stability boundaries are depicted in **c** and **d**. A particular segment of **d** is magnified in **e** to demonstrate **f** chatter-free motion, **g** frictional chatter, and **h** regenerative chatter

Potential energy is:

$$V = \frac{1}{2}\int_{0}^{L}\left(GI_o\left(\frac{\mathrm{d}^2\phi_\mathrm{w}(t,S)}{\mathrm{d}S^2}\right)^2 + EI_x\left(\frac{\mathrm{d}^2X_\mathrm{w}(t,S)}{\mathrm{d}S^2}\right)^2 + EI_y\left(\frac{\mathrm{d}^2Y_\mathrm{w}(t,S)}{\mathrm{d}S^2}\right)^2\right)\mathrm{d}S + \frac{1}{2}k_\mathrm{g}\left(\frac{\mathrm{d}X_\mathrm{g}(t)}{\mathrm{d}t}\right)^2, \tag{6.25}$$

where G and E represent the shear modulus and Young's modulus of the workpiece material, $I_o = \pi r_\mathrm{w}^4/2$, $I_\mathrm{x} = c\pi r_\mathrm{w}^4/4$, and $I_y = \pi r_\mathrm{w}^4/4$, are the cross-sectional area's moment of inertia for the workpiece. Given all the deformations, the kinetic energy of the system is:

$$T = \int_0^L \frac{1}{2}\rho\pi r_{\mathrm{W}}^2\left(\left(\frac{\mathrm{d}X_{\mathrm{W}}(t,S)}{\mathrm{d}t}\right)^2 + \left(\frac{\mathrm{d}Y_{\mathrm{W}}(t,S)}{\mathrm{d}t}\right)^2\right)\mathrm{d}S$$
$$+ \frac{1}{2}\rho\pi\left(r_p^2 - r_{\mathrm{w}}^2\right)\left(\left(\frac{\mathrm{d}X_{\mathrm{W}}(t,P)}{\mathrm{d}t}\right)^2 + \left(\frac{\mathrm{d}Y_{\mathrm{W}}(t,P)}{\mathrm{d}t}\right)^2\right)W$$
$$+ \int_0^L \frac{1}{4}\rho\pi r_{\mathrm{w}}^4\left(\frac{\mathrm{d}\phi_{\mathrm{W}}(t,S)}{\mathrm{d}t}\right)^2\mathrm{d}S$$
$$+ \frac{1}{4}\rho\pi\left(r_p^4 - r_{\mathrm{w}}^4\right)\left(\frac{\mathrm{d}\phi_{\mathrm{w}}(t,P)}{\mathrm{d}t}\right)^2 W + \frac{1}{2}m_{\mathrm{g}}\left(\frac{\mathrm{d}x_{\mathrm{g}}(t)}{\mathrm{d}t}\right)^2, \tag{6.26}$$

The dissipation is:

$$D = \frac{1}{2}\left(c_{\mathrm{w}}\left(\frac{\mathrm{d}x_{\mathrm{p}}}{\mathrm{d}t}\right)^2 + c_{\mathrm{w}}\left(\frac{\mathrm{d}Y_{\mathrm{p}}}{\mathrm{d}t}\right)^2 + c_{\mathrm{t}}\left(\frac{\mathrm{d}\phi_{\mathrm{p}}}{\mathrm{d}t}\right)^2 + c_{\mathrm{g}}\left(\frac{\mathrm{d}x_{\mathrm{g}}}{\mathrm{d}t}\right)^2\right), \tag{6.27}$$

where c_{w}, c_{t}, and c_{g} are the equivalent viscous damping coefficients.

To discretize this model, the bending displacement field of the workpiece is defined as

$$X_{\mathrm{w}}(t,S) = \begin{cases} a_{x0}(t) + a_{x1}(t)S + a_{x2}(t)S^2 + a_{x3}(t)S^3, \text{ if S} \le \text{P}, \\ b_{x0}(t) + b_{x1}(t)S + b_{x2}(t)S^2 + b_{x3}(t)S^3, \text{ if P} \le \text{S}, \end{cases} \tag{6.28}$$

where a_{xi} and b_{xi} $(i = 0, 1, 2, 3)$ are functions of time. In addition, the boundary conditions are

$$\begin{cases} X_{\mathrm{W}}(t,0) = 0, \\ \frac{\mathrm{d}X_{\mathrm{W}}(t,0)}{\mathrm{d}S} = 0, \\ X_{\mathrm{w}}(t,S)|_{S\to P^-} = X_p(t), \\ X_{\mathrm{W}}(t,S)|_{S\to P^+} = X_p(t), \\ \left.\frac{\mathrm{d}X_{\mathrm{W}}(t,S)}{\mathrm{d}S}\right|_{S\to P^-} = \left.\frac{\mathrm{d}X_{\mathrm{W}}(t,S)}{\mathrm{d}S}\right|_{S\to P^+}, \\ \left.\frac{\mathrm{d}^2X_{\mathrm{W}}(t,S)}{\mathrm{d}S^2}\right|_{S\to P^-} = \left.\frac{\mathrm{d}^2X_{\mathrm{W}}(t,S)}{\mathrm{d}S^2}\right|_{S\to P^+}, \\ X_{\mathrm{W}}(t,L) = 0, \\ \frac{\mathrm{d}^2X_{\mathrm{W}}(t,0)}{\mathrm{d}S^2} = 0, \end{cases} \tag{6.29}$$

where $X_{\mathrm{p}}(t) = X_{\mathrm{w}}(t,P)$ represents the horizontal displacement of the disc. It yields

$$X_{\mathrm{w}}(t,S) = \begin{cases} \frac{3L(2L-P)PS^2+\left(P^2-2LP-2L^2\right)S^3}{P^3(4L-P)(L-P)}X_{\mathrm{p}}(t), & \text{if S} \le \text{P}, \\ \frac{-2L^3P+6L^3S+3L(P-3L)S^2+(3L-P)S^3}{(L-P)^2P(4L-P)}X_{\mathrm{p}}(t), & \text{if P} \le \text{S}. \end{cases} \tag{6.30}$$

The vertical deformation is

$$Y_w(t, S) = \begin{cases} \frac{3L(2L-P)PS^2+(P^2-2LP-2L^2)S^3}{P^3(4L-P)(L-P)} Y_p(t), & \text{if } S \leq P, \\ \frac{-2L^3P+6L^3S+3L(P-3L)S^2+(3L-P)S^3}{(L-P)^2P(4L-P)} Y_p(t), & \text{if } P \leq S, \end{cases} \quad (6.31)$$

where $Y_p(t) = Y_w(t, P)$. The torsional displacement $\phi(t, S)$ is

$$\phi_w(t, S) = \begin{cases} a_{\phi 0}(t) + a_{\phi 1}(t)S, & \text{if } S \leq P, \\ b_{\phi 0}(t) + b_{\phi 1}(t)S, & \text{if } P \leq S. \end{cases} \quad (6.32)$$

With the boundary conditions:

$$\begin{cases} \phi(t, 0) = \Omega_w t, \\ \phi(t, S)|_{S \to P^-} = \phi_p(t), \\ \phi(t, S)|_{S \to P^+} = \phi_p(t), \\ \left.\frac{d\phi(t,S)}{dS}\right|_{S \to P^-} = \left.\frac{d\phi(t,S)}{dS}\right|_{S \to P^+}, \end{cases} \quad (6.33)$$

it yields

$$\phi_w(t, S) = t\Omega_w + \frac{S\left(\phi_p(t) - \Omega_w t\right)}{P}, \ (0 \leq S \leq L). \quad (6.34)$$

Lagrange's equations are then used to obtain the governing equation as follows

$$\frac{d}{dt}\frac{\partial L}{\partial (dq_i/dt)} - \frac{\partial L}{\partial q_i} + \frac{\partial D}{\partial (dq_i/dt)} = Q_{q_i}, \quad (6.35)$$

where the Largrange L is $L = T - V$, q_i represents generalized coordinates $X_g(t)$, $X_p(t)$, $Y_p(t)$, and $\phi_p(t)$. Q_{q_i} is the corresponding generalized force. The discrete model is:

$$\begin{aligned} m_g\frac{d^2X_g(t)}{dt^2} + c_g\frac{dX_g(t)}{dt} + k_g X_g(t) &= F_x(t), \\ m_w\frac{d^2X_p(t)}{dt^2} + c_w\frac{dX_p(t)}{dt} + k_w X_p(t) &= -F_x(t), \\ m_w\frac{d^2Y_p(t)}{dt^2} + c_w\frac{dY_p(t)}{dt} + k_w Y_p(t) &= F_y(t), \\ J_t\frac{d^2\phi_p(t)}{dt^2} + c_t\frac{d\phi_p(t)}{dt} + k_t\left(\phi_p(t) - \Omega_w t\right) &= M(t), \end{aligned} \quad (6.36)$$

where the coefficients are

$$
\begin{aligned}
k_{\mathrm{w}} &= \frac{12EI_xL^3}{(L-P)^2(4L-P)P^3},\\
k_{\mathrm{t}} &= \frac{GI_OL}{P^2},\\
m_{\mathrm{W}} &= \frac{70\pi\rho P^2W\left(R_{\mathrm{p}}^2-R_{\mathrm{W}}^2\right)\left(4L^2-5LP+P^2\right)^2}{35(L-P)^2P^2(4L-P)^2}\\
&\quad+\frac{\pi\rho L^3R_{\mathrm{W}}^2\left(24L^4-24L^3P-4L^2P^2+8LP^3-P^4\right)}{35(L-P)^2P^2(4L-P)^2},
\end{aligned}
\tag{6.37}
$$

$$
J_{\mathrm{t}} = \frac{\pi\rho\left(6P^2WR_{\mathrm{p}}^4+\left(L^3-6P^2W\right)R_{\mathrm{W}}^4\right)}{6P^2}.
$$

Here, $F_x(t)$ and $F_y(t)$ are the grinding forces in the horizontal and vertical directions, and $M(t)$ is the moment. With the torsional angle of the disc $\theta(t) = \phi_{\mathrm{p}}(t) - \Omega_{\mathrm{w}}t$ used, the equation is changed into:

$$
\begin{aligned}
m_{\mathrm{g}}\frac{\mathrm{d}^2X_{\mathrm{g}}(t)}{\mathrm{d}t^2} + c_{\mathrm{g}}\frac{\mathrm{d}X_{\mathrm{g}}(t)}{\mathrm{d}t} + k_{\mathrm{g}}X_{\mathrm{g}}(t) &= F_x(t),\\
m_{\mathrm{w}}\frac{\mathrm{d}^2X_{\mathrm{p}}(t)}{\mathrm{d}t^2} + c_{\mathrm{w}}\frac{\mathrm{d}x_{\mathrm{p}}(t)}{\mathrm{d}t} + k_{\mathrm{w}}X_{\mathrm{p}}(t) &= -F_x(t),\\
m_{\mathrm{w}}\frac{\mathrm{d}^2Y_{\mathrm{p}}(t)}{\mathrm{d}t^2} + c_{\mathrm{w}}\frac{\mathrm{d}Y_{\mathrm{p}}(t)}{\mathrm{d}t} + k_{\mathrm{w}}Y_{\mathrm{p}}(t) &= F_y(t),
\end{aligned}
\tag{6.38}
$$

$$
J_{\mathrm{t}}\frac{\mathrm{d}^2\theta_{\mathrm{p}}(t)}{\mathrm{d}t^2} + c_{\mathrm{t}}\frac{\mathrm{d}\theta_{\mathrm{p}}(t)}{\mathrm{d}t} + k_{\mathrm{t}}\theta_{\mathrm{p}}(t) = c_{\mathrm{t}}\Omega_{\mathrm{w}} + M(t).
$$

Considering the contact γ, the forces are expressed as:

$$
\begin{aligned}
F_x &= F_{\mathrm{n}}\cos\gamma + F_{\mathrm{t}}\sin\gamma,\\
F_y &= F_{\mathrm{n}}\sin\gamma - F_{\mathrm{t}}\cos\gamma,\\
M &= F_{\mathrm{t}}r_{\mathrm{w}},
\end{aligned}
\tag{6.39}
$$

where F_{n} and F_{t} are the normal and the tangential grinding forces, respectively.

The grinding force exhibits a direct proportionality to the grinding depth [22]. Considering the workpiece's torsional and vertical motions, the grinding force is:

$$F_{\mathrm{n}} = \begin{cases} k_{\mu}\left(\frac{\Omega_{\mathrm{w}}+\frac{d\theta_{\mathrm{p}}}{dt}-\frac{d\gamma}{dt}}{\Omega_{\mathrm{g}}+\frac{d\gamma}{dt}}\right)^{2\mu-1} D_{\mathrm{g}}^{\mu}, & \text{if } \mathrm{D_g} > 0, \\ 0, & \text{if } \mathrm{D_g} \le 0, \end{cases} \tag{6.40}$$

where k_{μ}, D_{g}, and $\mu \in [0.5, 1]$ are grinding stiffness, depth, and dimensionless exponential coefficient, respectively.

Given workpiece motion, regeneration, and wheel wear, one obtains the grinding depth as follows:

$$\begin{aligned} D_{\mathrm{g}} = {} & f\cos(\gamma(t)) + \big(X_{\mathrm{p}}(t)\cos(\gamma(t)) - X_{\mathrm{g}}(t)\cos(\gamma(t)) - Y_{\mathrm{p}}(t)\sin(\gamma(t))\big) \\ & - \big(X_{\mathrm{p}}(t-T_{\mathrm{w}})\cos(\gamma(t-T_{\mathrm{w}})) - X_{\mathrm{g}}(t-T_{\mathrm{w}})\cos(\gamma(t-T_{\mathrm{w}})) \\ & - Y_{\mathrm{p}}(t-T_{\mathrm{w}})\sin(\gamma(t-T_{\mathrm{w}}))\big) \\ & - g\big(X_{\mathrm{p}}(t-T_{\mathrm{g}})\cos(\gamma(t-T_{\mathrm{g}})) - X_{\mathrm{g}}(t-T_{\mathrm{g}})\cos(\gamma(t-T_{\mathrm{g}})) \\ & - Y_{\mathrm{p}}(t-T_{\mathrm{g}})\sin(\gamma(t-T_{\mathrm{g}}))\big), \end{aligned} \tag{6.41}$$

where g is a small dimensionless parameter [22]. Here, T_{w} stands for the rotating period of the workpiece, while T_{g} represents the grinding wheel's rotation period. The delays are state-dependent as follows:

$$\begin{aligned} 2\pi &= \phi(t) - \gamma(t) - \phi(t-T_{\mathrm{w}}) + \gamma(t-T_{\mathrm{w}}) \\ &= \theta(t) - \gamma(t) - \theta(t-T_{\mathrm{w}}) + \gamma(t-T_{\mathrm{w}}) + T_{\mathrm{w}}\Omega_{\mathrm{w}}, \\ 2\pi &= \gamma(t) - \gamma(t-T_{\mathrm{g}}) + T_{\mathrm{g}}\Omega_{\mathrm{g}}. \end{aligned} \tag{6.42}$$

The tangent frictional force is:

$$F_{\mathrm{t}} = k_{\mathrm{f}}(V_{\mathrm{f}})F_{\mathrm{n}}, \tag{6.43}$$

where the frictional velocity V_{f} corresponds to the relative tangential speed between the wheel and workpiece at the contact zone, while k_{f} represents the frictional coefficient depending on V_{f}, which is given by:

$$V_{\mathrm{f}} = V_{\mathrm{fg}} - V_{\mathrm{fw}}, \tag{6.44}$$

where V_{fg} and V_{fw} refer to the tangential speeds of the wheel and workpiece at the contact zone. As seen, V_{fg} and V_{fw} are given by:

$$\begin{aligned} V_{\mathrm{fg}} &= \Omega_{\mathrm{g}}R_{\mathrm{g}} - \frac{\mathrm{d}x_{\mathrm{g}}(t)}{\mathrm{d}t}\sin(\gamma), \\ V_{\mathrm{fw}} &= \frac{\mathrm{d}\phi_{\mathrm{p}}(t)}{\mathrm{d}t}R_{\mathrm{p}} - \frac{\mathrm{d}Y_{\mathrm{p}}(t)}{\mathrm{d}t}\cos(\gamma) - \frac{\mathrm{d}x_{\mathrm{p}}(t)}{\mathrm{d}t}\sin(\gamma). \end{aligned} \tag{6.45}$$

Therefore, the relative speed V_f is:

$$
\begin{aligned}
V_\mathrm{f} &= \Omega_\mathrm{g} R_\mathrm{g} - \frac{\mathrm{d}\phi_\mathrm{p}(t)}{\mathrm{d}t} R_\mathrm{p} + \frac{\mathrm{d}Y_\mathrm{p}(t)}{\mathrm{d}t} \cos(\gamma) - \left(\frac{\mathrm{d}x_\mathrm{g}(t)}{\mathrm{d}t} + \frac{\mathrm{d}X_\mathrm{p}(t)}{\mathrm{d}t}\right) \sin(\gamma) \\
&= \Omega_\mathrm{g} R_\mathrm{g} - \Omega_\mathrm{w} R_\mathrm{p} - \frac{\mathrm{d}\theta_\mathrm{p}(t)}{\mathrm{d}t} R_\mathrm{p} + \frac{\mathrm{d}Y_\mathrm{p}(t)}{\mathrm{d}t} \cos(\gamma) \\
&\quad - \left(\frac{\mathrm{d}x_\mathrm{g}(t)}{\mathrm{d}t} - \frac{\mathrm{d}X_\mathrm{p}(t)}{\mathrm{d}t}\right) \sin(\gamma).
\end{aligned} \tag{6.46}
$$

The velocity-soften friction coefficient μ is:

$$
\mu = \mathrm{sign}(V_\mathrm{f}) \left(k_\mathrm{d} + (k_\mathrm{s} - k_\mathrm{d}) \exp\left(-\frac{|V_\mathrm{f}|}{V_\mathrm{s}}\right)\right). \tag{6.47}
$$

We introduce the following dimensionless parameters:

$$
\begin{aligned}
\xi_\mathrm{g} &= \frac{c_\mathrm{g}}{\sqrt{m_\mathrm{g} k_\mathrm{g}}}, \ \xi_\mathrm{w} = \frac{c_\mathrm{w}\sqrt{m_\mathrm{g}}}{m_\mathrm{w}\sqrt{k_\mathrm{g}}}, \ \xi_\mathrm{t} = \frac{c_\mathrm{t}\sqrt{m_\mathrm{g}}}{J_\mathrm{t}\sqrt{k_\mathrm{g}}}, \\
\kappa_\mu &= \frac{k_\mu}{k_\mathrm{g}} \left(R_\mathrm{g} + R_\mathrm{p}\right)^{\mu-1}, \ \kappa_\mathrm{w} = \frac{k_\mathrm{w} m_\mathrm{g}}{k_\mathrm{g} m_\mathrm{w}}, \ \kappa_\mathrm{t} = \frac{k_\mathrm{t} m_\mathrm{g}}{J_\mathrm{t} k_\mathrm{g}}, \\
\gamma_\mathrm{t} &= \frac{m_\mathrm{g} R_\mathrm{p}\left(R_\mathrm{g} + R_\mathrm{p}\right)}{J_\mathrm{t}}, \ \gamma_\mathrm{w} = \frac{m_\mathrm{g}}{m_\mathrm{w}}, \ v| = \frac{f}{R_\mathrm{p} + R_\mathrm{g}}, \\
r_\mathrm{p} &= \frac{R_\mathrm{p}}{R_\mathrm{p} + R_\mathrm{g}}, \ r_\mathrm{g} = \frac{R_\mathrm{g}}{R_\mathrm{p} + R_\mathrm{g}}, \ p = \frac{P}{L}, \\
\omega_\mathrm{w} &= \Omega_\mathrm{w}\sqrt{\frac{m_\mathrm{g}}{k_\mathrm{g}}}, \ \omega_\mathrm{g} = \Omega_\mathrm{g}\sqrt{\frac{m_\mathrm{g}}{k_\mathrm{g}}},
\end{aligned} \tag{6.48}
$$

and dimensionless variables:

$$
\begin{aligned}
\tau &= t\sqrt{\frac{k_\mathrm{g}}{m_\mathrm{g}}}, \ \tau_\mathrm{w} = T_\mathrm{w}\sqrt{\frac{k_\mathrm{g}}{m_\mathrm{g}}}, \ \tau_\mathrm{g} = T_\mathrm{g}\sqrt{\frac{k_\mathrm{g}}{m_\mathrm{g}}}, \ x_\mathrm{g}(\tau) = \frac{X_\mathrm{g}(t)}{R_\mathrm{g} + R_\mathrm{p}}, \\
\chi_\mathrm{w}(\tau) &= \frac{X_\mathrm{W}(t)}{R_\mathrm{g} + R_\mathrm{p}}, \ y_\mathrm{w}(\tau) = \frac{Y_\mathrm{w}(t)}{R_\mathrm{g} + R_\mathrm{p}}, \ \theta_\mathrm{p}(t) = \theta_\mathrm{p}(\tau), \\
\gamma &= \tan^{-1}\left(\frac{y_\mathrm{p}(\tau)}{1 - v - x_\mathrm{p}(\tau) + x_\mathrm{g}(\tau)}\right), \\
d_\mathrm{g} &= \left(v + x_\mathrm{p}(\tau) - x_\mathrm{g}(\tau)\right)\cos(\gamma(\tau)) - y_\mathrm{p}(t)\sin(\gamma(t)) \\
&\quad - \left(x_\mathrm{p}(\tau - \tau_\mathrm{w}) - x_\mathrm{g}(\tau - \tau_\mathrm{w})\right)\cos(\gamma(\tau - \tau_\mathrm{w})) \\
&\quad + y_\mathrm{p}(\tau - \tau_\mathrm{w})\sin(\gamma(\tau - \tau_\mathrm{w})) \\
&\quad - g\left(x_\mathrm{p}(\tau - \tau_\mathrm{g}) - x_\mathrm{g}(\tau - \tau_\mathrm{g})\right)\cos\left(\gamma\left(\tau - \tau_\mathrm{g}\right)\right)
\end{aligned}
$$

$$+ g y_p(\tau - \tau_g)\sin(\gamma(\tau - \tau_g)),$$
$$v_f = \omega_g r_g - \omega_w r_p - \frac{d\theta_p(\tau)}{d\tau} r_p + \frac{dy_p(\tau)}{d\tau}\cos(\gamma) - \left(\frac{dx_g(\tau)}{d\tau} - \frac{dx_p(\tau)}{d\tau}\right)\sin(\gamma). \tag{6.49}$$

This changes the model into

$$\begin{aligned}
&\frac{d^2x_g(\tau)}{d\tau^2} + \xi_g \frac{dx_g(\tau)}{d\tau} + \kappa_g x_g(\tau) = (\cos(\gamma) + k_f \sin(\gamma))\kappa_\mu \left(\frac{\tau_g}{\tau_w}\right)^{2\mu-1} d_g^\mu,\\
&\frac{d^2x_p(\tau)}{d\tau^2} + \xi_w \frac{dx_p(\tau)}{d\tau} + \kappa_w x_p(\tau) = -\gamma_w(\cos(\gamma) + k_f \sin(\gamma))\kappa_\mu \left(\frac{\tau_g}{\tau_w}\right)^{2\mu-1} d_g^\mu,\\
&\frac{d^2y_p(\tau)}{d\tau^2} + \xi_w \frac{dy_p(\tau)}{d\tau} + \kappa_w y_p(\tau) = \gamma_w(\sin(\gamma) - k_f \cos(\gamma))\kappa_\mu \left(\frac{\tau_g}{\tau_w}\right)^{2\mu-1} d_g^\mu,\\
&\frac{d^2\theta_p(\tau)}{d\tau^2} + \xi_t \frac{d\theta_p(\tau)}{d\tau} + \kappa_t \theta_p(\tau) = \gamma_t k_f \kappa_\mu \left(\frac{\tau_g}{\tau_w}\right)^{2\mu-1} d_g^\mu.
\end{aligned} \tag{6.50}$$

6.1.2.2 Grinding Stability

The parameter values are selected as follows:

$$\begin{aligned}
&L = 1\,\text{m},\ R_w = 0.1\,\text{m},\ R_p = 0.25\,\text{m},\ R_g = 0.25\,\text{m},\\
&m_g = 20\,\text{kg},\ c_g = 20 \times 10^5 \frac{\text{Ns}}{\text{m}},\ k_g = 3 \times 10^8\,\text{N/m},\\
&\rho = 7850\,\text{kg/m}^3,\ c_w = 1.2 \times 10^6\,\text{Ns/m}^2,\ E = 2.06 \times 10^{11}\,\text{Pa},\\
&c_t = 822 \frac{\text{Nms}}{\text{rad}},\ G = 7.93 \times 10^{10}\text{Pa},\ W = 0.03\,\text{m}.
\end{aligned} \tag{6.51}$$

The moment of inertia for the cross-sectional area of the workpiece is. Moreover, it is obtained that:

$$\begin{aligned}
&k_w = 6.1 \times 10^{10} \frac{\text{N}}{\text{m}},\quad k_t = 1.2 \times 10^9\text{Nm/rad},\\
&m_w = 1311.2\ \text{kg},\quad J_t = 43.9\ \text{kgm}^2.
\end{aligned} \tag{6.52}$$

Subsequently, the dimensionless parameters are calculated as follows:

$$\begin{aligned}
&\xi_g = 2.58199,\ \xi_w = 0.236301,\ \xi_t = 0.00483257,\\
&\kappa_w = 3.12484,\ \kappa_t = 1.89084,\ \gamma_t = 0.0569236,\\
&\gamma_w = 0.0152532,\ r_p = 0.5,\ r_g = 0.5.
\end{aligned} \tag{6.53}$$

Moreover,

$$g = \frac{1}{50}, v_s = 0.1, \mu = 0.7, \tau_{g0} = 20, k_s = 1, k_d = \frac{1}{3}, \tag{6.54}$$

while other parameters, τ_{w0}, κ_μ, ν and p, are left undefined.

Corresponding stable grinding, frictional chatter, and regenerative chatter are displayed in Fig. 6.5f–h respectively.

To illustrate the impact of friction, eigenvalue analyses were conducted with and without considering the Stribeck effect, as presented in Fig. 6.5. In Fig. 6.5c, the Coulomb model results in a stability boundary, represented by a black boundary, indicating regenerative instability only. The Stricbeck effect yields the stability in Fig. 6.5d, where red lines are introduced. Figure 6.5f–h illustrate stable grinding, frictional chatter, and regenerative chatter, respectively.

6.1.3 *Cutting Dynamics with Three Time Delays—Parallel Grinding*

6.1.3.1 Modeling of the Grinding Process

Figure 6.6 depicts a schematic of the parallel grinding process, involving the placement of two grinding wheels on either side of the workpiece [23]. The normal grinding forces are then partitioned into

$$\begin{aligned} \mathrm{F}_l(\mathrm{t}) &= \mathrm{F}_{nl}(\mathrm{t}) + \mathrm{F}_{ol}(\mathrm{t}), \\ \mathrm{F}_r(\mathrm{t}) &= \mathrm{F}_{nr}(\mathrm{t}) + \mathrm{F}_{or}(\mathrm{t}), \end{aligned} \tag{6.55}$$

where $F_{nl}(t)$ and $F_{nr}(t)$ are the grinding forces generated in the left and right non-overlapped regions, and $F_{ol}(t)$ and $F_{or}(t)$ are the forces in the overlapped regions. As discussed above, the grinding forces are proportional to its corresponding grinding depth

$$\begin{aligned} F_{nl}(t) &= W_l k_g \left(\frac{r_d \Omega_W}{V_l} D_{nl}(t) \right)^\mu, \\ F_{ol}(t) &= W_o k_g \left(\frac{r_d \Omega_w}{V_l} D_{ol}(t) \right)^\mu, \\ F_{nr}(t) &= W_r k_g \left(\frac{r_d \Omega_W}{V_r} D_{nr}(t) \right)^\mu, \\ F_{or}(t) &= W_o k_g \left(\frac{r_d \Omega_w}{v_r} D_{or}(t) \right)^\mu, \end{aligned} \tag{6.56}$$

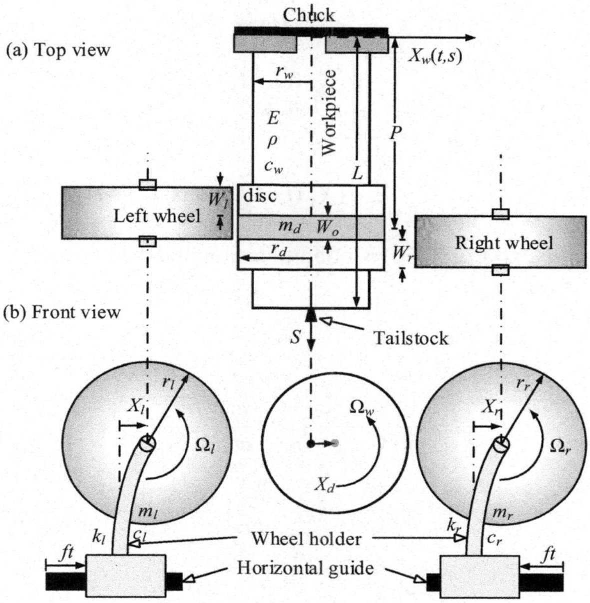

Fig. 6.6 Schematic of the parallel grinding process

where μ is an exponential coefficient. The grinding depths in left non-overlapped, left overlapped, right non-overlapped, and right overlapped zones are represented by $D_{nl}(t)$, $D_{ol}(t)$, $D_{nr}(t)$ and $D_{or}(t)$, respectively.

The grinding depth within the non-overlapping right zone is given by:

$$\begin{aligned} D_{nr}(t) = f + X_p(t) - X_r(t) - X_p(t - T_w) + X_r(t - T_w) \\ - g\big(X_p(t - T_r) - X_r(t - T_r)\big). \end{aligned} \tag{6.57}$$

The depth of grinding within the uncoordinated left zone is as follows

$$\begin{aligned} D_{nl}(t) = f - X_p(t) + X_l(t) + X_p(t - T_w) - X_l(t - T_w) \\ + g\big(X_p(t - T_l) - X_l(t - T_l)\big). \end{aligned} \tag{6.58}$$

The rotational periods are given by

$$T_w = \frac{2\pi}{\Omega_w},\ T_l = \frac{2\pi}{\Omega_l} = \frac{2\pi r_l}{V_l},\ T_r = \frac{2\pi}{\Omega_r} = \frac{2\pi r_r}{V_r}. \tag{6.59}$$

The grinding depths in the overlapped zones are

$$D_{or} = \frac{f}{2} + X_p(t) - X_r(t) + X_p\left(t - \frac{T_W}{2}\right) - X_l\left(t - \frac{T_W}{2}\right)$$
$$- g\big(X_p(t - T_r) - X_r(t - T_r)\big),$$
$$D_{ol}(t) = \frac{f}{2} - X_p(t) + X_l(t) - X_p\left(t - \frac{T_W}{2}\right) + X_r\left(t - \frac{T_W}{2}\right)$$
$$+ g\big(X_p(t - T_l) - X_l(t - T_l)\big). \tag{6.60}$$

Next, Eq. (6.60) is nondimensionalized with the following dimensionless parameters:

$$\gamma_{wl} = \frac{m_W}{m_l}, \gamma_{wr} = \frac{m_w}{m_r}, \kappa_l = \frac{k_l m_w}{k_w m_l}, \kappa_r = \frac{k_r m_w}{k_w m_r},$$
$$\xi_l = \frac{c_l}{m_l}\sqrt{\frac{m_w}{k_w}}, \xi_r = \frac{c_r}{m_r}\sqrt{\frac{m_W}{k_W}}, \xi_w = \frac{c_W}{m_w}\sqrt{\frac{m_w}{k_w}},$$
$$\gamma_r = \frac{r_d + r_l}{r_d + r_r}, \kappa_g = \frac{k_g}{k_w}\left(\frac{r_d(r_d + r_l)}{v_g}\sqrt{\frac{k_w}{m_w}}\right)^{\mu}, v = \frac{f}{r_d + r_l},$$
$$\tau_w = T_w\sqrt{\frac{k_w}{m_w}}, \omega_W = \Omega_w\sqrt{\frac{m_w}{k_w}} = \frac{2\pi}{\tau_W}, \tau_r = T_r\sqrt{\frac{k_W}{m_W}},$$
$$\tau_r = T_r\sqrt{\frac{k_W}{m_W}}, w_l = \frac{w_l}{r_d + r_l}, w_r = \frac{w_r}{r_d + r_l}, w_o = \frac{w_o}{r_d + r_l}, \tag{6.61}$$

and variables:

$$\tau = t\sqrt{\frac{k_w}{m_w}}, x_w(\tau) = \frac{x_w(t)}{r_d + r_l}, x_l(\tau) = \frac{x_l(t)}{r_d + r_l}, x_r(\tau) = \frac{X_r(t)}{r_d + r_l},$$
$$d_{nl}(\tau) = \frac{D_{nl}(t)}{r_d + r_l}, d_{ol}(\tau) = \frac{D_{ol}(t)}{r_d + r_l}, d_{nr}(\tau) = \frac{D_{nr}(t)}{r_d + r_l}, d_{or}(\tau) = \frac{D_{or}(t)}{r_d + r_l}, \tag{6.62}$$

which yields the dimensionless governing equation

$$x_l''(\tau) + \xi_l x_l'(\tau) + \kappa_l x_l(\tau) = -\gamma_{wl}\kappa_g\omega_w^{\mu}\big(w_l d_{nl}^{\mu}(\tau) + w_o d_{ol}^{\mu}(\tau)\big),$$
$$x_r''(\tau) + \xi_r x_r'(\tau) + \kappa_r x_r(\tau) = \gamma_{wr}\kappa_g\omega_w^{\mu}\big(w_r d_{nr}^{\mu}(\tau) + w_o d_{or}^{\mu}(\tau)\big),$$
$$x_p''(\tau) + \xi_w x_p'(\tau) + x_p(\tau) = \kappa_g\omega_w^{\mu}\big(w_l d_{nl}^{\mu}(\tau) + w_o d_{ol}^{\mu}(\tau) - w_r d_{nr}^{\mu}(\tau) - w_o d_{or}^{\mu}(\tau)\big). \tag{6.63}$$

6.1.3.2 Static Deflection and Linear Stability

Subjecting $x_l(\tau) \equiv x_{l0}$, $x_r(\tau) \equiv x_{r0}$ and $x_p(\tau) \equiv x_{p0}$ into Eq. (6.63) yields:

$$\begin{aligned} \kappa_l x_{l0} &= -\gamma_{wl}\kappa_g\omega_w^{\mu}\left(w_l d_{nl0} + w_o d_{ol0}^{\mu}\right), \\ \kappa_r x_{r0} &= \gamma_{wr}\kappa_g\omega_w^{\mu}\left(w_r d_{nr0}^{\mu} + w_o d_{or}^{\mu}\right), \\ x_{p0} &= \kappa_g\omega_w^{\mu}\left(w_l d_{nl0}^{\mu} + w_o d_{ol0}^{\mu} - w_r d_{nr0}^{\mu} - w_o d_{or0}^{\mu}\right), \end{aligned} \tag{6.64}$$

where

$$\begin{aligned} d_{nl0} &= v + g\left(x_{p0} - x_{l0}\right), \\ d_{ol0} &= \frac{v}{2} + (g-2)x_{p0} + (1-g)x_{l0} + x_{r0}, \\ d_{nr0} &= v - g\left(x_{p0} - x_{r0}\right), \\ d_{\mathrm{or}\,0} &= \frac{v}{2} + (2-g)x_{p0} + (g-1)x_{r0} - x_{l0}. \end{aligned} \tag{6.65}$$

Parameter values are selected as:

Mass density $\rho = 7850\,\mathrm{kgm}^2$
Mass of the right wheel $m_r = 30\,\mathrm{kg}$
Mass of the left wheel $m_l = 30\,\mathrm{kg}$
Young's modulus $E = 2.06 \times 10^{11}\,\mathrm{Pa}$
Stiffness of the right wheel $k_r = 7 \times 10^8\,\mathrm{Nm}^{-1}$
Stiffness of the left wheel $k_l = 7 \times 10^8\,\mathrm{Nm}^{-1}$
Specific cutting stiffness $k_g = 7.84 \times 10^7\,\mathrm{Nm}^{-1-\mu}$
Workpiece length $L = 0.25\,\mathrm{m}$
Workpiece radius $r_w = 0.00635\,\mathrm{m}$
Radius of the right wheel $r_r = 0.0889\,\mathrm{m}$
Radius of the left wheel $r_l = 0.0889\,\mathrm{m}$
Disc radius $r_d = 0.0381\,\mathrm{m}$
Disc position $P = 0.03\,\mathrm{m}$
Workpiece damping $c_w = 274.126\,\mathrm{Nsm}^{-1}$
Damping of the right wheel $c_r = 1.0 \times 10^4\,\mathrm{Nsm}^{-1}$
Damping of the left wheel $c_l = 1.0 \times 10^4\,\mathrm{Nsm}^{-1}$
Feed $f = 2.0 \times 10^{-6}\,\mathrm{mrev}^{-1}$
Surface speed of the right wheel $V_r = 35\,\mathrm{ms}^{-1}$
Surface speed of the left wheel $V_l = 35\,\mathrm{ms}^{-1}$
Exponential coefficient $\mu = 0.842$
Dimensionless ratio $g = 0.1$

Substituting $x_l(\tau) = x_{l0} + x_1(\tau)$, $x_r(\tau) = x_{r0} + x_2(\tau)$, and $x_p(\tau) = x_{p0} + x_3(\tau)$ in Eq. (6.65), one can study the grinding stability by studying $x_1(\tau)$, $x_2(\tau)$, and $x_3(\tau)$. The linear part is

$$\begin{aligned} Mx''(\tau) + Cx'(\tau) + Kx(\tau) &= Dx(\tau) + D_l x(\tau - \tau_l) + D_r x(\tau - \tau_r) \\ &+ D_w x(\tau - \tau_w) + D_h x\left(\tau - \frac{\tau_w}{2}\right), \end{aligned} \tag{6.66}$$

where $x(\tau) = (x_1(\tau), x_2(\tau), x_3(\tau))^{\mathrm{T}}$ is the state vector, with

$$\mathbf{M} = \begin{pmatrix} 1 & 0 & 0 \\ 0 & 1 & 0 \\ 0 & 0 & 1 \end{pmatrix}, \mathbf{C} = \begin{pmatrix} \xi_l & 0 & 0 \\ 0 & \xi_r & 0 \\ 0 & 0 & \xi_w \end{pmatrix}, \mathbf{K} = \begin{pmatrix} \kappa_l & 0 & 0 \\ 0 & \kappa_r & 0 \\ 0 & 0 & \kappa_w \end{pmatrix}. \tag{6.67}$$

In addition, with $c_1 = \mu \kappa_g \omega_w^\mu w_l d_{nl0}^{\mu-1}$, $c_2 = \mu \kappa_g \omega_w^\mu w_r d_{nr0}^{\mu-1}$, $c_3 = \mu \kappa_g \omega_w^\mu w_o d_{ol0}^{\mu-1}$, and $c_4 = \mu \kappa_g \omega_w^\mu w_o d_{nr0}^{\mu-1}$, the matrices are

$$\mathbf{D} = \begin{pmatrix} -\gamma_{wl}(c_1 + c_3) & 0 & \gamma_{wl}(c_1 + c_3) \\ 0 & -\gamma_{wr}(c_2 + c_4) & \gamma_{wr}(c_2 + c_4) \\ c_1 + c_3 & c_2 + c_4 & -(c_1 + c_2 + c_3 + c_4) \end{pmatrix}, \tag{6.68}$$

$$\mathbf{D}_l = \begin{pmatrix} \gamma_{wl}(c_1 + c_3) & 0 & -\gamma_{wl}(c_1 + c_3) \\ 0 & 0 & 0 \\ -(c_1 + c_3) & 0 & c_1 + c_3 \end{pmatrix}, \tag{6.69}$$

$$\mathbf{D}_r = \begin{pmatrix} 0 & 0 & 0 \\ 0 & \gamma_{wr}(c_2 + c_4) & -\gamma_{wr}(c_2 + c_4) \\ 0 & -(c_2 + c_4) & c_2 + c_4 \end{pmatrix}, \tag{6.70}$$

$$\mathbf{D}_w = \begin{pmatrix} \gamma_{wl}c_1 & 0 & -\gamma_{wl}c_1 \\ 0 & \gamma_{wr}c_2 & -\gamma_{wr}c_2 \\ -c_1 & -c_2 & c_1 + c_2 \end{pmatrix}, \tag{6.71}$$

$$\mathbf{D}_h = \begin{pmatrix} 0 & -\gamma_{wl}c_3 & \gamma_{wl}c_3 \\ -\gamma_{wr}c_4 & 0 & \gamma_{wr}c_4 \\ c_4 & c_3 & -(c_3 + c_4) \end{pmatrix}. \tag{6.72}$$

The stability boundary is determined through eigenvalue analysis, revealing the presence of two distinct "lobe" curves: Type I and Type II. As the overlap ratio (OR) increases, Type I curves rapidly ascend and vanish when OR surpasses 40%. Conversely, Type II curves show minimal movement. Consequently, the boundary exclusively features Type II curves, which are pushed downwards as the overlapped cutting zones expand. Especially for OR = 100%, the grinding stability can become worse than in the conventional case at certain workpiece speeds. Therefore, the most effective strategy for parallel grinding involves partial overlapping of the cutting zones of the two wheels. For instance, an OR of 60% results in a significantly larger stable region exclusively featuring Type II curves. Additionally, the number of "lobe" curves significantly widens the marginally stable area (Fig. 6.7).

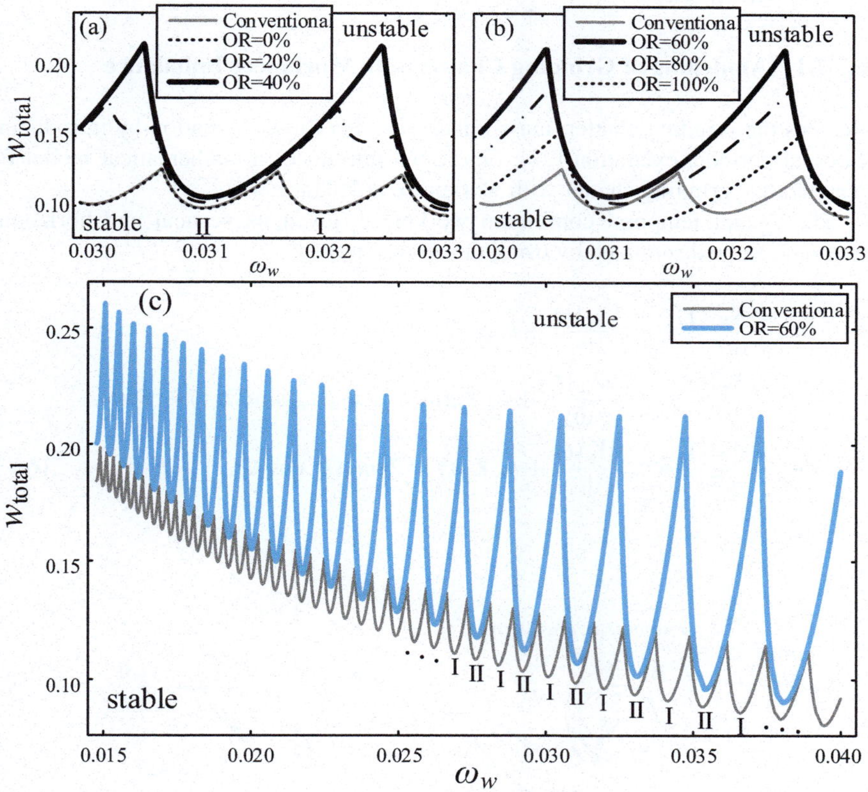

Fig. 6.7 Stability boundaries are. Increasing OR **a** lifts the "lobe" curves up for OR < 50%, but **b** pushes it down for OR > 50%. **c** An overlap ratio of 60% results in the highest stability boundary

6.2 Analysis of Nonlinear Cutting Chatter Based on Time Delay Model

6.2.1 Grinding Chatter with External Excitation—Workpiece Imbalance

6.2.1.1 Modelling of Grinding Chatter with Workpiece Imbalance

Besides the grinding chatter, mass imbalance also causes vibration by introducing external periodic excitation [24]. Figure 6.8 introduces a mathematical model for regenerative grinding chatter with workpiece imbalance.

The spatial–temporal continuum workpiece, given its vertical and horizontal motion, can be discretized by using the above method

$$
\begin{aligned}
&m_{\mathrm{g}}\frac{\mathrm{d}^2X_{\mathrm{g}}(t)}{\mathrm{d}t^2}+c_{\mathrm{g}}\frac{\mathrm{d}X_{\mathrm{g}}(t)}{\mathrm{d}t}+k_{\mathrm{g}}X_{\mathrm{g}}(t)=F_x(t),\\
&m_{\mathrm{w}}\frac{\mathrm{d}^2X_{\mathrm{p}}(t)}{\mathrm{d}t^2}+c_{\mathrm{W}}\frac{\mathrm{d}X_{\mathrm{p}}(t)}{\mathrm{d}t}+k_{\mathrm{w}}X_{\mathrm{p}}(t)=2em_{\mathrm{d}}\Omega_{\mathrm{W}}^2\sin(\Omega_{\mathrm{w}}t)-F_x(t),\\
&m_{\mathrm{w}}\frac{\mathrm{d}^2Y_{\mathrm{p}}(t)}{\mathrm{d}t^2}+c_{\mathrm{w}}\frac{\mathrm{d}Y_{\mathrm{p}}(t)}{\mathrm{d}t}+k_{\mathrm{w}}Y_{\mathrm{p}}(t)=2em_{\mathrm{d}}\Omega_{\mathrm{w}}^2\cos(\Omega_{\mathrm{w}}t)+F_y(t),
\end{aligned}
\tag{6.73}
$$

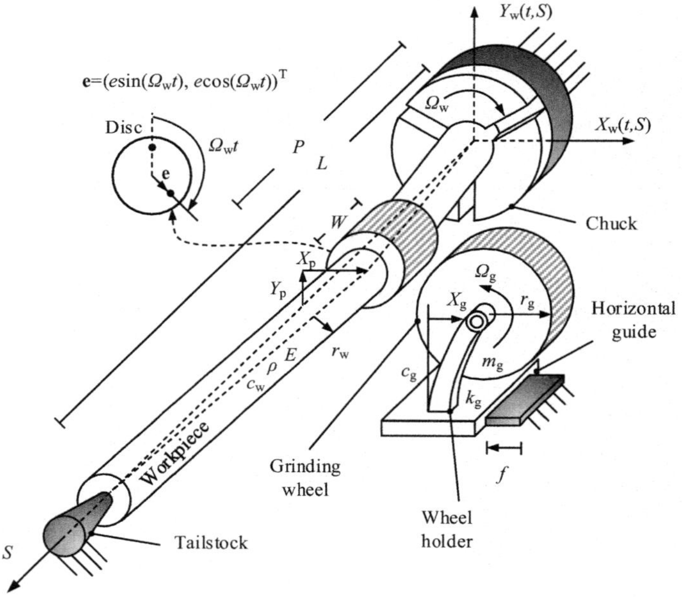

Fig. 6.8 Schematic of a plunge grinding process, which is of an eccentricity e

where m_{w} [kg] and k_{w} [N/m] are equivalent mass and stiffness. Moreover, it should be noticed that the terms, $2em_{\mathrm{d}}\Omega_{\mathrm{w}}^2\sin(\Omega_{\mathrm{w}}t)$ and $2em_{\mathrm{d}}\Omega_{\mathrm{w}}^2\cos(\Omega_{\mathrm{w}}t)$, represent the impact of the mass imbalance.

In addition to the excitation, another factor contributing to instability in Eq. (6.73) is the presence of $F_x(t)$ and $F_y(t)$, which represent the interactive grinding forces. The X- and Y-components of grinding forces, $F_x(t)$ and $F_y(t)$ are

$$\begin{aligned} F_x &= F_{\mathrm{n}}\cos\gamma + F_{\mathrm{t}}\sin\gamma, \\ F_y &= F_{\mathrm{n}}\sin\gamma - F_{\mathrm{t}}\cos\gamma, \end{aligned} \tag{6.74}$$

where F_{n} [N] and F_{n} [N] represent normal and tangential forces.

Each of the grinding wheel particles on the wheel engages with the workpiece surface, resulting in the generation of both cutting and frictional forces in the vertical and horizontal directions [10, 19]:

$$\begin{aligned} f_{\mathrm{ct}} &= \psi f_{\mathrm{cn}}, \\ f_{\mathrm{ft}} &= \mu f_{\mathrm{fn}}. \end{aligned} \tag{6.75}$$

Total normal and tangential grinding forces are represented by ref. [13]

$$\begin{aligned} F_{\mathrm{n}} &= WK_{\mathrm{c}}\frac{\Omega_{\mathrm{w}}}{\Omega_{\mathrm{g}}}D_{\mathrm{g}}(t) + WK_{\mathrm{f}}\frac{\Omega_{\mathrm{w}}}{\Omega_{\mathrm{g}}}D_{\mathrm{g}}(t)^{\frac{1}{2}}, \\ F_{\mathrm{t}} &= \psi WK_{\mathrm{c}}\frac{\Omega_{\mathrm{w}}}{\Omega_{\mathrm{g}}}D_{\mathrm{g}}(t) + \mu WK_{\mathrm{f}}\frac{\Omega_{\mathrm{w}}}{\Omega_{\mathrm{g}}}D_{\mathrm{g}}(t)^{\frac{1}{2}}, \end{aligned} \tag{6.76}$$

where K_{c} N/m^2 and K_{f} N/m$^{-3/2}$ are proportional coefficients. The doubly regenerative effect yields the following depth

$$\begin{aligned} D_{\mathrm{g}} = {} & f\cos(\gamma(t)) + \big(X_{\mathrm{p}}(t)\cos(\gamma(t)) - X_{\mathrm{g}}(t)\cos(\gamma(t)) - Y_{\mathrm{p}}(t)\sin(\gamma(t))\big) \\ & - \big(X_{\mathrm{p}}(t-T_{\mathrm{w}})\cos(\gamma(t-T_{\mathrm{w}})) - X_{\mathrm{g}}(t-T_{\mathrm{w}})\cos(\gamma(t-T_{\mathrm{w}})) \\ & - Y_{\mathrm{p}}(t-T_{\mathrm{w}})\sin(\gamma(t-T_{\mathrm{w}}))\big) \\ & - g\big(X_{\mathrm{p}}\big(t-T_{\mathrm{g}}\big)\cos\big(\gamma\big(t-T_{\mathrm{g}}\big)\big) - X_{\mathrm{g}}\big(t-T_{\mathrm{g}}\big)\cos\big(\gamma\big(t-T_{\mathrm{g}}\big)\big) \\ & - Y_{\mathrm{p}}\big(t-T_{\mathrm{g}}\big)\sin\big(\gamma\big(t-T_{\mathrm{g}}\big)\big)\big), \end{aligned} \tag{6.77}$$

where g is a small dimensionless ratio. Moreover, the state-dependent delays are governed by

$$\begin{aligned} 2\pi &= -\gamma(t) + \gamma(t-T_{\mathrm{w}}) + T_{\mathrm{w}}\Omega_{\mathrm{w}}, \\ 2\pi &= \gamma(t) - \gamma\big(t-T_{\mathrm{g}}\big) + T_{\mathrm{g}}\Omega_{\mathrm{g}}. \end{aligned} \tag{6.78}$$

In addition to accounting for the regenerative effect, the model also incorporates friction using a basic Coulomb model, where the frictional coefficient μ is selected as

$$\mu = \mathrm{sign}(V_\mathrm{f})\mu_\mathrm{d}. \tag{6.79}$$

Before the analysis, Eq. (6.73) is nondimensionalized by introducing the following dimensionless parameters:

$$\begin{aligned}
\xi_\mathrm{g} &= \frac{c_\mathrm{g}\sqrt{m_\mathrm{w}}}{m_\mathrm{g}\sqrt{k_\mathrm{w}}},\ \xi_\mathrm{w} = \frac{c_\mathrm{w}\sqrt{m_\mathrm{w}}}{m_\mathrm{w}\sqrt{k_\mathrm{w}}},\\
\kappa_\mathrm{c} &= \frac{K_\mathrm{c}m_\mathrm{w}}{k_\mathrm{w}m_\mathrm{g}}(R_\mathrm{g}+R_\mathrm{p}),\ \kappa_\mathrm{f} = \frac{K_\mathrm{f}m_\mathrm{w}}{k_\mathrm{w}m_\mathrm{g}}\sqrt{R_\mathrm{g}+R_\mathrm{p}},\ \kappa_\mathrm{g} = \frac{k_\mathrm{g}m_\mathrm{w}}{k_\mathrm{w}m_\mathrm{g}},\\
\gamma_\mathrm{w} &= \frac{m_\mathrm{w}}{m_\mathrm{g}},\ \nu = \frac{f}{R_\mathrm{p}+R_\mathrm{g}},\ \delta = \frac{8e\pi^2 m_\mathrm{d}}{m_\mathrm{w}(R_\mathrm{p}+R_\mathrm{g})},\\
r_\mathrm{p} &= \frac{R_\mathrm{p}}{R_\mathrm{p}+R_\mathrm{g}},\ r_\mathrm{g} = \frac{R_\mathrm{g}}{R_\mathrm{p}+R_\mathrm{g}},\ p = \frac{P}{L},\\
\omega_\mathrm{w} &= \Omega_\mathrm{w}\sqrt{\frac{m_\mathrm{w}}{k_\mathrm{w}}},\ \omega_\mathrm{g} = \Omega_\mathrm{g}\sqrt{\frac{m_\mathrm{w}}{k_\mathrm{w}}},
\end{aligned} \tag{6.80}$$

and variables:

$$\begin{aligned}
\tau &= t\sqrt{\frac{k_\mathrm{w}}{m_\mathrm{w}}},\ \tau_\mathrm{w} = T_\mathrm{w}\sqrt{\frac{k_\mathrm{w}}{m_\mathrm{w}}},\ \tau_\mathrm{g} = T_\mathrm{g}\sqrt{\frac{k_\mathrm{w}}{m_\mathrm{w}}},\\
x_\mathrm{g}(\tau) &= \frac{X_\mathrm{g}(t)}{R_\mathrm{g}+R_\mathrm{p}},\ x_\mathrm{W}(\tau) = \frac{X_\mathrm{W}(t)}{R_\mathrm{g}+R_\mathrm{p}},\ y_\mathrm{W}(\tau) = \frac{Y_\mathrm{w}(t)}{R_\mathrm{g}+R_\mathrm{p}},\\
\gamma &= \tan^{-1}\left(\frac{y_\mathrm{p}(\tau)}{1-\nu-x_\mathrm{p}(\tau)+x_\mathrm{g}(\tau)}\right),\\
d_\mathrm{g} &= \big(\nu+x_\mathrm{p}(\tau)-x_\mathrm{g}(\tau)\big)\cos(\gamma(\tau)) - y_\mathrm{p}(t)\sin(\gamma(t))\\
&\quad - \big(x_\mathrm{p}(\tau-\tau_\mathrm{w})-x_\mathrm{g}(\tau-\tau_\mathrm{w})\big)\cos(\gamma(\tau-\tau_\mathrm{w}))\\
&\quad + y_\mathrm{p}(\tau-\tau_\mathrm{w})\sin(\gamma(\tau-\tau_\mathrm{w}))\\
&\quad - g\big(x_\mathrm{p}(\tau-\tau_\mathrm{g})-x_\mathrm{g}(\tau-\tau_\mathrm{g})\big)\cos\big(\gamma(\tau-\tau_\mathrm{g})\big)\\
&\quad + g y_\mathrm{p}(\tau-\tau_\mathrm{g})\sin\big(\gamma(\tau-\tau_\mathrm{g})\big),\\
v_\mathrm{f} &= \omega_\mathrm{g}r_\mathrm{g} - \omega_\mathrm{w}r_\mathrm{p} + \frac{\mathrm{d}y_\mathrm{p}(\tau)}{\mathrm{d}\tau}\cos(\gamma) - \left(\frac{\mathrm{d}x_\mathrm{g}(\tau)}{\mathrm{d}\tau} - \frac{\mathrm{d}x_\mathrm{p}(\tau)}{\mathrm{d}\tau}\right)\sin(\gamma).
\end{aligned} \tag{6.81}$$

As a result, Eq. (6.107) is transformed into

$$\begin{aligned}
\frac{\mathrm{d}^2x_\mathrm{g}(\tau)}{\mathrm{d}\tau^2} + \xi_\mathrm{g}\frac{\mathrm{d}x_\mathrm{g}(\tau)}{\mathrm{d}\tau} + \kappa_\mathrm{g}x_\mathrm{g}(\tau) &= w\kappa_\mathrm{f}\frac{\tau_\mathrm{g}}{\tau_\mathrm{w}}\sqrt{d_\mathrm{g}}(\cos(\gamma)+\mu\sin(\gamma))\\
&\quad + w\kappa_\mathrm{c}\frac{\tau_\mathrm{g}}{\tau_\mathrm{w}}d_\mathrm{g}(\cos(\gamma)+\psi\sin(\gamma)),
\end{aligned}$$

$$
\begin{aligned}
\frac{\mathrm{d}^2 x_{\mathrm{p}}(\tau)}{\mathrm{d}\tau^2} + \xi_{\mathrm{w}} \frac{\mathrm{d}x_{\mathrm{p}}(\tau)}{\mathrm{d}\tau} + x_{\mathrm{p}}(\tau) = &-\frac{w\kappa_{\mathrm{f}}}{\gamma_{\mathrm{w}}} \frac{\tau_{\mathrm{g}}}{\tau_{\mathrm{w}}} \sqrt{d_{\mathrm{g}}}(\cos(\gamma) + \mu \sin(\gamma)) \\
&- \frac{w\kappa_{\mathrm{c}}}{\gamma_{\mathrm{w}}} \frac{\tau_{\mathrm{g}}}{\tau_{\mathrm{w}}} d_{\mathrm{g}}(\cos(\gamma) + \psi \sin(\gamma)) \\
&+ \frac{\delta}{\tau_{\mathrm{w0}}^2} \sin\left(\frac{2\pi}{\tau_{\mathrm{w0}}}\tau\right), \\
\frac{\mathrm{d}^2 y_{\mathrm{p}}(\tau)}{\mathrm{d}\tau^2} + \xi_{\mathrm{w}} \frac{\mathrm{d}y_{\mathrm{p}}(\tau)}{\mathrm{d}\tau} + y_{\mathrm{p}}(\tau) = &\frac{w\kappa_{\mathrm{f}}}{\gamma_{\mathrm{w}}} \frac{\tau_{\mathrm{g}}}{\tau_{\mathrm{w}}} \sqrt{d_{\mathrm{g}}}(\sin(\gamma) - \mu \cos(\gamma)) \\
&+ \frac{w\kappa_{\mathrm{c}}}{\gamma_{\mathrm{w}}} \frac{\tau_{\mathrm{g}}}{\tau_{\mathrm{w}}} d_{\mathrm{g}}(\sin(\gamma) - \psi \cos(\gamma)) \\
&+ \frac{\delta}{\tau_{\mathrm{w0}}^2} \cos\left(\frac{2\pi}{\tau_{\mathrm{w0}}}\tau\right).
\end{aligned} \tag{6.82}
$$

6.2.1.2 Grinding Vibration

Figure 6.9 presents the stability lobes, i.e., which represent stability boundaries. When the workpiece mass is uniformly distributed ($\delta = 0$), forced vibration does not occur during grinding, and only self-induced chatter is observed. Conversely, when imbalance is introduced ($\delta > 0$), chatter becomes perturbed, leading to diverse grinding dynamics. To illustrate, system parameters are chosen along Arrow I to analyze nonlinear chatter behavior.

The associated bifurcation diagram is presented in Fig. 6.10a. The blue dots represent $\max(d_{\mathrm{g}})$ achieved through forward simulation. The red dots correspond to the backward simulation. The chatter is triggered by a sub-critical Hopf bifurcation [13].

Then, a small workpiece imbalance ($\delta = 0.1$) is introduced into the grinding along Arrow I, and the corresponding bifurcation diagram is illustrated in Fig. 6.12. The bifurcation pattern ($\max(d_{\mathrm{g}})$ varies with respect to τ_{w0}) in Fig. 6.12a is very similar to that in Fig. 6.11a, although the periodic chatter transitions into a quasi-periodic state, the stable grinding experiences forced vibrations.

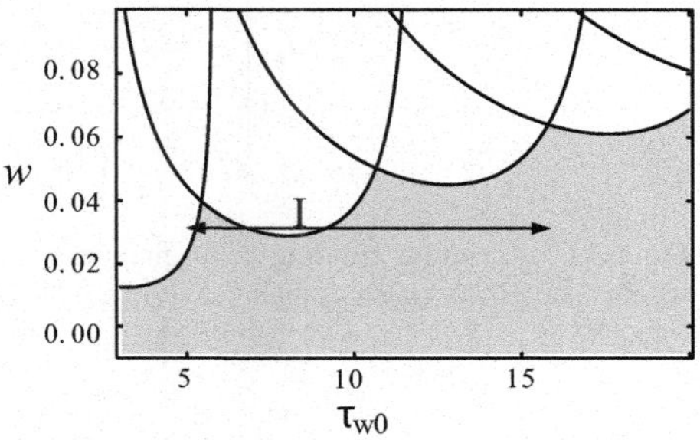

Fig. 6.9 Stability boundaries for the grinding process

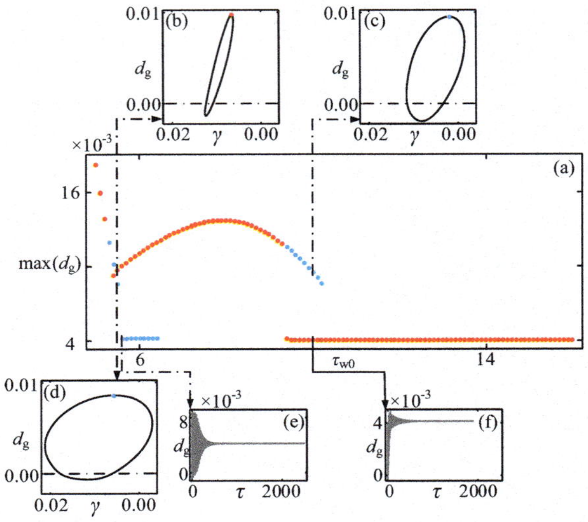

Fig. 6.10 Grinding depth varies with respect to τ_{w0}, which corresponds to Arrow II ($w = 0.033$, $\tau_{w0} \in [5, 16]$)

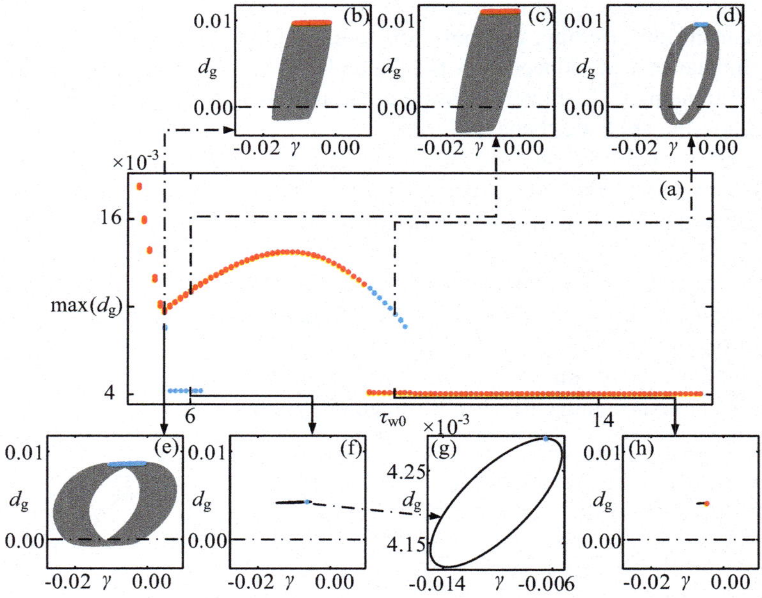

Fig. 6.11 Maximum grinding depth max(d_g) varies along Arrow II ($w = 0.033$, $\tau_{w0} \in [5, 16]$), where a small workpiece imbalance ($e = 0.1$) is considered

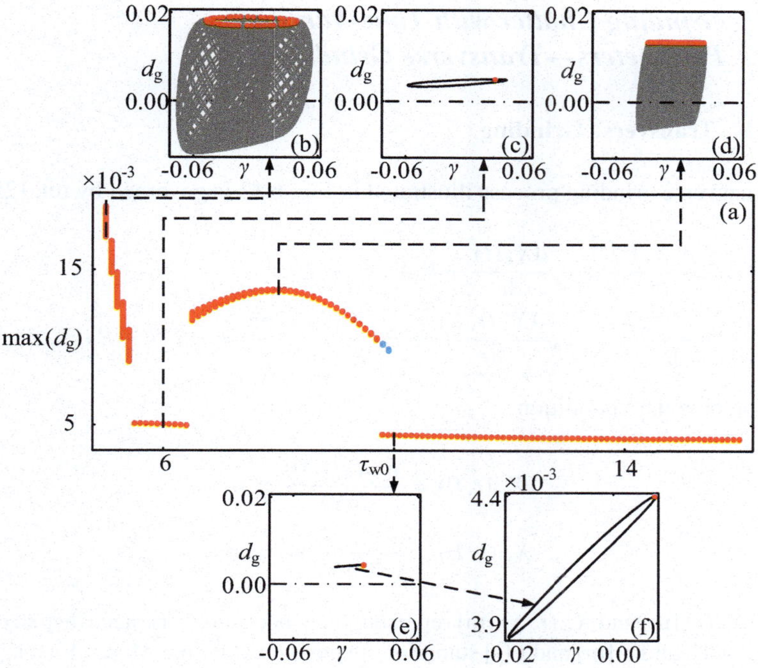

Fig. 6.12 Maximum grinding depth $\max(d_g)$ varies along Arrow II ($w = 0.033$, $\tau_{w0} \in [5, 16]$), where a large workpiece imbalance ($\delta = 1.0$) is considered

Next, we introduce a significant workpiece imbalance ($\delta = 1.0$) rather than a minor one. An apparent reduction in regions with coexisting dynamics is observed. Specifically, the coexistence between quasi-periodic chatter and forced vibration is restricted only for $\tau_{w0} \in [9.8, 9.9]$. Furthermore, it's evident that the substantial workpiece imbalance suppresses high-amplitude chatter within the τ_{w0} range of [5.5, 6.4]. Comparing this with Fig. 6.11c, there is a notable reduction in the fluctuation of the grinding depth, ensuring continuous contact between the wheel and the workpiece (d_g keeps positive), while the contact angle vibrates significantly.

6.2.2 *Grinding Chatter with Time-Varying Parameters—Transverse Grinding*

6.2.2.1 Transverse Grinding

The transverse grinding process illustrated in Fig. 6.13 is governed by ref. [25]

$$\begin{aligned} m_g\frac{\mathrm{d}^2X_g(t)}{\mathrm{d}t^2}+c_g\frac{\mathrm{d}X_g(t)}{\mathrm{d}t}+k_gX_g(t)&=F_g,\\ \rho A\frac{\partial^2X_w(t,S)}{\partial t^2}+c_w\frac{\partial X_w(t,S)}{\partial t}+EI\frac{\partial^4X_w(t,S)}{\partial S^4}&=-\delta(S-P)F_g, \end{aligned} \tag{6.83}$$

with the boundary condition

$$\begin{aligned} X_w(t,0)&=\frac{\partial^2X_w(t,0)}{\partial S^2}=0,\\ X_w(t,L)&=\frac{\partial^2X_w(t,L)}{\partial S^2}=0, \end{aligned} \tag{6.84}$$

where $X_g(t)$ [m] and $X_w(t,S)$ [m] represent the wheel and workpiece displacements respectively, and S [m] and t [s] stand for the position and time. Here, $A=\pi r_w^2$ [m^2] is the cross-sectional area of the workpiece, $I=\frac{\pi r_w^4}{4}$ [m^4] the inertia moment of the workpiece, $\delta(S-P)$ [m^{-1}] the Dirac delta function (Fig. 6.14).

Within Region I, the grinding wheel gradually wears down, grinding the new workpiece surface. The workpiece in Region II has already been ground in the previous period, and its surface is newly regenerated. Thus, D_g is

$$D_g=\begin{cases} D_{g1}=\Delta(t,P)-g\Delta\big(t-T_g,P\big) & \text{in Region I,}\\ D_{g2}=D_{g1}-\Delta(t-T_w,P) & \text{in Region II,}\end{cases} \tag{6.85}$$

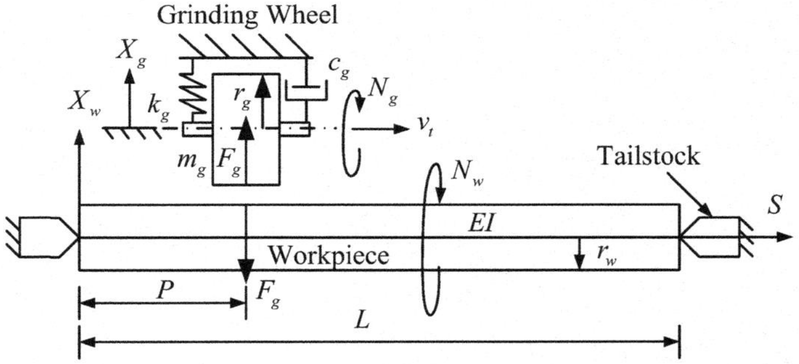

Fig. 6.13 Transverse grinding process

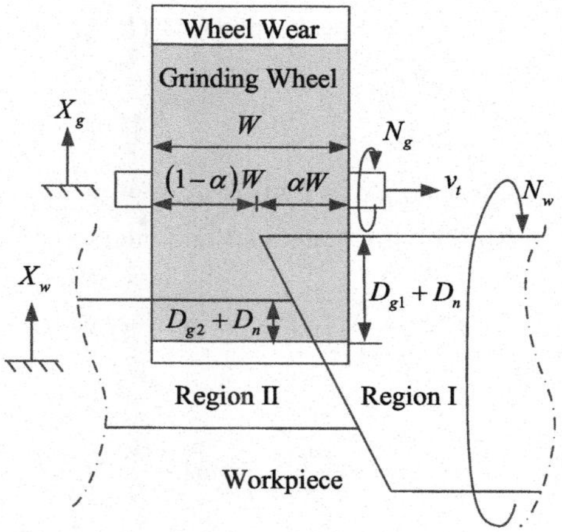

Fig. 6.14 Wheel-workpiece interactions

where T_g (s) and T_w (s) are the rotational periods. The contacting widths in Regions I and II are

$$W = \begin{cases} W_1 = \alpha W = v_t T_w & \text{in Region I,} \\ W_2 = (1-\alpha)W = W - v_t T_w & \text{in Region II,} \end{cases} \tag{6.86}$$

where v_t (ms^{-1}) is the transverse speed, and $\alpha = \frac{v_t T_w}{W}$ a dimensionless ratio.

The grinding forces are

$$F_{g1} = \begin{cases} k_c \alpha \left(D_n + D_{g1}\right)^{\mu} & \text{if } \mathrm{D_n} + \mathrm{D_{g1}} \geq 0, \\ 0 & \text{if } \mathrm{D_n} + \mathrm{D_{g1}} \leq 0, \end{cases} \tag{6.87}$$

and

$$F_{g2} = \begin{cases} k_c (1-\alpha)\left(D_n + D_{g2}\right)^{\mu} & \text{if } \mathrm{D_n} + \mathrm{D_{g2}} \geq 0, \\ 0 & \text{if } \mathrm{D_n} + \mathrm{D_{g2}} \leq 0, \end{cases} \tag{6.88}$$

where $k_c = WKC^{\nu}\left(\frac{r_w}{r_g}\right)^{2\mu-1}\left(\frac{\omega_w}{\omega_g}\right)^{2\mu-1} D_e^{1-\mu}$. In the aggregate, we have $F_g = F_{g1} + F_{g2}$.

The workpiece deformation is expanded as

$$X_w(t, S) = \sum_{i=1}^{n} X_i(t)\sin\left(\frac{iS\pi}{L}\right). \tag{6.89}$$

The static displacements are denoted as $X_g(t) = X_g^{(0)}$ and $X_w(t, S) = X_w^{(0)}(S) = \sum_{i=1}^{n} X_i^{(0)} \sin\left(\frac{iS\pi}{L}\right)$. Consequently, Eq. (6.83) becomes

$$
\begin{aligned}
k_g X_g^{(0)} &= k_c\Big(\alpha\big(D_n-(1-g)\Delta^{(0)}(P)\big)^{\mu}+(1-\alpha)\big(D_n+g\Delta^{(0)}(P)\big)^{\mu}\Big),\\
EI\frac{\partial^4 X_w^{(0)}(S)}{\partial S^4} &= -\delta(S-P)k_g X_g^{(0)},
\end{aligned}
\tag{6.90}
$$

where $\Delta^{(0)}(P)=X_w^{(0)}(P)-X_g^{(0)}$ is static relative deformation.

Multiplying both sides and integrating them yield

$$
\begin{aligned}
X_g^{(0)} &= \frac{k_c}{k_g}\Big(\alpha\big(D_n-(1-g)\Delta^{(0)}(P)\big)^{\mu}+(1-\alpha)\big(D_n+g\Delta^{(0)}(P)\big)^{\mu}\Big),\\
&\cdots,\\
X_i^{(0)} &= -\frac{2L^3k_g}{i^4\pi^4EI}\sin\left(\frac{iP\pi}{L}\right)X_g^{(0)},\\
&\cdots,
\end{aligned}
\tag{6.91}
$$

where $i=1,2,\cdots,n$. It yields

$$
\begin{aligned}
\Delta^{(0)}(P) = X_w^{(0)}(P)-X_g^{(0)} &= \sum_{i=1}^{n}X_i^{(0)}\sin\left(\frac{iP\pi}{L}\right)-X_g^{(0)}\\
&= -\left(\frac{k_c}{k_a}+\frac{2L^3k_c}{\pi^4EI}\sum_{i=1}^{n}\frac{\sin^2\left(\frac{iP\pi}{L}\right)}{i^4}\right)\Big(\alpha\big(D_n-(1-g)\Delta^{(0)}(P)\big)^{\mu}\\
&\quad +(1-\alpha)\big(D_n+g\Delta^{(0)}(P)\big)^{\mu}\Big).
\end{aligned}
\tag{6.92}
$$

For further analysis of the grinding stability, the equilibrium point is relocated as follows:

$$
\begin{aligned}
&X_g^*(t)=X_g(t)-X_g^{(0)},\ X_w^*(t,S)=X_w(t,S)-X_w^{(0)}(S),\\
&X_i^*(t)=X_i(t)-X_i^{(0)},\ \Delta^*(t,P)=\Delta(t,P)-\Delta^{(0)}(P),\\
&D_{g1}^*=\Delta^*(t,P)-g\Delta^*\big(t-T_g,P\big),\ D_{g2}^*=D_{g1}^*-\Delta^*(t-T_w,P),\\
&D_{g1}^{(0)}=(1-g)\Delta^{(0)}(P),\ D_{g2}^{(0)}=-g\Delta^{(0)}(P),
\end{aligned}
\tag{6.93}
$$

where $i=1,2,\cdots,n$. Thus, Eq. (6.83) becomes

$$
\begin{aligned}
&m_g\frac{\mathrm{d}^2X_g^*(t)}{\mathrm{d}t^2}+c_g\frac{\mathrm{d}X_g^*(t)}{\mathrm{d}t}+k_gX_g^*(t)=-k_gX_g^{(0)}\\
&\qquad +k_c\Big(\alpha\Big(D_n+D_{g1}^{(0)}+D_{g1}^*\Big)^{\mu}+(1-\alpha)\Big(D_n+D_{g2}^{(0)}+D_{g2}^*\Big)^{\mu}\Big),\\
&\rho A\frac{\partial^2X_w^*(t,S)}{\partial t^2}+c_w\frac{\partial X_w^*(t,S)}{\partial t}+EI\frac{\partial^4X_w^*(t,S)}{\partial S^4}=-EI\frac{\partial^4X_w^{(0)}(S)}{\partial S^4}\ g=0.1
\end{aligned}
$$

$$
\begin{aligned}
&- \delta(S - P)k_c\Big(\alpha\Big(D_n + D_{g1}^{(0)} + D_{g1}^{*}\Big)^{\mu} \\
&+(1-\alpha)\Big(D_n + D_{g2}^{(0)} + D_{g2}^{*}\Big)^{\mu}\Big).
\end{aligned} \tag{6.94}
$$

Next, the dimensionless variables and parameters are brought into consideration:

$$
\begin{aligned}
&\tau = t\sqrt{\frac{k_g}{m_g}},\ \tau_w = T_w\sqrt{\frac{k_g}{m_g}},\ \tau_g = T_g\sqrt{\frac{k_g}{m_g}},\\
&s = \frac{S}{D_n},\ y_g = \frac{X_g^{*}}{D_n},\ y_w = \frac{X_w^{*}}{D_n},\ d_{g1} = \frac{D_{g1}^{*}}{D_n},\\
&d_{g2} = \frac{D_{g2}^{*}}{D_n},\ d_{g1}^{(0)} = \frac{D_{g1}^{(0)}}{D_n},\ d_{g2}^{(0)} = \frac{D_{g2}^{(0)}}{D_n},\ l = \frac{L}{D_n},\ p = \frac{P}{D_n},\\
&y_g^{(0)} = \frac{X_g^{(0)}}{D_n},\ y_w^{(0)}(s) = \frac{x_w^{(0)}(S)}{D_n},\ \epsilon\xi_g = \frac{c_g}{k_g}\sqrt{\frac{k_g}{m_g}},\ \gamma = \frac{2m_g}{\rho A l D_n},\\
&\epsilon\xi_w = \frac{c_w}{\rho A}\sqrt{\frac{m_g}{k_g}},\ \kappa_w = \frac{m_g E I}{k_g \rho A D_n^4}\frac{\pi^4}{l^4},\ \epsilon\kappa_c = \frac{k_c}{k_g}D_n^{\mu-1}\left(\frac{\tau_g}{\tau_w}\right)^{2\mu-1}.
\end{aligned} \tag{6.95}
$$

Then Eq. (6.94) is transformed into

$$
\begin{aligned}
&\frac{\mathrm{d}^2 y_g(\tau)}{\mathrm{d}\tau^2} + y_g(\tau) = -y_g^{(0)} - \epsilon\xi_g\frac{\mathrm{d}y_g(\tau)}{\mathrm{d}\tau}\\
&\quad + \epsilon\kappa_c\Big(\alpha\Big(1 + d_{g1}^{(0)} + d_{g1}\Big)^{\mu} + (1-\alpha)\Big(1 + d_{g2}^{(0)} + d_{g2}\Big)^{\mu}\Big),\\
&\frac{\partial^2 y_w(\tau, s)}{\partial\tau^2} + \frac{l^4}{\pi^4}\kappa_w\frac{\partial^4 y_w(\tau, s)}{\partial s^4} = -\frac{l^4}{\pi^4}\kappa_w\frac{\partial^4 y_w^{(0)}}{\partial s^4} - \epsilon\xi_w\frac{\partial y_w(\tau, s)}{\partial\tau}\\
&\quad - \epsilon\left(\frac{lD_n}{2}\delta(s-p)\right)\gamma\kappa_c\Big(\alpha\Big(1 + d_{g1}^{(0)} + d_{g1}\Big)^{\mu}\\
&\quad +(1-\alpha)\Big(1 + d_{g2}^{(0)} + d_{g2}\Big)^{\mu}\Big).
\end{aligned} \tag{6.96}
$$

The boundary condition is changed into

$$
y_w(t, 0) = \frac{\partial^2 y_w(t, 0)}{\partial s^2} = 0,\ y_w(t, l) = \frac{\partial^2 y_w(t, l)}{\partial s^2} = 0. \tag{6.97}
$$

Subsequently, applying the Galerkin projection of Eq. (6.96) onto the basis outlined in Eq. (6.94) results in:

$$
\begin{aligned}
\frac{\mathrm{d}^2 y_g(\tau)}{\mathrm{d}\tau^2} + y_g(\tau) = {} & -y_g^{(0)} - \epsilon\xi_g \frac{\mathrm{d}y_g(\tau)}{\mathrm{d}\tau} \\
& + \epsilon\kappa_c\alpha\left(1 + d_{g1}^{(0)} + y_g(\tau) - \sum_{i=1}^{n} y_i(\tau)\sin\left(\frac{ip\pi}{l}\right)\right. \\
& \left. -g y_g(\tau - \tau_g) + g\sum_{i=1}^{n} y_i(\tau - \tau_g)\sin\left(\frac{ip\pi}{l}\right)\right)^{\mu} \\
& + \epsilon\kappa_c(1-\alpha)\left(1 + d_{g2}^{(0)} + y_g(\tau)\right. \\
& - \sum_{i=1}^{n} y_i(\tau)\sin\left(\frac{ip\pi}{l}\right) - g y_g(\tau - \tau_g) \\
& + g\sum_{i=1}^{n} y_i(\tau - \tau_g)\sin\left(\frac{ip\pi}{l}\right) - y_g(\tau - \tau_w) \\
& \left. + \sum_{i=1}^{n} y_i(\tau - \tau_w)\sin\left(\frac{ip\pi}{l}\right)\right)^{\mu}, \\
& \cdots, \\
\frac{\mathrm{d}^2 y_i(\tau)}{\mathrm{d}\tau^2} + i^4\kappa_w y_i(\tau) = {} & -\kappa_w y_w^{(0)} - \epsilon\xi_w \frac{\mathrm{d}y_i(\tau)}{\mathrm{d}\tau} \\
& - \gamma\sin\left(\frac{ip\pi}{l}\right)\kappa_c\alpha\left(1 + d_{g1}^{(0)} + y_g(\tau) - \sum_{i=1}^{n} y_i(\tau)\sin\left(\frac{ip\pi}{l}\right)\right. \\
& \left. -g y_g(\tau - \tau_g) + g\sum_{i=1}^{n} y_i(\tau - \tau_g)\sin\left(\frac{ip\pi}{l}\right)\right)^{\mu} \\
& - \gamma\sin\left(\frac{ip\pi}{l}\right)\epsilon\kappa_c(1-\alpha)\left(1 + d_{g2}^{(0)} + y_g(\tau)\right. \\
& - \sum_{i=1}^{n} y_i(\tau)\sin\left(\frac{ip\pi}{l}\right) - g y_g(\tau - \tau_g) \\
& + g\sum_{i=1}^{n} y_i(\tau - \tau_g)\sin\left(\frac{ip\pi}{l}\right) - y_g(\tau - \tau_w) \\
& \left. + \sum_{i=1}^{n} y_i(\tau - \tau_w)\sin\left(\frac{ip\pi}{l}\right)\right)^{\mu}, \\
& \cdots,
\end{aligned}
\tag{6.98}
$$

where $i = 1, 2, \cdots, n$. The position of the wheel, denoted as p, varies over time owing to the transverse movement of the wheel. Nevertheless, the transverse speed v_t of the wheel is typically quite low and can be regarded as quasi-static.

6.2.2.2 Grinding Stability

The physical parameters are selected as

$$\begin{aligned} m_g &= 20\text{ kg}, c_g = 200\text{ Nsm}^{-1}, k_g = 6.4\times10^8\text{ Nm}^{-1} \\ c_w &= 800\text{ Nsm}^{-2}, E = 2.06\times10^{11}\text{ Nm}^{-2} \\ r_w &= 0.05\text{ m}, L = 0.5\text{ m}, D_n = 10^{-5}\text{m}, r_g = 0.25\text{ m}, \\ v_t &= 10^{-3}\text{ms}^{-1}, W = 0.04\text{ m}, KC^{\upsilon} = 4.1\times10^6\text{Nm}^{-2}, \\ \rho &= 7850\text{ kgm}^{-3}, N_g = 5000\text{ rpm}, g = 0.002, \mu = 0.7. \end{aligned} \tag{6.99}$$

Thus, one has

$$\begin{aligned} A &= \pi r_w^2 = 7.85\times10^{-3}\left(\text{m}^2\right), I = \frac{\pi r_g^4}{4} = 4.91\times10^{-6}\left(\text{m}^4\right), \\ D_e &= 2\frac{r_w r_g}{r_w + r_g} = \frac{1}{12}(\text{m}). \end{aligned} \tag{6.100}$$

And

$$\begin{aligned} \kappa_w &= 0.8, \gamma = 1.3, \tau_g = 67.9, \alpha = 4.4\times10^{-6}\tau_w, \\ \epsilon\xi_g &= 0.0018, \epsilon\xi_w = 0.0023, \epsilon\kappa_c = \frac{0.0208}{\tau_w^{0.4}}. \end{aligned} \tag{6.101}$$

By using the previously employed eigenvalue analysis and continuation algorithm, we determine the stability of the grinding process incorporating the primary mode of the workpiece.

Figure 6.15 illustrates the relationship between the wheel position p and the time delay τ_w. It features chatter boundaries that distinguish chatter-free regions (grey) from chatter regions (white). Figure 6.15 identifies two distinct types of chatter regions: one related to the wheel and another to the workpiece. Figure 6.15 illustrates that the stability of the wheel is not affected by the wheel position p. Conversely, the workpiece stability is notably influenced by p. As depicted, the workpiece's chatter regions consistently cluster around $p = \frac{l}{2}$, indicating a tendency for the workpiece to vibrate when the wheel approaches the workpiece's center. With the identification of both chatter-free and chatter regions, further examination is required to delve into the diverse chatter behaviors.

6.2.2.3 Nonlinear Grinding Chatter

In the chatter analysis, we will treat the wheel position p as quasi-static because the transverse speed v_t is exceedingly small. To explore the chatter phenomenon, we will need to follow several subsequent steps. First, for a fixed p, we will determine the

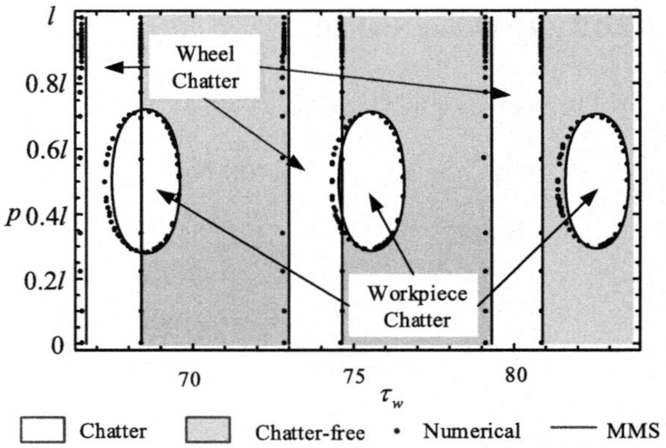

Fig. 6.15 Stability boundaries serve to demarcate regions where chatter occurs and regions that remain chatter-free

chatter amplitude in order to create the bifurcation diagrams. Subsequently, we will monitor the wheel position p throughout the grinding process. Taking into account both p and the bifurcation diagrams, we will proceed to predict the chatter motions. Finally, we will conduct simulations, capturing the time series of d_2 and incorporating it into the bifurcation diagrams to validate the quasi-static analysis.

During the grinding process, the wheel continually traverses across the workpiece in a transverse manner. Consequently, the wheel position p varies between 0 and l periodically. The wheel advances from one tailstock ($p = 0$) to the other ($p = l$) at a slow speed v_t. Upon reaching l, it reverses course and returns to the starting point. This cyclic process is then repeated until the conclusion of the grinding operation.

Corresponding to different chatter amplitude distributions, the following cases are selected:

$$\text{Case I}: \tau_w = 84, \text{Case II}: \tau_w = 80, \text{Case III}: \tau_w = 80.8,$$
$$\text{Case IV}: \tau_w = 82.6, \text{Case V}: \tau_w = 68, \text{Case VI}: \tau_w = 68.34.$$

Amplitude distributions concerning p are calculated and depicted in Fig. 6.16. In Fig. 6.16a, the grinding process remains free of chatter. Conversely, Fig. 6.16b consistently exhibits wheel chatter throughout the grinding process. In Fig. 6.16c, both stable grinding and wheel chatter coexist simultaneously. Figure 6.16d–f illustrate that workpiece chatter is only present near the center of the workpiece. To be more precise, Fig. 6.16d depicts the transition between workpiece chatter and stable grinding. Figure 6.16e demonstrates the coexistence of workpiece and wheel chatter without the presence of chatter-free grinding. In Fig. 6.16f, all three possible dynamics are concurrently presented in the same graph.

By means of the bifurcation diagrams, we aim to delve into the chattering phenomenon during transverse grinding. To achieve this, we estimate chatter amplitudes by monitoring the changes in p within the bifurcation diagram. Subsequently, we incorporate the resulting time series into the bifurcation diagrams, as displayed in Fig. 6.17. These diagrams reveal that the time series of the dynamic depth of

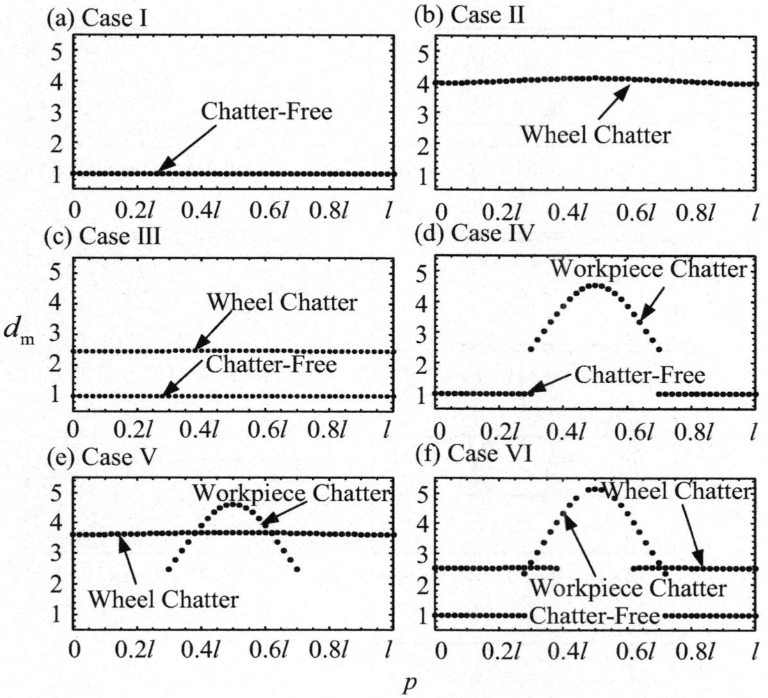

Fig. 6.16 Maximum depth of cut d_m with respect to wheel position p

cut, denoted as d_2, is encompassed within the bifurcation diagrams. In Fig. 6.17a, we observe the chatter-free transverse grinding process, while Fig. 6.17b showcases wheel chatter. Figure 6.17c and d reveal that both chatter-free grinding and wheel chatter manifest in identical scenarios. Figure 6.17e portrays the occurrence of workpiece chatter, which exhibits an intermittent pattern. In Fig. 6.17f, wheel chatter is observed, bearing a resemblance to the patterns seen in Fig. 6.17b and d, despite being associated with different bifurcation diagrams. Figure 6.17g is the most intricate, demonstrating transitions among chatter-free grinding, wheel chatter, and workpiece chatter.

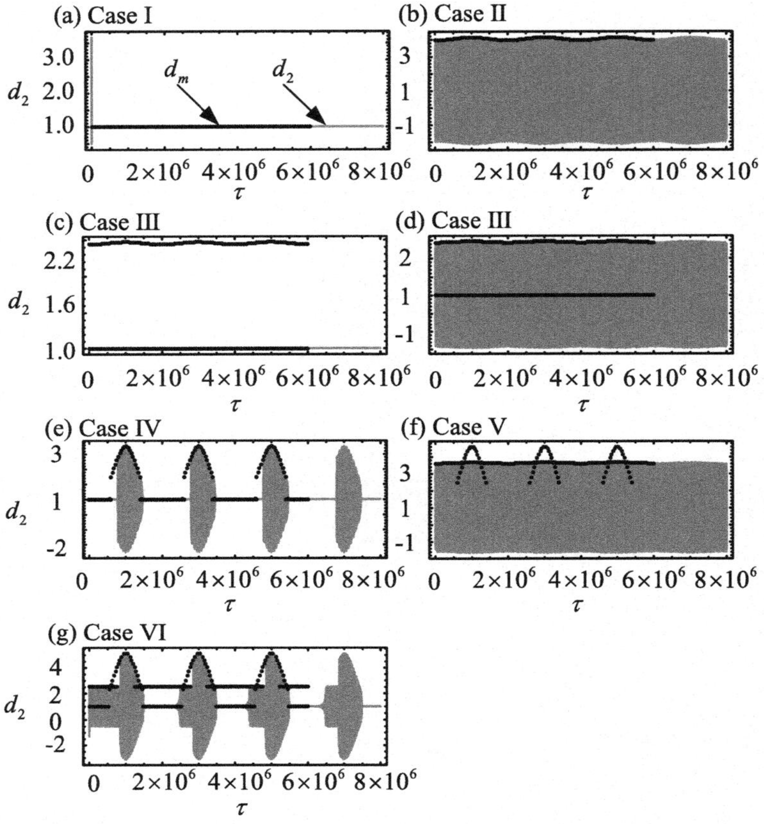

Fig. 6.17 Time series of d_2, with the chatter amplitude d_m

6.3 Estimate of Cutting Safety by Time Delay

6.3.1 Unsafe Cutting and Unsafe Zones

Apart from the discussion on linear cutting stability in Sect. 6.1.1, it has become increasingly evident over the past decade that nonlinearity plays a pivotal role in shaping cutting dynamics near the stability boundaries. This nonlinearity introduces the concept of multi-stability, which undermines linear cutting stability. Given the unpredictability associated with cutting dynamics in the presence of multi-stability, which is called unsafe cutting (UC), and the corresponding regions are referred to as unsafe zones (UZs) [26]. Typical UZs are introduced due to subcritical instability along the stability boundaries. This results in an unstable periodic orbit extending into the stable region until the cutting tool loses contact with the workpiece, leading to a chatter. Consequently, estimating UZs involves employing perturbation methods, tracing the trajectory of unstable periodic orbits, and assessing their non-smooth behavior [27].

The method of multiple scales (MMS) is utilized to approximate the unsafe zones (UZs) as shown in Fig. 6.18a. However, it is important to note that there is a discrepancy in Fig. 6.18b. To be specific, the analytical approach underestimates the UZ in the high-velocity zone but overestimates it in the low-velocity zone. As illustrated in the enlarged view in Fig. 6.18c, the analytical estimation is qualitatively accurate only within the shaded areas, but it falls short in predicting the UZs in other regions.

Bifurcation diagrams for $N = 3600$ rpm in Fig. 6.19 plots the minimum and maximum values of H and V_γ as functions of the depth of cut. Stationary cutting remains stable until a_p reaches 0.82 mm, where a significant amplitude chatter, indicated by red squares, emerges, coexists with the stable cutting. Upon further increasing a_p to 0.86 mm, as observed in the small magnified view, another form of small-amplitude chatter emerges, resulting in tristability. For $a_\mathrm{p} \geq 0.89$ mm, stable cutting ceases to exist, marking the entry of the cutting process into the unstable region. In Fig. 6.19c, it is evident that the large-amplitude chatter exhibits a sticking phase ($V_\gamma = 0$) for frictional chatter.

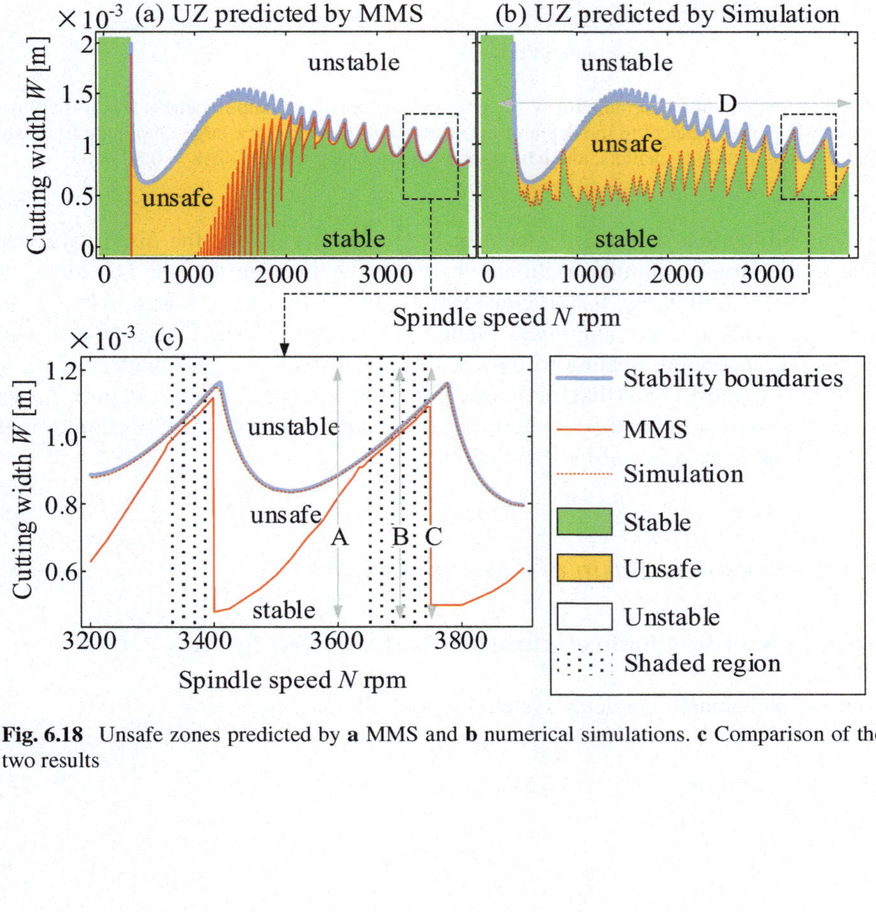

Fig. 6.18 Unsafe zones predicted by **a** MMS and **b** numerical simulations. **c** Comparison of the two results

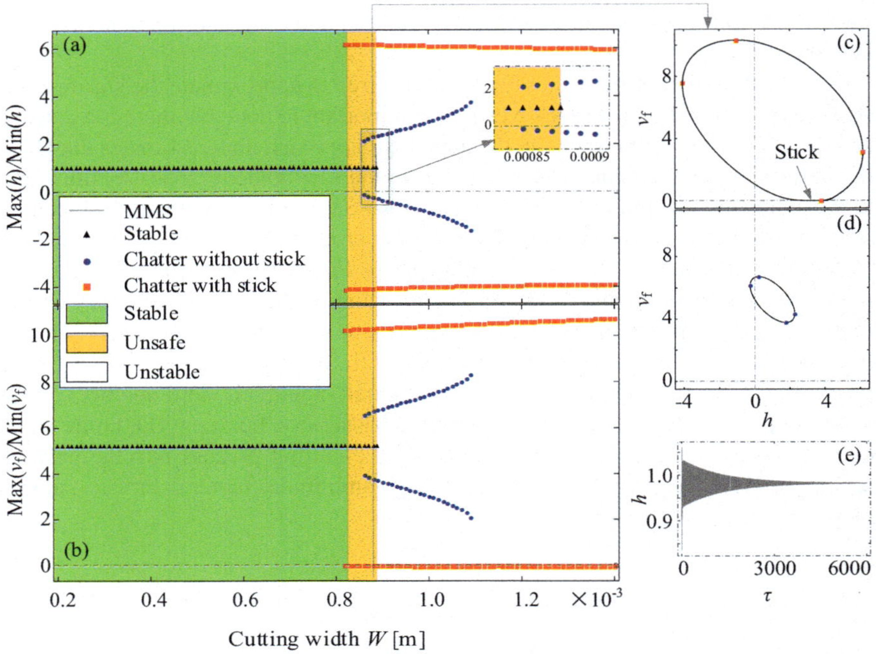

Fig. 6.19 Bifurcation diagrams for $N = 3600$ rpm, where the **a** chip thickness and **b** frictional velocity vary with respect to the depth of cut, with phase portraits of **c** large-amplitude frictional and **d** small-amplitude chatters, and **e** time series of stationary cutting for $W = 0.88$ mm

The bifurcation diagram depicted in Fig. 6.19 reveals that the disparity arises due to the large-amplitude frictional chatter. It's noticeable that the UZ for $a_\text{p} \in [1.01, 1.03]$ [mm] in Fig. 6.20 is considerably smaller compared to that in Fig. 6.19. The occurrence of large-amplitude chatter with sticking phases ($V_\gamma = 0$) is absent until a_p is increased, leading to the coexistence of two types of chatter for $a_\text{p} \in [1.09, 1.34]$ [mm]. Given that frictional chatter does not impact the size of the UZ, this scenario aligns with the theory put forth by Insperger et al. [13], thereby confirming the qualitative accuracy of the estimation.

6.3.2 *Statistical Basin of Attraction (SBoA)*

6.3.2.1 Statistical Basin of Attraction for Time Delay Systems

The n-dimensional time delay system has the following form [28]

$$\frac{\text{d}\mathbf{y}}{\text{d}t} = \mathbf{F}(t, \mathbf{y}(t), \mathbf{y}(t-\tau_1), \mathbf{y}(t-\tau_2), \cdots, \mathbf{y}(t-\tau_m)), \mathbf{y} \in \mathbb{R}^{n\times 1}, \quad (6.102)$$

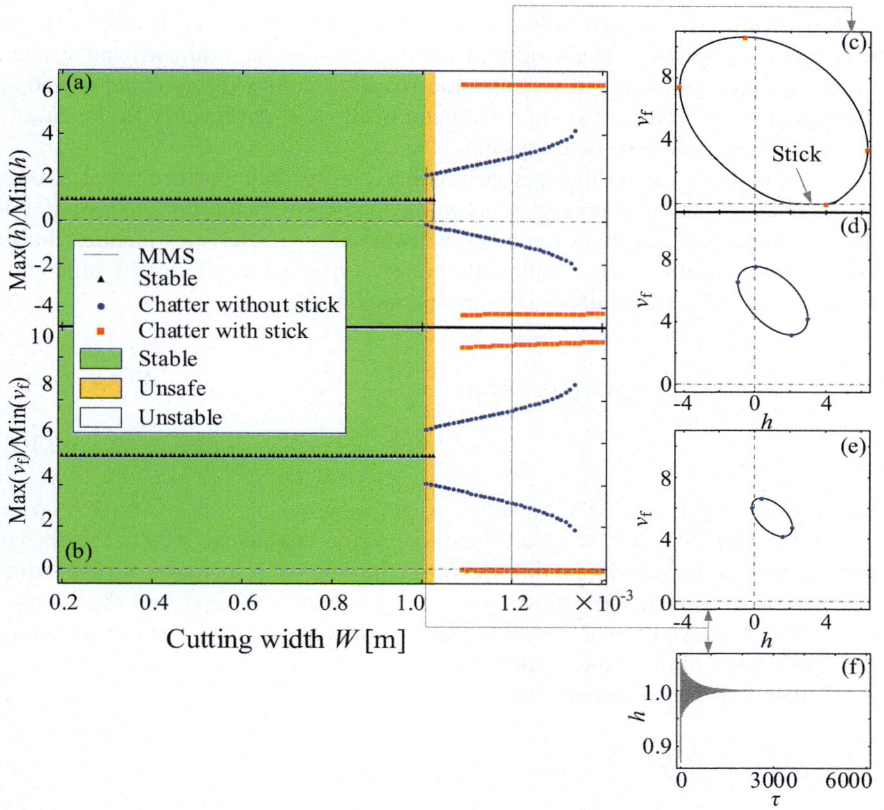

Fig. 6.20 Bifurcation diagrams for $N = 3700$ rpm, where the **a** chip thickness and **b** frictional velocity vary with respect to the depth of cut, with phase portraits of **c** large-amplitude frictional and **d** small-amplitude chatters for $W = 1.2$ mm, and **e** phase portrait of small-amplitude chatter and f time series of stationary cutting for $W = 1.01$ mm

where $0 < \tau_1 < \tau_2 < \cdots < \tau_m = \tau$ are m time delays. One can arrange $\mathbf{y}(t - \tau_i)$ $(i = 1, 2, \cdots, m)$ as follows

$$\mathbf{y}(t-\tau_i) = \begin{pmatrix} \mathbf{y}_1(t-\tau_i) \\ \mathbf{y}_2(t) \end{pmatrix}, \mathbf{y}_1 = \begin{pmatrix} y_1 \\ y_2 \\ \vdots \\ y_l \end{pmatrix} \in \mathbb{R}^{l\times 1}, \mathbf{y}_2 = \begin{pmatrix} y_{l+1} \\ y_{l+2} \\ \vdots \\ y_n \end{pmatrix} \in \mathbb{R}^{(n-l)\times 1}. \tag{6.103}$$

When $l = 0$, Eq. (6.102) becomes ordinary differential equations (ODEs).

In the presence of multi-stability in a nonlinear system, the long-term dynamics are determined by the initial conditions (ICs), and all ICs that converge to the same attractor are categorized within a single basin of attraction (BoA). However, it becomes increasingly challenging when dealing with high-dimensional systems in terms of computation and visualization. To address this, it's essential to find an appropriate low-dimensional approximation of the ICs, particularly in the case of infinite-dimensional time delay systems.

In this context, we use the approach outlined in Ref. [9], as it necessitates fewer coefficients to achieve a precise and continuous approximation. Furthermore, this method remains applicable even when non-smooth characteristics manifest in the cutting force. It effectively captures the time-varying delayed terms by utilizing a Fourier sine series superimposed on linear components

$$\mathbf{y}_1(t-s) = \mathbf{f}(t,s) = \mathbf{a}_{-1}(t)\frac{s}{\tau} + \mathbf{a}_0(t)\frac{\tau - s}{\tau} + \sum_{i=1}^{N} \mathbf{a}_i(t)\sin\left(i\pi\frac{s}{\tau}\right), s \in [0,\tau], \tag{6.104}$$

where $\mathbf{f} = (f_1, f_2, \cdots, f_l)^T \in \mathbb{R}^{l\times 1}$, $\mathbf{a}_i = (a_{1,i}, a_{2,i}, \cdots, a_{l,i})^T \in \mathbb{R}^{l\times 1}$ ($i = -1, 0, \cdots, N$), N and $\tau = \tau_m$ are function vector, coefficient vectors, number of basis functions, and maximum time delay. As the value of N increases, an expanding number of sinusoidal basis functions are incorporated to represent the infinite-dimensional delayed terms within a space of dimension $(N+2)$. Consequently, the significance of the residuals diminishes progressively.

Element-wise representation is:

$$\begin{aligned} y_j(t-s) = f_j(t,s) = a_{j,-1}(t)\frac{s}{\tau} \\ + a_{j,0}(t)\frac{\tau - s}{\tau} + \sum_{i=1}^{N} a_{j,i}(t)\sin\left(i\pi\frac{s}{\tau}\right), \quad s \in [0,\tau], \end{aligned} \tag{6.105}$$

where $j = 1, 2, \cdots, l$. Figure 6.21 shows that $f_j(t,0)$ and $f_j(t,\tau)$ respectively correspond to $y_j(t) = a_{j,0}(t)$ and $y_j(t-\tau) = a_{j,-1}(t)$, and the sum of the Fourier sine series is represented by the vertical separation between the two curves.

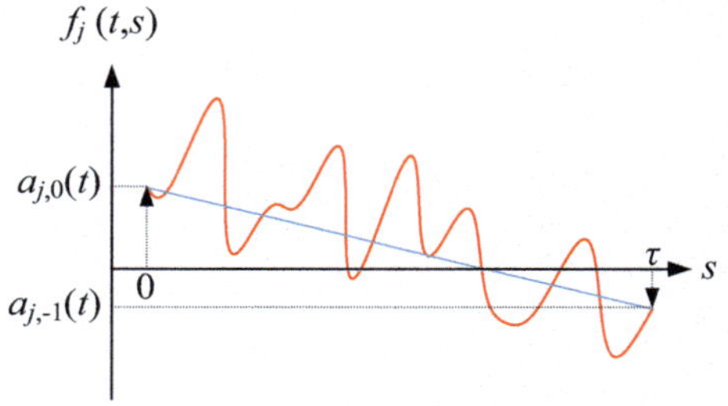

Fig. 6.21 Approximation of $x(t-s) = f(t,s)$ $(s \in [0,\tau])$

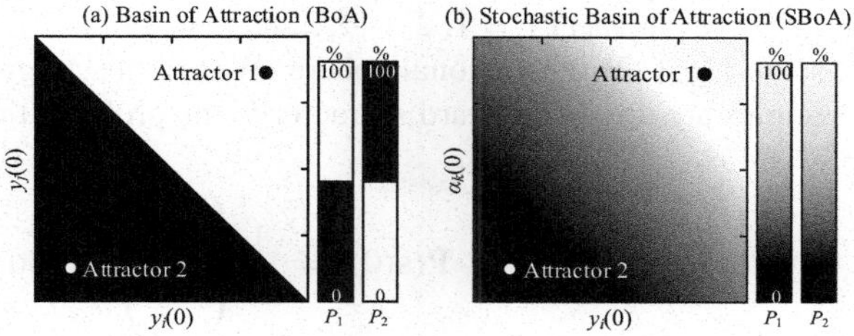

Fig. 6.22 **a** Conventional and **b** Statistical basin of attraction

We introduce $\alpha_j > 0$ satisfying

$$\alpha_j^2(t) = \sum_{i=-1}^{N} a_{j,i}^2(t), \tag{6.106}$$

to further compress the dimension. Equation (6.106) introduces a new problem the IC is not unique.

In line with this feature, we extend the traditional BoA concept to create what we refer to as a Statistical Basin of Attraction (SBoA). As depicted in Fig. 6.22a, the conventional BoA is marked by a distinct boundary that separates the basins of different attractors. For instance, ICs within the white region consistently lead to Attractor 1 (black dot). If we use

$$\mathbf{P}(\mathbf{s}(0)) = \begin{pmatrix} P_1(\mathbf{s}(0)) \\ P_2(\mathbf{s}(0)) \end{pmatrix} \tag{6.107}$$

to denote of the probabilities of the initial condition, $\mathbf{s}(0) = (y_i(0), y_j(0))^T$ $(i \neq j$ and $i, j = l+1, l+2, \cdots, n)$, going to the coexisting attractors, it has only binary values, which is either $\mathbf{P} = (100\%, 0\%)^T$ for Attractor 1 or $\mathbf{P} = (0\%, 100\%)^T$ for Attractor 2. In contrast, the boundary in Fig. 6.22b becomes soft, and the elements of $\mathbf{P}$ varies continuously between 0 and 100%.

In order to create a uniform distribution on the sphere for estimating $\mathbf{P}$ by Monte Carlo method, we independently generate M sets of $b_{j,i}$ $(j = 1, 2, \cdots, l$ and $i = -1, 0, \cdots, N)$ based on the normal (Gaussian) density, as described below:

$$b_{j,i} \sim \mathcal{N}(0, 1), \quad j = 1, 2, \cdots, \text{ land } i = -1, 0, \cdots, N. \tag{6.108}$$

Before normalization as follows:

$$a_{j,i} = \alpha_j \frac{b_{j,i}}{\sqrt{\sum\limits_{i=-1}^{N} b_{j,i}^2}}, \quad j = 1, 2, \cdots, \text{ land } i = -1, 0, \cdots, N. \tag{6.109}$$

As a result, $a_{j,i}$ ($j = 1, 2, \cdots, l$ and $i = -1, 0, \cdots, N$) is uniformly distributed on the $(N + 2)$-dimensional sphere. Within the M groups, if M_{k} ($k = 1, 2, \cdots$) samples are directed toward Attractor k, the probabilities are

$$\mathbf{P}(\mathbf{s}(0)) = \frac{1}{M}\begin{pmatrix} M_1 \\ M_2 \\ \vdots \end{pmatrix} \times 100\%. \tag{6.110}$$

In the visualization of the SBoA, the greyscale represents the probabilities.

In order to guide the long-term dynamics toward a desired attractor, one can implement a state-dependent intermittent control mechanism. Correspondingly, one has

$$\frac{\mathrm{d}\mathbf{y}}{\mathrm{d}t} = \mathbf{F}(t, \mathbf{y}(t), \mathbf{y}(t-\tau_1), \mathbf{y}(t-\tau_2), \cdots, \mathbf{y}(t-\tau_m)) + \mathbf{U}(\mathbf{s}(\mathbf{t})), \mathbf{y} \in \mathbb{R}^{n\times 1}, \tag{6.111}$$

where $\mathbf{U}(\mathbf{s}(\mathbf{t})) = h(\mathbf{s}(\mathbf{t}))\mathbf{u}(t)$, and $\mathbf{u}(t)$ represents the control input. The Heaviside step function is:

$$h(\mathbf{s}(t)) = \begin{cases} 1 \text{ if } s(t) \notin \mathcal{B}_{\mathrm{s}} \\ 0 \text{ if } s(t) \in \mathcal{B}_{\mathrm{s}} \end{cases}, \tag{6.112}$$

where $\mathcal{B}_{\mathrm{s}} = \{\mathbf{s} | P_{\mathrm{t}}(\mathbf{s}) \geq P_{\mathrm{th}}\}$ is the targeted basin, in which the probability to the targeted attractor, $P_{\mathrm{t}}(\mathbf{s})$, is not less than a predefined threshold, P_{th}.

In contrast to managing multiple stability in ODEs, where transitioning between basins in an instant is adequate, DDEs demand a more extended duration to shift between these basins. This is due to the sustained influence of the delayed state, $y_1(t - s)$, which persists for a specific period, $s \in (0, \tau]$. Therefore, we establish a monitoring period, T_{m}, to represent the time span between two consecutive observations, namely $\mathbf{s}(t)$ and $\mathbf{s}(t + T_{\mathrm{m}})$. This ensures that the control initiated at time t can remain effective for at least the duration $[t, t + T_{\mathrm{m}})$.

6.3.2.2 SBoA of the Cutting Chatter

Considering the cutting dynamics described above, the approximation of the delayed term is formulated as follows:

$$y_1(t - s) = a_{-1}(t)\frac{s}{\tau} + a_0(t)\frac{\tau - s}{\tau} + \sum_{i=1}^{N} a_i(t)\sin\left(i\pi\frac{s}{\tau}\right), s \in [0, \tau]. \tag{6.113}$$

Definition of the state

$$\mathbf{s}(t) = \begin{pmatrix} y_2(t) \\ \alpha(t) \end{pmatrix} = \begin{pmatrix} y_2(t) \\ \sqrt{\sum_{i=-1}^{N} a_i(t)^2} \end{pmatrix}. \tag{6.114}$$

It is seen in Fig. 6.23c–g, that a_i ($i = -1, 0, \cdots, N$), decrease towards zero. On the contrary, in Fig. 6.23h–l, the occurrence of chatter is not associated with a reduction in all the coefficients; rather, certain coefficients exhibit an increase. The analysis indicates that the involvement of a_i ($i > 20$) in the approximation is minimal, whether the cutting process is stable or unstable.

Figure 6.24 demonstrates that this estimation steadily converges as M exceeds 1000. Furthermore, Fig. 6.24b–d depict the scenarios for $M = 500$, 2500 and 10000, highlighting the minimal impact of N on P_c when $N > 15$. For subsequent analyses, $N = 21$ and $M = 2500$ will be employed, considering a balance between computational efficiency and accuracy.

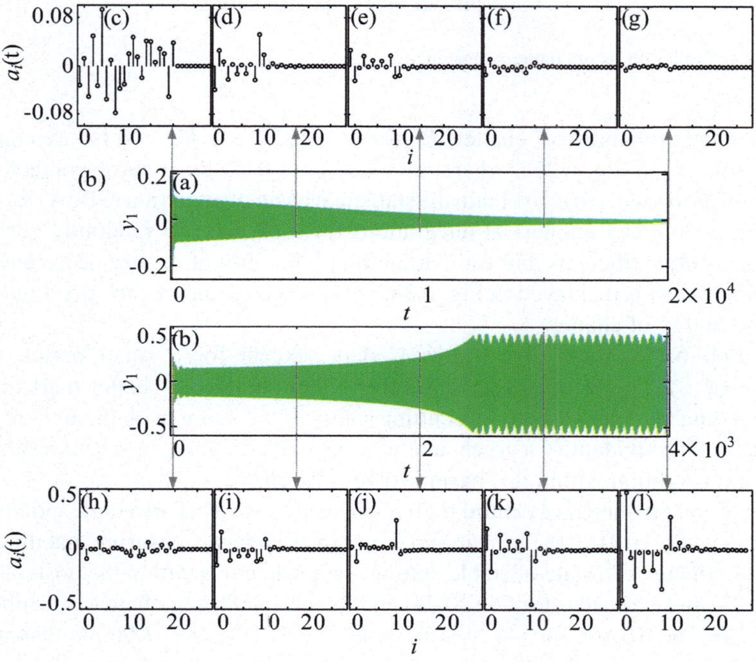

Fig. 6.23 **a** Time series of the stationary cutting with its coefficients in panels (**c**–**g**). **b** Time series of the coexisting chatter, with its coefficients in panels (**h**–**l**)

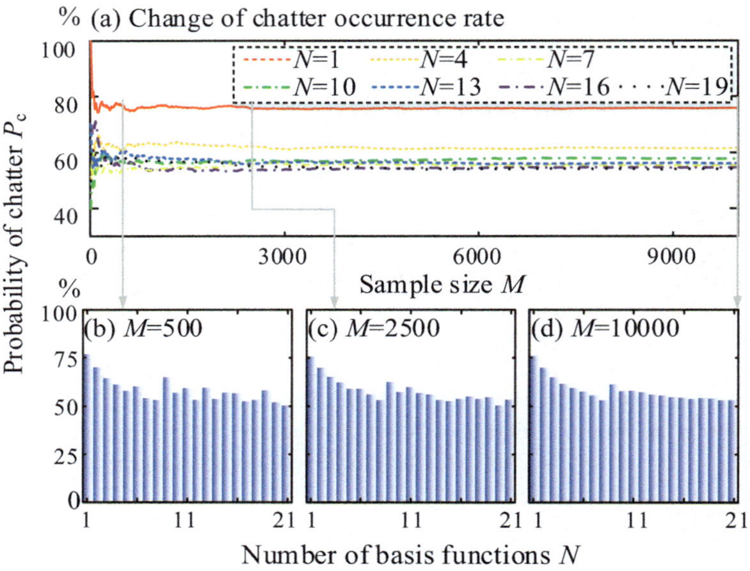

Fig. 6.24 Probabilities of chatter occurrence

Then the probability of chatter, P_c, for $\Omega = 0.24$ and $\frac{\eta_1}{\eta_{\min}} = 4.64$ are computed and displayed in Fig. 6.25a, where the location of each pixel corresponds to given values of $y_2(0)$ and $\alpha(0)$. To obtain the statistical basin of attraction (SBoA) for unsafe cutting, we initiated numerical integrations for each pixel by randomly generating M groups of coefficients and estimating the probability of chatter occurrence. The resulting SBoA is displayed in Fig. 6.25, with the grayscale of each pixel indicating the probability of chatter.

In Fig. 6.25a, most of the area is dark, except for a small region around $s(0) = (0, 0)$. This darkness indicates that either significant chatter marks or high tool bending velocity leads to low cutting safety. If we zoom in on the light region in Fig. 6.25, we can identify a green area with $P_c = 0\%$, signifying a safe basin where no initial condition within this basin evolves into chatter.

In Fig. 6.25d, there is a central region where $P_c = 0$. This area corresponds to the scenario where $\alpha(0) = 0.1$ transitions across the safe basin, ensuring that the cutting process remains in a state of stable cutting, even when it is within the unstable zone.

The SBoA estimated for $\Omega = 0.24$, and the probability of chatter is displayed in Fig. 6.26. The SBoAs for $\frac{\eta_1}{\eta_{\min}} = 4.62, 4.63, 4.64, 4.65$, and 4.66 are displayed in Fig. 6.26b–f, showing that the basin is gradually darkened for a higher probability of chatter. Meanwhile, the safe basin is uniformly shrunk when the cutting parameters approach the cutting instability.

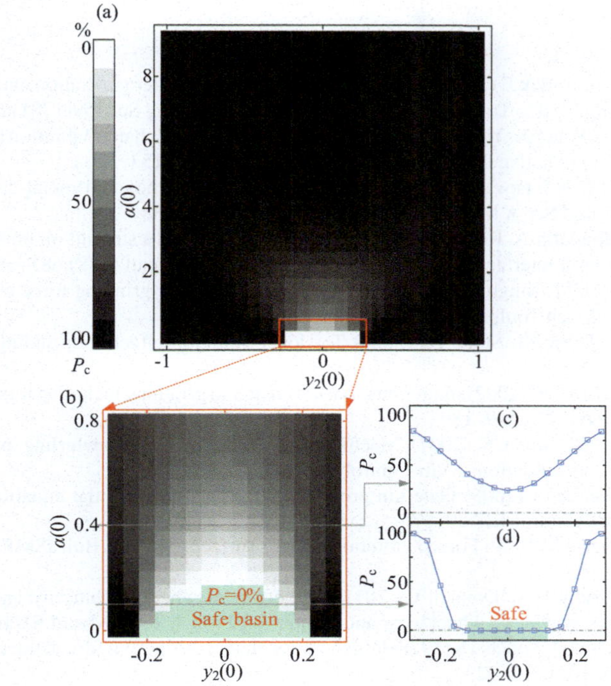

Fig. 6.25 Statistical basin of attraction for $\Omega = 0.24$ and $\frac{\eta_1}{\eta_{\min}} = 4.64$. **c** $\alpha(0) = 0.1$ and **d** 0.4 are displayed as functions of$y_2(0)$

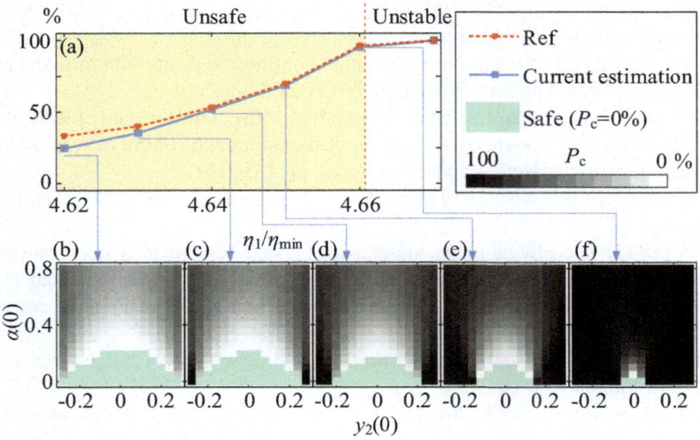

Fig. 6.26 Averaged probability of chatter as a function of $\frac{\eta_1}{\eta_{\min}}$

References

1. Wiercigroch M, Budak E (2001) Sources of nonlinearities, chatter generation and suppression in metal cutting. Philos Trans R Soc Lond Ser A Math Phys Eng Sci 359(1781):663–693
2. Fu Z, Zhang X, Wang X, Yang W (2014) Analytical modeling of chatter vibration in orthogonal cutting using a predictive force model. Int J Mech Sci 88:145–153
3. Wiercigroch M, Krivtsov AM (2001) Frictional chatter in orthogonal metal cutting. Philos Trans R Soc Lond Ser A Math Phys Eng Sci 359(1781):713–738
4. Claudin C, Mondelin A, Rech J, Fromentin G (2010) Effects of a straight oil on friction at the tool work material interface in machining. Int J Mach Tools Manuf 50(8):681–688
5. Durgumahanti USP, Singh V, Rao PV (2010) A new model for grinding force prediction and analysis. Int J Mach Tools Manuf 50(3):231–240
6. Mannan MA, Drew SJ, Stone BJ (2000) Torsional vibration effects in grinding. Ann CIRP 49:249–252
7. Hesterman D, Stone B (2002) Improved model of chatter in grinding, including torsional effects. J Multi-body Dyn 216:169–180
8. Dassanayake AV, Suh CS (2007) Machining dynamics involving whirling part I: model development and validation. J Vib Control 13(5):475–506
9. Wahi P, Chatterjee A (2005) Galerkin projections for delay differential equations. ASME J Dyn Syst Meas Control 127:80–87
10. Inazaki I, Yonetsu S (1969) Forced vibrations during surface grinding. Bull JSME 12(50):385–391
11. Badger J, Murphy S, O'Donnell G (2011) The effect of wheel eccentricity and run-out on grinding forces, waviness, wheel wear and chatter. Int J Mach Tools Manuf 51(10):766–774
12. Shimizu T, Inasaki I, Yonetsu S (1978) Regenerative chatter during cylindrical traverse grinding. Bull JSME 21(152):317–323
13. Insperger T, David AWB, Stépán G (2008) Criticality of Hopf bifurcation in state-dependent delay model of turning processes. Int J Non-Linear Mech 43(2):140–149
14. Ji JC (2015) Two families of super-harmonic resonances in a time-delayed nonlinear oscillator. J Sound Vib 349:299–314
15. Wang HL, Hu HY, Wang ZH (2004) Global dynamics of a duffing oscillator with delayed displacement feedback. Int J Bifurcat Chaos 14(08):2753–2775
16. Zheng YG, Sun JQ (2017) Attractive domain of nonlinear systems with time-delayed feedback control and time-delay effects. Procedia IUTAM 22:51–58
17. Hu HY (2005) Global dynamics of a duffing system with delayed velocity feedback. In: Rega G, Vestroni F (eds) IUTAM symposium on chaotic dynamics and control of systems and processes in mechanics. Springer Netherlands, Dordrecht, pp 335–344
18. Yan Y, Xu J, Wiercigroch M (2019) Modelling of regenerative and frictional cutting dynamics. Int J Mech Sci 156:86–93
19. Tustin A (1947) The effects of backlash and of speed-dependent friction on the stability of closed-cycle control systems. Electr Eng Part II A Autom Regul Servo Mechan 94(1):143–151
20. Das MK, Tobias SA (1967) The relation between the static and the dynamic cutting of metals. Int J Mach Tool Des Res 7(2):63–89
21. Eynian M (2010) Chatter stability of turning and milling with process damping. The University of British Columbia
22. Yan Y, Xu J, Wiercigroch M (2016) Regenerative and frictional chatter in plunge grinding. Nonlinear Dyn 86:283–307
23. Yan Y, Xu J, Wiercigroch M (2018) Stability and dynamics of parallel plunge grinding. Int J Adv Manuf Technol 99:881–895
24. Yan Y, Xu J, Wiercigroch M (2017) Regenerative chatter in a plunge grinding process with workpiece imbalance. Int J Adv Manuf Technol 89:2845–2862
25. Yan Y, Xu J, Wiercigroch M (2014) Chatter in a transverse grinding process. J Sound Vib 333:937–953

26. Yan Y, Xu J, Wiercigroch M (2017) Basins of attraction of the bistable region of time-delayed cutting dynamics. Phys Rev E 96(3):032205
27. Yan Y, Liu GJ, Wiercigroch M, Xu J (2021) Safety estimation for a new model of regenerative and frictional cutting dynamics. Int J Mech Sci 201:106468
28. Yan Y, Zhang S, Guo Q, Xu J, Kim KC (2020) Energy determines multiple stability in time-delayed systems. Nonlinear Dyn 102:2399–2416

Chapter 7
Effect of Time Delay on Network Dynamics

Network systems widely exist in human and nature society, such as the internet, electric power, transportation, and interpersonal relationships [1]. It can be said that any system composed of the same or different individuals can be represented by a network. When the individuals and the interactions between them are respectively abstracted as nodes and edges, these individuals can be represented by a network system. With the extensive usage of network dynamics in epidemic preparedness [2], internet congestion [3], the traffic system [4] and neural networks [5], etc., the dynamics and control of network systems have received widespread attention and in-depth research in the academic community [6].

Previous studies have shown that most complex dynamic behaviours of network systems are accompanied by changes in network topology and node dynamics [7, 8]. In particular, network systems with time delay usually exhibit unique behaviours such as synchronization, bifurcation, stability, oscillation, and chaos [9, 10]. Studying these dynamic behaviours provides essential theoretical bases for applying network systems in various fields. However, for a network system with time delay coupling, since its characteristic equation is a transcendental equation containing an exponential function, the characteristic roots cannot be accurately solved, which brings difficulties to the dynamic analysis of the network system [11]. So, it is extremely challenging to study the dynamics of time delay network systems. It has important scientific significance and practical application value for the development of human society.

The primary objectives of examining the dynamic behavior of a network system encompass two key facets: firstly, comprehending the influence of the network's topology on its dynamic tendencies; and secondly, devising suitable control strategies to guide the network toward anticipated dynamic states. In this chapter, we use the time-delayed neural network system and the internet congestion control system as two main objects to research the dynamic behaviour and control of the network system. For this reason, the chapter initiates with a scrutiny of the time-delayed oscillator within neural systems, elucidating its dynamical characteristics such as periodic

J. Xu, *Nonlinear Dynamics of Time Delay Systems*,
https://doi.org/10.1007/978-981-99-9907-1_7

oscillation, high-dimensional torus, and chaos. A distinctive three-dimensional torus and its bifurcation route to chaos are identified. Then, we chose a BAM (bidirectional associative memory) network with a double-layer structure. The dynamical classification near the codimension-twofold-Hopf bifurcation is studied. The approximate calculation of solutions is given when network nodes have synchronous periods. Furthermore, we consider a general n-dimensional time delay neural network to study the hyperchaos and synchronization problem of the system. Next, we investigate the adaptive synchronization problem of the drive-response system under uncertain parameters, and design a class of integral sliding-mode structure controllers to realize the chaotic projective synchronization problem with different time delays. Finally, we discuss the problem of congestion control in the internet systems, specifically highlighting the nonlinear dynamics stemming from variations in time delay within the congestion control model of elevated dimension. Our investigations on such topic can be extended to high-dimensional network systems with diverse topologies. In such scenarios, stabilization is achieved through meticulous parameter adjustments and the incorporation of controllers.

7.1 Chaotic Oscillation of Time Delay Coupled Wilson-Cowan Neural Oscillator System

As a new interdisciplinary subject, the neural network is a network structure constructed by abstract modeling of the human brain nervous system, which has powerful self-organization, powerful self-learning ability and high-speed parallel information processing ability [12, 13]. Due to the superior performance of the neural network, it has developed rapidly in many fields [14]. It can deal with problems which are not solved by traditional programs and mathematical computing power [15]. With the deepening of neural network research, in order to further simulate different characteristics of the human brain nervous system, different types of neural network systems have been discussed by many scholars, such as convolutional neural network [16], deep neural network [17] and probabilistic neural network [18]. Compared with traditional artificial neural networks, chaotic neural networks are more advanced systems for simulating intelligent information processing [19]. Studies have shown that human mind evolved on the boundary between chaos and order [20], so research on chaotic neural networks has high application value.

In this section, based on the time-delayed Wilson-Cowan oscillator system [21], we explore the complex dynamic behaviors, including periodic oscillation, quasi-periodic and chaotic motion, especially a special oscillation mode with a three-dimensional torus was discovered. In fact, the neural network system is usually composed of many interacting complex subsystems. To quantitatively describe the overall operation of this type of complex system, studying its fundamental structural unit system is very imperative.

7.1.1 Delay-Coupled Wilson-Cowan Neural Oscillator System

As we all know, time delay is an indispensable factor in lots of biological systems due to neuronal membrane capacitance, transmembrane resistance, and the limited propagation speed of potentials [22]. Furthermore, artificial neural networks with delay coupling can demonstrate complicated dynamical behaviours, which can be used for information storage [23], adaptive learning [24], secure communication [25], and so on. Therefore, the time-delay neural oscillator system we consider in this section is

$$\begin{cases} \dfrac{dx(t)}{dt} = -x(t) + a_1 F(x(t-\tau)) + a_2 F(y(t-\tau)) + P, \\ \dfrac{dy(t)}{dt} = -y(t) + a_3 F(x(t-\tau)) + a_4 F(y(t-\tau)) + Q, \end{cases} \tag{7.1}$$

where a_1, a_4 represent coupling strength of self-connection, respectively. a_2, a_3 represent coupling strength of other connection, respectively. τ represents the coupling delay, here we set time delay of self-connection and other connection is equal. P, Q are the external excitation, $F(u) = 1/(1 + e^{-ku})$ is the activation function, and k is called the slope rate of activation function.

7.1.2 Periodic Oscillation Under the Effect of Time Delay

The equilibrium points (x_0, y_0) in the Eq. (7.1) satisfies

$$\begin{cases} x_0 = a_1 F(x_0) + a_2 F(y_0) + P, \\ y_0 = a_3 F(x_0) + a_4 F(y_0) + Q. \end{cases} \tag{7.2}$$

In order to translate (x_0, y_0) to the trivial equilibrium point $(0, 0)$, use the transformation as shown below:

$$x \to x - x_0, y \to y - y_0, \tag{7.3}$$

then the linearization equation of the Eq. (7.1) is obtained

$$\begin{cases} \dfrac{dx(t)}{dt} = a_1 \alpha x(t-\tau) + a_2 \beta y(t-\tau) - x(t), \\ \dfrac{dy(t)}{dt} = a_3 \alpha x(t-\tau) + a_4 \beta y(t-\tau) - y(t), \end{cases} \tag{7.4}$$

where $\alpha = F'(x_0)$, $\beta = F'(y_0)$, (x_0, y_0) is the equilibrium point of the system. The corresponding characteristic equation is

$$\begin{aligned}\Delta(\lambda,\tau) = 1 - (a_1\alpha + a_4\beta)e^{-\lambda\tau} + (a_1a_4 - a_2a_3)\alpha\beta e^{-2\lambda\tau} + \\ (2 - (a_1\alpha + a_4\beta)e^{-\lambda\tau})\lambda + \lambda^2 = 0.\end{aligned} \tag{7.5}$$

Assuming that the characteristic Eq. (7.5) has pure imaginary roots, its form is $\lambda = i\omega$, $\omega > 0$, we can get

$$\begin{aligned}1 - (a_1\alpha + a_4\beta)e^{-i\omega\tau} + (a_1a_4 - a_2a_3)\alpha\beta e^{-2i\omega\tau} + \\ (2 - (a_1\alpha + a_4\beta)e^{-i\omega\tau})(i\omega) + (i\omega)^2 = 0.\end{aligned} \tag{7.6}$$

Separating the real part and the imaginary part, we get

$$\begin{cases}(1 - \omega^2 + a_1a_4\alpha\beta - a_2a_3\alpha\beta)\cos\omega\tau - 2\omega\sin\omega\tau - (a_1\alpha + a_4\beta) = 0, \\ (1 - \nu^2 + a_1a_4\alpha\beta - a_2a_3\alpha\beta)\sin\omega\tau + 2\omega\cos\omega\tau - (a_1\alpha + a_4\beta)\omega = 0.\end{cases} \tag{7.7}$$

Eliminate the harmonic term in (7.7), that is

$$G(\omega) = \omega^4 + n_2\omega^2 + n_0 = 0, \tag{7.8}$$

where

$$\begin{aligned}n_2 &= 2 - (a_1^2\alpha^2 + 2a_2a_3\alpha\beta + a_4^2\beta^2), \\ n_0 &= 1 - (a_1^2\alpha^2 + 2a_2a_3\alpha\beta + a_4^2\beta^2 - \alpha^2\beta^2(a_2a_3 - a_1a_4)^2).\end{aligned}$$

Thus, the root of (7.8) is

$$\omega_\pm = \left[\left(-n_2 \pm \sqrt{n_2^2 - 4n_0}\right)/2\right]^{1/2} \tag{7.9}$$

Choose the system parameters as $a_1 = -6$, $a_2 = 2.5$, $a_3 = 2.5$, $a_4 = -6$, $P = 0.4$, $Q = 0.4$, $k = 15.8$, $\tau = 0.1$. The system has a stable periodic oscillation. The phase diagram and time history diagram are shown in Fig. 7.1.

7.1.3 Torus and Chaotic Oscillations

For the chosen system parameters $a_1 = -6$, $a_2 = 2.5$, $a_3 = 2.5$, $a_4 = -6$, $P = 0.4$, $Q = 0.4$, we continue to study the impact of time delay and slope rate on the dynamic behaviours. In order to express different oscillation modes, we choose the Poincaré section as $\dot{x}(t - \tau) = 0$. That is to say, if the system has a single-period or multi-period oscillation mode, the Poincaré cross-section has one or more discrete points. When the crossing points on the section can form a closed loop, the system must have a quasi-periodic oscillation mode. For the chaotic motion, the Poincaré

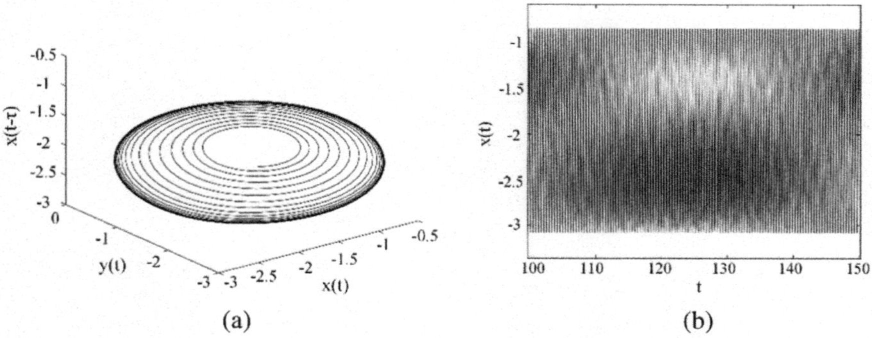

Fig. 7.1 The stable periodic oscillation mode of the system

cross-section has an irregular pattern composed of countless points, and the system presents strange attractors.

Fixed the time delay $\tau = 0.1$, the different Poincaré maps are shown in Fig. 7.2 with changing the slope rate k. For further analyzing dynamic behavior of these different modes, we give the system's phase diagram in the (x, y) plane and use fast Fourier transform (FFT) to calculate the power spectrum of $x(t)$ as shown in Figs. 7.3 and 7.4. We can see that when the slope rate $k = 15.8$, the system presents a periodic solution, as shown in Fig. 7.3a, and the Poincaré section only has a unique point $(-0.1226, -0.1203)$ as shown in Fig. 7.2a. The power spectrum analysis of $x(t)$ only has a unique fundamental frequency $f_1 \approx 2.48$Hz, as shown in Fig. 7.4a. With the increase of the slope rate, when $k = 22.2$, the periodic solution of the system generates a quasi-periodic oscillation through Neimark-Sacker bifurcation, and its Poincaré diagram is a closed loop as Fig. 7.2b. At this time, there are two fundamental frequencies in the system, namely $f_1 \approx 1.90$Hz and $f_2 \approx 1.24$Hz, as shown in Fig. 7.4b. Due to the mutual coupling of two fundamental frequencies, the system has a quasi-periodic oscillation which is usually called a two-dimensional torus motion (2-torus).

When the slope rate k is increased to 22.4, a two-dimensional torus attractor appears on the Poincaré section, as Fig. 7.2c. That is to say, there are two fundamental frequencies on the Poincaré section, and the corresponding system motion has three different fundamental frequencies, the power spectrum analysis is shown in Fig. 7.4c. Although the third fundamental frequency cannot be distinguished very well, compared with the two-dimensional quasi-periodic oscillation, the frequency spectrum has more discrete peak values, and the phase diagram is more complicated than the two-dimensional torus motion, as Fig. 7.3c. Finally, when the parameter k changes to 29, the system enters into the chaotic region by through multiple bifurcations. At this time, the system presents a chaotic oscillation, as Fig. 7.3d, and the Poincaré section diagram is an irregular pattern, as shown in Fig. 7.2d. The corresponding power spectrum is a continuous broadened, which is characteristic of chaos attractor, as shown in Fig. 7.4d.

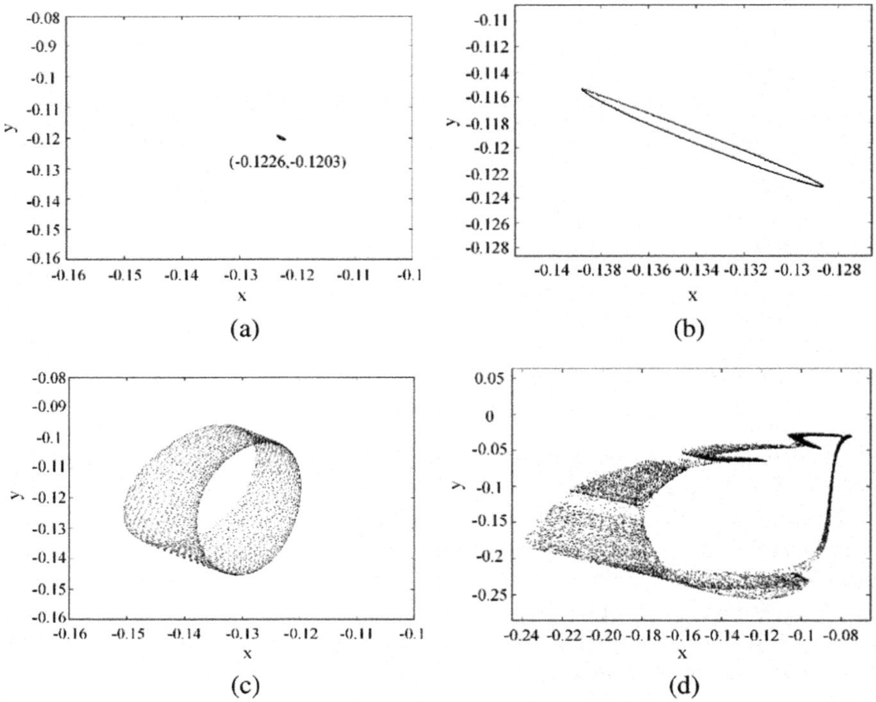

Fig. 7.2 The path of the system going into chaos when the time delay is $\tau = 0.1$: **a** periodic motion: $k = 15.8$, **b** almost periodic motion: $k = 22.2$, **c** 3D torus: $k = 22.4$, **d** chaotic attractor $k = 29$

7.2 Fold-Hopf Bifurcation and Approximated Computation of the Synchronous Periodic Solutions in the BAM Network System

The bidirectional associative memory (BAM) network was initially expanded from the Hopfield single-layer structure to a two-layer network by the professor Bart Kosko in University of Southern California [26, 27]. The neurons in the same layer are not connected, and they are fully connected in different layers. Each layer can be used as both input and output, so that information can flow bidirectionally between the two layers [28]. Because of the bidirectional structure, BAM neural networks can search forward and backward directions for desired patterns, and have practical applications in storing paired patterns or memories [29]. The BAM network overcomes the limitation of the Hopfield neural network that can only carry out self-association, which realizes the mutual association between different modes [30]. At the same time, BAM network system has good robustness [31]. When the information in the stored pattern pair is polluted or missing (it can be regarded as this memory has been forgotten), it can also reach a steady state to link the missing data and restore the original data

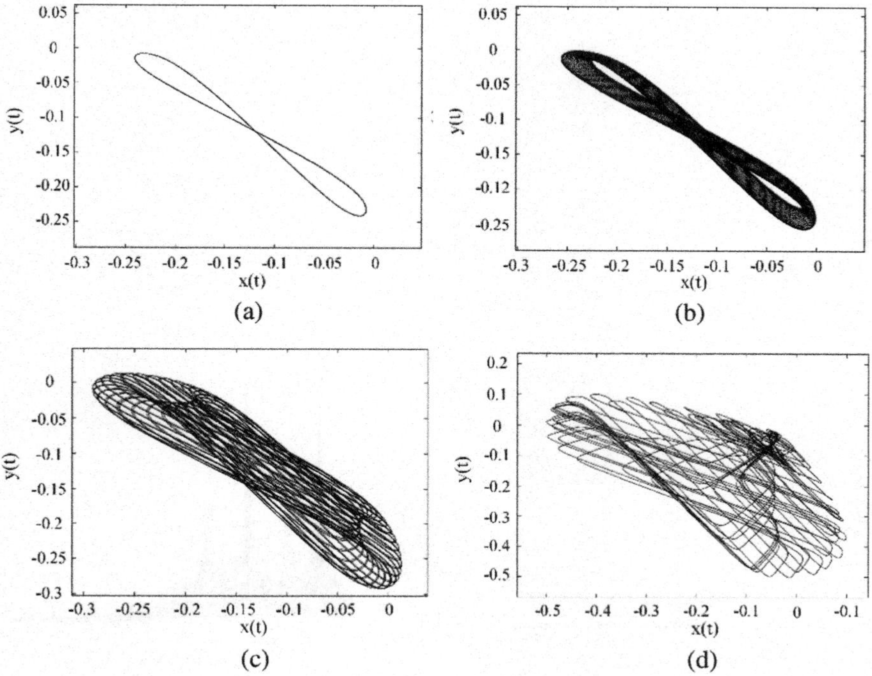

Fig. 7.3 The corresponding phase diagram for the time delay $\tau = 0.1$

through the dynamic addressing in the network [32, 33]. BAM networks provide the possibility for the mutual association of multi-modal information, mutual cooperation and completion, and the simulation of human cognitive process. At present, the BAM network system has been applied in many fields, such as biology, geology, physics, etc., and has been widely used in associative memory, pattern recognition, intelligent control, optimization problems and so on [34–36]. Therefore, it is very meaningful to discuss the dynamic behaviour of this widely used network system.

7.2.1 System Model

In this section, the BAM network system has 4 neurons and 2 time delays, which is

$$\begin{cases} \dot{x}_1(t) = -\mu_1 x_1(t) + c_{21} f_1(x_2(t-\tau_2)) + c_{31} f_1(x_3(t-\tau_2)) + c_{41} f_1(x_4(t-\tau_2)), \\ \dot{x}_2(t) = -\mu_2 x_2(t) + c_{12} f_2(x_1(t-\tau_1)), \\ \dot{x}_3(t) = -\mu_3 x_3(t) + c_{13} f_3(x_1(t-\tau_1)), \\ \dot{x}_4(t) = -\mu_4 x_4(t) + c_{14} f_4(x_1(t-\tau_1)), \end{cases} \tag{7.10}$$

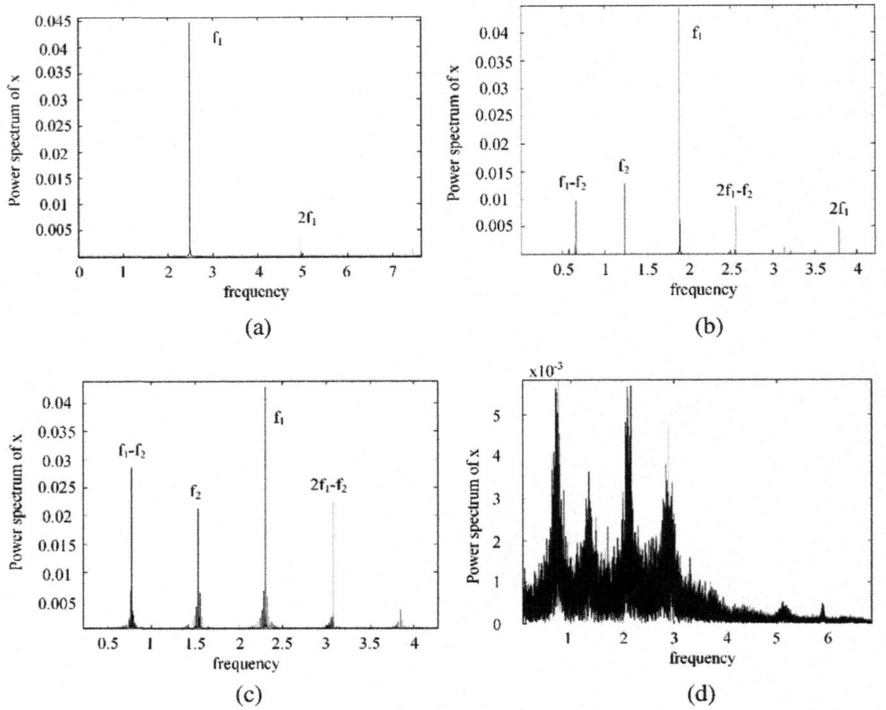

Fig. 7.4 The corresponding power spectrum diagram when the time delay $\tau = 0.1$

where $x_i (i = 1, 2, 3, 4)$ represents the state of the i-th neuron, $f_i(\bullet)$ is the activation function $f(u) = \tanh(u)$, $\mu_i > 0$ $(i = 1, 2, 3, 4)$ represents the decay rate inside each neuron. The non-negative constants τ_1 and τ_2 are the time delays because of the limited transmission speed between signals, c_{k1} and $c_{1k} (k = 2, 3, 4)$ are the neuron connection weights.

7.2.2 Fold-Hopf Bifurcation

Linearized the Eq. (7.10) at the trivial equilibrium point (0, 0, 0, 0), the characteristic equation of the linear system is obtained as

$$\lambda^4 + c_3\lambda^3 + c_2\lambda^2 + c_1\lambda + \left(d_2\lambda^2 + d_1\lambda + d_0\right)e^{-\lambda\tau} + c_0 = 0 \tag{7.11}$$

where $c_0 = \mu_1\mu_2\mu_3\mu_4$, $c_1 = \mu_1\mu_2\mu_3 + \mu_1\mu_2\mu_4 + \mu_1\mu_3\mu_4 + \mu_2\mu_3\mu_4$, $c_3 = \mu_1 + \mu_2 + \mu_3 + \mu_4$, $c_2 = \mu_1\mu_2 + \mu_1\mu_3 + \mu_1\mu_4 + \mu_2\mu_3 + \mu_2\mu_4 +$

$\mu_3\mu_4$, $d_0 = -c_{21}c_{12}\mu_3\mu_4 - c_{31}c_{13}\mu_2\mu_4 - c_{41}c_{14}\mu_2\mu_3$, $d_1 = -c_{21}c_{12}(\mu_3 + \mu_4) - c_{31}c_{13}(\mu_2 + \mu_4) - c_{41}c_{14}(\mu_2 + \mu_3)$, $d_2 = -c_{21}c_{12} - c_{31}c_{13} - c_{41}c_{14}$.

Based on the analysis of the characteristic roots, we get the parameter conditions when the Eq. (7.10) has a Fold-Hopf bifurcation, i.e.$c_0 + d_0 = 0$, and

$$\tau_j \overset{\text{def}}{=} \begin{cases} \eta_1[\eta_2 + 2(j+1)\pi], & P < 0, Q > 0 \\ \eta_1[\eta_2 + (j+1)\pi], & Q < 0 \\ \eta_1[\eta_2 + 2j\pi], & P > 0, Q > 0 \end{cases} \tag{7.12}$$

where $\eta_1 = 1/\omega$, $\eta_2 = \arctan(P/Q)$, $P = (c_1d_0 - c_0d_1)\omega - (c_3d_0 - c_2d_1 + c_1d_2)\omega^3 - (d_1 - c_3d_2)\omega^5$, $Q = d_2\omega^6 + (c_3d_1 - c_2d_2 - d_0)\omega^4 + (c_0d_2 + c_2d_0 - c_1d_1)\omega^2 - c_0d_0$.

We choose the connection weight c_{12} and coupling delay τ as two bifurcation parameters. Fixed system parameters $\mu_1 = 2$, $\mu_2 = 0.75$, $\mu_3 = \mu_4 = 1.8$, $c_{21} = c_{31} = c_{41} = 1$, $c_{13} = 2$, $c_{14} = 4$. The critical values of c_{12} and τ can be calculated as $\alpha_c = -1$, $\tau_0 = 5.7528$. At this time, the characteristic Eq. (7.11) has a pair of simple pure imaginary roots and one zero root, other eigenvalues all have negative real parts. Further, based on the central manifold theorem and normal form method, we get the normal form of the system

$$\begin{cases} \dot{r} = a_1 r + a_2 r^3 + a_3 z_3^2 r + h.o.t, \\ \dot{z}_3 = b_1 z_3 + b_2 z_3^3 + b_3 r^2 z_3 + h.o.t, \\ \dot{\nu} = -\omega\tau_0 - \mathrm{Im}(k_1)\varepsilon\delta_1 - \mathrm{Im}(k_2)\varepsilon\delta_2 + h.o.t, \end{cases} \tag{7.13}$$

where $a_1 = 0.277291\varepsilon\delta_1 + 0.0346338\varepsilon\delta_2$, $a_2 = -3.9698 < 0$, $a_3 = -4.59405 < 0$, $b_1 = l_1\varepsilon\delta_1 + l_2\varepsilon\delta_2 = 0.609746\varepsilon\delta_1$, $b_2 = l_3 = -1.82547 < 0$, $b_3 = l_4 = -9.81031 < 0$. By classifying the equilibrium points of (7.13), the dynamic behaviour near the Fold-Hopf bifurcation point (α_c, τ_0) in original Eq. (7.10) can be obtained, as Fig. 7.5, where the line marked with "Hopf" is $\tau = -2.25357 - 8.00637c_{12}$, while the line labeled "pithfork" is $c_{12} = -1$.

The theoretical analysis is verified by through numerical simulation of the original Eq. (7.10), as shown Fig. 7.6. It follows that the numerical simulations agree well with theoretical results. Furthermore, the stable periodic solutions in regions II and III are verified quantitatively. We consider the effect of time delay τ on the stable periodic solution when the connection strength c_{12} is fixed, as shown in Fig. 7.7. When time delay τ is near the critical value τ_0, the numerical results which obtained by normal form theory and central manifold theorem have high precision in both qualitatively and quantitatively, but when time delay τ is far from the critical value τ_0, the accuracy of the theory method is getting worse.

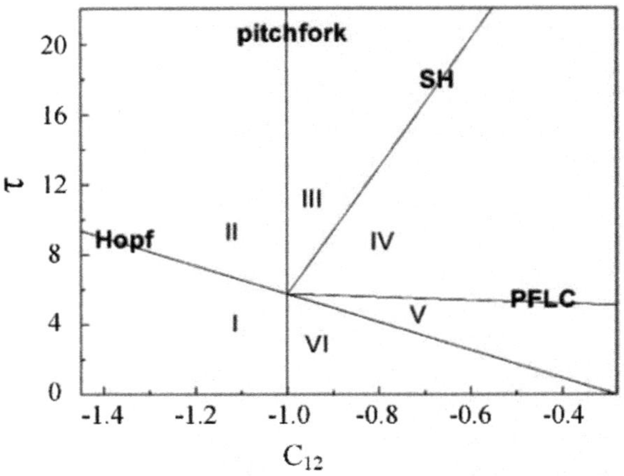

Fig. 7.5 Dynamic behaviors near the fold-hopf bifurcation point on the $c_{12} - \tau$ plane

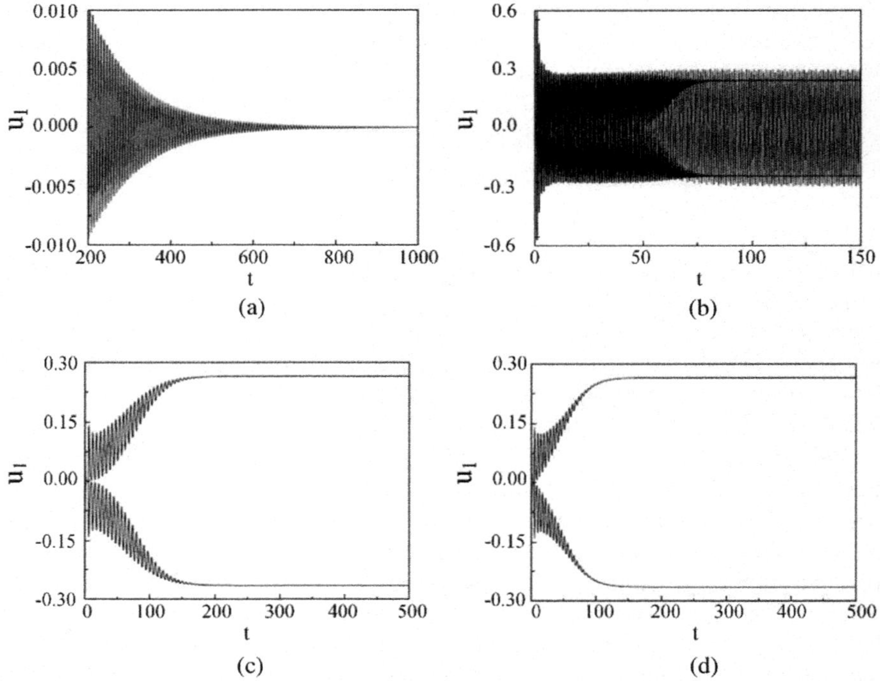

Fig. 7.6 **X** **a** Stable trival equilibrium point in region I where $(c_{12}, \tau) = (-1.2, 6)$. **b** Multi-stable state in region IV (two stable equilibrium points and a stable periodic solution) where $(c_{12}, \tau) = (-0.8, 8)$. **c** Two stable equilibrium points in V where $(c_{12}, \tau) = (-0.8, 5)$. **d** Two stable equilibrium points in VI where $(c_{12}, \tau) = (-0.8, 4)$

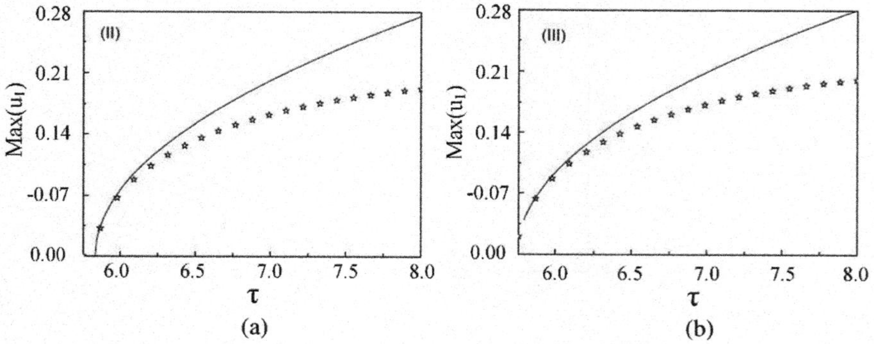

Fig. 7.7 An approximate solution (solid line) and numerical simulation (asterisk) of the (c_{12}, τ) periodic solution in regions II and III, where **a** $c_{12} = -1.01$, **b** $c_{12} = -0.999$

7.2.3 Delay-Induced Approximated Computation of Synchronous Periodic Solutions

Transform Eq. (7.10) with $u_1(t) = x_1(t + (\tau_2 - \tau_1)/2)$, $u_2(t) = x_2(t)$, $u_3(t) = x_3(t)$, $u_4(t) = x_4(t)$. Letting $\tau = (\tau_1 + \tau_2)/2$ and $\mu_k = \mu(k = 1, 2, 3, 4)$, then Eq. (7.10) is equivalent to the following system

$$\begin{cases} \dot{u}_1(t) = -\mu u_1(t) + c_{21} f(u_2(t-\tau)) + c_{31} f(u_3(t-\tau)) + c_{41} f(u_4(t-\tau)), \\ \dot{u}_2(t) = -\mu u_2(t) + c_{12} f(u_1(t-\tau)), \\ \dot{u}_3(t) = -\mu u_3(t) + c_{13} f(u_1(t-\tau)), \\ \dot{u}_4(t) = -\mu u_4(t) + c_{14} f(u_1(t-\tau)). \end{cases} \tag{7.14}$$

For Eq. (7.14), the necessary and sufficient conditions for the delay-induced synchronous periodic solution are $c_{12} < -\mu$, $c_{12} = c_{21} + c_{31} + c_{41}$, $c_{12} = c_{13} = c_{14}$, $\tau_c = [1/(c_{12}^2 - \mu^2)^{1/2}] \arccos(\mu/c_{12})$.

Now, we want to approximate the completely synchronous periodic solution $\mathbf{Z}(t)$ of Eq. (7.14) when $\varepsilon \neq 0$.

$$\mathbf{Z}(t) = \begin{Bmatrix} \varepsilon r \cos\big((\omega + \varepsilon^2\sigma)t + \theta\big) \\ \varepsilon r \cos\big((\omega + \varepsilon^2\sigma)t + \theta\big) \\ \varepsilon r \cos\big((\omega + \varepsilon^2\sigma)t + \theta\big) \\ \varepsilon r \cos\big((\omega + \varepsilon^2\sigma)t + \theta\big) \end{Bmatrix} \tag{7.15}$$

Using the PIS (perturbation-incremental scheme) method, the approximate expressions of periodic solution's amplitude and frequency is as follow

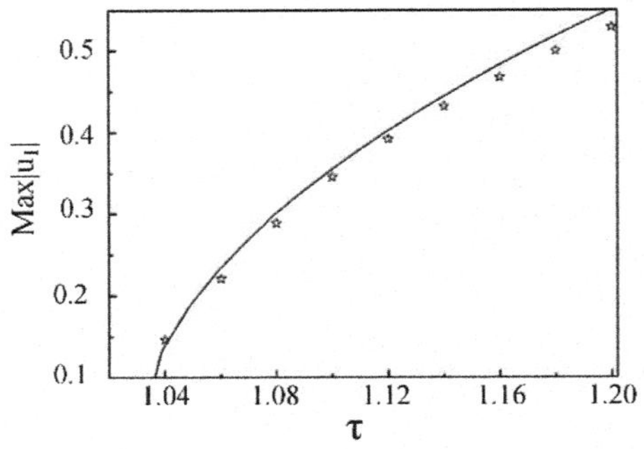

Fig. 7.8 Delay-induced Hopf bifurcations, where the solid line represents the perturbed solution (7.17), and the asterisk represents the numerical simulation

$$\varepsilon r = 2\sqrt{2}\omega\sqrt{-\frac{(c_{13}c_{31}+c_{14}c_{41}+c_{12}(2c_{21}+c_{31}+c_{41}))}{c_{12}}}$$

$$\frac{\sqrt{\tau-\tau_c}}{\sqrt{\frac{(\mu+(\mu^2+\omega^2)\tau_c)f^{(3)}(0)\left(c_{12}\sum\limits_{k=2}^{4}c_{k1}+\sum\limits_{k=2}^{4}c_{1k}c_{k1}\right)}{c_{12}}}} \tag{7.16}$$

$$\varepsilon^2\sigma = \frac{-\mu^2\omega-\omega^3}{\mu+\mu^2\tau_c+\omega^2\tau_c}(\tau-\tau_c).$$

Choose $\mu = 2$, $c_{12} = -3$, $c_{21} = c_{31} = c_{41} = -1$, substitute into (7.16), we get

$$\varepsilon r = 1.33277\sqrt{\tau - 1.02883}, \quad \varepsilon^2\sigma = -1.78736(\tau - 1.02883). \tag{7.17}$$

From Figs. 7.8 and 7.9, it is obvious that the analytical solution (7.17) obtained by the PIS method agrees well with the numerical simulation results, which not only demonstrate that the PIS method has high precision, but also verifies the correctness of the results in this section.

7.3 Hyperchaos and Synchronous Control of Neural Network System with Time Delay

In the above research, we focus on the complex dynamic behavior of the low-dimensional neural network system, including equilibrium point, periodic solution, bifurcation and chaos. In particular, based on the PIS method, the approximate calculation of the synchronous periodic solution was given. However, the generated chaotic signal in low-dimensional chaotic system is easy to extract which resulting in poor

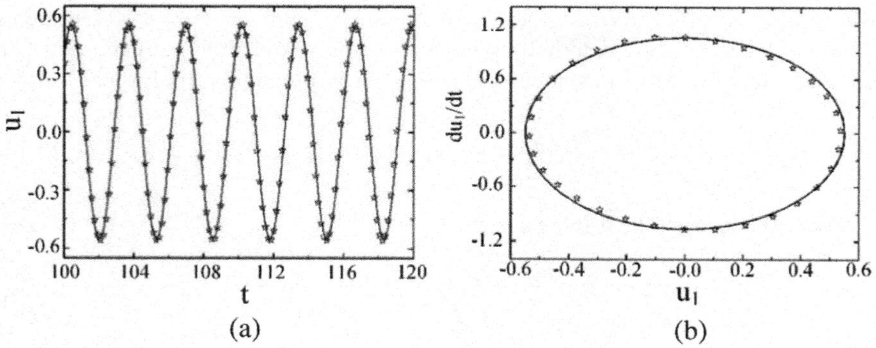

Fig. 7.9 Comparing perturbed with numerical solutions on the time history diagram (**a**) and phase diagram (**b**), where $\tau = 1.2$

security of the system [37, 38]. In order to increase safety, scholars have reconstructed high-dimensional chaotic systems by using nonlinear dynamical prediction methods [39] and neural network methods [40]. The time-delayed chaotic system can generate multiple positive Lyapunov exponents to become a hyperchaotic system even when it only has a simple structure [41, 42]. Applying this system to the confidential communication improves safety and reliability. Whether real or artificial, many neural networks are time-delayed systems that can produce chaotic or even hyperchaotic behavior [43, 44]. Therefore, in this section, we focus on a group of n-dimensional time-delayed neural network systems. We not only study the chaos, hyperchaos and synchronization of the system, but also consider the self-adapting synchronous control of the drive responses in neural network systems with uncertain parameters. Furthermore, we design a class of integral sliding mode controllers to realize the chaotic synchronization in neural networks with different delays [45, 46].

7.3.1 Hyperchaotic Synchronization in Neural Networks with Time Delay

In this section, we propose a class of time-delayed neural network systems

$$
\begin{aligned}
\frac{dx_i(t)}{dt} &= -a_i x_i(t) + \sum_{j=1}^{n} b_{ij} f\left(x_j(t)\right) \\
&+ \sum_{j=1}^{n} c_{ij} f\left(x_j(t-\tau)\right) + I_i, \\
& i = 1, 2, \cdots, n,
\end{aligned}
\tag{7.18}
$$

where all the coefficients a_i, b_{ij}, c_{ij} are real numbers, and $a_i > 0$. $\tau \geq 0$ represents the time delay, I_i is the external input. The activation function $f(x)$ is a constant nonlinear function. For convenience, Eq. (7.18) is written in matrix form as follows

$$\frac{dx(t)}{dt} = -Ax(t) + Bf(x(t)) + Cf(x(t-\tau)) + I. \tag{7.19}$$

We first construct the driving system

$$\begin{aligned} \dot{x} &= -Ax + Bf(x) + Cf(x(t-\tau)) + I, \\ x(t) &= \Phi_1(t), \ t \in [-\tau, 0]. \end{aligned} \tag{7.20}$$

The corresponding response system is

$$\begin{aligned} \dot{y} &= -Ay + Bf(y) + Cf(y(t-\tau)) + I + U(m(x, x(t-\tau)), y, y(t-\tau)), \\ y(t) &= \Phi_2(t), \ t \in [-\tau, 0]. \end{aligned} \tag{7.21}$$

To realize synchronization of Eqs. (7.20) and (7.21), a suitable nonlinear controller needs to be designed. Set

$$U(m(x, x(t-\tau)), y, y(t-\tau)) = m(x, x(t-\tau)) - m(y, y(t-\tau)), \tag{7.22}$$

where

$$m(x, x(t-\tau)) = Bf(x) + Cf(x(t-\tau)) - K_1x - K_2x(t-\tau). \tag{7.23}$$

$m(x, x(t-\tau))$ is the transmission signal, also known as the synchronization signal, K_1 and K_2 are the constant control matrices.

By using theoretical analysis, we get that if there exists K_1 and K_2 which satisfy

$$(K_1 - A)^T + (K_1 - A) + K_2^T K_2 + I < 0, \tag{7.24}$$

Equations (7.20) and (7.21) are asymptotically synchronized, where I represents the unit matrix.

Further, if there are positive definite matrices K_1 and K_2, satisfying

$$\lambda_{\max}(K_1) + \lambda_{\max}(K_2) < \lambda_{\min}(A), \tag{7.25}$$

the Eqs. (7.20) and (7.21) are exponentially synchronized, where A is the coefficient matrix in the Eqs. (7.20) and (7.21).

As a numerical simulation example of the theoretical analysis in this section, we construct the following neural network system

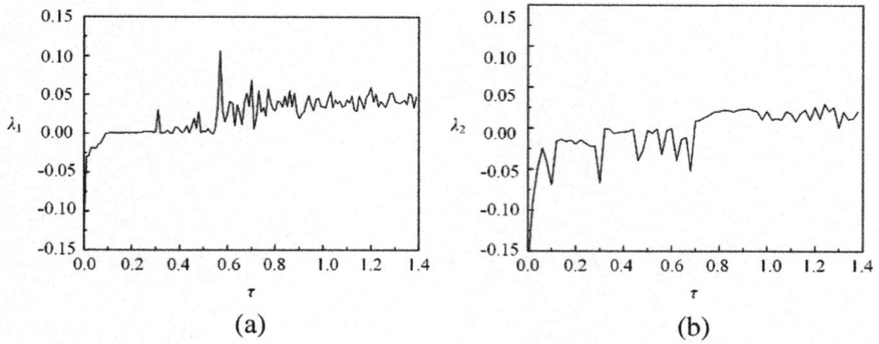

Fig. 7.10 Largest and second largest Lyapunov exponents

$$\begin{cases} \dot{x}_1(t) = pf(x_2(t)) + qf(x_2(t-\tau)) - x_1(t), \\ \dot{x}_2(t) = qf(x_1(t)) + pf(x_1(t-\tau)) - x_2(t). \end{cases} \tag{7.26}$$

The parameters are taken as $p = 3, q = 5$. Activation function $f(x) = \sin(x)$. The initial condition is set to $x(t) \equiv [2, 2]^T,\ t \in [-\tau, 0]$.

We first need to determine the parameter values to make the system have complex hyperchaotic attractor. For this purpose, we calculate the largest and second largest Lyapunov exponents of the system, as shown in Fig. 7.10. It follows that when time delay $\tau = 0$, $\lambda_1 < 0$ and $\lambda_2 < 0$, Eq. (7.26) has an equilibrium point. When time delay τ changes from 0 to 0.1, λ_1 gradually approaches 0 from a negative value and λ_2 is still less than 0, the system will produce a stable periodic orbit. When time delay τ varies from 0.55 to 0.7, $\lambda_1 > 0$ and $\lambda_2 < 0$ or $\lambda_2 = 0$, the system has a chaotic attractor, where it only has one positive Lyapunov exponent. When time delay $0.75 < \tau < 1.4$, $\lambda_1 > 0$ and $\lambda_2 > 0$, more than one positive Lyapunov exponent exist in the system. The system has a hyperchaotic oscillation, and its corresponding phase diagram is shown in Fig. 7.11.

In order to verify the chaotic and hyperchaotic oscillations in the system, we give the Poincaré diagram of the corresponding attractor, where we take the Poincaré section as $x_1(t-\tau) = 0$, as Fig. 7.12. It follows that the Poincaré diagram is neither a few countable points nor a closed curve, which indicates that the dynamical behaviour of the original system is neither a periodic solution nor a quasi-periodic oscillation. Lots of discretely distributed points indicates the presence of the chaotic attractor. And obviously, Fig. 7.12b is much more complicated than Fig. 7.12a.

Further, for numerical simulations of the drive-response system, we fix time delay $\tau = 1.0$. At this time, if we take $K_1 = diag(0.3, 0.3)$ and $K_2 = diag(-0.5, -0.5)$ which satisfy Eq. (7.24), the drive-response system exhibits the asymptotic synchronization, as shown in Fig. 7.13a. Further, when we take $K_1 = diag(0.2, 0.2)$, $K_2 = diag(0.3, 0.3)$, which satisfy the condition (7.25), the drive-response system presents the exponential synchronization, as shown in Fig. 7.13b.

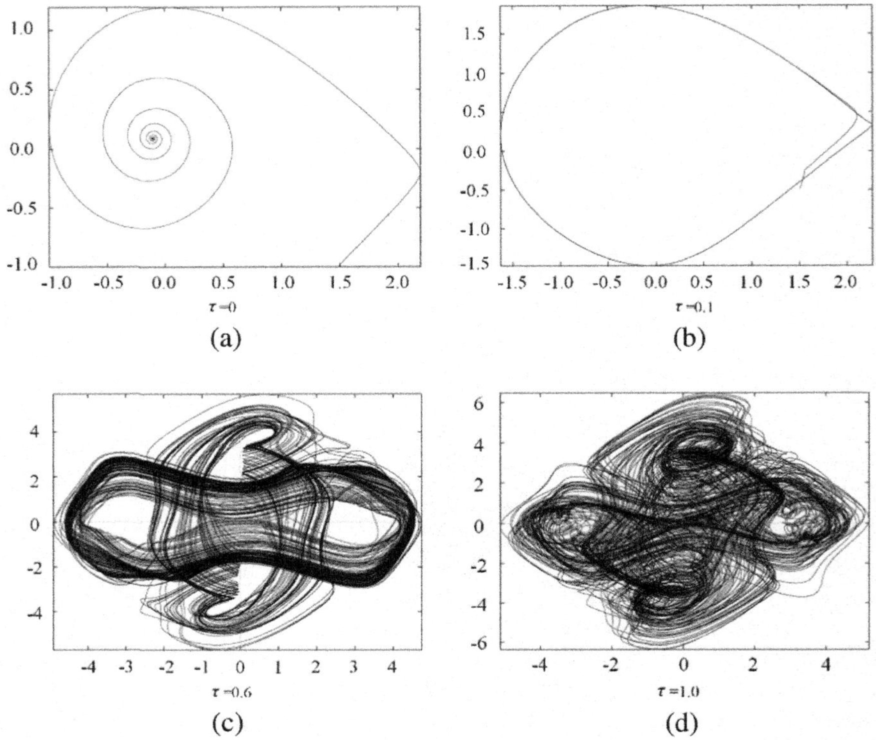

Fig. 7.11 The attractor projection of Eq. (7.26) on the plane $x_1(t) - x_2(t)$

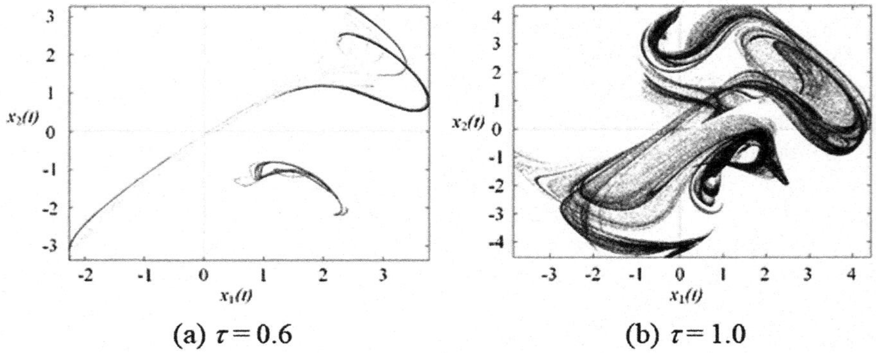

Fig. 7.12 Poincaré crossp section of attractor

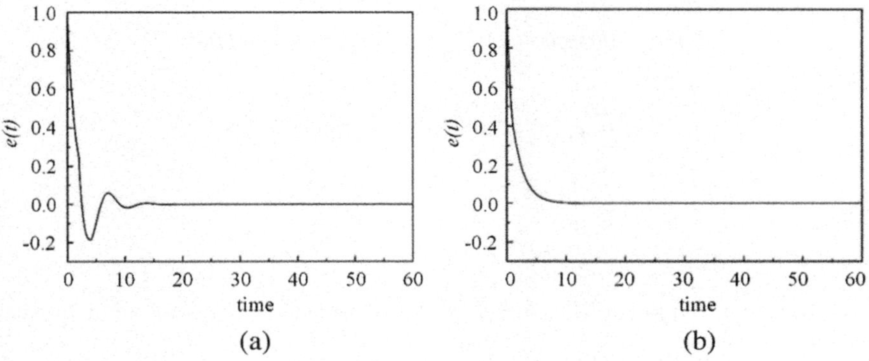

Fig. 7.13 **a** Asymptotic convergence and **b** exponential convergence of $e(t)$ illustrate the asymptotic and exponential synchronization of the drive-response system.

7.3.2 Adaptive Synchronization with Uncertain Parameters

In the real world, due to the influence of surrounding environment and external disturbances, some parameter values of the system cannot be accurately determined. The synchronization is likely to be destroyed by these uncertainties. The adaptive synchronization method can resolve this problem very well. In this section, we study the problem of parameter identification and adaptive synchronization for the time-delayed chaotic neural network systems with uncertain parameters. The driving system is the time-delayed chaotic neural network studied in the previous section

$$\dot{x}_i(t) = \sum_{j=1}^{n} b_{ij} f_j\big(x_j(t)\big) + \sum_{j=1}^{n} c_{ij} f_j\big(x_j(t-\tau)\big) - a_i x_i(t) + I_i\,, \quad i = 1, 2, \cdots, n. \tag{7.27}$$

The response system can be represented by the following delay differential equation

$$\dot{y}_i(t) = \sum_{j=1}^{n} b_{ij} f_j\big(y_j(t)\big) + \sum_{j=1}^{n} c_{ij} f_j\big(y_j(t-\tau)\big) - a_i y_i(t) + I_i + u_i(t), \quad i = 1, 2, \cdots, n, \tag{7.28}$$

where $y_i(t)$ represents the state variable of the i-th neuron, $u_i(t)$ is an external control input designed to achieve a certain control. Subtracting the driving Eq. (7.27) from the response Eq. (7.28) yields the following error dynamical system

$$\begin{aligned} \dot{e}_i(t) = u_i(t) - a_i e_i(t) + \sum_{j=1}^{n} b_{ij}[f_j(y_j(t)) - f_j(x_j(t))] \\ + \sum_{j=1}^{n} c_{ij}[f_j(y_j(t-\tau)) - f_j(x_j(t-\tau))], \end{aligned} \tag{7.29}$$

where $i, j = 1, 2, \ldots, n$.

Case 1:

If all parameters in Eqs. (7.27) and (7.28) are unknown, the Lyapunov function can be constructed as

$$V(t) = \frac{1}{2}\left(\sum_{i=1}^{n} e_i^2(t) + \sum_{i=1}^{n} \tilde{a}_i^2(t) + \sum_{i=1}^{n}\sum_{j=1}^{n} \tilde{b}_{ij}^2(t) + \sum_{i=1}^{n}\sum_{j=1}^{n} \tilde{c}_{ij}^2(t)\right), \tag{7.30}$$

where $\tilde{a}_i(t) = a_i - \hat{a}_i(t)$, $\tilde{b}_{ij}(t) = b_{ij} - \hat{b}_{ij}(t)$, and $\tilde{c}_{ij}(t) = c_{ij} - \hat{c}_{ij}(t)$. $\hat{a}_i(t)$, $\hat{b}_{ij}(t)$ and $\hat{c}_{ij}(t)$ are respectively the estimated values of the unknown parameters a_i, b_{ij}, and c_{ij}. We designed the adaptive controller and the update rule for the estimated parameters as follows

$$\begin{aligned} u_i(t) = \hat{a}_i(t)e_i(t) + \sum_{j=1}^{n} \hat{b}_{ij}(t)[f_j(x_j(t)) - f_j(y_j(t))] \\ + \sum_{j=1}^{n} \hat{c}_{ij}(t)[f_j(x_j(t-\tau)) - f_j(y_j(t-\tau))] - k_i e_i(t), \end{aligned} \tag{7.31}$$

where

$$\begin{cases} \dot{\hat{a}}_i(t) = -e_i(t)e_i(t), \\ \dot{\hat{b}}_{ij}(t) = e_i(t)\big[f_j\big(y_j(t)\big) - f_j\big(x_j(t)\big)\big], \quad i, j = 1, 2, \ldots, n, k_i > 0 \\ \dot{\hat{c}}_{ij}(t) = e_i(t)\big[f_j\big(y_j(t-\tau)\big) - f_j\big(x_j(t-\tau)\big)\big], \end{cases}$$

are the given parameters.

Case 2:

If the parameters b_{ij} and c_{ij} in Eqs. (7.27) and (7.28) are known, or either of them is known, we need to firstly assume the activation function $f_j : R \to R$, $j \in \{1, 2, \ldots n\}$ to satisfy the Lipschitz condition, i.e. there is always a Lipschitz constant L_p, making $\left|f_j(\xi) - f_j(\eta)\right| \le L_p|\xi - \eta|(\forall \xi, \eta \in R$, and $\xi \ne \eta)$. Moreover, the state variables in the error system are $|e_i(t)| = |y_i(t) - x_i(t)| \le 2B_d$, $i \in \{1, 2, \ldots, n\}$. Let's assume b_{ij} and c_{ij} are known parameters in the Eqs. (7.27) and (7.28) respectively, a_i is an unknown parameter, then the Lyapunov function, adaptive controller and update rule for estimated parameters are designed as

$$V(t) = \frac{1}{2}\left(\sum_{i=1}^{n} e_i^2(t) + \sum_{i=1}^{n} \tilde{a}_i^2(t)\right). \tag{7.32}$$

$$\begin{aligned} u_i(t) &= \hat{a}_i(t)e_i(t) - 2nB_dL_P(\beta_i + \gamma_i)g(e_i) - k_ie_i(t), \\ \dot{\hat{a}}_i(t) &= -e_i(t)e_i(t). \end{aligned} \tag{7.33}$$

When some parameters in Eqs. (7.27) and (7.28) are unknown, synchronization can be achieved under the action of adaptive controller and the update rules (7.33) for the estimated parameters.

As a simulation example in this section, we take the hyperchaotic neural network system proposed in the previous section as an example, assuming that the following systems are the driving system and the response system respectively

$$\begin{cases} \dot{x}_1(t) = -a_1x_1(t) + b_{12}\sin(x_2(t)) + c_{12}\sin(x_2(t-1)), \\ \dot{x}_2(t) = -a_2x_2(t) + b_{21}\sin(x_1(t)) + c_{21}\sin(x_1(t-1)), \end{cases} \tag{7.34}$$

and

$$\begin{cases} \dot{y}_1(t) = -a_1y_1(t) + b_{12}\sin(y_2(t)) \\ \quad +c_{12}\sin(y_2(t-1)) + u_1(t), \\ \dot{y}_2(t) = -a_2y_2(t) + b_{21}\sin(y_1(t)) \\ \quad +c_{21}\sin(y_1(t-1)) + u_2(t). \end{cases} \tag{7.35}$$

For case 1, all parameters in Eqs. (7.34) and (7.35) are unknown. The initial conditions of the driving system and the response system are selected as $x(t) \equiv (2, 2)^T$ and $y(t) \equiv (-2, -2)^T$, where $t \in [-1, 0]$. The controller parameters $k_i(i = 1, 2)$ set to be $k_1 = k_2 = 3.0$, and the unknown parameters' initial estimates are taken as $\hat{a}_1(0) = \hat{a}_2(0) = 3.6$, $\hat{b}_{12}(0) = 2.5$, $\hat{b}_{21}(0) = 4.5$, $\hat{c}_{12}(0) = 4.5$, and $\hat{c}_{21}(0) = 2.0$. We use Simulink module of Matlab for simulation, and the result is shown in Fig. 7.14.

In case 2, we hypothesize that only parameter $a_i(i = 1, 2)$ is unknown. The initial states of the Eqs. (7.34) and (7.35) are respectively set as $x(t) \equiv (2, 2)^T$ and $y(t) \equiv (1, 3)^T$, where $t \in [-1, 0]$. And the initial estimates of the unknown parameters are taken separately $\hat{a}_1(0) = 2.2$ and $\hat{a}_2(0) = 0.4$. According to theoretical analysis and numerical simulations, the controller parameters are set to the following values: $n = 2$, $B_d = 7$, $L_p = 1$, $\beta_1 = 3$, $\beta_2 = 5$, $\gamma_1 = 5$, $\gamma_2 = 3$ and $k_1 = k_2 = 1.0$. The corresponding simulation results are shown in Fig. 7.15. It follows from Figs. 7.14 and 7.15 that the error variables and converge rapidly to zero respectively, namely the drive and response systems obtain synchronization. At the same time we can see the estimate values of "unknown" parameters tend to the true values quickly. This illustrative example verifies the effectiveness of the adaptive controllers.

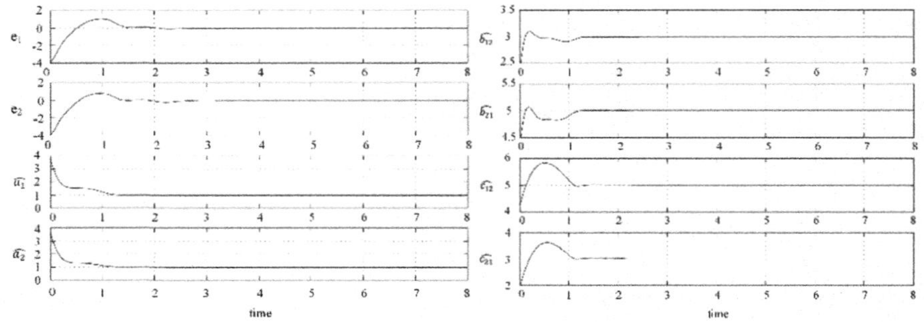

(a) (b)

Fig. 7.14 Time evolution of **a** error variables and **b** estimated parameters (all parameters unknown)

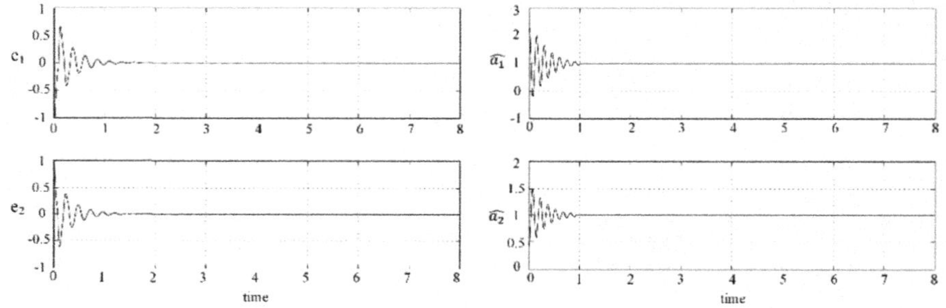

(a) (b)

Fig. 7.15 Time evolution of **a** error variable and **b** the estimated parameters (parameter $a_i\,(i = 1, 2)$ are unknown)

7.3.3 *Projective Synchronization Based on Sliding Mode Variable Structure Control*

We assume that the driving system is

$$\dot{x}(t) = Ag(x(t)) + Bg(x(t-\tau)) - Cx(t) + H\,. \tag{7.36}$$

The measured output of the Eq. (7.36) depends on the instantaneous state and the state before a certain period of time, it can be expressed as the following equation

$$y(t) = J_1 x(t) + J_2 x(t-\tau)\,, \tag{7.37}$$

where $y(t) \in R^m$, $J_1, J_2 \in R^{m\times n}$ are known constant matrices.

The response system is represented as follows

$$\dot{z}(t) = Pf(z(t)) + Qf(z(t-\tau)) - Dz(t) + W + u(t)\,, \tag{7.38}$$

where $z(t) \in R^n$ is the state vector in the response system, $u(t)$ is the control input to be designed. Activation function $g(\cdot)$ and $f(\cdot)$ are Lipschitz continuous and monotonically nondecreasing. Thus, there are constants $\lambda_{1j} > 0$, $\lambda_{2j} > 0$, $(j = 1, 2, \ldots n)$, which satisfy:$0 \le [g_j(\theta) - g_j(\varphi)]/(\theta - \varphi) \le \lambda_{1j}$ and $0 \le [f_j(\theta) - f_j(\varphi)]/(\theta - \varphi) \le \lambda_{2j}$ for any $\theta \ne \varphi \in R$. For convenience, we denote as $\Lambda_1 = diag(\lambda_{11}, \lambda_{12}, \ldots \lambda_{1n})$ and $\Lambda_2 = diag(\lambda_{21}, \lambda_{22}, \cdots \lambda_{2n})$.

In this section, we propose a suitable controller $u(t)$ to make

$$\lim_{t\to\infty} ||e(t)|| = \lim_{t\to\infty} ||z(t) - \alpha x(t)|| = 0 , \tag{7.39}$$

here $|| \cdot ||$ represents the Euclidean norm of a vector. If Eq. (7.39) is satisfied, we believe that Eqs. (7.36) and (7.38) have achieved the projective synchronization.

We define the projection synchronization deviation between Eqs. (7.36) and (7.38) as $e(t) = z(t) - \alpha x(t)$, where $\alpha \ne 0$. Then we can get the error system from (7.36) and (7.38) as follows

$$\begin{aligned}\dot{e}(t) = &-De(t) + P\psi(e(t)) + Q\psi(e(t-\tau)) + u(t)\\ &+\alpha(C - D)x(t) - \alpha Ag(x(t)) - \alpha Bg(x(t-\tau))\\ &+Pf(\alpha x(t)) + Qf(\alpha x(t-\tau)) + W - \alpha H ,\end{aligned} \tag{7.40}$$

where $\psi(e(t)) = f(z(t)) - f(\alpha x(t))$.

Construct a suitable sliding surface

$$\begin{aligned}S(t) = e(t) + \int_0^t &[Ce(s) - P\psi(e(s)) - Q\psi(e(s-\tau))\\ &-K(\alpha y(s) - J_1 z(s) - J_2 z(s-\tau))]\, ds ,\end{aligned} \tag{7.41}$$

where the gain matrix $K \in R^{n\times m}$ needs to be chosen appropriately.

Thus, from (7.37) and (7.40), we can get

$$\begin{aligned}S(t) = e(0) - \int_0^t &[(D - C - KJ_1)e(s) - KJ_2 e(s-\tau) - u(s)\\ &+ \alpha(D - C)x(s) + \alpha Ag(x(s)) + \alpha Bg(x(s-\tau))\\ &- Pf(\alpha x(s)) - Qf(\alpha x(s-\tau)) - W + \alpha H]ds ,\end{aligned} \tag{7.42}$$

where $e(0)$ is the initial condition of the error Eq. (7.40).

According to the sliding mode control theory, there will be $S(t) = 0$ and $\dot{S}(t) = 0$ when the state trajectory of the error Eq. (7.40) enters the sliding mode. So, an equivalent control law from Eq. (7.42) and $\dot{S}(t) = 0$ can be designed as follows

$$\begin{aligned}u_{eq}(t) = -KJ_1e(t) - KJ_2e(t-\tau) + (D-C)z(t) + \alpha Ag(x(t)) \\ +\alpha Bg(x(t-\tau)) - Pf(\alpha x(t)) - Qf(\alpha x(t-\tau)) - W + \alpha H.\end{aligned} \tag{7.43}$$

Substituting Eq. (7.43) into Eq. (7.40), we can get the sliding mode dynamics

$$\dot{e}(t) = -(C + KJ_1)e(t) - KJ_2e(t-\tau) + P\psi(e(t)) + Q\psi(e(t-\tau)). \tag{7.44}$$

In this section, the approach control law is designed as $u_r(t) = k_w u_w(t)$, thus the total control $u(t)$ is determined as

$$u(t) = u_{eq}(t) + u_r(t) = u_{eq}(t) + k_w u_w(t) , \tag{7.45}$$

where k_w is the positive conversion gain, and $u_w(t)$ can be obtained

$$u_w(t) = \text{sgn}(S(t)) , \tag{7.46}$$

here $\text{sgn}(\cdot)$ represents a sign function. However, the sign function will cause the control input to produce the chattering phenomenon. To reduce the chattering on the simulation results, we use the saturation function $\text{sat}(S(t)/\delta)$ which is defined as follows to replace the symbolic function $\text{sgn}(S(t))$,

$$\text{sat}\left(\frac{S(t)}{\delta}\right) = \left[\text{sat}\left(\frac{S_1(t)}{\delta}\right), \text{sat}\left(\frac{S_2(t)}{\delta}\right), \ldots, \text{sat}\left(\frac{S_n(t)}{\delta}\right)\right]^T, \tag{7.47}$$

where

$$\text{sat}\left(\frac{S_i(t)}{\delta}\right) = \begin{cases} +1 , & S_i(t) > \delta , \\ \kappa S_i(t) , & |S_i(t)| \le \delta , \quad i = 1, 2, \ldots, n \\ -1 , & S_i(t) < -\delta , \end{cases} \tag{7.48}$$

with the boundary layer's thickness $\delta > 0$, the gain constant $\kappa > 0$ and $S(t) = [S_1(t), S_2(t), \ldots, S_n(t)]^T$.

Theorem 7.1 *If there are matrices $M = M^T > 0$, $N_i = N_i^T > 0$ $(i = 1, 2)$, $R_j = R_j^T > 0$ $(j = 1, 2, 3, 4)$ and diagonal $\Gamma_1 = diag(\gamma_{11}, \gamma_{12}, \ldots, \gamma_{1n}) > 0$, $\Gamma_2 = diag(\gamma_{21}, \gamma_{22}, \ldots, \gamma_{2n}) > 0$ such that the linear matrix inequality* (7.49) *hold, the sliding mode dynamical Eq.* (7.44) *is globally asymptotically stable for any constant delay τ satisfying $0 < \tau \le \tau_m$.*

$$\Pi = \begin{bmatrix} \Pi_{11} & \Pi_{12} & \Pi_{13} & \Pi_{14} \\ \Pi_{12}^T & \Pi_{22} & \Pi_{23} & \Pi_{24} \\ \Pi_{13}^T & \Pi_{23}^T & \Pi_{33} & \Pi_{34} \\ \Pi_{14}^T & \Pi_{24}^T & \Pi_{34}^T & \Pi_{44} \end{bmatrix} < 0 , \tag{7.49}$$

where

$$\Pi_{11} = -M(C + KJ_1 + KJ_2) - (C + KJ_1 + KJ_2)^T M + N_1$$
$$+ \tau_m (C + KJ_1)^T R_1 (C + KJ_1) + \tau_m M K J_2 \sum_{j=1}^{4} R_j^{-1} J_2^T K^T M,$$
$$\Pi_{12} = 0, \Pi_{13} = MP + \Gamma_1 \Lambda_2, \Pi_{14} = MQ, \Pi_{22} = -N_1 + \tau_m J_2^T K^T R_2 K J_2$$
$$\Pi_{23} = 0, \Pi_{24} = \Gamma_2 \Lambda_2, \Pi_{33} = N_2 + \tau_m P^T R_3 P - 2\Gamma_1, \Pi_{34} = 0,$$
$$\Pi_{44} = -N_2 + \tau_m Q^T R_4 Q - 2\Gamma_2. \tag{7.50}$$

Further, as a numerical example, we hypothesize that the Hopfield neural network with time-delay chaos is a driving system

$$\begin{bmatrix} \dot{x}_1(t) \\ \dot{x}_2(t) \end{bmatrix} = -C \begin{bmatrix} x_1(t) \\ x_2(t) \end{bmatrix} + A \begin{bmatrix} g_1(x_1(t)) \\ g_2(x_2(t)) \end{bmatrix} + B \begin{bmatrix} g_1(x_1(t-\tau)) \\ g_2(x_2(t-\tau)) \end{bmatrix}, \tag{7.51}$$

where $g_j(x_j(t)) = \tanh(x_j(t)),\ j = 1, 2$.

The response system is the time-delayed chaotic cellular neural network

$$\begin{bmatrix} \dot{z}_1(t) \\ \dot{z}_2(t) \end{bmatrix} = -D \begin{bmatrix} z_1(t) \\ z_2(t) \end{bmatrix} + P \begin{bmatrix} f_1(z_1(t)) \\ f_2(z_2(t)) \end{bmatrix} + Q \begin{bmatrix} f_1(z_1(t-\tau)) \\ f_2(z_2(t-\tau)) \end{bmatrix} + u(t)\ , \tag{7.52}$$

where $f_j(z_j(t)) = 0.5(|z_j(t) + 1| - |z_j(t) - 1|),\ j = 1, 2$. The parameters of the driving Eq. (7.51) and the response Eq. (7.52) are respectively taken as

$$C = \mathrm{diag}(1, 1), A = \begin{bmatrix} 2 & -0.1 \\ -5 & 4.5 \end{bmatrix}, B = \begin{bmatrix} -1.5 & -0.1 \\ -0.2 & -4 \end{bmatrix}, D = \mathrm{diag}(1, 1),$$

$$P = \begin{bmatrix} 1 + \pi/4 & 20 \\ 0.1 & 1 + \pi/4 \end{bmatrix}, Q = \begin{bmatrix} -1.3\sqrt{2}\pi/4 & 0.1 \\ 0.1 & -1.3\sqrt{2}\pi/4 \end{bmatrix}, \tau = 1.0.$$

The initial conditions are $[x_1(\varphi), x_2(\varphi)] = [0.4, 0.6]$ and $[z_1(\varphi), z_2(\varphi)] = [0.01, 0.1]$, where $\varphi \in [-1, 0]$. The chaotic behaviour of the driving Eq. (7.51) and the responding Eq. (7.52) without controller $u(t)$ is shown in Figs. 7.16 and 7.17, respectively.

Obviously, the activation functions $g(\cdot)$ and $f(\cdot)$ satisfy Theorem 7.1 and $\Lambda_1 = \Lambda_2 = I$, where I is the identity matrix. The parameters of the measured output (7.37) are fixed as $J_1 = (0, 0.1)$, $J_2 = (0, 1)$. The gain matrix can be chosen as $K = (6,\ 6)^T$. According to Theorem 7.1, there are 26 variables need to be calculated. The maximum of time delay τ_m can be obtained by solving the linear matrix inequality (7.49) through the LMI toolbox of Matlab, which is 1.1683. The conversion gain k_w and the boundary layer thickness δ are chosen as $k_w = 0.5$, $\delta = 0.05$. For scale factor $\alpha = 1.5$, the simulation results when $\tau = 1.0$ are displayed in Fig. 7.18 and Fig. 7.19.

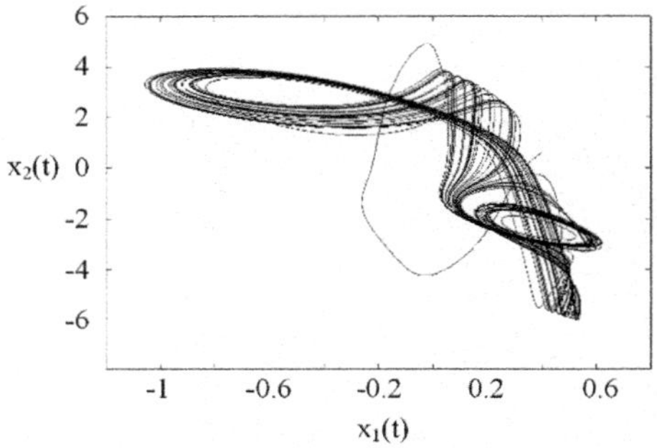

Fig. 7.16 Chaotic attractors for driving systems

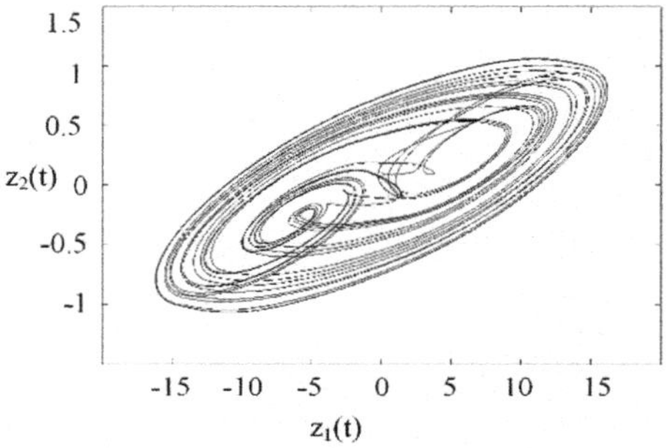

Fig. 7.17 Chaotic attractors for uncontrolled response systems

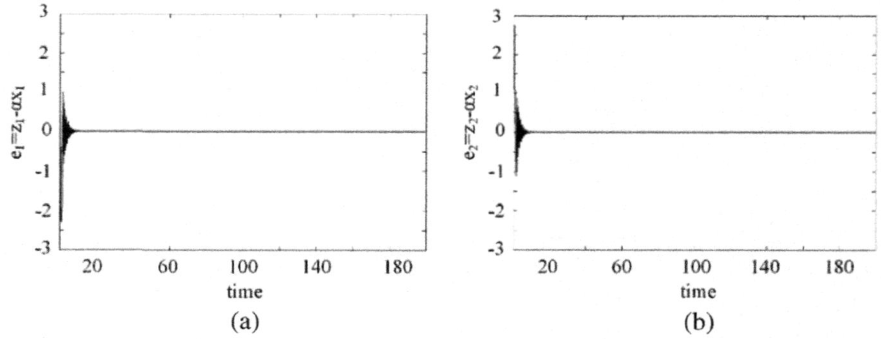

Fig. 7.18 Time history diagram of error system ($\alpha = 1.5$)

It can be seen from Fig. 7.18 that the projective synchronization errors $e_1(t)$ and $e_2(t)$ between Eqs. (7.51) and (7.52) are stabilized to zero after a short period of time. That is, the driving Eq. (7.51) and the responding Eq. (7.52) have reached a state of projective synchronization. After the projective synchronization state is generated when $\alpha = 1.5$, the time series of $x(t)$ and $z(t)$ are shown in Fig. 7.19a and c. The corresponding projective synchronization manifolds are shown in Fig. 7.19b and d,

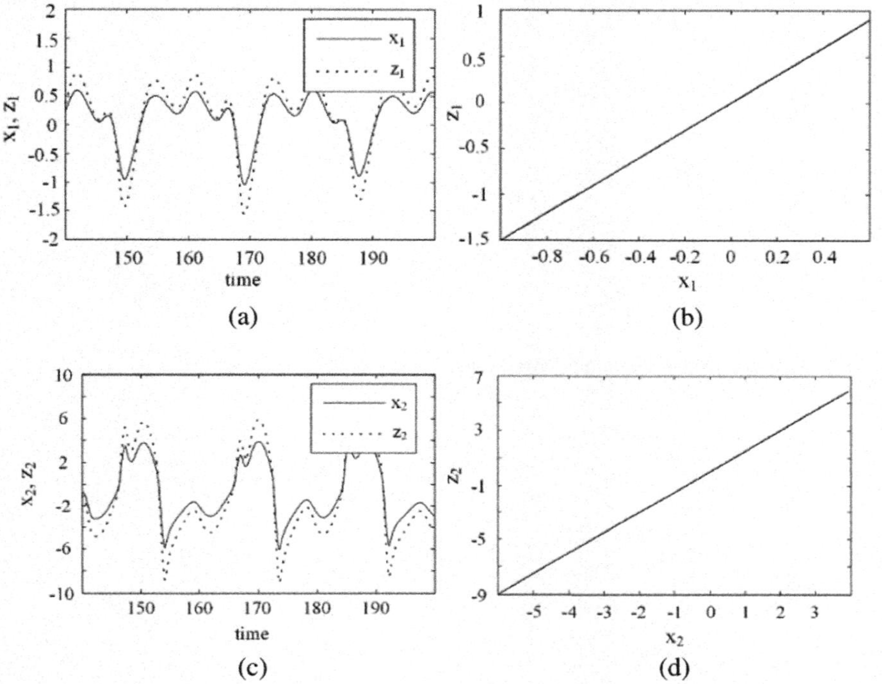

Fig. 7.19 **a** and **c** Projective synchronization of $x(t)$ and $z(t)$ when $\alpha = 1.5$, **b** and **d** Corresponding projective synchronization manifold.

respectively. Figure 7.19a and c shows that the driving state $x(t)$ and the response state $z(t)$ are in a proportional relationship $z(t) = 1.5x(t)$. It can be seen from Fig. 7.19b and d that the two trajectories shaped line has the same slope of 1.5.

For the scale factor with a negative sign, we can see the anti-phase projective synchronization phenomenon in Figs. 7.20 and 7.21. At this time, time delay τ is still stabilized at 1.0. The state trajectories of the anti-phase projective synchronization are shown in Fig. 7.20. After synchronization occurs, the time histories and the anti-phase projective synchronization manifolds of $x(t)$ and $z(t)$ are illustrated in Fig. 7.21. The amplitude of response state $z(t)$ has been scaled by − 0.5.

7.4 Nonlinear Dynamics of Internet Congestion Control Model with Time Delay

Since the pioneer work of Kelly [47, 48], researches on the mathematics of Internet congestion control have become an important area in the theoretical research on the Internet. To study the behavior of the real network system both qualitatively and

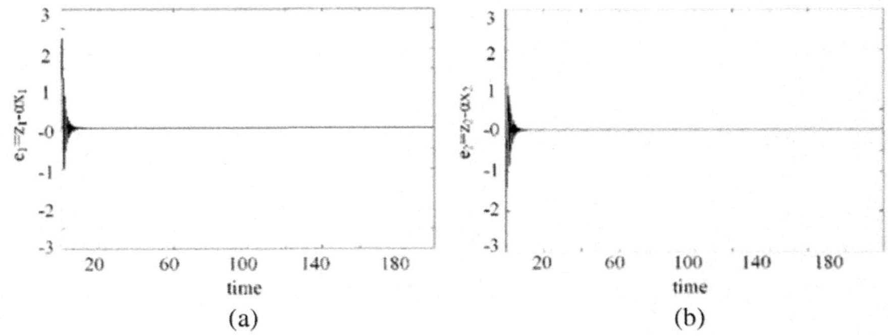

Fig. 7.20 Time history diagram of error system ($\alpha = -0.5$)

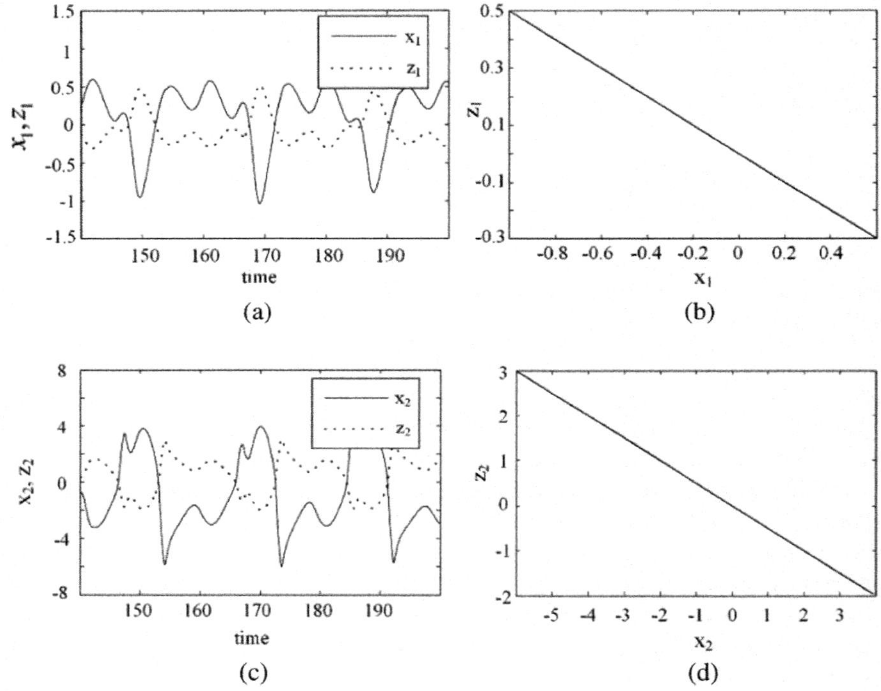

Fig. 7.21 **a** and **c** Antiphase projective synchronization of $x(t)$ and $z(t)$ for $\alpha = -0.5$, **b** and **d** Corresponding antiphase projective synchronization manifold.

quantitatively, several mathematical models have been founded by means of maps or differential equations[49]. Based on these researches, from the point of view of control theory, many authors have paid much attention to the stability of the equilibrium for different congestion control models, as in Refs. [50–53]. Control theory has yielded certain sufficient conditions to maintain equilibrium stability. However, these conditions often prove excessively stringent for real network systems. Consequently, one may ask: What outcomes arise for a network system when the stability of the equilibrium does not persist? The resolution to this inquiry is intricately linked to the nonlinear dynamics inherent in the congestion control model. For example, in Ref. [54], the Hopf bifurcation in the model given by Kelly[48] has been studied by center manifold reduction and normal form computation. The problem of controlling the bifurcation and chaos has been investigated in Refs. [55, 56].

The equilibrium of the Internet congestion control model holds significant physical implications, because there is direct connection between congestion and equilibrium instability. Consequently, ensuring the stability of the congestion control model stands out as a primary concern in the formulation of congestion control algorithms. This section presents our research findings, specifically addressing the mitigation of oscillations to stabilize the equilibrium of the congestion control system.

7.4.1 Time-Varying Delayed Feedback Control of an Internet Congestion Control Model

In Ref. [48], a proportionally-fair controller with a single link and a single user has been proposed and given by

$$\dot{y}(t) = k(w - y(t - \tau)p(y(t - \tau))), \tag{7.53}$$

where y, k, w and τ represent the data sending rate of the user, the positive gain parameter, the positive target and the RTT or time delay, respectively. p is an increasing function in its variable and non-negative. In the real network system, p is considered as the marking probability for a data packet.

It has been widely accepted that the network system is free of congestion if the equilibrium of Eq. (7.53) is stable. To suppress the oscillation that may arise in Eq. (7.53), we impose a periodic perturbation to time delay for the purpose of stabilization. We come to the result that Eq. (7.53) may be stabilized after undergoing a Hopf bifurcation if the perturbation parameters are properly chosen.

Assume that the perturbation imposed on the delay is periodic with small amplitude, then the method of multiple scales could be employed in the same procedure as in the case with constant time delay. Then it is found that the coefficients in normal form equations are functions of time and thus the stationary solution of the amplitude equation may not exist. However, given the specific form the amplitude equation, one can direct solve it to obtain the analytic solution, based on which the effectiveness of the proposed control method can be discussed.

7.4.2 Oscillation Suppression Through Periodic Delay

Imposing a periodic perturbation to the delay, the system under consideration now read

$$\frac{\mathrm{d}y(t)}{\mathrm{d}t} = k\,(w - y(t - \tau(t))\,p(y(t - \tau(t)))). \tag{7.54}$$

Assume that $k = 1$, $w = 1$, $p(x) = x/(20 - 3x)$ [54], and

$$\tau(t) = \tau_c + \epsilon\tau_\epsilon = \tau_c + \tau_0 + B\sin(\Omega t),$$

where τ_c is the critical value of time delay for the occurrence of the Hopf bifurcation for the case with constant delay. In the case that B is 0.35, τ_0 0.1 and Ω 0.02, a bursting-like phenomenon is then observed as shown in Fig. 7.22.

Researchers have accomplished some qualitative studies [57, 58] on periodic delayed differential systems. The present subsection will follow the idea originates from [58]. Given restrictions on τ_0 and B so that $\tau_0 + B$ is relatively small comparing to τ_c, $\tau_0 + B\sin(\Omega\,t)$ could be viewed as a perturbation to τ around τ_c as is suggested in [58]. Then, the linearization of Eq. (7.54) has the same form as that of Eq. (7.53). Furthermore, we point out that the employment of MMS should be added some small modifications so that time-varying delayed system could be handled within the same procedure by noticing the following relations:

$$\begin{aligned}\tau(t) &= \tau_c + \epsilon\,\tau_\epsilon = \tau_c + \tau_0 + B\sin(\Omega\,t) = \tau_c + \tau_0 + B\sin\left(\frac{\Omega}{\epsilon}\epsilon\,t\right)\\ &= \tau_c + \tau_0 + B\sin\left(\frac{\Omega}{\epsilon}T_1\right) = \tau_c + \epsilon\,\tau_\epsilon(T_1),\end{aligned} \tag{7.55}$$

which means time delay should be treated as a slow variable of time in the application of MMS. Then, by the method of MMS we obtain the corresponding form of equations for time-varying delay:

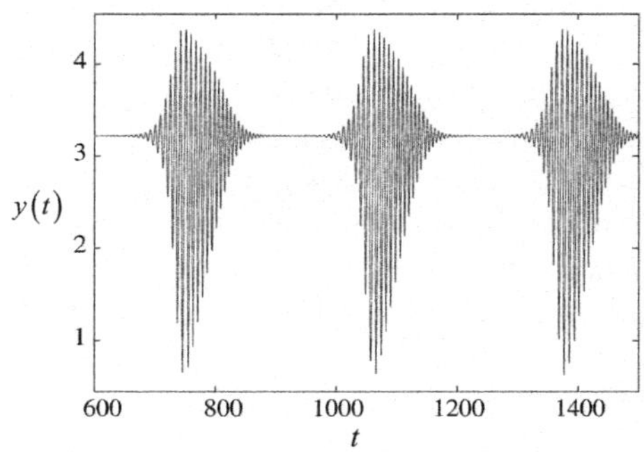

Fig. 7.22 Bursting-like phenomenon while time delay is periodically perturbed

$$\begin{cases} \dot{R}(t) = r_1(t)\, R(t) + r_3(t)\ R(t)^3 + r_5(t)\, R(t)^5, \\ \dot{\varphi}(t) = f_1(t) + f_2(t)\ R(t)^2 + f_4(t)\, R(t)^4. \end{cases} \tag{7.56}$$

7.4.3 *Analytical Solution of Time-Varying Delayed System (7.56)*

By means of MMS for which only some details should be noticed due to the time-dependent perturbation to the delay, we obtain that:

$$\begin{gathered} r_1(t) = g_0 + g_1 \sin(\Omega\, t) + g_2 \cos(\Omega\, t) + g_3 \sin(2\,\Omega\, t) + g_4 \cos(2\,\Omega\, t) \\ + g_5 \sin(3\,\Omega\, t) + g_6 \cos(3\,\Omega\, t) + g_7 \sin(4\,\Omega\, t) + g_8 \cos(4\,\Omega\, t), \\ r_3(t) = h_0 + h_1 \sin(\Omega\, t) + h_2 \cos(\Omega\, t) + h_3 \sin(2\,\Omega\, t) + h_4 \cos(2\,\Omega\, t), \\ r_5(t) = i_0, \end{gathered} \tag{7.57}$$

and

$$\begin{gathered} f_0(t) = m_0 + m_1 \sin(\Omega\, t) + m_2 \cos(\Omega\, t) + m_3 \sin(2\,\Omega\, t) + m_4 \cos(2\,\Omega\, t) \\ + m_5 \sin(3\,\Omega\, t) + m_6 \cos(3\,\Omega\, t) + m_7 \sin(4\,\Omega\, t) + m_8 \cos(4\,\Omega\, t), \\ f_2(t) = n_0 + h_1 \sin(\Omega\, t) + n_2 \cos(\Omega\, t) + n_3 \sin(2\,\Omega\, t) + n_4 \cos(2\,\Omega\, t), \\ f_4(t) = l_0, \end{gathered} \tag{7.58}$$

where

$$\begin{aligned} g_0 &= -0.102769B^2 - 0.0354582B^4 + 0.239655\tau_0 + 0.216587B^2\tau_0 - 0.205538\tau_0^2 \\ &\quad - 0.283665B^2\tau_0^2 + 0.144392\tau_0^3 - 0.0945551\tau_0^4 \\ &\quad + 0.0712306B^2\Omega^2, \\ g_1 &= 0.239655B + 0.108294B^3 - 0.411077B\tau_0 - 0.283665B^3\tau_0 \\ &\quad + 0.433175B\tau_0^2 - 0.378221B\tau_0^3 + 0.168791B\Omega^2 + 0.171999B\tau_0\Omega^2 \\ g_2 &= 0.293864B\Omega + 0.000136399B^3\Omega + 0.12891B\tau_0\Omega \\ &\quad + 0.000545596B\tau_0^2\Omega - 0.0727134B\Omega^3 \\ g_3 &= 0.0644549B^2\Omega + 0.000545596B^2\tau_0\Omega \\ g_4 &= 0.102769B^2 + 0.0472776B^4 - 0.216587B^2\tau_0 \\ &\quad + 0.283665B^2\tau_0^2 - 0.100768B^2\Omega^2 \\ g_5 &= -0.0360979B^3 + 0.0945551B^3\tau_0 \\ g_6 &= -0.000136399B^3\Omega, \quad g_7 = 0, \quad g_8 = -0.0118194B^4 \end{aligned}$$

$$h_0 = -0.0274087 - 0.0460022B^2 + 0.0675705\tau_0 - 0.0920045\tau_0^2$$
$$h_1 = 0.0675705B - 0.184009B\tau_0, \quad h_2 = 0.00493831B\Omega,$$
$$h_3 = 0, \quad h_4 = 0.0460022B^2, \quad i_0 = -0.00386815, \tag{7.59}$$

and

$$\begin{aligned}
m_0 &= 0.0753104B^2 + 0.00402557B^4 - 0.376449\tau_0 - 0.078343B^2\tau_0 \\
&\quad + 0.150621\tau_0^2 + 0.0322046B^2\tau_0^2 - 0.0522287\tau_0^3 + 0.0107349\tau_0^4 \\
&\quad - 0.0510767B^2\Omega^2, \\
m_1 &= -0.376449B - 0.0391715B^3 + 0.301242B\tau_0 + 0.0322046B^3\tau_0 \\
&\quad - 0.156686B\tau_0^2 + 0.0429395B\tau_0^3 - 0.0788404B\Omega^2 - 0.111519B\tau_0\Omega^2, \\
m_3 &= -0.0572482B^2\Omega - 0.00205197B^2\tau_0\Omega, \\
m_4 &= -0.0753104B^2 - 0.00536743B^4 + 0.078343B^2\tau_0 \\
&\quad - 0.0322046B^2\tau_0^2 + 0.0604419B^2\Omega^2 \\
m_5 &= 0.0130572B^3 - 0.0107349B^3\tau_0, \\
m_6 &= 0.000512992B^3\Omega, \quad m_7 = 0, \quad m_8 = 0.00134186B^4, \\
n_0 &= -0.0356391 - 0.0566331B^2 + 0.0858149\tau_0 - 0.113266\tau_0^2, \\
n_1 &= 0.0858149B - 0.226532B\tau_0, \quad n_2 = -0.00452029B\Omega, \\
n_3 &= 0, \quad n_4 = 0.0566331B^2, \quad l_0 = -0.00400312.
\end{aligned} \tag{7.60}$$

For simplicity, we neglect the term $(\epsilon R(t))^5$ in the right-hand side of Eq. (7.56). Then we're able to derive that

$$\dot{R}(t) = r_1(t)\, R(t) + r_3(t)\, R(t)^3 \,. \tag{7.61}$$

We have the following theorem:

Theorem 7.2 [59]. *The analytical solution of non-autonomous Eq.* (7.61) *could be solved explicitly, i.e.,*

$$\epsilon R(t) = 1/\sqrt{C\, e^{\int -2r_1(t)\,dt} + e^{\int -2r_1(t)\,dt} \int 2\,r_3(t)\, e^{\int 2r_1(t)\,dt}\, dt}, \tag{7.62}$$

where C appears to meet the initial condition.

After obtaining the expression of $R(t)$, we can give the solution of Eq. (7.54). However, one should notice that this analytical solution cannot be expressed in the form of elementary functions. Then series expansion is employed to transform the solution of $\dot{u} = -2r_1(t)u(t) - 2r_3(t)$ into an elementary form. While C is chosen to be zero, we focus on the term $e^{\int -2r_1(t)\,dt} \times \int -2\,r_3(t)\, e^{\int 2r_1(t)\,dt}\, dt$. Treat $e^{\int 2r_1(t)\,dt}$

as the product of two parts: $e^{g_0(t)}$ and

$$\exp\left(-\frac{\cos(t\Omega)g_1}{\Omega}+\frac{\sin(t\Omega)g_2}{\Omega}-\frac{\cos(2t\Omega)g_3}{2\Omega}+\frac{\sin(2t\Omega)g_4}{2\Omega}\right.$$
$$\left.-\frac{\cos(3t\Omega)g_5}{3\Omega}+\frac{\sin(3t\Omega)g_6}{3\Omega}-\frac{\cos(4t\Omega)g_7}{4\Omega}+\frac{\sin(4t\Omega)g_8}{4\Omega}\right), \tag{7.63}$$

where "exp" represents the exponential function. Then hold the former part and expand the latter part into a trigonometric series:

$$\exp\left(-\frac{\cos(t\Omega)g_1}{\Omega}+\frac{\sin(t\Omega)g_2}{\Omega}-\frac{\cos(2t\Omega)g_3}{2\Omega}+\frac{\sin(2t\Omega)g_4}{2\Omega}\right.$$
$$\left.-\frac{\cos(3t\Omega)g_5}{3\Omega}+\frac{\sin(3t\Omega)g_6}{3\Omega}-\frac{\cos(4t\Omega)g_7}{4\Omega}+\frac{\sin(4t\Omega)g_8}{4\Omega}\right)$$
$$=e^{-\frac{\cos(t\Omega)g_1}{\Omega}}\times e^{\frac{\sin(t\Omega)g_2}{\Omega}}\times\cdots \tag{7.64}$$
$$=\left(1-\frac{2\cos(t\Omega)g_1}{\Omega}+\frac{2\cos(t\Omega)^2g_1^2}{\Omega^2}-\frac{4\cos(t\Omega)^3g_1^3}{3\Omega^3}+\cdots\right)$$
$$\times\left(1+\frac{\sin(t\Omega)g_2}{\Omega}+\frac{\sin(t\Omega)^2g_2^2}{2\Omega^2}+\frac{\sin(t\Omega)^3g_2^3}{6\Omega^3}+\cdots\right)\times\cdots.$$

Applying the same procedure to $e^{\int -2r_1(t)\,dt}$ and using the formula

$$\int e^{\alpha t}\sin(\beta t)\,dt=\frac{e^{\alpha t}\,(-\beta\cos(\beta t)+\alpha\sin(\beta t))}{\alpha^2+\beta^2},$$
$$\int e^{\alpha t}\cos(\beta t)\,dt=\frac{e^{\alpha t}\,(\alpha\cos(\beta t)+\beta\sin(\beta t))}{\alpha^2+\beta^2}, \tag{7.65}$$

we obtain a solution in the form of elementary functions. Here we notice that unless

$$g_1\sin(\Omega t)+g_2\cos(\Omega t)+g_3\sin(2\Omega t)+g_4\cos(2\Omega t)$$
$$+g_5\sin(3\Omega t)+g_6\cos(3\Omega t)+g_7\sin(4\Omega t)+g_8\cos(4\Omega t)$$

is close to zero, one has to include many terms into the expansion until it is accurate enough. In the next, a comparison between solutions of Eq. (7.61) through Theorem 7.2 with expanding the exponential function above to different orders is given, as shown in Fig. 7.23.

In spite of this, expression (7.62) does have its meaning. While the value of $g_1, g_2, \cdots, g_8$ is close enough to zero the approximation of the expression based on Taylor expansion could trace the solution of Eq. (7.61) accurately.

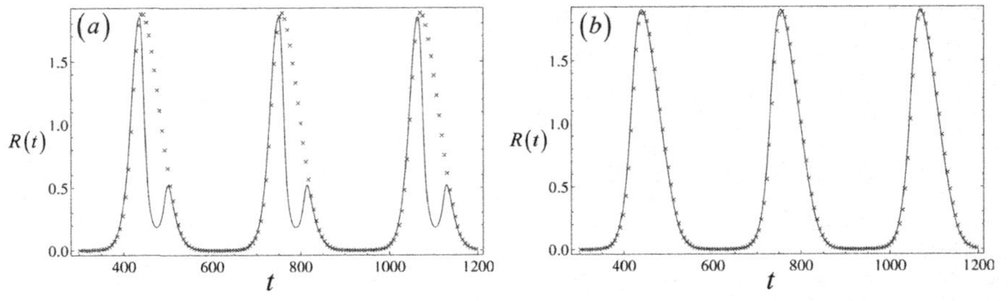

Fig. 7.23 Comparison between solutions of Eq. (7.61) through Theorem 7.2 using Taylor expansion up to different orders, i.e., expanding $e^{g_1 \sin(\Omega t)+g_2 \cos(\Omega t)+\cdots}$ to **a** $(E_1(t) + o(\sin(\Omega t)^{30}))$ $(E_2(t) + o(\cos(2\Omega t)^5))$ $(E_3(t) + o(\sin(3\Omega t)))$ $(E_4(t) + o(1))\cdots$; **b** $(F_1(t) + o(\sin(\Omega t)^{20}))$ $(F_2(t) + o(\cos(2\Omega t)^3))$ $(F_3(t) + o(\sin(3\Omega t)))$ $(F_4(t) + o(1))\cdots$. Solid lines for solutions of Eq. (7.61) with Theorem 7.2, "×" for numerical method

7.4.4 Discussion on the Oscillation Control of System (7.54)

In this part one necessary and sufficient condition for decay of the periodic solution of Eq. (7.61) where $r_1(t)$ and $r_3(t)$ take the form of (7.57) is obtained. We have one theorem as follows:

Theorem 7.3 [60]. *The sufficient and necessary condition for decay of the periodic solution of Eq.* (7.61) *is that* $g_0 < 0$. *Note that* g_0 *is expressed in* (7.61).

While a critical boundary is given by a set of (B, τ_0), *a comparison could be drawn between the theoretical result and numerical simulation. For example, while* $\tau_0 = 0.05$, *the critical value of* B *is* 0.3444 *by solving the equation* $g_0(\tau_0, B, 0.02) = 0$ *and plots of time history for* $B = 0.335$ *and* $B = 0.355$ *exhibit qualitatively distinct performances, as shown in Fig.* 7.24. *In the case that* $\tau_0 = 0.1$, *the critical value of* B *by analytical method is* 0.4885 *as is shown in Fig.* 7.25. *Figure* 7.26 *shows some numerical results while* τ_0 *takes the value of* 0.15, *for which the theoretical analysis predicts the critical value of* B *to be* 0.5966.

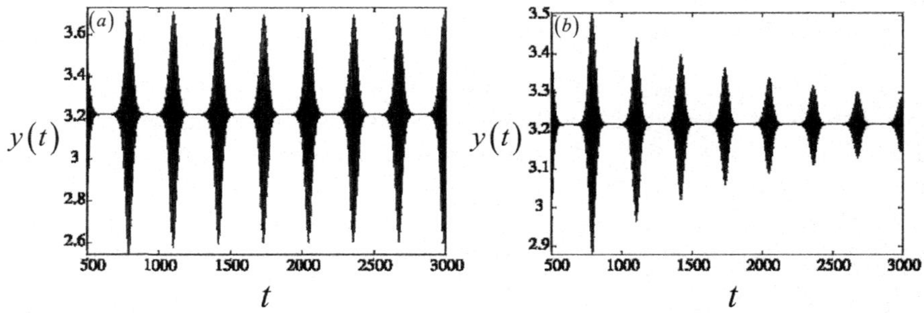

Fig. 7.24 Time history around the critical value of B for $\tau_0 = 0.05$, $\Omega = 0.02$ while the analytical result predicts a critical value of 0.3444: **a** $B = 0.335$ and **b** $B = 0.355$

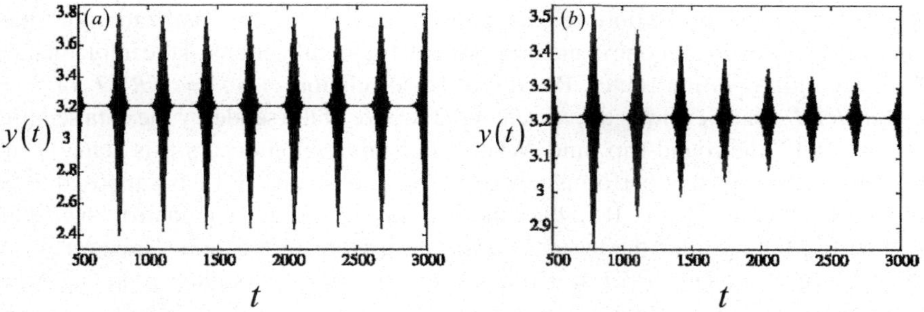

Fig. 7.25 Time history around the critical value of B for $\tau_0 = 0.1$, $\Omega = 0.02$ while the analytical result predicts a critical value of 0.4885: **a** $B = 0.48$ and **b** $B = 0.5$

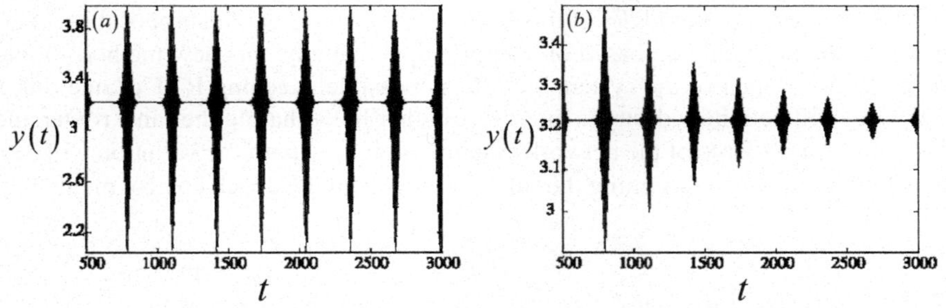

Fig. 7.26 Time history around the critical value of B for $\tau_0 = 0.15$, $\Omega = 0.02$ while the analytical result predicts a critical value of 0.5966: **a** $B = 0.59$ and **b** $B = 0.62$

Remark 7.1 In theory, the realization of this time-varying delay feedback control method depends on whether the perturbation to time delay or RTT is considerable in comparison with RTT itself. Note that RTT is the summation of two parts, namely, propagation delay for data transmission in the link and queueing delay for data packets queueing and processing by the link or router. Under the consideration that periodic perturbation to RTT is only to be exerted within the router and therefore only queueing delay could be changed through adjustments of the algorithm, this kind of control could be realized in the following cases:

a. The proposed method is proven effective when suppressing congestion within a local-area network. In such cases, the range of propagation is relatively limited, and the ratio of queueing delay to the total delay, i.e., the round-trip time (RTT) is noteworthy.
b. The method is applicable in scenarios where congestion has already manifested in a crucial router, or if the router's buffer design is suboptimal, resulting in a persistent accumulation of excessive data packets awaiting processing. In these situations, the queueing time may significantly surpass typical levels, leading to a notable increase in both queueing delay and the amplitude of perturbations.

Remark 7.2 It should be noticed that in technological layer the realization of time-varying delay feedback control method may call for modifications of the implemented TCP algorithms in the router. Recall the basic relation $cwnd = y\,RTT$ or $y = cwnd/RTT$ where $cwnd$ is the current window size of the sender, y the data sending rate and RTT the round-trip time. When Hopf bifurcation occurs it is actually the window size $cwnd$ that performs a periodic motion because RTT is supposed to be constant guaranteed by TCP. In the time-varying delay feedback control $cwnd$ will still be oscillatory after the system undergoes a Hopf bifurcation. However, from the relation $y = cwnd/RTT$ it is possible to stabilize y through introducing some proper mechanism in which the oscillation of RTT is allowed. Surely that this idea demands modifications of current TCP algorithms which used to keep a constant value of RTT.

Remark 7.3 Some may argue that since it is possible to bring down RTT in one period then why not hold RTT at this level. We point out that the proposal of a new control scheme should be based on one principal, namely, this new method should increase the burden of the system as little as possible. Keeping RTT of one link a low level will definitely destroys interests of other links sharing the same router and take up extra resource of the network system. Therefore, periodic variation will be a feasible choice when designing the time-varying delay feedback control method.

7.4.5 Discussions on Cases for Different Positive Gain Parameter k in System (7.54)

For a different value of k, the critical value of time delay will change to $\pi/2\alpha k$. The above procedure can also be employed to obtain a critical boundary of stabilizing Eq. (7.54), given by

$$\begin{aligned} G_0(B, \tau_0, \Omega, k) = &-0.102769B^2k^3 - 0.0354582B^4k^5 + 0.0712306B^2k\Omega^2 \\ &+0.239655k^2\tau_0 + 0.216587B^2k^4\tau_0 - 0.205538k^3\tau_0^2 \\ &-0.283665B^2k^5\tau_0^2 + 0.144392k^4\tau_0^3 - 0.0945551k^5\tau_0^4 \\ = &\ 0. \end{aligned}$$

Then, as is shown in Fig. 7.27a, when Ω is fixed at 0.02, it seems that k is preferably to be fixed at a large value, because for certain value of τ_0, it demands a smaller value of B in the case with a large k than a small one to control the oscillation of Eq. (7.54). However, it should be noticed that the larger the value of k, the smaller the value of τ_c. Therefore, it is more reasonable to measure the real expense of applying the variable delay control method to stabilize the system by introducing a new parameter which is defined as $H = B_c/\tau_c$ where B_c is the value of B on the critical boundary of stabilization. As is shown in Fig. 7.27b, it is more advisable to adopt a small value of k. Basically B_c and H could be solved explicitly in terms of τ_0 and k by letting

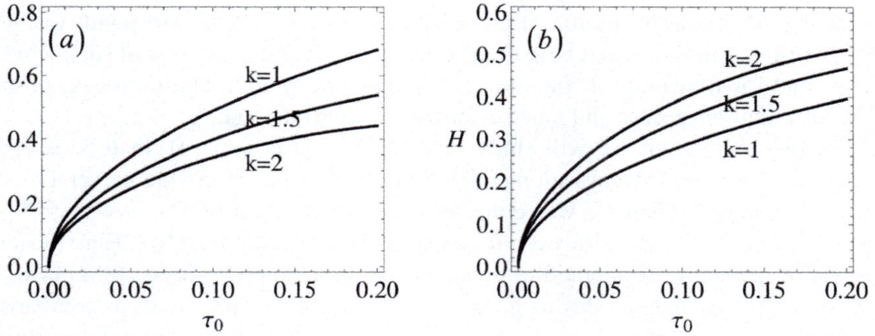

Fig. 7.27 Critical boundaries for eliminating the oscillation of Eq. (7.54) for different values of k in **a** (τ_0, B) plane and **b**(τ_0, H) plane where H is defined by $H = B_c/\tau_c$, B_c the value of B on the critical boundary and is a function of τ_0 and k. In both cases Ω takes a constant value of 0.05

Fig. 7.28 Derivative of H with respect to k. $\partial H/\partial k$ is negative for a large range of k, which means the expense of feedback control increases with the increase of k

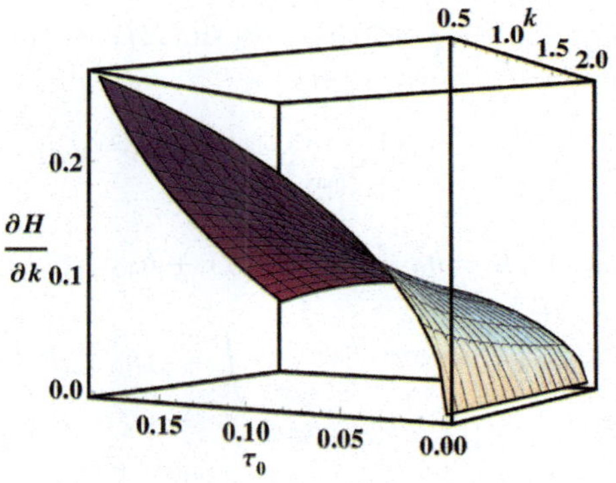

$G_0 = 0$. Then, as is demonstrated in Fig. 7.28 where Ω is fixed at 0.02, the derivative of H with respect to k is always positive, indicating the monotonous increase of H with k. That means a small value of k will bring low expenses for stabilization.

7.4.6 Utilization of the Bursting-Like Phenomenon

The condition $g_0 < 0$ that requires a large value of B for a fixed τ_0. Sometimes this demand may not be necessary in some specific situations. Besides, although the oscillation of Eq. (7.53) could be stabilized as some specific conditions hold for perturbation parameters, extra restrictions on parameters will be imposed on the system. In other words, the relation $g_0 < 0$ may sometimes not be allowable by the real application. On the other hand, it is observed in Fig. 7.22 that on the curve there are pieces of almost straight segments. On these segments it is possible to send data packets at a steady rate, leading to the idea that these segments could be utilized as windows of data transmission. If the value of τ_c, i.e., 1.72316 corresponds to a RTT of 20 milliseconds, it could be estimated from Fig. 7.22 that the size of such window is around 150 time units or equally 1.76 s in real time system. This period is not short for information process and may be utilized to transmit data.

In this subsection we will study in detail the case $g_0 \geq 0$. In this case, the term $C\,\mathrm{e}^{-2g_0 t}\,G(t)$ decays along with time thus it is reasonable to set $C = 0$ for convenience. Then we will concentrate on the study of $H(t) = \mathrm{e}^{\int -2r_1(t)\,dt} \times \int -2r_3(t)\,\mathrm{e}^{\int 2r_1(t)\,dt}\,dt$. However, it is very difficult to simplify $H(t)$. Here the technique of inequality is adopted to give an estimation of $H(t)$. In the first place, we observe that the magnitudes of $g_2, g_3, \cdots, g_8$ and h_2, h_3 are small in comparison with that of g_0, g_1 and h_0, h_1, h_4, respectively, thus these terms are to be neglected for simplicity. We have:

$$\begin{aligned}
&\int -2\,(h_0 + h_1 \sin(\Omega\,t) + h_4 \cos(2\,\Omega\,t))\,\mathrm{e}^{\int 2\,(g_0 + g_1 \sin(\Omega\,t))\,dt}\,dt \\
&= \int -2\,(h_0 + h_1 \sin(\Omega\,t) + h_4 \cos(2\,\Omega\,t))\,\mathrm{e}^{2\,g_0\,t - \frac{2\,g_1 \cos(\Omega\,t)}{\Omega}}\,dt \\
&\leq 2\left|\tilde{h}\right|_{\max} \mathrm{e}^{2\,|g_1/\Omega|} \int e^{2\,g_0\,t}dt = \frac{2\left|\tilde{h}\right|_{\max} \mathrm{e}^{2\,|g_1/\Omega|}}{2\,g_0}\,e^{2\,g_0\,t},
\end{aligned} \tag{7.66}$$

where $\tilde{h} = h_0 + h_1 \sin(\Omega\,t) + h_4 \cos(2\,\Omega\,t)$. Then the following relation holds

$$\begin{aligned}
&e^{\int -2\,(g_0 + g_1 \sin(\Omega\,t))dt} \int -2\,(h_0 + h_1 \sin(\Omega\,t))\,\mathrm{e}^{\int 2\,(g_0 + g_1 \sin(\Omega\,t))\,dt}\,dt \\
&\leq \frac{\left|\tilde{h}\right|_{\max} \mathrm{e}^{2\,|g_1/\Omega|}}{g_0}\,\mathrm{e}^{2\,g_1 \cos(\Omega\,t)/\Omega}.
\end{aligned} \tag{7.67}$$

Thus, we have

$$1\Bigg/\sqrt{\frac{\left|\tilde{h}\right|_{\max} \mathrm{e}^{2\,|g_1/\Omega|}}{g_0}\,\mathrm{e}^{2\,g_1 \cos(\Omega\,t)\Omega}} = \sqrt{\frac{g_0}{\left|\tilde{h}\right|_{\max}}}\,\mathrm{e}^{-|g_1/\Omega| - g_1 \cos(\Omega\,t)\Omega} \leq R(t). \tag{7.68}$$

Due to lack of more advisable analytical tools, the expression:

$$\hat{R}(t) = \sqrt{\frac{g_0}{\left|\tilde{h}\right|_{\max}}}\,\mathrm{e}^{-|g_1/\Omega| - g_1 \cos(\Omega\,t)\Omega} \tag{7.69}$$

is employed to simulate the solution of Eq. (7.61), where $\hat{R}(t)$ is the estimate of $R(t)$. It should be noticed that the inequality used in (7.67) is somewhat coarse and results in quantitative deviation from the accurate solution of (7.61), as shown in Fig. 7.29.

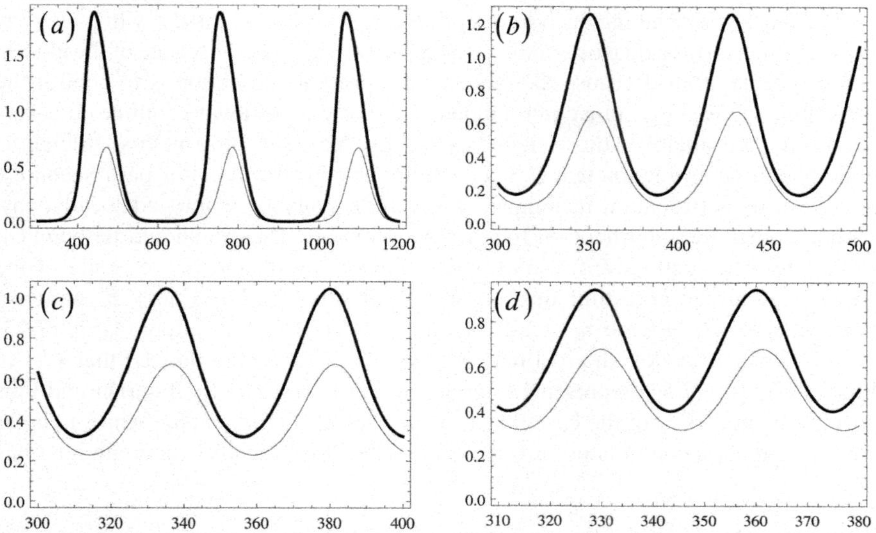

Fig. 7.29 Solutions of Eq. (7.61) obtained numerically and $\hat{R}(t)$ given by (7.69) for different values of Ω: **a** $\Omega = 0.02$; **b** $\Omega = 0.08$; **c** $\Omega = 0.14$ and **d** $\Omega = 0.2$. Here $\tau_0 = 0.1$ and $B = 0.35$. Thick solid lines for numerical method, thin solid lines for (7.69)

However, it is observed in the same figure that although $\hat{R}(t)$ deviates from accurate solution of (7.61) in quantity, this approximation describes the qualitative property of the solution curve of (7.61). In other words, the shape of $\hat{R}(t)$ is nearly the same as the solution curve of (7.61). This leads to the idea that $\hat{R}(t)$ is a preferable replacement of the solution of (7.61) in the study of the effect of the parameters (B, τ_0, Ω) on the behavior of Eq. (7.61). Next, we give a definition:

Definition Let $R_{\max}$ and $R_{\min}$ be the maximum and minimum value of $R(t)$, respectively, and $A = \{\tilde{t} | \tilde{t} \in \mathbb{R}, R(\tilde{t}) \leq R_{\min} + \epsilon(R_{\max} - R_{\min})\}$ be a subset of $\mathbb{R}$. Then we define l_{abs} and l_{rel} as the absolute and relative length of interval with appointed distance to zero indicated by ϵ as follows:

$$l_{abs} = \mathrm{m}(A), \quad l_{rel} = \mathrm{m}(A) \Big/ \frac{2\pi}{\Omega}, \tag{7.70}$$

where m(A) represents the Lebesgue measure of A.

It could be seen that basically l_{abs} and l_{rel} mainly depends on the configuration of the solution of (7.61) rather than its specific values. Again, this point of view supports us to study $\hat{R}(t)$ as an appropriate replacement of the solution of Eq. (7.61) when parameters vary. In order to simplify the problem, we make some assumptions of parameter B and τ_0:

$$\tau_0 \leq B, \quad \tau_0 \leq 0.3, \quad \tau_0 + B \leq 0.6. \tag{7.71}$$

The reasons of making such assumptions are as follows. Firstly, while $B \leq \tau_0$ Eq. (7.54) performs a quasi-periodic motion and the state of quiescence of the motion is not notable at all. Because the target of the present subsection is to control the oscillation of the congestion control model and thus we can always confine ourselves to consider the situations those enable us to apply the controlling method efficiently, consequently the situation that $B \leq \tau_0$ will not be considered in this part. Secondly, as a kind of perturbation technique, MMS could only be employed locally, say, within a small neighborhood of the bifurcation point. Thus τ_0 should be bounded in the region around the bifurcation point while being treated as the center of the periodic perturbation to time delay. Thirdly, for small τ_0, large $\tau_0 + B$ implies a large value of B. In large τ_0 cases, large $\tau_0 + B$ may lead to failure of Eq. (7.53) or (7.54) which models the real network system. It should be noticed that $y(t)$ in Eqs. (7.53) and (7.54) represents sending rate of a source in the Internet and thus will not be negative. In the case that $\tau_0 + B$ exceeds definite value while τ_0 is far from τ_c, numerical simulation reveals that $y(t)$ will be negative for some intervals of time.

Theorem 7.4 *Let $\hat{l}_{rel}$ be the estimation of l_{rel} and ϵ a small value. Note that $\hat{l}_{rel}$ could be calculated by replacing $R(t)$ by $\hat{R}(t)$ in our definition. Then we have.*

$$\hat{l}_{rel} = \frac{1}{\pi} \arccos\left(-\frac{\Omega}{g_1} \log\left((1-\epsilon)\, \mathrm{e}^{-\frac{2g_1}{\Omega}} + \epsilon \right) - 1 \right), \tag{7.72}$$

and (7.72) *is monotonically decreasing with* Ω.

Proof Firstly, it should be noticed that for condition (7.71) and the problem under consideration, $g_0 > 0$, $g_1 > 0$, $h_0 < 0$, $h_1 > 0$ are valid and thus it is easily verified that $\left|\tilde{h}\right|_{\max} = h_1 + h_4 - h_0$. Then $\hat{R}(t)$ is written as

$$\hat{R}(t) = \sqrt{\frac{g_0}{(h_1 + h_4 - h_0)}\, \mathrm{e}^{-(g_1 + g_1 \cos(\Omega t))/\Omega}}. \tag{7.73}$$

It is easily proved that $\hat{R}(t) \leq \hat{R}_{\min} + \epsilon \left(\hat{R}_{\max} - \hat{R}_{\min} \right)$ if and only if $\cos(\Omega t) \geq -\frac{2\Omega}{g_1} \log\left((1-\epsilon)\, \mathrm{e}^{-\frac{g_1}{\Omega}} + \epsilon \right) - 1$ where $\hat{R}(t)$ takes the form in (7.73). Thus

$$\hat{l}_{abs} = \arccos\left(-\frac{\Omega}{g_1} \log\left((1-\epsilon)\, \mathrm{e}^{-2g_1/\Omega} + \epsilon \right) - 1 \right), \tag{7.74}$$

and

$$\hat{l}_{rel} = \frac{1}{\pi} \arccos\left(-\frac{\Omega}{g_1} \log\left((1-\epsilon)\, \mathrm{e}^{-2g_1/\Omega} + \epsilon \right) - 1 \right). \tag{7.75}$$

This finishes the proof of the first part of the theorem. Let $X = \Omega/2\,g_1$. We want to verify that while $X > 0$, $\arccos\big(-2\,X\log\big((1-\epsilon)\,\mathrm{e}^{-1/X}+\epsilon\big)-1\big)$ is monotonically decreasing with X thus is monotonically decreasing with Ω. To this end, consider the expression

$$X\log\big((1-\epsilon)\,\mathrm{e}^{-1/X}+\epsilon\big) \tag{7.76}$$

Differentiate (7.76) for once, yields

$$\frac{\mathrm{d}}{\mathrm{d}X}\Big(X\log\Big((1-\epsilon)\,\mathrm{e}^{-\frac{1}{X}}+\epsilon\Big)\Big)=\frac{(1-\epsilon)\,\mathrm{e}^{-\frac{1}{X}}}{X\Big((1-\epsilon)\,\mathrm{e}^{-\frac{1}{X}}+\epsilon\Big)}+\log\Big((1-\epsilon)\,\mathrm{e}^{-\frac{1}{X}}+\epsilon\Big). \tag{7.77}$$

The right-hand side of (7.77) could be written as

$$\frac{1}{(1-\epsilon)\,\mathrm{e}^{-1/X}+\epsilon}\left[(1-\epsilon)\,\frac{\mathrm{e}^{-1/X}}{X}+\big((1-\epsilon)\,\mathrm{e}^{-1/X}+\epsilon\big)\log\big((1-\epsilon)\,\mathrm{e}^{-1/X}+\epsilon\big)\right] \tag{7.78}$$

Next, we will demonstrate that

$$(1-\epsilon)\,\frac{\mathrm{e}^{-1/X}}{X}+\big((1-\epsilon)\,\mathrm{e}^{-1/X}+\epsilon\big)\log\big((1-\epsilon)\,\mathrm{e}^{-1/X}+\epsilon\big) \tag{7.79}$$

is negative for $X > 0$. Differentiate (7.79):

$$\begin{aligned}&\frac{d}{dx}\left((1-\epsilon)\,\frac{\mathrm{e}^{-1/X}}{X}+\big((1-\epsilon)\,\mathrm{e}^{-1/X}+\epsilon\big)\log\big((1-\epsilon)\,\mathrm{e}^{-1/X}+\epsilon\big)\right)\\&=\frac{\mathrm{e}^{-1/X}\,(1-\epsilon)}{X^2}\left(\frac{1}{X}+\log\big((1-\epsilon)\,\mathrm{e}^{-1/X}+\epsilon\big)\right)\end{aligned} \tag{7.80}$$

Note that $\frac{\mathrm{e}^{-1/X}\,(1-\epsilon)}{X^2} > 0$ for small ϵ. Differentiating the following expression

$$\frac{1}{X}+\log\big((1-\epsilon)\,\mathrm{e}^{-1/X}+\epsilon\big), \tag{7.81}$$

yields

$$\frac{\mathrm{d}}{\mathrm{d}X}\left(\frac{1}{X}+\log\big((1-\epsilon)\,\mathrm{e}^{-1/X}+\epsilon\big)\right)=-\frac{1}{X^2}+\frac{(1-\epsilon)\,\mathrm{e}^{-1/X}}{X^2\,\big((1-\epsilon)\,\mathrm{e}^{-1/X}+\epsilon\big)}. \tag{7.82}$$

It is easily proved that for small ϵ,

$$-\frac{1}{X^2}+\frac{(1-\epsilon)\,\mathrm{e}^{-1/X}}{X^2\left((1-\epsilon)\,\mathrm{e}^{-1/X}+\epsilon\right)}=-\frac{\epsilon}{X^2\left((1-\epsilon)\,\mathrm{e}^{-1/X}+\epsilon\right)}<0 \tag{7.83}$$

which means (7.81) is decreasing with X for $X > 0$. While $X \to \epsilon f t y$, (7.81)$\to 0$, indicating that (7.81) is always positive for $X > 0$ and therefore (7.79) is increasing with X for $X > 0$. While $X \to \infty$ one can easily obtain that (7.79)$\to 0$ and thus (7.79)< 0 for all $X > 0$. Finally, (7.76) is decreasing with X leading that $\arccos\left(-2\,X\log\left((1-\epsilon)\,\mathrm{e}^{-1/X}+\epsilon\right)-1\right)$ is monotonically decreasing with X and Ω. This completes the proof of the second part of the theorem.

Furthermore, providing restrictions (7.71) on the perturbation parameters, one can derive that $g_1 = g_1(B, \tau_0, \Omega)$ is monotonically increasing with B and decreasing with τ_0. Therefore, by performing the same procedure as the proof of Theorem 7.4 one can easily obtain the similar results:

Theorem 7.5 *Given* (7.71), $\hat{l}_{rel} = \frac{1}{\pi}\arccos\left(-\frac{\Omega}{g_1}\log\left((1-\epsilon)\,\mathrm{e}^{-2g_1/\Omega}+\epsilon\right)-1\right)$ *is monotonically increasing with* B, *decreasing with* τ_0.

Proof See the proof of Theorem 7.4.

Surely, we want to apply (7.72) as an estimation of l_{rel}. However, as is pointed out above, the approximation (7.69) deviates somewhat from the solution of Eq. (7.61), thus numerical method should be employed to verify the results given by Theorems 7.4 and 7.5, as shown in Figs. 7.30, 7.31, and Fig. 7.32.

Once more we emphasize that (7.69) is just used to study the varying tendency of the solution of Eq. (7.61) while one changes the values of parameters of the perturbation to time delay. In other words, no attempt to trace exactly the solution curve of the time-varying delayed system is made here. It is observed that though there is some deviation in the quantitative description by (7.72) in comparison with the direct numerical simulation, (7.72) does reveal the rule of variation of $R(t)$ while parameters are changed. Therefore, (7.72) is proved to be an effective expression to predict the varying rules of the amplitude of $y(t)$ in (7.54).

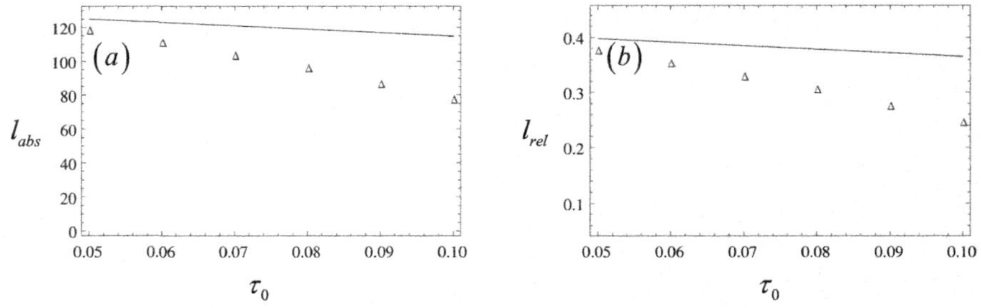

Fig. 7.30 Varying tendency of **a** l_{abs} and **b** l_{rel} while $\tau_0 \in [0.05, 0.10]$ for $B = 0.3$ and $\Omega = 0.02$. Solid lines for (7.72) in (**a**) and (7.74) in (**b**), $\triangle$ for numerical method

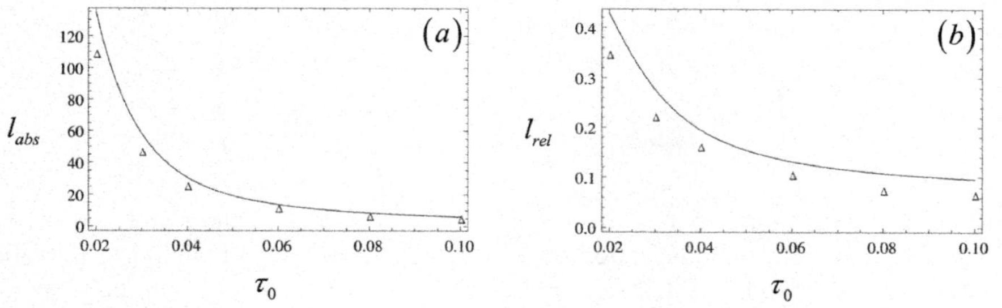

Fig. 7.31 Varying tendency of **a** l_{abs} and **b** l_{rel} while $\Omega \in [0.02, 0.10]$ for $B = 0.35$ and $\tau_0 = 0.1$. Solid lines for (7.72) in (**a**) and (7.74) in (**b**), $\triangle$ for numerical method

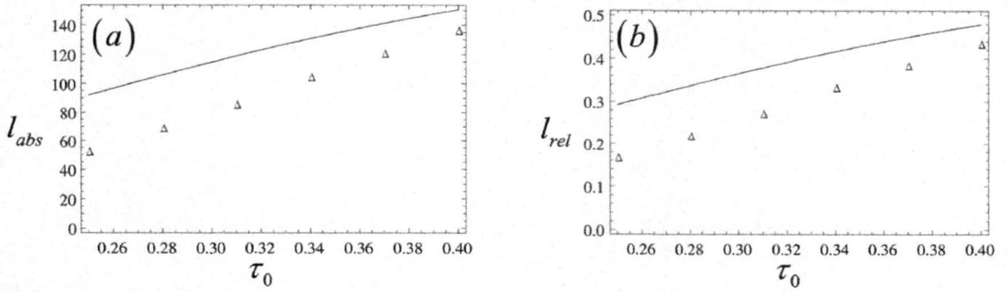

Fig. 7.32 Varying tendency of **a** l_{abs} and **b** l_{rel} while $B \in [0.25, 0.40]$ for $\tau_0 = 0.1$ and $\Omega = 0.02$. Solid lines for (7.72) in (**a**) and (7.74) in (**b**), $\triangle$ for numerical method

7.4.7 Generalization to the Case with n Users

In this subsection, we study a generalization of (7.53), namely, the congestion control model for the case of a single link shared by n users. We assume that p and the RTT are the same for all the users. This is understandable because the users are sharing a common link. Then the model takes the form

$$\dot{y}_i(t) = k_i(w_i - y_i(t-\tau)p(y_1(t-\tau), \cdots, y_n(t-\tau))), \tag{7.84}$$

for $i = 1, 2, \ldots n$. We assume that there is only one positive equilibrium, denoted by $\mathbf{Y}^* = \left\{y_1^*, y_2^*, \ldots y_n^*\right\}^T$, in Eq. (7.84). Through the linearization of Eq. (7.84) around $\mathbf{Y}^*$ and letting $\mathbf{X} = \mathbf{Y} - \mathbf{Y}^*$ where $\mathbf{X} = \{x_1, x_2, \cdots, x_n\}^T$ and $\mathbf{Y} = \{y_1, y_2, \ldots y_n\}^T$, a differential equation for new state variable $\mathbf{X}$ is obtained:

$$\dot{\mathbf{X}} = \mathbf{F}(\mathbf{X}(t-\tau)). \tag{7.85}$$

Denote the Jacobian matrix of $\mathbf{F}(\mathbf{X})$ by $\mathbf{J}$. Let $\mathbf{E} = \lambda\mathbf{I} - \mathbf{J}e^{-\lambda\tau}$. Then $\mathbf{E}$ takes the form

$$\begin{pmatrix} a_{1,1}e^{-\lambda\tau}+\lambda & a_{1,2}e^{-\lambda\tau} & \cdots & a_{1,3}e^{-\lambda\tau} \\ a_{2,1}e^{-\lambda\tau} & a_{2,2}e^{-\lambda\tau}+\lambda & \cdots & a_{2,3}e^{-\lambda\tau} \\ \vdots & \vdots & \ddots & \vdots \\ a_{n,1}e^{-\lambda\tau} & a_{n,2}e^{-\lambda\tau} & \cdots & a_{n,n}e^{-\lambda\tau}+\lambda \end{pmatrix},$$

where $a_{i,j} = a_{i,k}$ for $1 \le j \le n,$, $1 \le k \le n,$, $j \ne i,$, $k \ne i$, and $i = 1, 2, \cdots, n$. Noticing that from the congestion control model (7.84), one obtains $a_{i,j} > 0$ for $i = 1, 2, \cdots, n$, and $j = 1, 2, \ldots, n$. Let $a_{i,j} = a_i$, where $1 \le j \le n$ and $j \ne i,$. By letting $|\mathbf{E}|$ be zero, one may obtain the following equation

$$e^{-n\lambda\tau}\begin{vmatrix} a_{1,1}+\lambda e^{\lambda\tau} & a_{1,2} & \cdots & a_{1,3} \\ a_{2,1} & a_{2,2}+\lambda e^{\lambda\tau} & \cdots & a_{2,3} \\ \vdots & \vdots & \ddots & \vdots \\ a_{n,1} & a_{n,2} & \cdots & a_{n,n}+\lambda e^{\lambda\tau} \end{vmatrix} = 0,$$

or equivalently

$$\begin{vmatrix} a_{1,1}+\lambda e^{\lambda\tau} & a_{1,2} & \cdots & a_{1,3} \\ a_{2,1} & a_{2,2}+\lambda e^{\lambda\tau} & \cdots & a_{2,3} \\ \vdots & \vdots & \ddots & \vdots \\ a_{n,1} & a_{n,2} & \cdots & a_{n,n}+\lambda e^{\lambda\tau} \end{vmatrix} = 0. \tag{7.86}$$

Let $\rho = \lambda e^{\lambda\tau}$. Equation (7.86) can be rewritten as

$$\begin{vmatrix} \frac{a_{1,1}}{a_1}+\frac{1}{a_1}\rho & 1 & \cdots & 1 \\ 1 & \frac{a_{2,2}}{a_2}+\frac{1}{a_2}\rho & \cdots & 1 \\ \vdots & \vdots & \ddots & \vdots \\ 1 & 1 & \cdots & \frac{a_{n,n}}{a_n}+\frac{1}{a_n}\rho \end{vmatrix} = 0. \tag{7.87}$$

Then, we have the following result:
Lemma [61]. *Let*

$$\mathbf{T} = \begin{pmatrix} \chi_1(\rho) & 1 & \cdots & 1 \\ 1 & \chi_2(\rho) & \cdots & 1 \\ \vdots & \vdots & \ddots & \vdots \\ 1 & 1 & \cdots & \chi_n(\rho) \end{pmatrix}, \tag{7.88}$$

where $\chi_i(\rho) = \frac{a_{i,i}}{a_i} + \frac{1}{a_i}\rho, i = 1, 2, \cdots, n.$

Assuming that the zeros of $\chi_i = (\rho)$ for $i = 1, 2, \cdots, n$ are different from each other, we can conclude that all roots of $|\mathbf{T}| = 0$ are real numbers.

Now suppose that the assumption in Lemma above is satisfied. Then there are n different real roots for Eq. (7.87), denoted by $\gamma_1, \gamma_2, \cdots, \gamma_n$. In order to determine the critical delay for the Hopf bifurcation, we have the following theorem:

Theorem 7.6 [62]. *Let* $\omega = \max |\gamma_i|, i = 1, 2, \cdots, n$. *Then* $\tau_c = \pi/2\omega$, *where* τ_c *represents the critical delay for the Hopf bifurcation.*

Based on the above results, the nonlinear analysis for the case with periodic perturbation to the delay can be performed in the same way as discussed in previous subsections and as a result the critical value of the perturbation parameter for the decay of the oscillation can be determined. The details will not be discussed here and the readers are referred to [61].

7.5 Oscillatory Dynamics Induced by Time Delay in an Internet Congestion Control Model with a Ring Topology

Typically, there are several kinds of physical topologies in the Internet, namely, point-to-point, bus, star, ring, mesh and tree [63]. This section concentrates on the ring topology. From a complex network perspective, the ring topology serves as a straightforward model that facilitates theoretical analysis. Moreover, the ring topology is prevalent in the Internet, notably within local area networks and campus area networks. Examples include the Fiber Distributed Data Interface (FDDI) network, which embodies a physical ring topology. Additionally, the evolving Ethernet ring is recognized as another manifestation of a ring network.

This section explores an -dimensional congestion control model with a ring-type physical topology to investigate network oscillations. Among the various parameters under consideration, particular emphasis is placed on time delay. The Hopf bifurcation theory [64, 65] and the method of multiple scales are employed to analyze the periodic motion induced by the delay. The findings reveal that significant time delays can trigger periodic motion within the ring network system. Additionally, the oscillation of data sending rates in the network may be attributed to factors such as extended transmission distances and limited link capacities.

7.5.1 *Model of Congestion Control for a Ring Network*

In this subsection, an n-dimensional congestion control model is established to study the oscillation in the network whose physical topology is a ring. As to the logical topology, we consider a scenario where there are n users (or transmissions) and n links, and every transmission utilizes m consecutive links to transmit data packets, and consequently the logical topology can be illustrated by Fig. 7.33 [59].

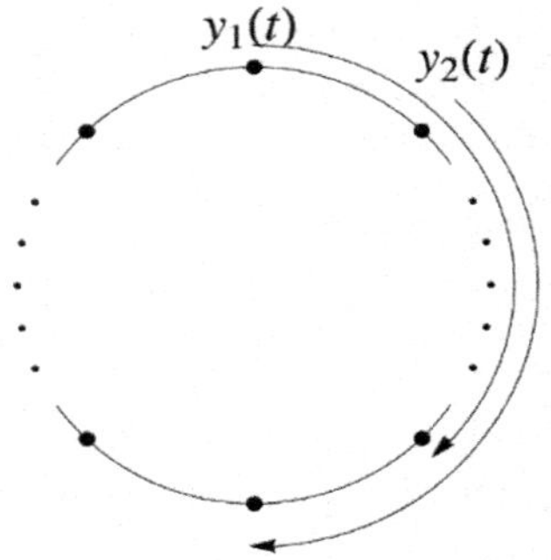

Fig. 7.33 Logical topology for the transmission

Assume that $m \leq n-1$. In other words, we exclude the case that a user sends a connection request to itself. Then the model reads.

$$
\begin{aligned}
&\dot{y}_1(t) = k(w - y_1(t-2m\tau))\cdot \\
&\big(p\big[y_1(t-2m\tau) + y_n(t-(2m+1)\tau) + \ldots + y_{n-(m-2)}(t-(3m-1)\tau)\big] \\
&+p\big[y_2(t-(2m-1)\tau) + y_1(t-2m\tau) + \ldots + y_{n-(m-3)}(t-(3m-2)\tau)\big] \\
&\vdots \\
&+p\big[y_m(t-(m+1)\tau) + y_{m-1}(t-(m+2)\tau) + \ldots + y_1(t-2m\tau)\big]\big), \\
&\dot{y}_2(t) = k(w - y_2(t-2m\tau))\cdot \\
&\big(p\big[y_2(t-(2m-1)\tau) + y_1(t-2m\tau) + \ldots + y_{n-(m-3)}(t-(3m-1)\tau)\big] \\
&+p\big[y_3(t-(2m-1)\tau) + y_2(t-2m\tau) + \ldots + y_{n-(m-3)}(t-(3m-2)\tau)\big] \\
&\vdots \\
&+p\big[y_{m+1}(t-(m+1)\tau) + y_m(t-(m+2)\tau) + \ldots + y_1(t-2m\tau)\big]\big), \\
&\vdots \\
&\dot{y}_2(t) = k(w - y_n(t-2m\tau))\cdot \\
&\big(p\big[y_n(t-(2m-1)\tau) + y_{n-1}(t-2m\tau) + \ldots + y_{n-(m-1)}(t-(3m-1)\tau)\big] \\
&+p\big[y_1(t-(2m-1)\tau) + y_n(t-2m\tau) + \ldots + y_{n-(m-2)}(t-(3m-2)\tau)\big] \\
&\vdots \\
&+p\big[y_{m-1}(t-(m+1)\tau) + y_{m-2}(t-(m+2)\tau) + \ldots + y_1(t-2m\tau)\big]\big),
\end{aligned}
\tag{7.89}
$$

where k is the positive gain parameter, w is the target set in advance, p is the probability of a certain mark, τ is the time delay from a node to the one next to it. In other words, τ can be viewed as half of the round-trip time for a single link. In this work, we define $p(x)$ to be $\theta\sigma^2 x/\big(\theta\sigma^2 x + 2(c-x)\big)$, as suggested in [66] and [67], where c is the link capacity, θ is a positive constant and σ^2 is the variance of the packets in the link. $\theta\sigma^2$ is usually fixed to be $1/2$. The system can be viewed as an extension of the classical model proposed in [67]. To provide a further understanding of the model, consider the network consists of routers. Then n represents the number of users (or transmissions) and the links, m describes the times that the data packets be transferred by the routers from the source to the destination. As to the rules of establishing the delayed terms in each equation, we follow the suggestions of [68].

7.5.2 Analysis on the Stability of the Equilibrium

It should be noticed that there may be quite a lot of equilibriums for Eq. (7.89). However, it can be observed that Eq. (7.89) has some symmetric appearance, which indicates critical information about the system's dynamics. Enlightened by this observation, we seek the equilibrium which satisfies $y_1 = y_2 = \cdots = y_n = y^*$. Then we have

$$w = y^*\big(mp(my^*)\big).$$

It can be verified that the equilibrium is stable for $\tau = 0$. Let $x_i = y_i - y^*$ for $i = 1, 2, \cdots, n$ then we can formulate the linearized system around the equilibrium. It is clear that for the stability switch, $\lambda = \omega i$ should be one of the eigenvalues of the linearized system. Through a direct analysis, one obtains that

$$\tau_c = \frac{\pi}{4\omega m}, \tag{7.90}$$

where

$$\omega = -\frac{1}{4}\csc\left(\frac{\pi}{8m}\right)^2\left(\frac{8\left(-2+\sqrt{2}\right)\left(-3w+\sqrt{w(16c+9w)}\right)ck}{m\left(8c+9w-3\sqrt{w(16c+9w)}\right)} - \frac{4\left(3w-\sqrt{w(16c+9w)}\right)\sin\left(\frac{\pi}{8m}\right)km}{-8c-9w+3\sqrt{w(16c+9w)}}\right). \tag{7.91}$$

Afterwards, we can compute τ_c based on (7.90). We stress that it is difficult to prove the validity of (7.90) and (7.91) from the theoretical level. But the numeral simulations show that the values of τ_c and ω determined by (7.90) and (7.91) are the critical ones which we pursue.

7.5.3 Study on the Periodic Motion Induced by the Delay Through the Method of Multiple Scales

In this section, we use an example to demonstrate how to compute the periodic motion arises from the Hopf bifurcation by the method of multiple scales, based on the work in [69] and [70]. Let $n = 8$, $m = 3$, $k = 1$, $w = 1$, $c = 5$ and

$$T = \begin{pmatrix} 1&1&1&0&0&0&1&1 \\ 1&1&1&1&0&0&0&1 \\ 1&1&1&1&1&0&0&0 \\ 0&1&1&1&1&1&0&0 \\ 0&0&1&1&1&1&1&0 \\ 0&0&0&1&1&1&1&1 \\ 1&0&0&0&1&1&1&1 \\ 1&1&0&0&0&1&1&1 \end{pmatrix}.$$

From (7.90) and (7.91), one obtains that $\tau_c = 0.09865$ and $\omega = 2.6538$. Perturb the delay, namely, let $\tau = \tau_c + \varepsilon\tau_\varepsilon$. Linearize the system around the equilibrium and let $\mathrm{X}(t) = \mathrm{Y}(t) - \mathrm{Y}^*$ where $\mathrm{X} = \{x_1(t), x_2(t), \cdots, x_8(t)\}^T$ and $\mathrm{Y} = \{y_1(t), y_2(t), \cdots, y_8(t)\}^T$. Therefore, the original system can be rewritten as

$$\dot{\mathrm{X}}(t) = \mathrm{F}(\mathrm{X}(t-4\tau), \mathrm{X}(t-5\tau), \cdots, \mathrm{X}(t-8\tau)). \tag{7.92}$$

Assume that the bifurcating periodic solution takes the form

$$\mathrm{X}(t) = \mathrm{X}(T_0, T_1, T_2, \cdots) = \sum_{i=1} \varepsilon^i \mathrm{X}_i(T_0, T_1, T_2, \cdots), \tag{7.93}$$

where $\mathrm{X}_i = \{X_{1,i}, X_{2,i}, \cdots, X_{8,i}\}^T$ for $i = 1, 2, \cdots$ and $T_k = \varepsilon^k t$ for $k = 0, 1, 2, \cdots$. Substitute Eq. (7.93) into Eq. (7.92). The delayed terms should be handled specially. Namely, we expand $x_{i,j}(T_0 - \tau_c - \varepsilon\tau_\varepsilon, T_1 - \varepsilon(\tau_c - \varepsilon\tau_\varepsilon), \cdots)$ at $(T_0 - \tau_c, T_1, T_2, \cdots)$ to series of ε. According to the standard process of applying the method of multiple scales for systems with time delay, we have

$$\begin{cases} \dot{R}(t) = R(t)\big(r_1 + r_3 R(t)^2\big), \\ \dot{\varphi}(t) = r_0 + r_2 R(t)^2, \end{cases} \tag{7.94}$$

where $R(t)$ is the amplitude of the periodic solution and φ is the phase angle. For this example, we compute the coefficients in (7.94) as

$$r_0 = (-18.96 - 132.1\tau_\varepsilon)\tau_\varepsilon, r_1 = (11.77 - 318.8\tau_\varepsilon)\tau_\varepsilon, r_2 = -0.855, r_3 = -0.6427.$$

7.5.4 Effects of M*: Long Transmission Distance Will Induce Oscillation*

Based on (7.90) in Sect. 7.3, we obtain the relation of τ, m and c while the system starts to oscillate. Though the theory is not strict, the validity of the method can be proven with numerical simulations. For $k = 1$ and $w = 1$, we obtain from (7.90) and (7.91) that

$$\tau_c = \left(\left(9 + 8c - 3\sqrt{9+16c}\right)^2 \pi \sin\left(\frac{\pi}{8m}\right)^2\right) / \left(8\left(\left(-2+\sqrt{2}\right)c\left(-3+\sqrt{9+16c}\right) + \left(-4c\left(-9+\sqrt{9+16c}\right) - 9\left(-3+\sqrt{9+16c}\right)\right)m^2 \sin\left(\frac{\pi}{8m}\right)^2\right)\right) \tag{7.95}$$

Now we can show the relation between τ_c and m for different values of c, as is shown in Fig. 7.34.

It can be observed from the figure that for any c, the critical value of the delay decreases while m's value increases. Remind that m is the number of links a transmission passes by. In essence, m indicates of the transmission distance when the delay of each link remains constant. Consequently, the figure illustrates a crucial observation: the network's transition from a stable state to an oscillatory state is apparent when the system permits transmissions that traverse multiple links. However, one may argue that in this instance, the underlying factor to drive the system into the unstable state is m, τ since this multiplication indicates the total transmission distance. Now we employ (7.95) to check which factor has more obvious effects on the stability of the system. Let $\Delta = \mathrm{m}\,\tau_\mathrm{c}$. Then Δ measures the total delay for the transmission from the source to the destination while the system loses the stability. The relation of Δ and m can be seen from Fig. 7.35.

From Figs. 7.34 and 7.35, both τ_c and Δ decreases while m increases. Therefore, m is decisive to inducing the oscillatory dynamics of the network.

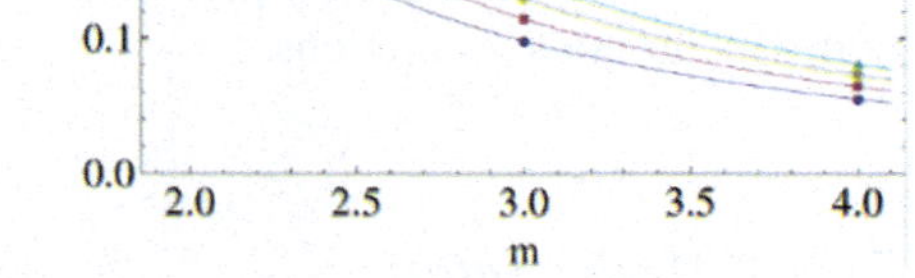

Fig. 7.34 Critical boundary for the Hopf bifurcation in $m - \tau_c$ plane based on Eq. (7.95) while $c = 5, 6, 7, 8$ respectively

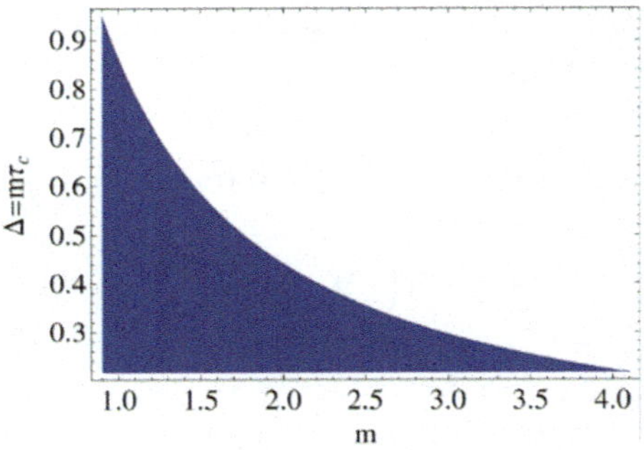

Fig. 7.35 Critical boundary for the Hopf bifurcation in $m - \Delta$ plane based on Eq. (7.95) while $c = 5$. Equilibrium is stable in coloured area

7.6 Desynchronization-Based Congestion Suppression for a Star-Type Internet System

For the system discussed in this section, the occurrence of synchronization is attributed to the symmetric structure of the system setup. Consequently, the state feedback controller must be formulated to introduce symmetry-breaking to the original system. To ascertain the conditions for the effective operation of the desynchronization scheme, a quantitative analysis of the dynamics around the unstable steady state is necessary. This analysis is conducted on the mathematical model of Internet congestion control, expressed in terms of a system of delayed differential equations (DDE).

For the quantitative investigation of DDE especially DDE with symmetry breaking phenomenon, researchers have developed many theoretical tools, such as center manifold reduction [65, 71], theory of equivariant degree [72], method of multiple scales (MMS) [69, 73] and so on. In this section, we utilize the Method of Multiple Scales (MMS) to derive the truncated normal form system associated with Hopf and equivariant Hopf bifurcations. These bifurcations have the potential to induce desynchronous oscillations, as deduced from the analysis of the linearized system around the equilibrium. The stability of the bifurcated periodic solutions is assessed through an examination of the truncated normal form system. This analysis allows us to establish criteria for the parameters of the controller, ensuring the occurrence of desynchronous oscillations.

7.6.1 Model Setup

This section is dedicated to elucidating the model under examination. Subsequently, we will present both the original model and the model augmented by the desynchronization mechanism. The base model assumes the same form as last subsection.

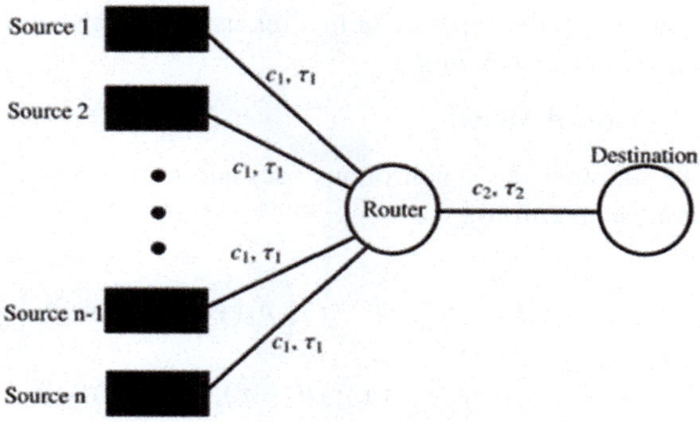

Fig. 7.36 Topology of the network system under investigation

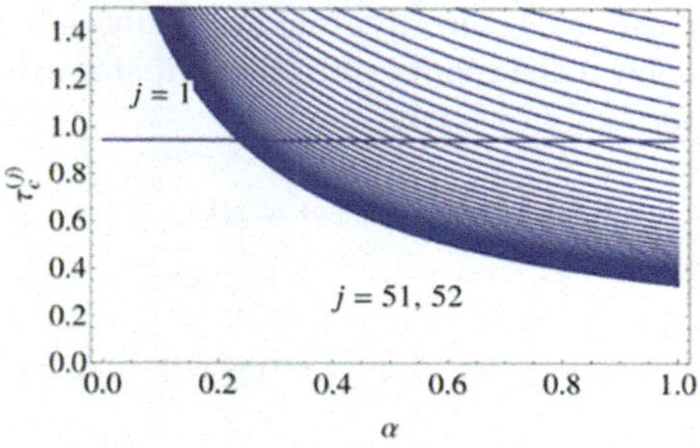

Fig. 7.37 Solution curves of the characteristic equation in terms of α and $\tau_c^{(j)}$

We now examine the congestion control system within a star-type topology, as depicted in Fig. 7.36 [74]. It is essential to clarify that our reference to "topology" aligns with the terminology in computer science rather than that in mathematics. In the context of graph theory, the topology illustrated in Fig. 7.37 is categorized as a fully connected network.

This network can be viewed as a simplification of the system investigated by Liu. et al. in [75]. The corresponding mathematical model can be written as

$$\dot{y}_i(t) = k\left(w - y_i(t-\tau)\left(p_1(y(t-\tau_1)) + p_2\left(\sum_{j=1}^{n} y_j(t-\tau)\right)\right)\right), \tag{7.96}$$
$$i = 1, 2, \cdots, n,$$

where $\tau = 2\tau_1 + 2\tau_2$ is the total round trip time and n is the number of sources (or equivalently, connections). Throughout the section, we have the following assumption:

$$n \geq 3, \quad p_1(x) = \frac{\theta x}{\theta x + 2(c_1 - x)}, \quad p_2(x) = \frac{\theta x}{\theta x + 2(c_2 - x)}$$

where c_1 is the capacity of the link used by each source alone and c_2 is the capacities of the bottleneck link.

Controlled Model

To suppress the synchronous oscillation of (7.96), we introduce the state feedback control as follows

$$\dot{y}_i(t) = k\left(w - y_i(t-\tau)\left(p_1(y(t-\tau)) + p_2\left(\sum_{k=1}^{n} y_k(t-\tau)\right)\right)\right) \\ + f_i(y_1(t-\tau), y_2(t-\tau), \cdots, y_n(t-\tau)), \quad i = 1, 2, \cdots, n, \tag{7.97}$$

where f_i represents the control signal provided by the router. The inclusion of delayed terms arises directly from the feedback loop, and we make the assumption that the delay in the control corresponds to that in the original system. We consider a specific form of f_i, namely,

$$f_i(y_1, y_2, \cdots, y_n) = \alpha(y_{i+1} + y_{i-1} - 2y_i) + \beta(y_{i+1} + y_{i-1} - 2y_i)^2 \\ + \gamma(y_{i+1} + y_{i-1} - 2y_i)^3, \quad \text{for all } i (\bmod n). \tag{7.98}$$

Given (7.98), Eq. (7.96) is rewritten as

$$\dot{y}_i(t) = k\left(w - y_{i,\tau}\left(p_1(y_{i,\tau}) + p_2\left(\sum_{k=1}^{n} y_{k,\tau}\right)\right)\right) + \alpha\left(y_{i+1,\tau} + y_{i-1,\tau} - 2y_{i,\tau}\right) \\ + \beta\left(y_{i+1,\tau} + y_{i-1,\tau} - 2y_{i,\tau}\right)^2 + \gamma\left(y_{i+1,\tau} + y_{i-1,\tau} - 2y_{i,\tau}\right)^3, i = 1, 2, \cdots, n, \tag{7.99}$$

where $y_{i,\tau} = y_i(t-\tau)$. The rest of the subsection is devoted to the study of (7.99). It should be noticed that (7.99) has multiple equilibria. We will focus our attention on the one that has strongest symmetry, namely, $Y^* = \{y^*, y^*, \cdots, y^*\}^T$. We will start from determining the critical conditions for the stability switch around Y^*.

7.6.2 *Critical Conditions for Stability Switch and Oscillatory Patterns*

Our primary focus centers on mitigating desynchronization of oscillations in instances of system instability. Drawing upon bifurcation theory, we identify small-amplitude oscillations as solutions arising from the linearized system around equilibrium. The characteristics of oscillation patterns are intricately connected to the eigenvectors of the linearized system. In this subsection, our objective is to pinpoint critical conditions for stability transitions and the corresponding eigenvectors linked to these critical eigenvalues.

Letting $y_i(t) = y^* + x_i(t)$, linearization around Y^* as

$$\dot{X}(t) = M'X(t - \tau), \tag{7.100}$$

where $X = \{x_1, x_2, \cdots, x_n\}^T$ and

$$M' = \begin{pmatrix} g_1 & g_2 & g_3 & \cdots & g_3 & g_2 \\ g_2 & g_1 & g_2 & \cdots & g_3 & g_3 \\ \vdots & \vdots & \vdots & \ddots & \vdots & \vdots \\ g_2 & g_3 & g_3 & \cdots & g_2 & g_1 \end{pmatrix},$$

here $g_1 = -k(m_1 + m_2 + m_3) - 2\alpha$, $g_2 = \alpha - km_3$, $g_3 = -km_3k$ and $m_1 = p_1(y^*) + p_2(ny^*)$, $m_2 = y^* p_1{}'(y^*)$, $m_3 = y^* p_2{}'(ny^*)$. Apparently, m_1, m_2 and m_3 are all positive.

The characteristic matrix can be written as

$$M = \lambda I - e^{-\lambda\tau} M', \tag{7.101}$$

where I is the identity matrix of size n, λ represents the eigenvalue of the linearized Eq. (7.100). Following the idea of [76], we rewrite M' as

$$M' = (-km_1 - km_2 - 2\alpha)I + \alpha K_1 - km_3 K_2, \tag{7.102}$$

where

$$K_1 = \begin{pmatrix} 0 & 1 & 0 & \cdots & 0 & 1 \\ 1 & 0 & 1 & \cdots & 0 & 0 \\ \vdots & \vdots & \vdots & \ddots & \vdots & \vdots \\ 1 & 0 & 0 & \cdots & 1 & 0 \end{pmatrix},$$

and

$$K_2 = \begin{pmatrix} 1 & 1 & 1 & \cdots & 1 & 1 \\ 1 & 1 & 1 & \cdots & 1 & 1 \\ \vdots & \vdots & \vdots & \ddots & \vdots & \vdots \\ 1 & 1 & 1 & \cdots & 1 & 1 \end{pmatrix}.$$

Let $\mathrm{v}_j = \left\{1, \chi^{(j-1)}, \chi^{2(j-1)}, \cdots, \chi^{(n-1)(j-1)}\right\}^T$, where $\chi = e^{2\pi \mathrm{i}/n}$. Clearly $\chi^{(n-1)(j-1)} = \chi^{-(j-1)}$, $\chi^{(n-2)(j-1)} = \chi^{-2(j-1)}$, ..., Therefore,

$$K_1 \mathrm{v}_j = \begin{pmatrix} \chi^{(n-1)(j-1)} + \chi^{j-1} \\ 1 + \chi^{2(j-1)} \\ \chi^{j-1} + \chi^{3(j-1)} \\ \vdots \\ 1 + \chi^{(n-2)(j-1)} \end{pmatrix} = (\chi^{(j-1)} + \chi^{-(j-1)})\mathrm{v}_j = 2\cos\left(\frac{2\pi(j-1)}{n}\right)\mathrm{v}_j,$$

$$K_2 \mathrm{v}_j = \begin{cases} n\mathrm{v}_j & j = 1, \\ 0 & j \neq 1. \end{cases}$$

Let $\lambda_1 = ke^{-\lambda\tau}(-m_1 - m_2 - n\,m_3)$, $\lambda_j = e^{-\lambda\tau}\left(-km_1 - km_2 - 2\alpha + 2\alpha\cos\left(\frac{2\pi(j-1)}{n}\right)\right)$, $j = 2, 3, \cdots, n$. Then we have

$$M \cdot \{\mathrm{v}_1, \mathrm{v}_2, \cdots, \mathrm{v}_n\} = \Lambda\{\lambda - \lambda_1, \lambda - \lambda_2, \cdots, \lambda - \lambda_n\} \cdot \{\mathrm{v}_1, \mathrm{v}_2, \cdots, \mathrm{v}_n\}, \tag{7.103}$$

where $\Lambda\{\cdot\}$ represents the diagonal matrix consisting of the elements in the brackets in sequence. It can be obtained directly from (7.103) that

$$\det M \cdot \det\{\mathrm{v}_1, \mathrm{v}_2, \cdots, \mathrm{v}_n\} = \det \Lambda\{\lambda - \lambda_1, \lambda - \lambda_2, \cdots, \lambda - \lambda_n\} \cdot \det\{\mathrm{v}_1, \mathrm{v}_2, \cdots, \mathrm{v}_n\}. \tag{7.104}$$

Note that $\{\mathrm{v}_1, \mathrm{v}_2, \cdots, \mathrm{v}_n\} \neq 0$, Eq. (7.104) implies that

$$\det M = \det \Lambda\{\lambda - \lambda_1, \lambda - \lambda_2, \cdots, \lambda - \lambda_n\} = \prod_{i=1}^{n} (\lambda - \lambda_i). \tag{7.105}$$

Therefore, all the eigenvalues of the system can be derived by solving $\lambda - \lambda_i = 0$ where λ_i is function of λ and the corresponding eigenvector is v_i, $i = 1, 2, \cdots, n$.

Note that $\mathrm{v}_1 = \{1, 1, \cdots, 1\}^T$ represents the pattern of complete synchronization (CS). It can also be proved that $\sum_{i=1}^{n} v_{j,i} = 0$, where $v_{j,i}$ is the i-th component of v_j for $j = 2, 3, \cdots, n$. Furthermore, we have the following proposition.

Proposition 7.1 *If* $j \not\equiv 0 \bmod n$, *then* $\sum_{k=0}^{n-1} \chi^{jk} = 0$.

We point out that χ^{jk} *represents the relation of sources during the oscillation. Therefore, Proposition 1 lays down the basis of the desynchronization of the oscillation.* v_j *for* $j = 2, 3, \cdots, n$ *corresponds the desynchronous patterns of oscillation we are seeking. The following discussion in this section will be divided into two parts depending on the parity of n. In what follows, the time delay is considered as the bifurcation parameter that induces the stability switch. This consideration makes sense in the real-world engineering because the time delay is determined by neither the hardware nor the software and thus should be treated as a crucial parameter that introduces uncertainty to the system. Before moving on, we provide the following result addressing the stability of the system with zero delay.*

Case 1: n is odd

For this case, the characteristic polynomial (7.105) can be rewritten as

$$\det M = (\lambda - \lambda_1) \prod_{i=2}^{(n+1)/2} (\lambda - \lambda_i)^2, \tag{7.106}$$

because

$$\cos\left(\frac{2\pi(j-1)}{n}\right) = \cos\left(\frac{2\pi(n-j+1)}{n}\right), \Rightarrow \lambda_j = \lambda_{n+2-j}.$$

To find the critical value of the parameter, i.e., the time delay, we will attempt to find all the eigenvalues with zero real part through solving $\det M = 0$, which is equivalent to

$$\lambda + \kappa_j e^{-\lambda\tau} = 0, \quad \text{for some } i. \tag{7.107}$$

Substituting $\lambda = \omega\mathrm{i}$ into (7.107) yields

$$\omega i + \kappa_j e^{-\omega\tau_c i} = 0, \quad \text{for some } j. \tag{7.108}$$

It can be verified that $\kappa_j > 0$ for all j as long as $\alpha > \alpha_c$. Separating the real and imaginary parts of (7.108) gives

$$\kappa_j \cos(\omega\tau_c) = 0, \quad \omega - \kappa_j \sin(\omega\tau_c) = 0, \tag{7.109}$$

which implies

$$\omega = \kappa_j, \quad \tau_c = \pi/(2\kappa_j) \tag{7.110}$$

Define $\tau_c^{(i)} = \pi/(2\kappa_j)$. Then we're able to determine if the transversality condition is satisfied, namely, the sign of the speed of the real part of the eigenvalue with respect to the delay evaluated at $\lambda = \kappa_j\mathrm{i}$, $\tau = \tau_c^{(j)}$. To this end, we handle λ as a function of τ, namely, $\lambda = \lambda(\tau) = a(\tau) + \omega(\tau)\mathrm{i}$ and differentiate both sides of $\det M = 0$. For $j = 1$, we have

$$\left.\frac{\mathrm{d}(\det M)}{\mathrm{d}\tau}\right|_{\lambda=\kappa_1\mathrm{i},\tau=\tau_c^{(j)}} = \left.\frac{\mathrm{d}\Big(\big(\lambda + \kappa_1 e^{\lambda\tau}\big)\tilde{M}^1(\lambda(\tau),\tau)\Big)}{\mathrm{d}\tau}\right|_{\lambda=\kappa_1\mathrm{i},\tau=\tau_c^{(1)}} = 0, \tag{7.111}$$

where

$$\tilde{M}^1 = \prod_{i=2}^{(n+1)/2} (\lambda - \lambda_i)^2.$$

Calculating the left side of (7.111) gives

$$\begin{aligned}&\left(\frac{\mathrm{d}\lambda(\tau)}{\mathrm{d}\tau}+\kappa_1 e^{-\lambda(\tau)\tau}\left(\lambda(\tau)-\tau\frac{\mathrm{d}\lambda(\tau)}{\mathrm{d}\tau}\right)\right)\tilde{M}^1(\lambda,\tau)\Big|_{\lambda=\kappa_1\mathrm{i},\tau=\tau_c^{(1)}}\\&+\left(\lambda(\tau)+\kappa_1 e^{-\lambda(\tau)\tau}\right)\frac{\mathrm{d}\tilde{M}^1(\lambda(\tau),\tau)}{\mathrm{d}\tau}\Big|_{\lambda=\kappa_1\mathrm{i},\tau=\tau_c^{(1)}}\\&=\left(\frac{\mathrm{d}\lambda(\tau)}{\mathrm{d}\tau}+\kappa_1 e^{-\lambda(\tau)\tau}\left(\lambda(\tau)-\tau\frac{\mathrm{d}\lambda(\tau)}{\mathrm{d}\tau}\right)\right)\tilde{M}^1(\lambda(\tau),\tau)\Big|_{\lambda=\kappa_1\mathrm{i},\tau=\tau_c^{(1)}}=0,\end{aligned} \tag{7.112}$$

which implies

$$\begin{aligned}&\left(\frac{\mathrm{d}\lambda(\tau)}{\mathrm{d}\tau}+\kappa_1 e^{-\lambda(\tau)\tau}\left(\lambda(\tau)-\tau\frac{\mathrm{d}\lambda(\tau)}{\mathrm{d}\tau}\right)\right)\Big|_{\lambda=\kappa_1\mathrm{i},\tau=\tau_c^{(1)}}\\&=\left(1+\frac{\pi\mathrm{i}}{2}\right)\frac{\mathrm{d}\lambda(\tau)}{\mathrm{d}\tau}\Big|_{\lambda=\kappa_1\mathrm{i},\tau=\tau_c^{(1)}}-\kappa_1^2\\&=\left(1+\frac{\pi\mathrm{i}}{2}\right)\left(\frac{\mathrm{d}a(\tau)}{\mathrm{d}\tau}+\mathrm{i}\frac{\mathrm{d}\omega(\tau)}{\mathrm{d}\tau}\right)\Big|_{\lambda=\kappa_1\mathrm{i},\tau=\tau_c^{(1)}}-\kappa_1^2=0.\end{aligned} \tag{7.113}$$

Thus we have

$$\frac{\mathrm{d}a(\tau)}{\mathrm{d}\tau}\Big|_{\lambda=\kappa_1\mathrm{i},\tau=\tau_c^{(1)}}=\frac{\kappa_1^2}{1+\pi^2/4}>0. \tag{7.114}$$

For the case that $j>1$, one needs to differentiate λ with respect to τ twice, due to the multiplicity of λ_j, and then solve for $\mathrm{d}a(\lambda)/\mathrm{d}\tau$ in a similar way. The expression of $\mathrm{d}a(\lambda)/\mathrm{d}\tau$ is exactly the same as (7.114). This result together with the fact that the equilibrium is stable for $\tau=0$ indicates that the system will become unstable when the delay increases over $\tau_c^{(\tilde{j})}=\min_j \tau_c^{(j)}$ and remains unstable as τ keeps increasing. In other words, $\tilde{j}$ is the index of the delay for which the stability switch occurs. The pattern of oscillation is governed by the eigenvector corresponding to $\lambda=\kappa_{\tilde{j}}\mathrm{i}$. In order to remove the CS of (7.99), we only need to find the control parameters such that $\min\limits_{j>1}\tau_c^{(j)}<\tau_c^{(1)}$.

According to (7.110), this condition is equivalent to

$$\max_{j>1}\kappa_j>\kappa_1. \tag{7.115}$$

Since n is odd, it can be seen that

$$\kappa_{\frac{n+1}{2}}=\kappa_{\frac{n+3}{2}}=\max_{j>1}\kappa_j. \tag{7.116}$$

With (7.115), we have

$$\alpha > \alpha_1^c := \frac{knm_3}{2(1-\cos(\pi(n-1)/n))}, \tag{7.117}$$

which represents the condition that CS induced by the delay does not take place when the system becomes unstable.

From (7.116), we find that

$$\tilde{j} = \frac{n+1}{2}. \tag{7.118}$$

It should be noticed that if (7.117) is satisfied, multiple purely imaginary roots for the characteristic equation will arise. More precisely, (7.99) is a D_n-equivariant system where D_n is the dihedral group of n-th order [77]. Roughly speaking, D_n-equivariant system is the system that is invariant under the transformations $y_j \to y_{j+1 \text{ Mod } n}$, for $j = 1, 2, \cdot, n$ and $y_j \leftrightarrow y_{n+2-j}$ for $j = 2, 3, \cdots, n$. Such system is featured by the interaction of different modes of oscillation modes when the equilibrium loses the stability. We will address this problem in the next section.

Case 2: *n* is even

For this case, the characteristic polynomial (7.105) can be rewritten as

$$\det M = (\lambda - \lambda_1)(\lambda - \lambda_{\frac{n}{2}+1}) \prod_{i=2}^{n/2} (\lambda - \lambda_i)^2, \tag{7.119}$$

then the eigenvalue can be obtained by solving

$$\lambda + \kappa_j e^{-\lambda\tau} = 0, \quad \text{for some } i, \tag{7.120}$$

and the critical value of the delay is given by

$$\tau_c^j = \frac{\pi}{2\kappa_j}. \tag{7.121}$$

Following the same procedure as the case that n is even, we claim that $da(\lambda)/d\tau > 0$, evaluated at $\lambda = \kappa_j \mathrm{i}$ and $\tau = \tau_c^{(j)}$. We also have that $\kappa_{n/2+1} = \max_{j>1} \kappa_j$ and thus $\tau_c^{(\tilde{j})} = \tau_c^{(n+2)/2} = \min_{j>1} \tau_c^{(j)}$. That is,

$$\tilde{j} = \frac{n}{2} + 1. \tag{7.122}$$

Furthermore, $\tau_c^{((n+2)/2)} = \min_{j \geq 1} \tau_c^{(j)}$ if

$$\alpha > \alpha_2^c := \frac{knm_3}{4}. \tag{7.123}$$

(7.123) represents the condition that CS induced by the delay does not take place when the system becomes unstable. Unlike the case that n is odd, a single pair of purely imaginary roots for the characteristic equation will arise and corresponding eigenvector is $\mathrm{v}_{n/2-1} = \left\{1, e^{\frac{2\pi}{n}\frac{n}{2}}, e^{\frac{4\pi}{n}\frac{n}{2}}, \cdots, e^{\frac{(n-1)\pi}{n}\frac{n}{2}}\right\}^T = \{1, -1, 1, \cdots, -1\}^T$. Therefore, Hopf bifurcation may occurs as the equilibrium is about to be unstable for n being even. We will study this problem in the next section.

7.6.3 Method of Multiple Scales

In last section, the condition of stability switch is obtained and the possible modes of oscillation are determined. In this section, we will employ the method of multiple scales to calculate the normal form and analyze the nonlinear dynamics when the equilibrium becomes unstable, especially the stability of the oscillations. In the following, we focus on the case that n is odd and the analysis can be performed in a similar way for the case that n is even.

Assume that (7.117) is satisfied. Denote $\tau_c^{\left(\frac{n+1}{2}\right)}$ by τ_c. To investigate the dynamical behaviour of (7.99) when τ is close to τ_c, we write τ as $\tau = \tau_c + \varepsilon\tau_\varepsilon$ where ε is a small quantity and τ_ε a detuning parameter. Following similar procedure as in previous discussions,

$$x_{i,1}(T_0, T_1, T_2) = A(T_1, T_2)\chi_1^{\bar{j}(i-1)} e^{\mathrm{i}\omega T_0} + B(T_1, T_2)\chi_2^{\bar{j}(i-1)} e^{\mathrm{i}\omega T_0} + c.c., \quad i = 1, 2, \cdots, n, \tag{7.124}$$

where $x_{i,1}(T_0, T_1, T_2)$ is the approximated solution of $x_i(t)$ at ε, $c.c.$ represents the conjugate of the previous terms and $T_j = \varepsilon^j t$ for $i = 1, 2, \cdots, n$ and $j = 0, 1, 2$. Following the method of multiple scales and solving the equations of higher order of ε, we have

$$\begin{aligned}\dot{A} &= \varepsilon D_1 A + \varepsilon^2 D_2 A = A\big(\varepsilon s + \varepsilon^2 r_1 + \varepsilon^2 (r_2 A\overline{A} + r_3 B\overline{B})\big),\\ \dot{B} &= \varepsilon D_1 B + \varepsilon^2 D_2 B = B\big(\varepsilon s + \varepsilon^2 r_1 + \varepsilon^2 (r_2 A\overline{A} + r_3 B\overline{B})\big).\end{aligned}$$

Let

$$A = z_1(t) + iz_2(t), \quad B = z_3(t) + iz_4(t), \tag{7.125}$$

and

$$\begin{aligned} z_1(t) &= \rho_1(t)\cos(\phi_1(t)), \quad z_2(t) = \rho_1(t)\sin(\phi_1(t)),\\ z_3(t) &= \rho_2(t)\cos(\phi_2(t)), \quad z_4(t) = \rho_2(t)\sin(\phi_2(t)).\end{aligned} \tag{7.126}$$

Then it's easy to obtain that

$$\begin{aligned}
\dot{\rho}_1 &= \rho_1\left(\mathrm{Re}\left(\varepsilon s+\varepsilon^2 r_1\right)+\varepsilon^2\left(\mathrm{Re}(r_2)\rho_1^2+\mathrm{Re}(r_3)\rho_2^2\right)\right),\\
\dot{\rho}_2 &= \rho_2\left(\mathrm{Re}\left(\varepsilon s+\varepsilon^2 r_1\right)+\varepsilon^2\left(\mathrm{Re}(r_3)\rho_1^2+\mathrm{Re}(r_2)\rho_2^2\right)\right),\\
\rho_1\dot{\phi}_1 &= \rho_1\left(\mathrm{Im}\left(\varepsilon s+\varepsilon^2 r_1\right)+\varepsilon^2\left(\mathrm{Im}(r_2)\rho_1^2+\mathrm{Im}(r_3)\rho_2^2\right)\right),\\
\rho_2\dot{\phi}_2 &= \rho_2\left(\mathrm{Im}\left(\varepsilon s+\varepsilon^2 r_1\right)+\varepsilon^2\left(\mathrm{Im}(r_3)\rho_1^2+\mathrm{Im}(r_2)\rho_2^2\right)\right).
\end{aligned} \tag{7.127}$$

The normal form system of D_n- equivariant Hopf bifurcation in real coordinates can be utilized to analyze the stability of the bifurcated periodic solution.

According to equivariant bifurcation theory [77, 78], D_n- equivariant Hopf bifurcation will generate three types of periodic solutions, namely, discrete waves (DW), standing waves (SW) and mirror-reflecting waves (MRW), which corresponds to $\rho_2 = 0$ (or $\rho_1 = 0$), $\rho_1 = -\rho_2$ and $\rho_1 = \rho_2$, respectively.

It can be verified from (7.124) that $\sum_{j=1}^{n} y_i(t) \rightarrow$ constant. Only when $A(T_1, T_2)\overline{B}(T_1, T_2)$ and $B(T_1, T_2)\overline{A}(T_1, T_2)$ are both zero, i.e., $\rho_1 = 0$ or $\rho_2 = 0$. In other words, the perfect desynchronization is realized only when DW arises. A direct computation shows that the conditions that DW exists and is stable are

$$\mathrm{Re}\left(\varepsilon s+\varepsilon^2 r_1\right) > 0, \quad \mathrm{Re}\left(\varepsilon^2 r_2\right) < 0, \quad \frac{\mathrm{Re}(r_3)}{\mathrm{Re}(r_2)} > 1. \tag{7.128}$$

It should be noticed that solutions corresponding to $\rho_1 \neq 0$, $\rho_2 = 0$ and $\rho_1 = 0$, $\rho_2 \neq 0$ are both stable, namely, multi-periodic solutions coexist.

7.6.4 Examples

Case: $n = 101$.

For this case, assume $c_1 = 5$, $c_2 = 250$. Then a simple calculation shows that $y^* = 1.836$. The relation between τ_c^j and α is depicted in Fig. 7.37. Based on (7.117), $\alpha_1^c = 0.2356$. Let $\alpha = 0.5$. Then $\omega = 2.719$ and $\tau_c = \tau_c^{(51)} = 0.5777$. The conditions on the controller parameters β and γ such that DW exists and is stable as shown in Fig. 7.38. We select a set of parameter values from the shaded region, for example, $\beta = -0.1$ and $\gamma = 0$. Then the numerical results for $\tau = 0.6$ are shown in Fig. 7.39. μ is calculated to be 0.033.

Remark. We claimed that perfect DW arises if the number of connections is odd. One possible reason that perfect DW is not observed in this case is that characteristic equation's solution is dense around $\alpha = 0.5$, $\tau = 0.6$, implying multiple modes of oscillation and that interaction among different modes may break the DW.

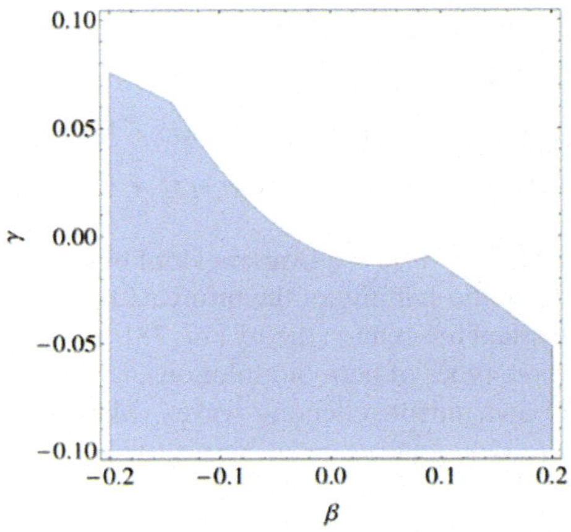

Fig. 7.38 Region of β and γ such that DW exists and is stable (shaded)

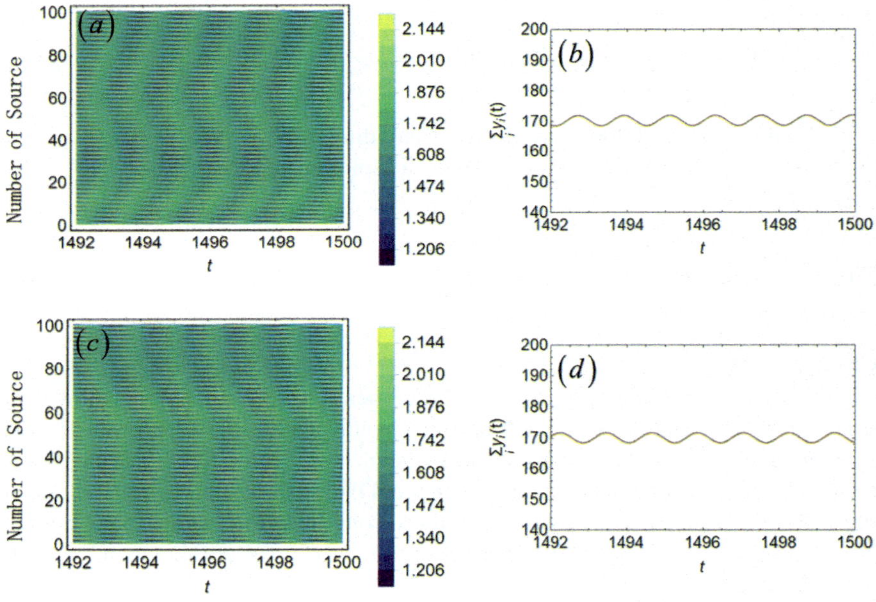

Fig. 7.39 Numerical simulations for $\alpha = 0.5$, $\beta = -0.1$, $\gamma = 0$ and $\tau = 0.6$ with initial conditions $y_i(t) = 1.83 + 0.2\sin(2(i-1)\pi/101)$ for **a** and **b**, $y_i(t) = 1.83 + 0.2\sin(2(i-1)\pi/101)$ for **c** and **d**, $-\tau \leq t \leq 0$

References

1. Bullo F (2020) Lectures on network systems, vol 3. Kindle Direct Publishing Seattle, DC, USA
2. Kamp C (2010) Untangling the interplay between epidemic spread and transmission network dynamics. PLoS Comput Biol 6(11):e1000984. https://doi.org/10.1371/journal.pcbi.1000984
3. Agrawala AK, Sanghi D (1995) Network dynamics: an experimental study of the Internet. [Conference Record] GLOBECOM'92-Communications for Global Users: IEEE: IEEE; 1992. pp. 782–786
4. Liu WR, Qin GR, He Y, Jiang F (2017) Distributed cooperative reinforcement learning-based traffic signal control that integrates V2X networks' dynamic clustering. IEEE Trans Veh Technol 66(10):8667–8681. https://doi.org/10.1109/Tvt.2017.2702388
5. Nagabandi A, Kahn G, Fearing RS, Levine S (2018) Neural network dynamics for model-based deep reinforcement learning with model-free fine-tuning. In: 2018 IEEE international conference on robotics and automation (ICRA): IEEE, pp 7559–7566
6. Barzel B, Barabasi AL (2013) Universality in network dynamics. Nat Phys 9(10):673–681. https://doi.org/10.1038/nphys2741
7. Wang XF (2002) Complex networks: topology, dynamics and synchronization. Int J Bifurcat Chaos 12(5):885–916. https://doi.org/10.1142/S0218127402004802
8. Tang LK, Lu JA, Wu XQ, Lü JH (2013) Impact of node dynamics parameters on topology identification of complex dynamical networks. Nonlinear Dyn 73(1–2):1081–1097. https://doi.org/10.1007/s11071-013-0854-3
9. Michiels W, Nijmeijer H (2009) Synchronization of delay-coupled nonlinear oscillators: an approach based on the stability analysis of synchronized equilibria. Chaos 19(3):033110. https://doi.org/10.1063/1.3187792
10. Tang LK, Lu JA, Lü JH, Yu XH (2012) Bifurcation analysis of synchronized regions in complex dynamical networks. Int J Bifurcat Chaos 22(11):1250282. Artn 1250282 https://doi.org/10.1142/S0218127412502823
11. Lakshmanan M, Senthilkumar DV (2011) Dynamics of nonlinear time-delay systems. Springer Science & Business Media
12. Yuste R (2015) From the neuron doctrine to neural networks. Nat Rev Neurosci 16(8):487–497. https://doi.org/10.1038/nrn3962
13. Aggarwal CC (2018) Neural networks and deep learning. Springer 10(978):3
14. Wu YC, Feng JW (2018) Development and application of artificial neural network. Wireless Pers Commun 102(2):1645–1656. https://doi.org/10.1007/s11277-017-5224-x
15. Abiodun OI, Jantan A, Omolara AE, Dada KV, Mohamed NA, Arshad H (2018) State-of-the-art in artificial neural network applications: a survey. Heliyon 4(11):e00938. https://doi.org/10.1016/j.heliyon.2018.e00938
16. Bernal J, Kushibar K, Asfaw DS, Valverde S, Oliver A, Marti R, Llado X (2019) Deep convolutional neural networks for brain image analysis on magnetic resonance imaging: a review. Artif Intell Med 95:64–81. https://doi.org/10.1016/j.artmed.2018.08.008
17. Kim HC, Bandettini PA, Lee JH (2019) Deep neural network predicts emotional responses of the human brain from functional magnetic resonance imaging. Neuroimage 186:607–627. https://doi.org/10.1016/j.neuroimage.2018.10.054
18. Varuna Shree N, Kumar TNR (2018) Identification and classification of brain tumor MRI images with feature extraction using DWT and probabilistic neural network. Brain Inform. 5(1):23–30. https://doi.org/10.1007/s40708-017-0075-5
19. Aihara K, Takabe T, Toyoda M (1990) Chaotic neural networks. Phys Lett A 144(6–7):333–340. https://doi.org/10.1016/0375-9601(90)90136-C
20. Goerner S (2014) Chaos, evolution, and deep ecology. Chaos theory in psychology and the life sciences. Psychology Press, pp 17–38
21. Wilson HR, Cowan JD (1972) Excitatory and inhibitory interactions in localized populations of model neurons. Biophys J 12(1):1–24. https://doi.org/10.1016/S0006-3495(72)86068-5
22. Sherratt JA, Murray JD (1990) Models of epidermal wound healing. Proc Biol Sci 241(1300):29–36. https://doi.org/10.1098/rspb.1990.0061

23. Nakagawa M (1998) Chaos associative memory with a periodic activation function. J Phys Soc Jpn 67(7):2281–2293. https://doi.org/10.1143/Jpsj.67.2281
24. Song Y, Yu X, Chen G, Xu J-X, Tian Y-P (2002) Time delayed repetitive learning control for chaotic systems. Int J Bifurcat Chaos 12(05):1057–1065
25. Li CD, Liao XF, Wong KW (2004) Chaotic lag synchronization of coupled time-delayed systems and its applications in secure communication. Phys D-Nonlinear Phenomena 194(3–4):187–202. https://doi.org/10.1016/j.physd.2004.02.005
26. Kosko B (1987) Adaptive bidirectional associative memories. Appl Opt 26(23):4947–4960. https://doi.org/10.1364/AO.26.004947
27. Bidirectional KB, Memories A (1988) IEEE Trans Syst Man Cybern 18(1):49–60. https://doi.org/10.1109/21.87054
28. Ge J, Xu J (2010) Computation of synchronized periodic solution in a BAM network with two delays. IEEE Trans Neural Netw 21(3):439–450. https://doi.org/10.1109/TNN.2009.2038911
29. Ge JH, Xu J (2010) Fold-Hopf bifurcation in a simplified four-neuron BAM (bidirectional associative memory) neural network with two delays. Sci China-Technol Sci 53(3):633–644. https://doi.org/10.1007/s11431-010-0079-3
30. Wang L, Jiang M, Liu R, Tang X (2008) Comparison bam and discrete hopfield networks with CPN for processing of noisy data. In: 2008 9th international conference on signal processing: IEEE, pp 1708–1711
31. Liao X, Wong KW (2004) Robust stability of interval bidirectional associative memory neural network with time delays. IEEE Trans Syst Man Cybern B Cybern 34(2):1142–1154. https://doi.org/10.1109/tsmcb.2003.821455
32. Li YY, Li JR, Li J, Duan SK, Wang LD, Guo MJ (2021) A reconfigurable bidirectional associative memory network with memristor bridge. Neurocomputing 454:382–391. https://doi.org/10.1016/j.neucom.2021.04.077
33. Du S, Zhang Z, Li J, Sun C, Sun J, Hong Q (2023) Multidirectional associative memory neural network circuit based on memristor. IEEE Trans Biomed Circuits Syst 17(3):433–445. https://doi.org/10.1109/TBCAS.2023.3276077
34. Wu AL, Zeng ZG, Song XG (2016) Global Mittag-Leffler stabilization of fractional-order bidirectional associative memory neural networks. Neurocomputing 177:489–496. https://doi.org/10.1016/j.neucom.2015.11.055
35. Sakthivel R, Vadivel P, Mathiyalagan K, Arunkumar A, Sivachitra M (2015) Design of state estimator for bidirectional associative memory neural networks with leakage delays. Inf Sci 296:263–274. https://doi.org/10.1016/j.ins.2014.10.063
36. Feng C, Plamondon R (2003) Stability analysis of bidirectional associative memory networks with time delays. IEEE Trans Neural Netw 14(6):1560–1565. https://doi.org/10.1109/TNN.2003.820829
37. Xu X, Hu HY, Wang HL (2006) Stability switches, HOPF bifurcation and chaos of a neuron model with delay-dependent parameters. Phys Lett A 354(1–2):126–136. https://doi.org/10.1016/j.physleta.2006.01.034
38. Qiu W, Yan S (2019) An image encryption algorithm based on the combination of low-dimensional chaos and high-dimensional chaos. In: 2019 3rd International conference on electronic information technology and computer engineering (EITCE): IEEE, pp 684–687
39. Perez G, Cerdeira HA (1995) Extracting messages masked by chaos. Phys Rev Lett 74(11):1970–1973. https://doi.org/10.1103/PhysRevLett.74.1970
40. Yang T, Yang LB, Yang CM (1998) Application of neural networks to unmasking chaotic secure communication. Physica D 124(1–3):248–257. https://doi.org/10.1016/S0167-2789(98)00199-7
41. Shen CW, Yu SM, Lu JH, Chen GR (2014) Designing hyperchaotic systems with any desired number of positive lyapunov exponents via a simple model. IEEE Trans Circ Syst I-Regular Pap 61(8):2380–2389. https://doi.org/10.1109/Tcsi.2014.2304655
42. Yang QG, Zhu DY, Yang LB (2018) A new 7D hyperchaotic system with five positive lyapunov exponents coined. Int J Bifurcat Chaos. 28(5):1850057. Artn 1850057 https://doi.org/10.1142/S0218127418500578

43. Le W, Li G, Guo H (2004) Application on classification of one dimension sequence using nonlinear model based on olfactory system. J Syst Simul 16(3):564–569
44. Zhou L, Hu G (2008) Global exponential periodicity and stability of cellular neural networks with variable and distributed delays. Appl Math Comput 195(2):402–411. https://doi.org/10.1016/j.amc.2007.04.114
45. Zhang D (2010) Adaptive synchronization of a class of chaotic time-delayed neural networks with uncertain parameters. In: 2010 2nd IEEE international conference on information management and engineering. IEEE, pp 374–377
46. Dong Z, Xu J (2010) Projective synchronization of different chaotic time-delayed neural networks based on integral sliding mode controller. Appl Math Comput
47. Jacobson V (1988) Congestion avoidance and control. ACM SIGCOMM Comput Commun Rev 18(4):314–329
48. Floyd S, Jacobson V (1993) Random early detection gateways for congestion avoidance. IEEE-ACM Trans Netw 1(4):397–413. https://doi.org/10.1109/90.251892
49. Alpcan T, Basar T (2003) Global stability analysis of an end-to-end congestion control scheme for general topology networks with delay. In: IEEE Conference on Decision and Control
50. Kelly FP (2000) Models for a self-managed Internet. Phil Trans Royal Soc a-Math Phys Eng Sci 358(1773):2335–2348. https://doi.org/10.1098/rsta.2000.0651
51. Kelly FP, Maulloo AK, Tan DKH (1998) Rate control for communication networks: shadow prices, proportional fairness and stability. J Operat Res Soc 49(3):237–252. https://doi.org/10.2307/3010473
52. Jiang K, Wang X, Xi Y (2004) Bifurcation analysis of an Internet congestion control model. In: ICARCV 2004 8th control, automation, robotics and vision conference, 2004: IEEE, pp. 590–594
53. Liu F, Guan ZH, Wang HO (2010) Controlling bifurcations and chaos in TCP-UDP-RED. Nonl Anal-Real World Appl 11(3):1491–1501. https://doi.org/10.1016/j.nonrwa.2009.03.005
54. Li CG, Chen GR, Liao XF, Yu JB (2004) HOPF bifurcation in an Internet congestion control model. Chaos, Solitons Fractals 19(4):853–862. https://doi.org/10.1016/S0960-0779(03)00269-8
55. Xiao M, Cao JD (2007) Delayed feedback-based bifurcation control in an Internet congestion model. J Math Anal Appl 332(2):1010–1027. https://doi.org/10.1016/j.jmaa.2006.10.062
56. Wang ZF, Chu TG (2006) Delay induced HOPF bifurcation in a simplified network congestion control model. Chaos, Solitons Fractals 28(1):161–172. https://doi.org/10.1016/j.chaos.2005.05.047
57. Brunner H, Maset S (2009) Time transformations for delay differential equations. Discrete Contin Dynam Syst 25(3):751–775. https://doi.org/10.3934/dcds.2009.25.751
58. Choi Y (2005) Periodic delay effects on cutting dynamics. J Dyn Diff Equat 17:353–389
59. Zhang S, Xu J (2012) Oscillatory dynamics induced by time delay in an internet congestion control model with ring topology. Appl Math Comput 218(22):11033–11041
60. Ding D-W, Jie Z, Luo X-S, Huang L-S, Hu Y-J (2009) Nonlinear dynamics in Internet congestion control model with TCP Westwood under RED. J China Univ Posts and Telecommun 16(4):53–58
61. Zhang S, Xu J (2011) Oscillation control for-dimensional congestion control model via time-varying delay. Sci China-Technol Sci 54(8):2044–2053. https://doi.org/10.1007/s11431-011-4488-8
62. Zhang S, Xu J (2011) Oscillation control for n-dimensional congestion control model via time-varying delay. Science China Technol Sci 54:2044–2053
63. TamaraDean. Network+ guide to networks. Network+ guide to networks
64. Kuznetsov YA (2004) Elements of applied bifurcation theory. Appl Math Sci 288(2):715–730
65. Hale JK (1977) Theory of functional differential equation. Theo Funct Diff Equat
66. Srikant R (2006) The mathematics of internet congestion control. Birkhäuser
67. Kelly FP (2000) Models for a self-managed Internet. Philosoph Trans Royal Soc: A Math Phys Eng Sci 358(1773):2335–2348. https://doi.org/10.1098/rsta.2000.0651

68. Kunniyur SS, Srikant R (2003) Stable, scalable, fair congestion control and aqm schemes that achieve high utilization in the internet. Automat Cont IEEE Trans 48(11):2024–2028
69. Das SL, Chatterjee A (2002) Multiple scales without center manifold reductions for delay differential equations near HOPF bifurcations. Nonlinear Dyn 30(4):323–335. https://doi.org/10.1023/A:1021220117746
70. Nayfeh AH (2008) Order reduction of retarded nonlinear systems - the method of multiple scales versus center-manifold reduction. Nonlinear Dyn 51(4):483–500. https://doi.org/10.1007/s11071-007-9237-y
71. Song Y, Xu J, Zhang T (2011) Bifurcation, amplitude death and oscillation patterns in a system of three coupled van der Pol oscillators with diffusively delayed velocity coupling. Chaos 21(2):023111. https://doi.org/10.1063/1.3578046
72. Wu JH (1998) Symmetric functional differential equations and neural networks with memory. Trans Am Math Soc 350(12):4799–4838. https://doi.org/10.1090/S0002-9947-98-02083-2
73. Guo YX, Jiang WH, Niu B (2013) Multiple scales and normal forms in a ring of delay coupled oscillators with application to chaotic Hindmarsh-Rose neurons. Nonlinear Dyn 71(3):515–529. https://doi.org/10.1007/s11071-012-0678-6
74. Zhang S, Xu J, Chung K-W (2017) Desynchronization-based congestion suppression for a star-type internet system with arbitrary dimension. Neurocomputing 266:42–55
75. Qiu LL, Zhang Y, Keshav S (2001) Understanding the performance of many TCP flows. Comput Netw-the Int J Comput Telecommun Network 37(3–4):277–306. https://doi.org/10.1016/S1389-1286(01)00203-1
76. Yuan Y, Campbell SA (2004) Stability and synchronization of a ring of identical cells with delayed coupling. J Dyn Diff Equat 16:709–744
77. Grob K, Biedermann M (1985) Singularities and groups in bifurcation theory. Singularities and groups in bifurcation theory
78. Guo SJ, Huang LH (2007) Stability of nonlinear waves in a ring of neurons with delays. J Diff Eqn 236(2):343–374. https://doi.org/10.1016/j.jde.2007.01.027

Chapter 8
Delay Effect in Biology

In fact, life phenomena are far more complicated than physical and mechanical phenomena. In the study of life science, dynamical systems have been developed and successfully used in a variety of subjects, including physics, mechanics, and life science [1–3]. Especially since the late 1970s, the discipline has developed rapidly, and it has penetrated almost every branch of life science, which promotes the development of life science from micro to macro.

Time delays exist widely in biological systems. It is very necessary to consider the impact of time delay when we analyze the dynamic behaviour of biological systems. In immunology, the immune system produces corresponding effector cells to kill tumor cells. However, the effector cells are accompanied by delay effects in the processes of activation, maturation, and transportation. Medical research shows that implanting an artificial pancreas in diabetic patients is one of the methods used to treat diabetes. Nevertheless, the blood glucose-insulin system in the human body will experience a technical delay caused by artificial pancreas and a physiological delay caused by the liver, which will cause blood glucose levels to fluctuate and raise the risk of diabetic complications. Therefore, it is crucial to research and evaluate how time delays affect the dynamics of blood glucose concentration. In the process of neuron signal transmission of the biological nervous system, the chemical and electrical signals are also taken time to encode, transmit, and decode. Additionally, the human body maintains static standing through the coordination of vestibular organs, muscles, tendons, proprioceptors in joints, and the central nervous system, among which the closed-loop control system composed of the neuromusculoskeletal system needs to consider the influence of time delay. Similarly, there are incubation delays and reaction delays in population dynamics, which require us to establish different types of delay differential equations.

In this chapter, we introduce time-delayed factor into the biological systems and focus on its influence on the dynamic behaviour of the system. First, the dynamics of the neuronal time-delay coupling system, as well as the dynamics of the blood glucose-insulin system with treatment delay and liver response delay, are taken into

J. Xu, *Nonlinear Dynamics of Time Delay Systems*,
https://doi.org/10.1007/978-981-99-9907-1_8

consideration. Secondly, the problem of humans standing under the action of randomness and time delay is discussed. Under the condition of a non-uniform spatial distribution of biological populations, the influence between population diffusion in space and the reaction delay of intraspecific interaction on population distribution and population density is studied. This chapter studies several common problems in application of biology systems. We attempt to explain to readers how to construct the appropriate mathematical model under the action of time delay and how to carry out the mathematical analysis of the model. The enlightenment and significance of the research results for life science problems are the contents that need to be paid attention to.

8.1 Dynamic Analysis of a Coupled FitzHugh-Nagumo Neural System with Time Delay

Neurons are the basic units of nervous system to obtain structures and functions. The large-scale network formed by their interconnection determines the perception, learning, emotion, and memory in organisms. Neurons produce action potentials in response to environmental stimulation, and then process and transmit information in the form of electrical or chemical pulses. With the in-depth study of the biological neuron system, it is increasingly found that the nervous system exhibits extremely complex nonlinear characteristics. The study of the nonlinear dynamic behaviour of the biological neuronal systems is not only helpful to the mechanism researches of neurophysiological phenomena, but also to guide the experimental researches of neurophysiology and the application of neuromedicine and intelligent robots. The analysis of neural dynamics promotes the in-depth development of the theory and methods of nonlinear dynamical systems.

The effect of time delay on signal transmission of neuronal information cannot be avoided in the real brain nervous system due to the finite speed of neuron signal propagation and the presence of synaptic gaps, which will have a certain impact on the stability of neural network and the response of information transmission. Experiments show that the transmission speed of nerve signals in unmyelinated axial fibers is 1 m/s, which leads to the propagation delay through the cerebral cortex network reaching 80 ms [4]. Therefore, the study of the dynamic behaviour for the time-delayed neuronal system is one of the important contents in the neural dynamical fields.

The FitzHugh-Nagumo (FHN) model can describe physiological phenomena [5–7] similar to those corresponding to the Hodgkin-Huxley model [8]. The FHN model is easy to analyze by mathematical methods, and it can simulate most of the spiking and bursting behaviour of biological neurons, so it has attracted widespread attention by researchers [9, 10]. In this section, we focus on the nonlinear dynamical behaviour of the delayed FHN neural system by using the center manifold theorem and normal form theory. The generation mechanism of periodic oscillations of the system is expounded theoretically to obtain the parameter conditions. Moreover, the effect of time delay on static bifurcation is studied. The dynamical classification in the parameter region near the bifurcation point is given in detail to obtain the system dynamical behaviours [11].

8.1.1 A Single FHN Neuron Model and Delay Coupled System

A single FHN neuron model is taken the following form [7]

$$\begin{cases} \dot{u}_1 = -u_1(u_1 - 1)(u_1 - a) - u_2, \\ \dot{u}_2 = b(u_1 - \gamma u_2), \end{cases} \tag{8.1}$$

where u_1 and u_2 indicate the membrane potential and recovery variable of the neuron, respectively. Further, the parameters in Eq. (8.1) satisfy the following conditions

$$0 < a < 0.5, b > 0, r > 0, \frac{1}{r} - \frac{1 - a + a^2}{3} > 0, \frac{1 - a + a^2}{3} - br > 0.$$

In this section, we consider the following delayed FHN neural system, which is

$$\begin{cases} \dot{u}_1 = -u_1(u_1 - 1)(u_1 - a) - u_2 + c\tanh(u_3(t - \tau)), \\ \dot{u}_2 = b(u_1 - \gamma u_2), \\ \dot{u}_3 = -u_3(u_3 - 1)(u_3 - a) - u_4 + c\tanh(u_1(t - \tau)), \\ \dot{u}_4 = b(u_3 - \gamma u_4), \end{cases} \tag{8.2}$$

where $\tau > 0$ reflects time delay in signal transmission, c represents coupling strength. Further, c is used as the bifurcation parameter to discuss the dynamics behaviour of Eq. (8.2).

8.1.2 Analysis on Eigenvalues

The characteristic equation corresponding to (8.2) at the trivial equilibrium is

$$D(\lambda, c) = \lambda^4 + d_1\lambda^3 + d_2\lambda^2 + d_3\lambda + d_4 + \left(d_5\lambda^2 + d_6\lambda + d_7\right)e^{-2\lambda\tau} = 0, \tag{8.3}$$

where

$$d_1 = 2(a + b\gamma), d_2 = (a + b\gamma)^2 + 2b(a\gamma + 1), d_3 = 2b(a + b\gamma)(a\gamma + 1),$$
$$d_4 = (a\gamma + 1)^2b^2, d_5 = -c^2, d_6 = -2c^2b\gamma, d_7 = -c^2b^2\gamma^2.$$

First, we consider the conditions for the occurrence of single zero eigenvalues of Eq. (8.2). It is easy to verify that λ=0 is a root of Eq. (8.3) when $c = \pm(a + 1/\gamma)$. Differentiating Eq. (8.3) with respect to the parameter λ gives

$$\frac{dD(\lambda, c)}{d\lambda}\bigg|_{\lambda=0, c=\pm(a+1/\gamma)} = 2b^2(a\gamma + 1)^2\left[\tau - \frac{1 - b\gamma^2}{b\gamma(a\gamma + 1)}\right].$$

It indicates that $\lambda = 0$ is a single root of Eq. (8.3) if and only if $c = \pm(a + 1/\gamma)$ and $\tau \neq \left(1 - b\gamma^2\right)/(b\gamma(a\gamma + 1))$.

Let

$$D_1(\lambda) = D(\lambda, \pm(a + 1/\gamma))/\lambda,$$

then we have

$$\lim_{\lambda \to 0} D_1(\lambda) = 2b^2(a\gamma + 1)^2\left(\tau - \frac{1 - b\gamma^2}{b\gamma(a\gamma + 1)}\right).$$

As can be observed, $D_1(\lambda) < 0$ when $\tau < \left(1 - b\gamma^2\right)/(b\gamma(a\gamma + 1))$ and $\lambda \to 0^+$. Since $\tau > 0$, there exists a positive number λ_0 satisfied with $D_1(\lambda_0) = 0$. It indicates that Eq. (8.3) has at least one positive root when $c = \pm(a + 1/\gamma)$ and $\tau \neq \left(1 - b\gamma^2\right)/(b\gamma(a\gamma + 1))$.

Further, we consider the conditions that Eq. (8.2) has a pair of purely imaginary roots. Substituting $\lambda = i\omega$ into Eq. (8.3) and separating the real and imaginary parts, we obtain

$$\begin{cases} \omega^4 - d_2\omega^2 + d_4 + \left(-d_5\omega^2 + d_7\right)\cos(2\omega\tau) + d_6\omega\sin(2\omega\tau) = 0, \\ -d_1\omega^3 + d_3\omega + d_6\omega\cos(2\omega\tau) - \left(-d_5\omega^2 + d_7\right)\sin(2\omega\tau) = 0. \end{cases} \tag{8.4}$$

Eliminating τ from (8.4) and letting $v = \omega^2$, one has

$$v\left(v^3 + r_1v^2 + r_2v + r_3\right) = 0, \tag{8.5}$$

where

$$\begin{aligned} r_1 &= 2\left((a + b\gamma)^2 - 2b(a\gamma + 1)\right), \\ r_2 &= -c^4 + 2(a\gamma + 1)^2b^2 + \left((a + b\gamma)^2 - 2b(a\gamma + 1)\right)^2, \\ r_3 &= -2c^4b^2\gamma^2 + 2b^2\left((a + b\gamma)^2 - 2b(a\gamma + 1)\right)(a\gamma + 1)^2. \end{aligned}$$

Let

$$N_1(v) = v^3 + r_1v^2 + r_2v + r_3 = 0. \tag{8.6}$$

Define

$$\Delta = \frac{q^2}{4} + \frac{p^3}{27}, \tag{8.7}$$

where $p = (2r_2 - r_1^2)/3$ and $q = (2r_1^3 - 9r_1r_2 + 27r_3)/27$. Clearly, if $\Delta > 0$, Eq. (8.6) has one real root and a pair of conjugate complex roots. Therefore, if $\Delta > 0$ and $r_3 > 0$, Eq. (8.6) only has one negative real root; if $\Delta > 0$ and $r_3 < 0$, Eq. (8.6) only has one positive real root; if $\Delta > 0$ and $r_3 = 0$, Eq. (8.6) only has one zero root. Furthermore, if $\Delta \leq 0$, Eq. (8.6) has three real roots. Assume that $v = v_0$ is one real root of Eq. (8.6), then Eq. (8.6) can be rewritten as

$$N_1(v) = (v - v_0)\left(v^2 + p_1 v + p_2\right) = 0, \tag{8.8}$$

where p_1 and p_2 are non-zero real numbers.

Comparing the coefficients in Eq. (8.6) with those in Eq. (8.8), we get

$$-v_0 + p_1 = r_1, -v_0 p_1 + p_2 = r_2, -v_0 p_2 = r_3. \tag{8.9}$$

If $\Delta \leq 0$ and $r_3 > 0$, we get $v_0 < 0$ and $p_2 > 0$. Thus, it is easy to see that Eq. (8.6) has two positive real roots and one negative real root, or has three negative real roots. According to the Roth-Hurwitz criterion, the necessary and sufficient conditions for all real roots of Eq. (8.6) to be negative give are

$$r_1 > 0, r_2 > 0, r_3 > 0, r_1r_2 > r_3. \tag{8.10}$$

So, Eq. (8.6) has two positive real roots if $\Delta \leq 0$ and the conditions in (8.10) are not met. Then, if $\Delta \leq 0$ and $r_3 < 0$, we get $v_0 > 0$ and $p_2 > 0$. Then, Eq. (8.6) has two negative real roots and one positive real root, or has three positive real roots. Substituting $-\overline{v}$ for v in Eq. (8.6) gives

$$N_2(\overline{v}) = \overline{v}^3 - r_1\overline{v}^2 + r_2\overline{v} - r_3 = 0. \tag{8.11}$$

According to the Roth-Hurwitz criterion, the necessary and sufficient conditions for (8.11) all real roots to be negative real roots are as follows

$$r_1 < 0, r_2 > 0, r_3 < 0, r_1r_2 < r_3. \tag{8.12}$$

In summary, Eq. (8.6) has three positive real roots if $\Delta \leq 0$ and (8.12) are satisfied, Eq. (8.6) has two negative real roots and one positive real root if $\Delta \leq 0$ and (8.12) are not satisfied. Further, we assume that $\lambda = \pm i\omega_0(\omega_0 > 0)$ are the roots of Eq. (8.3) with $c = \pm(a + 1/\gamma)$ and $\tau \neq \left(1 - b\gamma^2\right)/(b\gamma(a\gamma + 1))$, then we get

$$\tau_j = \frac{1}{2\omega_0}\left\{\sin^{-1}\left(\frac{A\omega_0}{B}\right) + 2j\pi\right\}, \quad j = 0, 1, 2, \ldots \tag{8.13}$$

where

$$A = (d_1d_5 - d_6)\omega_0^4 + (-d_1d_7 + d_2d_6 - d_3d_5)\omega_0^2 - d_4d_6 + d_3d_7,$$
$$B = d_5^2\omega_0^4 + \left(d_6^2 - 2d_5d_7\right)\omega_0^2 + d_7^2.$$

Then, if Eq. (8.3) with $c = \pm(a + 1/\gamma)$ and $\tau \neq (1 - b\gamma^2)/(b\gamma(a\gamma + 1))$ has a simple pair of purely imaginary roots and satisfies $\mathrm{Re}\big(d\lambda(\tau_j)/(d\tau)\big) \neq 0$, Eq. (8.2) undergoes a zero-pair bifurcation at the trivial equilibrium. According to the center manifold theorem, there exists a central manifold that makes Eq. (8.2) and its reduction equations on the manifold are topologically equivalent. The reduction equation is as follows

$$\begin{cases} \dot{z}_1 = k_{a1}\mu_1 z_1 + k_{a2}z_1^2 + k_{a3}z_2z_3 + k_{a4}z_1^3 + k_{a5}z_1z_2z_3, \\ \dot{z}_2 = i\omega_0\tau_0 z_2 + (k_{b1}\mu_1 + k_{b2}\mu_2)z_2 + k_{b3}z_1z_2 + k_{b4}z_1^2z_2 + k_{b5}z_2^2z_3, \\ \dot{z}_3 = -i\omega_0\tau_0 z_3 + \left(\overline{k}_{b1}\mu_1 + \overline{k}_{b2}\mu_2\right)z_3 + \overline{k}_{b3}z_1z_3 + \overline{k}_{b4}z_1^2z_3 + \overline{k}_{b5}z_2z_3^2, \end{cases} \tag{8.14}$$

where

$$k_{a1} = 2Hb_{\tau_0}\gamma^2, k_{a2} = 2Hb_{\tau_0}\gamma^3(a{+}1), k_{a2} = 2Hb_{\tau_0}\gamma(a{+}1)(1 + q_2\overline{q}_2),$$
$$k_{b1} = \rho\tau_0(q_2 + \overline{q}_2)e^{-\mathrm{i}\omega_0\tau_0},$$
$$k_{b2} = \rho\big(-(a + \overline{q}_1)(1 + q_2\overline{q}_2) + (\gamma\overline{q}_1 - 1)(q_1 + \overline{q}_2q_3) + c_0(\overline{q}_2 + q_2)e^{-\mathrm{i}\omega_0\tau_0}\big),$$
$$k_{b3} = 2\rho\tau_0\gamma(a{+}1)(1 + q_2\overline{q}_2),$$

with

$$q_1 = \frac{b}{\mathrm{i}\omega_0 + b\gamma}, q_2 = \frac{\left(-\omega_0^2 + \mathrm{i}\omega_0(b\gamma + a) + b(a\gamma + 1)\right)\gamma e^{\mathrm{i}\omega_0\tau_0}}{\mathrm{i}\omega_0 + b\gamma},$$
$$q_3 = q_1q_2, \frac{1}{H} = 2c_0b\gamma^2(\tau - \tau_{00}), c_0 = c - \mu_1, \tau_0 = \tau - \mu_2,$$
$$\tau_{00} = \frac{1 - b\gamma^2}{b\gamma(a\gamma + 1)}, \frac{1}{\overline{\rho}} = 1 + bq_1\overline{q}_1 + q_2\overline{q}_2 + bq_3\overline{q}_3 + c_0(q_2 + \overline{q}_2)e^{\mathrm{i}\omega_0\tau_0}.$$

8.1.3 Numerical Simulations

Firstly, we choose the parameter values as $a = 0.1$, $b = 0.09$, $\gamma = 3.2$, one can calculate that $c_0 = 0.4125$, $\left(1 - b\gamma^2\right)/(b\gamma(a\gamma + 1)) = 0.206$, $\Delta = 1.3\mathrm{e}-5 > 0$, $r_3 = -0.0073 < 0$. Let $\tau_0 = 5.3243$, we get $\mathrm{Re}(d\lambda(c_0, \tau_0)/(d\tau)) > 0$.

From the above analysis, we know that $k_{ai} > 0$, $\mathrm{Re}(k_{bi}) > 0 (i = 1, 2, 3)$. The bifurcation diagram of (8.14) in the parameter space (c, τ) for this case is given in Fig. 8.1. Stationary solutions on the w_1-axis of the planar Eq. (8.14) corresponds to equilibria of Eq. (8.2), whereas stationary solution off this axis corresponds to a periodic solution. As is shown, E_3 is a saddle and bifurcation cannot occur here, which means (8.2) has no stable limit cycle near the origin. The time history diagrams corresponding to Fig. 8.1 are shown in Figs. 8.2 and 8.3, respectively.

Then, we choose the parameter values as $a = 0.1$, $b = 0.01$, $\gamma = 3.2$, one can calculate that $p = (2r_2 - r_1^2)/3$, $c_0' = 0.4125$, $\left(1 - b\gamma^2\right)/(b\gamma(a\gamma + 1)) = 21.25$, $\Delta = -1.05\mathrm{e} - 6 < 0$, $r_2 = -0.046 < 0$, $r_3 = -6.24e - 5 < 0$. Let $\tau_0' = 14.2564$, we can calculate $\mathrm{Re}\left(d\lambda(c_0', \tau_0')/(d\tau)\right) < 0$. From the above analysis, we know that $k_{ai} < 0$, $\mathrm{Re}(k_{bi}) > 0 (i = 1, 2, 3)$, $p_1 \neq 0$, $l_1 > 0$. The bifurcation diagram of (8.14) in the parameter space (c, τ) is given in Fig. 8.4. It can be seen that the limit

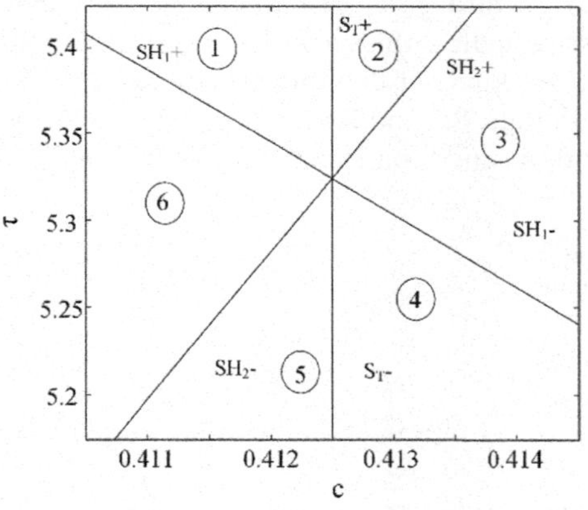

Fig. 8.1 The bifurcation diagram of Eq. (8.14) in the parameter plane (c, τ) for $a = 0.1$, $b = 0.09$, $\gamma = 3.2$, $c_0 = 0.4125$, and $\tau_0 = 5.3243$

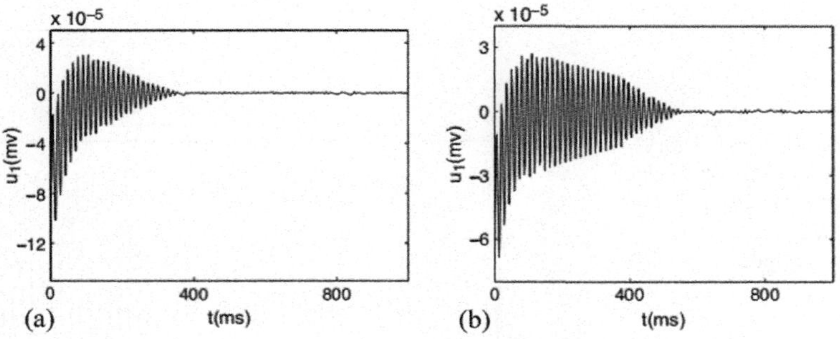

Fig. 8.2 The time-dependent curves of u_1 in Eq. (8.2) with **a** $c = 0.36$ and $\tau = 4.8$, lying in region ⑤ in Fig. 8.1; **b** $c = 0.35$ and $\tau = 5.3$, lying in region ⑥ in Fig. 8.1, which show that E_1 is the stable trivial equilibrium in regions ⑤ and ⑥

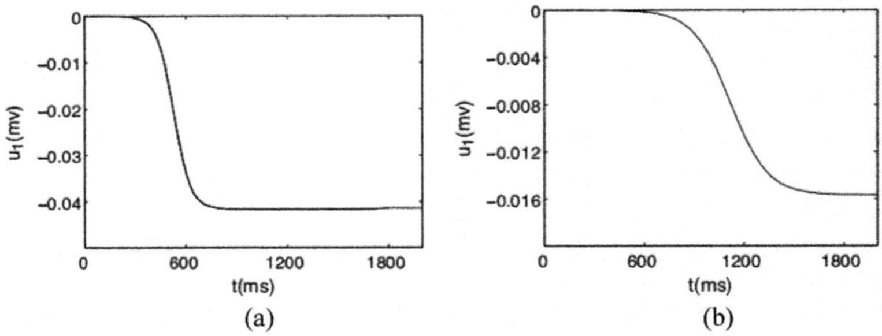

Fig. 8.3 The time-dependent curves of u_1 in Eq. (8.2) with **a** $c = 0.43$ and $\tau = 5.3$, lying in region ③ in Fig. 8..8.1; **b** $c = 0.43$ and $\tau = 4.8$, lying in region ④ in Fig. 8.1, which show that E_2 is the stable trivial equilibrium in regions ③ and ④

cycle bifurcated from E_3 through Hopf bifurcation is unstable, which is equivalent to an unstable quasi-periodic solution in Eq. (8.2). According to the bifurcation diagram in Fig. 8.4, two kinds of the time history curves of u_1 in Eq. (8.2) can be observed, as shown in Figs. 8.5 and 8.6, respectively, in which we can also observe the quasi-periodic motion and limit cycle in regions ④ and ⑧.

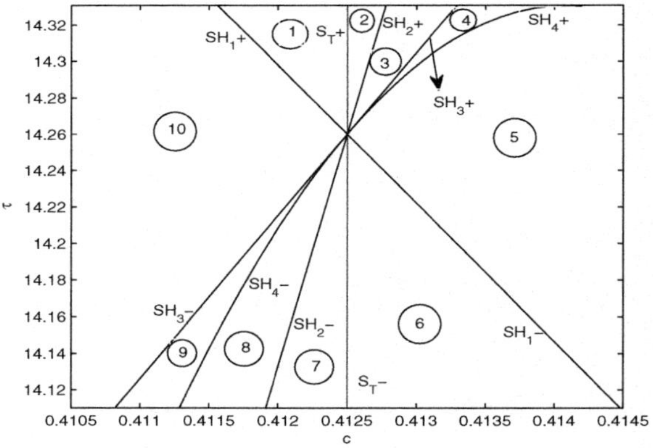

Fig. 8.4 The bifurcation diagram of Eq. (8.14) in the parameter plane (c, τ) for $a = 0.1$, $b = 0.01$, $\gamma = 3.2$, $c_0' = 0.4125$, and $\tau_0' = 14.2564$

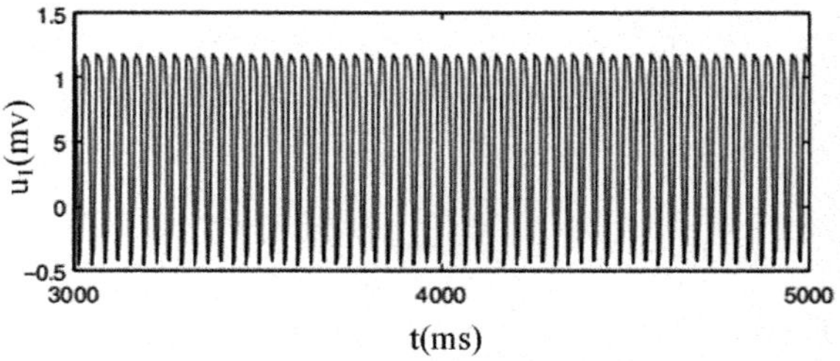

Fig. 8.5 The time history curve of u_1 in Eq. (8.2) when $c = 0.46$, $\tau = 14.8$, which shows the quasi-periodic motion in region ④

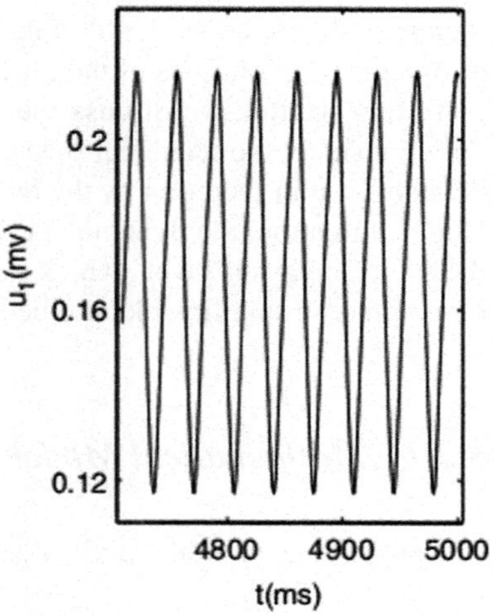

Fig. 8.6 The time history curve of u_1 in Eq. (8.2) when $c = 0.4$, $\tau = 13.9$, which shows that a stable limit cycle exists in region ⑧

8.2 Effects of the Technological Delay and Physiological Delay on the Insulin and Blood Glucose System

The concentration of blood glucose (BG) in the human body is controlled by the regulation of the endocrine system. However, there will be many diseases in the human body if the endocrine system is disordered, one of which is diabetes [12]. Diabetes is a disease that can cause a series of complications, such as retinopathy, nephropathy, peripheral neuropathy, and blindness [13]. Medical research shows that implanting an artificial pancreas (AP) in diabetic patients is one of the most effective methods to treat diabetes. The AP, however, takes time to measure and regulate blood sugar levels. The presence of inherent technological delay perturbs the system and can lead to instability in the glucose concentration, i.e., fluctuations in the BG. The existence of glycemia during glucose fluctuation can lead to many secondary complications of diabetes.

In order to seek advantages and avoid disadvantages, it is crucial to investigate the effect of the technical delay of AP on the dynamics of the system and give the setting range of parameters from the perspective of theoretical analysis. Up until now, researchers have established many related delayed mathematical models to study the mechanism of blood glucose-insulin interactions [14–22]. For example, Al-Hussein et al. proposed a new time delay mathematical model for the glucose-insulin endocrine metabolic regulatory feedback system [19]. The results showed that the time delay in insulin secretion to blood glucose level and the delay in glucose drop due to increased insulin concentration can give rise to complex dynamics. Mohabati et al. [20] considered a delay differential model for glucose-insulin interaction. The simulations revealed that the sustained ultradian oscillations of glucose and insulin are stable. Additionally, when the time delay increases, the amplitude of oscillation increases simultaneously. All of these dynamical behaviours indicate different kinds of dynamical evolutions of the glucose fluctuations or concentrations.

In this section, we discuss the blood glucose-insulin physiological system not only considers the technical delay of the AP, but also considers the physiological delay of insulin secretion by the liver. Based on the mathematical analysis of double Hopf bifurcation, the dynamical classification of the system under the combined of these time delays is given, which expounds the parameter intervals corresponding to blood glucose stability, blood glucose periodic fluctuation, and complex fluctuation.

8.2.1 Mathematical Model

The physiological model under consideration is given by

$$\begin{cases} \dot{I}(t) = \alpha f_1(G) - \frac{1}{t_1} + (1-\alpha) f_1(G(t-\tau_1)), \\ \dot{G}(t) = Eg + f_5(I(t-\tau_2)) - (f_2(G) + f_3(G) f_4(I)), \end{cases} \tag{8.15}$$

where $I(t)$ represents the insulin quantity, $G(t)$ represents the glucose quantity (mg) in the body, Eg indicates the glucose quantity supplied by the external medium corresponding to food intake, α reveals the infection level of patients. The time delays τ_1 and τ_2 in the model (8.15) represent the technological time delays of AP and the reaction time of the liver to produce the hepatic glucose, respectively. The functions $f_i (i = 1, 2, \cdots 5)$ are

$$\begin{cases} f_1(G) = \frac{209}{1+e^{6.6-(G/300V_3)}}, \; f_2(G) = 72\left(1 - e^{-(G/144V_3)}\right), \\ f_3(G) = \frac{0.01G}{V_3}, \; f_4(I) = \frac{90}{1+e^{-1.772\times\log(I/V_1)+7.76}} + 4, \\ f_5(I) = \frac{180}{1+e^{(0.29I/V_1)-7.5}}, \end{cases} \tag{8.16}$$

where V_1 is the volume of the insulin compartment, V_3 is the volume of the glucose compartment. The reference values of the parameters are

$$\begin{array}{c} V_1 = 3(l), \tau_1 = 6(\text{min}), V_3 = 10(l), Eg = 180(\text{mg/min}) \\ V_3 = 10(l), \tau_2 = 50(\text{min}). \end{array} \tag{8.17}$$

In this section, we investigate the effects of the technological delay τ_1 and the infection degree α of the patient on the dynamics of Eq. (8.15).

8.2.2 Double Hopf Bifurcation Analysis

Equation (8.15) has a unique equilibrium point $E^* = (I^*, G^*) = (81.078, 11784.46)$. It is found that (8.15) has multiple double Hopf bifurcation points in the characteristic equation of linear approximate system near E^*. We choose one of the double Hopf bifurcation points $(\alpha_c, \tau_{1c}) = (0.68365, 224.4)$ as the research object. At (α_c, τ_{1c}), the characteristic equation of Eq. (8.15) has two pairs of pure imaginary roots $\pm i\omega_1$ and $\pm i\omega_2$, where i is an imaginary unit, $\omega_1 = 0.039708$, $\omega_2 = 0.050707$, and $\omega_1 : \omega_2 = \sqrt{2} : \sqrt{3}$.

Let $\tau_1 = \tau_{1c} + \Delta\tau_1$ and $\alpha = \alpha_c + \Delta\alpha$, the dynamic equation of Eq. (8.15) on the central manifold can be obtained by using the center manifold theorem. The amplitude form of the normal form can be obtained by using Nayfeh's normal form method

$$\begin{cases} \dot{r}_1 = \mu_1 r_1 + a_{11} r_1^3 + a_{12} r_1 r_2^2, \\ \dot{r}_2 = \mu_2 r_2 + a_{21} r_1^2 r_2 + a_{22} r_2^3, \end{cases} \tag{8.18}$$

where

$$\begin{aligned} &\mu_1 = -0.036565\Delta\alpha - 0.000131\Delta\tau_1, \mu_2 = 0.005451\Delta\alpha + 0.000215\Delta\tau_1, \\ &\Delta\tau_1 = \varepsilon\tau_{1\varepsilon}, \Delta\alpha = \varepsilon\alpha_\varepsilon, a_{11} = -2.244366 \times 10^{-6}, \\ &a_{12} = -4.45535 \times 10^{-6}, a_{21} = -0.000021968, a_{22} = -0.0000109703. \end{aligned}$$

It follows from (8.18) that the classification of the double Hopf bifurcation are obtained which is shown in Fig. 8.7. In region I, Eq. (8.15) has a stable focus. In regions II and III, there are stable periodic solutions only related to the frequency ω_2 in Eq. (8.15). It is worth noting that an unstable periodic solution also appears in region III. In region IV, Eq. (8.15) has a bistable periodic solution related to the frequency ω_1 and an unstable quasi-periodic solution. In regions V and VI, there are only stable periodic solutions with frequency ω_1 in Eq. (8.15). Furthermore, there is an unstable periodic solution in region V.

To verify the above classification results, we employ the fourth order Runge–Kutta method to give numerical simulations. In region I, we choose the parameter values as $\Delta\alpha = 0.06$ and $\Delta\tau_1 = -4$, there is a stable focus E^*(see Fig. 8.8a) in Eq. (8.15). In regions II and III, for $\Delta\alpha = 0.07$, $\Delta\tau_1 = 6$ and $\Delta\alpha = -0.05$, $\Delta\tau_1 = 13$, there are stable periodic solutions with the almost frequency of ω_2 in Eq. (8.15) (see Fig. 8.8b, c). In region IV, for $\Delta\alpha = -0.26365$, $\Delta\tau_1 = 55$, there appear two stable periodic motions in (8.15) (see Fig. 8.8d). According to our result, it should be noted that for $\Delta\alpha = 0.42$, $\Delta\tau_1 = 280$ in region IV, there also exist the above mentioned co-existing periodic solutions (cf. Fig. 8.10 in [12]). Similarly, in regions V and VI, for $\Delta\alpha = -0.26365$, $\Delta\tau_1 = 16$ and $\Delta\alpha = -0.26365$, $\Delta\tau_1 = 6$, there are stable periodic solutions with frequency ω_1 in Eq. (8.15) (see Fig. 8.8e, f). The above results of the numerical simulation qualitatively illustrate the correctness of

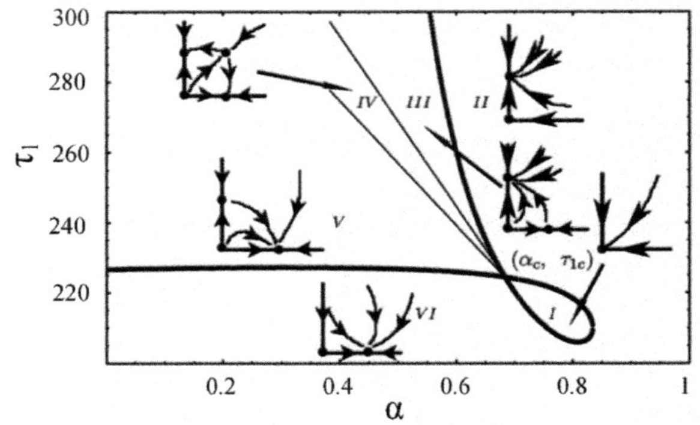

Fig. 8.7 The bifurcation sets of Eq. (8.18) near the $\sqrt{2} : \sqrt{3}$ double Hopf bifurcation point (α_c, τ_{1c})

the classification results, which reveal the complexity and richness of the dynamics of this physiological Eq. (8.15).

The classification in Fig. 8.7 contains profound physiological significance. For $(\alpha, \tau_1) \in$ I, it follows from the analytical prediction that the BG concentration of the diabetic patient will be stable, i.e., the patient will recover, which is the desired result. When $(\alpha, \tau_1) \in$ II, $(\alpha, \tau_1) \in$ III, $(\alpha, \tau_1) \in$ V, $(\alpha, \tau_1) \in$ VI, there is the BG fluctuation in the patient's body, i.e., the patient will not recover. In the above cases, the patient should be vigilant and monitored during the treatment. For $(\alpha, \tau_1) \in$ IV, there will be a complex BG fluctuation in the patient, which implies that the patient will take a turn for the worse, and we must avoid this situation.

In this section, the physiological Eq. (8.15) of using external auxiliary equipment to treat diabetes is studied, in which the technological delay τ_1 of external auxiliary and the physiological delay τ_2 of liver are considered. It is found that there are resonant and non-resonant double Hopf bifurcations in Eq. (8.15) due to the technological delay τ_1. The dynamic behaviours of the resulting non-resonant bifurcation are classified by applying nonlinear dynamics theory. The results show that with the change of technological delay τ_1 and the infection degree α of diabetic patients, different medical outcomes can be predicted by using this system, such as BG stability, simple and complex BG fluctuations. The research results have potential application value in analyzing, predicting, and optimizing the medical results of a diabetes treatment scheme and evaluating the medical risk and feasibility of the scheme. The significance of the results of this section lies in the fact that the technological delay of the auxiliary equipment can be qualitatively adjusted according to the different degrees of diabetes to achieve a better therapeutic effect.

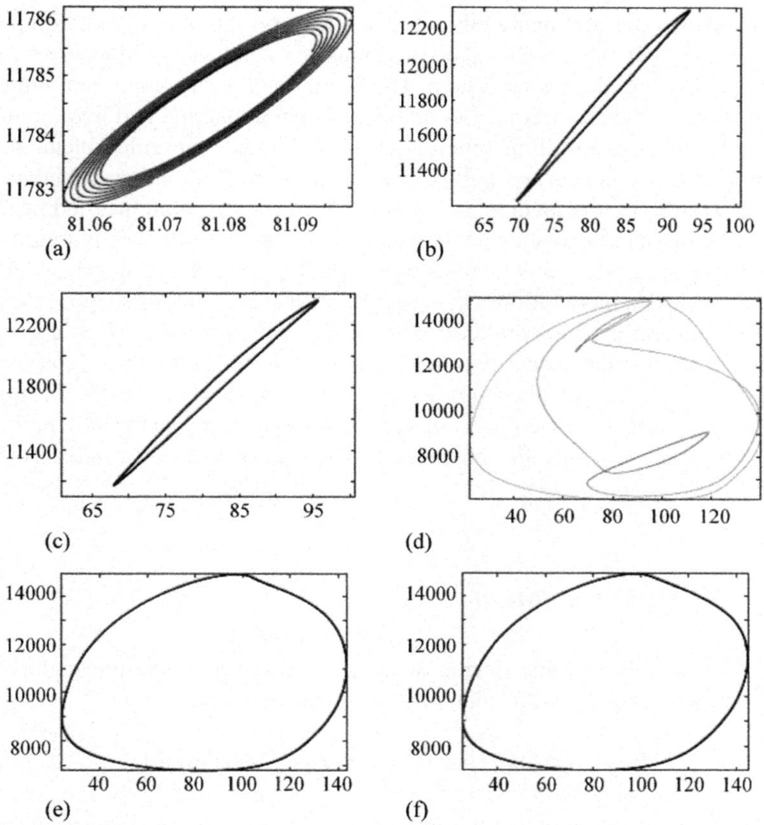

Fig. 8.8 The numerical simulations in plane (I, G) of Eq. (8.15) for bifurcations sets shown in Fig. 8.7, from Region I to VI, respectively. **a** $\Delta\alpha = 0.06$, $\Delta\tau_1 = -4$; **b** $\Delta\alpha = 0.07$, $\Delta\tau_1 = 6$; **c** $\Delta\alpha = -0.05$, $\Delta\tau_1 = 13$; **d** $\Delta\alpha = -0.26365$, $\Delta\tau_1 = 55$; **e** $\Delta\alpha = -0.26365$, $\Delta\tau_1 = 16$; **f** $\Delta\alpha = -0.26365$, $\Delta\tau_1 = 6$

8.3 Effects of Time Delay and Noise on Asymptotic Stability in Human Quiet Standing Model

In the previous research, we paid attention to the influence of time delay on the dynamic behaviour, especially the different dynamic behaviours corresponding to different values of parameters. In fact, biological systems in nature are more or less affected by external environmental noise and internal uncertainties [23–28]. In this part, we study a biomechanical model for the human in quiet standing. Considering the joint influence of time delay and random, the asymptotic stability of the system is studied by the Lyapunov exponent [29].

In fact, the human body maintains a standing posture under the influence of physiological, psychological and other internal factors and environmental noise (including light intensity, odor and many other external factors), through the coordination of various aspects of the vestibular organs, muscles, tendons, proprioceptors in the joints, and the central nervous system. This involves two very important influencing factors, namely, the random factors of environmental external and human internal, and the feedback controlling time-delay of the human musculoskeletal nervous system. For convenience, we only consider the role of the ankle in human quiet standing to simplify the modeling as a single degree of freedom inverted pendulum model. The random factors inside the human body and outside the environment are abstracted as multiplicative white noise perturbations, which are introduced into the inverted pendulum model for the human in quiet standing. In this section, the center manifold theorem is applied to reduce an infinite-dimensional stochastic system of the human standing model to a two-dimensional ordinary differential stochastic equation. Then, the stochastic averaging method is employed to obtain the Itô equation. The necessary and sufficient conditions of the asymptotically stable with probability one are obtained to study the influence of time delay and noise intensity on the stability of the system.

8.3.1 Model Formulation

In this sub section, a single degree of freedom inverted pendulum model for the human in quiet standing with stochastic perturbation is studied

$$I\ddot{\theta} + \gamma\dot{\theta} - mgL\sin\theta = \tilde{f}(\theta(t-\tau)) + \tilde{c}\eta(t)\theta(t), \tag{8.19}$$

where I represents the moment of inertia of human body around the ankle; θ represents the inclination angle of the human body relative to the vertical position; g indicates the gravity acceleration; m represents the body mass; L represents the distance from the ankle joint to the center of mass of the human body; γ represents the damping coefficient at the joint; $\eta(t)$ represents for Gauss white noise with zero mean value; $\tilde{f}(\theta(t-\tau))$ is a feedback function describing the control of the nervous system in the human body, where τ represents the time delay.

Let $q = L\sin\theta$, where q is the transverse displacement of the gravity center, then Eq. (8.19) can be rewritten as follows

$$\ddot{q} = \frac{mgL}{I}q - \frac{r}{I}\dot{q} + f + \tilde{\tilde{b}}\eta(t)q(t). \tag{8.20}$$

Rewrite Eq. (8.20) as follows

$$\dot{x}(t) = ax(t) + bf(x(t-\tau)) + c\eta(t)x(t), \tag{8.21}$$

where

$$x(t) = q + d\dot{q},\ a = \frac{mgL}{I}\left(\frac{r}{I} + \sqrt{\frac{r^2}{I^2} + 4\frac{mgL}{I}}\right) > 0,\ d = \frac{r}{I} + \sqrt{\frac{r^2}{I^2} + 4\frac{mgL}{I}},$$

and $b < 0$ represents the feedback coefficients. Here, we take the feedback function as $f(x(t-\tau)) = \tanh(x(t-\tau))$.

8.3.2 Asymptotically Stable Analysis

In this section, we analyze the stability of the periodic solution of Eq. (8.21) near $\tau = \tau_0$ through the center manifold theorem and stochastic averaging method. The ordinary differential equation on the central manifold can be obtained as follows

$$\dot{y}(t) = By(t) + \varepsilon\overline{\psi}(0)g(\mu, y(t)), \tag{8.22}$$

namely

$$\begin{cases} \dot{y}_1(t) = -wy_2 - \varepsilon p_1(a\mu y_1(t) + b\mu(y_1(t)\cos w\theta + y_2(t)\sin w\theta) \\ \quad -\frac{1}{3}b\tau(y_1(t)\cos w\theta - y_2(t)\sin w\theta)^3 + \varepsilon^{-1/2}\tau c\eta(t)y_1(t)\big), \\ \dot{y}_2(t) = wy_1 - \varepsilon p_2(a\mu y_1 + b\mu(y_1(t)\cos w\theta + y_2(t)\sin w\theta) \\ \quad -\frac{1}{3}b\tau(y_1(t)\cos w\theta + y_2(t)\sin w\theta)^3 + \varepsilon^{-1/2}\tau c\eta(t)y_1(t)\big), \end{cases} \tag{8.23}$$

where

$$p_1 = \frac{-\psi_{22}}{\psi_{11}\psi_{22} - \psi_{12}^2},\ p_2 = \frac{\psi_{12}}{\psi_{11}\psi_{22} - \psi_{12}^2}.$$

Carrying out a change of variables from (y_1, y_2) to (z_1, z_2) in Eq. (8.23), namely

$$z_1 = y_1 - \frac{p_1}{p_2}y_2,\ z_2 = y_2 - \frac{p_1}{p_2}wy_1. \tag{8.24}$$

Write Eq. (8.24) as a functional expression about (y_1, y_2) as follows

$$\begin{cases} y_1 = \frac{p_2^2}{p_2^2 + p_1^2w^2}\left(z_1 + \frac{p_1}{p_2}wz_2\right), \\ y_2 = \frac{p_2^2}{p_2^2 + p_1^2w^2}\left(z_2 - \frac{p_1}{p_2}wz_1\right). \end{cases} \tag{8.25}$$

Then the stochastic ordinary differential equations in Eq. (8.23) on the central manifold become

$$
\begin{aligned}
\dot{z}_1(t) &= -wz_2, \\
\dot{z}_2(t) &= wz_1 - \varepsilon\frac{p_2^2 + p_1^2 w}{p_2}\Big(-\frac{1}{3}b\tau\,(p_{11}z_1(t) + p_{12}z_2(t))^3 \\
&\quad +\mu(q_{11}z_1(t) + q_{12}z_2(t)) + \varepsilon^{-1/2}\tau c\eta(t)(g_{11}z_1(t) + g_{12}z_2(t))\Big),
\end{aligned} \tag{8.26}
$$

where

$$
\begin{aligned}
p_{11} &= \frac{p_2^2}{p_2^2 + p_1^2 w^2}\left(\cos w - \frac{p_1}{p_2}w\sin w\right),\ p_{12} = \frac{p_2^2}{p_2^2 + p_1^2 w^2}\left(\sin w + \frac{p_1}{p_2}w\cos w\right), \\
q_{11} &= \frac{p_2^2}{p_2^2 + p_1^2 w^2}\left(a + b\cos w - \frac{bp_1}{p_2}w\sin w\right),\ g_{11} = \frac{p_2^2}{p_2^2 + p_1^2 w^2}, \\
q_{12} &= \frac{p_2^2}{p_2^2 + p_1^2 w^2}\left(\frac{p_1}{p_2}(a + bw\cos w) + \sin w\right),\ g_{12} = \frac{p_2 p_1 w}{p_2^2 + p_1^2 w^2}.
\end{aligned}
$$

Then using the relations $z_1 = \beta(t)\sin\Theta$ and $\Theta = wt + \varphi(t)$, Eq. (8.26) is transformed into the following system with polar coordinates

$$
\begin{aligned}
\dot{\beta}(t) &= \varepsilon\frac{p_2^2 + p_1^2 w}{p_2^2}\beta\Big(-\frac{1}{3}b\beta^2\tau\,(p_{11}\sin\Theta - p_{12}\cos\Theta)^3 \\
&\quad +\mu(q_{11}\sin\Theta - q_{12}\cos\Theta) + \varepsilon^{-1/2}\tau c\eta(t)(g_{11}\sin\Theta - g_{12}\cos\Theta)\Big)\cos\Theta, \\
\dot{\varphi}(t) &= -\varepsilon\frac{p_2^2 + p_1^2 w}{p_2^2}\Big(-\frac{1}{3}b\tau\beta^2\,(p_{11}\sin\Theta - p_{12}\cos\Theta)^3 \\
&\quad +\mu(q_{11}\sin\Theta - q_{12}\cos\Theta) + \varepsilon^{-1/2}\tau c\eta(t)(g_{11}\sin\Theta - g_{12}\cos\Theta)\Big)\sin\Theta.
\end{aligned} \tag{8.27}
$$

According to the Khasminskii limit theorem and the stochastic averaging method, we get the averaging Itô stochastic equations of Eq. (8.27) as follows

$$
\begin{aligned}
\mathrm{d}\beta(t) &= \frac{p_2^2 + p_1^2 w}{p_2^2}\varepsilon\beta\Big(\frac{1}{8}b\tau(p_{11}^2 p_{12} + p_{12}^3)\beta^2 - \frac{1}{2}q_{12}\mu + \frac{p_2^2 + p_1^2 w}{4p_2^2}\tau^2 \\
&\quad (g_{11}^2 + g_{12}^2)c^2\Big)\mathrm{d}t \ + \varepsilon^{1/2}\frac{\sqrt{2(g_{11}^2 + 3g_{12}^2)(p_2^2 + p_1^2 w)}}{4p_2^2}\tau\beta c\mathrm{d}w_1,
\end{aligned} \tag{8.28a}
$$

$$
\begin{aligned}
\mathrm{d}\varphi(t) &= \varepsilon\frac{p_2^2 + p_1^2 w}{p_2^2}\left(\frac{1}{8}b\tau(p_{11}^3 + p_{11}p_{12}^2)\beta^2 - \frac{1}{2}q_{11}\mu\right)\mathrm{d}t \\
&\quad + \varepsilon^{1/2}\frac{\sqrt{2(3g_{11}^2 + g_{12}^2)(p_2^2 + p_1^2 w)}}{4p_2^2}\tau c\mathrm{d}w_2,
\end{aligned} \tag{8.28b}
$$

where w_1 and w_2 are independent standard Wiener processes. The averaged amplitude Eq. (8.28a) and phase Eq. (8.28b) are decoupled, so we study stability using the averaged amplitude Eq. (8.28a). Let $\beta(t) = \beta_0(t) + r(t)$ and $\rho(t) = \ln r(t)$, where $r(t)$ represents a small disturbance around $\beta_0(t)$. Using the stochastic differential rule to Itô, we obtain the linearized Itô equation with regard to $r(t)$ and $\rho(t)$, namely,

$$\mathrm{d}r(t) = \frac{p_2^2 + p_1^2 w}{p_2^2}\varepsilon r\left(\frac{3}{8}b\tau(p_{11}^2 p_{12} + p_{12}^3)\beta_0^2 - \frac{1}{2}q_{12}\mu + \frac{p_2^2 + p_1^2 w}{4p_2^2}\tau^2\right.$$
$$\left.(g_{11}^2 + g_{12}^2)c^2\right)\mathrm{d}t + \varepsilon^{1/2}\frac{\sqrt{2(g_{11}^2 + 3g_{12}^2)(p_2^2 + p_1^2 w)}}{4p_2^2}\tau r c \mathrm{d}w_1, \tag{8.29}$$

$$\mathrm{d}\rho(t) = \frac{p_2^2 + p_1^2 w}{p_2^2}\varepsilon\left(\frac{3}{8}b\tau(p_{11}^2 p_{12} + p_{12}^3)\beta_0^2 - \frac{1}{2}q_{12}\mu + \frac{p_2^2 + p_1^2 w}{4p_2^2}\tau^2\right.$$
$$\left.(g_{11}^2 + g_{12}^2)c^2\right)\mathrm{d}t + \varepsilon^{1/2}\frac{\sqrt{2(g_{11}^2 + 3g_{12}^2)(p_2^2 + p_1^2 w)}}{4p_2^2}\tau c \mathrm{d}w_1. \tag{8.30}$$

The sample stability of the stochastic dynamical system is determined by the qualitative evaluation of the Lyapunov exponents. According to the multiplicative ergodic theorem [30], the max Lyapunov exponent λ of the amplitude process is obtained as

$$\lambda = \frac{p_2^2 + p_1^2 w}{p_2^2}\varepsilon\left[\frac{3}{8}b\tau(p_{11}^2 p_{12} + p_{12}^3)E[\beta_0^2] - \frac{1}{2}q_{12}\mu\right.$$
$$\left. + \frac{p_2^2 + p_1^2 w}{4p_2^2}\tau^2(g_{11}^2 + g_{12}^2)c^2\right], \tag{8.31}$$

where $E[\beta_0^2]$ is the expected value of β_0^2. For the trivial solution $\beta_0^2 = 0$, Eq. (8.23) gives

$$\lambda = -\frac{1}{2}q_{12}\mu + \frac{p_2^2 + p_1^2 w}{4p_2^2}\tau^2(g_{11}^2 + g_{12}^2)c^2 = 0.$$

So, if $\lambda < 0$, namely

$$\mu < (g_{11}^2 + g_{12}^2)\tau^2 c^2 \frac{p_2^2 + p_1^2 w}{2p_2^2 q_{12}},$$

and $q_{12} > 0$, the trivial solution is asymptotically stable with probability one (w.p.1), otherwise, the trivial solution is unstable when $q_{12} < 0$.

From Fig. 8.9, we can observe that the max Lyapunov exponent λ increases as the excitation intensity c or time delay τ increases. From the judgment conditions of asymptotically stable, we know that the original Eq. (8.19) is stable if $\lambda < 0$

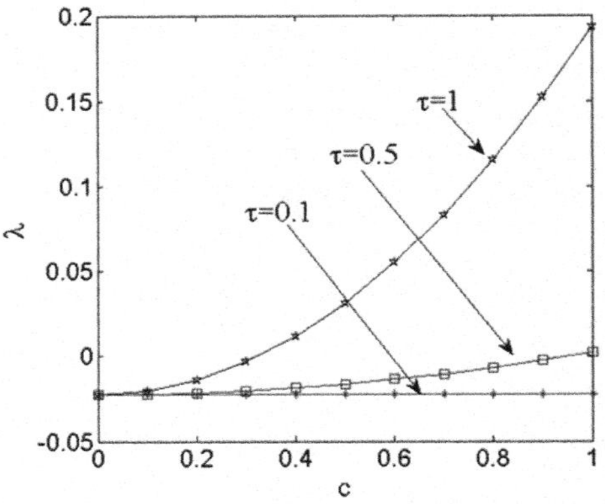

Fig. 8.9 The max Lyapunov exponent λ of Eq. (8.19) with the change of excitation intensity c for different values of time delay τ, where the other parameters are $a = 1, b = -2, \varepsilon = 0.1$, $\mu = 0.1$

and unstable if $\lambda > 0$. It follows that when $\tau = 1.0$, Eq. (8.19) becomes stable at $c = 0.447$ and unstable at $c = 0.8$. When $\tau = 0.10$, the noise intensity has little effect on the balance of the human in quiet standing. However, when τ becomes larger, the noise intensity has a greater impact on the system. Compared with small noise intensity, the influence of time delay on system stability is more sensitive at large noise intensity. Furthermore, a small change in the strength of noise may destabilize the quiet standing for a large delay.

8.4 Pattern Dynamics of Population Reaction–Diffusion Models with Spatiotemporal Delay

In research involving biological sciences, scientists always focus on two aspects: one is to investigate the microscopic behaviour of organisms, which includes the nervous system, insulin-blood system, and human standing system shown in the above chapters; on the other hand, the organism is regarded as the macroscopic research subject, and the competition, predation, and relationship with the surrounding environment are emphatically studied, which is population dynamics in ecology. Up until now, the development of population dynamics has been extremely rapid. The dynamical models and methods not only directly promote the development of ecology, but also have a significant impact on the related fields of biomathematics and dynamics [31].

Among the population dynamics researches in the ecosystem, the life processes of animals such as the gestation period [32–34], maturity period [35–37], and food transformation period [32] all go through a period of time; that is, there is a time delay effect. The existence of time delay may lead to more complex dynamic properties of the studied system. Under certain conditions, the time delay will destabilize the population coexistence equilibrium, causing a periodic fluctuation in population density and even population extinction [38, 39]. The mathematical model of the time delay differential equation plays a very important role in describing biodynamic

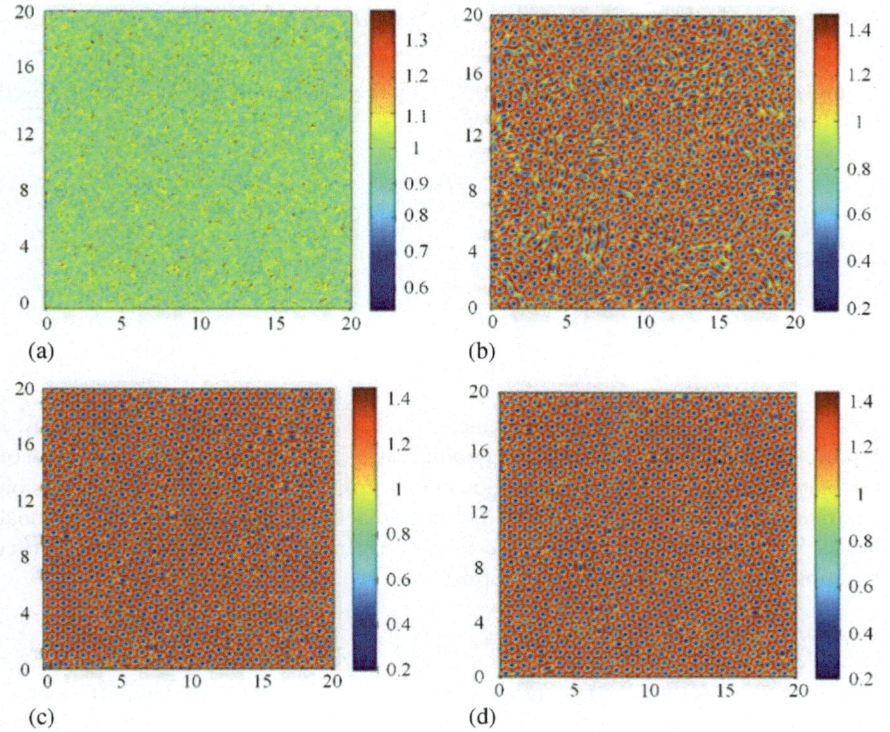

Fig. 8.10 Snapshots of contour pictures of the time evolution of u at the different times with the parameter values $\beta = 1.0$ and $\tau = 0.05$. **a** $t = 0$, **b** $t = 10$, **c** $t = 1000$, **d** $t = 10{,}000$

behaviour. It explains the dynamic behaviour between many populations or between populations and environment from a mathematical point of view, which is helpful for people to understand population dynamics so as to control the interaction between population and environment. Furthermore, the establishment of an appropriate time delay system has an important mathematical guiding role in maintaining the function of the ecosystem, guiding agricultural production practices, adjusting the direction of energy flow, and promoting material recycling.

In this section, we focus on the spatiotemporal dynamics under the effect of time delays, aiming at single population system and multi-population reaction–diffusion dynamic system. The main research contents include: the dynamical behaviour of a single species reaction–diffusion system model with spatiotemporal delay is studied. Time delay can induce the spatiotemporal distributions of the density including spots pattern, stripes pattern, and fluctuation. Further, the conditions of different distributions of predator and prey populations under the spatiotemporal delay are extended. It is found that the time delay has an important influence on the evolution of the distribution pattern of predator–prey populations, which enriches the pattern formation of predator–prey models.

8.4.1 Single Species Reaction–Diffusion Model

A single species reaction–diffusion model with the spatiotemporal delay is expressed as follows

$$u_t = d\Delta u + ru\big(1 + \alpha u - \beta u^2 - (1 + \alpha - \beta)(f * *u)\big), \tag{8.32}$$

where

$$(f * *u)(x, t) = \int_{-\infty}^{t} \int_{\mathrm{R}} f(x - y, t - s)u(y, s)dyds,$$

for $(x, t) \in \mathbb{R} \times (0, +\infty)$. The parameters d, r, α, and β are positive constants. r represents the intrinsic growth rate of population, α indicates the aggregation effect of population, β represents the competition rate of population, d represents the diffusion coefficient of population, $\Delta = \partial^2/\partial x^2$ is Laplacian operator in one dimensional space. And we further assume $1 + \alpha - \beta > 0$. In the above equation, the kernel function $f(x, t)$ has the following forms

$$f(x,t) = \frac{1}{\sqrt{4\pi t}}e^{-\frac{|x|^2}{4t}}\frac{1}{\tau}e^{-\frac{t}{\tau}},\ f(x,t) = \frac{1}{\sqrt{4\pi t}}e^{-\frac{|x|^2}{4t}}\frac{t}{\tau^2}e^{-\frac{t}{\tau}},$$
$$f(x,t) = ae^{-a|x|}be^{-bt},\ f(x,t) = \frac{1}{\sqrt{4\pi t}}e^{-\frac{|x|^2}{4t}}\delta(t-\tau),\ f(x,t) = \delta(x)\delta(t-\tau),$$
$$f(x,t) = \delta(x)be^{-bt},\ f(x,t) = ae^{-a|x|}\delta(t),\ f(x,t) = \delta(x)\delta(t),$$

and so on. The terms in Eq. (8.32) have the following biological interpretations: αu is an index to measure the aggregation of population; $-\beta u^2$ indicates the competition of population for space; the integral term $-(1 + \alpha - \beta)(f * *u)$ indicates the competition of food resources within the population. Moreover, the kernel function $f(x, t)$ satisfies the following properties:

(H1) $f \in L^1(\mathbb{R} \times (0, \infty))$ and $tf \in L^1(\mathbb{R} \times (0, \infty))$, where $f \in L^1(\mathbb{R} \times (0, \infty))$ indicates that convolution $f * *u$ is an integrable function with respect to time and space.

(H2) f satisfies equation $f * *1 = 1$, that is,

$$\int_{R}\int_{0}^{\infty} f(x, t)dtdx = 1.$$

(H3) Since f is a weight function, the kernel function satisfies $f(x, t) \geq 0$.

(H4) The kernel function f depicts the non-local effect of $u(y, s)(s \leq t)$ from x to y.

In this subsection, we choose the kernel function as

$$f(x,t) = \frac{1}{\sqrt{4\pi t}} e^{-\frac{|x|^2}{4t}} \frac{1}{\tau} e^{-\frac{t}{\tau}}.$$

Given

$$\begin{aligned} v(x,t) &= f ** u \\ &= \int_{-\infty}^{t} \int_{R} f(x-y, t-s) u(y,s) dy ds \\ &= \int_{-\infty}^{t} \int_{-\infty}^{+\infty} \frac{1}{\sqrt{4\pi t}} e^{-\frac{(x-y)^2}{4(t-s)}} \frac{1}{\tau} e^{-\frac{t-s}{\tau}} u(y,s) dy ds. \end{aligned}$$

Let $z = x - y, \theta = t - s$, we get

$$v(x,t) = \int_{0}^{+\infty} \int_{-\infty}^{+\infty} \frac{1}{\sqrt{4\pi\theta}} e^{-\frac{z^2}{4\theta}} \frac{1}{\tau} e^{-\frac{\theta}{\tau}} u(x-z, t-\theta) dz d\theta.$$

Thus, Eq. (8.32) is transformed into the following equations

$$\begin{cases} u_t = d\Delta u + f_1(u,v), \\ v_t = \Delta v + f_2(u,v), \end{cases} \tag{8.33}$$

where

$$\begin{aligned} f_1(u,v) &= ru\big(1 + \alpha u - \beta u^2 - (1+\alpha-\beta)v\big), \\ f_2(u,v) &= \tfrac{1}{\tau}(u - v). \end{aligned}$$

Equation (8.33) has three equilibria $(0,0)$, $(-1/\beta, -1/\beta)$, and $(1,1)$. From the biological point of view, we are mainly focusing on the third equilibrium $(1,1)$ since this equilibrium corresponds to the equilibrium state of the carrying capacity of the population living environment. Linearize Eq. (8.33) at the equilibrium $(1,1)$ to obtain the following linear system

$$\begin{cases} u_t = d\Delta u + a_{11}u + a_{12}v, \\ v_t = \Delta v + a_{21}u + a_{22}v, \end{cases} \tag{8.34}$$

where

$$a_{11} = r(\alpha - 2\beta), a_{12} = -r(1+\alpha-\beta), a_{21} = \frac{1}{\tau}, a_{22} = -\frac{1}{\tau}.$$

For research convenience, we assume that Eq. (8.33) satisfies the following initial conditions

$$u(0, x) = \phi_1(t, x) \geq 0, v(0, x) = \phi_2(t, x) \geq 0, x \in \mathbb{R},$$

and with smooth boundary conditions. It is further assumed that the boundary condition of Eq. (8.33) is the Newman boundary condition, i.e.,

$$\frac{\partial u(t, x)}{\partial n} = \frac{\partial v(t, x)}{\partial n} == 0, t \geq 0, x \in \partial\mathbb{R},$$

where n represents the outer normal vector of the boundary $\partial\Omega$.

According to the bifurcation theory of reaction diffusion systems, the conditions of Turing bifurcation refer to a uniform steady state of the reaction–diffusion system that is stable for the corresponding ordinary differential equations, but it is unstable in the partial differential equations with diffusion terms. To achieve this end, we need to obtain the characteristic equation determined by Eq. (8.34). So, we conducted stability analysis on the state variables u and v in the temporal direction, as well as Fourier expansion in spatial direction

$$\begin{pmatrix} u \\ v \end{pmatrix} = \begin{pmatrix} \alpha_1 \\ \alpha_2 \end{pmatrix} \cos(k_1 X) \cos(k_2 Y) e^{\lambda t}, \tag{8.35}$$

where λ represents the growth rate in the time direction, k_1 and k_2 represent wave vector components in spatial directions X and Y, respectively. Substituting Eq. (8.35) into Eq. (8.34), one gets

$$\lambda \begin{pmatrix} \alpha_1 \\ \alpha_2 \end{pmatrix} = \begin{pmatrix} a_{11} - dk^2 & a_{12} \\ a_{21} & a_{22} - k^2 \end{pmatrix} \begin{pmatrix} \alpha_1 \\ \alpha_2 \end{pmatrix}.$$

We can obtain the characteristic equation for the growth rate λ as the determinant $\det A = 0$, where

$$\det A = \begin{vmatrix} a_{11} - dk^2 - \lambda & a_{12} \\ a_{21} & a_{22} - k^2 - \lambda \end{vmatrix}, k^2 = k_1^2 + k_2^2.$$

That is equivalent to the following equation

$$\lambda^2 - Tr_k \lambda + \Delta_k = 0, \tag{8.36}$$

where

$$\begin{aligned} Tr_k &= a_{11} + a_{22} - (d+1)k^2 = Tr_0 - (d+1)k^2, \\ \Delta_k &= a_{11}a_{22} - a_{12}a_{21} - (a_{11} + a_{22}d)k^2 + dk^4. \end{aligned}$$

Further, we get roots of Eq. (8.36) are

$$\lambda_{k1} = \frac{Tr_k + \sqrt{Tr_k^2 - 4\Delta_k}}{2}, \lambda_{k2} == \frac{Tr_k - \sqrt{Tr_k^2 - 4\Delta_k}}{2}. \tag{8.37}$$

If $\tau = \tau_H = 1/(r\alpha - 2\beta\alpha)$ and $l = 2dr(\alpha - 2\beta) + 4rd(1 + \beta)$, then Eq. (8.33) will undergo the spatially uniform Hopf bifurcation. Moreover, when

$$\tau = \tau_T = \sqrt{l^2 - 4r^2d^2(\alpha - 2\beta)^2}/2r^2(\alpha - 2\beta)^2,$$

the Turing bifurcation will occur in Eq. (8.33) at the critical wave number

$$k_T = \sqrt{\frac{r\tau_T(\alpha - 2\beta) - d}{2d\tau_T}}.$$

By the method of multiple scales, the amplitude equation of the pattern induced by the system is described as

$$\begin{aligned}
s_0\frac{\partial A_1}{\partial t} &= \mu A_1 + h\overline{A}_2\overline{A}_3 - \left[g_1|A_1|^2 + g_2\left(|A_2|^2 + |A_3|^2\right)\right]A_1,\\
s_0\frac{\partial A_2}{\partial t} &= \mu A_2 + h\overline{A}_1\overline{A}_3 - \left[g_1|A_2|^2 + g_2\left(|A_1|^2 + |A_3|^2\right)\right]A_2,\\
s_0\frac{\partial A_3}{\partial t} &= \mu A_3 + h\overline{A}_2\overline{A}_1 - \left[g_1|A_3|^2 + g_2\left(|A_2|^2 + |A_1|^2\right)\right]A_3,
\end{aligned}$$

where

$$\begin{aligned}
s_0 &= \frac{(1-d)l}{\tau_T[lb_{11} + b_{12} - dl(lb_{21} + b_{22})]},\\
\mu &= \frac{\tau_T - \tau}{\tau_T}, h = \frac{2l^2(\alpha - 3\beta)r + 2lr(\beta - \alpha - 1)}{\tau_T[lb_{11} + b_{12} - dl(lb_{21} + b_{22})]},\\
g_1 &= \frac{G_1}{\tau_T[lb_{11} + b_{12} - dl(lb_{21} + b_{22})]},\\
g_2 &= \frac{G_2}{\tau_T[lb_{11} + b_{12} - dl(lb_{21} + b_{22})]}.
\end{aligned}$$

In this way, we get the amplitude equation that determines different Turing patterns and also gives the coefficient expression of the amplitude equation. The classical finite difference method is used to calculate Eq. (8.33) to get the pattern of the system in two-dimensional space. All our numerical simulations use the Neumann boundary conditions with the size of 20×20. The space step is $\Delta X = \Delta Y = 0.1$, and the time step is $\Delta t = 0.002$. We fix $d = 0.02$, $r = 5.0$, $\alpha = 3.0$. If we choose $\beta = 1.0$ and $\tau = 0.05$, we can get the spatial distribution of the state variable u at time $t = 0$,

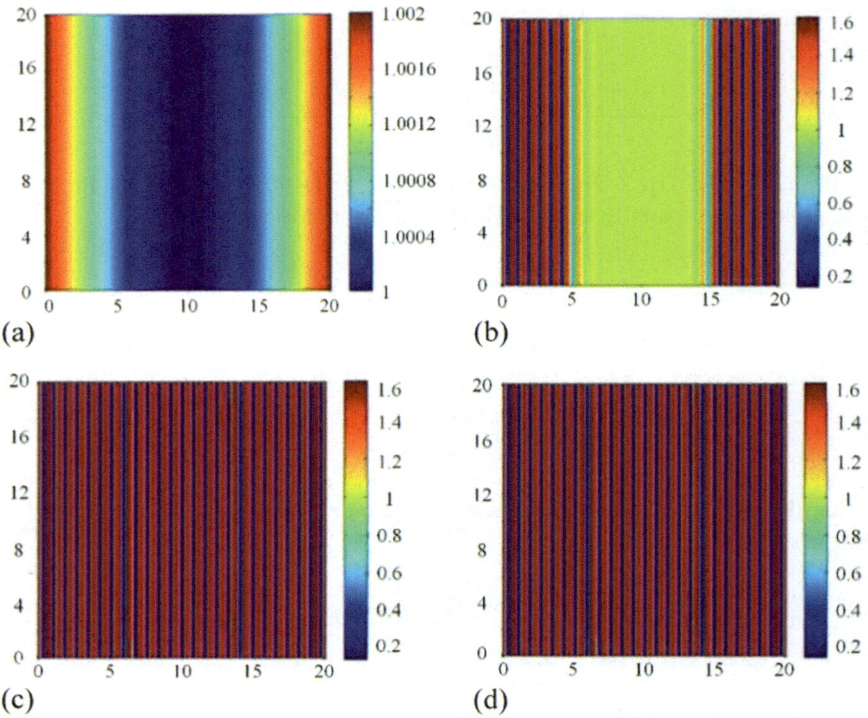

Fig. 8.11 Snapshots of contour pictures of the time evolution of u at the different times with the parameter values $\beta = 1.0$ and $\tau = 0.1$. **a** $t = 0$, **b** $t = 10$, **c** $t = 1000$, **d** $t = 10000$

$t = 10$, $t = 1000$ and $t = 10{,}000$, as shown in Fig. 8.10. Similarly, if we choose $\beta = 1.0$ and $\tau = 0.1$, we can get the spatial distribution of the state variable u at time $t = 0$, $t = 10$, $t = 1000$ and $t = 10{,}000$, as shown in Fig. 8.11. From the above numerical simulation results, we obtain the spots pattern and the stripes pattern in this single species reaction–diffusion model with spatiotemporal delay. It is difficult to find these similar patterns in our realistic world.

8.4.2 Predator–Prey Reaction–Diffusion Model

In order to explore the spatial distribution pattern of predator and prey population density, we consider the following predator–prey system with spatiotemporal effect

$$\begin{cases} u_t = ru\big(1 + \alpha u - \beta u^2 - (1 + \alpha - \beta) f{**}u\big) - uv + d\Delta u, \\ v_t = av(u - b) + \Delta v, \end{cases} \tag{8.38}$$

where

$$(f{**}u)(x,t) = \int_{-\infty}^{t}\int_{R^2} f(x-y,t-s)u(y,s)dyds,$$

and

$$f(x,t) = \frac{1}{4\pi t}e^{-\frac{|x|^2}{4t}}\frac{1}{\tau}e^{-\frac{t}{\tau}}$$

for $(x,t) \in \mathbb{R}^2 \times (0,+\infty)$. Here, $u(x,t)$ is the prey population density, $v(x,t)$ is the predator population density. Parameters a and b represent predatory effect and mortality rate of the predator, respectively. The parameters r, α, β, d, τ are positive constants, and their biological meanings are consistent with subsection 8.4.1. $\Delta = \partial^2/\left(\partial X^2 + \partial Y^2\right)$ is Laplacian operator in two-dimensional space, and we further assume $1+\alpha-\beta > 0$. The corresponding theoretical analysis can determine the conditions that the time delay parameters and predation rate should meet when the system produces a Turing bifurcation. Finally, the finite difference method is used to numerical calculations on the system, and the calculation results show that the spatial distribution pattern of Eq. (8.38) will obviously change in the presence of predators.

All our numerical simulations use the Neumann boundary conditions with the size of 20×20. The space step is $\Delta X = \Delta Y = 0.1$, and the time step is $\Delta t = 0.0025$. We fix $d = 0.001$, $r = 3.0$, $\alpha = 0.4$, $\beta = 0.1$, $b = 0.2$, $a = 1.0$, $\tau = 1.0$, and the initial value is a random disturbance in a spatially uniform state. Then we can get the spatial distribution of the state variable u at time $t = 0$, $t = 10$, $t = 1000$ and $t = 10000$, as shown in Fig. 8.12. Similarly, if we fix $d = 0.001$, $r = 3.0$, $\alpha = 0.4$, $\beta = 0.1$, $b = 0.2$, $a = 1.0$, $\tau = 1.0$ and the periodic perturbations

$$u_0 = 0.2 + 0.01\cos(4.58X)\cos(4.58Y)$$

and

$$v_0 = 2.44 + 0.01 \times \cos(4.58X)\cos(4.58Y)$$

of the spatially uniform state are taken as initial conditions, then we can get the spatial distribution of the state variable u at time $t = 0$, $t = 10$, $t = 1000$ and $t = 10000$, as shown in Fig. 8.13.

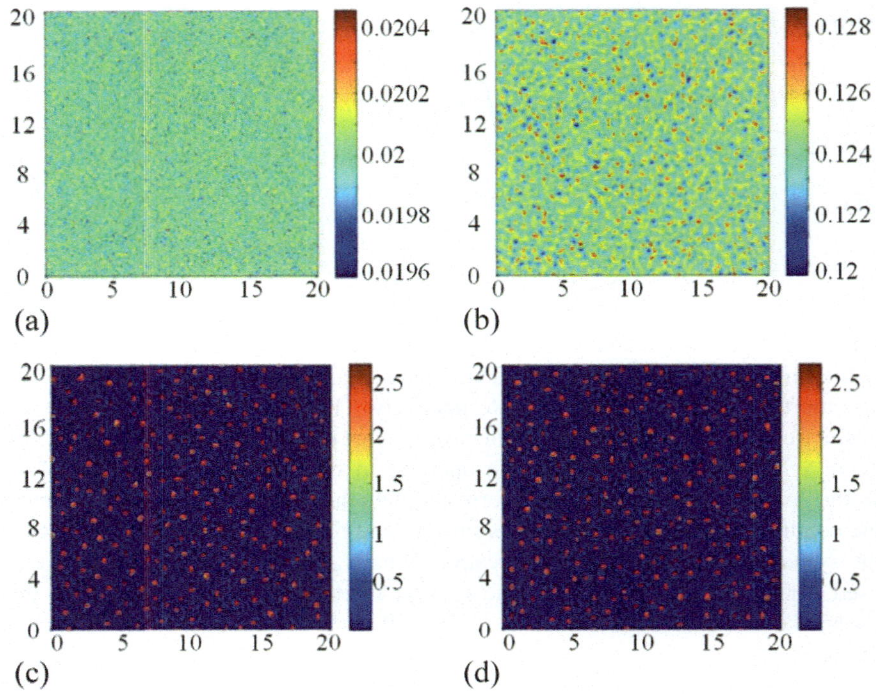

Fig. 8.12 Snapshots of contour pictures of the time evolution of u at the different times with the parameter values in the Turing space. **a** $t = 0$; **b** $t = 10$; **c** $t = 1000$; **d** $t = 10{,}000$

8.4.3 Three-Species Food Chain System and Chaotic Behaviour

Chaos refers to the random and irregular motion that occurs in a certain system. The concept of chaos was introduced into ecology to explain the widespread fluctuations of population. Tens of thousands of species are present in an ecosystem. The vast and complex systems are easily exhibited to chaos. Nonlinear dynamics is a necessary condition of chaos, which is also common in ecological time series, and many driving factors of population dynamics are also chaotic [40]. In 2022, Rogers et al. applied many detection methods of the chaos to the global database containing time series of 172 populations. They found evidence of the existence of chaos in more than 30% of the data. These results show that chaos in natural populations is much more common than previously thought, and chaos is not uncommon in natural populations [41]. In the last part of this chapter, we consider a three-species food chain system with time delay for food digestion and discuss the chaos of the population.

The three-species food chain system is as follows

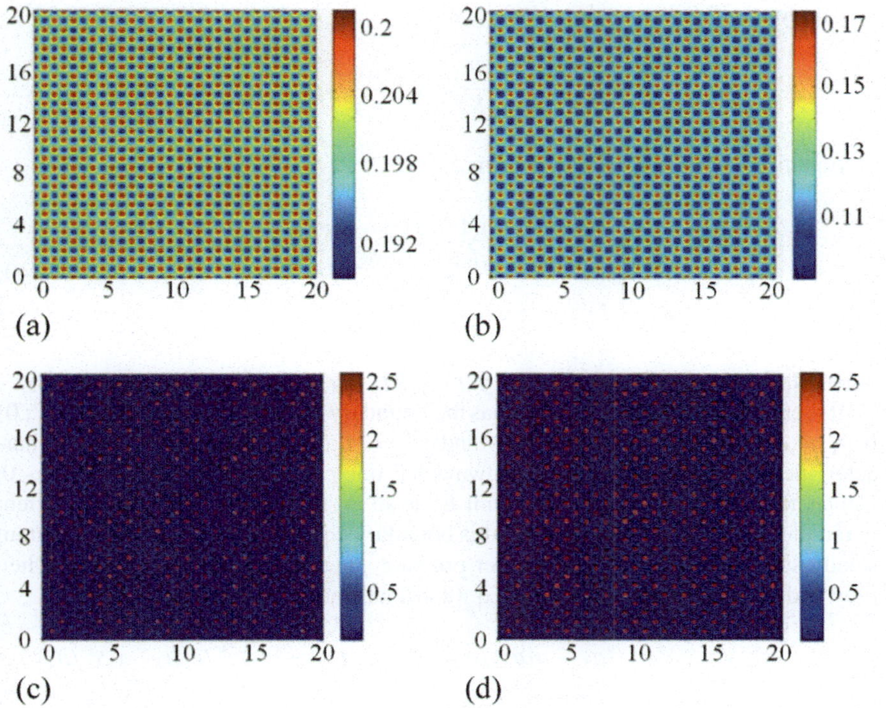

Fig. 8.13 Snapshots of contour pictures of the time evolution of u at the different times with the parameter values in the Turing space. **a** $t = 0$, **b** $t = 10$, **c** $t = 1000$, **d** $t = 10000$

$$\begin{cases} \frac{dX(T)}{dT} = r_1 X(T)\Big(1 - \frac{X(T)}{K}\Big) - \alpha_1 Y(T), \\ \frac{dY(T)}{dT} = r_2 Y(T)\Big(1 - \frac{Y(T)}{c_1 X(T-\theta_1)}\Big) - \alpha_2 Z(T), \\ \frac{dZ(T)}{dT} = r_3 Z(T)\Big(1 - \frac{Z(T)}{c_2 Y(T-\theta_2)}\Big). \end{cases} \tag{8.39}$$

The variables X, Y, and Z in Eq. (8.39) represent resource, consumer and super predator, respectively. All parameters in the system are positive. Here $r_i (i = 1, \ldots 3)$ represent the intrinsic growth rate of species X, Y, Z, respectively; K is the environment carrying capacity of X; c_1 and c_2 are the measures of the food quality that the corresponding prey provides for conversion into the births of consumer and super predator, respectively; α_1 and α_2 are the predation rates of Y and Z. Two time delays θ_1 and θ_2 are introduced to the functional response term involved with the growth equation of consumer and super predator population. It is a digestion period corresponding to the consumer-eat-resource and super predator-eat-consumer, which are called the resource digestion delay (RDD) and consumer digestion delay (CDD), respectively.

Employing the time scale transformation

$$x = \frac{X}{K}, y = \frac{\alpha_1 Y}{r_1 K}, z = \frac{\alpha_1 \alpha_2 Z}{r_1^2 K}, t = r_1 T,$$

Equation (8.39) thus takes the form

$$\begin{cases} \frac{dx(t)}{dt} = x(t)(1 - x(t)) - y(t), \\ \frac{dy(t)}{dt} = y(t)\left(a_1 - \frac{a_2 y(t)}{x(t-\tau_1)}\right) - z(t), \\ \frac{dz(t)}{dt} = z(t)\left(a_3 - \frac{a_4 z(t)}{y(t-\tau_2)}\right), \end{cases} \tag{8.40}$$

where $a_1 = r_2/r_1, a_2 = r_2/\alpha_1 c_1, a_3 = r_3/r_1, a_4 = r_3/\alpha_2 c_2, \tau_1 = r_1\theta_1, \tau_2 = r_1\theta_{2.}$.

It is easy to verify that Eq. (8.40) has the boundary equilibrium point $E^0(x^0, y^0, 0)$ if $a_2 > a_1$, where $x^0 = (a_2 - a_1)a_2$ and $y^0 = a_1 x^0/a_2$. The corresponding characteristic equation evaluated at E^0 always has one positive eigenvalue $\lambda = a_3 > 0$. Hence, the boundary equilibrium point E^0 is an unstable node whatever the system parameter values are. It implies that this boundary equilibrium point is a nonexistent steady state for Eq. (8.40). The super predator population cannot go extinct when prey and predator exist. Moreover, Eq. (8.40) has the interior equilibrium point

$$E^c : \left(x^c, y^c, z^c\right) = \left((a_3 - a_1 a_4)/a_2 a_4 + 1,\ (a_1 a_4 - a_3)x^c/a_2 a_4,\ a_3 y^c/a_4\right)$$

if and only if condition $a_3 < a_1 a_4 < a_3 + a_2 a_4$. In what follows, we investigate the coupling effects of multiple delays (RDD and CDD) on the dynamical properties of Eq. (8.40) by analyzing the local stability of the equilibrium point E^c. The stable species coexistence and chaos behaviour are illustrated in detail. Larger values of the RDD and CDD can make populations go extinct more easily.

To demonstrate chaotic behaviour in Eq. (8.40), we choose parameter set $a_1 = 0.5$, $a_2 = 0.68$, $a_3 = 0.2$, $a_4 = 0.56$, whereas multiple delays τ_1 and τ_2 are considered as the variable parameters. The initial condition is given as $(x_1(t), x_2(t), x_3(t)) = (0.8, 0.2, 0.06)$ for $-h < t \le 0$, where $h = \max(\tau_1, \tau_2)$. To describe the combined influences of multiple delays τ_1 and τ_2 on species population dynamics, bifurcation diagrams as the function of τ_2 are shown in Fig. 8.14 for the different fixed τ_1 by using the Poincaré section $\dot{y}(t - \tau_1) = 0$. With τ_2 varying, the different states of Eq. (8.40) (periodic, multi-periodic and chaotic behaviour) create a bifurcation diagram. In fact, the Poincaré map is the intersection of a periodic orbit with a certain lower-dimensional subspace in dynamical systems. It can be interpreted as a discrete dynamical system with a state space that is one dimension smaller than the original continuous system. If the steady state of the original system is periodic or multi-periodic motion, there are only one or multiple points in the Poincaré section. The number of points, however, becomes limitless in the case of chaotic behaviour. An irregular pattern in Poincaré section indicates the existence of a chaos behaviour.

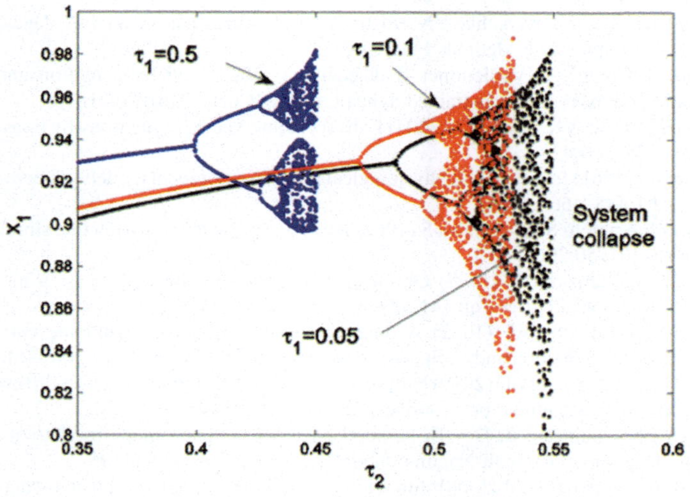

Fig. 8.14 Bifurcation diagram as a function of τ_2 varied in the range [0.35, 0.6]. Other parameters fixed as $a_1 = 0.5$, $a_2 = 0.68$, $a_3 = 0.2$, $a_4 = 0.56$ for the different τ_1 values

It follows from Fig. 8.14 that the food chain Eq. (8.40) exhibits period-1, period-2, period-4, and chaotic motion. It is a sequence of period-doubling bifurcations that eventually lead to chaos motion with τ_2 increasing. Further, the increase of τ_1 makes it easier for the system to emerge from period-doubling bifurcation and chaos motion. Moreover, delays τ_1 and τ_2 induce food chain systems to enter into system collapse due to species outbreaks. As a matter of fact, for the fixed digestion period of resource species τ_1, the large digestion period of predator species τ_2 implies the small needs of consumer species in the food chain system. The consumer species evolves into a population outbreak and enters a system collapse due to the predator's decreasing predation. On the other hand, when the predator digestion period is fixed, the large digestion period of consumer species induces small needs in resource species. The decline in consumer consumption may potentially cause the species population to go extinct. Of course, it should be noted that in actual ecological systems, a variety of factors, including age structure, natural death rate, capture rate, food consumption rate, and others, are likely to have an impact on population dynamics. It may be that the large food consumption rate of predator species will postpone or avoid the prey population outbreak in the predator–prey system.

References

1. Sriram K, Rodriguez-Fernandez M, Doyle FJ (2012) A detailed modular analysis of heat-shock protein dynamics under acute and chronic stress and its implication in anxiety disorders. PLoS ONE 7:e42958

2. Parmar K, Blyuss KB, Kyrychko YN, Hogan SJ (2015) Time-delayed models of gene regulatory networks. Comput Math Methods Med 2015:347273
3. Bonella S, Coretti A, Vuilleumier R, Ciccotti G (2020) Adiabatic motion and statistical mechanics via mass-zero constrained dynamics. Phys Chem 22:10775–10785
4. Dhamala M, Jirsa VK, Ding MZ (2004) Enhancement of neural synchrony by time delay. Phys Rev Lett 92:074104
5. FitzHugh R (1961) Impulses and physiological states in theoretical models of nerve membrane. Biophys J 1:445–466
6. Nagumo J, Arimoto S, Yoshizawa S (1962) An active pulse transmission line simulating nerve axon. Proc IRE 50:2061–2070
7. Kakiuchi N, Tchizawa K (1997) On an explicit duck solution and delay in the FitzHugh-Nagumo equation. J Differ Equ 141:327–339
8. Hodgkin AL, Huxley AF (1952) A quantitative description of membrane current and its application to conduction and excitation in nerve. J Physiol-London 117:500–544
9. Zhang CR, Ke A, Zheng BD (2019) Patterns of interaction of coupled reaction-diffusion systems of the FitzHugh-Nagumo type. Nonlinear Dyn 97:1451–1476
10. Davison EN, Aminzare Z, Dey B, Leonard NE (2019) Mixed mode oscillations and phase locking in coupled FitzHugh-Nagumo model neurons. Chaos 29:033105
11. Zhen B, Xu J (2010) Fold-Hopf bifurcation analysis for a coupled FitzHugh-Nagumo neural system with time delay. Int J Bifurcation Chaos 20(12):3919–3934
12. Engelborghs K, Lemaire V, Belair J, Roose D (2001) Numerical bifurcation analysis of delay differential equations arising from physiological modeling. J Math Biol 42:361–385
13. Derouich M, Boutayeb A (2002) The effect of physical exercise on the dynamics of glucose and insulin. J Biomech 35:911–917
14. Huang MZ, Li JX, Song XY, Guo XJ (2012) Modeling impulsive injections of insulin: towards artificial pancreas. SIAM J Appl Math 72:1524–1578
15. Xu J, Pei LJ (2010) Effects oftechnologicaldelayoninsulinandbloodglucose in aphysiologicalmodel. Int J NonLin Mech 45:628–633
16. Huard B, Easton JF, Angelova M (2015) Investigation of stability in a two-delay model of the ultradian oscillations in glucose-insulin regulation. Commun Nonlinear Sci Numer Simul 26:211–222
17. Piemonte V, Capocelli M, De Santis L, Maurizi AR, Pozzilli P (2017) A novel three-compartmental model for artificial pancreas: development and validation. Artif Organs 41:e326–e336
18. Murillo AL, Li JX, Castillo-Chavez C (2019) Modeling the dynamics of glucose, insulin, and free fatty acids with time delay: the impact of bariatric surgery on type 2 diabetes mellitus. Math Biosci Eng 16:5765–5787
19. Al-Hussein ABA, Rahma F, Jafari S (2020) Hopf bifurcation and chaos in time-delay model of glucose-insulin regulatory system. Chaos Solitons Fractals 137:109845
20. Mohabati F, Molaei M (2020) Bifurcation analysis in a delay model of IVGTT glucose-insulin interaction. Theory Biosci 139:9–20
21. Subramanian V, Bagger JI, Holst JJ, Vilsboll T (2022) A glucose-insulin-glucagon coupled model of the isoglycemic intravenous glucose infusion experiment. Front Physiol 13:911616
22. Golestani F, Tavazoei MS (2022) Delay-Independent regulation of blood glucose for type-1 diabetes mellitus patients via an observer-based predictor feedback approach by considering quantization constraints. Eur J Control 63:240–252
23. Boulet J, Balasubramaniam R, Daffertshofer A, Longtin A (2010) Stochastic two-delay differential model of delayed visual feedback effects on postural dynamics. Philos Trans R Soc A-Math Phys Eng Sci 368:423–438
24. Wang WQ, Zhong SM (2012) Stochastic stability analysis of uncertain genetic regulatory networks with mixed time-varying delays. Neurocomputing 82:143–156
25. Guo Q, Sun ZK, Xu W (2018) Stochastic bifurcations in a birhythmic biological model with time-delayed feedbacks. Int J Bifurcation Chaos 28:1850048

26. Tuerxun N, Abdurahman X, Teng ZD (2020) Global dynamics and optimal harvesting in a stochastic two-predators one-prey system with distributed delays and Levy noise. J Biol Dyn 14:32–56
27. Pan LJ, Song Q, Cao JD, Ragulskis M (2021) Pinning impulsive synchronization of stochastic delayed neural networks via uniformly stable function. IEEE Trans Neural Netw Learn Syst 33:4491–4501
28. Zhang Y, Gao SJ, Chen SH (2022) A stochastic predator-prey eco-epidemiological model with the fear effect. Appl Math Lett 134:108300
29. Wang CH, Xu J (2010) Effects of time delay and noise on asymptotic stability in human quiet standing model. 2010:829484
30. Oseledec VI (1968) A multiplicative ergodic theorem, Lyapunov characteristic numbers for dynamical systems. Trans Moscow Math Soc 19:197–231
31. Chen LS (1991) Mathematical ecological model and research method. Science Press
32. Mondal S, Samanta GP (2019) Dynamics of an additional food provided predator-prey system with prey refuge dependent on both species and constant harvest in predator. Physica A 534:122301
33. Thakur, NK. Ojha A, Jana D, Upadhyay RK (2020) Modeling the plankton-fish dynamics with top predator interference and multiple gestation delays. Nonlinear Dyn 100:4003–4029
34. Luo YT, Zhang L, Teng ZD, Zheng TT (2021) Stability and bifurcation for a stochastic differential algebraic Holling-II predator-prey model with nonlinear harvesting and delay. Int J Biomath 14:2150019
35. Onana M, Mewoli B, Tewa JJ (2020) Hopf bifurcation analysis in a delayed Leslie-Gower predator-prey model incorporating additional food for predators, refuge and threshold harvesting of preys. Nonlinear Dyn 100:3007–3028
36. Anacleto M, Vidal C, Tewa JJ (2020) Dynamics of a delayed predator-prey model with Allee effect and Holling type II functional response. Math Meth Appl Sci 43:5708–5728
37. Jiao XB, Li XD, Yang YP (2022) Dynamics and bifurcations of a Filippov Leslie-Gower predator-prey model with group defense and time delay. Chaos Solitons Fractals 162:112436
38. Yang GX, Xu J (2015) Analysis of spatiotemporal patterns in a single species reaction–diffusion model with spatiotemporal delay. Nonlinear Anal-Real World Appl 22:54–65
39. Xu J, Yang GX, Xi HG, Su JZ (2015) Pattern dynamics of a predator-prey reaction–diffusion model with spatiotemporal delay. Nonlinear Dyn 81:2155–2163
40. Song ZG, Zhen B, Xu J (2014) Species coexistence and chaotic behaviour induced by multiple delays in a food chain system. Ecol Complex 19:9–17
41. Rogers T, Johnson B, Munch S (2022) Chaos is not rare in natural ecosystems. Nat Ecol Evol 6:1105–1111

Chapter 9
Impact of Time Delay on Traffic Flow

Since the 1930s, there has been a consistent increase in human's demand for transportation. A substantial number of vehicles have been manufactured and integrated into road transportation operations. However, due to limited transportation resources, an imbalance between transportation supply and demand has arisen, leading to the emergence of issues in traffic flow. Among these issues, traffic jam causes concern. Up to now, a comprehensive resolution for this problem remains elusive. Numerous experts and engineers from various fields have dedicated themselves to the investigation of traffic flow issues. They have developed traffic flow models within their respective domains of research, with the purpose of comprehending the fundamental mechanisms underlying the formation and evolution of traffic jam.

Existing traffic flow models can be broadly categorized into three types. The first type comprises continuous models based on macroscopic approaches. These models conceptualize traffic flow as a compressible continuous fluid consisting of a large number of vehicles, allowing for the study of the collective behaviour of vehicles. The second type consists of gas kinetic theory models established through a mesoscopic approach. In these models, traffic flow is treated as interacting particles. The third type comprises vehicle following models and cellular automaton models developed through a microscopic approach [1–6], which primarily focus on the study of behaviour of individual vehicles during their interaction.

Specifically, the research on microscopic vehicle following models provides valuable insights into the characteristics of individual drivers. This research has significant importance for the simulation, evaluation, and control of modern transportation systems, such as vehicle automatic cruise systems and intelligent traffic systems.

Vehicle following theory is used to describe the state where individual vehicles on a single lane follow one another. The behaviour of following drivers is influenced by various factors, with the most significant factors originating from the actual road design, the mechanical properties of vehicles, and the individual attributes of the drivers. Additionally, these behaviours may also be affected by random elements

J. Xu, *Nonlinear Dynamics of Time Delay Systems*,
https://doi.org/10.1007/978-981-99-9907-1_9

such as weather conditions and pedestrians crossing the street. In our study, the individual attribute on which we mainly focus is the reaction time delay [7–13].

Due to various influencing factors, the identification of reaction time delay can be challenging. Davis [14] conducted numerical simulations and measured driver reaction time delays ranging from 0.75 to 1.0 s after receiving a stimulus signal. Green [8] classified driver reaction time delays based on the strength of the stimulus signals, and analyzed a vast amount of data to determine the size of driver reaction time delays under different signal strengths. Mehmood et al. [15] categorized driver responses into braking reaction time delays and acceleration (deceleration) reaction time delays. They estimated the impact of individual driver characteristics (such as age and gender) and signal types (sudden or anticipated situations) on driver reaction time delays.

Given these researches, there still exists a lack of systematic summary on vehicle-following models that include driver reaction time delays and their bifurcation phenomena [16–24]. In this chapter, we first provide a review of research on time delay vehicle-following models and their bifurcated dynamics in Sect. 9.1. Section 9.2 is devoted to the study on bistable traffic modes induced by the variation of reaction time delay. In Sect. 9.3, we introduced the stabilization method for oscillatory traffic modes by imposing periodic perturbation to the parameter.

9.1 Full Velocity Difference Model and Traffic Patterns

This section includes three parts: the process of building the time delay full velocity difference (FVD) model, the definitions of several traffic modes involved in this section, and the effectiveness of the method of multiple scales and numerical continuation method by means of DDE-BIFTOOL [25]. Here, these three elements run parallel to one another.

9.1.1 Time Delay Full Velocity Difference Model

In this part, we introduce the time delay full velocity difference (FVD) model, and discuss how to treat boundary conditions, the driver's reaction time delay, and the nonlinear function of inter-vehicle distance in such model.

9.1.1.1 Model Setup

In this subsection, we will develop a time delay FVD model based on the microscopic car-following model. The FVD model encodes the driver's response to changes during the car-following process by monitoring the vehicle ahead, assessing its speed, the spacing between vehicles, and the relative velocity. The model is named as such because it considers the reaction time experienced by the driver while operating the vehicle. Specifically, a driver cannot instantaneously perceive changes in the spacing and relative speed of the vehicle ahead. The introduction of reaction time delay is essential, given the physiological or psychological factors inherent in the driver's perception process. Hence, we present the development of the time delay FVD model to provide a detailed account of its construction.

Figure 9.1 shows a typical car-following process, with n vehicles following one another counterclockwise on a circular road. Here, we make some basic assumptions—(i) each car has the same mechanical characteristics, or sensitivity coefficients, (ii) all drivers process information within the same time delay, and (iii) lane changes and overtaking are not allowed. The i th vehicle's position is represented by $x_i(t)(i = 1, 2, ..., n)$, and the position of the adjacent vehicle ahead is denoted by $x_{i+1}(t)(i = 1, 2, ..., n)$. The i th vehicle's velocity at time t is denoted as $v_i(t)$. The relative displacement of the $(i + 1)$ th vehicle to the i th vehicle is marked as the i th vehicle's inter-vehicle distance $h_i(t)$, namely $h_i(t) = x_{i+1}(t) - x_i(t), i = 1, 2, ..., n - 1$.

Differentiating $h_i(t)$ with respect to time yields the relative velocity between the $(i + 1)$th vehicle and the ith vehicle, termed as the velocity difference of the ith vehicle and is expressed as:

$$\dot{h}_i(t) = v_{i+1}(t) - v_i(t), \; i = 1, 2, \ldots, n - 1. \tag{9.1}$$

In order to formulate $h_n(t)$, assume the length of the road is L and the length of each vehicle is neglected, then we have $h_n = L - \sum_{i=1}^{n-1} h_i$, and consequently

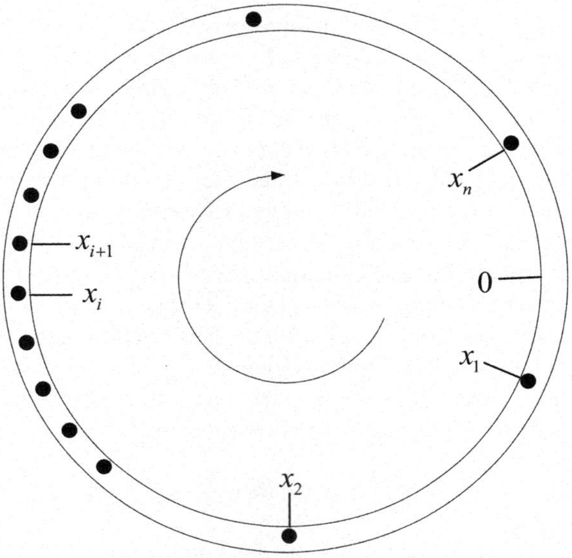

Fig. 9.1 Schematic of the position of vehicles on a circular road

$$\dot{h}_n = v_1 - v_n. \tag{9.2}$$

The original optimized velocity (OV) model was proposed in 1995 by Bando et al. [26]. This model operates under the assumption that each vehicle possesses an ideal following velocity which is a function of the inter-vehicle distance. The driver's objective is to modify the velocity to match the ideal velocity, which can be expressed by

$$\dot{v}_i(t) = \alpha(V(h_i(t)) - v_i(t)), \tag{9.3}$$

where α is the sensitivity coefficient, and $V(h_i)$ is the so called optimized velocity function.

The OV model does not consider the impact of variations in the velocity difference between the leading and following vehicles on the decision-making process of the following driver. However, driving decisions are influenced by changes in both the distance and velocity difference between vehicles [27]. Shamoto [28] observed that the acceleration of each vehicle is positively correlated with the change in its velocity difference. Consequently, while following a vehicle, drivers need to pay attention to the evolving velocity difference between their vehicle and the car ahead. Jiang [29] also proposed the FVD model in 2001, which took the impact of velocity difference into account, that is

$$\dot{v}_i(t) = \alpha(V(h_i(t)) - v_i(t)) + \beta\dot{h}_i(t), \tag{9.4}$$

where $\alpha > 0$ and $\beta > 0$ are sensitivity coefficients, and $V(h_i)$ is the optimized velocity function.

The driver's reaction time delay is not taken into consideration in the classic FVD model. Due to the driver's physiological aspect, there is a delay between the driver's absorbing information about changes in the car ahead and responding to changes. Even with automatic cruise control, there will still be a reaction time delay as the information is sensed, the control signal is calculated, and then the vehicle is eventually actuated. Therefore, time delay should be a necessary part of the mathematical model of the car-following model in traffic flow theory. Driver reaction time delay has been observed in Greenshields's [12, 13] early traffic assessments to highlight the significance of individual driver's feature in the driving process. As a result, this subsection proposes a FVD model that considers the driver's reaction time delay. This model, which builds on the FVD model, accounts for the driver's reaction time delay to the three stimulus signals—inter-vehicle distance, own velocity, and velocity difference, which are denoted as τ_1, τ_2 and τ_3, respectively. Consequently, the ith vehicle's acceleration can be written as,

$$\dot{v}_i(t) = \alpha\ (\ V(h_i(t - \tau_1)) - v_i(t - \tau_2)) + \beta\ \dot{h}_i(t - \tau_3). \tag{9.5}$$

Given the non-zero time delay in the driver's reaction, Eq. (9.5) is designated as the time delay Full Velocity Difference (FVD) model.

9.1.1.2 Model Processing

This section makes the assumption that $\tau_1 = \tau_3 = \tau \neq 0$, $\tau_2 = 0$, which indicates that the driver can perceive his own velocity instantly without a reaction time delay, but there is a reaction time delay when detecting the inter-vehicle spacing and velocity difference [30]. Here, we assume that the driver's reaction time delay to the inter-vehicle spacing is the same as that of the velocity difference. In accordance with this assumption, we combine Eqs. (9.1), (9.2) and (9.5) to construct the dynamical system that will be used in this subsection, which consists of a time delay differential equation with $2n - 1$ dimension given by

$$\begin{cases} \dot{h}_i(t) = v_{i+1}(t) - v_i(t), \\ \dot{v}_i(t) = \alpha\,(V(h_i(t-\tau)) - v_i(t)) \\ \qquad + \beta\,(v_{i+1}(t-\tau) - v_i(t-\tau)), \qquad\qquad i = 1, ..., n-1, \\ \dot{v}_n(t) = \alpha\,(V(L - \sum_{i=1}^{n-1} h_i(t-\tau)) - v_n(t)) \\ \qquad + \beta\,(v_1(t-\tau) - v_n(t-\tau)), \end{cases} \tag{9.6}$$

where $\alpha > 0$ and $\beta > 0$ are sensitivity coefficients, and $V(h_i)$ is the optimized velocity function. Both the time delay of the driver's response to the velocity difference and inter-vehicle distance are represented by τ. This system suggests that a driver must not only adjust his own velocity to maintain the optimal speed $V(h_i)$, but also keep close attention to the change of velocity difference between his vehicle and the vehicle ahead to avoid collisions.

The optimized velocity function in Eq. (9.6) has many forms for selection. Batista [31] has systematically reviewed various ways of choosing the optimized velocity function $V(h_i)$. Orosz [32] proposed a dimensionless optimized velocity function in the following form

$$V(h) = \begin{cases} 0, & if\ h \in [0, h_{stop}], \\ v^0 \dfrac{[(h - h_{stop})]^3}{h_{stop}^3 + (h - h_{stop})^3}, & if\ h \in [h_{stop}, \infty), \end{cases} \tag{9.7}$$

where $h_{\mathrm{stop}} > 0$ denotes the minimum inter-vehicle distance at which the vehicle should stop and $v^0 > 0$ denotes the desired velocity. This optimized velocity function is a piecewise smooth nonlinear function whose configuration is shown in Fig. 9.2. Furthermore, The optimized velocity function of (9.7) has the following properties:

(1) $V(h)$ is a continuous, non-negative and monotonically increasing function, which means that the optimized velocity function increases continuously with the corresponding increase of the inter-distance between vehicles;
(2) As the distance between vehicles increases infinitely, the optimized velocity function converges to the optimal velocity, i.e. $V(h) \to v^0$, $h \to \infty$;

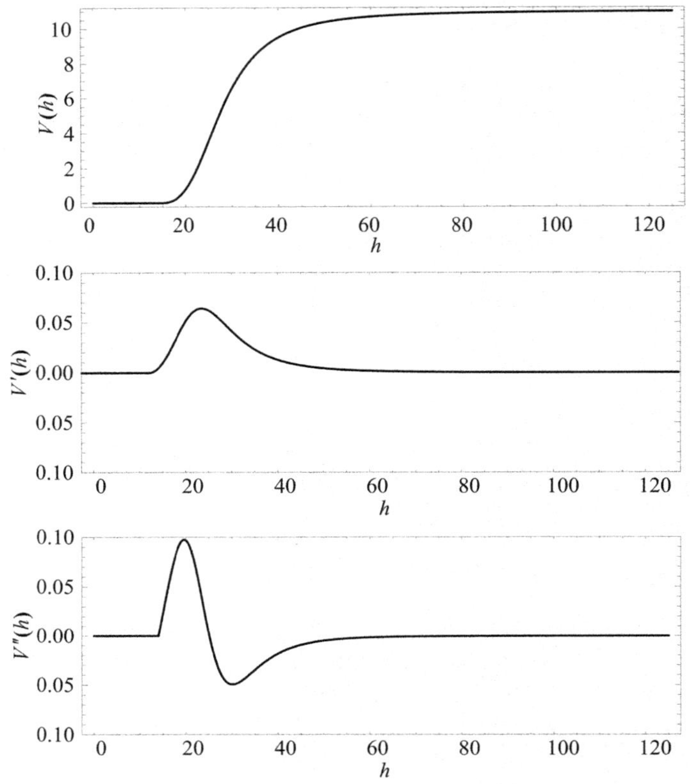

Fig. 9.2 Optimized velocity function and its derivative

(3) When the inter-vehicle distance is small, the optimized velocity function is zero, i.e. $V(h) = 0,\ h \in [0, h_{stop}]$.

Property (1) benefits the smooth implementation of the numerical continuation in our subsequent work, while Properties (2) and (3) are based on the actual traffic situation. Property (2) considers the mechanical properties of the vehicle so that there must be a finite maximum velocity, and Property (3) is to avoid the occurrence of collision accidents in the driving process. In Fig. 9.2, we choose $h_{stop} = 14(m)$, $v^0 = 11(m/s)$ to draw the images of the function $V(h)$ and its derivative with respect to the inter-vehicle distance h, from which we can clearly see that $V(h)$ satisfies the above three properties.

9.1.1.3 Reasonability of the Model

The time delay FVD model concept is reasonable for the following reasons. First, if we set the driver's reaction time delay τ_1, τ_2 and τ_3 to zero, the model will degenerate into the form of Jiang's [29] FVD model. Secondly, the time delay FVD model will exclude irrational events like vehicle reversing or negative speed within a reasonable range of value of the reaction time delay. Finally, the time delay FVD model not only explains the relationship between the driver's reaction time delay and traffic jam but also uncovers new traffic patterns.

A further issue that needs to be clarified is the choice of a single-lane circular road as the background for the time delay car-following model in this subsection. Despite the fact that the assumption of the road's condition is simplistic, it is still important because it may enlighten the adaptive cruise control on highways and other intelligent transportation systems.

We choose the following parameters for the case study shown in Fig. 9.3:

$$\begin{aligned} n &= 33\,(vehicles), \alpha = 0.9\,(\mathrm{s}^{-1}), v^0 = 11\,(\mathrm{m/s}), \\ h_{stop} &= 14\,(\mathrm{m}), h^* = 34\,(\mathrm{m}), \tau = 0.8(\mathrm{s}). \end{aligned} \tag{9.8}$$

Then, for cases that $\beta = 0$, $\beta = 0.1$, and $\beta = 2$, DDE-BIFTOOL is used to numerically calculate all the roots of the characteristic equation of the uniform flow in Eq. (9.6). The Argand diagram which indicates the distribution of the real and imaginary parts of these characteristic roots is plotted to show that the uniform flow solution's stability can be preserved in the time delay FVD model by properly accounting for the velocity difference. As observed in Fig. 9.3a, the uniform flow solution is unstable when $\beta = 0$, that is, the influence of velocity difference is not taken into consideration. As a result, the characteristic root with a positive real part is observed. As illustrated in Fig. 9.3b, the uniform flow solution is stable when $\beta = 0.1$, which means that all the real parts of the characteristic roots are negative. Further, the uniform flow solution becomes more unstable if the model is over sensitive to the velocity difference, as is the case for $\beta = 2$ which is depicted in Fig. 9.3c.

9.1.2 *Criteria for Traffic Jams and Traffic Patterns*

In this part, the criteria for distinguishing traffic jams is provided and then the various traffic patterns that will be studied in this subsection for the one-lane circular road condition are defined.

The time delay car-following model has a number of solutions when model parameters are changed, and we can use bifurcation theory and numerical continuation to determine the relationship between these solutions and the driver's reaction time delay. However, to establish the relation between the solutions and actual traffic patterns, we must first define the criteria for determining various traffic patterns as well as their classifications.

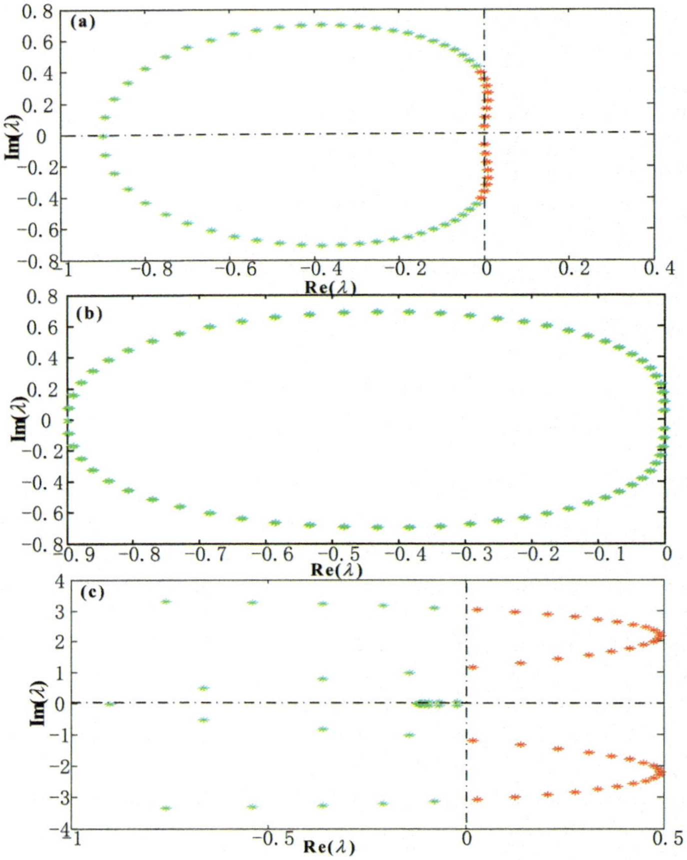

Fig. 9.3 Argand diagrams for the uniform flow solution of Eq. (9.6) when **a** $\beta = 0$, **b** $\beta = 0.1$, **c** $\beta = 2$

9.1.2.1 Establishing the Jam Criteria

In the microscopic car-following model, the judgment of whether or not the traffic jam occurs may depend on the average velocity of the vehicle over a certain period, the amplitude of the velocity of the vehicle, the inter-vehicle distance, and other indexes. Using the hysteresis loop, spatiotemporal pattern, and other visualization techniques, one can examine the onset and severity of traffic jam. In this subsection, we use the distribution of vehicle velocity extracted from the time series graph as an indicator of whether a traffic jam exists, and the distribution of jam clusters on the spatiotemporal pattern as an indicator of distinguishing the severity of traffic jams. We now run the numerical simulation of Eq. (9.6) starting from two initial conditions as an illustration to show the rationality of the proposed indicators of traffic jam.

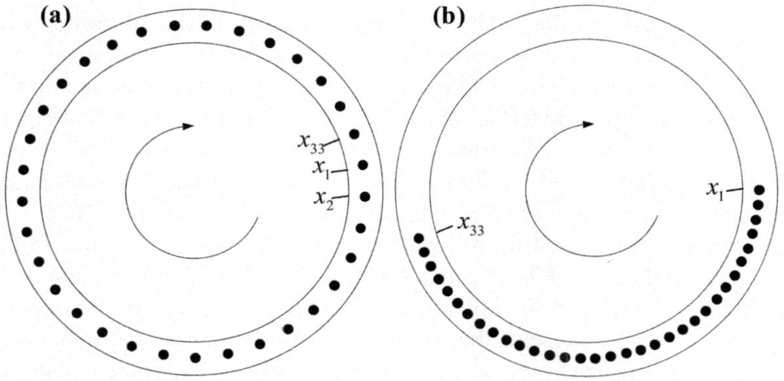

Fig. 9.4 Initial positions of 33 vehicles on a circular road. **a** Scattered initial condition; **b** jammed initial condition

In Fig. 9.4, two initial states are given for 33 vehicles traveling counterclockwise along a circular road of length L. The initial state of the vehicles on the road under the scattered initial condition is shown in Fig. 9.4a, i.e.

$$\begin{aligned} &h_1(t) = 35, v_1(t) = 5, h_i(t) = 30, v_i(t) = 6.5871, i = 2, n-1, \\ &h_{33}(t) = 25, v_{33}(t) = 6.5871, t \in [-\tau, 0]. \end{aligned} \tag{9.9}$$

Equation (9.9) depicts the scatter initial condition where the first vehicle slows down, which increases the inter-vehicle distance, while the last vehicle speeds up, which decreases the inter-vehicle distance, and the remaining vehicles continue to maintain the same inter-vehicle distance and velocity in the uniform flow condition at the initial moment. In Fig. 9.4b, we show the initial state of each car on the road with the initial jam conditions as follows

$$h_i(t) = 14, v_i(t) = 0, t \in [-\tau, 0]. \tag{9.10}$$

According to Eq. (9.10), all vehicles are placed together in a wide jam cluster with the initial condition of traffic jam. The cars are all separated by safe inter-vehicle distances when stopped, and the velocities are all zero. The parameters that we choose for the numerical simulation of several jam indicators are as follows

$$n = 33, \alpha = 0.9, \beta = 0.1, v^0 = 11, h_{stop} = 14, h^* = 30. \tag{9.11}$$

Velocity and Jam Cluster Criteria

We begin by demonstrating the velocity and jam cluster criteria. In this subsection, we choose each vehicle's velocity and the speed distribution across all vehicles to determine whether a traffic jam occurs, and distinguish traffic jams of different levels by the structure, distribution, and width of jam clusters in the macroscopic spatiotemporal pattern. It is easy to distinguish different traffic patterns under different driver reaction time delays, as well as different traffic jam modes with different initial conditions and the same driver reaction time delays, by using the velocity and jam cluster as the criteria of traffic jam. For instance, in Fig. 9.5, we chose $\tau = 1.8$ (s) and plot the time histories of velocity and inter-vehicle distance, spatiotemporal pattern, as well as the statistical graph of velocities of vehicles under the two initial conditions.

The time history plots of the velocity and the inter-vehicle distance of the first vehicle are shown in Fig. 9.5a_1, b_1 under the two initial conditions within 0–2000 s. In these figures, the black solid curve represents the time history of the velocity, and the red dashed curve represents the time history of the inter-vehicle distance. From these two figures, we can figure out two points—firstly, the temporal evolution of the velocity and the inter-vehicle distance stay synchronized; the only difference is in the amplitude. This suggests that we can use either the amplitude of the velocity or that of the inter-vehicle distance as the criterion of traffic jam because the inter-vehicle distance remains at the minimal value for a period of time when the velocity is zero. Secondly, the time history with the scatter initial conditions has a completely different evolution compared with that of the time history with the jam initial conditions: the

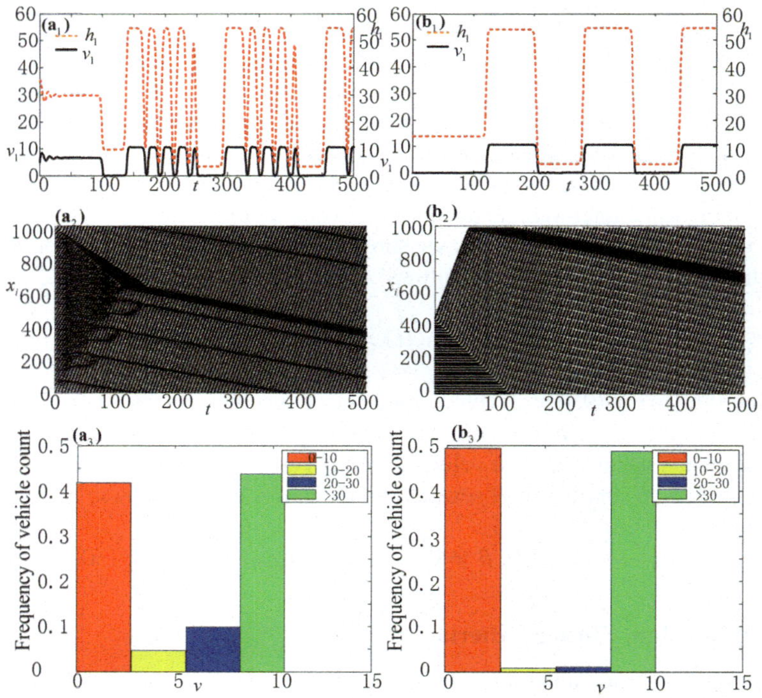

Fig. 9.5 Various plots with two initial conditions for $\tau = 1.8$. $\mathbf{a_1}$–$\mathbf{a_3}$ are the plots with the scattered initial condition, $\mathbf{b_1}$–$\mathbf{b_3}$ are the plots with the jammed initial condition

former has four different small-period oscillations within one large period, while the latter oscillates only once within one period.

The time history plot, however, can only reflect the driving conditions of an individual vehicle on the road, and it can only detect whether there is a traffic jam on the road, but is unable to show the conditions of all the vehicles on the entire road when there is a traffic jam. Therefore, we need to use the spatiotemporal pattern to visualize the severity of traffic jam on the road during a given period. According to Fig. 9.5a_2, b_2, the spatiotemporal pattern of the location distribution of all vehicles within 0–500 s with the scatter and jammed initial conditions shows that four jam clusters with different widths appear, which means that as long as the vehicle encounters a traffic jam, it will have to pass through four different levels of traffic jam to exit the congested area. On the contrary, under the initial condition of a jam, only one cluster is seen in the spatiotemporal pattern, indicating that the vehicle only needs to wait for a while to exit the congested area instead of continuously entering it. We can see from the spatiotemporal pattern that different initial conditions, even with the same reaction time delay, will result in different traffic jam modes.

When the spatiotemporal pattern is unable to clearly show the difference among various traffic jam modes, we can use the histogram of velocity statistics to identify different jam modes and the severities of the traffic jam. As seen in Fig. 9.5a_3, b_3, for the scatter initial condition and the jammed initial condition, we divide the velocities of all vehicles within 0–2000 s into four periods, i.e. [0, 10), [10, 20), [20, 30), and [30, 39.6] in kilometers per hour. The figures show that the frequency of the velocities of vehicles fall into [0, 10) under the scatter initial condition is much lower than they do under the jammed initial condition, indicating that the traffic jam is more severe under the jammed initial condition than it is under the scatter initial condition.

Other Criteria

We also present a summary of other criteria to explain why criteria such as the amplitude of the vehicle velocity, the average velocity of the vehicle, and the hysteresis loop are not selected.

First, we demonstrate that the amplitude of the vehicle velocity is not ideal as a criterion for determining the degree of traffic jam. As shown in Fig. 9.6, the DDE-BIFTOOL is used to plot the maximum of the first vehicle's inter-vehicle distance as the time delay of the driver's reaction varies. In both initial conditions, the change in amplitude is the same, indicating that the amplitude of the vehicle velocity does not distinguish the traffic patterns under different initial conditions.

Now we explain why the following indexes, the average velocity and vehicle occupancy ratio in the congested area are not selected as indicators of different traffic jam patterns. For either of the two initial conditions, we plot the change in the average velocity along with the variation of the driver reaction time delay in Fig. 9.7a. Although we already know that different traffic patterns exist under the two initial conditions from the previous subsection, it is not obvious from Fig. 9.7a that the traffic patterns will be different. Figure 9.7b shows the relationship between the percentage of vehicles in congested area to the total number of vehicles and the driver reaction time delay under the two initial conditions. Although the emergence

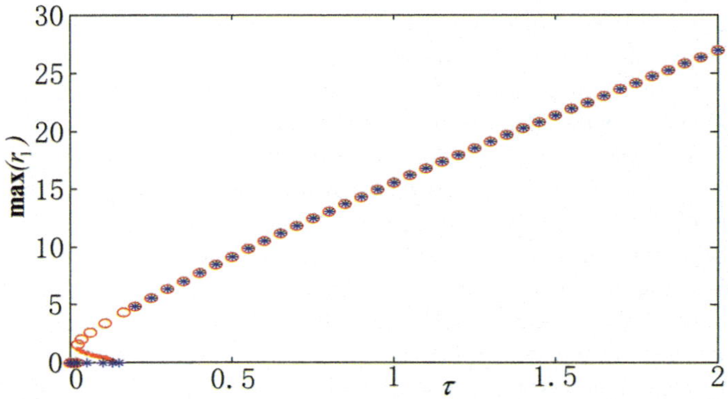

Fig. 9.6 Bifurcation plots of the amplitude of the inter-vehicle distance of the first vehicle, as the time delay varies, with the blue and red circles denoting the scattered initial condition and the jammed initial condition, respectively, and the red dotted curve representing the unstable periodic solutions that arise with both initial conditions

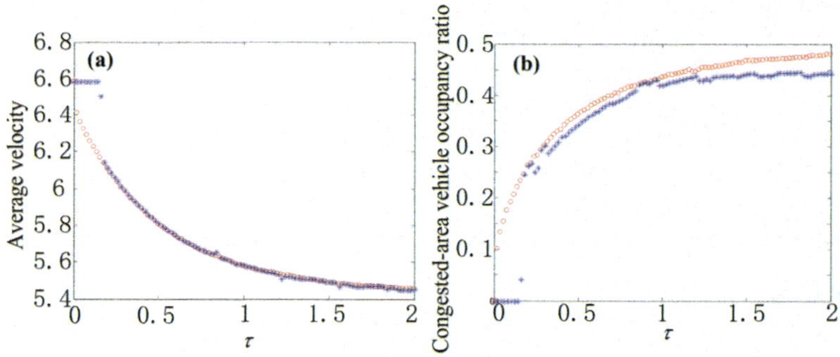

Fig. 9.7 **a** Change in average velocity of vehicles with time delay; **b** vehicle occupancy ratio in the congested area

of traffic jams and different traffic jam modes of different initial conditions can be observed qualitatively, it is not possible to know the emergence moment or the location of the traffic jam, while the spatiotemporal pattern can help us access such information.

Finally, the hysteresis loop criterion is also analyzed. In Fig. 9.8, we show the trajectories of the vehicle in the plane of velocity and inter-vehicle distance with two different initial conditions and identical reaction time delays, to which we refer as hysteresis loop. It can be seen from Fig. 9.8 that the hysteresis loop has a number of merits, one of which is that we can easily detect various traffic patterns with different initial conditions. In Fig. 9.8a, the vehicle maintains a constant velocity and inter-vehicle distance, or equivalently the uniform flow state, under the scatter initial condition for $\tau = 0.1$, but under the jammed initial condition, the vehicle's

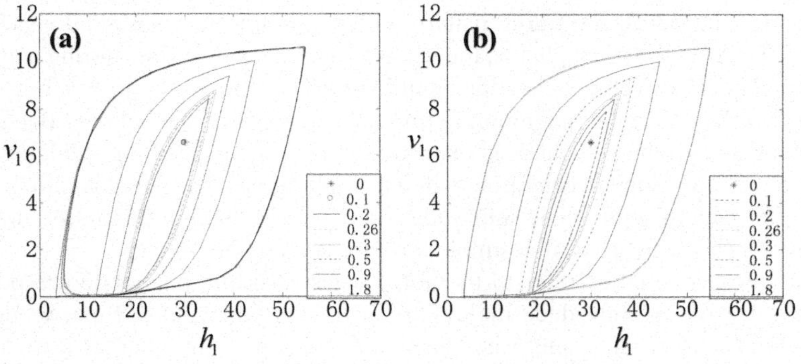

Fig. 9.8 Hysteresis loops with different driver reaction time delays with **a** scattered initial conditions; **b** jammed initial conditions

velocity and inter-vehicle distance start to oscillate irregularly for the same time delay. Another example indicates that, when $\tau = 1.8$, the hysteresis loop under the scatter initial condition consists of multiple loops while consisting of just one loop in the jammed initial condition, indicating the presence of various jam patterns. Secondly, information such as the inter-vehicle distance when the jam arises, as well as whether a jam occurs, can be inferred from the hysteresis loop. Despite these advantages, the hysteresis loop is still unable to determine the severity of the jam, the emergence time and the location of the jam.

9.1.2.2 Definitions of Various Traffic Patterns

For ease of use in the following discussion, we provide definitions for a number of traffic patterns that will be mentioned. We select amplitude, frequency, and maximum and minimum velocities extracted from time series of velocity to judge whether a traffic jam occurs. For instance, we assume that there is no traffic jam if the amplitude of velocity is zero, i.e., the maximum and minimum velocities are equal, and no frequency component is present. Another example is that if the minimum velocity is significantly greater than zero and the amplitude is not zero but still relatively small, we also believe that the instability of the uniform flow state does not result in traffic jams. When the minimum velocity is close or equal to zero and the amplitude is not zero, we say that a traffic jam has occurred. Different oscillatory patterns suggest the emergence of various traffic jam modes. To distinguish these jam modes, we take into account the structure and width of the jam clusters on the spatiotemporal pattern. Here, we go into further details about these traffic patterns.

We start by defining the uniform flow pattern. If the vehicle velocity is non-zero constant, as shown in the time history of velocity when $v = c(c \neq 0)$ and $v_{amp} = 0$, we say that there is no traffic jam at this moment and refer to this traffic pattern as the uniform flow pattern. The velocity and the inter-vehicle distance are both constant under this pattern.

We now introduce the traffic pattern with small-amplitude oscillations. If $v = c(t)(c(t) \neq 0)$, $t \in [0, T]$, that is, the velocity fluctuates over time during a cycle of the time history of velocity, and the amplitude of the velocity $v_{amp} \neq 0$, but v_{amp} is much less that $v_{\max}$, the maximum permissible velocity of the vehicle. Because of this, even though the vehicle does not travel at a constant speed, the vehicle simply keeps accelerating and decelerating without stopping. Such a traffic pattern is called the traffic pattern with small-amplitude oscillation. Although this mode does not directly cause traffic jams, it is simple for it to develop into the jam.

The above two cases refer to the non-congested traffic patterns corresponding to $v \neq 0$. Now we introduce the definitions of the jam modes. First we need to understand what a jam pattern is. When $v = 0$, namely the vehicle's velocity is decelerated to zero, we consider that the traffic jam occurs and such mode is classified as a jammed one because vehicles must have encountered various levels of traffic jam in such situation. We distinguish these different jam patterns based on the features of the jammed road segments on the circular road and the frequency components of time history of the vehicle's velocity. Furthermore, we will describe the features of such jammed road sections using the spatiotemporal pattern of positions of vehicles.

Now we define three types of traffic jam modes, namely, the almost jammed mode, equal-width jammed mode, and wide-narrow alternative jammed mode. For each jammed mode, we create a quantity T_{jam} to describe the level of the jam. T_{jam} represents the time a vehicle spends on a congested road. If $T_{jam} = 0$, the vehicle velocity reduces to zero but then immediately increases, i.e., the vehicle enters the congested region but soon exits it without halting. This jam mode is then referred to an almost jammed mode. We define a traffic jam as an equal-width jammed mode if $T_{jam} = c(c \neq 0)$, which means that vehicles enter the congested section and stay in such state for a while. For the wide-narrow alternative jammed mode, we mean that the vehicles enter the congested roadway and stop, but the resting duration varies in each congested roadway, namely, $T_{jam} = c(t)(c(t) \neq 0)$.

Both the almost jammed mode and equal-width jammed mode have only one frequency component. There are two or more frequencies present in the wide-narrow alternative jammed mode. Using the spectral analysis in the frequency domain, it is also possible to identify to which type the jammed mode belongs. The traffic mode with no frequency component represents a uniform flow mode. The case that the traffic mode has one frequency component is a bit complicated: it can be a traffic mode with small-amplitude oscillations (non-congested), almost jammed mode, or equal-width jammed mode.

9.1.3 Example

In this part, we provide a simple example with the following two purposes: firstly, to demonstrate that the instability of a uniform flow can be induced by the time delay in driver's reaction, and consequently lay the foundation for subsequent studies on the delay-induced bifurcation in the car-following model; secondly, to demonstrate that the method of multiple scales can be applied to analyze the nonlinear dynamics caused by the time delay in the car-following model with satisfactory accuracy.

Focusing on the time delay car-following model with three vehicles, we study the impact of driver's reaction time delay on the stability of uniform flow mode and the behaviour of delay-induced dynamics when the uniform flow becomes unstable. It is shown that driver's reaction time delay promotes the instability of the uniform flow mode and results in a traffic pattern with small-amplitude oscillations through linear analysis and the method of multiple scales. The theoretical result is in good agreement with the results of numerical continuation. In Fig. 9.9, we show the results of the time delay car-following model with three cars, where $\alpha = 0.1$, $\beta = 0.5$, $h^* = 25$. We plot the bifurcation diagram of the uniform flow solution obtained by the method of multiple scales, represented by the solid curve, which agrees with the results obtained by running DDE-BIFTOOL.

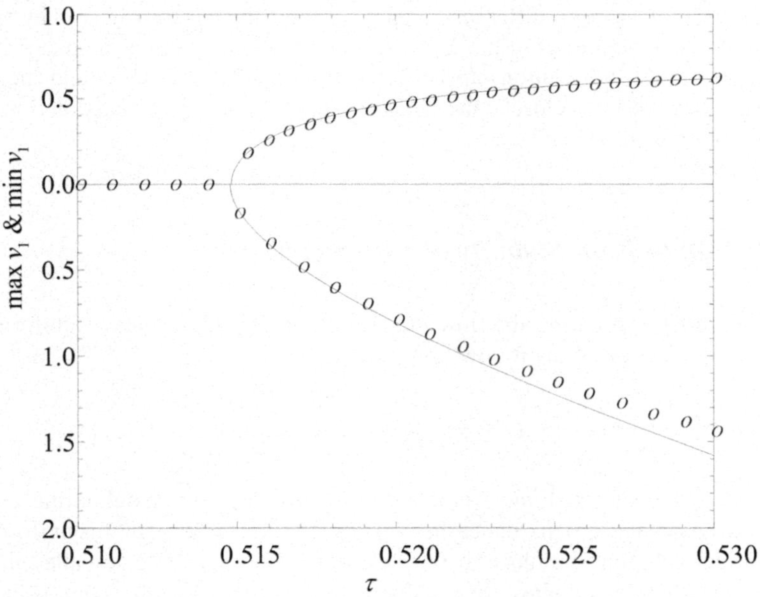

Fig. 9.9 Comparison of bifurcation diagrams by theoretical analysis (the method of multiple scales) and numerical methods (bifurcation continuation by DDE-BIFTOOL) with the reaction time delay being treated as the bifurcation parameter

9.2 Bistable Traffic Patterns Induced by Reaction Time Delay

This section focuses on the bifurcated dynamics when the parameter of driver reaction time delay is close to the critical value for the occurrence of bifurcation. There have been various studies on the local bifurcation behaviour regarding a single parameter in the time delay car-following model. Igarashi [33] discussed the uniform flow solution's bifurcation phenomenon using the average inter-vehicle distance as the bifurcation parameter. It has been discovered that the most important mechanism contributing to the destabilization of the uniform flow solution is the subcritical Hopf bifurcation. Using the theoretical analysis which accounts for the translational symmetry, Orosz [34] also chose the average inter-vehicle distance as the bifurcation parameter and analytically determined the relationship between the inter-vehicle distance and traffic modes. However, none of the current researches has considered the influence of velocity difference signals in their models, nor have they specifically addressed the impact of the driver reaction time delay on traffic patterns. This section examines the relationship between the reaction time delay and the traffic mode when the delay is close to the critical value, by treating the reaction time delay as a bifurcation parameter.

9.2.1 Analysis on Stability of Uniform Flow

All OV models, including the time delay FVD model (9.5), have a uniform flow equilibrium solution of the following form:

$$h_i(t) = h^* = \frac{L}{n},\ v_i(t) = v* = V(h^*),\ \dot{h}_i(t) \equiv 0,\ i = 1, ..., n, \tag{9.12}$$

where h^* denotes a fixed inter-vehicle distance between two neighboring vehicles and v^* is a constant velocity under the uniform flow mode. We call this uniform flow equilibrium solution, the uniform flow without traffic jam. The so-called uniform flow means that the vehicles are equally spaced and travel at the same velocity. It should be pointed out that the distance between vehicles in uniform flow should be between 0–125 m to ensure the interaction between vehicles on the road [35], and to ensure the effectiveness of using the vehicle-following model. In the following, we will investigate the stability of the uniform flow when the driver' reaction time delay is used as a bifurcation parameter.

Let us first define a perturbation variable of the uniform flow (9.12) as follows:

$$r_i(t) = h_i(t) - h^*, u_i(t) = v_i(t) - v^*, i = 1, 2, ..., n. \tag{9.13}$$

The velocity function $V(h)$ is expanded to Taylor series at $h_i(t) = h^*$ up to the fifth term, so that Eq. (9.6) becomes:

$$\begin{cases} \dot{r}_i(t) = u_{i+1}(t) - u_i(t), \\ \dot{u}_i(t) = -\alpha u_i(t) + \alpha \sum_{j=1}^{5} b_j r_i^j(t-\tau) + \beta(u_{i+1}(t-\tau) - u_i(t-\tau)), \\ \qquad i = 1, ..., n-1, \\ \dot{u}_n(t) = -\alpha u_n(t) + \alpha \sum_{j=1}^{5} b_j(-\sum_{i=1}^{n-1} r_i(t-\tau))^j + \beta(u_1(t-\tau) - u_n(t-\tau)), \end{cases} \tag{9.14}$$

where

$$b_j = \frac{1}{j!} V^{(j)}(h^*),\ j = 1, 2, 3, 4, 5. \tag{9.15}$$

Therefore, the trivial equilibrium of Eq. (9.14) is equivalent to the uniform flow solution of Eq. (9.6). Discussion on the stability of the uniform flow solution of Eq. (9.6) is now transformed into the discussion on the stability of the trivial equilibrium of Eq. (9.14). In the following, we will focus on the stability of the trivial equilibrium of Eq. (9.14). Firstly, substitute the solution in the form of $e^{\lambda t}(\lambda \in \mathbb{C})$ into the linear part of Eq. (9.14) to obtain the characteristic equation which is given by:

$$(\alpha + \lambda) \prod_{k=1}^{n-1} \left(\lambda^2 + \alpha\lambda + e^{-\lambda\tau}(b_1\alpha + \beta\lambda) - e^{-\lambda\tau}(b_1\alpha + \beta\lambda) e^{\frac{2\pi k}{n} \mathrm{i}} \right) = 0. \tag{9.16}$$

Next, we substitute the critical eigenvalue $\lambda = \mathrm{i}\omega$, $\omega > 0$ into the characteristic equation and separate the real and imaginary parts to obtain the following system of equations:

$$\begin{cases} \sin(\tau\omega) = p - q \cot \dfrac{k\pi}{n} \\ \cos(\tau\omega) = p \cot \dfrac{k\pi}{n} + q \end{cases},\ k = 1, \ldots, n-1, \tag{9.17}$$

where $k = 1, 2, \ldots n-1$ is introduced to represent the n roots on the unit circle, denoting the discrete wave number. Here

$$p = \frac{\omega(\beta\omega^2 + \alpha^2 b_1)}{2(\beta^2\omega^2 + \alpha^2 b_1^2)},\ q = \frac{\alpha\omega^2(-\beta + b_1)}{2(\beta^2\omega^2 + \alpha^2 b_1^2)}.$$

Using the trigonometric identity $\sin^2(\tau\omega) + \cos^2(\tau\omega) = 1$ yields

$$\omega_j = \sqrt{-\frac{\alpha^2}{2} + 2\beta^2 \sin^2\frac{k\pi}{n} + \frac{1}{2}\sqrt{16b^2\alpha^2\csc^2\frac{k\pi}{n} + \left(-4\beta^2 + \alpha^2\csc^2\frac{k\pi}{n}\right)^2 \sin^2\frac{k\pi}{n}}}.$$

Solving Eq. (9.17), we get

$$\tau_k^{(j)} = \frac{1}{\omega_k}\arccos(p\cot\frac{k\pi}{n} + q) + \frac{2j\pi}{\omega_k},$$

where $k = 1, 2, ..., n-1$, $j = 0, 1, 2, ...$, and $\pm i\omega_k$ is a pair of purely imaginary roots of Eq. (9.1).

We define the critical time delay as follows:

$$\tau_c = \tau_{k_0}^{(0)} = \min_{k_0 \in \{1,...,n-1\}} \{\tau_k^{(0)}\},\ \omega_c = \omega_{k_0}. \tag{9.18}$$

Next, we will check the transversality condition with respect to the reaction time delay τ, which is a necessary condition for the occurrence of the Hopf bifurcation, namely

$$\text{Re}\left(\frac{d\lambda}{d\tau}|_{\tau_c}\right) = \frac{c_0 + c_1\tau_c + c_2\tau_c^2}{d_0[\alpha^2 b_1^2(\alpha^2 + 3\omega^2 + d_1^2)] + \beta^2\omega^2 d_1 - 2\alpha b_1\omega^2(\alpha + \alpha^2\tau_c + \omega^2\tau_c)} > 0. \tag{9.19}$$

Here the expressions for c_i, d_i are given in Appendix 9.1.

When the transversality condition (9.19) is satisfied, it follows from the Hopf bifurcation theorem [36] that a Hopf bifurcation may occur for Eq. (9.14) at the critical value $\tau = \tau_c$. That is, the uniform flow solution of Eq. (9.6) becomes unstable, and a bifurcation solution exhibiting periodic oscillations arises at the critical time delay $\tau = \tau_c$.

In order to verify the reliability of above analytical results, we show an example for numerical verification. We consider the case with 33 vehicles, and the other parameters in the time delay FVD model are chosen as follows:

$$\alpha = 0.9\ (\text{s}^{-1}),\ \beta = 0.1\ (\text{s}^{-1}),\ v^0 = 11\ (\text{m/s}),\ h_{stop} = 14\ (\text{m}),\ L = 1122\ (\text{m}). \tag{9.20}$$

It is important to note that the selection of these parameters [32] is reasonable regarding the real traffic conditions. Using Eq. (9.18), we get $\tau_c = 0.84$ (s), $\omega_c = 0.06$. Then it can be verified that the transversality condition is satisfied. We also plot the distribution of the real and imaginary parts of leading characteristic roots of the characteristic equation with the aid of the numerical continuation software

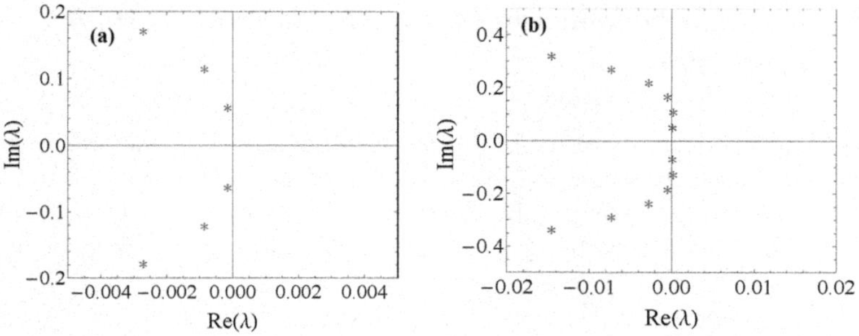

Fig. 9.10 Distribution of real and imaginary parts of eigenvalues of the uniform flow solution of Eq. (9.6)

package DDE-BIFTOOL. The horizontal axis in Fig. 9.10 represents the real part of the characteristic roots, while the vertical axis represents the imaginary part. In Fig. 9.10a, we choose $\tau = 0.83 < \tau_c$, from which we see that all characteristic roots, or equivalently eigenvalues, have negative real parts. Therefore, the trivial equilibrium of Eq. (9.14) is stable, i.e., the uniform flow solution of Eq. (9.6) is stable. In Fig. 9.10b, we choose $\tau = 0.85 > \tau_c$, from which we can see that some pairs of conjugate eigenvalues cross the imaginary axis, i.e., the characteristic equation has roots with positive real parts. This indicates that the uniform flow solution of Eq. (9.6) has become unstable. So far, we see that the analytical results are in agreement with the numerical results.

It is discovered that the time delay of a driver's reaction can make the uniform flow unstable, and it is possible to determine the interval of the delay on which the uniform flow solution is stable. The uniform flow is stable when the critical time delay $\tau < \tau_c$ and the traffic jam may not occur. When $\tau > \tau_c$, the Hopf bifurcation occurs, the uniform flow becomes unstable due to the Hopf bifurcation and the bifurcated solution in the form of periodic oscillation arises. It should be noted that a locally stable uniform flow solution does not exclude traffic jams. For instance, the subcritical Hopf bifurcation can result in traffic jam, even at $\tau < \tau_c$ when the initial values are not in the basin of attraction of the uniform flow solution. Similar to this, we also think that unstable uniform flows do not necessarily imply traffic jams. Depending on the direction of the bifurcation, the instability of the uniform flow may or may not lead to traffic jam. In particular, for instance, the occurrence of a supercritical Hopf bifurcation only results in the periodic solution with small amplitude, and such bifurcation does not always result in the traffic jam. However, subcritical Hopf bifurcation causes periodic solutions with large amplitudes, which significantly increases the likelihood of traffic jams.

9.2.2 Bistable Phenomenon Induced by Subcritical Hopf Bifurcation

To investigate the stability of the bifurcated solution as well as the direction of the bifurcation, in this subsection, we investigate the dynamical behaviour of Eq. (9.14) when the uniform flow solution becomes unstable, using the method of multiple scales [37]. In order to rewrite Eq. (9.14) in matrix form, we introduce the notation

$$\mathrm{y}(t) = (r_1(t), u_1(t), ..., r_{n-1}(t), u_{n-1}(t), u_n(t))^T. \tag{9.21}$$

Then Eq. (9.14) can be written as

$$\dot{\mathrm{y}}(t) = \mathrm{C}y(t) + \mathrm{D}y(t-\tau) + \mathrm{F}(\mathrm{y}(t-\tau)), \tag{9.22}$$

where $y : \mathbb{R} \to \mathbb{R}^{2n-1}$, and

$$\mathrm{C} = \begin{pmatrix} 0 & -1 & 0 & 1 & 0 & 0 & \cdots & 0 & 0 \\ 0 & -\alpha & 0 & 0 & 0 & 0 & \cdots & 0 & 0 \\ 0 & 0 & 0 & -1 & 0 & 1 & \cdots & 0 & 0 \\ 0 & 0 & 0 & -\alpha & 0 & 0 & \cdots & 0 & 0 \\ \vdots & \vdots & \vdots & \vdots & \vdots & \vdots & \ddots & \vdots & \vdots \\ 0 & 0 & 0 & 0 & 0 & 0 & \cdots & 0 & 0 \\ 0 & 0 & 0 & 0 & 0 & 0 & \cdots & -1 & 1 \\ 0 & 0 & 0 & 0 & 0 & 0 & \cdots & -\alpha & 0 \\ 0 & 0 & 0 & 0 & 0 & 0 & \cdots & 0 & -\alpha \end{pmatrix},$$

$$\mathrm{D} = \begin{pmatrix} 0 & 0 & 0 & 0 & 0 & 0 & \cdots & 0 & 0 \\ \alpha b & -\beta & 0 & \beta & 0 & 0 & \cdots & 0 & 0 \\ 0 & 0 & 0 & 0 & 0 & 0 & \cdots & 0 & 0 \\ 0 & 0 & \alpha b & -\beta & 0 & \beta & \cdots & 0 & 0 \\ \vdots & \vdots & \vdots & \vdots & \vdots & \vdots & \ddots & \vdots & \vdots \\ 0 & 0 & 0 & 0 & 0 & 0 & \cdots & \beta & 0 \\ 0 & 0 & 0 & 0 & 0 & 0 & \cdots & 0 & 0 \\ 0 & 0 & 0 & 0 & 0 & 0 & \cdots & -\beta & \beta \\ -\alpha b & \beta & -\alpha b & 0 & -\alpha b & 0 & \cdots & 0 & -\beta \end{pmatrix},$$

$$\mathrm{F}(\mathrm{y}(t-\tau)) = \left(0, \alpha \sum_{i=2}^{5} b_i r_1^i(t-\tau), 0, \alpha \sum_{i=2}^{5} b_i r_2^i(t-\tau), ..., \right.$$
$$\left. \alpha \sum_{i=2}^{5} b_i r_{n-1}^i(t-\tau), \alpha \sum_{i=2}^{5} b_i \Big(-\sum_{k=1}^{n-1} r_k(t-\tau)\Big)^i \right)^T. \tag{9.23}$$

In order to study the Hopf bifurcation at $\tau = \tau_c$, we perform the perturbation analysis on the reaction time delay by letting:

$$\tau = \tau_c + \varepsilon\tau_\varepsilon, \tag{9.24}$$

here τ_c denotes the critical value of the parameter and $\varepsilon\tau_\varepsilon$ denotes a small perturbation to the parameter near its critical value.

Before using the method of multiple scales to analyze the dynamical behaviour of the system, it is necessary to introduce the time scale as $T_i(t) = \varepsilon^i t$, $i = 0, 1, 2, 3, 4$, where ε is a dimensionless parameter much smaller than 1. Then the solution of Eq. (9.14) is written as $y(t) = Y(T_0, T_1, T_2, T_3, T_4)$ which can be further expanded in terms of powers of ε as:

$$Y(T_0, T_1, T_2, T_3, T_4) = \sum_{i=1}^{5} \varepsilon^i Y_i(T_0, T_1, T_2, T_3, T_4) + ..., \tag{9.25}$$

where $Y_i = (Y_{i,1}, Y_{i,2}, ..., Y_{i,2n-1})^T$. Accordingly,

$$\dot{y}(t) = \sum_{i=1}^{5} \varepsilon^i \frac{\partial Y_i(T_0, T_1, T_2, T_3, T_4)}{\partial T_{i-1}} + \cdots. \tag{9.26}$$

Now $\mathbf{y}(t-\tau) = \mathbf{Y}(T_0 - \tau, T_1 - \varepsilon\tau, T_2 - \varepsilon^2\tau, T_3 - \varepsilon^3\tau, T_4 - \varepsilon^4\tau)$ is expanded to Taylor series at $(T_0 - \tau_c, T_1, T_2, T_3, T_4)$ and the lower order terms of ε are retained. Then combining Eqs. (9.23)–(9.26), substituting all these expressions into Eq. (9.22) so as to balance the coefficient of the same power of ε, we obtain a series of equations. The equation for the lowest power of ε is

$$D_0\mathbf{Y}_1 - \mathbf{C}\mathbf{Y}_1 - \mathbf{D}\mathbf{Y}_{1\tau_c} = 0. \tag{9.27}$$

The general form of the equations of higher powers of ε is as follows

$$\begin{aligned}
D_0\mathbf{Y}_i - \mathbf{C}\mathbf{Y}_i - \mathbf{D}\mathbf{Y}_{i\tau_c} &= -\sum_{j=1}^{i-1} D_j\mathbf{Y}_{i-j} - \mathbf{D}\{(\tau_c D_1 + \tau_\varepsilon D_0)\mathbf{Y}_{i-1,\tau_c} + (\tau_c D_2 \\
&+ \tau_\varepsilon D_1 - \frac{(\tau_c D_1 + \tau_\varepsilon D_0)^2}{2})\mathbf{Y}_{i-2,\tau_c} + [\tau_c D_3 + \tau_\varepsilon D_2 - (\tau_c D_1 + \tau_\varepsilon D_0)(\tau_c D_2 \\
&+ \tau_\varepsilon D_1) + \frac{(\tau_c D_1 + \tau_\varepsilon D_0)^3}{6}]\mathbf{Y}_{i-2,\tau_c}\} + [\tau_c D_4 + \tau_\varepsilon D_3 + (\tau_c D_3 + \tau_\varepsilon D_2)(\tau_c D_2 \\
&+ \tau_\varepsilon D_1) - (\tau_c D_1 + \tau_\varepsilon D_0)^2 \tau_c D_2 + \tau_\varepsilon D_1) + \frac{(\tau_c D_1 + \tau_\varepsilon D_0)^4}{24}]\mathbf{Y}_{i-4,\tau_c} \\
&+ \mathbf{F}_i, \ i = 2, \ldots, 5,
\end{aligned} \tag{9.28}$$

where $D_i = \partial D/\partial T_i$ denotes the partial derivative with respect to the ith time scale, and the exact analytic representation of $\mathbf{F}_i (i = 2, 3, 4, 5)$ is given in Appendix 9.1. The definitions of $\mathbf{Y}_i$ and $\mathbf{Y}_{i,\tau_c}$ in Eq. (9.28) are as follows

$$\mathbf{Y}_i = \begin{cases} \mathbf{Y}_i(T_0, T_1, T_2, T_3, T_4), & i = 1, ..., 5, \\ 0 & else, \end{cases} ;$$

$$\mathbf{Y}_{i,\tau_c} = \begin{cases} \mathbf{Y}_{i,\tau_c}(T_0, T_1, T_2, T_3, T_4), & i = 1, ..., 5, \\ 0 & else. \end{cases}$$

The solution of Eq. (9.27) is assumed as

$$\mathbf{Y}_1 = \mathbf{A}_1 \sin(\omega_c T_0) + \mathbf{B}_1 \cos(\omega_c T_0), \tag{9.29}$$

here $\mathbf{A}_1 = \mathbf{A}_1(T_1, T_2, T_3, T_4), \mathbf{B}_1 = \mathbf{B}_1(T_1, T_2, T_3, T_4)$. Substitute (9.29) into Eq. (9.27) so that Eq. (9.27) becomes an algebraic equation as follows

$$\begin{pmatrix} (s\mathbf{D} + \omega\mathbf{I})^T & -(\mathbf{C} + c\mathbf{D})^T \\ -(\mathbf{C} + c\mathbf{D})^T & -(s\mathbf{D} + \omega\mathbf{I})^T \end{pmatrix} \begin{pmatrix} \mathbf{A}_1 \\ \mathbf{B}_1 \end{pmatrix} = \mathbf{0}, \tag{9.30}$$

where $c = \cos(\omega_c T_c), s = \sin(\omega_c T_c)$ is known constants. Solving Eq. (9.30), one can observe that all the undetermined coefficients in (9.29) can be expressed in terms of $A_{1,1}$ and $B_{1,1}$, the first components of $\mathbf{A}_1$ and $\mathbf{B}_1$, respectively, namely

$$A_{1,i} = \alpha_{1,i}(A_{1,1}, B_{1,1}),\ B_{1,i} = \beta_{1,i}(A_{1,1}, B_{1,1}), i = 2, 3, \ldots, 2n - 1, \tag{9.31}$$

here $A_{1,i}, B_{1,i}$ denotes the ith component of $\mathbf{A}_1, \mathbf{B}_1$, respectively; $\alpha_{1,i}$ denotes the expression of $A_{1,i}$ with respect to $A_{1,1}, B_{1,1}$, and $\beta_{1,i}$ denotes the expression of $B_{1,i}$ with respect to $A_{1,1}, B_{1,1}$.

Noting that $\alpha_{1,i}$ and $\beta_{1,i}$ are linear functions in their variables, we have

$$D_1 A_{1,i} = \alpha_{1,i}(D_1 A_{1,1}, D_1 B_{1,1}),\ D_1 B_{1,i} = \beta_{1,i}(D_1 A_{1,1}, D_1 B_{1,1}), \\ i = 2, 3, \ldots, 2n - 1. \tag{9.32}$$

It follows from Eq. (9.32) that $D_1 A_{1,i}, D_1 B_{1,i}$ can be calculated in terms of the two quantities $D_1 A_{1,1}, D_1 B_{1,1}$. Then the solution of Eq. (9.28) has the following form

$$\mathbf{Y}_i = \mathbf{Q}_i(T_1, T_2, T_3, T_4) + \sum_{k=1}^{i} \mathbf{A}_i^{(k)}(T_1, T_2, T_3, T_4) \sin(k\omega_c T_0) \\ + \mathbf{B}_i^{(k)}(T_1, T_2, T_3, T_4) \cos(k\omega_c T_0), \tag{9.33}$$

where $\mathbf{Y}_2$ is given as an example,

$$\begin{aligned}\mathbf{Y}_2 &= \mathbf{Q}_2(T_1, T_2, T_3, T_4) + \mathbf{A}_2^{(1)}(T_1, T_2, T_3, T_4)\sin(\omega_c T_0)\\ &+ \mathbf{B}_2^{(1)}(T_1, T_2, T_3, T_4)\cos(\omega_c T_0)\\ &+ \mathbf{A}_2^{(2)}(T_1, T_2, T_3, T_4)\sin(2\omega_c T_0) + \mathbf{B}_2^{(2)}(T_1, T_2, T_3, T_4)\cos(2\omega_c T_0).\end{aligned}$$

Note that the existence of the term $\mathbf{Q}_i$ in the solution (9.33) is a consequence of the existence of the square nonlinearity in Eq. (9.28).

Starting with $i = 2$, we substitute $\mathbf{Y}_i$ from Eq. (9.33) into Eq. (9.28) and eliminate the secular terms of $\sin(\omega_c T_0)$ and $\cos(\omega_c T_0)$ contained in the equation. This computational process is repeated in an iterative manner until $\mathbf{Y}_5$, the solution of the equation at the highest power of ε under our truncation is calculated. The solvability conditions (i.e., the system of equations obtained by eliminating the secular terms) obtained during this computation process can be expressed by the following algebraic equation,

$$\begin{pmatrix} (s\mathbf{D} + \omega\mathbf{I})^T & -(\mathbf{C} + c\mathbf{D})^T \\ -(\mathbf{C} + c\mathbf{D})^T & -(s\mathbf{D} + \omega\mathbf{I})^T \end{pmatrix} \begin{pmatrix} \mathbf{A}_i^{(1)} \\ \mathbf{B}_i^{(1)} \end{pmatrix} = \mathbf{U}_i, \tag{9.34}$$

where $\mathbf{I}$ denotes the identity matrix of order $2n - 1$; $\mathbf{A}_i^{(1)}$, $\mathbf{B}_i^{(1)}$ denotes the partial derivatives of $\mathbf{A}_i(T_1, T_2, T_3, T_4)$, $\mathbf{B_i}(T_1, T_2, T_3, T_4)$ with respect to the scale T_1, respectively. The right hand term $\mathbf{U}_i$ is expressed as follows

$$\begin{aligned}\mathbf{U}_i = \mathbf{D}\sum_{j=1}^{2}\Bigg[&\begin{pmatrix} -S_{6j-5} & S_{6j-4} \\ -S_{6j-4} & -S_{6j-5} \end{pmatrix}\begin{pmatrix} s\mathbf{B}_{i+2-3j}^{(1)} + c\mathbf{A}_{i+2-3j}^{(1)} \\ s\mathbf{A}_{i+2-3j}^{(1)} + c\mathbf{B}_{i+2-3j}^{(1)} \end{pmatrix} - \begin{pmatrix} S_{2j+1} & S_{2j+2} \\ S_{2j+2} & -S_{2j+1} \end{pmatrix}\\ &\begin{pmatrix} s\mathbf{A}_{i-j-1}^{(1)} + c\mathbf{B}_{i-j-1}^{(1)} \\ s\mathbf{B}_{i-j-1}^{(1)} + c\mathbf{A}_{i-j-1}^{(1)} \end{pmatrix}\Bigg] - \sum_{j=1}^{i-2}(\tau_c D_{j+1} + \tau_\varepsilon D_j)\begin{pmatrix} s\mathbf{A}_{i-j-1}^{(1)} + c\mathbf{B}_{i-j-1}^{(1)} \\ s\mathbf{B}_{i-j-1}^{(1)} + c\mathbf{A}_{i-j-1}^{(1)} \end{pmatrix}\\ &-\sum_{j=1}^{i-1} D_j \begin{pmatrix} \mathbf{B}_{i-j}^{(1)} \\ \mathbf{A}_{i-j}^{(1)} \end{pmatrix} + \begin{pmatrix} \mathbf{F}_i^c \\ \mathbf{F}_i^s \end{pmatrix}, i = 2, \ldots, 5,\end{aligned} \tag{9.35}$$

where $\mathbf{F}_i^c$, $\mathbf{F}_i^s$ denotes the coefficient of $\sin(\omega_c T_0)$ and $\cos(\omega_c T_0)$ in $\mathbf{F}_i$, respectively, whose expression is given in Appendix 9.1. Here S_j denotes the jth component of $\mathbf{S}$. The specific expression of $\mathbf{S}$ is as follows

$$\begin{aligned}\mathbf{S} = \Bigg(&\varepsilon\tau_\varepsilon, \tau_c D_1, \frac{\tau_c^2 D_1^2 - \varepsilon^2\tau_\varepsilon^2}{2}, -\varepsilon\tau_\varepsilon\tau_c D_1, \tau_c D_1 \frac{6(\tau_c D_2 + \tau_\varepsilon D_1) - \tau_c^2 D_1^2 + 3\varepsilon^2\tau_\varepsilon^2}{6},\\ &-\frac{\varepsilon\tau_\varepsilon(\varepsilon^2\tau_\varepsilon^2 - 3\tau_c^2 D_1^2)}{6} - \varepsilon\tau_\varepsilon(\tau_c D_2 + \tau_\varepsilon D_1),\\ &\varepsilon\tau_\varepsilon(\tau_c D_3 + \tau_\varepsilon D_2 - 2\tau_c D_1(\tau_c D_2 + \tau_\varepsilon D_1)\end{aligned}$$

$$+\frac{\tau_c D_1(\tau_c^2 D_1^2-\varepsilon^2\tau_\varepsilon^2)}{6}), \tau_c D_1(\tau_c D_3+\tau_\varepsilon D_2)+(\varepsilon^2\tau_\varepsilon^2-\tau_c^2 D_1^2)(\tau_c D_2+\tau_\varepsilon D_1)$$
$$+\frac{\varepsilon^4\tau_\varepsilon^4-6\varepsilon^2\tau_\varepsilon^2\tau_c^2 D_1^2+\tau_c^4 D_1^4}{24})^T.$$

Let us summarize the process of solving Eq. (9.28). First, we need to solve the algebraic Eq. (9.34), whose solution provides the information of the solvability condition of Eq. (9.28). With this solvability condition, the solution of the Eq. (9.28) can be found. Therefore, the main goal becomes to solve the algebraic equation. According to Fredholm Alternative Theorem [38], it follows that Eq. (9.34) has a solution if and only if the right hand term $\mathbf{U}_i$ of (9.34) is orthogonal to each solution of the conjugate homogeneous equation (adjoint equation) of Eq. (9.34). The homogeneous conjugate equation of Eq. (9.34) is as follows

$$\begin{pmatrix} (s\mathbf{D}+\omega\mathbf{I})^T & -(\mathbf{C}+c\mathbf{D})^T \\ -(\mathbf{C}+c\mathbf{D})^T & -(s\mathbf{D}+\omega\mathbf{I})^T \end{pmatrix}\mathbf{X}=0. \tag{9.36}$$

Using the orthogonality of $\mathbf{U}_i$ and $\mathbf{X}$, the following equation can be obtained

$$\mathbf{U}_i^T\mathbf{X}=0, i=2,3,4,5. \tag{9.37}$$

Starting from $i=2$ until working up until $i=5$, we solve Eq. (9.37) and finally obtain

$$D_i A_{1,1}=M_i(A_{1,1},B_{1,1}),\ \ D_i B_{11}=N_i(A_{1,1},B_{1,1}), i=1,2,3,4. \tag{9.38}$$

Based on Eq. (9.38), we get

$$\begin{cases} \dot{A}_{1,1}(t)=\varepsilon D_1 A_{1,1}+\varepsilon^2 D_2 A_{1,1}+\varepsilon^3 D_3 A_{1,1}+\varepsilon^4 D_4 A_{1,1}, \\ \dot{B}_{1,1}(t)=\varepsilon D_1 B_{1,1}+\varepsilon^2 D_2 B_{1,1}+\varepsilon^3 D_3 B_{1,1}+\varepsilon^4 D_4 B_{1,1}, \end{cases} \tag{9.39}$$

where "$\cdot$" denotes the derivative with respect to t.

The following transformations are introduced to simplify Eq. (9.39)

$$A_{1,1}(t)=R(t)\cos(\varphi(t)),\ B_{1,1}(t)=R(t)\sin(\varphi(t)), \tag{9.40}$$

and Eq. (9.39) can be transformed to

$$\begin{cases} \dot{R}(t)=f_1(\varepsilon,\tau_\varepsilon)R(t)+f_3(\varepsilon,\tau_\varepsilon)R(t)^3+f_5(\varepsilon,\tau_\varepsilon)R(t)^5, \\ \dot{\varphi}(t)=g_0(\varepsilon,\tau_\varepsilon)+g_2(\varepsilon,\tau_\varepsilon)R(t)^2+g_4(\varepsilon,\tau_\varepsilon)R(t)^4. \end{cases} \tag{9.41}$$

Now the problem of studying the bifurcation direction and stability of the bifurcated periodic solution of Eq. (9.14) is transformed into studying the stability of the equilibrium of Eq. (9.41). It can be shown that when $f_3(\varepsilon,\tau_\varepsilon)<0$, then the

Hopf bifurcation of Eq. (9.14) is a supercritical one; when $f_3(\varepsilon, \tau_\varepsilon) > 0$, the Hopf bifurcation is a subcritical one.

With the aid of Eq. (9.41) obtained by the method of multiple scales, we can easily determine the bifurcation direction and stability of the bifurcated periodic solution of Eq. (9.14). Together with the visualization tools such as time history plot, spatiotemporal pattern, and density plot of vehicle distribution in the congested area, we are able to investigate various traffic patterns induced by drivers' reaction time delay. In the following discussions, we will provide two examples to better explain this procedure.

9.2.3 *Examples*

9.2.3.1 Example: 3 Vehicles

Example 1 is provided to demonstrate the agreement between the theoretical prediction by the method of multiple scales and the numerical continuation resutls, thereby showing the viability of the theoretical analysis. Furthermore, we provide time history plots of the bifurcated solution caused by subcritical Hopf bifurcation, allowing us to observe the traffic patterns associated with this bistable scenario. Besides, we make use of the concept of attraction domain to quantitatively demonstrate the distribution of various traffic patterns.

Let $n = 3$, $\alpha = 0.41$/s, and $\beta = 0.5$/s. From (9.18), we have $\tau_c = 0.3382$ s and $\omega_c = 1.3275$. The uniform flow of Eq. (9.6) is obtained as

$$\mathbf{Y}^* = \{16\,\text{m}, 19.041\,\text{m/s}, 16\,\text{m}, 19.041\,\text{m/s}, 19.041\,\text{m/s}\}^T. \tag{9.42}$$

Based on the method of multiple scales introduced in last subsection, we obtain the following equation

$$\begin{cases} \epsilon \dot{R}(t) = \left(1.1272\epsilon\tau_\epsilon - 0.5351\epsilon^2\tau_\epsilon^2 - 2.1945\epsilon^3\tau_\epsilon^3 + 3.6683\epsilon^4\tau_\epsilon^4\right)\epsilon R[t] \\ \qquad + \left(0.0027 - 0.0022\epsilon\tau_\epsilon - 0.0145\epsilon^2\tau_\epsilon^2 + 0.0007\epsilon^6\tau_\epsilon^6\right)(\epsilon R[t])^3 \\ \qquad - 0.00008(\epsilon R[t])^5, \\ \epsilon\dot{\varphi}(t) = \left(-0.1801\epsilon\tau_\epsilon - 1.4729\epsilon^2\tau_\epsilon^2 + 1.4473\epsilon^3\tau_\epsilon^3 + 3.2565\epsilon^4\tau_\epsilon^4\right) \\ \qquad + \left(0.0026 - 0.0076\epsilon\tau_\epsilon - 0.0031\epsilon^2\tau_\epsilon^2 + 0.0002\epsilon^6\tau_\epsilon^6\right)(\epsilon R[t])^2 \\ \qquad - 0.00003(\epsilon R[t])^4. \end{cases} \tag{9.43}$$

By equating $\varepsilon\dot{R}(t)$ to zero, we obtain the following theoretical solutions for the stable and unstable non-zero equilibria of Eq. (9.43)

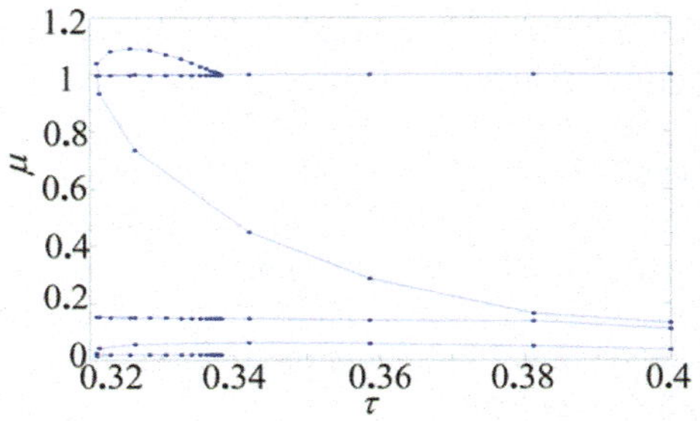

Fig. 9.11 Distribution of Floquet multipliers with the variation of τ

$$\varepsilon R_{st} = \begin{cases} 0.5\sqrt{-2\mathrm{E}_1 + 2.\sqrt{-49883.9\mathrm{E}_2\epsilon\tau_\epsilon + \mathrm{E}_1^2}}, \epsilon\tau_\epsilon \geq 0.3183, \\ 0.5\sqrt{-2E_1 - 2.\sqrt{-49883.9\mathrm{E}_2\epsilon\tau_\epsilon + \mathrm{E}_1^2}}, 0.3183 \leq \epsilon\tau_\epsilon \leq 0.3383. \end{cases}$$
$$\mathrm{E}_1 = -33.156 + 28.026\epsilon\tau_\epsilon + 180.56\epsilon^2\tau_\epsilon^2 - 8.5\epsilon^6\tau_\epsilon^6,$$
$$\mathrm{E}_2 = -1.127 + 0.535\epsilon\tau_\epsilon + 2.194\epsilon^2\tau_\epsilon^2 - 3.668\epsilon^3\tau_\epsilon^3. \tag{9.44}$$

By analyzing the eigenvalues of linearized system around the equilibrium, we can conclude that the nontrivial equilibrium εR_{st} is unstable when $0.3183 \leq \epsilon\tau_\epsilon \leq 0.3383$; when $\epsilon\tau_\epsilon \geq 0.3183$, the nontrivial equilibrium εR_{st} is stable.

The above theoretical results are verified by numerical methods. Figure 9.11 shows the distribution of Floquet multipliers of the bifurcated periodic solution when the time delay τ changes, by using the numerical continuation software package DDE-BIFTOOL. According to the theory of Floquet multipliers [39, 40], it can be seen that the interval of the delay on which unstable periodic solution exists is (0.3183, 0.3382) and the stable periodic solution exists for $\tau > 0.3382$. This is consistent with theoretical predictions.

Substituting (9.44) into the second equation of Eq. (9.43) yields the expression of $\varepsilon\varphi(t)$. This allows us to determine the approximation of the stable periodic solution of Eq. (9.22) as follows

$$\mathbf{y}(t) = (\varepsilon R)^2\mathbf{Y}_0 + \varepsilon R\sin(\Psi(t))\mathbf{Y}_{11} + \varepsilon R\cos(\Psi(t))\mathbf{Y}_{12} + (\varepsilon R)^2(\sin(2\Psi(t))$$
$$\mathbf{Y}_{21} + \cos(2\Psi(t))\mathbf{Y}_{22}) + +(\varepsilon R)^3(\sin(3\Psi(t))\mathbf{Y}_{31} + \cos(3\Psi(t))\mathbf{Y}_{32}), \tag{9.45}$$

where

$$\mathbf{Y}_0 = (-0.1173 + 0.0022(\varepsilon R[t])^2)\{0, 1, 0, 1, 1\}^T,$$
$$\mathbf{Y}_{11} = \{1, 0.38 - 0.62\varepsilon\tau_\varepsilon - 0.16(\varepsilon\tau_\varepsilon)^2 + 1.52(\varepsilon\tau_\varepsilon)^3, -0.5, 0.38 + 0.51\varepsilon\tau_\varepsilon$$
$$- 0.7\varepsilon\tau_\varepsilon - 0.68(\varepsilon\tau_\varepsilon)^3, -0.77 + 0.1\varepsilon\tau_\varepsilon + 0.85(\varepsilon\tau_\varepsilon)^2 - 0.84(\varepsilon\tau_\varepsilon)^3\}^T,$$
$$\mathbf{Y}_{12} = \{0, -0.66 - 0.24\varepsilon\tau_\varepsilon + 0.89(\varepsilon\tau_\varepsilon)^2 - 0.09(\varepsilon\tau_\varepsilon)^3, 0.87, 0.66 - 0.42\varepsilon\tau_\varepsilon$$

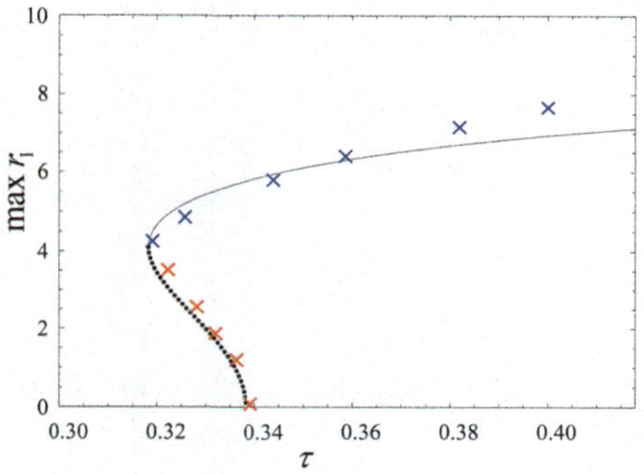

Fig. 9.12 Comparison of results obtained by the method of multiple scales and numerical continuation

$$
\begin{aligned}
&- 0.58(\varepsilon\tau_\varepsilon)^2 + 1.36(\varepsilon\tau_\varepsilon)^3, 0.65\varepsilon\tau_\varepsilon - 1.27(\varepsilon\tau_\varepsilon)^3\}^T,\\
\mathbf{Y}_{21} &= \{0.02\varepsilon\tau_\varepsilon + 0.01(\varepsilon\tau_\varepsilon)^2, 0.02 - 0.02\varepsilon\tau_\varepsilon - 0.09(\varepsilon\tau_\varepsilon)^2, 0.01,\\
&\quad + \epsilon\tau_\epsilon(-0.01 - 0.03\epsilon\tau_\epsilon) - 0.02 + \epsilon\tau_\epsilon(-0.03 + 0.09\epsilon\tau_\epsilon), 0.05\tau_\epsilon\epsilon\}^T,\\
\mathbf{Y}_{22} &= \{0.01 + \epsilon\tau_\epsilon(-0.03\epsilon\tau_\epsilon), -0.01 + \epsilon\tau_\epsilon(-0.04 + 0.05\epsilon\tau_\epsilon), -0.01 + \epsilon\tau_\epsilon\\
&\quad (-0.02 + 0.01\epsilon\tau_\epsilon), -0.01 + \epsilon\tau_\epsilon(0.04 + 0.05\epsilon\tau_\epsilon), 0.02 + \epsilon\tau_\epsilon(-0.1\epsilon\tau_\epsilon)\}^T,\\
\mathbf{Y}_{31} &= (0.0006 + 0.0003\epsilon\tau_\epsilon)\{0, 1, 0, 1, 1\},\\
\mathbf{Y}_{32} &= (0.0002 - 0.0024\epsilon\tau_\epsilon)\{0, 1, 0, 1, 0, 1\}.
\end{aligned}
\tag{9.46}
$$

Based on Eq. (9.46), the bifurcation diagram can be plotted as shown in Fig. 9.12, where with the vertical axis represents the amplitude of velocity of the first vehicle and the horizontal axis represents the reaction time delay τ. The theoretical predictions of stable and unstable periodic solutions are represented by the black solid and dashed curves, respectively while the stable periodic solution is represented by the black solid line. Numerical continuation results of the stable and unstable periodic solutions are represented by the blue and "$\times$" symbols, respectively. The comparison further demonstrates the effectiveness of the method of multiple scales for studying the Hopf bifurcation of Eq. (9.22) when the delay is close to its critical value.

From Fig. 9.12, one can observe the bistable phenomenon when $\tau \in (0.3211, 0.3382)$. In the following, we choose $\tau = 0.33$, and perform numerical simulations with different initial values. We find that change in initial conditions may result in different stationary solutions. For instance, the evolution of traffic flow ends up with almost jammed mode in Fig. 9.13a$_1$, while the traffic eventually evolves into a uniform flow mode in Fig. 9.13b$_1$.

9.2.3.2 Example: 33 Vehicles

For this example, we choose $n = 33$ and the other parameters are fixed as in (9.20). With such parameters, the theoretical calculation by means of the method of multiple scales provides

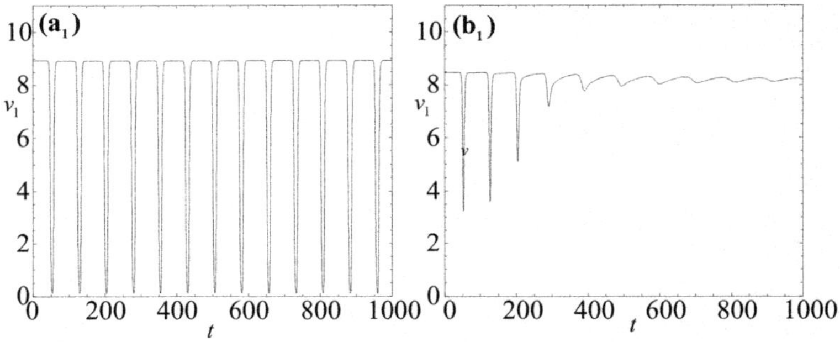

Fig. 9.13 Time history plot of the velocity of the first vehicle when $\tau = 0.33$ for different initial conditions

$$\begin{cases} \epsilon \dot{R}(t) = \left(0.0035\epsilon\tau_\epsilon - 0.000068\epsilon^2\tau_\epsilon^2 - 0.00003\epsilon^3\tau_\epsilon^3\right)\epsilon R[t] \\ \qquad + (0.001 + 0.0544\epsilon\tau_\epsilon)(\epsilon R[t])^3 - 0.01859(\epsilon R[t])^5, \\ \epsilon\dot{\varphi}(t) = -0.01 - 0.008\epsilon\tau_\epsilon(\epsilon R[t])^2 + 0.0156(\epsilon R[t])^4. \end{cases} \tag{9.47}$$

From Eq. (9.47), we figure that $f_3(\varepsilon, \tau_\varepsilon) > 0$, implying that the bifurcation of Eq. (9.14) is subcritical.

In order to demonstrate the bistable phenomenon induced by the subcritical Hopf bifurcation, we choose the following initial conditions: all vehicles are in a uniform flow mode, and only one of them have its velocity decreased by v_{per} and its inter-vehicle distance increased by h_{per}. i.e., let this vehicle have a deceleration $a_{per} < 0$ in the time period of T_{per} so that $v_{per} = |a_{per}|T_{per}$, $h_{per} = (|a_{per}|^2 T_{per}^2)/2$. In Fig. 9.14a, we set $a_{per} = -0.125$ and $T_{per} = 4$, and consequently $v_{per} = 0.5$ and $h_{per} = 1$, then it can be observed that such perturbation to the initial state eventually decays and the traffic flow eventually returns to the uniform flow mode. In Fig. 9.14b, we let $a_{per} = -2$ and $T_{per} = 4$, implying $v_{per} = 8$ and $h_{per} = 16$, then it can be observed that this perturbation is eventually amplified, and the traffic flow eventually evolves toward the almost jammed mode.

Figure 9.15 shows the spatiotemporal pattern of each vehicle's position at each instant under various initial conditions. The black curve represents the first vehicle's displacement. It is clear from Fig. 9.15a that each vehicle moves in a uniform flow mode because there are not any curved lines in the spatiotemporal pattern. In Fig. 9.15b, the sparsely and densely distributed and curved lines in the spatiotemporal pattern indicate that the vehicle's velocity is changing as it travels. There is also a section of almost flat segment at the bend of the line, showing that the car is moving at velocity of almost zero during that time, which is referred to as an almost jammed mode. As can be seen in Fig. 9.15b, the vehicle will enter the almost jammed roadway about once or twice in one cycle.

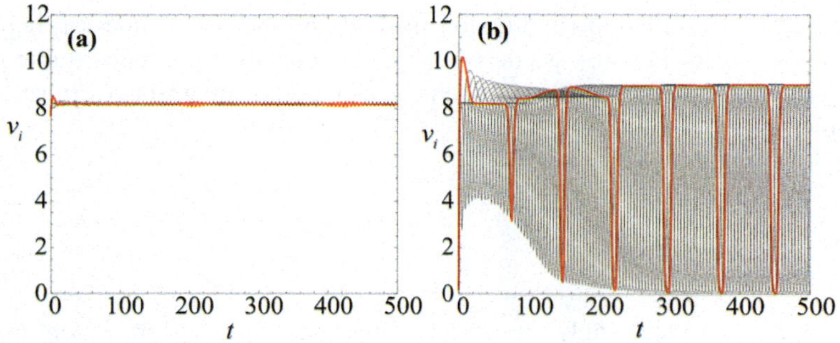

Fig. 9.14 Time history plots of the velocity of each vehicle in the region of bistability for different initial conditions for **a** $a_{br} = -0.125$ and **b** $a_{br} = -2$, with the red curve representing the time history of the velocity of the first vehicle

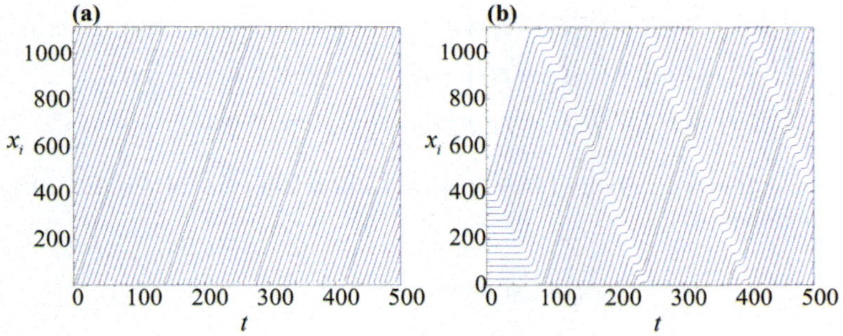

Fig. 9.15 Spatiotemporal patterns with two initial conditions

9.3 Control of Traffic Jam by Time-Varying Delay

In this section, we aim to explore how to suppress the traffic jam and maintain the stability of uniform flow mode. With the background of the current adaptive cruise control system, we continue to employ the time-delay FVD model. In this section, however, the driver's reaction time delay originates from the perception, decision-making, and actuation of the cruise equipment. Our control strategy involves imposing a time-varying time delay feedback signal to enlarge the uniform flow's stable region, and ultimately suppress the traffic jam.

Currently, there is no existing research on utilizing time-varying delay for traffic jam control. Sipahi [41] and Niculescu [42] have noted that time-varying delay in the car-following models impacts the stability of the uniform traffic flow. This insight motivates our research in this section, where we attempt to suppress the traffic jam induced by the loss of stability of the uniform flow, by using variable time delay. Shampint [43] provided a methodology on using the software package ddesd to carry out numerical simulations of differential equations with variable delays.

Enlightened by oscillation suppression methods by means of time-varying delay from other systems [44–46], we derive the conditions for suppressing traffic jam in a car-following system composed of vehicles equipped with adaptive cruise control equipment.

9.3.1 Model Setup

In this section, we still study the model represented by Eq. (9.6), except that this section is based on the implementation of an adaptive cruise control system. The driver reaction time delay represents the reaction time of the device during the process of perception, decision making, and finally actuation. We consider the time delay car-following model consisting of three cars as follows

$$\begin{cases} \dot{v}_1(t) = \alpha\ [\ V(h_1(t-\tau_1)) - v_1(t-\tau_2)] + \beta\ \dot{h}_1(t-\tau_3), \\ \dot{v}_2(t) = \alpha\ [\ V(h_2(t-\tau_1)) - v_2(t-\tau_2)] + \beta\ \dot{h}_2(t-\tau_3), \\ \dot{v}_3(t) = \alpha\ [\ V(h_3(t-\tau_1)) - v_3(t-\tau_2)] + \beta\ \dot{h}_3(t-\tau_3), \end{cases} \tag{9.48}$$

where the optimized velocity function $V(h_i)$ still retains the form in Eq. (9.7).

Assume that [47]

$$\tau_1 = \tau_2 = \tau_3 > 0. \tag{9.49}$$

In other words, the adaptive cruise control device has the same reaction time in processing the distance and velocity difference between the neighboring vehicles, and the vehicle's own velocity, which is different from what we assume for human drivers. With this assumption and the vehicle distance condition $h_n(t) = L - h_1(t) - h_2(t)$, Eq. (9.48) is now written as

$$\begin{cases} \dot{h}_1(t) = v_2(t) - v_1(t), \\ \dot{v}_1(t) = \alpha\ [\ V(h_1(t-\tau)) - v_1(t-\tau)] + \beta\ \dot{h}_1(t-\tau), \\ \dot{h}_2(t) = v_3(t) - v_2(t), \\ \dot{v}_2(t) = \alpha\ [\ V(h_2(t-\tau)) - v_2(t-\tau)] + \beta\ \dot{h}_2(t-\tau) \\ \dot{v}_3(t) = \alpha\ (V(L - h_1(t-\tau) - h_2(t-\tau)) - v_n(t)) + \beta\ (v_1(t-\tau) - v_3(t-\tau)). \end{cases} \tag{9.50}$$

The values of parameters of the three vehicles in the time delay FVD model are assumed as

$$\alpha = 0.9/\text{s},\ \beta = 0.1/\text{s},\ v^0 = 11\ \text{m/s},\ h_{stop} = 14\ \text{m}. \tag{9.51}$$

Under the operation in previous sections, the nontrivial equilibrium of Eq. (9.51) can be translated to the trivial equilibrium, and consequently Eq. (9.50) is transformed to

$$\begin{cases} \dot{r}_1(t) = u_2(t) - u_1(t), \\ \dot{u}_1(t) = \alpha\,[\,V(r_1(t-\tau)+h^*) - u_1(t-\tau)] + \beta\,\dot{r}_1(t-\tau) - \alpha\,v^*, \\ \dot{r}_2(t) = u_3(t) - u_2(t), \\ \dot{u}_2(t) = \alpha\,[\,V(r_2(t-\tau)+h^*) - u_2(t-\tau)] + \beta\,\dot{r}_2(t-\tau) - \alpha\,v^*, \\ \qquad \dot{u}_3(t) = \alpha\,(V(h^* - r_1(t-\tau) - r_2(t-\tau)) \\ \quad -u_3(t)) + \beta\,(u_1(t-\tau) - u_3(t-\tau)) - \alpha\,v^*. \end{cases} \tag{9.52}$$

In the next subsection, we will analyse the stability of the equilibrium of Eq. (9.52), which actually represents the stability of the uniform flow solution of Eq. (9.50), as well as the pattern of traffic jam that results from instability of the uniform flow.

9.3.2 *Traffic Jam Mode Under Constant Delay*

9.3.2.1 Stability of Uniform Flow

In this part, we analyze the stability of the uniform flow solution of Eq. (9.50). Since the uniform flow of Eq. (9.50) corresponds to the trivial solution of Eq. (9.52), we discuss the stability of the trivial equilibrium of Eq. (9.52) to obtain the stability of the uniform flow solution.

Firstly, the nonlinear optimized velocity function $V(h_i)$ that appears on the right hand side of Eq. (9.50) is expanded to Taylor series at $h = h^*$, namely

$$V(h) = v^* + b_1(h-h^*) + b_2(h-h^*)^2 + b_3(h-h^*)^3 + \cdots, \tag{9.53}$$

where

$$v^* = V(h^*), b_1 = V'(h^*), b_2 = \frac{V''(h^*)}{2}, b_3 = \frac{V'''(h^*)}{6}. \tag{9.54}$$

Substituting Eq. (9.53) into Eq. (9.52) yields

$$\begin{cases} \dot{r}_1(t) = u_2(t) - u_1(t), \\ \dot{u}_1(t) = \alpha \left[\sum_{i=1}^{3} b_i r_1^i(t-\tau) - u_1(t-\tau) \right] + \beta\, \dot{r}_1(t-\tau), \\ \dot{r}_2(t) = u_3(t) - u_2(t), \\ \dot{u}_2(t) = \alpha \left[\sum_{i=1}^{3} b_i r_2^i(t-\tau) - u_2(t-\tau) \right] + \beta\, \dot{r}_2(t-\tau), \\ \quad \dot{u}_3(t) = \alpha \left(\sum_{i=1}^{3} b_i(-r_1(t-\tau) - r_2(t-\tau))^i - u_3(t) \right) \\ \qquad\qquad +\beta\, (u_1(t-\tau) - u_3(t-\tau)). \end{cases} \tag{9.55}$$

The characteristic equation is derived out of the linearization of Eq. (9.55), as follows

$$(\lambda + \alpha e^{-\lambda\tau}) \prod_{k=1}^{2} ((\lambda^2 + e^{-\lambda\tau}(b_1\alpha + (\alpha+\beta)\lambda)) - e^{-\lambda\tau + \frac{i(2k\pi)}{3}}(b_1\alpha + \beta\lambda)) = 0. \tag{9.56}$$

To obtain the Hopf bifurcation curve in the parameter plane consisting of (h^*, τ), substituting $\lambda = i\omega$ into the characteristic Eq. (9.56) and separating the real and imaginary parts yield

$$\begin{cases} \sin(\tau\omega) = \dfrac{\omega^2(\omega(\alpha+\beta-\beta\cos\frac{k\pi}{3}) - \alpha b_1 \sin\frac{k\pi}{3})}{\omega^2\left((\alpha+\beta)^2 - 2\beta(\alpha+\beta)\cos\dfrac{k\pi}{3}\right) + 2\alpha^2 b_1\left(-\omega\sin\dfrac{k\pi}{3} + b_1(1-\cos\dfrac{k\pi}{3})\right)} \\ \cos(\tau\omega) = \dfrac{\omega^2(\beta\omega\sin\frac{k\pi}{3} + 2\alpha b_1 \sin\frac{k\pi}{3})}{\omega^2\left((\alpha+\beta)^2 - 2\beta(\alpha+\beta)\cos\dfrac{k\pi}{3}\right) + 2\alpha^2 b_1\left(-\omega\sin\dfrac{k\pi}{3} + b_1(1-\cos\dfrac{k\pi}{3})\right)} \end{cases}, k = 1, 2, 3. \tag{9.57}$$

Utilizing the trigonometric identity yields

$$\begin{cases} b_1 = \dfrac{\omega(2\alpha\cos\frac{k\pi}{3} - \sqrt{2}\omega P)}{4\alpha\sin\frac{k\pi}{3}}, \\ \tau = \dfrac{1}{\omega}\arccos\dfrac{(\alpha+2\beta)\sin\frac{2k\pi}{3} - \sqrt{2}P}{\omega}, \end{cases} \quad k = 1, 2, \tag{9.58}$$

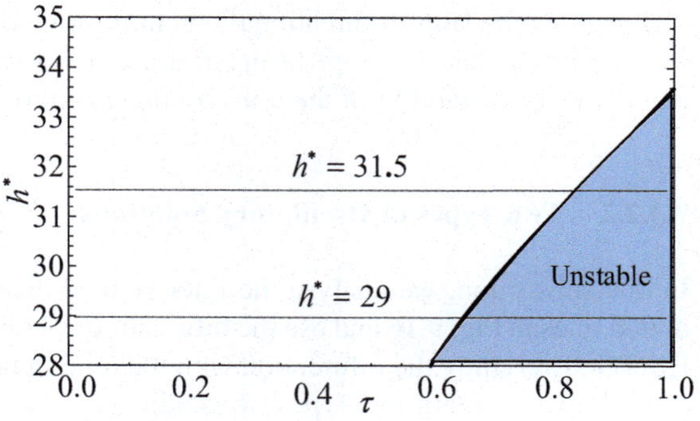

Fig. 9.16 Stability boundary lines in the parameter plane (h^*, τ)

where $P = \sqrt{\omega^2 - 2(\alpha + 2\beta)^2 \sin^2 \frac{k\pi}{3}}$.

Here (b_1, τ) are both functions of ω, so the stability switch boundary curve in the plane of (h^*, τ) can be obtained as shown in Fig. 9.16. Since the curve for $k = 1$ marks the initial loss of stability of the uniform flow, which may correspond to the Hopf bifurcation curve that we aim to discover, we only present this curve in Fig. 9.16.

The stability boundary curve of the uniform flow in the parameter plane (h^*, τ) is shown in Fig. 9.16, where the blue shaded region indicates the region for which the uniform flow is stable. The two dashed lines indicate vehicle distances $h^* = 31.5$ and $h^* = 29$, respectively. It is found that the uniform flow is gradually destabilized as the reaction time delay increases, while it becomes stable again as the vehicle distance increases. This suggests that the stability of the uniform flow can be maintained either by reducing the reaction time delay or increasing the length of the road. However, when neither of these can be realized, variable time delay control can be used to guarantee the stability of the uniform flow. By choosing the reaction delay τ as the bifurcation parameter and fixing the vehicle distance, the transversality condition for the occurrence of the Hopf bifurcation with respect to τ can be derived as follows

$$
\begin{aligned}
&\mathrm{Re}(\frac{d\lambda}{d\tau})\Big|_{\tau_c} \\
&= \frac{1}{2\omega_c^2} \frac{\left(d\omega_c^2 + \alpha^2 b_1\left(-\sqrt{3}\omega_c + 3b_1\right)\right)\left(4d\omega_c^2 - 2\omega_c^4 + \alpha^2 b_1\left(-5\sqrt{3}\omega_c + 18b_1\right)\right)}{d\omega_c^2(1+\tau_c^2\omega_c^2) + \alpha b_1\left(-\omega_c\left(6\beta\tau_c\omega_c + \alpha\left(2\sqrt{3} + \tau_c\omega_c\left(3 + \sqrt{3}\tau_c\omega_c\right)\right)\right) + 3\alpha(4+\tau_c^2\omega_c^2)b_1\right)}.
\end{aligned}
\tag{9.59}
$$

Substituting (9.51) and $h^* = 31.5$, $h^* = 29$ into Eq. (9.59) yields

$$
\mathrm{Re}(\frac{d\lambda}{d\tau})\Big|_{\tau_c} = 0.806 > 0, \mathrm{Re}(\frac{d\lambda}{d\tau})\Big|_{\tau_c} = 0.814 > 0. \tag{9.60}
$$

Based on the Hopf bifurcation theorem for time delay differential equations [48, 49], we know that the Hopf bifurcation occurs when $\tau = \tau_c$, and the interval on which the local stability of the uniform flow solution can be guaranteed is $[0, \tau_c)$.

9.3.2.2 Two Types of Oscillatory Solutions

In this subsection, we analyse the inter-vehicle distance corresponding to the two dotted lines in Fig. 9.16 and use the bifurcation continuation software package DDE-BIFTOOL to study the bifurcation with the time-delay being the bifurcation parameter. Then we obtain two types of oscillatory solutions under different bifurcation mechanisms after the uniform flow solution becomes unstable, from a numerical perspective. In Fig. 9.17, we present the bifurcation diagrams for $h^* = 31.5$ and $h^* = 29$, respectively. Additionally, to illustrate the oscillatory solutions under these two distinct bifurcation mechanisms, we provide the corresponding time histories in Fig. 9.18.

From Fig. 9.17a, it can be seen that the uniform flow is destabilized as a consequence of the supercritical Hopf bifurcation. Although the destabilization caused by this bifurcation mechanism only leads to a periodic oscillation with small amplitude which does not immediately result in traffic jams, it still needs to be prevented. The reason is that with the increase of time delay, the amplitude of the oscillation will also

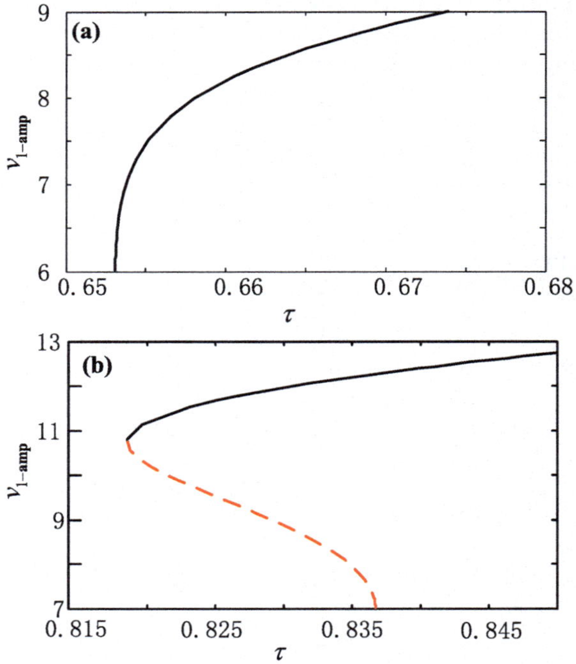

Fig. 9.17 **a** Bifurcation diagram for $h^* = 29$, indicating a supercritical case; **b** bifurcation diagram for $h^* = 31.5$, indicating a subcritical case

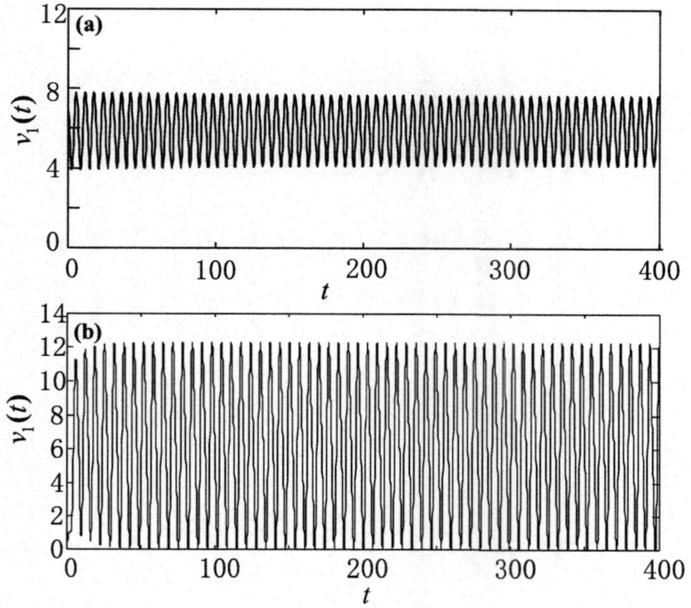

Fig. 9.18 Time history plots for **a** $h^* = 29$ with the critical value of τ for the Hopf bifurcation is $\tau_c = 0.654$, and **b** $h^* = 31.5$ with the critical value of τ for the Hopf bifurcation is $\tau_c = 0.835$

increase, which gradually increase the risk of traffic jam. As shown in Fig. 9.17b, the uniform flow destabilization caused by the subcritical Hopf bifurcation leads to a periodic solution with large amplitude, which rapidly evolves into a traffic jam. Therefore, this situation is far more dangerous than the destabilization caused by the supercritical Hopf bifurcation. In order to demonstrate the oscillatory solutions resulting from these two mechanisms, the time history plots of the velocity of the first vehicle for these two cases are shown in Fig. 9.18a, b, respectively.

9.3.2.3 Traffic Jam Mode

In order to present the jammed modes corresponding to the above two oscillatory solutions, this subsection provides the spatiotemporal patterns as shown in Fig. 9.19.

Figure 9.19 presents a 3D spatiotemporal pattern of time-position-velocity, where each point in the figure represents the position of a certain vehicle on the road at a given moment, and the spatiotemporal pattern of the first vehicle is represented by the blue curve. Figure 9.19a depicts the periodic solution with large-amplitude induced by the subcritical Hopf bifurcation, corresponding to the stop-and-go jammed mode; Fig. 9.19b depicts the periodic solution with small-amplitude induced by the supercritical Hopf bifurcation. In such mode, the vehicle just goes through the process of acceleration and deceleration constantly, but the vehicle does not virtually come to a

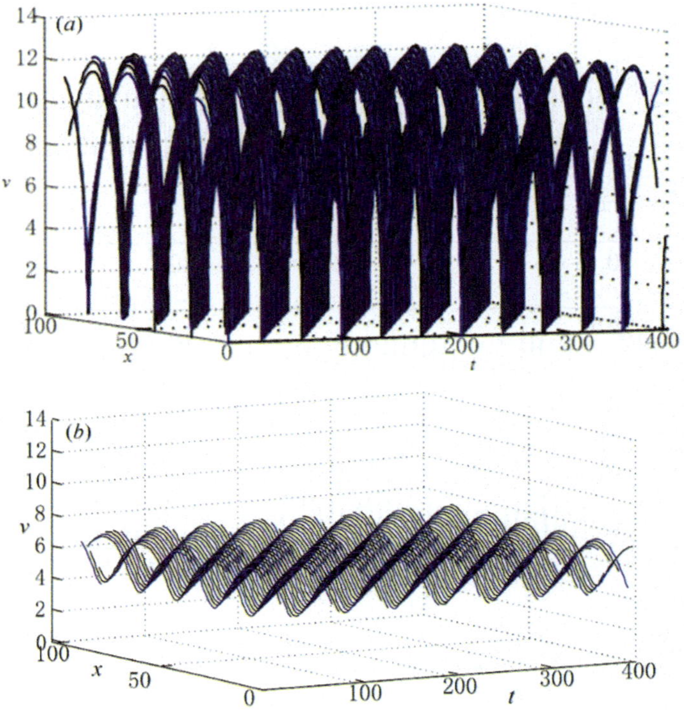

Fig. 9.19 Spatiotemporal pattern of **a** stop-and-go jammed mode and **b** oscillations with small amplitude

resting state. Although this state will not cause traffic jam, it is easy to develop into the traffic jam, as can be verified by Fig. 9.17a. Therefore, regardless of the type of mechanism that leads to the instability of the uniform flow, it is necessary to apply the control to guarantee the stability of the uniform flow as possible as we can.

9.3.3 Suppress Traffic Jam Through Time-Varying Delay

To suppress the two jammed modes discussed in the preceding part, this subsection employs the control with time-varying delay. The specific strategy is to add a periodic perturbation on top of the constant time delay, transforming it into a time-varying quantity. The stable region of the uniform flow is enlarged by modifying the perturbation amplitude of the time-varying delay, and thus the goal of jam suppression is achieved.

9.3.3.1 Form of Time-Varying Delay

In order to suppress the traffic jam after the uniform flow loses its stability when $\tau \geq \tau_c$, the control strategy with variable delay is adopted, and the form of variable delay is given by

$$\tau(t) = (\tau_c + \varepsilon\tau_\varepsilon) + \varepsilon B \sin(\varepsilon\Omega t),\ \varepsilon\tau_\varepsilon > 0, \tag{9.61}$$

where Ω is the frequency of periodic perturbation to the delay. If $B = 0$, i.e. there is no periodic perturbation to the delay, Eq. (9.61) becomes $\tau(t) = \tau_c + \varepsilon\tau_\varepsilon$. From the bifurcation diagram in Fig. 9.17, it can be seen that the uniform flow is destabilized and traffic jam occurs. If $B \neq 0$, that is, the periodic perturbation is applied to the constant delay, the uniform flow can still maintain its stability and the traffic jam is avoided. Such phenomenon suggests that a proper periodic perturbation to a constant delay may expand the interval of delay on which the uniform flow is stable, for example, from $[0, \tau_c)$ to $[0, \tau_c + \varepsilon\tau_\varepsilon)$. The key issue now is to find such suitable value of the perturbation parameter B.

9.3.3.2 Conditions of Traffic Jam Suppression

Recall that all vehicles are assumed to be identical. Then we apply to each vehicle the time-varying perturbation of time delay with the same amplitude and frequency, so that the time delay FVD model containing the variable delay is given by

$$\begin{cases} \dot{r}_1(t) = u_2(t) - u_1(t), \\ \dot{u}_1(t) = \alpha\left(\sum_{i=1}^{3} b_i r_1^i(t-\tau(t)) - u_1(t-\tau(t))\right) + \beta\, \dot{r}_1(t-\tau(t)), \\ \dot{r}_2(t) = u_3(t) - u_2(t), \\ \dot{u}_2(t) = \alpha\left(\sum_{i=1}^{3} b_i r_2^i(t-\tau(t)) - u_2(t-\tau(t))\right) + \beta\, \dot{r}_2(t-\tau(t)), \\ \dot{u}_3(t) = \alpha\left(\sum_{i=1}^{3} b_i(-r_1(t-\tau(t)) - r_2(t-\tau(t)))^i - u_3(t-\tau(t))\right) \\ \qquad +\beta\, (u_1(t-\tau(t)) - u_3(t-\tau(t))), \end{cases} \tag{9.62}$$

where $\tau(t) = \tau_c + \varepsilon\tau_\varepsilon + \varepsilon B \sin(\varepsilon\Omega t)$. ε is a small dimensionless quantity, so that εB and $\varepsilon\Omega$ represent small amplitude and frequency, respectively. To facilitate the following study, Eq. (9.62) is rewritten in matrix form as

$$\dot{\mathbf{y}}(t) = \mathbf{C}\mathbf{y}(t) + \mathbf{D}\mathbf{y}(t-\tau(t)) + \mathbf{F}(\mathbf{y}(t-\tau(t))), \tag{9.63}$$

where $\mathbf{y}(t) = (y_1(t), y_2(t), y_3(t), y_4(t), y_5(t))^T = (r_1(t), u_1(t), r_2(t), u_2(t), , u_3(t))^T$, and

$$\mathrm{C} = \begin{pmatrix} 0 & -1 & 0 & 1 & 0 \\ 0 & 0 & 0 & 0 & 0 \\ 0 & 0 & 0 & -1 & 1 \\ 0 & 0 & 0 & 0 & 0 \\ 0 & 0 & 0 & 0 & 0 \end{pmatrix}, \quad \mathrm{D} = \begin{pmatrix} 0 & 0 & 0 & 0 & 0 \\ \alpha b_1 & -\beta-\alpha & 0 & \beta & 0 \\ 0 & 0 & 0 & 0 & 0 \\ 0 & 0 & \alpha b_1 & -\beta-\alpha & \beta \\ -\alpha b_1 & \beta & -\alpha b_1 & 0 & -\beta-\alpha \end{pmatrix},$$

$$\mathbf{F}(\mathbf{y}(t-\tau(t))) = \left(0, \alpha\sum_{i=2}^{3} b_i r_1^i(t-\tau(t)), 0, \alpha\sum_{i=2}^{3} b_i r_2^i(t-\tau(t)), \alpha\sum_{i=2}^{3} b_i(-\sum_{k=1}^{2} r_k(t-\tau(t)))^i\right)^T.$$

Introducing the time scales $T_j = \varepsilon^j t$, $j = 0, 1, 2$, we have

$$\mathbf{y}(t) = \mathbf{Y}(T_0, T_1, T_2) = \sum_{i=1}^{3} \varepsilon^i \mathbf{Y}_i(T_0, T_1, T_2), \; \dot{\mathbf{y}}(t) = \varepsilon\frac{\partial \mathbf{Y}(T_0, T_1, T_2)}{\partial T_0} + \varepsilon^2\frac{\partial \mathbf{Y}(T_0, T_1, T_2)}{\partial T_1} + \varepsilon^3\frac{\partial \mathbf{Y}(T_0, T_1, T_2)}{\partial T_2},$$

$$\begin{aligned} y_i(t-\tau(t)) &= \varepsilon y_{i,1} + \varepsilon^2\left(y_{i,2} - \tau_c\frac{\partial y_{i,1}}{\partial T_1} - (B\sin(\Omega T_1) + \tau_\varepsilon)\frac{\partial y_{i,1}}{\partial T_0}\right) + \frac{1}{2}\varepsilon^3 \\ &\left(2y_{i,3} - 2\tau_c\left(\frac{\partial y_{i,1}}{\partial T_2} + \frac{\partial y_{i,2}}{\partial T_1}\right) - 2(B\sin(\Omega T_1) + \tau_\varepsilon)\left(\frac{\partial y_{i,1}}{\partial T_1} + \frac{\partial y_{i,2}}{\partial T_0}\right)\right. \\ &\left. + \left(\tau_c\frac{\partial y_{i,1}}{\partial T_1} + (B\sin(\Omega T_1) + \tau_\varepsilon)\frac{\partial y_{i,1}}{\partial T_0}\right)^2\right), \end{aligned} \tag{9.64}$$

where $y_{i,j,\tau_c} = y_{i,j}(T_0 - \tau_c, T_1, T_2)$. Substituting the result in (9.64) into Eq. (9.63) and equating the coefficients of ε, ε^2 and ε^3 to zero yield

$$D_0\mathbf{Y}_1 - \mathbf{C}\mathbf{Y}_1 - \mathbf{D}\mathbf{Y}_{1\tau_c} = 0, \tag{9.65}$$

$$\begin{aligned} &D_0\mathbf{Y}_2 - \mathbf{C}\mathbf{Y}_2 - \mathbf{D}\mathbf{Y}_{2,\tau_c} \\ &= \left(-\frac{\partial y_{1,1}}{\partial T_1}, -\frac{\partial y_{2,1}}{\partial T_1} + \alpha b_2 y_{1,1,\tau_c}^2 - \sum_{i=1}^{5}\left(\tau_c\frac{\partial y_{i,1,\tau_c}}{\partial T_1} + (\tau_\varepsilon + B\right.\right. \\ &\sin(\Omega T_1))\frac{\partial y_{i,1,\tau_c}}{\partial T_0} D(2,i), -\frac{\partial y_{3,1}}{\partial T_1}, -\frac{\partial y_{4,1}}{\partial T_1} +, \alpha b_2 y_{3,1,\tau_c}^2 \end{aligned}$$

$$
\begin{aligned}
&-\sum_{i=1}^{5} D(4,i)\left(\tau_c \frac{\partial y_{i,1,\tau_c}}{\partial T_1} + (\tau_\varepsilon + B\sin(\Omega T_1))\frac{\partial y_{i,1,\tau_c}}{\partial T_0}\right) - \frac{\partial y_{5,1}}{\partial T_1}\\
&+\alpha b_2 (y_{1,1,\tau_c} + y_{3,1,\tau_c})^2 - \sum_{i=1}^{5} D(5,i)\\
&\left(\tau_c \frac{\partial y_{i,1,\tau_c}}{\partial T_1} + (\tau_\varepsilon + B\sin(\Omega T_1))\frac{\partial y_{i,1,\tau_c}}{\partial T_0}\right)\Bigg)\Bigg), \qquad (9.66)
\end{aligned}
$$

$$
\begin{aligned}
&\mathbf{D}_0\mathbf{Y}_3 - \mathbf{C}\mathbf{Y}_3 - \mathbf{D}\mathbf{Y}_{3,\tau_c}\\
&= \left(-\frac{\partial y_{1,2}}{\partial T_1} - \frac{\partial y_{1,1}}{\partial T_2}, -\frac{\partial y_{2,2}}{\partial T_1} - \frac{\partial y_{2,1}}{\partial T_2} + \alpha b_3 y_{1,1,\tau_c}^3 + \alpha b_2 y_{1,1,\tau_c}\right.\\
&(y_{1,2,\tau_c} - \tau_c \frac{\partial y_{1,1,\tau_c}}{\partial T_1} - (\tau_\varepsilon + B\sin(\Omega T_1))\frac{\partial y_{1,1,\tau_c}}{\partial T_0}) - \sum_{i=1}^{5} D(2,i)(\tau_c \frac{\partial y_{i,1,\tau_c}}{\partial T_2}\\
&+ (\tau_\varepsilon + B\sin(\Omega T_1))\frac{\partial y_{i,1,\tau_c}}{\partial T_1} + \tau_c \frac{\partial y_{i,2,\tau_c}}{\partial T_1} + (\tau_\varepsilon + B\sin(\Omega T_1))\frac{\partial y_{i,2,\tau_c}}{\partial T_0} + \frac{1}{2}\\
&\left(\tau_c \frac{\partial y_{i,1,\tau_c}}{\partial T_1} + (\tau_\varepsilon + B\sin(\Omega T_1))\frac{\partial y_{i,1,\tau_c}}{\partial T_0}\right)^2, -\frac{\partial y_{3,2}}{\partial T_1} - \frac{\partial y_{3,1}}{\partial T_2}, -\frac{\partial y_{4,2}}{\partial T_1} - \frac{\partial y_{4,1}}{\partial T_2}\\
&+ \alpha b_2 y_{3,1,\tau_c}(y_{3,2,\tau_c} - \tau_c \frac{\partial y_{3,1,\tau_c}}{\partial T_1} - (\tau_\varepsilon + B\sin(\Omega T_1))\frac{\partial y_{3,1,\tau_c}}{\partial T_0}) + \alpha b_3 y_{3,1,\tau_c}^3\\
&- \sum_{i=1}^{5} D(4,i)(\tau_c \frac{\partial y_{i,1,\tau_c}}{\partial T_2} + (\tau_\varepsilon + B\sin(\Omega T_1))\frac{\partial y_{i,1,\tau_c}}{\partial T_1} + \tau_c \frac{\partial y_{i,2,\tau_c}}{\partial T_1}\\
&+ (\tau_\varepsilon + B\sin(\Omega T_1))\frac{\partial y_{i,2,\tau_c}}{\partial T_0} + \frac{1}{2}\left(\tau_c \frac{\partial y_{i,1,\tau_c}}{\partial T_1} + (\tau_\varepsilon + B\sin(\Omega T_1))\frac{\partial y_{i,1,\tau_c}}{\partial T_0}\right)^2, \qquad (9.67)\\
&-\frac{\partial y_{5,2}}{\partial T_1} - \frac{\partial y_{5,1}}{\partial T_2} + \alpha b_3 (y_{1,1,\tau_c} + y_{3,1,\tau_c})^3 + \alpha b_2 (y_{1,1,\tau_c} + y_{3,1,\tau_c})\\
&\sum_{i=1}^{2} (y_{2i-1,2,\tau_c} - \tau_c \frac{\partial y_{2i-1,1,\tau_c}}{\partial T_1} - (\tau_\varepsilon + B\sin(\Omega T_1))\frac{\partial y_{2i-1,1,\tau_c}}{\partial T_0})\\
&- \sum_{i=1}^{5} D(5,i)(\tau_c \frac{\partial y_{i,1,\tau_c}}{\partial T_2} + (\tau_\varepsilon + B\sin(\Omega T_1))\frac{\partial y_{i,1,\tau_c}}{\partial T_1} + \tau_c \frac{\partial y_{i,2,\tau_c}}{\partial T_1}\\
&+ (\tau_\varepsilon + B\sin(\Omega T_1))\frac{\partial y_{i,2,\tau_c}}{\partial T_0} + \frac{1}{2}\left(\tau_c \frac{\partial y_{i,1,\tau_c}}{\partial T_1} + (\tau_\varepsilon + B\sin(\Omega T_1))\frac{\partial y_{i,1,\tau_c}}{\partial T_0}\right)^2\Bigg).
\end{aligned}
$$

The procedure of the method of multiple scales is briefly presented below. The solution of Eq. (9.65) can be formulated as

$$
\mathbf{Y}_1 = \mathbf{A}_1(T_1, T_2)\sin(\omega_c T_0) + \mathbf{B}_1(T_1, T_2)\cos(\omega_c T_0). \qquad (9.68)
$$

Then substituting (9.68) into Eq. (9.65), we obtain the following algebraic equation

$$\begin{pmatrix} (s\mathbf{D}+\omega\mathbf{I})^T & -(\mathbf{C}+c\mathbf{D})^T \\ -(\mathbf{C}+c\mathbf{D})^T & -(s\mathbf{D}+\omega\mathbf{I})^T \end{pmatrix}\begin{pmatrix} \mathbf{A}_1 \\ \mathbf{B}_1 \end{pmatrix} = 0, \tag{9.69}$$

where $c = \cos(\omega_c T_0)$ and $s = \sin(\omega_c T_0)$. Solving the algebraic equation yields

$$A_{1,i} = \alpha_{1,i}(A_{1,1}, B_{1,1}),\ \ B_{1,i} = \beta_{1,i}(A_{1,1}, B_{1,1}), i = 2, 3, 4, 5. \tag{9.70}$$

Substituting Eqs. (9.68) and (9.70) into Eq. (9.66), one finds that the solution of Eq. (9.66) should take the following form

$$\mathbf{Y_2} = \mathbf{Q}_2(T_1, T_2) + \sum_{k=1}^{2} \mathbf{A}_{2,k}(T_1, T_2)\sin(k\omega_c T_0) + \mathbf{B}_{2,k}(T_1, T_2)\cos(k\omega_c T_0). \tag{9.71}$$

To eliminate the secular term in Eq. (9.71), let

$$\mathbf{A}_{2,1} = 0, \mathbf{B}_{2,1} = 0. \tag{9.72}$$

The Fredholm Alternative Theorem can be used to obtain $\mathrm{D}_1 A_{1,1}$ and $\mathrm{D}_1 B_{1,1}$ from Eq. (9.72) and $\mathrm{D}_2 A_{1,1}$ and $\mathrm{D}_2 B_{1,1}$ from Eq. (9.67). Then we have

$$\begin{cases} \dot{A}_{1,1} = \varepsilon \mathrm{D}_1 A_{1,1} + \varepsilon^2 \mathrm{D}_2 A_{1,1} + \cdots, \\ \dot{B}_{1,1} = \varepsilon \mathrm{D}_1 B_{1,1} + \varepsilon^2 \mathrm{D}_2 B_{1,1} + \cdots. \end{cases} \tag{9.73}$$

Apply the following polar coordinate transformations to Eq. (9.73)

$$A_{1,1}(t) = R(t)\cos(\varphi(t)), B_{1,1}(t) = R(t)\sin(\varphi(t)), \tag{9.74}$$

then we have

$$\begin{cases} \dot{R}(t) = f_1(t)R(t) + f_3(t)R(t)^3, \\ \dot{\varphi}(t) = g_0(t) + g_2(t)R(t)^2, \end{cases} \tag{9.75}$$

where $f_1(t) = p_0 + p_1(t)$. Note that $p_0 = p_0(\varepsilon\tau_\varepsilon, B, \Omega)$ is a constant, and $p_1(t)$ is a function of time. Based on the results from [50], the solution of the first equation in Eq. (9.75) is obtained as

$$
\begin{aligned}
\varepsilon R(t) &= 1/\sqrt{E_0 e^{-2\int f_1(t)dt} - 2e^{-2\int f_1(t)dt} \int f_3(t) e^{2\int f_1(t)dt} dt} \\
&= \frac{1}{\sqrt{E_0 e^{-2p_0 t} G(t) + H(t)}}.
\end{aligned}
\tag{9.76}
$$

The critical condition under which the bifurcated periodic solution of Eq. (9.62) can be suppressed is that $p_0(\varepsilon\tau_\varepsilon, B, \Omega) = 0$. If $p_0 < 0$, the periodic oscillations are suppressed, which means the traffic jam is avoided.

9.3.3.3 Numerical Examples of Jam Suppression

Numerical Example 1: Suppression of Stop-and-go Jammed Mode

We choose $h^* = 31.5$ and the rest of the parameters are chosen as in (9.51). According to Eq. (9.58), the critical value of delay for the occurrence of subcritical Hopf bifurcation is $\tau_c = 0.835$ and the corresponding frequency $\omega_c = 0.978$. It can be observed from Fig. 9.19a that there arises a stop-and-go jammed mode. Then the methods in the previous subsection are used to suppress this traffic jam. The coefficients of the first equation in Eq. (9.75) are obtained as

$$
\begin{aligned}
f_1(t) &= -0.337B^2 + 0.814246\tau_\varepsilon - 0.673\tau_\varepsilon^2 + 0.337B^2\cos(2\Omega t) \\
&\quad + 0.814B\sin(\Omega t) - 1.347B\sin(\Omega t)\tau_\varepsilon, \\
f_3(t) &= 0.0005.
\end{aligned}
\tag{9.77}
$$

The critical condition for suppressing such mode is $p_0 = 0$, i.e.,

$$
p_0 = -0.337B^2 + 0.814246\tau_\varepsilon - 0.673\tau_\varepsilon^2 = 0. \tag{9.78}
$$

In Fig. 9.20, the critical boundary curve for the suppression of oscillation is plotted based on Eq. (9.78) in the parameter plane $(\varepsilon\tau_\varepsilon, B)$. In the shaded region, i.e., $p_0 < 0$, the stop-and-go traffic jam can be suppressed. To show the results of the theoretical analysis here, we fix $\varepsilon\tau_\varepsilon = 0.002$ and the value of the critical of B is obtained as $B_c = 0.1$ by Eq. (9.78). We select values of B on both sides of the critical curve respectively, namely, $B = 0.09$ from the white region and $B = 0.11$ in the shaded region. Then numerical simulation of Eq. (9.62) is carried out by using the numerical simulation software ddesd [40] and the results are shown in Fig. 9.21.

If $B < B_c$, the variable time delay control cannot suppress the oscillations with large amplitude, as shown in Fig. 9.21a, which is consistent with the theoretical prediction. Meanwhile, if $B > B_c$, oscillations with large amplitude are effectively suppressed as shown in Fig. 9.21b and the traffic jam is avoided. Thus, the variable time delay control is effective in controlling the traffic jam induced by subcritical Hopf bifurcation.

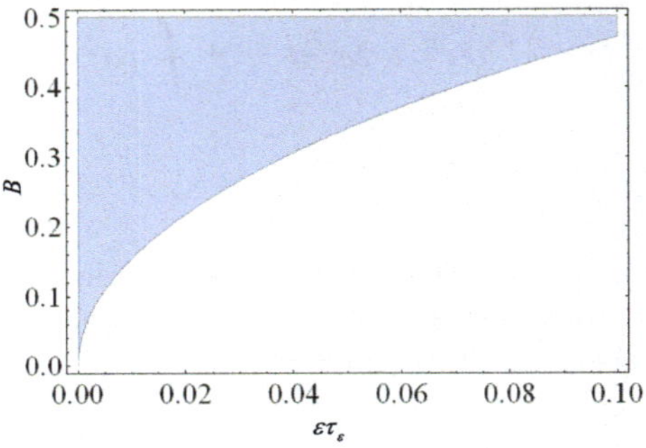

Fig. 9.20 Critical boundary curve for the suppression of oscillation in the plane of B and $\varepsilon\tau_\varepsilon$ for $h^* = 31.5$, where the shaded region corresponds to the set of parameters for which the traffic jam can be suppressed

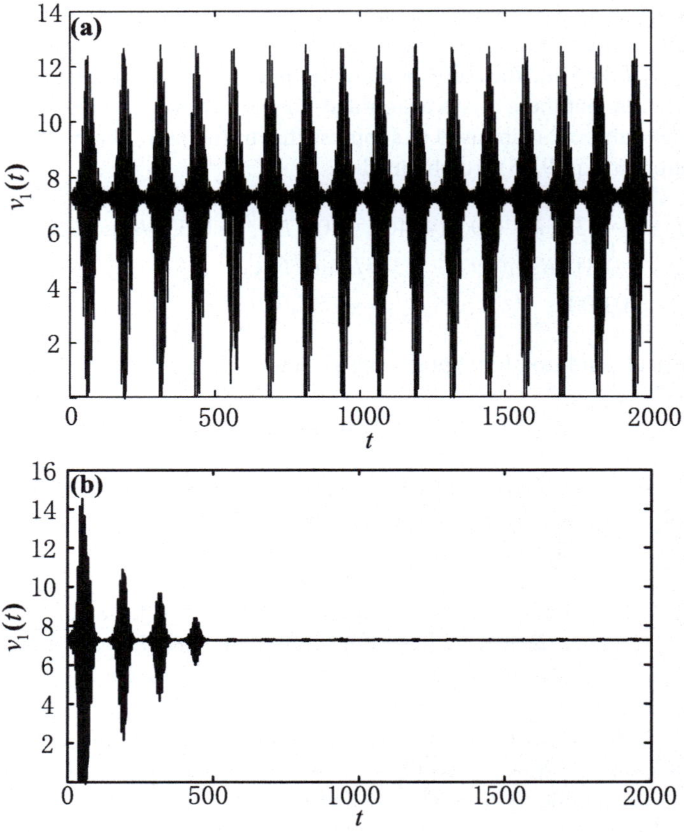

Fig. 9.21 Time history plots of the first vehicle for **a** $B = 0.09 (B < B_c)$, and **b** $B = 0.11 (B > B_c)$

Numerical Example 2: Suppression of Oscillatory Traffic Modes with Small Amplitude

If $h^* = 29$ and the rest of the parameters are the same as above, the value of the critical time delay for a supercritical Hopf bifurcation to occur is $\tau_c = 0.654$, according to Eq. (9.58), and the frequency $\omega_c = 1.025$. For such parameters, the oscillatory traffic mode with small amplitude is observed and shown in Fig. 9.19b. Although this traffic pattern does not directly cause traffic jam, it can potentially lead to jammed state, and therefore should be avoided. The theoretical analysis of the previous subsection is used to suppress the emergence of this mode.

The coefficients of the first equation in Eq. (9.75) are calculated as

$$\begin{aligned} f_1(t) &= -0.0399675B^2 + 0.0399675B^2\cos(2t\Omega) + 0.806099B\sin(t\Omega) \\ &\quad + 0.806099\tau_0 - 0.15987B\sin(t\Omega)\tau_0 - 0.079935\tau_0^2, \\ f_3(t) &= -0.00015. \end{aligned} \tag{9.79}$$

From which one can obtain the condition under which this mode can be suppressed as follows:

$$p_0 = -0.04B^2 + 0.806\tau_\varepsilon - 0.08\tau_\varepsilon^2 = 0. \tag{9.80}$$

We select values of B on both sides of the critical curve respectively, namely, $B = 0.09$ from the white region and $B = 0.11$ in the shaded region. Then numerical simulation of Eq. (9.62) is carried out by using the numerical simulation software ddesd [40] and the results are shown in Fig. 9.21.

In Fig. 9.22, the critical boundary curve for the suppression of oscillation is plotted based on Eq. (9.80) in the parameter plane $(\varepsilon\tau_\varepsilon, B)$. In the shaded region, i.e., $p_0 < 0$, the stop-and-go traffic jam can be suppressed. To show the results of the theoretical analysis here, we fix $\varepsilon\tau_\varepsilon = 0.002$ and the value of the critical of B is obtained as $B_c = 0.2$ by Eq. (9.80).

In Fig. 9.22, the critical curve for oscillation suppression is plotted in the parameter plane $(\varepsilon\tau_\varepsilon, B)$ based on Eq. (9.80), where the shaded area, i.e., $p_0 < 0$, can suppress small-amplitude oscillating traffic pattern. Similarly, to show the results of the theoretical analysis here, we fix $\varepsilon\tau_\varepsilon = 0.002$ and the value of the critical of B is obtained as $B_c = 0.2$. We select values of B on both sides of the critical curve respectively, namely, $B = 0.19$ from the white region and $B = 0.21$ from the shaded region to run numerical simulations of Eq. (9.62), and the results are shown in Fig. 9.23.

As shown in Fig. 9.21a, the variable time delay control fails to suppress the oscillation if $B < B_c$. Meanwhile, if $B > B_c$, the oscillation is effectively suppressed as shown in Fig. 9.21b.

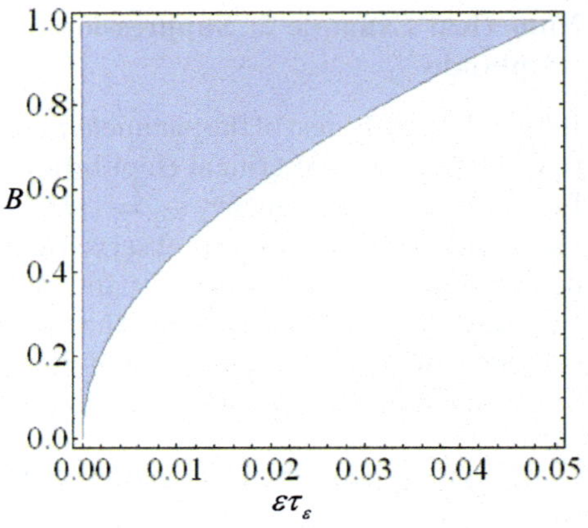

Fig. 9.22 Critical boundary curve for the suppression of oscillation in the plane of B and $\varepsilon\tau_\varepsilon$ for $h^* = 29$, where the shaded region corresponds to the set of parameters for which the traffic jam can be suppressed

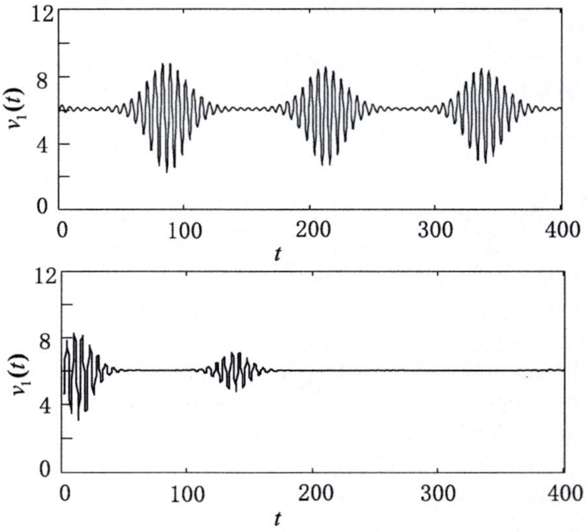

Fig. 9.23 Time history plots of the first vehicle for **a** $B = 0.19(B < B_c)$, and **b** $B = 0.21(B > B_c)$

Appendix

$$
\begin{aligned}
c_0 &= \{2\alpha^2 b_1^2(\alpha^2 - 3\beta^2 + 6\beta b_1 + 2\omega_c^2) \\
&\quad + \beta\omega_c^2[-2b_1(2\alpha^2 - 4\beta^2 + \omega_c^2) + \beta(\alpha^2 - 2\beta^2 + \omega_c^2)] \\
&\quad + \cos\frac{2k_0\pi}{n}[6\alpha^2\beta(\beta - 2b_1)b_1^2 + 2\beta^3(\beta - 4b_1)\omega_c^2]\}\alpha\omega_c^2(\alpha^2 b_1^2 + \beta^2\omega_c^2) \\
d_0 &= 2(\alpha^3 b_1^2 + \alpha b_1\beta\omega_c^2 - \alpha^2 b_1^2\omega_c^2\tau_c - \beta^2\omega_c^4\tau_c) \\
d_1 &= \omega_c^2(1 + \alpha\tau_c)^2 + \omega_c^2\tau_c^2 \\
c_1 &= 4\alpha^2\omega_c^2 b_1\cos\frac{2k_0\pi}{n}(\alpha^2 b_1^2 + \beta^2\omega_c^2)[(3b_1 - 2\beta)(\alpha^3 b_1^2 + \beta^2\omega_c^2) + 2b_1\beta^2\omega_c^2]
\end{aligned}
$$

$$+ 2\omega_c^2(\alpha^2 b_1^2 + \beta^2\omega_c^2)[4\alpha^4\beta b_1^3 - 6\alpha^4 b_1^4 - 6\alpha^2\beta^2 b_1^2\omega_c^2$$
$$+ \omega_c^2(2\alpha^2 - 4\beta^2 + \omega_c^2)(\beta^2\omega_c^2 - \alpha^2\beta b_1 + \alpha^2 b_1^2)]$$
$$c_2 = 2\alpha(\alpha^2 b_1^2\omega_c + \beta^2\omega_c^3)^2\left[\alpha^2\omega_c^2 + \omega_c^4 + 2(\alpha^2 b_1^2 + \beta^2\omega_c^2)\left(\cos\frac{2k_0\pi}{n} - 1\right)\right]$$

$$\begin{aligned} F_{i,2j-1} = {} & \{\alpha b_i Y_1^i + (i-1)\alpha b_{i-1}[Y_2 - (\tau_c D_1 + \tau_\varepsilon D_0)Y_1]Y_1^{i-2} \\ & + (i-2)\alpha b_{i-2}[Y_3 - (\tau_c D_2 + \tau_\varepsilon D_1)Y_1 \\ & - 2(\tau_c D_1 + \tau_\varepsilon D_0)Y_2 + (\tau_c D_1 + \tau_\varepsilon D_0)^2 Y_1 \\ & + \frac{1}{2Y_1}Y_2^2]Y_1^{i-3} + (i-3)\alpha b_{i-3}[Y_4 - \tau_c D_3 - \tau_\varepsilon D_2 \\ & + (\tau_c D_1 + \tau_\varepsilon D_0)(\tau_c D_2 + \tau_\varepsilon D_1) - \frac{1}{6}(\tau_c D_1 + \tau_\varepsilon D_0)^3]Y_1^{i-4}\}_{2j-1,\tau_c} \\ F_{i,2j} = {} & 0,\ j = 1, \ldots, n;\ i = 2, \ldots, 5 \end{aligned}$$

$$F_{i,2n-1} = \sum_{i=1}^{n-1} F_{i,2j-1}.$$
$$F_2^c = 0,\ F_2^s = 0$$
$$F_{j,2i-1}^c = 0, = F_{j,2i-1}^s = 0,\ j = 3, 4, 5$$

$$\begin{pmatrix} F_{3,2i}^c \\ F_{3,2i}^s \end{pmatrix} = \frac{3\alpha b_3(A_{1,1,2i-1}^2 + B_{1,1,2i-1}^2)}{4}\begin{pmatrix} cB_{1,1,2i-1} - sA_{1,1,2i-1} \\ cA_{1,1,2i-1} + sB_{1,1,2i-1} \end{pmatrix}$$

$$\begin{aligned} \begin{pmatrix} F_{4,2i}^c \\ F_{4,2i}^s \end{pmatrix} = {} & \frac{A_{1,1,2i-1}^2 + B_{1,1,2i-1}^2}{4}\left[\begin{pmatrix} cB_{2,1,2i-1} - sA_{2,1,2i-1} \\ cA_{2,1,2i-1} + sB_{2,1,2i-1} \end{pmatrix}\right. \\ & \left. - \begin{pmatrix} S_1 & 3S_2 \\ 3S_2 & -S_1 \end{pmatrix}\begin{pmatrix} cA_{1,1,2i-1} + sB_{1,1,2i-1} \\ cB_{1,1,2i-1} - sA_{1,1,2i-1} \end{pmatrix}\right] \\ & \alpha b_3 + 2(B_{1,1,2i-1}B_{2,1,2i-1} + A_{1,1,2i-1}A_{2,1,2i-1}) \\ & \begin{pmatrix} cB_{1,1,2i-1} - sA_{1,1,2i-1} \\ cA_{1,1,2i-1} + sB_{1,1,2i-1} \end{pmatrix} \end{aligned}$$

$$\begin{aligned} \begin{pmatrix} F_{5,2i}^c \\ F_{5,2i}^s \end{pmatrix} = {} & \frac{3\left(A_{1,1,2i-1}^2 + B_{1,1,2i-1}^2\right)}{4} \\ & \left[\begin{pmatrix} cB_{3,1,2i-1} - sA_{3,1,2i-1} \\ cA_{3,1,2i-1} + sB_{3,1,2i-1} \end{pmatrix} - 2\begin{pmatrix} S_1 & S_2 \\ S_2 & -S_1 \end{pmatrix}\begin{pmatrix} cA_{2,1,2i-1} + sB_{2,1,2i-1} \\ cB_{2,1,2i-1} - sA_{2,1,2i-1} \end{pmatrix}\right. \\ & \alpha b_3 + \begin{pmatrix} 3\left(S_2^2 - S_1^2 - \tau_\varepsilon D_1 - \tau_c D_2\right) & 2S_1S_2 \\ -2S_1S_2 & 3\left(S_2^2 - S_1^2 - \tau_\varepsilon D_1 - \tau_c D_2\right) \end{pmatrix} \end{aligned}$$

$$\begin{pmatrix} cB_{1,1,2i-1} - sA_{1,1,2i-1} \\ cA_{1,1,2i-1} + sB_{1,1,2i-1} \end{pmatrix} \Bigg] + \frac{3\alpha b_3}{2} \begin{pmatrix} cB_{1,1,2i-1} - sA_{1,1,2i-1} \\ cA_{1,1,2i-1} + sB_{1,1,2i-1} \end{pmatrix}$$

$$[B_{3,1,2i-1}B_{1,1,2i-1} + A_{3,1,2i-1}A_{1,1,2i-1} + A^2_{2,1,2i-1} + B^2_{2,1,2i-1}$$
$$+ 2S_1(B_{2,1,2i-1}A_{1,1,2i-1} - A_{2,1,2i-1}B_{1,1,2i-1})]$$
$$+ 3\alpha b_3(B_{2,1,2i-1}B_{1,1,2i-1} + A_{2,1,2i-1}A_{1,1,2i-1})$$
$$\begin{pmatrix} cB_{2,1,2i-1} - sA_{2,1,2i-1} \\ cA_{2,1,2i-1} + sB_{2,1,2i-1} \end{pmatrix} + \frac{5\alpha b_5}{8}(A^2_{1,1,2i-1} + B^2_{1,1,2i-1})^2$$
$$\begin{pmatrix} cB_{1,1,2i-1} - sA_{1,1,2i-1} \\ cA_{1,1,2i-1} + sB_{1,1,2i-1} \end{pmatrix}$$

References

1. Chowdhury D, Santen L, Schadschneider A (2000) Statistical physics of vehicular traffic and some related systems. Phys Rep 329:199–329
2. Helbing D (2001) Traffic and related self-driven many-particle systems. Rev Mod Phys 73(4):1067–1141
3. Bellomo N, Delitala M (2002) On the mathematical theory of vehicular traffic flow I. Fluid dynamics and kinetic modeling. Math Model Method Appl Sci 12(12):1801–1843
4. Shvetsov VI (2003) Mathematical modeling of traffic flows. Autom Remote Control 64(11):1651–1689
5. Nagatani T (2002) The physics of traffic jams. Rep Prog Phys 65:1331–1386
6. Nagel K (2003) Still following: approaches to traffic flow and traffic jam modeling. Oper Res 51(5):685–710
7. Bando M, Hasebe K, Nakanishi K et al (2000) Delay of vehicle motion in traffic dynamics. Japan J Indust Appl Math 17:275–294
8. Green M (2000) "How long does it take to stop?" methodological analysis of driver perception-brake times. Transp Hum Factors 2(3):95–216
9. Mahmassani H (2005) Transportation and traffic theory: flow, dynamics and human interaction. Elsevier, The Netherlands, pp 245–266
10. Sipahi R, Niculescu SI (2006) Analytical stability study of a deterministic car following model under multiple delay interactions. In: At invited session traffic dynamics under presence of time delays. IFAC Time Delay Systems Workshop, Italy
11. Sipahi R, Niculescu SI (2006) Some remarks on the characterization of delay interactions in deterministic car following models. MTNS, Kyoto
12. Greenshields BD (1936) Reaction time and automobile driving. J Appl Psychol 20:353–358
13. Greenshields BD (1935) Reaction time and traffic behavior. Civ Eng 7(6):384–386
14. Davis LC (2003) Modifications of the optimal velocity traffic model to include delay due to driver reaction time. Physica A 319:557–567
15. Mehmood A, Easa SM (2009) Modeling reaction time in carfollowing behaviour based on human factors. Int J Appl Sci Eng Technol 5(2):93–101
16. Rothery RE (1998) Traffic flow theory, 2nd edn. Transportation Research Board Special Report 165
17. Subramanian H (1996) Estimation of car following models. Massachusetts Institute of Technology, pp 1–93

18. Brackstone M, McDonald M (1999) Car-following: a historical review. Transp Res Part F 2(4):181–196
19. Hoogendoorn SP, Bovy PHL (2001) State-of-the-art of vehicular traffic flow modelling. J Syst Control Eng 25(4):283–304
20. Tampère C (2004) Human-kinetic multiclass trafficflow theory and modelling: With application to advanced driver assistance systems in congestion. Thesis Series, The Netherlands
21. Toledo T (2007) Driving behaviour: models and challenges. Transp Rev 27(1):65–84
22. Yanlin W, Tiejun W (2006) Car-following models of vehicular traffic. J Zhejiang Univ Sci 3(4):412–417
23. Baogui C, Zhaosheng Y (2009) Car-following models study progress. In: Proceedings of the second international symposium on knowledge acquisition and modeling, pp 190–193
24. Atay FM (2010) Complex time delay systems. Springer, Berlin, pp 297–320
25. Engelborghs K, Luzyanina T, Samaey G, Roose D & Verheyden K (2007) DDE-BIFTOOL v. 2.03: a Matlab package for bifurcation analysis of delay differential equations, available at http://twr.cs.kuleuven.be/research/software/delay/ddebiftool.shtml
26. Bando M, Hasebe K, Nakayama A, Shibata A, Sugiyama Y (1995) Dynamical model of traffic congestion and numerical simulation. Phys Rev E, Statis Phys Plasmas Fluids Relat Interdiscipl Top 51(2):1035–1042
27. Wagner P (2010) Fluid-dynamical and microscopic description of traffic flow: a data-driven comparison. Philos Trans R Soc A: Math Phys Eng Sci 368(1928):4481–4495
28. Shamoto D, Tomoeda A, Nishi R et al (2011) Car-following model with relative-velocity effect and its experimental verification. Phys Rev E 83(4):046105
29. Jiang R, Wu Q, Zhu Z (2001) Full velocity difference model for a car-following theory. Phys Rev E 64(1):017101
30. Orosz G, Wilson RE, Stépán G (2010) Traffic jams: dynamics and control. Philos Trans R Soc A Math Phys Eng Sci 368(1928):4455–4479
31. Batista M, Twrdy E (2010) Optimal velocity functions for car-following models. J Zhejiang Univ Sci A 11(7):520–529
32. Orosz G, Wilson RE, Krauskopf B (2004) Global bifurcation investigation of an optimal velocity traffic model with driver reaction time. Phys Rev E 70(2):26207
33. Igarashi Y, Itoh K, Nakanishi K et al (2001) Bifurcation phenomena in the optimal velocity model for traffic flow. Phys Rev E 64(4):47102
34. Orosz G, Stépán G (2006) Subcritical Hopf bifurcations in a car-following model with reaction-time delay. Proc R Soc A Math Phys Eng Sci 462(2073):2643–2670
35. Rothery RE, Traffic Flow Theory (1998) Transportation research board special report 165, 2nd edn.
36. Ruan S, Wei J (2003) On the zeros of transcendental functions with applications to stability of delay differential equations with two delays. Dyn Contin Discr Impuls Syst Ser A 10:863–874
37. Nayfeh AH (2008) Order reduction of retarded nonlinear systems–the method of multiple scales versus center-manifold reduction. Nonlinear Dyn 51(4):483–500
38. Nayfeh AH (2011) Introduction to perturbation techniques. Wiley
39. Kuznetsov IA (1998) Elements of applied bifurcation theory. Springer Science & Business Media
40. Guckenheimer J, Holmes P (1983) Nonlinear oscillations, dynamical systems, and bifurcations of vector fields. Springer Science & Business Media
41. Sipahi R, Niculescu S (2006) Slow time-varying delay effects-robust stability characterization of deterministic car following models. IEEE
42. Niculescu S (2001) Delay effects on stability: a robust control approach. Springer Science & Business Media
43. Shampine LF (2005) Solving ODEs and DDEs with residual control. Appl Numer Math 52(1):113–127
44. Zhang S, Xu J (2011) Oscillation control for n-dimensional congestion control model via time-varying delay. Sci China Technol Sci 54(8):2044–2053

45. Yan Y, Xu J (2013) Suppression of regenerative chatter in a plunge-grinding process by spindle speed. J Manuf Sci Eng 135(4):41019
46. Michiels W, Van Assche V, Niculescu S (2005) Stabilization of time-delay systems with a controlled time-varying delay and applications. IEEE Trans Autom Control 50(4):493–504
47. Stépán G (1989) Retarded dynamical systems: stability and characteristic functions. Longman Scientific & Technical
48. Hale JK (1971) Functional differential equations. Springer
49. Zhen B, Xu J (2010) Bautin bifurcation analysis for synchronous solution of a coupled FHN neural system with delay. Commun Nonlinear Sci Numer Simul 15(2):442–458
50. Song ZG, Xu J (2013) Stability switches and double Hopf bifurcation in a two-neural network system with multiple delays. Cogn Neurodyn 7(6):505–521

Chapter 10
Nonlinear Dynamics of Car-Following Model Induced by Time Delay and Other Parameters

Chapter 9 explores the bistable solution induced by bifurcation and presents the corresponding bistable traffic mode by studying the subcritical Hopf bifurcation when the driver's reaction delay is near its critical value. The research is conducted when other parameters such as sensitivity coefficient and inter-vehicle distance are fixed, with only driver's reaction delay being allowed to vary near the critical value. The study finds that the mechanisms of stability switch of uniform flow are dependent on the road length as well as the sensitivity coefficient of the driver. Therefore, this chapter proposes to analyze the dynamic behaviour induced by the bifurcation and the corresponding traffic modes under the joint influence of driver's reaction time delay and road length/sensitivity coefficient.

10.1 Traffic Modes Induced by Reaction Delay and Road Length

Zhao and Orosz [1] investigated the Bautin bifurcation in a dynamic traffic model and fixed the time delay while using two weight parameters as bifurcation parameters. They observed a bistable phenomenon in the dynamic traffic model. However, previous studies have not adequately considered the bifurcation with codimension two that occurs when using the driver's reaction delay as a bifurcation parameter alongside other parameters. When analyzing the Bautin bifurcation induced by the joint action of the driver's reaction time delay and road length, we follow similar ideas and research approaches as those detailed in literature [2, 3].

This section investigates the car-following model with full velocity difference, including the driver's reaction delay, for vehicles placed on a circular road. The average inter-vehicle distance and driver's reaction delay are selected as the bifurcation parameters. Firstly, through analytical analysis, the stability boundary curve of the uniform flow solution in the parameter plane is obtained. Numerical simulations

J. Xu, *Nonlinear Dynamics of Time Delay Systems*,
https://doi.org/10.1007/978-981-99-9907-1_10

with parameter values along the boundary curve reveal that the induced instability of the uniform flow solution across the stability boundary is dependent on whether the instability is caused by supercritical Hopf bifurcation or subcritical Hopf bifurcation. Secondly, secondary bifurcation curves of the aforementioned Hopf bifurcation is plotted using the numerical continuation software package DDE-BIFTOOL [4]. Thirdly, the parameter plane consisting of the average inter-vehicle distance and the driver's reaction time delay is classified into three regions: one containing only stable uniform flow solutions, one containing periodic solutions with small amplitude, and one containing both uniform flow solutions and periodic solutions with large amplitude. With the aid of spatiotemporal pattern of traffic flow, different solutions of the micro model are transformed into corresponding macro traffic modes. Finally, the influence of average inter-vehicle distance and reaction time delay on the macroscopic traffic mode is investigated, making it possible to clearly observe the degree of traffic jam in different regions of the parameter plane.

10.1.1 Stability in the Parameter Plane Defined by the Average Inter-Vehicle Distance and Driver's Response Time Delay

10.1.1.1 Stability of the Uniform Flow in a Two-Parameter Plane

The state of uniform flow in which all vehicles in a roadway travel at the same inter-vehicle distance and at the same velocity is called an equilibrium state of the vehicle system. In this part we analyze the stability of the equilibrium state in the parameter plane (h^*, τ).

The uniform flow equilibrium of Eq. (9.6) is given in (9.12). In order to study the stability of the equilibrium, we assume that $r_i(t)$, $u_i(t)$ is a perturbation to h^*, v^* respectively, i.e., $h_i(t) = h^* + r_i(t)$, $v_i(t) = v^* + u_i(t)$, and consider the linearized system of Eq. (9.6)

$$\begin{cases} \dot{r}_i(t) = u_{i+1}(t) - u_i(t), \\ \dot{u}_i(t) = -\alpha u_i(t) + \alpha V'(h^*) r_i(t-\tau) + \beta(u_{i+1}(t-\tau) - u_i(t-\tau)), \\ \qquad i = 1, \ldots, n-1, \\ \dot{u}_n(t) = -\alpha u_n(t) + \alpha V'(h^*) \sum\limits_{i=1}^{n-1} (-r_i(t-\tau)) + \beta(u_1(t-\tau) - u_n(t-\tau)), \end{cases} \tag{10.1}$$

here, $V'(h^*)$ is the first order derivative of $V(h)$ at h^*.

Setting $r_i(t) = A_i e^{\lambda t}, \lambda, A_i \in \mathbb{C}$ and substituting into (10.1), we obtain the characteristic equation of Eq. (10.1) as follows

$$(\alpha+\lambda)\prod_{k=1}^{n-1}\left(\lambda^2+\alpha\lambda+\alpha e^{-\lambda\tau}(V'(h^*)+\beta\lambda)-e^{-\lambda\tau}(\alpha V'(h^*)+\beta\lambda)e^{\frac{2\pi k}{n}\mathrm{i}}\right)=0. \tag{10.2}$$

In order to determine the stability of the uniform flow in the parameter plane (h^*, τ), we substitute the critical eigenvalue $\lambda = \mathrm{i}\omega, \omega > 0$ into the characteristic Eq. (10.2) and separate the real and imaginary parts to obtain the critical curve for stability switch in the parameter plane $\left(V'(h^*), \tau\right)$

$$\begin{cases} V'(h^*) = \dfrac{\sqrt{\omega^2\left(-4\beta^2+\left(\alpha^2+\omega^2\right)\csc\frac{k\pi}{n}^2\right)}}{2\alpha} \\ \tau = \dfrac{1}{\omega}\arccos\left(\dfrac{\omega\left(\alpha(b-\beta)\omega+\left(b\alpha^2+\beta\omega^2\right)\cot\frac{k\pi}{n}\right)}{2\left(b^2\alpha^2+\beta^2\omega^2\right)}\right) \end{cases}, k=1,\ldots,n-1, \tag{10.3}$$

where $V'(h^*)$, τ are all functions of ω. From the first equation of Eq. (10.3), we can find the relation of h^* with respect to ω. Then we can transform the critical curve for the stability switch of the uniform flow from plane $(V'(h^*), \tau)$ into plane (h^*, τ). We choose the parameter values as follows

$$\alpha = 0.9 \text{ (/s)}, \beta = 0.1 \text{ (/s)}, v^0 = 11 \text{ (m/s)}, h_{\text{stop}} = 14 \text{ (m)}. \tag{10.4}$$

Here we use the Routh-Hurwitz criterion [5] to prove the stability of the uniform flow on the upper and lower sides of the critical curve. We plot the boundary curves for stability switch of the uniform flow in the parameter plane (h^*, τ) by taking $n = 3, 5, 9, 13$ respectively, as shown in Fig. 10.1.

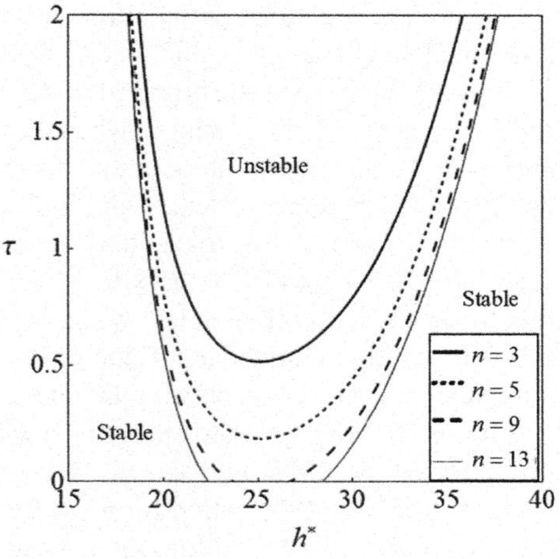

Fig. 10.1 The stability switch boundaries of the uniform flow in plane (h^*, τ)

In Fig. 10.1, we show the boundary curves of stability switch of the uniform flow for the case where the number of vehicles is taken as $n = 3, 5, 9, 13$ respectively. The horizontal axis denotes the mean vehicle distance h^*, and the vertical axis denotes the driver reaction time delay τ. Figure 10.1 illustrates three main points. Firstly, in the parameter plane (h^*, τ) defined by the average inter-vehicle distance and the driver's reaction time delay, the stable region of the uniform flow mode expands as the number n of vehicles on the road increases. This implies that reducing the number of vehicles on the road can, to some extent, stabilize the uniform flow. Secondly, when the average inter-vehicle distance remains constant, decreasing the number of vehicles leads to an expansion of the stable region of the parameter. This suggests that the instability of the uniform flow induced by crowded vehicles can be suppressed by reducing the reaction time delay. Thirdly, when the driver's reaction time delay is fixed, the uniform flow is not necessarily more stable for large inter-vehicle distance in comparison with small inter-vehicle distance (i.e., larger inter-vehicle distances are not always better).

10.1.1.2 Numerical Investigation of the Mechanism of Instability of Uniform Flow

The critical curve obtained from the linear analysis shows that as the two parameters (the average inter-vehicle distance and the driver's reaction time delay) enter the unstable region from the stable region, the uniform flow becomes unstable, and the velocity and inter-vehicle distance of the vehicles start to oscillate over time. To investigate this instability, we conduct numerical analysis along the lines corresponding to $h^* = 28(m)$ and $h^* = 34(m)$. The inter-vehicle distance selected for this analysis conforms to the assumption of inter-vehicle spacing [6]. The number of vehicles n is selected as 3, 5, 9, 13, respectively, and other parameters are set to the same values as in (10.4). The analysis then turns out to be a bifurcation problem with a single parameter, i.e., the driver's reaction time delay. We use DDE-BIFTOOL to carry out bifurcation continuation with respect to the driver's reaction delay. The numerical results are presented in Fig. 10.2.

Figure 10.2 presents the bifurcation diagram of the maximum and minimum velocities of each vehicle as τ changes when the number of vehicles n is taken respectively as 3, 5, 9, 13 under two inter-vehicle distances. Figure 10.2 (a_1), (b_1), (c_1) and (d_1) correspond to the supercritical Hopf bifurcation with $h^* = 28(\mathrm{m})$ and $n = 3, 5, 9, 13$, while Fig. 10.2 (a_2), (b_2), (c_2) and (d_2) correspond to the subcritical Hopf bifurcation with $h^* = 34(\mathrm{m})$ and $n = 3, 5, 9, 13$. The horizontal axis in Fig. 10.2 represents the driver's reaction time delay, while the vertical axis indicates the maximum and minimum velocities of the first vehicle. The "•" symbols represent the stable solutions while the "o" symbols represent the unstable solutions.

Figure 10.2 demonstrates that the stability switch mechanisms of uniform flow differ for $h^* = 28$ m and $h^* = 34$ m. The uniform flow is destabilized by the supercritical Hopf bifurcation for $h^* = 28$ m, while for $h^* = 34$ m it is destabilized by the subcritical Hopf bifurcation. Distinct mechanisms can lead to completely

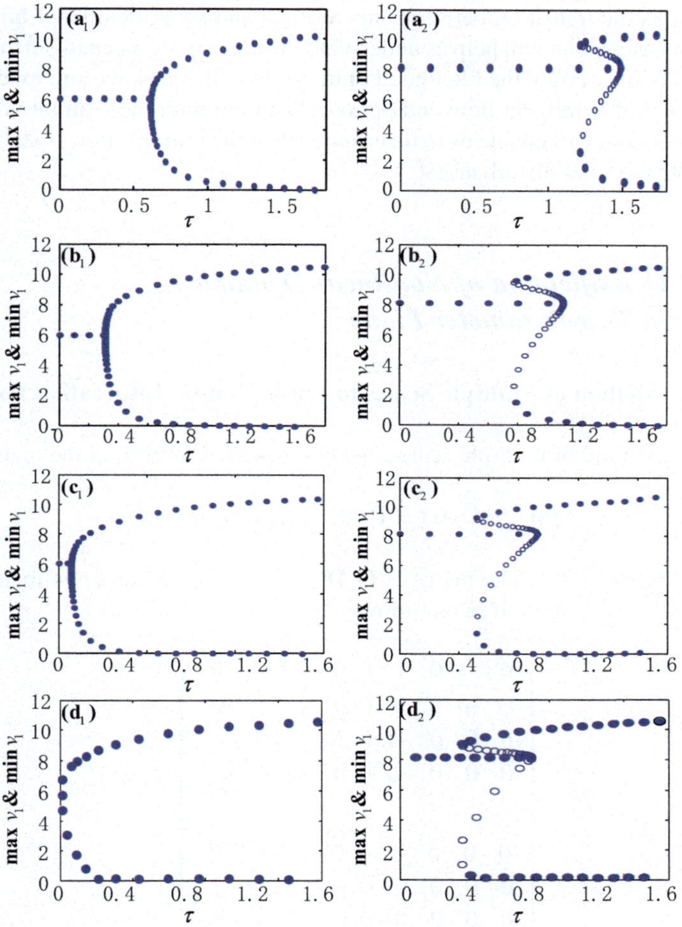

Fig. 10.2 Supercritical Hopf bifurcation diagrams ($h^* = 28$): $(a_1)n = 3$, $(b_1)n = 5$, $(c_1)n = 9$, and $(d_1)n = 13$. Subcritical Hopf bifurcation diagrams ($h^* = 34$): $(a_2)n = 3$, $(b_2)n = 5$, $(c_2)n = 9$, and $(d_2)n = 13$. The circle "o" and the dot "•" indicate unstable and stable solutions, respectively

different traffic modes. For instance, a supercritical Hopf bifurcation may result in a vibration-free traffic mode, while a subcritical one can induce stop-and-go traffic jams. Additionally, under the subcritical Hopf bifurcation, there exists a bistable area where two different traffic modes coexist. In this area, whether traffic jams occur or not depends on the driver's initial velocity and the inter-vehicle distance.

According to Fig. 10.2 it should be noted that the amplitude of the periodic solution induced by the supercritical Hopf bifurcation is small, while that induced by the subcritical Hopf bifurcation solution is large. This large amplitude can easily lead to a stop-and-go traffic mode. Therefore, in the following discussion we aim to determine the Bautin bifurcation point on the stability switch boundary of the uniform flow, which marks the transition between supercritical and subcritical Hopf bifurcations. Identifying such point will help us understand the underlying mechanism that induces instability and ascertain the likelihood of traffic jam. Besides, we aim to identify the region where the uniform flow and periodic solution with large amplitude coexist. This information will enable us to determine when the uniform flow is stable and not susceptible to initial disturbances.

10.1.2 Classification of Nonlinear Dynamics in Two-Parameter Plane

10.1.2.1 Method of Multiple Scales to Locate Bautin Bifurcation Point

To use the method of multiple scales, we first rewrite Eq. (9.6) in the matrix form

$$\dot{\mathbf{y}}(t) = \mathbf{C}\mathbf{y}(t) + \mathbf{D}\mathbf{y}(t-\tau) + \mathbf{F}(\mathbf{y}(t-\tau)), \tag{10.5}$$

where $\mathbf{y} : \mathbb{R} \to \mathbb{R}^{2n-1}$. The matrices $\mathbf{C}, \mathbf{D} \in \mathbb{R}^{(2n-1)\times(2n-1)}$ and nonlinear function $\mathbf{F} : \mathbb{R}^{2n-1} \to \mathbb{R}^{2n-1}$ are given as follows

$$\boldsymbol{C} = \begin{pmatrix} 0 & -1 & 0 & 1 & 0 & 0 & \cdots & 0 & 0 \\ 0 & -\alpha & 0 & 0 & 0 & 0 & \cdots & 0 & 0 \\ 0 & 0 & 0 & -1 & 0 & 1 & \cdots & 0 & 0 \\ 0 & 0 & 0 & -\alpha & 0 & 0 & \cdots & 0 & 0 \\ \vdots & \vdots & \vdots & \vdots & \vdots & \vdots & \ddots & \vdots & \vdots \\ 0 & 0 & 0 & 0 & 0 & 0 & \cdots & 0 & 0 \\ 0 & 0 & 0 & 0 & 0 & 0 & \cdots & -1 & 1 \\ 0 & 0 & 0 & 0 & 0 & 0 & \cdots & -\alpha & 0 \\ 0 & 0 & 0 & 0 & 0 & 0 & \cdots & 0 & -\alpha \end{pmatrix},$$

$$\boldsymbol{D} = \begin{pmatrix} 0 & 0 & 0 & 0 & 0 & 0 & \cdots & 0 & 0 \\ \alpha b & -\beta & 0 & \beta & 0 & 0 & \cdots & 0 & 0 \\ 0 & 0 & 0 & 0 & 0 & 0 & \cdots & 0 & 0 \\ 0 & 0 & \alpha b & -\beta & 0 & \beta & \cdots & 0 & 0 \\ \vdots & \vdots & \vdots & \vdots & \vdots & \vdots & \ddots & \vdots & \vdots \\ 0 & 0 & 0 & 0 & 0 & 0 & \cdots & \beta & 0 \\ 0 & 0 & 0 & 0 & 0 & 0 & \cdots & 0 & 0 \\ 0 & 0 & 0 & 0 & 0 & 0 & \cdots & -\beta & \beta \\ -\alpha b & \beta & -\alpha b & 0 & -\alpha b & 0 & \cdots & 0 & -\beta \end{pmatrix}$$

$$\mathbf{F}(\mathbf{y}(t-\tau)) = \left(0, \alpha \sum_{i=2}^{3} b_i y_1^i(t-\tau), 0, \alpha \sum_{i=2}^{3} b_i y_3^i(t-\tau), \ldots, \right.$$
$$\left. \alpha \sum_{i=2}^{3} b_i y_{2n-3}^i(t-\tau), \alpha \sum_{j=2}^{3} b_j \left(-\sum_{i=1}^{2n-3} y_i(t-\tau) \right)^j \right)^T. \quad (10.6)$$

The eigenmatrix $\mathbf{M} = \lambda\mathbf{I}-\mathbf{C}-\mathbf{D}e^{-\mathrm{i}\omega\tau_c}$ of the linearized system of (10.6) can be rewritten as

$$\mathbf{M} = \mathbf{M_R} + \mathrm{i}\mathbf{M_I}, \quad (10.7)$$

where $\mathbf{M_R}, \mathbf{M_I} \in \mathbb{R}^{(2n-1)\times(2n-1)}$ denotes the real and imaginary parts of $\mathbf{M}$, respectively. For each eigenvalue λ there exist the left and right eigenvectors

$$\mathbf{r} + \mathrm{i}\mathbf{s} = \begin{pmatrix} r_1 + \mathrm{i}s_1 \\ r_2 + \mathrm{i}s_2 \\ \vdots \\ r_{2n-1} + \mathrm{i}s_{2n-1} \end{pmatrix}, \mathbf{p} + \mathrm{i}\mathbf{q} = \begin{pmatrix} p_1 + \mathrm{i}q_1 \\ p_2 + \mathrm{i}q_2 \\ \vdots \\ p_{2n-1} + \mathrm{i}q_{2n-1} \end{pmatrix}, \quad (10.8)$$

which satisfy the following relations:

$$\begin{cases} r\mathrm{M_R} = s\mathrm{M_I}, \\ s\mathrm{M_R} = -r\mathrm{M_I}, \end{cases} \quad (10.9)$$

and

$$\begin{cases} \mathrm{M_R}p = \mathrm{M_I}q, \\ \mathrm{M_R}q = -\mathrm{M_I}p. \end{cases} \quad (10.10)$$

In general, there are infinitely many solutions of the eigenvectors satisfying Eqs. (10.9) and (10.10), and here we choose a particular solution as the eigenvector of the eigenmatrix $\mathbf{M}$. For this purpose, we let $r_1 = 1, s_1 = 0, p_1 = 1, q_1 = 0$. In this way we find the unique solution satisfying Eqs. (10.9) and (10.10).

When the mean inter-vehicle distance and the reaction time delay lie exactly on the critical curve of stability switch of the uniform flow, we have $\lambda = i\omega$. In this case, the bifurcated dynamical behaviour of the system near the critical curve can be analyzed by the method of multiple scales. To this end, we need to introduce different time scales as follows

$$T_0 = t, T_1 = \varepsilon t, T_2 = \varepsilon^2 t, \quad (10.11)$$

where $\varepsilon \ll 1$ is a dimensionless parameter. Accordingly, the time derivative of the state variable $\mathbf{y}(t)$ becomes:

$$\frac{d\mathbf{y}(t)}{dt} = \varepsilon\frac{\partial \mathbf{y}(T_0, T_1, T_2)}{\partial T_0} + \varepsilon^2\frac{\partial \mathbf{y}(T_0, T_1, T_2)}{\partial T_1} + \varepsilon^3\frac{\partial \mathbf{y}(T_0, T_1, T_2)}{\partial T_2}, \tag{10.12}$$

where

$$\mathbf{y}(t) = \mathbf{y}(T_0, T_1, T_2) = \varepsilon\begin{pmatrix} y_{1,1}(T_0, T_1, T_2) \\ y_{2,1}(T_0, T_1, T_2) \\ \vdots \\ y_{2n-1,1}(T_0, T_1, T_2) \end{pmatrix} + \varepsilon^2\begin{pmatrix} y_{1,2}(T_0, T_1, T_2) \\ y_{2,2}(T_0, T_1, T_2) \\ \vdots \\ y_{2n-1,2}(T_0, T_1, T_2) \end{pmatrix}$$
$$+\varepsilon^3\begin{pmatrix} y_{1,3}(T_0, T_1, T_2) \\ y_{2,3}(T_0, T_1, T_2) \\ \vdots \\ y_{2n-1,3}(T_0, T_1, T_2) \end{pmatrix}. \tag{10.13}$$

The delayed term in (10.13) can be written as

$$\mathbf{y}(t-\tau) = \mathbf{y}\big(T_0 - \varepsilon\tau, T_1 - \varepsilon^2\tau, T_2 - \varepsilon^3\tau\big). \tag{10.14}$$

We perform the perturbation analysis in a very small region near the critical parameter τ_c, h_c^*. Namely, we let

$$\tau = \tau_c + \varepsilon\tau_\varepsilon, h^* = h_c^* + \varepsilon h_\varepsilon^*. \tag{10.15}$$

In the following we expand the delayed term at $\mathbf{y}(T_0 - \tau_c, 0, 0)$ and truncate it to the cubic term of ε:

$$\begin{aligned}\mathbf{y_i}(t-\tau) = \mathbf{y_i}(T_0 - \tau_c, T_1, T_2) &= \varepsilon y_{i,1}(T_0 - \tau_c, T_1, T_2) \\ &+ \varepsilon^2\left(y_{i,2}(T_0 - \tau_c, T_1, T_2) - \tau_c\frac{\partial y_{i,1}(T_0 - \tau_c, T_1, T_2)}{\partial T_1}\right) \\ &+ \varepsilon^3\left(y_{i,3}(T_0 - \tau_c, T_1, T_2) - \tau_c\frac{\partial y_{i,1}(T_0 - \tau_c, T_1, T_2)}{\partial T_2}\right. \\ &\left.-\tau_c\frac{\partial y_{i,2}(T_0 - \tau_c, T_1, T_2)}{\partial T_1} + \frac{1}{2}\tau_c^2\frac{\partial^2 y_{i,1}(T_0 - \tau_c, T_1, T_2)}{\partial T_1^2}\right), \\ i &= 1, 2, \ldots, 2n-1. \end{aligned} \tag{10.16}$$

Clearly, at the Bautin bifurcation point we have

$$\tau = \tau_c, h^* = h_c^*. \tag{10.17}$$

Now we substitute (10.12), (10.13), (10.16), (10.17) into the matrix form of Eq. (10.5), regroup the equation according to the order of ε and equate the coefficients of $\varepsilon^1, \varepsilon^2, \varepsilon^3$ to zero to obtain a series of equations:

The equation obtained from the coefficient of ε^1 is given by

$$\frac{\partial y_{2i-1,1}}{\partial T_0}(T_0, T_1, T_2) + y_{2i,1}(T_0, T_1, T_2) - y_{2i+2,1}(T_0, T_1, T_2) = 0, i = 1, 2, \ldots, n-1, \tag{10.18}$$

$$\begin{aligned} &\frac{\partial y_{2i,1}}{\partial T_0}(T_0, T_1, T_2) + \alpha y_{2i,1}(T_0, T_1, T_2) - \alpha b_{1c} y_{2i-1,1}(T_0 - \tau_c, T_1, T_2) \\ &\quad + \beta y_{2i,1}(T_0 - \tau_c, T_1, T_2) - \beta y_{2i+2,1}(T_0 - \tau_c, T_1, T_2) = 0, \\ &\quad i = 1, 2, \ldots, n-1, \end{aligned} \tag{10.19}$$

$$\begin{aligned} &\frac{\partial y_{2n-1,1}}{\partial T_0}(T_0, T_1, T_2) + \alpha b_{1c} \sum_{i=1}^{n-1} y_{2i-1,1}(T_0 - \tau_c, T_1, T_2) \\ &\quad + \alpha y_{2n-1,1}((T_0, T_1, T_2)) + \beta y_{2n-1,1}(T_0 - \tau_c, T_1, T_2) \\ &\quad - \beta y_{2,1}(T_0 - \tau_c, T_1, T_2) = 0. \end{aligned} \tag{10.20}$$

Likewise, from the coefficient of ε^2 we obtain

$$\begin{aligned} &\frac{\partial y_{2i-1,2}}{\partial T_0}(T_0, T_1, T_2) + y_{2i,2}(T_0, T_1, T_2) - y_{2i+2,2}(T_0, T_1, T_2) \\ &\quad = -\frac{\partial y_{2i-1,1}}{\partial T_1}(T_0, T_1, T_2), i = 1, 2, \ldots, n-1, \end{aligned} \tag{10.21}$$

$$\begin{aligned} &\frac{\partial y_{2i,2}}{\partial T_0}(T_0, T_1, T_2) + \alpha y_{2i,2}(T_0, T_1, T_2) - \alpha b_{1c} y_{2i-1,2}(T_0 - \tau_c, T_1, T_2) \\ &\quad + \beta\big(y_{2i,2}(T_0 - \tau_c, T_1, T_2) - y_{2i+2,2}(T_0 - \tau_c, T_1, T_2)\big) = 0, \\ &\quad i = 1, 2, \ldots, n-1, \end{aligned} \tag{10.22}$$

$$\begin{aligned} &\frac{\partial y_{2n-1,2}}{\partial T_0}(T_0, T_1, T_2) + \alpha b_{1c} \sum_{i=1}^{n-1} y_{2i-1,2}(T_0 - \tau_c, T_1, T_2) + \alpha y_{2n-1,2}(T_0, T_1, T_2) \\ &\quad + \beta y_{2n-1,2}(T_0 - \tau_c, T_1, T_2) - \beta y_{2,2}(T_0 - \tau_c, T_1, T_2) \\ &\quad = -\frac{\partial y_{2n-1,1}}{\partial T_1}(T_0, T_1, T_2) \\ &\quad + \alpha b_{2c}\left(\sum_{i=1}^{n-1} y_{2i-1,1}(T_0 - \tau_c, T_1, T_2)\right)^2 \end{aligned}$$

$$+\alpha b_{1c}\tau_c\sum_{i=1}^{n-1}\frac{\partial y_{2i-1,1}}{\partial T_1}(T_0-\tau_c,T_1,T_2)$$
$$-\beta\tau_c\left(\frac{\partial y_{2,1}}{\partial T_1}(T_0-\tau_c,T_1,T_2)-\frac{\partial y_{2n-1,1}}{\partial T_1}(T_0-\tau_c,T_1,T_2)\right). \tag{10.23}$$

The equation for the coefficients of ε^3 is

$$\frac{\partial y_{2i-1,3}}{\partial T_0}(T_0,T_1,T_2)+y_{2i,3}(T_0,T_1,T_2)-y_{2i+2,3}(T_0,T_1,T_2)$$
$$=-\frac{\partial y_{2i-1,1}}{\partial T_2}(T_0,T_1,T_2)-\frac{\partial y_{2i-1,2}}{\partial T_1}(T_0,T_1,T_2),$$
$$i=1,2,\ldots,n-1, \tag{10.24}$$

$$\frac{\partial y_{2i,3}}{\partial T_0}(T_0,T_1,T_2)+\alpha y_{2i,3}(T_0,T_1,T_2)-\alpha b_{1c}y_{2i-1,3}(T_0-\tau_c,T_1,T_2)$$
$$+\beta\big(y_{2i,3}(T_0-\tau_c,T_1,T_2)-y_{2i+2,3}(T_0-\tau_c,T_1,T_2)\big)$$
$$=-\frac{\partial y_{2i,1}}{\partial T_2}(T_0,T_1,T_2)-\frac{\partial y_{2i,2}}{\partial T_1}(T_0,T_1,T_2)$$
$$+\alpha b_{3c}y_{2i-1,1}^3(T_0-\tau_c,T_1,T_2)$$
$$+2\alpha b_{2c}y_{2i-1,1}(T_0-\tau_c,T_1,T_2)\big(y_{2i-1,2}(T_0-\tau_c,T_1,T_2)$$
$$-\tau_c\frac{\partial y_{2i-1,1}}{\partial T_1}(T_0-\tau_c,T_1,T_2))$$
$$-\alpha b_{1c}\tau_c\left(\frac{\partial y_{2i-1,1}}{\partial T_2}(T_0-\tau_c,T_1,T_2)\right.$$
$$\left.+\frac{\partial y_{2i-1,2}}{\partial T_1}(T_0-\tau_c,T_1,T_2)\right)+\beta\tau_c\left(\frac{\partial y_{2i,1}}{\partial T_2}(T_0-\tau_c,T_1,T_2)\right.$$
$$+\frac{\partial y_{2i,2}}{\partial T_1}(T_0-\tau_c,T_1,T_2)-\frac{\partial y_{2i+2,1}}{\partial T_2}(T_0-\tau_c,T_1,T_2)$$
$$\left.-\frac{\partial y_{2i+2,2}}{\partial T_1}(T_0-\tau_c,T_1,T_2)\right)$$
$$+\frac{\alpha b_{1c}\tau_c^2}{2}\frac{\partial^2 y_{2i-1,1}}{\partial T_1^2}(T_0-\tau_c,T_1,T_2)$$
$$-\frac{\beta\tau_c^2}{2}\left(\frac{\partial^2 y_{2i,1}}{\partial T_1^2}(T_0-\tau_c,T_1,T_2)\right.$$
$$\left.-\frac{\partial^2 y_{2i+2,1}}{\partial T_1^2}(T_0-\tau_c,T_1,T_2)\right)=0,$$
$$i=1,2,\ldots,n-1, \tag{10.25}$$

$$
\begin{aligned}
&\frac{\partial y_{2n-1,3}}{\partial T_0}(T_0, T_1, T_2) + \alpha b_{1c} \sum_{i=1}^{n-1} y_{2i-1,3}(T_0 - \tau_c, T_1, T_2) + \alpha y_{2n-1,3}(T_0, T_1, T_2) \\
&\quad + \beta y_{2n-1,3}(T_0 - \tau_c, T_1, T_2) - \beta y_{2,3}(T_0 - \tau_c, T_1, T_2) \\
&\quad = -\frac{\partial y_{2n-1,1}}{\partial T_2}(T_0, T_1, T_2) - \frac{\partial y_{2n-1,2}}{\partial T_1}(T_0, T_1, T_2) \\
&\quad + \alpha b_{3c} \left(\sum_{i=1}^{n-1} -y_{2i-1,1}(T_0 - \tau_c, T_1, T_2) \right)^3 \\
&\quad + 2\alpha b_{2c} \left(\sum_{i=1}^{n-1} -y_{2i-1,1}(T_0 - \tau_c, T_1, T_2) \right) \\
&\quad \left(\sum_{i=1}^{n-1} \tau_c \frac{\begin{array}{c}(-y_{2i-1,2}(T_0 - \tau_c, T_1, T_2) \\ +\partial y_{2i-1,1}(T_0 - \tau_c, T_1, T_2))\end{array}}{\partial T_1} \right) \\
&\quad + \alpha b_{1c} \tau_c \left(\sum_{i=1}^{n-1} \frac{\partial y_{2i-1,1}}{\partial T_2}(T_0 - \tau_c, T_1, T_2) \right. \\
&\quad \left. + \frac{\partial y_{2i-1,2}}{\partial T_1}(T_0 - \tau_c, T_1, T_2) \right) \\
&\quad - \beta \tau_c \left(\frac{\partial y_{2,1}}{\partial T_2}(T_0 - \tau_c, T_1, T_2) + \frac{\partial y_{2,2}}{\partial T_1}(T_0 - \tau_c, T_1, T_2) \right. \\
&\quad \left. - \frac{\partial y_{2n-1,1}}{\partial T_2}(T_0 - \tau_c, T_1, T_2) - \frac{\partial y_{2n-1,2}}{\partial T_1}(T_0 - \tau_c, T_1, T_2) \right) \\
&\quad - \frac{\alpha b_{1c} \tau_c^2}{2} \sum_{i=1}^{n-1} \frac{\partial^2 y_{2i-1,1}}{\partial T_1^2}(T_0 - \tau_c, T_1, T_2) \\
&\quad + \frac{\beta \tau_c^2}{2} \left(\begin{array}{c} \frac{\partial^2 y_{2,1}}{\partial T1^2}(T_0 - \tau_c, T_1, T_2) \\ -\frac{\partial^2 y_{2n-1,1}}{\partial T_1^2}(T_0 - \tau_c, T_1, T_2) \end{array} \right),
\end{aligned} \tag{10.26}
$$

where

$$
b_{1c} = V'(h_c^*), b_{2c} = \frac{1}{2} V''(h_c^*), b_{3c} = \frac{1}{6} V'''(h_c^*). \tag{10.27}
$$

Combined with Eq. (10.10), it can be seen that the solutions of Eqs. (10.18), (10.19) and (10.20) take the following form

$$\mathbf{y}_1(T_0, T_1, T_2) = \begin{pmatrix} y_{1,1}(T_0, T_1, T_2) \\ y_{2,1}(T_0, T_1, T_2) \\ \vdots \\ y_{2n-1,1}(T_0, T_1, T_2) \end{pmatrix} = B(T_1, T_2)(\mathrm{p} + \mathrm{i}q)e^{\mathrm{i}\omega T_0} + \text{c.c.}, \quad (10.28)$$

where c.c. denotes the complex conjugate of all the terms left to it, and ω is from the imaginary part of the critical eigenvalue $\lambda = \mathrm{i}\omega$.

By substituting (10.28) into Eqs. (10.21), (10.22) and (10.23), we obtain:

$$\frac{\partial \mathbf{y}_2(T_0, T_1, T_2)}{\partial T_0} - \mathbf{C}\mathbf{y}_2(T_0, T_1, T_2) - \mathbf{D}\mathbf{y}_2(T_0 - \tau_c, T_1, T_2) = \mathrm{ST}_1 + \text{c.c.} + \mathrm{NST}_1, \quad (10.29)$$

where ST_1 represents the resonance term (solvability condition) and NST_1 the remaining non-resonance terms. The resonance term ST_1 is given by

$$\mathrm{ST}_1 = \begin{pmatrix} -p_1 - \mathrm{i}q_1 \\ -p_2 - \mathrm{i}q_2 + \Delta_{1,1} \\ \vdots \\ -p_{2i-1} - \mathrm{i}q_{2i-1} \\ -p_{2i} - \mathrm{i}q_{2i} + \Delta_{1,i} \\ \vdots \\ -p_{2n-3} - \mathrm{i}q_{2n-3} \\ -p_{2n-2} - \mathrm{i}q_{2n-2} + \Delta_{1,n-1} \\ -p_{2n-1} - \mathrm{i}q_{2n-1} + \Delta_{1,n} \end{pmatrix} e^{\mathrm{i}\omega T_0} \frac{\partial B(T_1, T_2)}{\partial T_1}, \quad (10.30)$$

where

$$\begin{aligned} \Delta_{1,i} &= -\alpha b_{1c}\tau_c(p_{2i-1} + \mathrm{i}q_{2i-1})e^{-\mathrm{i}\omega\tau_c} + \beta\tau_c(p_{2i} + \mathrm{i}q_{2i} - p_{2i+2} - \mathrm{i}q_{2i+2})e^{-\mathrm{i}\omega\tau_c}, \\ i &= 1, 2, \cdots, n-2, \\ \Delta_{1,n-1} &= -\alpha b_{1c}\tau_c(p_{2n-3} + \mathrm{i}q_{2n-3})e^{-\mathrm{i}\omega\tau_c} \\ &\quad + \beta\tau_c(p_{2n-2} + \mathrm{i}q_{2n-2} - p_{2n-1} - \mathrm{i}q_{2n-1})e^{-\mathrm{i}\omega\tau_c}, \\ \Delta_{1,n} &= -\alpha b_{1c}\tau_c \sum_{i=1}^{n-1} (p_{2i-1} + \mathrm{i}q_{2i-1})e^{-\mathrm{i}\omega\tau_c} \\ &\quad - \beta\tau_c(p_2 + \mathrm{i}q_2 - p_{2n-1} - \mathrm{i}q_{2n-1})e^{-\mathrm{i}\omega\tau_c}. \end{aligned}$$

According to Nayfeh [7, 8], in order to get rid of the secular term, we need to eliminate the resonance term ST_1 in Eq. (10.29). For this purpose, we assume that Eq. (10.29) has the particular solution of the following form

$$\mathbf{y}_2^*(T_0, T_1, T_2) = \varphi(T_1, T_2)e^{\mathrm{i}\omega T_0} = \begin{pmatrix} \varphi_1(T_1, T_2) \\ \varphi_2(T_1, T_2) \\ \vdots \\ \varphi_{2n-1}(T_1, T_2) \end{pmatrix} e^{\mathrm{i}\omega T_0}, \tag{10.31}$$

and the particular solution $\mathbf{y}_1^*$ needs to satisfy

$$(\mathbf{M_R} + \mathrm{i}\mathbf{M_I})\mathbf{y}_1^*(T_0, T_1, T_2) = \mathrm{ST}_1. \tag{10.32}$$

By Fredholm Alternative Theorem, a sufficient condition for the existence of $\mathbf{y}_1^*(T_0, T_1, T_2)$ is that

$$(\mathrm{r} + \mathrm{i}s).\mathrm{ST}_1 = 0. \tag{10.33}$$

So far, we have transformed the search for the solvability condition of Eq. (10.29) into the problem of solving Eq. (10.33). Then the solvability condition for Eq. (10.29) is obtained as

$$\frac{\partial B(T_1, T_2)}{\partial T_1} = 0. \tag{10.34}$$

Substituting the solution $\mathbf{y}_1$ in (10.28) and the solvability condition (10.34) into Eq. (10.29), we can obtain the following equations:

$$\frac{\partial \mathbf{y}_2(T_0, T_1, T_2)}{\partial T_0} - \mathbf{C}\mathbf{y}_2(T_0, T_1, T_2) - \mathbf{D}\mathbf{y}_2(T_0 - \tau_c, T_1, T_2)$$
$$= \begin{pmatrix} 0 \\ -\alpha b_{2c}(p_1 + \mathrm{i}q_1)^2 e^{-2\mathrm{i}\omega\tau_c} \\ \vdots \\ 0 \\ -\alpha b_{2c}\left(p_{2i-1} + \mathrm{i}q_{2i-1}\right)^2 e^{-2\mathrm{i}\omega\tau_c} \\ \vdots \\ -\alpha b_{2c}\left(p_{2n-3} + \mathrm{i}q_{2n-3}\right)^2 e^{-2\mathrm{i}\omega\tau_c} \\ -\alpha b_{2c}\left(\sum_{i=1}^{n-1}(p_{2i-1} + \mathrm{i}q_{2i-1})\right)^2 e^{-2\mathrm{i}\omega\tau_c} \end{pmatrix} B(T_1, T_2)^2 e^{2\mathrm{i}\omega T_0}$$

$$
+\begin{pmatrix} 0 \\ -2\alpha b_{2c}\left(p_1^2+q_1^2\right) \\ \vdots \\ 0 \\ -2\alpha b_{2c}\left(p_{2i-1}^2+q_{2i-1}^2\right) \\ \vdots \\ -2\alpha b_{2c}\left(p_{2n-3}^2+q_{2n-3}^2\right) \\ -2\alpha b_{2c}\left(\left(\sum_{i=1}^{n-1} p_{2i-1}\right)^2+\left(\sum_{i=1}^{n-1} q_{2i-1}\right)^2\right) \end{pmatrix} B(T_1,T_2)\overline{B}(T_1,T_2)
$$

$$
+\ \text{c.c.}, \tag{10.35}
$$

where $\overline{B}(T_1, T_2)$ denotes the complex conjugate of $B(T_1, T_2)$. Based on Eq. (10.35), we can assume that

$$
\mathbf{y}_2=\begin{pmatrix} \Delta_{2,1} \\ \Delta_{2,2} \\ \vdots \\ \Delta_{2,2n-1} \end{pmatrix} B(T_1,T_2)^2 e^{i2\omega T_0}+\begin{pmatrix} \Delta_{3,1} \\ \Delta_{3,2} \\ \vdots \\ \Delta_{3,2n-1} \end{pmatrix} B(T_1,T_2)\overline{B}(T_1,T_2)+\text{c.c.}. \tag{10.36}
$$

Substituting (10.36) into Eq. (10.35), we obtain the equations about $\Delta_{2,i}$ and $\Delta_{3,i}$, $i = 1, 2, \ldots, 2n - 1$, respectively. The equation for $\Delta_{3,i}$ is as follows

$$
\begin{aligned}
\Delta_{3,2i} &- \Delta_{3,2i+2} = 0, i = 1, 2, \ldots, n-2 \\
&- 2\alpha b_{2c}\left(p_{2i-1}^2+q_{2i-1}^2\right) + (\alpha+\beta)\Delta_{3,2i} - \beta\Delta_{3,2i+2} \\
&- \alpha b_{1c}\Delta_{3,2i-1} = 0, \\
i &= 1, 2, \ldots, n-2 \\
\Delta_{3,2n-2} &- \Delta_{3,2n-1} = 0, i = 1, 2, \ldots, n-2 \\
&- 2\alpha b_{2c}\left(p_{2n-3}^2+q_{2n-3}^2\right) + (\alpha+\beta)\Delta_{3,2n-2} - \beta\Delta_{3,2n-1} \\
&- \alpha b_{1c}\Delta_{3,2n-3} = 0, \\
&- 2\alpha b_{2c}\left(\left(\sum_{i=1}^{n-1} p_{2i-1}\right)^2+\left(\sum_{i=1}^{n-1} q_{2i-1}\right)^2\right) + (\alpha+\beta)\Delta_{3,2n-1} \\
&- \beta\Delta_{3,2n-1} - \alpha b_{1c}\sum_{i=1}^{n-1}\Delta_{3,2i-1} = 0.
\end{aligned} \tag{10.37}
$$

Solving Eq. (10.37) yields the following analytic form for $\Delta_{3,i}$

$$\Delta_{3,2i-1} = \frac{2b_{2c}}{9b_{1c}}\left(-9p_{2i-1}^2 - 9q_{2i-1}^2 + 2\sum_{i=1}^{n-1}\sum_{j=i}^{n-1} p_{2i-1}p_{2j-1},\right.$$
$$\left.+2\sum_{i=1}^{n-1}\sum_{j=i}^{n-1} q_{2i-1}q_{2j-1}\right),$$
$$\Delta_{3,2i} = \frac{4b_{2c}}{9}\left(\sum_{i=1}^{n-1}\sum_{j=i}^{n-1} p_{2i-1}p_{2j-1} + \sum_{i=1}^{n-1}\sum_{j=i}^{n-1} q_{2i-1}q_{2j-1}\right),$$
$$i = 1, \ldots, n-1,$$
$$\Delta_{3,2n-1} = \frac{4b_{2c}}{9}\left(\sum_{i=1}^{n-1}\sum_{j=i}^{n-1} p_{2i-1}p_{2j-1} + \sum_{i=1}^{n-1}\sum_{j=i}^{n-1} q_{2i-1}q_{2j-1}\right). \tag{10.38}$$

The $2n-1$ equations of $\Delta_{2,i}$ are as follows:

$$\begin{aligned}
&2\mathrm{i}\omega\Delta_{2,2i-1} + \Delta_{2,2i} - \Delta_{2,2i+2} = 0,\, i = 1, 2, \ldots, n \\
&\quad - 2\left(\alpha + 2\mathrm{i}\omega + \beta e^{-2\mathrm{i}\omega\tau_c}\right)\Delta_{2,2i} \\
&\quad - \beta e^{-2\mathrm{i}\omega\tau_c}\Delta_{2,2i+2} = \alpha b_{2c}(p_{2i-1} + \mathrm{i}q_{2i-1})^2 e^{-2\mathrm{i}\omega\tau_c} \\
&\quad + \alpha b_{1c}e^{-2\mathrm{i}\omega\tau_c}\Delta_{2,2i-1},\, i = 1, 2, \ldots, n-2 \\
&2\mathrm{i}\omega\Delta_{2,2n-3} + \Delta_{2,2n-2} - \Delta_{2,2n-1} = 0,\, \left(\alpha + 2\mathrm{i}\omega + \beta e^{-2\mathrm{i}\omega\tau_c}\right)\Delta_{2,2n-2} \\
&\quad - \beta e^{-2\mathrm{i}\omega\tau_c}\Delta_{2,2n-1} = \alpha b_{2c}(p_{2n-3} + \mathrm{i}q_{2n-3})^2 e^{-2\mathrm{i}\omega\tau_c} \\
&\quad + \alpha b_{1c}e^{-2\mathrm{i}\omega\tau_c}\Delta_{2,2n-3},\, \left(\alpha + 2\mathrm{i}\omega + \beta e^{-2\mathrm{i}\omega\tau_c}\right)\Delta_{2,2n-1} \\
&\quad - \beta e^{-2\mathrm{i}\omega\tau_c}\Delta_{2,2} = \alpha b_{2c}\left(\sum_{i=1}^{n-1} p_{2i-1} + \mathrm{i}q_{2i-1}\right)^2 e^{-2\mathrm{i}\omega\tau_c} \\
&\quad - \alpha b_{1c}e^{-2\mathrm{i}\omega\tau_c}\sum_{i=1}^{n-1}\Delta_{2i-1}.
\end{aligned} \tag{10.39}$$

Solving the first $2n-2$ Eq. (10.39) yields:

$$\Delta_{2,2i-1} = \frac{-\alpha b_{2c}(p_{2i-1} + \mathrm{i}q_{2i-1})^2 + (\alpha + 2\mathrm{i}\omega)e^{2\mathrm{i}\omega\tau_c}\Delta_{2,2i}}{\alpha b_{1c} + 2\mathrm{i}\omega\beta},$$

$$\Delta_{2,2i+2} = \frac{\begin{gathered}- 2\mathrm{i}\omega\alpha b_{2c}(p_{2i-1} + \mathrm{i}q_{2i-1})^2 \\ + \left(\alpha b_1 + 2\mathrm{i}\beta\omega + 2\mathrm{i}e^{2\mathrm{i}\tau_c\omega_1}(\alpha + 2i\omega)\omega\right)\Delta_{2,2i}\end{gathered}}{\alpha b_{1c} + 2\mathrm{i}\omega\beta},$$

$$\Delta_{2,2n-3} = \frac{-\alpha b_{2c}(p_{2n-3} + \mathrm{i}q_{2n-3})^2 + (\alpha + 2\mathrm{i}\omega)e^{2\mathrm{i}\omega\tau_c}\Delta_{2,2n-2}}{\alpha b_{1c} + 2\mathrm{i}\omega\beta},$$

$$\Delta_{2,2n-1} = \frac{\begin{array}{l} -\,2\mathrm{i}\omega\alpha b_{2c}(p_{2n-3} + \mathrm{i}q_{2n-3})^2 \\ +\,\left(\alpha b_1 + 2\mathrm{i}\beta\omega + 2\mathrm{i}e^{2\mathrm{i}\tau_c\omega_1}(\alpha + 2i\omega)\omega\right)\Delta_{2,2n-2} \end{array}}{\alpha b_{1c} + 2\mathrm{i}\omega\beta}, \tag{10.40}$$

where $i = 1, 2, \ldots, n-2$. From Eq. (10.40) we have

$$\Delta_{2,i} = f_i\left(\Delta_{2,2}\right), i = 1, 2, \ldots, 2n-1, \tag{10.41}$$

i.e., all $\Delta_{2,i}$ can be expressed in terms of $\Delta_{2,2}$. Consequently, we obtain the following equation of $\Delta_{2,2}$

$$\left(\alpha + 2\mathrm{i}\omega + \beta e^{-2\mathrm{i}\omega\tau_c}\right) f_{2n-1}\left(\Delta_{2,2}\right) - \beta e^{-2\mathrm{i}\omega\tau_c}\Delta_{2,2} = \alpha b_{2c}\left(\sum_{i=1}^{n-1} p_{2i-1} + \mathrm{i}q_{2i-1}\right)^2$$
$$e^{-2\mathrm{i}\omega\tau_c} - \alpha b_{1c} e^{-2\mathrm{i}\omega\tau_c}\sum_{i=1}^{n-1} f_{2i-1}\left(\Delta_{2,2}\right). \tag{10.42}$$

Solving Eq. (10.42), we can obtain

$$\Delta_{2,2} = f(p_1, q_1, \ldots, p_{2n-1}, q_{2n-1}), \tag{10.43}$$

that is, $\Delta_{2,2}$ is a function of the right eigenvector of $\mathbf{M}$. As a result, we now obtain the particular solution $\mathbf{y_2}$.

Following the similar process as above, we have

$$\frac{\partial \mathbf{y_3}(T_0, T_1, T_2)}{\partial T_0} - \mathbf{C}\mathbf{y_3}(T_0, T_1, T_2) - \mathbf{D}\mathbf{y_3}(T_0 - \tau_c, T_1, T_2) = \mathrm{ST}_2 + \mathrm{c.c.} + \mathrm{NST}_2, \tag{10.44}$$

where the resonance term ST_2 is given by

$$\mathrm{ST}_2 = \begin{pmatrix} \Delta_{4,1} \\ \Delta_{4,2} \\ \vdots \\ \Delta_{4,2n-1} \end{pmatrix} e^{\mathrm{i}\omega T_0} B(T_1, T_2)^2 \overline{B}(T_1, T_2) + \begin{pmatrix} \Delta_{5,1} \\ \Delta_{5,2} \\ \vdots \\ \Delta_{5,2n-1} \end{pmatrix} e^{\mathrm{i}\omega T_0} \frac{\partial B(T_1, T_2)}{\partial T_2}. \tag{10.45}$$

The coefficients are obtained as

$$\Delta_{4,2i-1} = 0, i = 1, 2, \ldots, n-1$$
$$\Delta_{4,2i} = 2\alpha e^{-\mathrm{i}\omega\tau_c} b_{2c}\big((p_{2i-1} - \mathrm{i}q_{2i-1})\Delta_{2,2i-1} + (p_{2i-1} + \mathrm{i}q_{2i-1})\Delta_{3,2i-1}\big) + 6\alpha e^{-\mathrm{i}\omega\tau_c} b_{2c}b_{3c}(p_{2i-1} - \mathrm{i}q_{2i-1})(p_{2i-1} + \mathrm{i}q_{2i-1})^2,$$
$$i = 1, 2, \ldots, n-2$$
$$\Delta_{4,2n-2} = 2\alpha e^{-\mathrm{i}\omega\tau_c} b_{2c}\big((p_{2n-3} - \mathrm{i}q_{2n-3})\Delta_{2,2n-3} + (p_{2n-3} + \mathrm{i}q_{2n-3})\Delta_{3,2n-3}\big) + 3\alpha e^{-\mathrm{i}\omega\tau_c} b_{3c}(p_{2n-3} - \mathrm{i}q_{2n-3})(p_{2n-3} + \mathrm{i}q_{2n-3})^2,$$
$$\Delta_{4,2n-1} = 2\alpha e^{-\mathrm{i}\omega\tau_c} b_{2c}\left(\sum_{i=1}^{n-1}(p_{2i-1} - \mathrm{i}q_{2i-1})\sum_{i=1}^{n-1}\Delta_{2,2i-1} + \sum_{i=1}^{n-1}(p_{2i-1} + \mathrm{i}q_{2i-1})\sum_{i=1}^{n-1}\Delta_{3,2i-1}\right) + 3\alpha e^{-\mathrm{i}\omega\tau_c} b_{3c}\sum_{i=1}^{n-1}(p_{2i-1} - \mathrm{i}q_{2i-1})\left(\sum_{i=1}^{n-1}(p_{2i-1} + \mathrm{i}q_{2i-1})\right)^2,$$

$$\Delta_{5,2i-1} = -p_{2i-1} - \mathrm{i}q_{2i-1}, i = 1, 2, \ldots, n-1,$$
$$\Delta_{5,2i} = -p_{2i} - \mathrm{i}q_{2i} - \alpha b_{1c}\tau_c(p_{2i-1} + \mathrm{i}q_{2i-1})e^{-\mathrm{i}\omega\tau_c} + \beta\tau_c(p_{2i} + \mathrm{i}q_{2i} - p_{2i+2} - \mathrm{i}q_{2i+2})e^{-\mathrm{i}\omega\tau_c}, i = 1, 2, \ldots, n-2,$$
$$\Delta_{5,2n-2} = -p_{2n-2} - \mathrm{i}q_{2n-2} - \alpha b_{1c}\tau_c(p_{2n-3} + \mathrm{i}q_{2n-3})e^{-\mathrm{i}\omega\tau_c} + \beta\tau_c(p_{2n-2} + \mathrm{i}q_{2n-2} - p_{2n-1} - \mathrm{i}q_{2n-1})e^{-\mathrm{i}\omega\tau_c},$$
$$\Delta_{5,2n-1} = -p_{2n-1} - \mathrm{i}q_{2n-1} - \alpha b_{1c}\tau_c\sum_{i=1}^{n-1}(p_{2i-1} + \mathrm{i}q_{2i-1})e^{-\mathrm{i}\omega\tau_c} - \beta\tau_c(p_2 + \mathrm{i}q_2 - p_{2n-1} - \mathrm{i}q_{2n-1})e^{-\mathrm{i}\omega\tau_c}.$$

To remove the secular term ST_2, we use the same method as above to obtain the following system of algebraic equations

$$(\mathrm{r} + \mathrm{is}) \cdot \mathrm{ST}_2 = 0. \tag{10.46}$$

Solving Eq. (10.46) leads to the following solvability condition of Eq. (10.44)

$$\frac{\partial B(T_1, T_2)}{\partial T_2} = \frac{\Lambda_1 + \mathrm{i}\Lambda_2}{\Lambda_3 + \mathrm{i}\Lambda_4} B(T_1, T_2)^2 \overline{B}(T_1, T_2), \tag{10.47}$$

where

$$\Lambda_1 = \mathrm{Re}\left(-\sum_{i=1}^{2n-1}(r_i + \mathrm{i}s_i)\Delta_{4,i}\right), \Lambda_2 = \mathrm{Im}\left(-\sum_{i=1}^{2n-1}(r_i + \mathrm{i}s_i)\Delta_{4,i}\right),$$

$$\Lambda_3 = \mathrm{Re}\left(\sum_{i=1}^{2n-1}(r_i + \mathrm{i}s_i)\Delta_{5,i}\right), \Lambda_4 = \mathrm{Im}\left(\sum_{i=1}^{2n-1}(r_i + \mathrm{i}s_i)\Delta_{5,i}\right).$$

Based on Eqs. (10.44) and (10.47), we obtain the following equation for $B(T_1, T_2)$:

$$\begin{aligned}\frac{dB}{dt} &= \varepsilon\frac{\partial B(T_1, T_2)}{\partial T_1} + \varepsilon^2\frac{\partial B(T_1, T_2)}{\partial T_2} \\ &= -\varepsilon^2\frac{\Lambda_1 + \mathrm{i}\Lambda_2}{\Lambda_3 + \mathrm{i}\Lambda_4}B(T_1, T_2)^2\overline{B}(T_1, T_2). \end{aligned} \tag{10.48}$$

Since we have obtained the normal form of Eq. (10.5), we can proceed to investigate the mechanism that induces the instability of the uniform flow and the resulting oscillatory modes by discussing the amplitude $B(T_1, T_2)$. For this purpose, the polar coordinates are introduced

$$B(T_1, T_2) = \frac{1}{2}R(T_1, T_2)e^{\mathrm{i}\phi(T_1,T_2)}. \tag{10.49}$$

Substituting Eq. (10.49) into Eq. (10.48) and separating the real and imaginary parts yield

$$\begin{aligned}\frac{\mathrm{d}R(T_1, T_2)}{\mathrm{d}t} &= \frac{1}{4}\frac{\Lambda_1\Lambda_3 + \Lambda_2\Lambda_4}{\Lambda_3^2 + \Lambda_4^2}\varepsilon^2 R(T_1, T_2)^3, \\ \frac{\mathrm{d}\phi(T_1, T_2)}{\mathrm{d}t} &= \frac{1}{4}\frac{\Lambda_2\Lambda_3 - \Lambda_1\Lambda_4}{\Lambda_3^2 + \Lambda_4^2}\varepsilon^2 R(T_1, T_2)^2. \end{aligned} \tag{10.50}$$

Equation (10.50) is the polar form of the normal form of the system. For the occurrence of the Bautin bifurcation, the first Lyapunov coefficient in the normal form should be zero, i.e., the coefficient of $R(T_1, T_2)^3$ in the first equation of Eq. (10.50) is zero, that is

$$\frac{\Lambda_1\Lambda_3 + \Lambda_2\Lambda_4}{\Lambda_3^2 + \Lambda_4^2} = 0. \tag{10.51}$$

10.1.2.2 Numerical Examples for Different Numbers of Vehicles

In this section, we follow the process of calculating the Bautin bifurcation point as outlined above, selecting the number of vehicles $n = 3, 5, 9, 13$ respectively. The calculated results for each component of the left and right eigenvectors ($\mathbf{r} + \mathrm{i}\mathbf{s}$ and $\mathbf{p} + \mathrm{i}\mathbf{q}$) of $\mathbf{M}$ for each case are presented in Tables 10.1, 10.2, 10.3 and 10.4.

When the number of vehicles $n = 3$, the critical parameter value at the Bautin point is obtained by using the above method as follows

Table 10.1 Left and right eigenvectors for $n = 3$

i	Left eigenvectors				Right eigenvectors			
	r_{2i}	s_{2i}	r_{2i+1}	s_{2i+1}	p_{2i}	q_{2i}	p_{2i+1}	q_{2i+1}
1	−0.765	0.437	0.500	−0.866	0.206	−0.356	−0.500	0.866
2	0.761	0.444	0.004	−0.881	0.206	0.356	−0.412	0

Table 10.2 Left and right eigenvectors for $n = 5$

i	Left eigenvectors				Right eigenvectors			
	r_{2i}	s_{2i}	r_{2i+1}	s_{2i+1}	p_{2i}	q_{2i}	p_{2i+1}	q_{2i+1}
1	−0.878	0.411	1.309	−0.951	0.389	−0.282	0.309	0.951
2	0.128	0.938	0.500	−1.539	0.389	0.282	−0.809	0.588
3	0.932	0.168	−0.309	−0.951	−0.148	0.457	−0.809	−0.588
4	0.448	−0.834	−0.655	−0.684	−0.480	0	−0.148	−0.457

Table 10.3 Left and right eigenvectors for $n = 9$

i	Left eigenvectors				Right eigenvectors			
	r_{2i}	s_{2i}	r_{2i+1}	s_{2i+1}	p_{2i}	q_{2i}	p_{2i+1}	q_{2i+1}
1	−1.074	0.179	2.014	−0.368	0.501	−0.204	0.871	0.356
2	−1.023	0.577	2.906	−1.115	0.500	0.317	0.659	0.552
3	−0.824	0.963	3.536	−2.202	0.442	0.142	0.399	0.689
4	−0.482	1.280	3.774	−3.535	0.341	0.264	0.130	0.727
5	−0.017	1.475	3.524	−4.975	0.213	0.338	−0.116	0.672
6	0.526	1.502	2.741	−6.343	0.079	0.362	−0.315	0.550
7	1.087	1.330	1.443	−7.440	−0.045	0.362	0.340	0.380
8	1.592	0.947	−0.993	−0.184	−0.147	0.362	0.282	0.199

$$h_c^* = 29.93(\mathrm{m}), \tau_c = 0.784(\mathrm{s}), \tag{10.52}$$

and the values of each component of the left eigenvector $\mathbf{r} + \mathbf{i}\mathbf{s}$ and the right eigenvector $\mathbf{p} + \mathbf{i}\mathbf{q}$ of $\mathbf{M}$ are shown in Table 10.1.

When the number of vehicles $n = 5$, the value of the critical parameter at the Bautin point is obtained as

$$h_c^* = 29.44(\mathrm{m}), \tau_c = 0.384(\mathrm{s}). \tag{10.53}$$

The values of the individual components of the left eigenvector $\mathbf{r} + \mathbf{i}\mathbf{s}$ and the right eigenvector $\mathbf{p} + \mathbf{i}\mathbf{q}$ are shown in Table 10.2.

As the third example, we choose $n = 9$. The values of the left and right eigenvectors are given in Table 10.3, and the critical value of parameters at the Bautin

Table 10.4 Left and right eigenvectors for $n = 13$

i	Left eigenvectors				Right eigenvectors			
	r_{2i}	s_{2i}	r_{2i+1}	s_{2i+1}	p_{2i}	q_{2i}	p_{2i+1}	q_{2i+1}
1	−1.041	0.244	1.885	−0.465	0.527	−0.130	0.885	0.465
2	−0.808	0.700	2.454	−1.288	0.527	0.130	0.568	0.823
3	−0.391	0.995	2.574	−2.280	0.407	0.360	0.121	0.993
4	0.117	1.062	2.220	−3.215	0.193	0.508	−0.355	0.935
5	0.600	0.887	1.471	−3.879	−0.065	0.539	−0.749	0.663
6	0.941	0.508	0.500	−4.118	−0.309	0.447	−0.971	−0.239
7	1.069	0.012	−0.471	−3.879	−0.481	0.252	−0.971	−0.239
8	0.952	−0.486	−1.220	−3.215	−0.543	0	−0.749	−0.663
9	0.617	−0.873	−1.574	−2.280	−0.481	−0.252	−0.355	−0.935
10	0.141	−1.060	−1.454	−1.288	−0.309	−0.447	0.121	−0.993
11	−0.367	−1.004	−0.885	−0.465	−0.065	−0.539	0.568	−0.823
12	−0.792	−0.718	−1.035	−0.268	0.193	−0.508	0.407	−0.360

bifurcation point is found to be

$$h_c^* = 29.15(\mathrm{m}), \tau_c = 0.158(\mathrm{s}). \tag{10.54}$$

As the fourth example, we choose $n = 13$ and find the critical parameter value at the Bautin point is

$$h_c^* = 28.56(\mathrm{m}), \tau_c = 0.044(\mathrm{s}). \tag{10.55}$$

The values of the components of the left eigenvector $\mathbf{r}+i\mathbf{s}$ and the right eigenvector $\mathbf{p}+i\mathbf{q}$ are tabulated in Table 10.4.

10.1.2.3 Numerical Continuation Around the Bautin Bifurcation Point

To precisely locate the region where bistable phenomenon exists, we use the DDE-BIFTOOL to perform the bifurcation continuation. Starting from the Bautin point, we will locate the fold bifurcation points on the branch of periodic solutions, record the critical parameter values at these fold bifurcation points and connect them into a curve in the parameter plane (h^*, τ). This process will enable us to obtain a classification of dynamics in the parameter plane, as shown in Fig. 10.3.

From Fig. 10.3, several observations can be made. Firstly, below the Bautin bifurcation curve, there is a region B where both uniform flow and periodic solutions coexist. In this region, the stability of the uniform flow depends on the initial conditions. Secondly, as the number of vehicles increases, the bistable region B will be enlarged, and the Bautin point moves closer to the Hopf curve, with the critical values

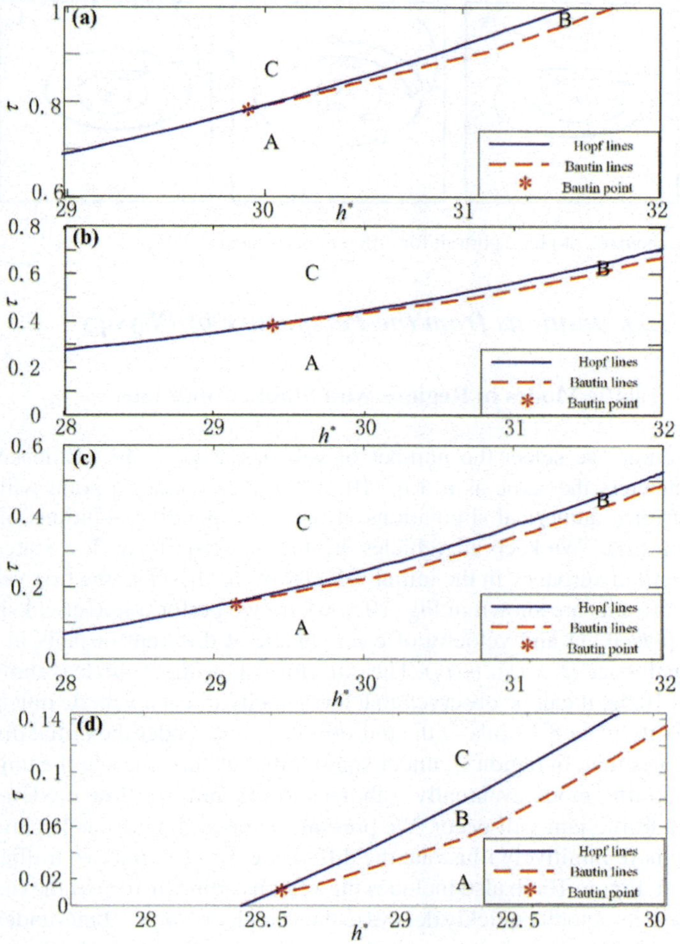

Fig. 10.3 Bifurcation diagram in the parameter plane for: **a** $n = 3$, **b** $n = 5$, **c** $n = 9$, and **d** $n = 13$

becoming smaller. This indicates that, with all other parameters fixed, increasing the number of vehicles leads to a decrease in the area of absolute stability for the uniform flow, indicating the possibility of the occurrence of bistable phenomenon. In other words, when there are more vehicles on the road, the traffic conditions become more susceptible to the influence of individual drivers.

Figure 10.4 shows schematic diagrams of the solutions in the three regions depicted in Fig. 10.3. The solid points in the figure represent the uniform flow, while the dashed ellipse denotes the unstable periodic solution and the solid ellipse represents the stable periodic solution. In Region A, there exists a stable uniform flow solution. In Region B, there exist a stable uniform flow solution, a stable periodic solution, and an unstable periodic solution. In Region C, there exists a stable periodic solution.

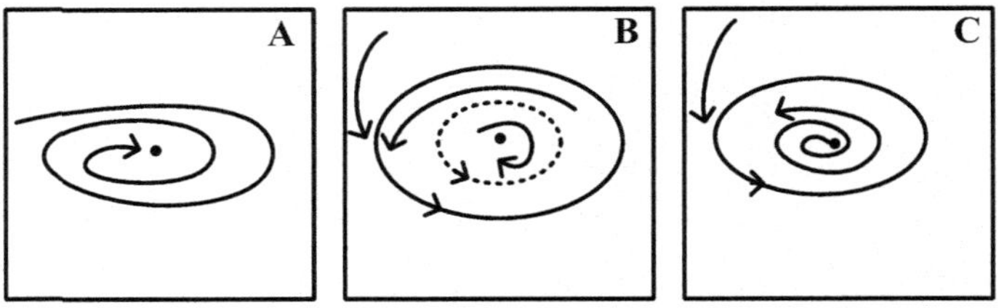

Fig. 10.4 Schematic of phase portrait for different parameters

10.1.3 Explanations from the Perspective of Physics

10.1.3.1 Traffic Modes in Regions with Stable Conditions

In this section, we select the number of vehicles as $n = 13$, while keeping the other parameters the same as in Eq. (10.4). We then choose a point within region B, and perform numerical simulations of the corresponding solutions of the original system (9.6). We keep all vehicles driving in the uniform flow state, and only allow a small disturbance in the initial velocity of the driver of the first vehicle. The simulation results are shown in Fig. 10.5, where we perform numerical simulations of the displacement and velocity of each vehicle at different instants in the three-dimensional space $(t, x_i(t), v_i(t))$. The trajectory of the first vehicle is shown in blue. From Fig. 10.5a, it can be observed that the velocity of each vehicle quickly decays to the velocity $v^* = 8.19$ m/s of the uniform flow state under the initial disturbance. This indicates that, in region B, under small initial disturbances in the uniform flow mode, the traffic mode eventually stabilizes to the uniform flow mode and consequently no traffic jam will occur. We present the three-dimensional spatiotemporal pattern to more intuitively illustrate the difference between the two traffic modes in such region. Figure 10.5b also includes the velocity–time history of the first vehicle, showing that its velocity quickly decays to the velocity $v^* = 8.19$ m/s under the small disturbance. In Fig. 10.5c, we provide the spatiotemporal pattern of the displacement of each vehicle at different instants in the plane $(t, x_i(t))$. In the spatiotemporal pattern of the uniform flow mode, the trajectories of each vehicle are parallel straight lines without any clusters of traffic jam.

In Fig. 10.6, we fix all the parameters the same as in Fig. 10.5, but change the initial condition to the initial condition of congestion, that is, the velocity of the first vehicle at the initial moment is zero, and the distance between vehicles is much smaller than the average distance between vehicles ($h^* = 34$) under uniform flow. The results of the numerical simulation are shown in Fig. 10.6. In Fig. 10.6a, we plot the spatiotemporal pattern of the displacement and velocity of each vehicle in 3D space $(t, x_i(t), v_i(t))$ at various instants. The stop-and-go traffic jammed mode is clearly observed. In Fig. 10.6b, we plot the velocity–time history of the first vehicle. The vehicle needs to experience a period of constant velocity driving and a period of stopping during the cycle of driving and stopping. In Fig. 10.6c, we draw the

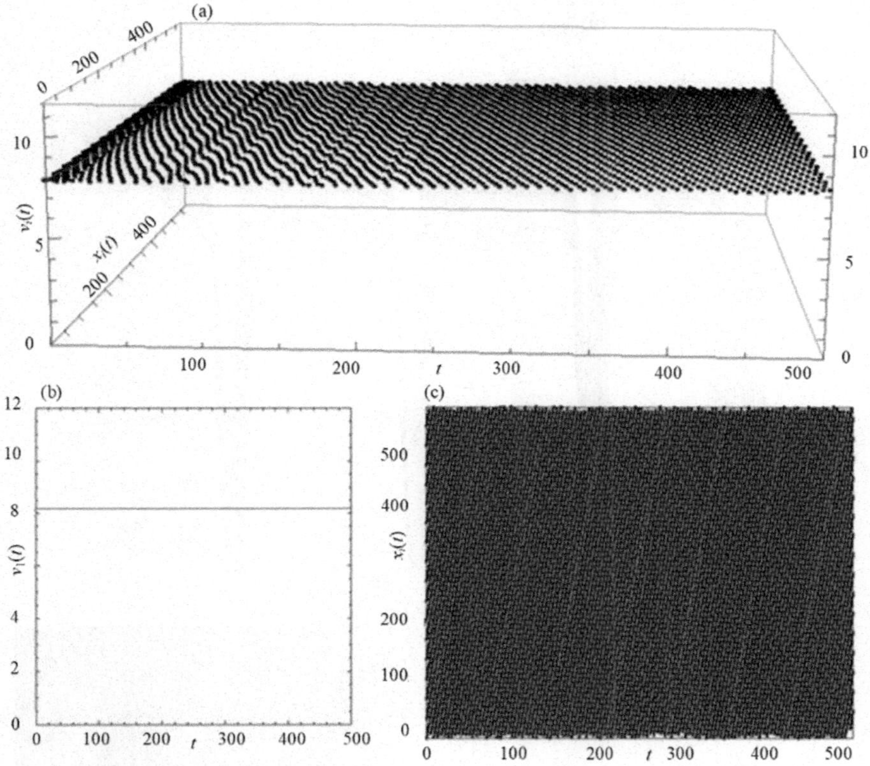

Fig. 10.5 When $n = 13$, $h^* = 34$, $\tau = 0.9$, the spatiotemporal pattern of the first vehicle under small perturbations in the velocity selected from region B. The blue line indicates the trajectory of the first vehicle. **a** Spatiotemporal pattern in 3D space $(t, x_i(t), v_i(t))$; **b** the velocity–time history of the first vehicle; **c** 2D spatiotemporal pattern in the plane $(t, x_i(t))$

two-dimensional spatiotemporal pattern of the vehicle's position trajectory on the road at each moment. It can be seen that there are equal-width traffic jam clusters with equal width on the road. We call this traffic mode the equal-width traffic jam mode.

10.1.3.2 Traffic Modes Under Different Bifurcation Mechanisms

From previous Section we learn that the destabilization mechanisms for uniform flows are different. The Hopf curve in Fig. 10.3 is divided into two parts by the Bautin point, with the Hopf curve below the Bautin point representing the supercritical Hopf bifurcation, and the Hopf curve above the Bautin point representing the subcritical Hopf bifurcation. In the parameter plane (h^*, τ) shown in Fig. 10.3, the uniform flow solution undergoes a supercritical Hopf bifurcation when crossing the supercritical

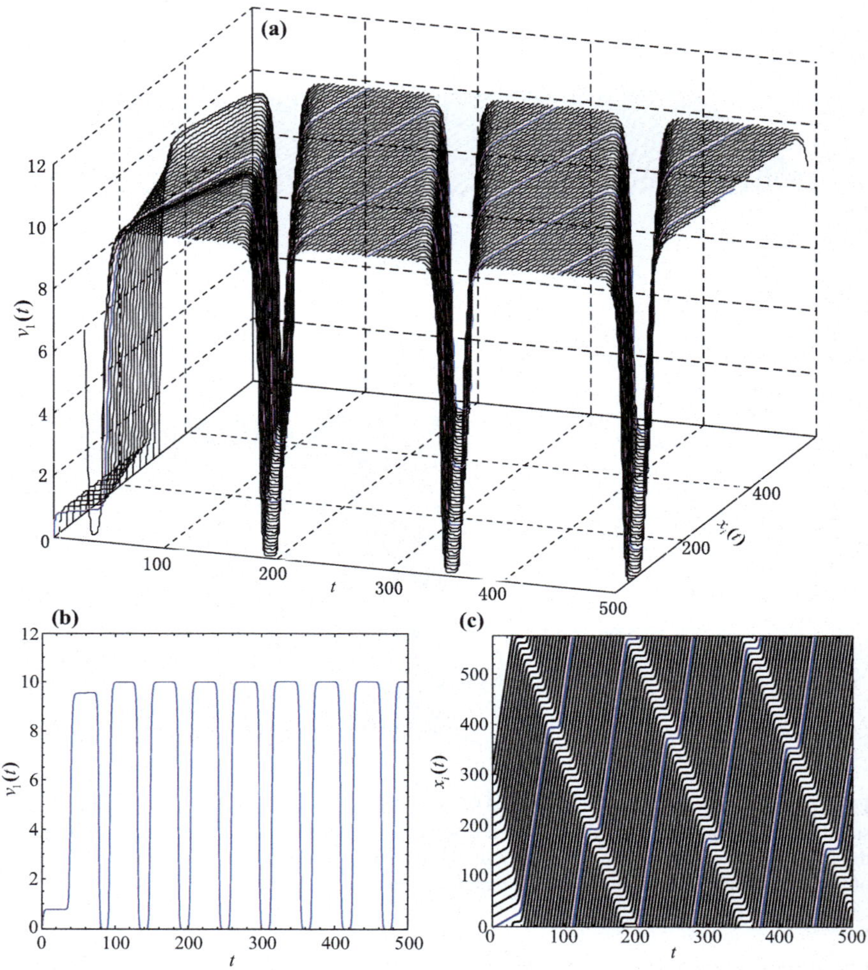

Fig. 10.6 When $n = 13$, $h^* = 34$, $\tau = 0.9$, the spatiotemporal pattern of velocity of the first vehicle. The blue line indicates the trajectory of the first vehicle. **a** Spatiotemporal pattern in 3D space $(t, x_i(t), v_i(t))$; **b** the velocity–time history of the first vehicle; **c** 2D spatiotemporal pattern in the plane $(t, x_i(t))$

Hopf curve, and a subcritical Hopf bifurcation when passing through the subcritical Hopf bifurcation curve. We propose that the periodic solutions resulting from these bifurcations are essentially distinct according to the corresponding traffic modes. The amplitude of the periodic solution resulting from the supercritical Hopf bifurcation is relatively small, and may only lead to a small-amplitude oscillatory traffic mode that may not induce traffic jam. In contrast, the subcritical Hopf bifurcation produces a large-amplitude periodic solution, which may correspond to the equal-width jammed

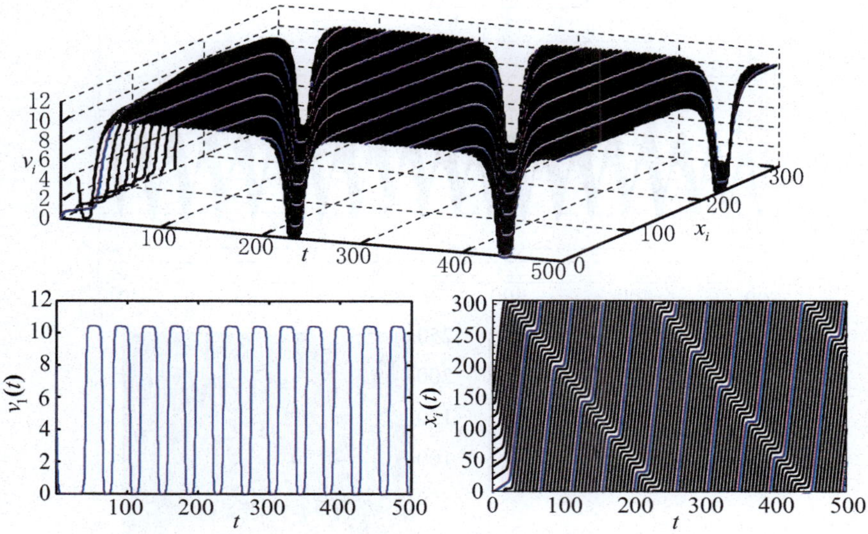

Fig. 10.7 Spatiotemporal pattern of equal-width traffic jam induced by the subcritical Hopf mechanism

mode. Therefore, we argue that a destabilization of uniform flow does not necessarily imply the occurrence of traffic jams. To demonstrate our viewpoint, we show the periodic solutions and corresponding traffic modes under the two bifurcation regimes with 9 vehicles in Fig. 10.7.

In Fig. 10.7, we select a set of parameters that cross the subcritical Hopf curve above the Bautin point in Fig. 10.3, and plot the three-dimensional spatial–temporal pattern of the vehicles, as well as two-dimensional spatial–temporal pattern in the time-displacement plane, and the velocity–time history of the first vehicle. It is observed that due to the subcritical Hopf mechanism the uniform flow evolves into an equal-width jammed mode. Additionally, we choose a set of parameter that cross the supercritical Hopf curve below the Bautin point in Fig. 10.3, and the results are presented in Fig. 10.8. From Fig. 10.8, it can be seen that the uniform flow evolves towards a small oscillatory traffic mode that may not cause traffic jam.

10.1.3.3 Evolution of Traffic Jam as the Number of Vehicles Increases

The aim of this study is to compare the level of traffic jam with an increase in the number of vehicles under the same inter-vehicle distance. As the periodic solution with large amplitude induced by the subcritical Hopf bifurcation corresponds to the traffic jam mode, we select an inter-vehicle distance $h^* = 32$, and then select the value of time delay corresponding to (a), (b), (c), (d) in region C of Fig. 10.3. We then

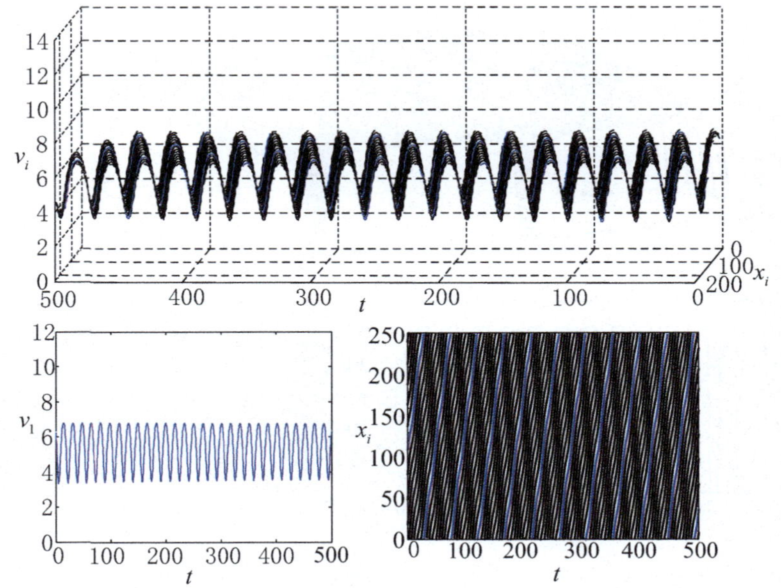

Fig. 10.8 Spatiotemporal pattern of small oscillatory traffic mode under the supercritical Hopf bifurcation mechanism

draw the spatial–temporal pattern for different numbers of vehicles ($n = 3, 5, 9, 13$), corresponding to in Fig. 10.9a–d, respectively.

From the width of jam cluster and the time it takes for vehicles to complete a full circle on the ring road, it is apparent that even at the same average inter-vehicle distance, an increase in the number of vehicles leads to a more severe traffic jam. When $n = 3$, each car takes approximately 16.7 s to complete a full circle, and there are almost no jam clusters formed on the road, meaning that traffic jam dissipates almost instantly. However, when $n = 13$, it can be observed that it takes about 71.4 s for the vehicles to complete a full circle, and there are significant jam clusters with a certain width on the road. As the number of vehicles gradually increases from to 3 to 13, the width of the jam cluster also increases gradually.

10.1.4 Conclusion

(1) A large average inter-vehicle distance will not necessarily lead to more stable state than a small average inter-vehicle distance (i.e., larger inter-vehicle distances are not necessarily more efficient). Additionally, the practice of maintaining the stability of the uniform flow by reducing the time delay is only effective when the average inter-vehicle distance takes an appropriate value.

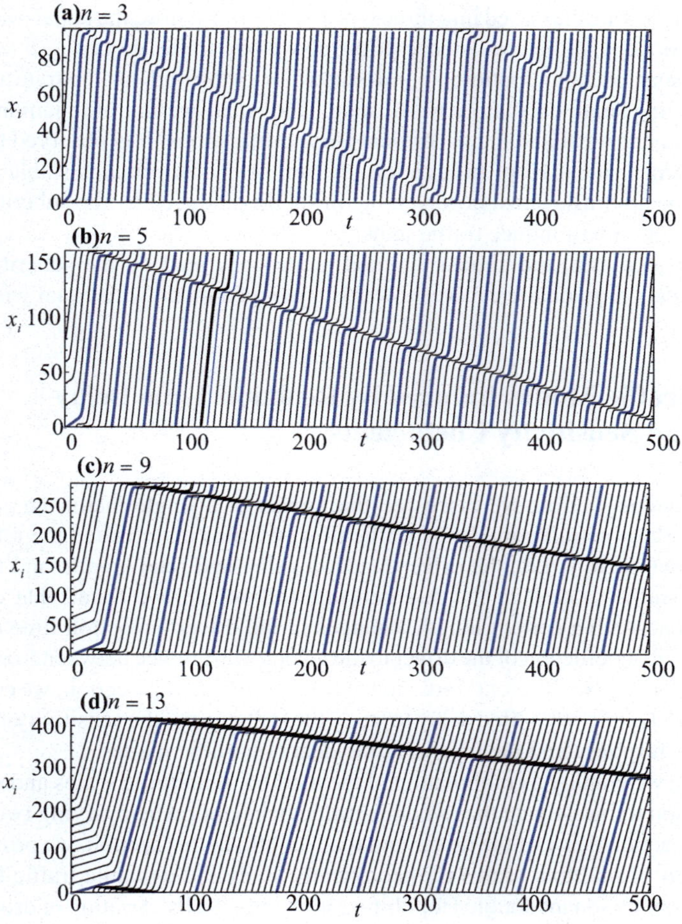

Fig. 10.9 Spatiotemporal pattern for **a** $n = 3$, **b** $n = 5$, **c** $n = 9$, and **d** $n = 13$

(2) The dynamic mechanism that leads to unstable uniform flow is discussed through the combination of the method of multiple scales and numerical continuation technique. Uniform flow loses stability through supercritical or subcritical Hopf bifurcation, but as the number of vehicles increases, the mechanism that induces instability is generally the subcritical Hopf bifurcation. The numerical continuation method is used to classify the dynamic behaviour induced by the Bautin bifurcation in the parameter plane, and the region where multiple stable solutions coexist is identified.

(3) The macroscopic spatiotemporal pattern is used to explain the traffic mode corresponding to the dynamic behaviour in each region of the parameter plane.

Dynamics are classified into three types: the uniform flow, the equal-width traffic jam mode, and the coexistence of the above two traffic modes.

(4) The macroscopic spatiotemporal pattern is employed to illustrate the corresponding small oscillatory traffic mode and equal-width traffic jam induced by the supercritical and the subcritical Hopf bifurcations. It is observed that under the supercritical Hopf bifurcation, the uniform flow remains stable and there will be no traffic jams; however, the instability induced by the subcritical Hopf bifurcation will induce traffic jams.

(5) As n increases, the almost jammed mode generated by the subcritical Hopf bifurcation gradually evolves towards a jammed mode with equal width.

10.2 Traffic Modes Induced by Reaction Time Delay and Sensitivity Coefficients

In the previous section, we discussed the impact of a single factor, i.e., the driver reaction delay, on traffic modes in traffic flow. We also investigated the relationship between two factors, i.e., the driver reaction delay and road length, and their joint effect on traffic modes. In the research that follows up, we found that even on a circular road with a certain length, changes in the driver's reaction time delay and the sensitivity coefficient of the driver to the velocity difference between front and rear vehicles, can lead to different traffic modes. Therefore, in this section, we continue to use the FVD model with time delay to study the influence of the sensitivity coefficient of driver's response and the reaction time delay on traffic modes.

In 10.1, we employ a combination of the method of multiple scales and numerical continuation to investigate the dynamic traffic behaviour when varying two parameters: the reaction delay of the driver and road length. In this section, we primarily use the method of multiple scales to theoretically study the dynamic traffic behaviour under the non-resonant double Hopf bifurcation induced by variation of driver's reaction delay and sensitivity coefficient. Carter et al. [9] introduced a feedback control term and establish a velocity model without delay, and investigated the double Hopf bifurcation induced by two bifurcation parameters from both theoretical and numerical perspectives. In this section, we also discuss the double Hopf bifurcation in traffic dynamics. The main difference between others' and our researches lies in two aspects. Firstly, the bifurcation parameters in our work, namely, the reaction delay and sensitivity coefficient, account for some natural features rather than control objectives. Secondly, we carefully investigate the dynamic behaviour in different regions, and seek their originalities rather than indiscriminately categorize the solutions when the uniform flow becomes unstable into traffic jams. We adopt the method presented in [10–16] for studying the non-resonant double Hopf bifurcation induced by the driver's reaction delay and sensitivity coefficient, and [12, 13] for the method of multiple scales.

10.2.1 Stability in the Plane of Sensitivity Coefficient and Reaction Delay

10.2.1.1 Stability of Uniform Flow in the Two-Parameter Plane

Let's consider the time delay FVD model in (9.6) for $n = 5$:

$$\begin{cases} \dot{h}_i(t) = v_{i+1}(t) - v_i(t), \\ \dot{v}_i(t) = \alpha(V(h_i(t-\tau)) - v_i(t)) \\ \qquad +\beta(v_{i+1}(t-\tau) - v_i(t-\tau)), \\ \dot{v}_5(t) = \alpha\left(V\left(L - \sum_{i=1}^{4} h_i(t-\tau)\right) - v_5(t)\right) \\ \qquad +\beta(v_1(t-\tau) - v_5(t-\tau)), \end{cases} \quad i = 1, \ldots, 4, \tag{10.56}$$

where $V(h)$ is referred to Eq. (9.7).

The uniform flow solution of Eq. (10.56) is given by

$$h_i(t) = h^* = \frac{L}{5},\ v_i(t) = v^* = V\left(h^*\right),\ \dot{h}_i(t) \equiv 0, \quad i = 1, \ldots, 5. \tag{10.57}$$

To analyze the stability of the uniform flow solution (10.57), a small perturbation to the uniform flow solution is introduced as follows

$$h_i(t) = h^* + r_i(t),\ v_i(t) = v^* + u_i(t). \tag{10.58}$$

Then we expand $V(h)$ to Taylor series at h^* and truncate at the third order:

$$\begin{cases} \dot{r}_i(t) = u_{i+1}(t) - u_i(t), \\ \dot{u}_i(t) = -\alpha u_i(t) + \alpha \sum_{j=1}^{3} b_j r_i^j(t-\tau) + \beta\left(u_{i+1}(t-\tau) - u_i(t-\tau)\right),\ i = 1, \ldots, 4, \\ \dot{u}_5(t) = -\alpha u_5(t) + \alpha \sum_{j=1}^{3} b_j \left(-\sum_{i=1}^{4} r_i(t-\tau)\right)^j + \beta(u_1(t-\tau) - u_5(t-\tau)), \end{cases} \tag{10.59}$$

where

$$b_j = \frac{1}{j!} V^{(j)}\left(h^*\right),\ j = 1, 2, 3. \tag{10.60}$$

The local stability of the uniform flow solution of Eq. (10.56) is equivalent to the stability of the trivial solution of Eq. (10.59). The characteristic equation of the linear part of Eq. (10.59) is given by

$$
\begin{aligned}
&\Big(e^{-2\lambda\tau}\Big((\beta\lambda+\alpha b_1)^2+\big(\lambda(\alpha+\beta+e^{\lambda\tau}\lambda)+\alpha b_1\big)^2\Big)-2\big(\lambda(\alpha+\beta+e^{\lambda\tau}\lambda)\\
&\quad+\alpha b_1)(\beta\lambda+\alpha b_1)e^{-2\lambda\tau}\cos\frac{2\pi}{5}\Big)\\
&\quad\Big(e^{-2\lambda\tau}\Big((\beta\lambda+\alpha b_1)^2+\big(\lambda(\alpha+\beta+e^{\lambda\tau}\lambda)+\alpha b_1\big)^2\Big)\\
&\quad-2e^{-2\lambda\tau}(\beta\lambda+\alpha b_1)\cos\frac{4\pi}{5}\big(\lambda(\alpha+\beta+e^{\lambda\tau}\lambda)+\alpha b_1\big)\Big)(\alpha+\lambda)=0. \qquad (10.61)
\end{aligned}
$$

Substituting $\lambda = \mathrm{i}\omega$ into Eq. (10.61) and separating the real and imaginary parts yield

$$
\begin{aligned}
&\omega^4\big(-(4\alpha+5\beta)\omega^3\cos(\tau\omega)+\big(4\alpha^3+15\alpha^2\beta+20\alpha\beta^2+10\beta^3\big)\omega\cos(3\tau\omega)\\
&\quad+\big(6\alpha^2+15\alpha\beta+10\beta^2\big)\omega^2\sin(2\tau\omega)\\
&\quad-\big(\alpha^4+5\alpha^3\beta+10\alpha^2\beta^2+10\alpha\beta^3+5\beta^4\big)\sin(4\tau\omega)\big)\\
&\quad+5\alpha b_1\big(\omega^3\,((3\alpha+4\beta)\omega^2\cos(2\tau\omega)\\
&\quad-(\alpha+2\beta)\big(\alpha^2+2\alpha\beta+2\beta^2\big)\cos(4\tau\omega)\\
&\quad+\omega^3\sin(\tau\omega)-\big(3\alpha^2+8\alpha\beta+6\beta^2\big)\omega\sin(3\tau\omega))\\
&\quad+\alpha b_1(2\omega^2(-(2\alpha+3\beta)\omega\cos(3\tau\omega)\\
&\quad-\omega^2\sin(2\tau\omega)+\big(\alpha^2+3\alpha\beta+3\beta^2\big)\sin(4\tau\omega))\\
&\quad+\alpha b_1(2\omega((\alpha+2\beta)\cos(4\tau\omega)\\
&\quad+\omega\sin(3\tau\omega)))-\alpha\sin(4\tau\omega)b_1))=0,\\
&\omega^4(\omega^4-\big(6\alpha^2+15\alpha\beta+10\beta^2\big)\omega^2\cos(2\tau\omega)\\
&\quad+\big(\alpha^4+5\alpha^3\beta+10\alpha^2\beta^2+10\alpha\beta^3+5\beta^4\big)\cos(4\tau\omega)\\
&\quad-(4\alpha+5\beta)\omega^3\sin(\tau\omega)+\big(4\alpha^3+15\alpha^2\beta+20\alpha\beta^2+10\beta^3\big)\omega\sin(3\tau\omega))\\
&\quad+5\alpha b_1\big(\omega^3\big(-\omega^3\cos(\tau\omega)+\big(3\alpha^2+8\alpha\beta+6\beta^2\big)\omega\cos(3\tau\omega)\\
&\quad+(3\alpha+4\beta)\omega^2\sin(2\tau\omega)-(\alpha+2\beta)\big(\alpha^2+2\alpha\beta+2\beta^2\big)\sin(4\tau\omega))\\
&\quad+\alpha b_1(2\omega^2(\omega^2\cos(2\tau\omega)-\big(\alpha^2+3\alpha\beta+3\beta^2\big)\cos(4\tau\omega)\\
&\quad-(2\alpha+3\beta)\omega\sin(3\tau\omega))+\alpha b_1(2\omega(-\omega\cos(3\tau\omega)\\
&\quad+(\alpha+2\beta)\sin(4\tau\omega))+\alpha\cos(4\tau\omega)b_1)))=0. \qquad (10.62)
\end{aligned}
$$

In the following discussion, values of parameters are fixed as

$$\alpha = 0.9(/\mathrm{s}),\ h^* = 28(\mathrm{m}),\ v^0 = 11(\mathrm{m/s}),\ h_{\mathrm{stop}} = 14(\mathrm{m}). \qquad (10.63)$$

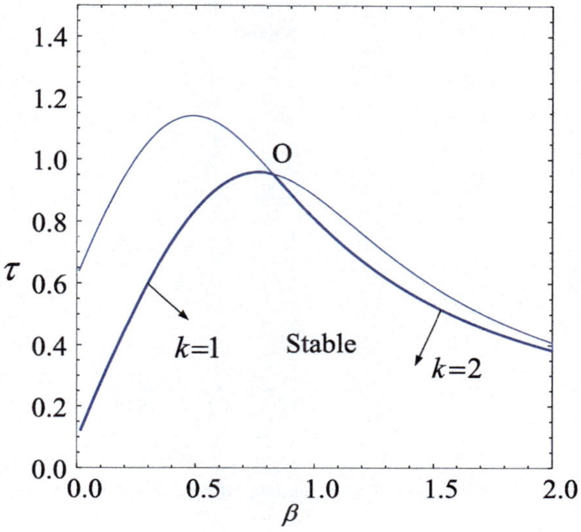

Fig. 10.10 Hopf bifurcation curves of the uniform flow in the parameter plane where O is the intersection of Hopf bifurcation curves

In Fig. 10.10, we plot two Hopf bifurcation curves of the uniform flow in the (β, τ) plane, denoted as $k = 1, k = 2$, respectively. When the sensitivity coefficient β is fixed, increasing the reaction delay will destabilize the equilibrium for parameters initially selected from the stable region. On the other hand, when the reaction time delay τ is fixed, the equilibrium will undergo "stable-unstable-stable" switch by increasing the value of. These results indicate that reducing the driver's reaction time delay can enhance the stability of the uniform flow. However, it is not always true that the larger the driver's sensitivity to velocity difference, the better the stability of the uniform flow. Too high and too low sensitivity both can lead to instability of the uniform flow.

It should be noticed that in Fig. 10.10 the two Hopf curves have an intersection O whose coordinates in the parameter plane is found to be $\beta_c = 0.8244$ and $\tau_c = 0.956$ with frequencies $\omega_1 = 0.8315, \omega_2 = 1.459$. The distribution of all characteristic roots at point O are shown in Fig. 10.11. There are two pairs of pure imaginary roots $\pm i\omega_1, \pm i\omega_2$, crossing the imaginary axis simultaneously. In the next, we apply the bifurcation continuation along horizontal lines $\tau = 0.955, \beta \in (0.8, 0.83)$ and $\tau = 0.959, \beta \in (0.8, 0.83)$ respectively, as shown in Figs. 10.12 and 10.13.

10.2.1.2 Bifurcation Continuation

Figure 10.12a is an enlargement of the corresponding part of Fig. 10.10, while Fig. 10.12b is the bifurcation diagram as β moves along the horizontal line $\tau = 0.955, \beta \in (0.8, 0.83)$, where B_1, B_2 represent the intersections of the horizontal line and the two Hopf bifurcation curves. According to the results of bifurcation continuation in Fig. 10.12b, we can deduce that the uniform flow solution is stable

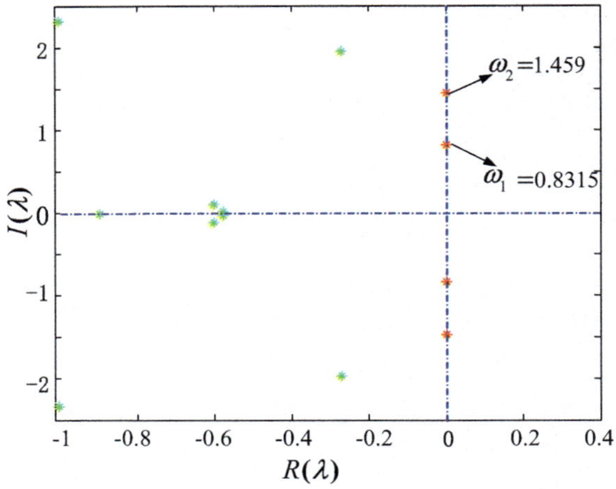

Fig. 10.11 Distribution of characteristic roots at point O

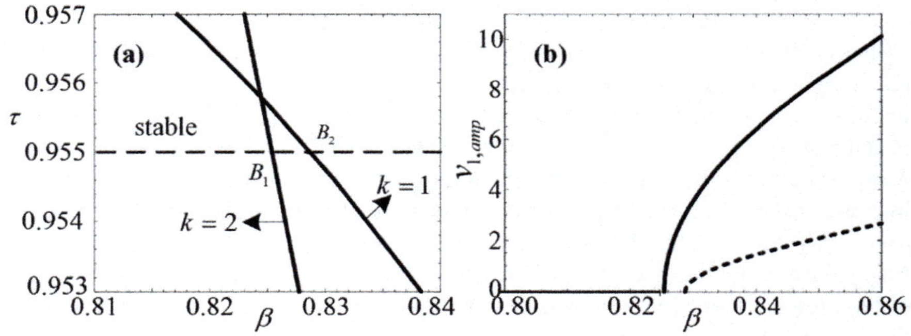

Fig. 10.12 Bifurcation analysis along the horizontal line where $\tau = 0.955$: **a** stable region obtained by linear analysis where B_1 and B_2 are the intersections of the horizontal line $\tau = 0.955$ with the two Hopf bifurcation curves; **b** bifurcation diagram where solid curve represents stable solutions and dashed curve unstable solutions

when β is located on the left side of B_1. When β passes through B_1, a stable periodic solution with period T_2 arises. As β is further increased to pass through B_2, the stable periodic solution with period T_2 still exists, but an unstable periodic solution with period T_1 emerges. Therefore, Fig. 10.12 indicates that there exists a region above point O where a stable periodic solution coexists with an unstable periodic solution.

Figure 10.13a is an enlargement of the corresponding part of Fig. 10.10, while Fig. 10.13b is the bifurcation diagram as β moves along the horizontal line $\tau = 0.959$, $\beta \in (0.8, 0.83)$, where A_1, A_2 represent the locations that the stable periodic solution with period T_1 and the unstable periodic solution with period T_2 appear, respectively. A_3, A_4 represent the locations where the stability of the two periodic solutions switch. According to the results of bifurcation continuation in Fig. 10.13b,

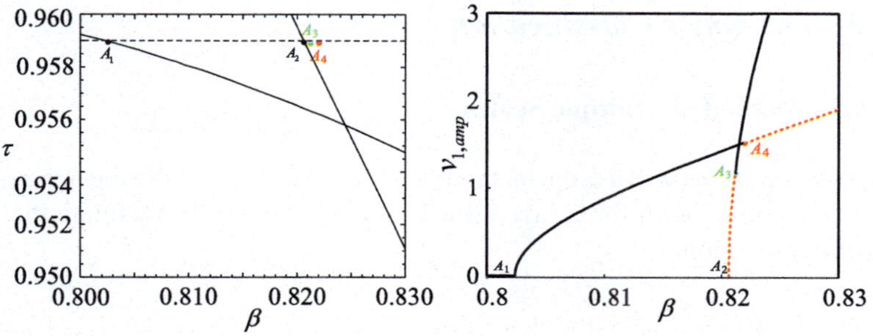

Fig. 10.13 Bifurcation analysis along the horizontal line where $\tau = 0.959$: **a** stable region obtained by linear analysis where A_1 and A_2 are the intersections of the horizontal line $\tau = 0.959$ and two Hopf curves, A_3 and A_4 the values of β for which the stability of the bifurcated period solutions switches; **b** bifurcation diagram where solid curve represents stable solutions and dashed curve unstable solutions

we can see that when β is on the left side of A_1, the uniform flow solution is stable. When β passes through A_1, a stable periodic solution with period T_1 arises. Furthermore, when β increases to A_2, an unstable periodic solution with period T_2 emerges. As β is further increased between A_3 and A_4, the unstable periodic solution becomes stable and coexists with the stable periodic solution with period T_1. Therefore, Fig. 10.13 reveals the presence of a region above point O where two periodic solutions coexist.

By combining the results shown in Figs. 10.12 and 10.13, we observe that the stability switch boundaris in Fig. 10.10, which is defined by the two Hopf curves, divides the region into four parts. Specifically, this region includes a zone where a stable periodic solution and an unstable periodic solution coexist, as well as a region where two stable periodic solutions coexist. Therefore, we can infer that point O is a double Hopf bifurcation point with codimension two. Based on the numerical continuation results and bifurcation theory, it is clear that the classification consisting of four regions obtained by the linear analysis is incomplete. Further subdivision of dynamic behaviour by nonlinear analysis is necessary to obtain a complete classification regarding the dynamic behaviour of the system in the parameter plane (β, τ). This comprehensive diagram can facilitate the study of classification of traffic mode in the parameter plane.

In the next subsection, we will use the method of multiple scales to classify the dynamis in the parameter plane (β, τ).

10.2.2 Dynamics Classification

10.2.2.1 Method of Multiple Scales

In this subsection, we will use the method of multiple scales to study the non-resonant double Hopf bifurcation for delayed differential equations. To facilitate the use of the method, we define

$$\mathbf{y}(t) = (r_1(t), u_1(t), \ldots, r_{2n-2}(t), u_{2n-2}(t), u_{2n-1}(t))^T,$$

then Eq. (10.59) is rewritten in matrix form as

$$\dot{y}\,(t) = \boldsymbol{C}\boldsymbol{y}(t) + \boldsymbol{D}\boldsymbol{y}(t-\tau) + \boldsymbol{F}(\boldsymbol{y}(t-\tau)), \tag{10.64}$$

where, $\mathbf{y} : \mathbb{R} \to \mathbb{R}^{2n-1}$. Matrices $\mathbf{C}, \mathbf{D} \in \mathbb{R}^{(2n-1)\times(2n-1)}$ and the nonzero nonlinear function $\mathbf{F} : \mathbb{R}^{2n-1} \to \mathbb{R}^{2n-1}$ are respectively given by

$$\mathbf{C} = \begin{pmatrix} 0 & -1 & 0 & 1 & 0 & 0 & \cdots & 0 & 0 \\ 0 & -\alpha & 0 & 0 & 0 & 0 & \cdots & 0 & 0 \\ 0 & 0 & 0 & -1 & 0 & 1 & \cdots & 0 & 0 \\ 0 & 0 & 0 & -\alpha & 0 & 0 & \cdots & 0 & 0 \\ \vdots & \vdots & \vdots & \vdots & \vdots & \vdots & \ddots & \vdots & \vdots \\ 0 & 0 & 0 & 0 & 0 & 0 & \cdots & 0 & 0 \\ 0 & 0 & 0 & 0 & 0 & 0 & \cdots & -1 & 1 \\ 0 & 0 & 0 & 0 & 0 & 0 & \cdots & -\alpha & 0 \\ 0 & 0 & 0 & 0 & 0 & 0 & \cdots & 0 & -\alpha \end{pmatrix},$$

$$\mathbf{D} = \begin{pmatrix} 0 & 0 & 0 & 0 & 0 & 0 & \cdots & 0 & 0 \\ \alpha b & -\beta & 0 & \beta & 0 & 0 & \cdots & 0 & 0 \\ 0 & 0 & 0 & 0 & 0 & 0 & \cdots & 0 & 0 \\ 0 & 0 & \alpha b & -\beta & 0 & \beta & \cdots & 0 & 0 \\ \vdots & \vdots & \vdots & \vdots & \vdots & \vdots & \ddots & \vdots & \vdots \\ 0 & 0 & 0 & 0 & 0 & 0 & \cdots & \beta & 0 \\ 0 & 0 & 0 & 0 & 0 & 0 & \cdots & 0 & 0 \\ 0 & 0 & 0 & 0 & 0 & 0 & \cdots & -\beta & \beta \\ -\alpha b & \beta & -\alpha b & 0 & -\alpha b & 0 & \cdots & 0 & -\beta \end{pmatrix},$$

$$\mathbf{F}(\mathbf{y}(t-\tau)) = \left(0, \alpha\sum_{i=2}^{3} b_i y_1^i(t-\tau), 0, \alpha\sum_{i=2}^{3} b_i y_3^i(t-\tau), \ldots, \right.$$

$$\left. \alpha\sum_{i=2}^{3} b_i y_{2n-3}^i(t-\tau), \alpha\sum_{j=2}^{3} b_j\left(-\sum_{i=1}^{2n-3} y_i(t-\tau)\right)^j\right)^T. \tag{10.65}$$

As in the previous discussions, multiple time scales, i.e., $T_0 = t$, $T_1 = \varepsilon t$, $T_2 = \varepsilon^2 t$, are introduced to apply the method of multiple scales. The solution of Eq. (10.64) is then expanded as

$$\mathbf{y}(t) = \mathbf{y}(T_0, T_1, T_2) = \varepsilon \begin{pmatrix} y_{1,1}(T_0, T_1, T_2) \\ y_{2,1}(T_0, T_1, T_2) \\ \vdots \\ y_{2n-1,1}(T_0, T_1, T_2) \end{pmatrix} + \varepsilon^2 \begin{pmatrix} y_{1,2}(T_0, T_1, T_2) \\ y_{2,2}(T_0, T_1, T_2) \\ \vdots \\ y_{2n-1,2}(T_0, T_1, T_2) \end{pmatrix} + \varepsilon^3 \begin{pmatrix} y_{1,3}(T_0, T_1, T_2) \\ y_{2,3}(T_0, T_1, T_2) \\ \vdots \\ y_{2n-1,3}(T_0, T_1, T_2) \end{pmatrix}. \tag{10.66}$$

The term with time delay in (10.64) can be written as

$$\mathbf{y}(t - \tau) = \mathbf{y}\big(T_0 - \varepsilon\tau, T_1 - \varepsilon^2\tau, T_2 - \varepsilon^3\tau\big). \tag{10.67}$$

Perturb the bifurcation parameters β, τ in a small neighborhood around their critical values, namely β_c, τ_c, yields

$$\beta = \beta_c + \varepsilon\beta_\varepsilon,\ \tau = \tau_c + \varepsilon\tau_\varepsilon, \tag{10.68}$$

where, $\varepsilon\beta_\varepsilon$, $\varepsilon\tau_\varepsilon$ denote small perturbations to the critical parameters β_c, τ_c, respectively.

Then we expand the term with time delay at $(T_0 - \tau_c, T_1, T_2)$ and truncate the terms up to ε^3 to arrive at

$$\begin{aligned} \mathbf{y_i}(t - \tau) = {} & \varepsilon y_{i,1}(T_0 - \tau_c, T_1, T_2) + \varepsilon^2 \\ & \left(y_{i,2}(T_0 - \tau_c, T_1, T_2) - \tau_c \frac{\partial y_{i,1}(T_0 - \tau_c, T_1, T_2)}{\partial T_1} \right) + \varepsilon^3 \\ & \left(y_{i,3}(T_0 - \tau_c, T_1, T_2) - \tau_c \frac{\partial y_{i,1}(T_0 - \tau_c, T_1, T_2)}{\partial T_2} \right. \\ & \left. - \tau_c \frac{\partial y_{i,2}(T_0 - \tau_c, T_1, T_2)}{\partial T_1} + \frac{1}{2}\tau_c^2 \frac{\partial^2 y_{i,1}(T_0 - \tau_c, T_1, T_2)}{\partial T_1^2} \right), \end{aligned} \tag{10.69}$$

where $i = 1, 2, \ldots, 2n - 1$.

Substituting Eqs. (10.65) and (10.66) into (10.64), regrouping terms in the equation, and equating coefficients of each power of ε to zero, we obtain the following equation from at ε^1:

$$\frac{\partial y_{2i-1,1}}{\partial T_0}(T_0, T_1, T_2) + y_{2i,1}(T_0, T_1, T_2) - y_{2i+2,1}(T_0, T_1, T_2) = 0,$$
$$i = 1, 2, \ldots, n-1, \tag{10.70}$$

$$\frac{\partial y_{2i,1}}{\partial T_0}(T_0, T_1, T_2) + \alpha y_{2i,1}(T_0, T_1, T_2) - \alpha b_1 y_{2i-1,1}(T_0 - \tau_c, T_1, T_2)$$
$$+ \beta_c y_{2i,1}(T_0 - \tau_c, T_1, T_2) - \beta_c y_{2i+2,1}(T_0 - \tau_c, T_1, T_2) = 0,$$
$$i = 1, 2, \ldots, n-1, \tag{10.71}$$

$$\frac{\partial y_{2n-1,1}}{\partial T_0}(T_0, T_1, T_2) + \alpha b_1 \sum_{i=1}^{n-1} y_{2i-1,1}(T_0 - \tau_c, T_1, T_2) + \alpha y_{2n-1,1}(T_0, T_1, T_2)$$
$$+ \beta_c y_{2n-1,1}(T_0 - \tau_c, T_1, T_2) - \beta_c y_{2,1}(T_0 - \tau_c, T_1, T_2) = 0. \tag{10.72}$$

From the coefficient of ε^2, we have

$$\frac{\partial y_{2i-1,2}}{\partial T_0}(T_0, T_1, T_2) + y_{2i,2}(T_0, T_1, T_2) - y_{2i+2,2}(T_0, T_1, T_2)$$
$$= -\frac{\partial y_{2i-1,1}}{\partial T_1}(T_0, T_1, T_2), i = 1, 2, \ldots, n-1, \tag{10.73}$$

$$\begin{aligned}\frac{\partial y_{2i,2}}{\partial T_0}(T_0, T_1, T_2) &+ \alpha y_{2i,2}(T_0, T_1, T_2) - \alpha b_1 y_{2i-1,2}(T_0 - \tau_c, T_1, T_2)\\
&+ \beta_c \big(y_{2i,2}(T_0 - \tau_c, T_1, T_2) - y_{2i+2,2}(T_0 - \tau_c, T_1, T_2)\big)\\
&= \alpha b_2 y_{2i-1,1}^2(T_0 - \tau_c, T_1, T_2)\\
&- \beta_\varepsilon \big(y_{2i,1}(T_0 - \tau_c, T_1, T_2) - y_{2i+2,1}(T_0 - \tau_c, T_1, T_2)\big)\\
&- \frac{\partial y_{2i,1}}{\partial T_1}(T_0 - \tau_c, T_1, T_2) - \alpha b_1 \tau_c \frac{\partial y_{2i-1,1}}{\partial T_1}(T_0 - \tau_c, T_1, T_2)\\
&+ \beta_c \tau_c \frac{\partial y_{2i,1}}{\partial T_1}(T_0 - \tau_c, T_1, T_2) - \beta_c \tau_c \frac{\partial y_{2i+2,1}}{\partial T_1}(T_0 - \tau_c, T_1, T_2)\\
&- \alpha b_1 \tau_\varepsilon \frac{\partial y_{2i-1,1}}{\partial T_0}(T_0 - \tau_c, T_1, T_2) + \beta_c \tau_\varepsilon\\
&\frac{\partial y_{2i,1}}{\partial T_0}(T_0 - \tau_c, T_1, T_2) - \beta_c \tau_\varepsilon \frac{\partial y_{2i+2,1}}{\partial T_0}(T_0 - \tau_c, T_1, T_2),\\
&i = 1, 2, \ldots, n-1,\end{aligned} \tag{10.74}$$

$$\frac{\partial y_{2n-1,2}}{\partial T_0}(T_0, T_1, T_2) + \alpha b_1 \sum_{i=1}^{n-1} y_{2i-1,2}(T_0 - \tau_c, T_1, T_2) + \alpha y_{2n-1,2}(T_0, T_1, T_2)$$
$$+ \beta_c y_{2n-1,2}(T_0 - \tau_c, T_1, T_2) - \beta_c y_{2,2}(T_0 - \tau_c, T_1, T_2)$$

$$
\begin{aligned}
&= -\frac{\partial y_{2n-1,1}}{\partial T_1}(T_0, T_1, T_2) + 2\alpha b_2 \sum_{i=1}^{n-2} y_{2i-1,1}(T_0 - \tau_c, T_1, T_2) \\
&\sum_{j=i}^{n-2} y_{2j+1,1}(T_0 - \tau_c, T_1, T_2) + \alpha b_1 \tau_c \\
&\sum_{i=1}^{n-1} \frac{\partial y_{2i-1,1}}{\partial T_1}(T_0 - \tau_c, T_1, T_2) + \alpha b_1 \tau_\varepsilon \\
&\sum_{i=1}^{n-1} \frac{\partial y_{2i-1,1}}{\partial T_0}(T_0 - \tau_c, T_1, T_2) \\
&- \beta\tau_c\left(\frac{\partial y_{2,1}}{\partial T_1}(T_0 - \tau_c, T_1, T_2) - \frac{\partial y_{2n-1,1}}{\partial T_1}(T_0 - \tau_c, T_1, T_2)\right) \\
&- \beta_c\tau_\varepsilon\left(\frac{\partial y_{2,1}}{\partial T_0}(T_0 - \tau_c, T_1, T_2) - \frac{\partial y_{2n-1,1}}{\partial T_0}(T_0 - \tau_c, T_1, T_2)\right) \\
&- \beta_\varepsilon\big(y_{2n-1,2}(T_0 - \tau_c, T_1, T_2) - y_{2,2}(T_0 - \tau_c, T_1, T_2)\big).
\end{aligned} \tag{10.75}
$$

The equation obtained from ε^3 is

$$
\begin{aligned}
&\frac{\partial y_{2i-1,3}}{\partial T_0}(T_0, T_1, T_2) + y_{2i,3}(T_0, T_1, T_2) - y_{2i+2,3}(T_0, T_1, T_2) \\
&\qquad = -\frac{\partial y_{2i-1,1}}{\partial T_2}(T_0, T_1, T_2) - \frac{\partial y_{2i-1,2}}{\partial T_1}(T_0, T_1, T_2), \\
&\qquad i = 1, 2, \ldots, n-1,
\end{aligned} \tag{10.76}
$$

$$
\begin{aligned}
&\frac{\partial y_{2i,3}}{\partial T_0}(T_0, T_1, T_2) + \alpha y_{2i,3}(T_0, T_1, T_2) - \alpha b_{1c} y_{2i-1,3}(T_0 - \tau_c, T_1, T_2 \\
&\quad + \beta_c(y_{2i,3}(T_0 - \tau_c, T_1, T_2) - y_{2i+2,3}(T_0 - \tau_c, T_1, T_2)) \\
&\quad = -\frac{\partial y_{2i,1}}{\partial T_2}(T_0, T_1, T_2) - \frac{\partial y_{2i,2}}{\partial T_1}(T_0, T_1, T_2) \\
&\quad + \alpha b_3 y_{2i-1,1}^3(T_0 - \tau_c, T_1, T_2) \\
&\quad + 2\alpha b_2 y_{2i-1,1}(T_0 - \tau_c, T_1, T_2)(y_{2i-1,2}(T_0 - \tau_c, T_1, T_2) \\
&\quad - \tau_c\frac{\partial y_{2i-1,1}}{\partial T_1}(T_0 - \tau_c, T_1, T_2) - \tau_\varepsilon\frac{\partial y_{2i-1,1}}{\partial T_0}(T_0 - \tau_c, T_1, T_2)) \\
&\quad - \beta_\varepsilon(y_{2i,2}(T_0 - \tau_c, T_1, T_2) - y_{2i+2,2}(T_0 - \tau_c, T_1, T_2)) \\
&\quad - \alpha b_1(\tau_c\frac{\partial y_{2i-1,1}}{\partial T_2}(T_0 - \tau_c, T_1, T_2) \\
&\quad + \tau_c\frac{\partial y_{2i-1,2}}{\partial T_1}(T_0 - \tau_c, T_1, T_2) + \tau_\varepsilon\frac{\partial y_{2i-1,1}}{\partial T_1}(T_0 - \tau_c, T_1, T_2)
\end{aligned}
$$

$$
\begin{aligned}
&+\tau_\varepsilon\frac{\partial y_{2i-1,2}}{\partial T_0}(T_0-\tau_c,T_1,T_2))+\beta_c\tau_c(\frac{\partial y_{2i,1}}{\partial T_2}(T_0-\tau_c,T_1,T_2)\\
&+\frac{\partial y_{2i,2}}{\partial T_1}(T_0-\tau_c,T_1,T_2)-\frac{\partial y_{2i+2,1}}{\partial T_2}(T_0-\tau_c,T_1,T_2)\\
&-\frac{\partial y_{2i+2,2}}{\partial T_1}(T_0-\tau_c,T_1,T_2))+\beta_c\tau_\varepsilon(\frac{\partial y_{2i,1}}{\partial T_1}(T_0-\tau_c,T_1,T_2)\\
&+\frac{\partial y_{2i,2}}{\partial T_0}(T_0-\tau_c,T_1,T_2)-\frac{\partial y_{2i+2,1}}{\partial T_1}(T_0-\tau_c,T_1,T_2)\\
&-\frac{\partial y_{2i+2,2}}{\partial T_0}(T_0-\tau_c,T_1,T_2))+\beta_\varepsilon\tau_c(\frac{\partial y_{2i,1}}{\partial T_1}(T_0-\tau_c,T_1,T_2),\\
&-\frac{\partial y_{2i+2,1}}{\partial T_1}(T_0-\tau_c,T_1,T_2))+\beta_\varepsilon\tau_\varepsilon(\frac{\partial y_{2i,1}}{\partial T_0}(T_0-\tau_c,T_1,T_2)\\
&-\frac{\partial y_{2i+2,1}}{\partial T_0}(T_0-\tau_c,T_1,T_2))+\frac{\alpha b_1}{2}(\tau_c\frac{\partial y_{2i-1,1}}{\partial T_1}(T_0-\tau_c,T_1,T_2)\\
&+\tau_\varepsilon\frac{\partial y_{2i-1,1}}{\partial T_0}(T_0-\tau_c,T_1,T_2))^2-\frac{\beta_c}{2}(\tau_c\frac{\partial y_{2i,1}}{\partial T_1}(T_0-\tau_c,T_1,T_2)\\
&+\tau_\varepsilon\frac{\partial y_{2i,1}}{\partial T_0}(T_0-\tau_c,T_1,T_2))^2+\frac{\beta_c}{2}(\tau_c\frac{\partial y_{2i+2,1}}{\partial T_1}(T_0-\tau_c,T_1,T_2)\\
&+\tau_\varepsilon\frac{\partial y_{2i+2,1}}{\partial T_0}(T_0-\tau_c,T_1,T_2))^2=0,\\
&i=1,2,...,n-1,
\end{aligned}
\tag{10.77}
$$

$$
\begin{aligned}
\frac{\partial y_{2n-1,3}}{\partial T_0}&(T_0,T_1,T_2)+\alpha b_1\sum_{i=1}^{n-1}y_{2i-1,3}(T_0-\tau_c,T_1,T_2)+\alpha y_{2n-1,3}(T_0,T_1,T_2)\\
&+\beta_c y_{2n-1,3}(T_0-\tau_c,T_1,T_2)-\beta_c y_{2,3}(T_0-\tau_c,T_1,T_2)\\
&=-\alpha b_3(\sum_{i=1}^{n-1}y_{2i-1,1}(T_0-\tau_c,T_1,T_2))^3\\
&+2\alpha b_2\sum_{i=1}^{n-1}(y_{2i-1,2}(T_0-\tau_c,T_1,T_2)\\
&-\tau_c\frac{\partial y_{2i-1,1}(T_0-\tau_c,T_1,T_2)}{\partial T_1}\\
&-\tau_\varepsilon\frac{\partial y_{2i-1,1}(T_0-\tau_c,T_1,T_2)}{\partial T_0})\sum_{i=1}^{n-1}y_{2i-1,1}(T_0-\tau_c,T_1,T_2)\\
&-\alpha b_1\sum_{i=1}^{n-1}(-\tau_c\frac{\partial y_{2i-1,1}(T_0-\tau_c,T_1,T_2)}{\partial T_2}\\
&-\tau_c\frac{\partial y_{2i-1,2}(T_0-\tau_c,T_1,T_2)}{\partial T_1}-\tau_\varepsilon\frac{\partial y_{2i-1,1}(T_0-\tau_c,T_1,T_2)}{\partial T_1}\\
&-\tau_\varepsilon\frac{\partial y_{2i-1,2}(T_0-\tau_c,T_1,T_2)}{\partial T_0}+\frac{1}{2}(\tau_c\frac{\partial y_{2i-1,1}(T_0-\tau_c,T_1,T_2)}{\partial T_1}
\end{aligned}
$$

$$
\begin{aligned}
&+\tau_\varepsilon\frac{\partial y_{2i-1,1}(T_0-\tau_c,T_1,T_2)}{\partial T_2})^2)+\beta_\varepsilon(y_{2,2}(T_0-\tau_c,T_1,T_2)\\
&-\tau_c\frac{\partial y_{2,1}(T_0-\tau_c,T_1,T_2)}{\partial T_1}-\tau_\varepsilon\frac{\partial y_{2,1}(T_0-\tau_c,T_1,T_2)}{\partial T_0})\\
&-\beta_\varepsilon(y_{2n-1,2}(T_0-\tau_c,T_1,T_2)-\tau_c\frac{\partial y_{2n-1,1}(T_0-\tau_c,T_1,T_2)}{\partial T_1}\\
&-\tau_\varepsilon\frac{\partial y_{2n-1,1}(T_0-\tau_c,T_1,T_2)}{\partial T_0})-\beta_c\tau_c(\frac{\partial y_{2,1}(T_0-\tau_c,T_1,T_2)}{\partial T_2}\\
&+\frac{\partial y_{2,2}(T_0-\tau_c,T_1,T_2)}{\partial T_1})-\beta_c\tau_\varepsilon(\frac{\partial y_{2,1}(T_0-\tau_c,T_1,T_2)}{\partial T_1}\\
&+\frac{\partial y_{2,2}(T_0-\tau_c,T_1,T_2)}{\partial T_0})+\frac{1}{2}\beta_c(\tau_c\frac{\partial y_{2,1}(T_0-\tau_c,T_1,T_2)}{\partial T_1}\\
&+\tau_\varepsilon\frac{\partial y_{2,1}(T_0-\tau_c,T_1,T_2)}{\partial T_0})^2+\beta_c\tau_c(\frac{\partial y_{2n-1,1}(T_0-\tau_c,T_1,T_2)}{\partial T_2}\\
&+\frac{\partial y_{2n-1,2}(T_0-\tau_c,T_1,T_2)}{\partial T_1})+\beta_c\tau_\varepsilon(\frac{\partial y_{2n-1,1}(T_0-\tau_c,T_1,T_2)}{\partial T_1}\\
&+\frac{\partial y_{2n-1,2}(T_0-\tau_c,T_1,T_2)}{\partial T_0})-\frac{1}{2}\beta_c(\tau_c\frac{\partial y_{2n-1,1}(T_0-\tau_c,T_1,T_2)}{\partial T_1}\\
&+\tau_\varepsilon\frac{\partial y_{2n-1,1}(T_0-\tau_c,T_1,T_2)}{\partial T_0})^2. \qquad (10.78)
\end{aligned}
$$

First let us solve Eqs. (10.70)–(10.72). Assume that the solution assumes the following form

$$
\begin{aligned}
\mathbf{Y}_1 &= \mathbf{A}_1(T_1,T_2)\sin(\omega_1T_0)+\mathbf{B}_1(T_1,T_2)\cos(\omega_1T_0)\\
&\quad+\mathbf{P}_1(T_1,T_2)\sin(\omega_2T_0)+\mathbf{Q}_1(T_1,T_2)\cos(\omega_2T_0). \qquad (10.79)
\end{aligned}
$$

Substituting the solution (10.79) into Eqs. (10.70)–(10.72) yields the algebraic equation as follows

$$
\begin{aligned}
&\begin{pmatrix}(s_1\mathbf{D}+\omega_1\mathbf{I})^T & -(\mathbf{C}+c_1\mathbf{D})^T\\ -(\mathbf{C}+c_1\mathbf{D})^T & -(s_1\mathbf{D}+\omega_1\mathbf{I})^T\end{pmatrix}\begin{pmatrix}\mathbf{A}_1\\ \mathbf{B}_1\end{pmatrix}=\mathbf{0},\\
&\begin{pmatrix}(s_2\mathbf{D}+\omega_2\mathbf{I})^T & -(\mathbf{C}+c_2\mathbf{D})^T\\ -(\mathbf{C}+c_2\mathbf{D})^T & -(s_2\mathbf{D}+\omega_2\mathbf{I})^T\end{pmatrix}\begin{pmatrix}\mathbf{P}_1\\ \mathbf{Q}_1\end{pmatrix}=\mathbf{0}, \qquad (10.80)
\end{aligned}
$$

where $c_i = \cos(\omega_i\tau_c), s_i = \sin(\omega_i\tau_c), i = 1, 2$, which are known constants. Note that β in $\mathbf{D}$ is actually β_c. Solving the algebraic Eq. (10.80) yields

$$
\begin{aligned}
&A_{1,i} = f^1_{1,i}(A_{1,1}, B_{1,1}),\ B_{1,i} = f^2_{1,i}(A_{1,1}, B_{1,1}),\\
&P_{1,i} = f^3_{1,i}(P_{1,1}, Q_{1,1}),\ Q_{1,i} = f^4_{1,i}(P_{1,1}, Q_{1,1}),\ i = 2, 3, \ldots, 2n-1, \qquad (10.81)
\end{aligned}
$$

where $A_{1,i}, B_{1,i}, P_{1,i}, Q_{1,i}$ indicate the i st component of $\mathbf{A}_1, \mathrm{B}_1, \mathbf{P}_1, \mathbf{Q}_1$ respectively.

From Eq. (10.81) we obtain

$$\begin{aligned} &D_1 A_{1,i} = f_{1,i}^1(D_1 A_{1,1}, D_1 B_{1,1}),\ D_1 B_{1,i} = f_{1,i}^2(D_1 A_{1,1}, D_1 B_{1,1}), \\ &D_1 P_{1,i} = f_{1,i}^3(D_1 P_{1,1}, D_1 Q_{1,1}),\ D_1 Q_{1,i} = f_{1,i}^4(D_1 P_{1,1}, D_1 Q_{1,1}), \\ &\quad i = 2, 3, \ldots, 2n-1. \end{aligned} \tag{10.82}$$

Thus, according to Eq. (10.82), $D_1 A_{1,i}, D_1 B_{1,i}$ can be represented analytically by $D_1 A_{1,1}$ and $D_1 B_{1,1}$, and $D_1 P_{1,i}, D_1 Q_{1,i}$ can be represented analytically by $D_1 P_{1,1}$ and $D_1 Q_{1,1}$. We conclude that (10.79) actually contains only four independent variables $A_{1,1}, B_{1,1}, P_{1,1}, Q_{1,1}$. The expression of Y_1 in terms of $A_{1,1}, B_{1,1}$ can be found by substituting Eq. (10.81) into (10.79). Based on the structure of the solutions, it can be assumed that the solutions to Eqs. (10.73)–(10.75) take the following form

$$\begin{aligned} \mathbf{Y}_2 = {} & \mathbf{E}_2(T_1, T_2) + \sum_{k=1}^{2} \mathbf{A}_2^{(k)}(T_1, T_2)\sin(k\omega_1 T_0) + \mathbf{B}_2^{(k)}(T_1, T_2)\cos(k\omega_1 T_0) \\ & + \sum_{k=1}^{2} \mathbf{P}_2^{(k)}(T_1, T_2)\sin(k\omega_2 T_0) + \mathbf{Q}_2^{(k)}(T_1, T_2)\cos(k\omega_2 T_0) \\ & + \mathbf{G}_2^1 \sin(\omega_1 T_0 - \omega_2 T_0) + \mathbf{G}_2^2 \sin(\omega_1 T_0 + \omega_2 T_0) \\ & + \mathbf{G}_2^3 \cos(\omega_1 T_0 - \omega_2 T_0) + \mathbf{G}_2^4 \cos(\omega_1 T_0 + \omega_2 T_0), \end{aligned} \tag{10.83}$$

where $\mathbf{A}_2^{(k)}, \mathbf{B}_2^{(k)}, \mathbf{P}_2^{(k)}, \mathbf{Q}_2^{(k)}$ respectively represents the coefficient of $\sin(k\omega_1 T_0)$, $\cos(k\omega_1 T_0)$, $\sin(k\omega_2 T_0)$ and $\cos(k\omega_2 T_0)$, $k = 1, 2$.

The existence of the constant term $\mathbf{E}_2$ in (10.83) is due to the quadratic nonlinearity of Eqs. (10.73)–(10.75). Substituting $\mathbf{Y}_2$ from (10.83) into Eqs. (10.73)–(10.75) and eliminating the secular term containing $\sin(\omega_1 T_0)$, $\cos(\omega_1 T_0)$ and $\sin(\omega_2 T_0)$, $\cos(\omega_2 T_0)$, the solvability condition can be expressed by a set of algebraic equations, namely

$$\begin{aligned} &\begin{pmatrix} (s_1\mathbf{D} + \omega_1\mathbf{I})^T & -(\mathbf{C} + c_1\mathbf{D})^T \\ -(\mathbf{C} + c_1\mathbf{D})^T & -(s_1\mathbf{D} + \omega_1\mathbf{I})^T \end{pmatrix} \begin{pmatrix} \mathbf{A}_2 \\ \mathbf{B}_2 \end{pmatrix} = \mathbf{U}_2^1, \\ &\begin{pmatrix} (s_2\mathbf{D} + \omega_2\mathbf{I})^T & -(\mathbf{C} + c_2\mathbf{D})^T \\ -(\mathbf{C} + c_2\mathbf{D})^T & -(s_2\mathbf{D} + \omega_2\mathbf{I})^T \end{pmatrix} \begin{pmatrix} \mathbf{P}_2 \\ \mathbf{Q}_2 \end{pmatrix} = \mathbf{U}_2^2, \end{aligned} \tag{10.84}$$

where $\mathbf{I}$ denotes the identity matrix with order $2n-1$, and $\mathbf{U}_2^1$, $\mathbf{U}_2^2$ are given in Appendix 1.

The major goal now becomes solving Eq. (10.84). According to the Fredholm Alternative Theorem, Eq. (10.84) admits a solution if and only if the term on the right hand side, namely, $\mathbf{U}_2^1, \mathbf{U}_2^2$ of is orthogonal to each solution of the adjoint equation of Eq. (10.84). In other words, the solutions to such adjoint equations need to be solved first.

$$\begin{pmatrix} (s_1\mathbf{D}+\omega_1\mathbf{I})^T & -(\mathbf{C}+c_1\mathbf{D})^T \\ -(\mathbf{C}+c_1\mathbf{D})^T & -(s_1\mathbf{D}+\omega_1\mathbf{I})^T \end{pmatrix}\mathbf{X}^1 = 0,$$
$$\begin{pmatrix} (s_2\mathbf{D}+\omega_2\mathbf{I})^T & -(\mathbf{C}+c_2\mathbf{D})^T \\ -(\mathbf{C}+c_2\mathbf{D})^T & -(s_2\mathbf{D}+\omega_2\mathbf{I})^T \end{pmatrix}\mathbf{X}^2 = 0. \tag{10.85}$$

The orthogonality of $\mathbf{U}_2^1$, $\mathbf{U}_2^2$ to solutions of the adjoint equation of Eq. (10.84) is expressed as

$$\left(\mathbf{U}_2^1\right)^T\mathbf{X}^1 = 0,$$
$$\left(\mathbf{U}_2^2\right)^T\mathbf{X}^2 = 0. \tag{10.86}$$

Solving Eq. (10.86) yields

$$D_1A_{1,1} = g^1(A_{1,1}, B_{1,1}),\ D_1B_{11} = g^2(A_{1,1}, B_{1,1});$$
$$D_1P_{1,1} = g^3(P_{1,1}, Q_{1,1}),\ D_1Q_{11} = g^4(P_{1,1}, Q_{1,1}); \tag{10.87}$$

where g^i, $i = 1, 2, 3, 4$ are linear function with respect to their variables.

The coefficients of each of the terms $\mathbf{A}_2^{(k)}, \mathbf{B}_2^{(k)}, \mathbf{P}_2^{(k)}, \mathbf{Q}_2^{(k)}$, $k = 1, 2$ in (10.83) as well as the constant term $\mathbf{E}_2$ can be obtained by using the solvability condition (10.87) and Eqs. (10.79), (10.81) and (10.82), and consequently $\mathbf{Y}_2$ can be solved. Substitute $\mathbf{Y}_1$, $\mathbf{Y}_2$ and the solvability condition into Eqs. (10.76)–(10.78). In order to remove the secular terms in Eqs. (10.76)–(10.78), a particular solution of $\mathbf{Y}_3$ is assumed as follows

$$\mathbf{Y}_3 = \mathbf{E}_3(T_1, T_2) + \sum_{k=1}^{3}\mathbf{A}_2^{(k)}(T_1, T_2)\sin(k\omega_1T_0) + \mathbf{B}_2^{(k)}(T_1, T_2)\cos(k\omega_1T_0)$$
$$+ \sum_{k=1}^{3}\mathbf{P}_2^{(k)}(T_1, T_2)\sin(k\omega_2T_0) + \mathbf{Q}_2^{(k)}(T_1, T_2)\cos(k\omega_2T_0)$$
$$+ \sum_{k=1}^{2}\mathbf{G}_2^{(k)}\sin(\omega_1T_0 + (-1)^k\omega_2T_0) + \mathbf{G}_2^{(k+2)}\cos(\omega_1T_0 + (-1)^k\omega_2T_0)$$
$$+ \sum_{k=1}^{2}\mathbf{G}_2^{(k+4)}\sin(2\omega_1T_0 + (-1)^k\omega_2T_0) + \mathbf{G}_2^{(k+6)}\cos(2\omega_1T_0 + (-1)^k\omega_2T_0)$$
$$+ \sum_{k=1}^{2}\mathbf{G}_2^{(k+8)}\sin(\omega_1T_0 + (-1)^k2\omega_2T_0) + \mathbf{G}_2^{(k+10)}\cos(\omega_1T_0 + (-1)^k2\omega_2T_0). \tag{10.88}$$

Substituting (10.88) into Eqs. (10.76)–(10.78) and eliminating the secular term yield the following algebraic equations

$$
\begin{pmatrix} (s_1\mathbf{D}+\omega_1\mathbf{I})^T & -(\mathbf{C}+c_1\mathbf{D})^T \\ -(\mathbf{C}+c_1\mathbf{D})^T & -(s_1\mathbf{D}+\omega_1\mathbf{I})^T \end{pmatrix}\begin{pmatrix} \mathbf{A}_3 \\ \mathbf{B}_3 \end{pmatrix} = \mathbf{U}_3^1,
$$

$$
\begin{pmatrix} (s_2\mathbf{D}+\omega_2\mathbf{I})^T & -(\mathbf{C}+c_2\mathbf{D})^T \\ -(\mathbf{C}+c_2\mathbf{D})^T & -(s_2\mathbf{D}+\omega_2\mathbf{I})^T \end{pmatrix}\begin{pmatrix} \mathbf{P}_3 \\ \mathbf{Q}_3 \end{pmatrix} = \mathbf{U}_3^2. \tag{10.89}
$$

Following the same idea as above, one obtains

$$
\begin{aligned}
(\mathbf{U}_3^1)^T\mathbf{X}^1 &= 0, \\
(\mathbf{U}_3^2)^T\mathbf{X}^2 &= 0,
\end{aligned} \tag{10.90}
$$

where $\mathbf{X}^1$, $\mathbf{X}^2$ are solutions to the adjoint equation of Eq. (10.89).

Solving Eq. (10.90) yields the solvability condition

$$
\begin{aligned}
D_2A_{1,1} &= h^1(A_{1,1}, B_{1,1}),\ D_2B_{11} = h^2(A_{1,1}, B_{1,1}), \\
D_2P_{1,1} &= h^3(P_{1,1}, Q_{1,1}),\ D_2Q_{11} = h^4(P_{1,1}, Q_{1,1}).
\end{aligned} \tag{10.91}
$$

Combining Eqs. (10.87) and (10.91) one can obtain that

$$
\begin{cases} \dot{A}_{1,1} = \varepsilon \mathrm{D}_1 A_{1,1} + \varepsilon^2 \mathrm{D}_2 A_{1,1}, \\ \dot{B}_{1,1} = \varepsilon \mathrm{D}_1 B_{1,1} + \varepsilon^2 \mathrm{D}_2 B_{1,1}, \end{cases} \tag{10.92}
$$

$$
\begin{cases} \dot{P}_{1,1} = \varepsilon \mathrm{D}_1 P_{1,1} + \varepsilon^2 \mathrm{D}_2 P_{1,1}, \\ \dot{Q}_{1,1} = \varepsilon \mathrm{D}_1 Q_{1,1} + \varepsilon^2 \mathrm{D}_2 Q_{1,1}, \end{cases} \tag{10.93}
$$

where "." denotes the derivative with respect to time t.

Introduce the following coordinate transformation to Eqs. (10.92) and (10.93)

$$
\begin{aligned}
A_{1,1}(t) &= R_1(t)\cos(\varphi_1(t)),\ B_{1,1}(t) = R_1(t)\sin(\varphi_1(t)), \\
P_{1,1}(t) &= R_2(t)\cos(\varphi_2(t)),\ Q_{1,1}(t) = R_2(t)\sin(\varphi_2(t)).
\end{aligned} \tag{10.94}
$$

Then Eqs. (10.92) and (10.93) now become

$$
\begin{cases} \varepsilon \dot{R}_1(t) = \varepsilon R_1(t)(\mu_1 + p_{11}(\varepsilon R_1(t))^2 + p_{12}(\varepsilon R_2(t))^2), \\ \varepsilon \dot{R}_2(t) = \varepsilon R_2(t)(\mu_2 + p_{21}(\varepsilon R_1(t))^2 + p_{22}(\varepsilon R_2(t))^2), \end{cases}
$$

$$
\begin{cases} \dot{\varphi}_1(t) = \gamma_1 + q_{11}(\varepsilon R_1(t))^2 + q_{12}(\varepsilon R_2(t))^2, \\ \dot{\varphi}_2(t) = \gamma_2 + q_{21}(\varepsilon R_1(t))^2 + q_{22}(\varepsilon R_2(t))^2. \end{cases} \tag{10.95}
$$

Equation (10.95) is the normal form of Eq. (10.59). Combined with the results of Eq. (10.95), it is possible to refine dynamics classification near the double Hopf bifurcation point in Fig. 10.10.

10.2.2.2 Example

In this section we provide an example where values of parameters are fixed as (10.63). Based on the previous theoretical derivation, the coefficients of the normal form equation are obtained as follows

$$\begin{aligned}
\mu_1 &= 0.076(\beta-\beta_c)+0.669(\beta-\beta_c)^2+(0.427+0.157(\beta-\beta_c))(\tau-\tau_c) \\
&\quad -0.516(\tau-\tau_c)^2, \\
\mu_2 &= 0.542(\beta-\beta_c)+0.174(\beta-\beta_c)^2+(0.639-0.669(\beta-\beta_c))(\tau-\tau_c) \\
&\quad -0.951(\tau-\tau_c)^2, \\
p_{11} &= -0.001,\, p_{12}=-0.002,\, p_{21}=-0.0006,\, p_{22}=-0.0005, \\
\gamma_1 &= 0.682(\beta-\beta_c)-0.057(\beta-\beta_c)^2+(-0.414-1.006(\beta-\beta_c))(\tau-\tau_c) \\
&\quad +0.106(\tau-\tau_c)^2, \\
\gamma_2 &= 0.77(\beta-\beta_c)-0.771(\beta-\beta_c)^2+(-1.06-1.079(\beta-\beta_c))(\tau-\tau_c) \\
&\quad +0.722(\tau-\tau_c)^2, \\
q_{11} &= -0.0002,\, q_{12}=-0.0015,\, q_{21}=0.0016,\, q_{22}=0.0003,
\end{aligned} \tag{10.96}$$

with $\beta_c = 0.8244$, $\tau_c = 0.9558$.

Then the four equilibria of the first group of equations, namely, the equations governing the evolution of $R_1(t)$ and $R_2(t)$ in Eq. (10.95), are calculated as

$$\begin{aligned}
Eq_1 &= (0,0), \\
Eq_2 &= (0, 46.588\sqrt{-2.335+0.894\beta+0.174\beta^2+3.009\tau-0.669\beta\tau-0.951\tau^2}), \\
Eq_3 &= (\sqrt{-6238.1+3251.41\beta+44.3769\beta^2+7502.9\tau-1958.16\beta\tau-2299.67\tau^2}, \\
&\quad \sqrt{3036.52-2283.83\beta+320.296\beta^2-3218.01\tau+1092.34\beta\tau+923.48\tau^2}). \\
Eq_4 &= (31.5083\sqrt{-0.364-1.177\beta+0.669\beta^2+1.284\tau+0.157\beta\tau-0.5163\tau^2}, 0).
\end{aligned} \tag{10.97}$$

In the next we analyse the nonlinear dynamic behaviour near the double Hopf bifurcation point (0.8244, 0.9558). A classification of the dynamical behaviour of Eq. (10.59) near the nonresonant double Hopf bifurcation is shown in Fig. 10.14, together with the number of equilibria in each region as well as their stability. By analyzing the equilibrium of the amplitude equation in Eq. (10.95) and their stability, the parameter plane (β, τ) is divided into six regions. In region A, the system admits a stable trivial equilibrium Eq_1, corresponding to the uniform flow solution of the original system, i.e., Eq. (10.59). When the set of parameters enters region B, the trivial equilibrium is destabilized and a stable equilibrium Eq_2 appears, which corresponds to the periodic solution of Eq. (10.59) with frequency close to ω_1. Namely, there exists a stable periodic solution with period T_1 in the original system. In region C, the unstable equilibrium Eq_1 and the stable equilibrium point Eq_2 still exist, but another unstable equilibrium Eq_4 arises, which corresponds to the unstable periodic solution of the original system with frequency close to ω_2. In this region, a stable periodic solution with period T_1 coexists with an unstable periodic solution with T_2 in the original system. In region D, the unstable equilibrium Eq_4 becomes stable and the stability of the original stable equilibrium Eq_2 persists. An unstable equilibrium Eq_3 appears, which corresponds to the unstable quasi-periodic solutions of Eq. (10.59). In such situation, two stable periodic solutions and one unstable quasi-periodic solution coexist in the original system. When the set of parameters enter region E, the unstable equilibrium Eq_3 disappears, and the equilibrium Eq_2 is destabilized, while the equilibrium Eq_4 maintains its stability. In this case, an unstable periodic solution with period T_1 coexists with a stable periodic solution with period T_2 in region E. In the last region F, the unstable equilibrium Eq_2 no longer exists, while the equilibrium Eq_4 remains stable. There is only one stable periodic solution in this region.

In order to verify the reliability of the theoretical classification of dynamics, we use DDE-BIFTOOL to carry out the bifurcation continuation, along the rectangle depicted in Fig. 10.15a. Starting from P_1 in Region A where the equilibrium of the system is stable, we plot the difference between the maximal and minimal velocities of the first vehicle, as a function of the driver's reaction delay τ or sensitivity coefficient β, depending on the rectangle edges on which the continuation is taken. As shown in Fig. 10.15b, the continuation results are in good agreement with the theoretical prediction by the method of multiple scales.

10.2.3 Modes of Traffic Flow Corresponding to the Solutions in Different Regions

In the last subsection, we used the method of multiple scales to classify the solutions of Eq. (10.59) in the parameter plane (β, τ) and numerically verified the reliability of the theoretical classification. We divided the parameter plane into six regions according to the feature of different solutions that may arise. In the following discussion, we will demonstrate the macroscopic traffic modes corresponding to the solutions in these regions using time history plot, spatiotemporal pattern, and histogram of velocity.

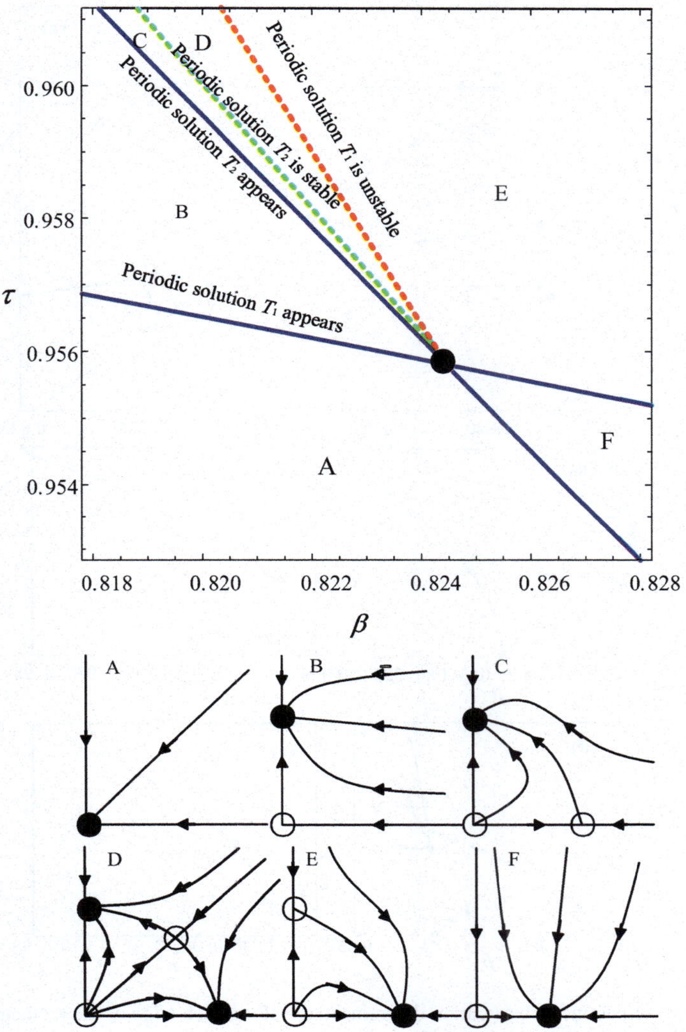

Fig. 10.14 Classification of dynamical behaviour near the double Hopf bifurcation point in the parameter plane consisting of the driver's reaction delay τ and sensitivity coefficient β. The equilibrium is stable in region A. Periodic solutions with frequency close to ω_1 exist in regions B and C. Bistable periodic solutions coexist in region D. Periodic solutions with frequency close to ω_2 exist in regions E and F. The solid line is the Hopf bifurcation line and the dashed line is the secondary bifurcation line. The stable and unstable equilibria are represented by solid and hollow circles, respectively

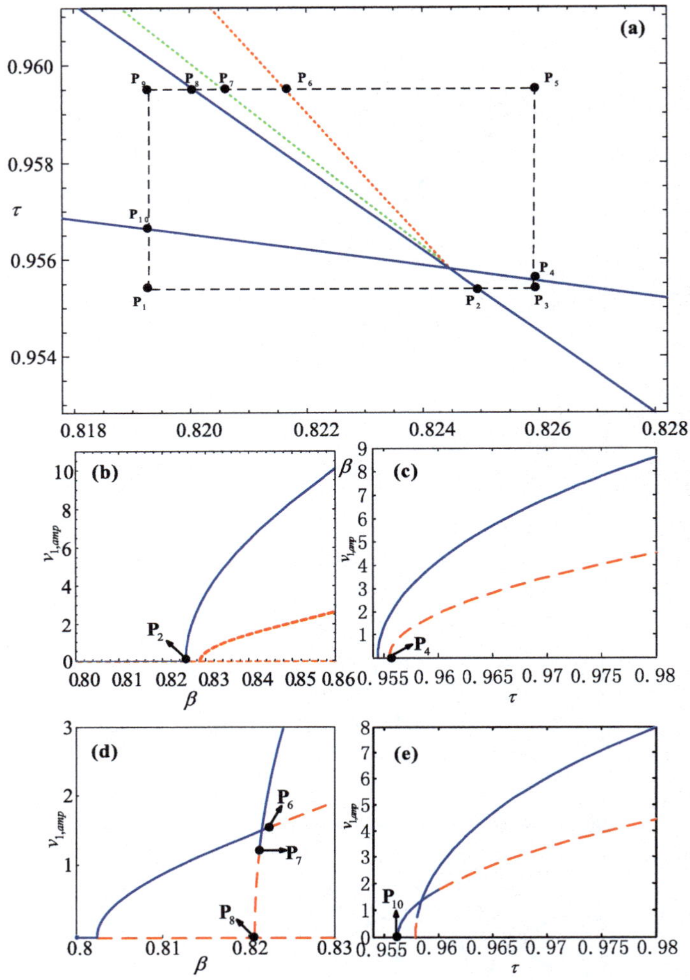

Fig. 10.15 One-parameter bifurcation diagrams made along each edge of the rectangular box in **a** i.e., **b** P_1P_3, **c** P_3P_5, **d** P_5P_9 and **e** P_9P_1

10.2.3.1 Traffic Modes of Uniform Flow

In order to demonstrate the macroscopic traffic mode of the uniform flow, we select a point P_1 in region A where the uniform flow is stable and conducted numerical simulations for Eq. (10.56). Figure 10.16a shows the evolution of the velocity of each vehicle over time, where the horizontal axis represents time and the vertical axis represents the velocity of each vehicle. The figure indicates that when the initial condition is disturbed (the first vehicle suddenly decelerates), the velocity of each vehicle deviates from the uniform flow velocity v^* at the initial moment. However,

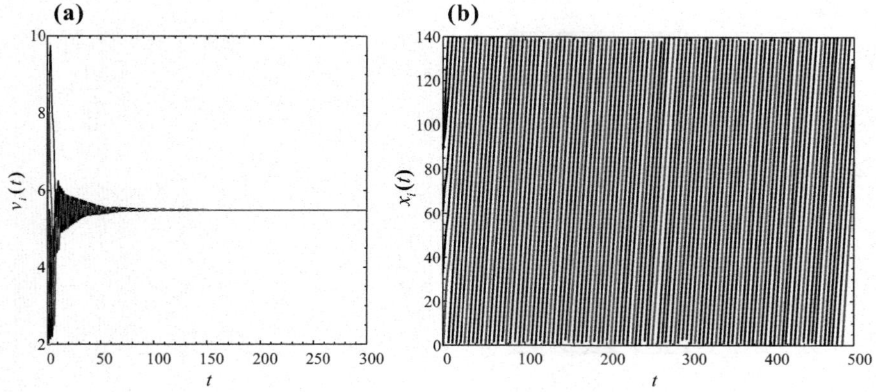

Fig. 10.16 **a** Time histories of the velocity of each vehicle for parameters corresponding to P_1 in the parameter plane, **b** Spatiotemporal pattern of each vehicle. The blue line indicates the evolution of velocity of the first vehicle in a and its traveling process in b

due to the stability of the uniform flow solution, the velocity of each vehicle quickly stabilizes to the uniform flow velocity. The vehicles then maintain a constant inter-vehicle distance h^* and travel at a constant velocity of v^*, i.e., the uniform flow velocity. To better understand the macroscopic traffic mode of the uniform flow solution at the microscopic level, we illustrate the spatiotemporal pattern of all vehicles in Fig. 10.16b. The figure shows that the uniform flow solution does not contain any traffic jam clusters, and the trajectory of each vehicle is a straight line, indicating that the vehicles always maintain the same velocity during their travel. Each straight line in the diagram is parallel to the others, and the distance between any two adjacent parallel lines is equal, which shows that the vehicles always maintain a fixed distance between them.

10.2.3.2 Almost Jammed Traffic Mode Corresponding to the Periodic Solution

This subsection is devoted to visualize the time history and spatiotemporal pattern for the periodic solutions with periods T_1 and T_2 in regions B and F in Fig. 10.14, respectively, for the purpose of identifying the macroscopic traffic modes corresponding to these solutions. To this end, we conduct numerical simulations for point P_9 in region B and point P_3 in region F, as shown in Fig. 10.15.

In Fig. 10.17a, the horizontal axis represents time t, and the vertical axis represents the velocity of each vehicle, where the blue line represents the time history of the velocity of the first vehicle. It can be seen that the period of the periodic solution is $T_1 = 20$ s. In each period, the vehicle has a duration of roughly 2 s within which the velocity is almost 0, and the maximal velocity in the whole period is 10.26 m/s. In Fig. 10.17b, the position of each vehicle is taken as the vertical axis, and the

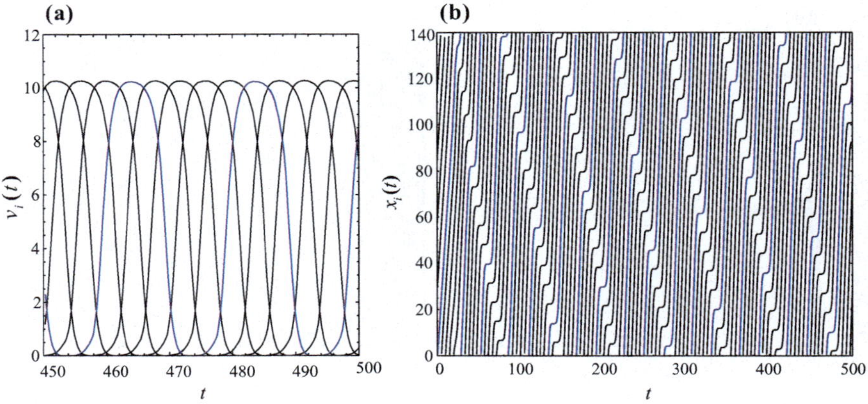

Fig. 10.17 **a** Time histories of the velocity of each vehicle for parameters corresponding to P_9 in the parameter plane, **b** Spatiotemporal pattern of each vehicle. The blue line indicates the evolution of velocity of the first vehicle in a and its traveling process in b

spatiotemporal pattern consisting positions of all vehicles for each moment is drawn. The trajectory of the first vehicle is marked by the blue line. One can observe in the spatiotemporal pattern many jam clusters with equal width. We call such phenomenon the equal-width jam mode. On a circular road placed with 5 vehicles, each vehicles will experience a traffic jam that will last for 2 s once every cycle of driving. But once they leave the jammed state, they will not experience traffic jam again within the same cycle. It takes about 23 s for each vehicle to drive through one cycle, which means that the time wasted in the jam within one cycle accounts for 8.7% of the total driving time.

Figure 10.18 shows the numerical simulations for the periodic solution with period T_2 for the set of parameters corresponding to the point P_3 in region F. In Fig. 10.18a, we provide the time history plot of the velocity of each vehicle. It can be seen that the period is $T_2 = 4$ s, and during each period, the maximal velocity is 10.4m/s while the minimal velocity is 0. Although the velocity of the vehicles decreases to zero at some instants, they do not come to a completely resting status but start to accelerate immediately. In other words, there is no actual stop-and-go situation during the driving. This can also be verified by the spatiotemporal pattern in Fig. 10.18b where no traffic jam clusters appear. However, the performance of the traffic flow is different from that of the uniform flow shown in Fig. 10.16b in that the trajectory of each vehicle in the spatiotemporal pattern is curled for multiple times due to constant acceleration and deceleration during the driving. In fact, each vehicle needs to go through 7 switches between acceleration and deceleration for every cycle in this case study. We call such phenomenon the almost jammed traffic mode. From Fig. 10.18b, it can be seen that the time it takes for each vehicle to travel for one cycle is approximately 28 s. Compared with the stop-and-go mode in Fig. 10.17b, this almost jammed traffic mode does not require the vehicle to rest for a long time,

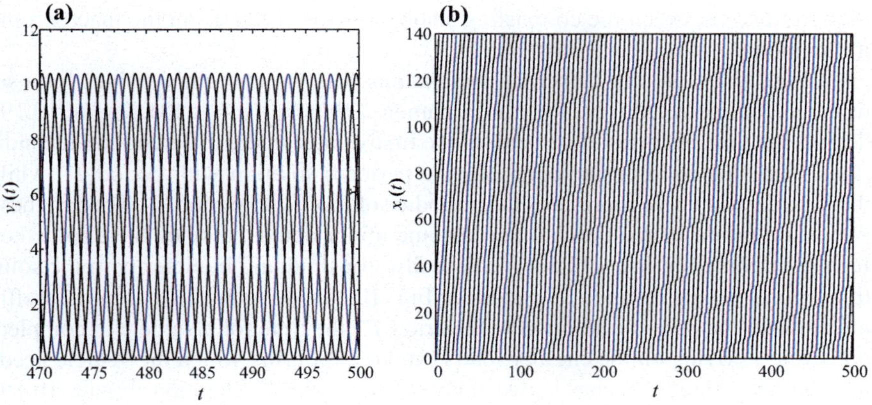

Fig. 10.18 **a** Time histories of the velocity of each vehicle for parameters corresponding to P_3 (periodic solution with T_2) in the parameter plane, **b** Spatiotemporal pattern of each vehicle. The blue line indicates the evolution of velocity of the first vehicle in a and its traveling process in b

but it takes longer for the vehicle to travel for one cycle. This is caused by the constant switch between acceleration and deceleration. This mode not only fails to improve the travel efficiency, but also exhausts the driver as a consequence of the constant acceleration and deceleration during the driving.

By inspecting the stop-and-go traffic mode with period T_1 in Fig. 10.17 and the almost jammed mode with period T_2 in Fig. 10.18, we see that different reaction delays and sensitivity coefficients of drivers may result in periodic solutions of (10.56) with different frequencies. Using the spatiotemporal pattern, we find that these solutions correspond to different traffic modes on a macro level, implying the significance of introducing nonlinear analysis to the study of traffic flow.

10.2.3.3 Jammed Modes Corresponding to the Coexisting Periodic Solutions

In this subsection, we focus on a particular point selected from region D in Fig. 10.14 where two stable periodic solutions coexist. We analyze the macroscopic traffic mode corresponding to these periodic solutions which arise under different initial conditions. To distinguish the traffic modes caused by different stable periodic solutions, we use not only the tool of the spatiotemporal pattern, but also the distribution curve of velocity to show the probability of a certain velocity detected within a given period of time under the two traffic modes. This provides us with another way to distinguish different traffic modes. Additionally, we use the histogram of the frequency distribution of vehicle velocity to estimate the probability that the velocity of the vehicle falls in four intervals within a period of time. From this, we can also observe the difference between the co-existing stable periodic solutions in the macroscopic traffic mode.

In Fig. 10.19, we run numerical simulations with various initial values for a set of parameters selected from Region D, namely, $(\beta, \tau) = (0.7, 1.13)$. Figure 10.19a depicts the time histories of velocity of the first vehicle under different initial conditions. The blue solid curve represents the periodic solution with period T_1, while the blue dashed curve represents the periodic solution with period T_2. Though these two solutions differ with each other in some minor details, they do have some key feature in common: the vehicles never really stops moving in either periodic solution. From the spatiotemporal pattern in Fig. 10.19b, we can see that in the traffic mode under the periodic solution with period T_1, the time for a vehicle to complete one cycle's driving on the circular road is about 28s during which the vehicle needs to experience three switches between acceleration and deceleration. Figure 10.19c shows that for the mode of the periodic solution with period T_2 it takes about 27 s for a vehicle to drive through one cycle within which the vehicle experiences five switches between acceleration and deceleration. Both traffic modes do not contain any traffic jam cluster and therefore can be called almost jammed modes.

An interesting phenomenon can be observed in this scenario. The vehicle experiences less switches between acceleration and deceleration in the first traffic mode than in the second, and the maximal and minimal velocities in the first mode are both greater than that in the second, implying that the traveling time of one cycle in the first mode should be less that that in the second. However, the spatiotemporal pattern suggests the opposite. The specific reason will then be explained in Fig. 10.20.

In Fig. 10.20, the probability of the velocity, i.e., the quantity on the vertical axis, is defined by

$$p_i = \frac{\text{total number that } v_i \text{ is detcted}}{n \times (\text{total steps})}. \tag{10.98}$$

From Fig. 10.20a, it can be seen that in the first traffic mode, about 1.4% of the vehicle velocities is distributed at the minimal value, and 0.9% is at the maximal value. Meantime, in the second traffic mode, from Fig. 10.20b it can be calculated that 1.4% of the vehicle velocities is at the minimal value, which is the same as the case of the first mode, however, 2.1% of the vehicle velocities reaches the maximal value and this percentage is much greater than that in the first traffic mode. Therefore, even the vehicle in the second mode suffers from more frequent acceleration and deceleration and the maximum velocity is not as large as that of the first mode, the vehicle can still travel faster than it can in the first mode.

The possible range of velocity of all vehicles from 200 to 500 s is divided into four intervals: velocities less than 10 km/h (severely jammed) correspond to the red bar; velocities between 10 km/h and 20 km/h (jammed) to the brown bar; velocities between 20 km/h and 30 km/h (mildly jammed) to the yellow bar; and velocities greater than 30 km/h (jam free) to the green bar. The comparison between Fig. 10.21a and b further explain the reason that vehicles travel faster in the first mode than they do in the second.

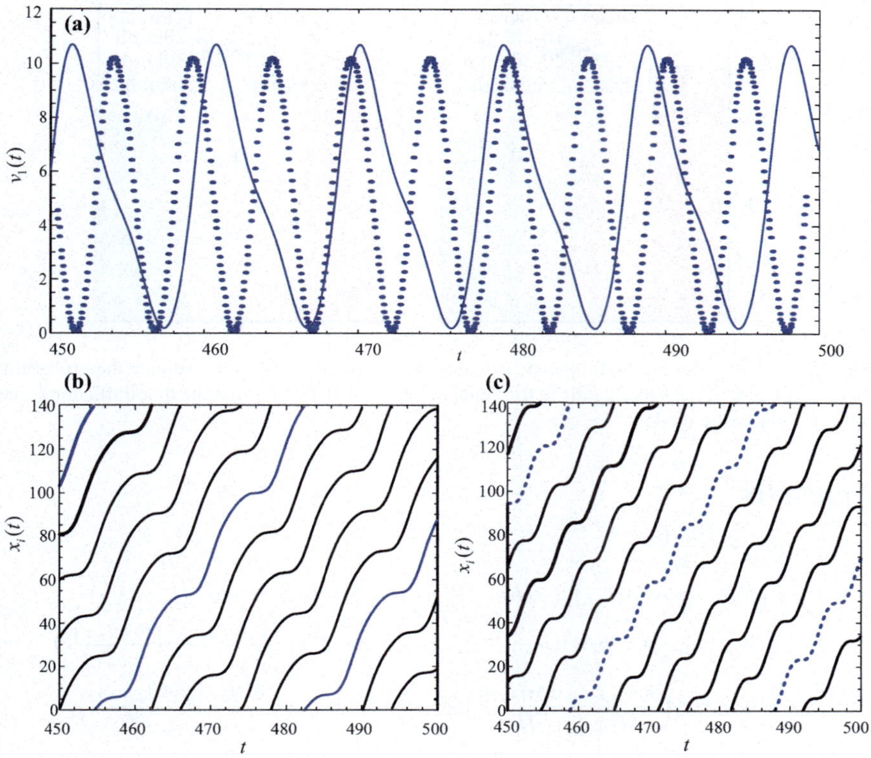

Fig. 10.19 **a** Coexisting periodic solutions for the first vehicle for parameters selected from region D in the parameter plane, starting from different initial values. Solid and dashed curves represent different periodic solutions. **b**, **c** Spatiotemporal patterns induced by different periodic solutions. In both figures, the blue curve indicates the position of the first vehicle

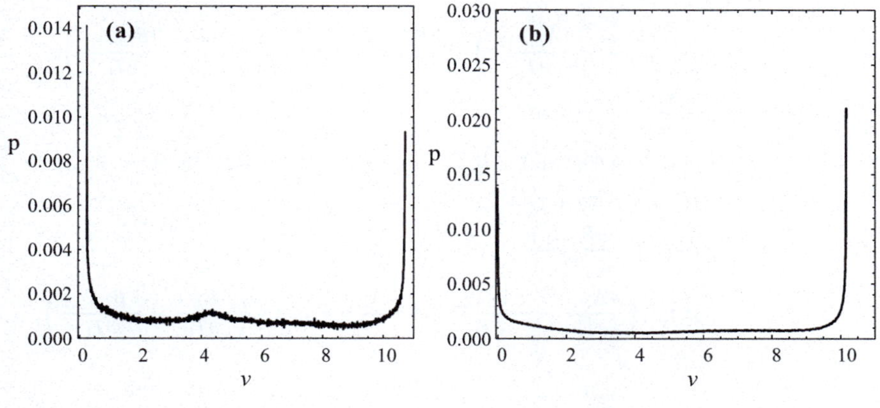

Fig. 10.20 Statistical data on the probability that the velocity arrives at some value, measured from 200 to 500 s, in **a** the first traffic mode and **b** the second traffic mode

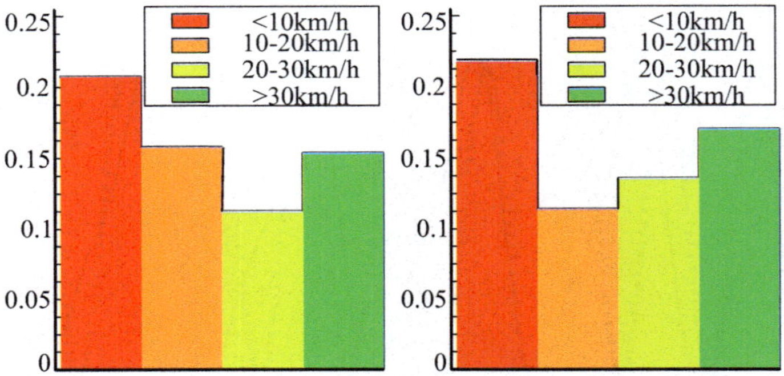

Fig. 10.21 Histogram of the frequency distribution of vehicle velocity to estimate the probability that the velocity of the vehicle falls in four intervals from 200 to 500 s in **a** the first traffic mode and **b** the second traffic mode

Appendix

$$U^1_{2i-1,2} = -\frac{\partial B_{2i-1,1}}{\partial T_1}$$

$$\begin{aligned} U^1_{2i,2} &= \alpha b_1 \tau_\varepsilon \omega_1 \left(c_1 A_{2i-1,1} + s_1 B_{2i-1,1}\right) + (\beta_\varepsilon s_1 + \beta_c \tau_\varepsilon \omega_1 c_1)\left(A_{2i+2,1} - A_{2i,1}\right) \\ &\quad + (\beta_\varepsilon c_1 - \beta_c \tau_\varepsilon \omega_1 s_1)\left(B_{2i,1} - B_{2i+2,1}\right) + \alpha b_1 \tau_c \left(c_1 \frac{\partial B_{2i-1,1}}{\partial T_1} - s_1 \frac{\partial A_{2i-1,1}}{\partial T_1}\right) \\ &\quad + \beta_c \tau_c s_1 \left(\frac{\partial A_{2i,1}}{\partial T_1} - \frac{\partial A_{2i+2,1}}{\partial T_1}\right) + \frac{\partial B_{2i,1}}{\partial T_1} - \beta_c \tau_c c_1 \left(\frac{\partial B_{2i,1}}{\partial T_1} - \frac{\partial B_{2i+2,1}}{\partial T_1}\right), \\ & i = 1, 2, \ldots, n-1 \end{aligned}$$

$$\begin{aligned} U^1_{2n-1,2} &= -\alpha b_1 \tau_\varepsilon \omega_1 \sum_{i=1}^{n-1} \left(c_1 A_{2i-1,1} + s_1 B_{2i-1,1}\right) - (\beta_\varepsilon s_1 + \beta_c \tau_\varepsilon \omega_1 c_1)\left(A_{2n-1,1} - A_{2,1}\right) \\ &\quad - (\beta_\varepsilon c_1 - \beta_c \tau_\varepsilon \omega_1 s_1)\left(B_{2,1} - B_{2n-1,1}\right) - \alpha b_1 \tau_c \sum_{i=1}^{n-1} \left(c_1 \frac{\partial B_{2i-1,1}}{\partial T_1} - s_1 \frac{\partial A_{2i-1,1}}{\partial T_1}\right) \\ &\quad - \beta_c \tau_c s_1 \left(\frac{\partial A_{2,1}}{\partial T_1} - \frac{\partial A_{2n-1,1}}{\partial T_1}\right) + \frac{\partial B_{2n-1,1}}{\partial T_1} + \beta_c \tau_c c_1 \left(\frac{\partial B_{2,1}}{\partial T_1} - \frac{\partial B_{2n-1,1}}{\partial T_1}\right), \end{aligned}$$

$$U^1_{2i-1+2n-1,2} = -\frac{\partial A_{2i-1,1}}{\partial T_1}$$

$$\begin{aligned} U^1_{2i+2n-1,2} &= \alpha b_1 \tau_\varepsilon \omega_1 \left(s_1 A_{2i-1,1} + c_1 B_{2i-1,1}\right) - (\beta_\varepsilon c_1 - \beta_c \tau_\varepsilon \omega_1 s_1)\left(A_{2i+2,1} - A_{2i,1}\right) \\ &\quad + (\beta_\varepsilon s_1 + \beta_c \tau_\varepsilon \omega_1 c_1)\left(B_{2i,1} - B_{2i+2,1}\right) \\ &\quad + \alpha b_1 \tau_c \left(s_1 \frac{\partial B_{2i-1,1}}{\partial T_1} + c_1 \frac{\partial A_{2i-1,1}}{\partial T_1}\right) \\ &\quad + \beta_c \tau_c s_1 \left(\frac{\partial A_{2i,1}}{\partial T_1} - \frac{\partial A_{2i+2,1}}{\partial T_1}\right) + \frac{\partial A_{2i,1}}{\partial T_1} - \beta_c \tau_c s_1 \left(\frac{\partial B_{2i,1}}{\partial T_1} - \frac{\partial B_{2i+2,1}}{\partial T_1}\right), \\ & i = 1, 2, \ldots, n-1, \end{aligned}$$

$$U^1_{4n-2,2} = -\alpha b_1 \tau_\varepsilon \omega_1 \sum_{i=1}^{n-1} \left(s_1 A_{2i-1,1} - c_1 B_{2i-1,1}\right)$$

$$
\begin{aligned}
&+(\beta_\varepsilon c_1-\beta_c\tau_\varepsilon\omega_1 s_1)\big(A_{2n-1,1}-A_{2,1}\big)\\
&-(\beta_\varepsilon s_1+\beta_c\tau_\varepsilon\omega_1 c_1)\big(B_{2,1}-B_{2n-1,1}\big)\\
&-\alpha b_1\tau_c\sum_{i=1}^{n-1}\left(s_1\frac{\partial B_{2i-1,1}}{\partial T_1}+c_1\frac{\partial A_{2i-1,1}}{\partial T_1}\right)\\
&+\beta_c\tau_c c_1\left(\frac{\partial A_{2,1}}{\partial T_1}-\frac{\partial A_{2n-1,1}}{\partial T_1}\right)+\frac{\partial A_{2n-1,1}}{\partial T_1}-\beta_c\tau_c s_1\left(\frac{\partial B_{2,1}}{\partial T_1}-\frac{\partial B_{2n-1,1}}{\partial T_1}\right),
\end{aligned}
$$

$i=1,2,\ldots,n-1,$

$$
U^2_{2i-1,2}=-\frac{\partial Q_{2i-1,1}}{\partial T_1},
$$

$$
\begin{aligned}
U^2_{2i,2}&=\alpha b_1\tau_\varepsilon\omega_2(c_2P_{2i-1,1}+s_2Q_{2i-1,1})+(\beta_\varepsilon s_2+\beta_c\tau_\varepsilon\omega_2c_2)(P_{2i+2,1}-Q_{2i,1})\\
&+(\beta_\varepsilon c_2-\beta_c\tau_\varepsilon\omega_2s_2)(Q_{2i,1}-Q_{2i+2,1})+\alpha b_1\tau_c\left(c_2\frac{\partial Q_{2i-1,1}}{\partial T_1}-s_2\frac{\partial P_{2i-1,1}}{\partial T_1}\right)\\
&+\beta_c\tau_c s_2\left(\frac{\partial P_{2i,1}}{\partial T_1}-\frac{\partial P_{2i+2,1}}{\partial T_1}\right)+\frac{\partial Q_{2i,1}}{\partial T_1}-\beta_c\tau_c c_2\left(\frac{\partial Q_{2i,1}}{\partial T_1}-\frac{\partial Q_{2i+2,1}}{\partial T_1}\right),
\end{aligned}
$$

$i=1,2,\ldots,n-1,$

$$
\begin{aligned}
U^2_{2n-1,2}&=-\alpha b_1\tau_\varepsilon\omega_2\sum_{i=1}^{n-1}(c_2P_{2i-1,1}+s_2Q_{2i-1,1})-(\beta_\varepsilon s_2+\beta_c\tau_\varepsilon\omega_2c_2)(P_{2n-1,1}-P_{2,1})\\
&-(\beta_\varepsilon c_2-\beta_c\tau_\varepsilon\omega_2s_2)(Q_{2,1}-Q_{2n-1,1})\\
&-\alpha b_1\tau_c\sum_{i=1}^{n-1}\left(c_2\frac{\partial Q_{2i-1,1}}{\partial T_1}-s_2\frac{\partial P_{2i-1,1}}{\partial T_1}\right)\\
&-\beta_c\tau_c s_2\left(\frac{\partial P_{2,1}}{\partial T_1}-\frac{\partial P_{2n-1,1}}{\partial T_1}\right)+\frac{\partial Q_{2n-1,1}}{\partial T_1}+\beta_c\tau_c c_2\left(\frac{\partial Q_{2,1}}{\partial T_1}-\frac{\partial Q_{2n-1,1}}{\partial T_1}\right),
\end{aligned}
$$

$$
\begin{aligned}
U^2_{2i-1+2n-1,2}&=-\frac{\partial P_{2i-1,1}}{\partial T_1},\\
U^2_{2i+2n-1,2}&=\alpha b_1\tau_\varepsilon\omega_2(s_1P_{2i-1,1}+c_1Q_{2i-1,1})\\
&-(\beta_\varepsilon c_2-\beta_c\tau_\varepsilon\omega_2s_2)(P_{2i+2,1}-P_{2i,1})\\
&+(\beta_\varepsilon s_2+\beta_c\tau_\varepsilon\omega_2c_2)(Q_{2i,1}-Q_{2i+2,1})\\
&+\alpha b_1\tau_c\left(s_2\frac{\partial Q_{2i-1,1}}{\partial T_1}+c_2\frac{\partial P_{2i-1,1}}{\partial T_1}\right)\\
&+\beta_c\tau_c s_2\Big(\frac{\partial P_{2i,1}}{\partial T_1}-\frac{\partial P_{2i+2,1}}{\partial T_1}\Big)+\frac{\partial P_{2i,1}}{\partial T_1}\\
&-\beta_c\tau_c s_2\left(\frac{\partial Q_{2i,1}}{\partial T_1}-\frac{\partial Q_{2i+2,1}}{\partial T_1}\right),
\end{aligned}
$$

$i=1,2,\ldots,n-1,$

$$
\begin{aligned}
U^2_{4n-2,2}&=-\alpha b_1\tau_\varepsilon\omega_2\sum_{i=1}^{n-1}(s_2P_{2i-1,1}-c_2Q_{2i-1,1})\\
&+(\beta_\varepsilon c_2-\beta_c\tau_\varepsilon\omega_2s_2)(P_{2n-1,1}-P_{2,1})
\end{aligned}
$$

$$
\begin{aligned}
&- (\beta_\varepsilon s_2 + \beta_c \tau_\varepsilon \omega_2 c_2)(Q_{2,1} - Q_{2n-1,1}) \\
&- \alpha b_1 \tau_c \sum_{i=1}^{n-1} \left(s_2 \frac{\partial Q_{2i-1,1}}{\partial T_1} + c_2 \frac{\partial P_{2i-1,1}}{\partial T_1} \right) \\
&+ \beta_c \tau_c c_2 (\frac{\partial P_{2,1}}{\partial T_1} - \frac{\partial P_{2n-1,1}}{\partial T_1}) + \frac{\partial P_{2n-1,1}}{\partial T_1} \\
&- \beta_c \tau_c s_2 \left(\frac{\partial Q_{2,1}}{\partial T_1} - \frac{\partial Q_{2n-1,1}}{\partial T_1} \right), \\
& i = 1, 2, \ldots, n-1,
\end{aligned}
$$

$$
\begin{aligned}
U^1_{2i-1,3} &= -\frac{\partial B_{2i-1,1}}{\partial T_2} - \frac{\partial B_{2i-1,2}}{\partial T_1}, \\
U^1_{2i,3} &= \frac{\alpha}{2}(s_1 A_{2i-1,1} - c_1 B_{2i-1,1})(b_1 \tau_\varepsilon^2 \omega_1^2 - 4 b_2 E_{2i-1,2} \\
&- \frac{3b_3}{2}(A^2_{2i-1,1} + B^2_{2i-1,1} + P^2_{2i-1,1} + Q^2_{2i-1,1})) \\
&- \alpha b_2 (A^{(2)}_{2i-1,2}(s_1 B_{2i-1,1} - c_1 A_{2i-1,1}) \\
&- B^{(2)}_{2i-1,2}(s_1 A_{2i-1,1} + c_1 B_{2i-1,1})) - \alpha b_1 \tau_\varepsilon \omega_1 (c_1 A^{(1)}_{2i-1,2} + s_1 B^{(1)}_{2i-1,2}) \\
&- \tau_\varepsilon \omega_1 ((A_{2i+2,1} - A_{2i,1})(c_1 \beta_\varepsilon - \frac{\beta_c \tau_\varepsilon \omega_1 s_1}{2}) \\
&- (B_{2i,1} - B_{2i+2,1})(s_1 \beta_\varepsilon + \frac{\beta_c \tau_\varepsilon \omega_1 c_1}{2})) \\
&- (s_1 \beta_\varepsilon + \beta_c \tau_\varepsilon \omega_1 c_1)(A^{(1)}_{2i+2,2} - A^{(1)}_{2i,2}) \\
&- (B^{(1)}_{2i+2,2} - B^{(1)}_{2i,2})(c_1 \beta_\varepsilon - \beta_c \tau_\varepsilon \omega_1 s_1) \\
&- \alpha b_2 (G^1_{2i-1,2}(c_1 P_{2i-1,1} + s_1 Q_{2i-1,1}) + G^2_{2i-1,2}(-c_1 P_{2i-1,1} + s_1 Q_{2i-1,1}) \\
&+ G^3_{2i-1,2}(s_1 P_{2i-1,1} + c_1 Q_{2i-1,1}) - G^4_{2i-1,2}(s_1 P_{2i-1,1} + c_1 Q_{2i-1,1})) \\
&- \alpha b_1 \tau_c (c_1 \frac{\partial B_{2i-1,1}}{\partial T_2} - s_1 \frac{\partial A_{2i-1,1}}{\partial T_2}) - \beta_c \tau_c (s_1 (\frac{\partial A_{2i,1}}{\partial T_2} - \frac{\partial A_{2i+2,1}}{\partial T_2}) \\
&- c_1 (\frac{\partial B_{2i,1}}{\partial T_2} - \frac{\partial B_{2i+2,1}}{\partial T_2})) - \frac{\partial B_{2i,1}}{\partial T_2} - \frac{\partial B^{(1)}_{2i,2}}{\partial T_1} \\
&- \alpha b_1 \tau_\varepsilon ((c_1 - \tau_c \omega_1 s_1) \frac{\partial B_{2i-1,1}}{\partial T_1} \\
&- (s_1 + \tau_c \omega_1 c_1) \frac{\partial A_{2i-1,1}}{\partial T_1}) - \alpha b_1 \tau_c (c_1 \frac{\partial B^{(1)}_{2i-1,2}}{\partial T_1} - s_1 \frac{\partial A^{(1)}_{2i-1,2}}{\partial T_1}) \\
&- (s_1 (\beta_c \tau_\varepsilon + \beta_\varepsilon \tau_c) + c_1 \beta_c \tau_c \tau_\varepsilon \omega_1)(\frac{\partial A_{2i,1}}{\partial T_1} - \frac{\partial A_{2i+2,1}}{\partial T_1}) \\
&- (c_1 (\beta_c \tau_\varepsilon + \beta_\varepsilon \tau_c) - s_1 \beta_c \tau_c \tau_\varepsilon \omega_1)(-\frac{\partial B_{2i,1}}{\partial T_1} + \frac{\partial B_{2i+2,1}}{\partial T_1}) \\
&- \beta_c \tau_c (s_1 (\frac{\partial A^{(1)}_{2i,2}}{\partial T_1} - \frac{\partial A^{(1)}_{2i+2,2}}{\partial T_1}) + c_1 (\frac{\partial B^{(1)}_{2i,2}}{\partial T_1} - \frac{\partial B^{(1)}_{2i+2,2}}{\partial T_1}))
\end{aligned}
$$

$$
-\frac{\alpha b_1\tau_c^2}{2}(s_1\frac{\partial^2 A_{2i-1,1}}{\partial T_1^2}-c_1\frac{\partial^2 B_{2i-1,1}}{\partial T_1^2})-\frac{\beta_c\tau_c^2}{2}(s_1(\frac{\partial^2 A_{2i+21}}{\partial T_1^2}-\frac{\partial^2 A_{2i,1}}{\partial T_1^2})
$$

$$
-c_1(\frac{\partial^2 B_{2i+21}}{\partial T_1^2}-\frac{\partial^2 B_{2i,1}}{\partial T_1^2})),i=1,2,\ldots,n-1,
$$

$$
\begin{aligned}
U^1_{2n-1,3} = &-(\frac{\alpha b_1\tau_\varepsilon^2\omega_1^2}{2}+2\alpha b_2\sum_{i=1}^{n-1}E_{2i-1,2}\\
&-\sum_{i=1}^{n-1}(P^2_{2i-1,1}+Q^2_{2i-1,1}))\sum_{i=1}^{n-1}(s_1A_{2i-1,1}-c_1B_{2i-1,1})\\
&+\frac{3\alpha b_3}{4}(\sum_{i=1}^{n-1}s_1A_{2i-1,1}-c_1B_{2i-1,1})^3\\
&+\alpha b_2\sum_{i=1}^{n-1}A^{(2)}_{2i-1,2}\sum_{i=1}^{n-1}(s_1B_{2i-1,1}-c_1A_{2i-1,1})\\
&-\alpha b_2\sum_{i=1}^{n-1}B^{(2)}_{2i-1,2}\sum_{i=1}^{n-1}(s_1A_{2i-1,1}+c_1B_{2i-1,1})\\
&-\alpha b_1\tau_\varepsilon\omega_1\sum_{i=1}^{n-1}(c_1A^{(1)}_{2i-1,2}+s_1B^{(1)}_{2i-1,2})\\
&+\tau_\varepsilon\omega_1((A_{2n-1,1}-A_{2,1})(c_1\beta_\varepsilon-\frac{\beta_c\tau_\varepsilon\omega_1 s_1}{2})\\
&-(B_{2,1}-B_{2n-1,1})(s_1\beta_\varepsilon+\frac{\beta_c\tau_\varepsilon\omega_1 c_1}{2}))\\
&+(s_1\beta_\varepsilon+\beta_c\tau_\varepsilon\omega_1 c_1)(A^{(1)}_{2n-1,2}-A^{(1)}_{2,2})\\
&-(B^{(1)}_{2n-1,2}-B^{(1)}_{2,2})(c_1\beta_\varepsilon-\beta_c\tau_\varepsilon\omega_1 s_1)\\
&+\alpha b_2(\sum_{i=1}^{n-1}G^1_{2i-1,2}\sum_{i=1}^{n-1}c_1P_{2i-1,1}+s_1Q_{2i-1,1}\\
&+\sum_{i=1}^{n-1}G^2_{2i-1,2}\sum_{i=1}^{n-1}(-c_1P_{2i-1,1}+s_1Q_{2i-1,1})\\
&+\sum_{i=1}^{n-1}G^3_{2i-1,2}\sum_{i=1}^{n-1}(s_1P_{2i-1,1}-c_1Q_{2i-1,1})\\
&+\sum_{i=1}^{n-1}G^4_{2i-1,2}\sum_{i=1}^{n-1}(s_1P_{2i-1,1}+c_1Q_{2i-1,1}))\\
&+\alpha b_1\tau_c\sum_{i=1}^{n-1}(c_1\frac{\partial B_{2i-1,1}}{\partial T_2}-s_1\frac{\partial A_{2i-1,1}}{\partial T_2})
\end{aligned}
$$

$$
\begin{aligned}
&+\beta_c\tau_c(s_1(\frac{\partial A_{2,1}}{\partial T_2}-\frac{\partial A_{2n-1,1}}{\partial T_2})-c_1(\frac{\partial B_{2,1}}{\partial T_2}-\frac{\partial B_{2n-1,1}}{\partial T_2}))\\
&-\frac{\partial B_{2n-1,1}}{\partial T_2}-\frac{\partial B^{(1)}_{2i,2}}{\partial T_1}-\alpha b_1\tau_\varepsilon\sum_{i=1}^{n-1}((c_1-\tau_c\omega_1 s_1)\frac{\partial B_{2i-1,1}}{\partial T_1}\\
&-(s_1+\tau_c\omega_1 c_1)\frac{\partial A_{2i-1,1}}{\partial T_1})\\
&-\alpha b_1\tau_c\sum_{i=1}^{n-1}(c_1\frac{\partial B^{(1)}_{2i-1,2}}{\partial T_1}-s_1\frac{\partial A^{(1)}_{2i-1,2}}{\partial T_1})+(s_1(\beta_c\tau_\varepsilon+\beta_\varepsilon\tau_c)\\
&+c_1\beta_c\tau_c\tau_\varepsilon\omega_1)(\frac{\partial A_{2,1}}{\partial T_2}-\frac{\partial A_{2n-1,1}}{\partial T_2})-(c_1(\beta_c\tau_\varepsilon+\beta_\varepsilon\tau_c)\\
&-s_1\beta_c\tau_c\tau_\varepsilon\omega_1)(\frac{\partial B_{2,1}}{\partial T_2}-\frac{\partial B_{2n-1,1}}{\partial T_2})\\
&+\beta_c\tau_c(s_1(\frac{\partial A^{(1)}_{2,2}}{\partial T_1}-\frac{\partial A^{(1)}_{2n-1,2}}{\partial T_1})-c_1(\frac{\partial B^{(1)}_{2,2}}{\partial T_1}-\frac{\partial B^{(1)}_{2n-1,2}}{\partial T_1}))\\
&-\frac{\alpha b_1\tau_c^2}{2}\sum_{i=1}^{n-1}(s_1\frac{\partial^2 A_{2i-1,1}}{\partial T_1^2}-c_1\frac{\partial^2 B_{2i-1,1}}{\partial T_1^2})\\
&-\frac{\beta_c\tau_c^2}{2}(s_1(\frac{\partial^2 A_{2n-1,1}}{\partial T_1^2}-\frac{\partial^2 A_{2,1}}{\partial T_1^2})\\
&-c_1(\frac{\partial^2 B_{2n-1,1}}{\partial T_1^2}-\frac{\partial^2 B_{2,1}}{\partial T_1^2})),
\end{aligned}
$$

$$
U^1_{2i-1+2n-1,3}=-\frac{\partial A_{2i-1,1}}{\partial T_2}-\frac{\partial A_{2i-1,2}}{\partial T_1},
$$

$$
\begin{aligned}
U^1_{2i+2n-1,3}=&-\frac{\alpha}{2}(c_1A_{2i-1,1}+s_1B_{2i-1,1})(b_1\tau_\varepsilon^2-4b_2E_{2i-1,2}\\
&-\frac{3b_3}{2}(A^2_{2i-1,1}+B^2_{2i-1,1}+P^2_{2i-1,1}+Q^2_{2i-1,1}))\\
&+\alpha b_2(A^{(2)}_{2i-1,2}(c_1B_{2i-1,1}+s_1A_{2i-1,1})\\
&-B^{(2)}_{2i-1,2}(c_1A_{2i-1,1}-s_1B_{2i-1,1}))\\
&-\alpha b_1\tau_\varepsilon\omega_1(s_1A^{(1)}_{2i-1,2}-c_1B^{(1)}_{2i-1,2})\\
&-\tau_\varepsilon\omega_1((A_{2i+2,1}-A_{2i,1})(s_1\beta_\varepsilon-\frac{\beta_c\tau_\varepsilon\omega_1c_1}{2})\\
&-(B_{2i,1}-B_{2i+2,1})(-c_1\beta_\varepsilon+\frac{\beta_c\tau_\varepsilon\omega_1s_1}{2}))\\
&-(-c_1\beta_\varepsilon+\beta_c\tau_\varepsilon\omega_1s_1)(A^{(1)}_{2i+2,2}-A^{(1)}_{2i,2})\\
&+(B^{(1)}_{2i+2,2}-B^{(1)}_{2i,2})(s_1\beta_\varepsilon+\beta_c\tau_\varepsilon\omega_1c_1)\\
&-\alpha b_2(G^1_{2i-1,2}(c_1P_{2i-1,1}-s_1Q_{2i-1,1})\\
&+G^2_{2i-1,2}(-c_1P_{2i-1,1}-s_1Q_{2i-1,1})+G^3_{2i-1,2}(s_1P_{2i-1,1}-c_1Q_{2i-1,1})\\
&+G^4_{2i-1,2}(s_1P_{2i-1,1}-c_1Q_{2i-1,1}))-\alpha b_1\tau_c(s_1\frac{\partial B_{2i-1,1}}{\partial T_2}+c_1\frac{\partial A_{2i-1,1}}{\partial T_2})
\end{aligned}
$$

$$
\begin{aligned}
&- \beta_c\tau_c(c_1(-\frac{\partial A_{2i,1}}{\partial T_2} + \frac{\partial A_{2i+2,1}}{\partial T_2}) - s_1(\frac{\partial B_{2i,1}}{\partial T_2} - \frac{\partial B_{2i+2,1}}{\partial T_2}))\\
&- \frac{\partial A_{2i,1}}{\partial T_2} - \frac{\partial A^{(1)}_{2i,2}}{\partial T_1} - \alpha b_1\tau_\varepsilon((s_1 + \tau_c\omega_1 c_1)\frac{\partial B_{2i-1,1}}{\partial T_1}\\
&- (-c_1 + \tau_c\omega_1 s_1)\frac{\partial A_{2i-1,1}}{\partial T_1}) - \alpha b_1\tau_c(s_1\frac{\partial B^{(1)}_{2i-1,2}}{\partial T_1} + c_1\frac{\partial A^{(1)}_{2i-1,2}}{\partial T_1})\\
&+ (c_1(\beta_c\tau_\varepsilon + \beta_\varepsilon\tau_c) + s_1\beta_c\tau_c\tau_\varepsilon\omega_1)(\frac{\partial A_{2i,1}}{\partial T_1} - \frac{\partial A_{2i+2,1}}{\partial T_1})\\
&+ (s_1(\beta_c\tau_\varepsilon + \beta_\varepsilon\tau_c) + c_1\beta_c\tau_c\tau_\varepsilon\omega_1)(\frac{\partial B_{2i,1}}{\partial T_1} - \frac{\partial B_{2i+2,1}}{\partial T_1})\\
&- \beta_c\tau_c(c_1(-\frac{\partial A^{(1)}_{2i,2}}{\partial T_1} + \frac{\partial A^{(1)}_{2i+2,2}}{\partial T_1}) - s_1(\frac{\partial B^{(1)}_{2i,2}}{\partial T_1} - \frac{\partial B^{(1)}_{2i+2,2}}{\partial T_1}))\\
&+ \frac{\alpha b_1\tau_c^2}{2}(c_1\frac{\partial^2 A_{2i-1,1}}{\partial T_1^2} + s_1\frac{\partial^2 B_{2i-1,1}}{\partial T_1^2}) + \frac{\beta_c\tau_c^2}{2}(c_1(\frac{\partial^2 A_{2i+21}}{\partial T_1^2} - \frac{\partial^2 A_{2i,1}}{\partial T_1^2})\\
&+ s_1(\frac{\partial^2 B_{2i+21}}{\partial T_1^2} - \frac{\partial^2 B_{2i,1}}{\partial T_1^2})), i = 1, 2, \ldots, n-1,
\end{aligned}
$$

$$
\begin{aligned}
U^1_{4n-2,3} = &-(\frac{\alpha b_1\tau_\varepsilon^2\omega_1^2}{2} + 2\alpha b_2\sum_{i=1}^{n-1} E_{2i-1,2} - \sum_{i=1}^{n-1}(P^2_{2i-1,1} + Q^2_{2i-1,1}))\\
&\sum_{i=1}^{n-1}(c_1 A_{2i-1,1} - s_1 B_{2i-1,1}) + \frac{3\alpha b_3}{4}(\sum_{i=1}^{n-1} c_1 A_{2i-1,1} - s_1 B_{2i-1,1})^3\\
&+ \alpha b_2\sum_{i=1}^{n-1} A^{(2)}_{2i-1,2}\sum_{i=1}^{n-1}(c_1 B_{2i-1,1} - s_1 A_{2i-1,1})\\
&- \alpha b_2\sum_{i=1}^{n-1} B^{(2)}_{2i-1,2}\sum_{i=1}^{n-1}(c_1 A_{2i-1,1} + s_1 B_{2i-1,1})\\
&- \alpha b_1\tau_\varepsilon\omega_1\sum_{i=1}^{n-1}(s_1 A^{(1)}_{2i-1,2} + c_1 B^{(1)}_{2i-1,2})\\
&+ \tau_\varepsilon\omega_1((A_{2n-1,1} - A_{2,1})(s_1\beta_\varepsilon - \frac{\beta_c\tau_\varepsilon\omega_1 c_1}{2})\\
&- (B_{2,1} - B_{2n-1,1})(c_1\beta_\varepsilon + \frac{\beta_c\tau_\varepsilon\omega_1 s_1}{2}))\\
&+ (c_1\beta_\varepsilon + \beta_c\tau_\varepsilon\omega_1 s_1)(A^{(1)}_{2n-1,2} - A^{(1)}_{2,2})\\
&- (B^{(1)}_{2n-1,2} - B^{(1)}_{2,2})(s_1\beta_\varepsilon - \beta_c\tau_\varepsilon\omega_1 c_1)\\
&+ \alpha b_2(\sum_{i=1}^{n-1} G^1_{2i-1,2}\sum_{i=1}^{n-1} s_1 P_{2i-1,1} + c_1 Q_{2i-1,1}\\
&+ \sum_{i=1}^{n-1} G^2_{2i-1,2}\sum_{i=1}^{n-1}(-s_1 P_{2i-1,1} + c_1 Q_{2i-1,1})
\end{aligned}
$$

$$
\begin{aligned}
&+\sum_{i=1}^{n-1} G_{2i-1,2}^{3} \sum_{i=1}^{n-1} (c_1 P_{2i-1,1} - s_1 Q_{2i-1,1}) \\
&+\sum_{i=1}^{n-1} G_{2i-1,2}^{4} \sum_{i=1}^{n-1} (c_1 P_{2i-1,1} + s_1 Q_{2i-1,1})) \\
&+\alpha b_1 \tau_c \sum_{i=1}^{n-1} (s_1 \frac{\partial B_{2i-1,1}}{\partial T_2} - c_1 \frac{\partial A_{2i-1,1}}{\partial T_2}) \\
&+\beta_c \tau_c (c_1 (\frac{\partial A_{2,1}}{\partial T_2} - \frac{\partial A_{2n-1,1}}{\partial T_2}) \\
&- s_1 (\frac{\partial B_{2,1}}{\partial T_2} - \frac{\partial B_{2n-1,1}}{\partial T_2})) - \frac{\partial B_{2n-1,1}}{\partial T_2} - \frac{\partial B_{2i,2}^{(1)}}{\partial T_1} \\
&- \alpha b_1 \tau_\varepsilon \sum_{i=1}^{n-1} ((s_1 - \tau_c \omega_1 c_1) \frac{\partial B_{2i-1,1}}{\partial T_1} \\
&- (c_1 + \tau_c \omega_1 s_1) \frac{\partial A_{2i-1,1}}{\partial T_1}) - \alpha b_1 \tau_c \sum_{i=1}^{n-1} (s_1 \frac{\partial B_{2i-1,2}^{(1)}}{\partial T_1} - c_1 \frac{\partial A_{2i-1,2}^{(1)}}{\partial T_1}) \\
&+ (c_1 (\beta_c \tau_\varepsilon + \beta_\varepsilon \tau_c) + s_1 \beta_c \tau_c \tau_\varepsilon \omega_1)(\frac{\partial A_{2,1}}{\partial T_2} - \frac{\partial A_{2n-1,1}}{\partial T_2}) \\
&- (s_1 (\beta_c \tau_\varepsilon + \beta_\varepsilon \tau_c) - c_1 \beta_c \tau_c \tau_\varepsilon \omega_1)(\frac{\partial B_{2,1}}{\partial T_2} - \frac{\partial B_{2n-1,1}}{\partial T_2}) \\
&+ \beta_c \tau_c (c_1 (\frac{\partial A_{2,2}^{(1)}}{\partial T_1} - \frac{\partial A_{2n-1,2}^{(1)}}{\partial T_1}) - s_1 (\frac{\partial B_{2,2}^{(1)}}{\partial T_1} - \frac{\partial B_{2n-1,2}^{(1)}}{\partial T_1})) \\
&- \frac{\alpha b_1 \tau_c^2}{2} \sum_{i=1}^{n-1} (c_1 \frac{\partial^2 A_{2i-1,1}}{\partial T_1^2} - s_1 \frac{\partial^2 B_{2i-1,1}}{\partial T_1^2}) \\
&- \frac{\beta_c \tau_c^2}{2} (c_1 (\frac{\partial^2 A_{2n-1,1}}{\partial T_1^2} - \frac{\partial^2 A_{2,1}}{\partial T_1^2}) - s_1 (\frac{\partial^2 B_{2n-1,1}}{\partial T_1^2} - \frac{\partial^2 B_{2,1}}{\partial T_1^2})).
\end{aligned}
$$

Note that the analytic form of U_3^2 is no longer provided here, since it can be obtained by replacing A_1, B_1 with P_1, Q_1, A_2, B_2 with P_2, Q_2, ω_1, c_1, s_1 with ω_2, c_2, s_2, and G_2^1 with $-G_2^1$ in U_3^1, respectively.

References

1. Zhao X, Orosz G (2014) Nonlinear day-to-day traffic dynamics with driver experience delay: modeling, stability and bifurcation analysis. Physica D 275:54–66
2. Zhen B, Xu J (2010) Bautin bifurcation analysis for synchronous solution of a coupled FHN neural system with delay. Commun Nonlinear Sci Numer Simul 15:442–458

3. Yan Y, Xu J, Wang W (2012) Nonlinear chatter with large amplitude in a cylindrical plunge grinding process. Nonlinear Dyn 69:1781–1793
4. Engelborghs K, Luzyanina T, Samaey G et al (2007) DDE-BIFTOOL v. 2.03: a Matlab package for bifur-cation analysis of delay differential equations, available at http://twr.cs.kuleuven.be/research/software/delay/ddebiftool.shtml
5. Stépán G (1989) Retarded dynamical systems: stability and characteristic functions. Longman Scientific & Technical
6. Rothery RE (1998) Traffic flow theory, 2nd edn. Transportation Research Board special report 165
7. Nayfeh AH (2005) Resolving controversies in the application of the method of multiple scales and the generalized method of averaging. Nonlinear Dyn 40:61–102
8. Nayfeh AH (2008) Order reduction of retarded nonlinear systems–the method of multiple scales versus center-manifold reduction. Nonlinear Dyn 51:483–500
9. Carter P, Christiansen PL, Gaididei YB et al (2014) Multijam solutions in traffic models with velocity-dependent driver strategies. SIAM J Appl Math 74:1895–1918
10. Xu J, Chung KN (2004) Delay reduced double Hopf bifurcation in a limit cycle oscillator: extension of a perturbation-incremental method. Dyn Continuous, Discrete Impulsive Syst, Ser B: Appl Algorithms 11:136–143
11. Xu J, Huang MS, Zhang YY (2007) Dynamics due to non-resonant double Hopf Bifurcationin in Van Del Pol-Duffing SYSTEM WITH DELAYED POSITION Feedback. Springer
12. Xu J, Chung K, Chan C (2007) An efficient method for studying weak resonant double Hopf bifurcation in nonlinear systems with delayed feedbacks. SIAM J Appl Dyn Syst 6:29–60
13. Zhang S, Xu J (2013) Quasiperiodic motion induced by heterogeneous delays in a simplified internet congestion control model. Nonlinear Anal Real World Appl 14:661–670
14. Song ZG, Xu J (2013) Stability switches and double Hopf bifurcation in a two-neural network system with multiple delays. Cogn Neurodyn 7:505–521
15. Wang W, Xu J, Sun X (2013) Strong and weak resonances in delayed differential systems. Int J Bifurcat Chaos 23:1350119
16. Ge J, Xu J (2012) Weak resonant double Hopf bifurcations in an inertial four-neuron model with time delay. Int J Neural Syst 22:63–75

SPRINGER NATURE

GPSR Compliance

The European Union's (EU) General Product Safety Regulation (GPSR) is a set of rules that requires consumer products to be safe and our obligations to ensure this.

If you have any concerns about our products, you can contact us on ProductSafety@springernature.com

In case Publisher is established outside the EU, the EU authorized representative is:

Springer Nature Customer Service Center GmbH
Europaplatz 3
69115 Heidelberg, Germany

The manufacturer's authorised representative in the EU is Springer Nature Customer Service Centre GmbH, Europaplatz 3, 69115 Heidelberg, Germany. If you have any concerns regarding our products, please contact ProductSafety@springernature.com

Printed and bound by CPI Group (UK) Ltd, Croydon, CR0 4YY

01/10/2026

02225868-0017